Corrosion Science and Technology

Mechanism, Mitigation and Monitoring

Corrosion Science and Technology

Mechanism, Mitigation and Monitoring

Editors

U. Kamachi Mudali

Baldev Raj

Alpha Science International Ltd.

Oxford, U.K.

Corrosion Science and Technology
Mechanism, Mitigation and Monitoring
598 pgs. | 86 tbs. | 346 figs.

Editors
U. Kamachi Mudali
Head, Corrosion Science and Technology Section - RPM
Corrosion Science and Technology Division
Indira Gandhi Centre for Atomic Research
Kalpakkam - 603 102, India

Baldev Raj
Distinguished Scientist and Director
Indira Gandhi Centre for Atomic Research
Kalpakkam - 603 102, India

ALPHA SCIENCE INTERNATIONAL LTD.

7200 The Quorum, Oxford Business Park North
Garsington Road, Oxford OX4 2JZ, U.K.

www.alphasci.com

ISBN 978-1-84265-508-5

Printed in India

Preface

The field of corrosion science and technology has opened up several avenues of research and development and challenges for the scientific community to produce new materials, surface modifications, coatings and other protection methodologies. Any material could be used for different environments if adequate protection methods and precaution to avoid corrosion are practised. Gold, platinum, many ceramic and carbon materials, withstand significant corrosion, while other metallic materials interact with the environment and produce substantial changes in their properties. For efficient corrosion control and prevention, it becomes necessary to know about the various manifestations of corrosion and their mechanisms. Corrosion can range from the most encountered general corrosion to the highly localized types, such as pitting corrosion, crevice corrosion, intergranular corrosion, stress corrosion cracking, etc. The success in combating corrosion attack lies in understanding the mechanisms of the processes, effects of various environmental factors such as pH, temperature etc. and susceptibility of different microstructures to the corrosion process. An understanding of the corrosion process vis-a-vis the material-environment combination helps in devising methods for combating corrosion. The possible ways include modifying the existing material or choosing a new material, controlling environmental factors, modification or choice of new fabrication processes, improving the design of the component, and applying chemical and electrochemical protection methods. Effective corrosion control measures could result from adequate corrosion management techniques. Apart from the understanding of corrosion processes, their mechanisms and effects of various parameters, good corrosion management would also involve effective monitoring of the health of the component during service, through electrochemical and non-destructive evaluation techniques and prediction of the remnant life of the components.

This edited volume is a collection of specialist articles on various aspects of mechanisms, mitigation and monitoring of corrosion processes provided by leading experts in the field. State of the art in mechanisms, mitigation and monitoring has been brought out under the categories of corrosion processes, corrosion protection, corrosion resistant materials and corrosion evaluation and monitoring. The articles are organised in these categories in order to provide recent information and views by authors who are well known in the respective field. Localised corrosion, high temperature corrosion, microbially influenced corrosion and stress corrosion cracking aspects are discussed in the category on corrosion processes; metallic and oxide coatings, electro- and electroless coatings, plasma based coatings, engineering coatings and vapour phase inhibitors are discussed in the category on corrosion protection; intermetallics, high nitrogen steels, biomaterials and nanomaterials are discussed in the category on corrosion resistant materials; and finally, nondestructive evaluation, scanning Kelvin probes, industrial corrosion monitoring and nuclear & Auger electron spectroscopy are discussed in the category on corrosion evaluation and monitoring. The difference that one can make in this edited volume is the wide spectrum of information on each topic of corrosion in an interdisciplinary manner.

The book is meant to be a treasure to all those pursuing corrosion science and technology as subject of study, research, profession, or in industry. It is a worthy collection for libraries from all over the world. In the recent times, no other edited book provides articles dealing with corrosion processes, corrosion control methods, and corrosion evaluation and monitoring in one place, tapping the expertise of distinguished specialists. The editors thank all the authors who have taken efforts to prepare such wonderful articles. Finally, we thank Narosa Publishing House personnel for their help in editing and production of a high quality book of this nature published in association with CRC Press, USA and Alpha Science International, Europe.

U. Kamachi Mudali
Baldev Raj

Contents

Corrosion Resistant Materials

Corrosion Evaluation and Monitoring

List of Contributors

V.S. Raja

Corrosion Science and Engineering, IIT Bombay, Mumbai - 400 076, India.

E-mail: *vsraja@iitb.ac.in*

P.K. Datta, H.L. Du and J.S. Burnell-Gray

Director, AMRI, Room E302 Ellison Building, School of Engineering, University of Northumbria at Newcastle NE1 8ST, United Kingdom.

E-mail: *psantu.datta@unn.ac.uk*

K.A. Natarajan

Indian Institute of Science, Bangalore - 560 012, India

E-mail: *kan@met.iisc.ernet.in*

R.K. Singh Raman

Departments of Chemical and Mechanical Engineering, Bldg 31, Monash University (Melbourne), Vic 3800, Australia

E-mail: *Raman.Singh@spme.monash.edu.au*

S. Ningshen and U. Kamachi Mudali

Corrosion Science and Technology Division, IGCAR, Kalpakkam - 603 102, India.

E-mail: *ning@igcar.gov.in*

T. Sankarayanan and S.K. Seshadri*

National Metallurgical Laboratory, Madras Centre, CSIR Complex, Chennai-600 113, India

*Dept of Metallurgical and Materials Engineering, Indian Institute of Technology Madras, Chennai-600 036, India.

E-mail: *tsnsn@rediffmail.com*

V.S. Raghunathan and P. Kuppusami

Metallurgy and Materials Group, IGCAR, Kalpakkam - 603 102, India.

E-mail: *pk@igcar.gov.in*

G. Sundararajan, L. Rama Krishna, Nitin P. Wasekar, G. Sivakumar and A. Jyothirmayi

International Advanced Research Centre for Powder Metallurgy & New Materials (ARCI), Balapur (P.O), R.R. District, Hyderabad –500 005, India.

E-mail: *gsundar@arci.res.in*

K.L. Vasanth

Carderock Division, Naval Surface Warfare Center, Code 613, 9500 MacArthur Blvd., West Bethesda, MD 20817, USA

E-mail: *VasanthKL@nswccd.naw.mil*

Shankar Rao

Department of Materials and Metallurgical Engineering, Indian Institute of Technology Kanpur, Kanpur-208 016, India

E-mail: *vshankarao@yahoo.co.in*

J. Foct

Laboratoire de Métallurgie Physique et Génie Matériaux, UMR CNRS 8517, Université de Lille I, 59655, Villeneuve d'Ascq Cedex, France.

E-mail: *jacques.foct@univ-lille1.fr*

Noam Eliaz

Department of Solid Mechanics, Materials and Systems, Tel-Aviv University, Ramat-Aviv, Tel-Aviv 69978, Israel.

E-mail: *neliaz@eng.tau.ac.il*

Anita Toppo, P. Shankar, H. Shaikh and A.K. Tyagi

Metallurgy and Materials Group, Corrosion Science and Technology Division, Indira Gandhi Centre for Atomic Research, Kalpkkam - 603 102, India.

E-mail: *anita@igcar.gov.in*

Baldev Raj, T. Jayakumar and G.K. Sharma

Metallurgy and Materials Group, Indira Gandhi Centre for Atomic Research, Kalpkkam - 603 102, India.

E-mail: *dirsec@igcar.gov.in*

M. Rohwerder

Max-Planck-Institut für Eisenforschung GmbH, Max-Planck-Straße 1, D-40237 Düsseldorf, Germany.

E-mail: *rohwerd2@mpie.de*

Alec Groysman

Oil Refineries Ltd., Haifa, Israel

E-mail: *GALEC@orl.co.il*

G. Amarendra and R. Govindaraj

Material Science Division, IGCAR, Kalpakkam - 603 102, India.

E-mail: *amar@igcar.gov.in*

CHAPTER 1

Localized Corrosion

V. S. Raja
Department of Metallurgical Engineering and Material Science, Indian Institute of Technology Bombay, Mumbai 400046, India

INTRODUCTION

Aqueous corrosion of metals/alloys can be broadly classified into two categories namely uniform and localized corrosion based on the way the anodic and cathodic reactions occur on the metal surface. What is common to all types of localized corrosion that differs from uniform corrosion is that the anodic and cathodic reactions of the former exhibit preference over some sites, microscopic/ macroscopic level, at which they occur. On the other hand both the anodic and cathodic reactions of uniform corrosion occur throughout the metallic surface with no preference to any site. Several factors can bias a metallic surface, at microscopic and/or macroscopic levels, into either an anodic site or cathodic site. They could be one or more of the following: (i) nature of alloy, (ii) nature of environment and (iii) external factors. As the alloys having passivity are known to suffer from severe localized corrosion, the above factors one way or the other become responsible for damaging the passivity at microscopic/macroscopic levels. Broadly, the following forms of corrosion viz., pitting corrosion, crevice corrosion, intergranular corrosion, selective leaching and stress corrosion cracking can be treated under this perspective. Hence this chapter discusses these forms of localized corrosion.

PITTING CORROSION

Pitting corrosion is one of the most insidious forms of attack that can severely damage engineering alloys with undesirable consequences. National Transportation Safety Board, USA implicates pitting corrosion for eight accidents that occurred on aircrafts/helicopters during the years 1994-96. Table-1 shows various components of aircrafts/helicopters that suffered pitting corrosion. Generally, pitting corrosion occurs on alloys when they are in the passive state. Some of the alloys/ metals such as stainless steels and Al, Ti, Zr and their alloys are

Table 1 Three years of pitting corrosion incidents of aircraft and helicopters (Data obtained from the website of National Transportation Safety Board, USA.)

Aircraft	Location of Failure	Cause	Incident Severity	Place	Year
DC-6	Engine, master connecting rod	Corrosion pitting	Fatal	AK, USA	1996
Piper PA-23	Engine, cylinder	Corrosion pitting	Fatal	AL, USA	1996
Boeing 75	Rudder Control	Corrosion pitting	Substantial damage to plane	WI, USA	1996
Embraer 120	Propeller Blade	Corrosion pitting	Fatal and serious, loss of plane	GA, USA	1995
Gulfstream GA-681	Hydraulic Line	Corrosion pitting	Loss of plane, no injuries	AZ, USA	1994
L-1011	Engine, compressor assembly disk	Corrosion pitting	Loss of plane, no injuries	AK, USA	1994
Embraer 120	Propeller Blade	Corrosion pitting	Damage to plane, no injuries	Canada	1994
Embraer 120	Propeller Blade	Corrosion pitting	Damage to plane, no injuries	Brazil	1994

known to passivate in a wide-ranging industrial environments. Fe and steel though do not normally passivate, can exhibit passivity in some environments and therefore can suffer pitting under these conditions. It should be pointed out that components can last longer in a non-passive state of a metal, with high uniform corrosion rate, than when the metal becomes susceptible to pitting corrosion under passivating conditions. For example, a pipeline made of type 304 SS can leak earlier than a carbon steel or cast iron pipeline when they need to be used for carrying seawater. Hence this topic assumes greater significance.

Pitting Process and Pitting Morphology

Pitting Process

Pitting corrosion of an alloy is caused by specific ions damaging the passive films at microscopic levels (Fig. 1). Pitting process follows two steps namely initiation and propagation. Between the two processes, propagation or growth is relatively better understood than the initiation process. This is because of the fact that growth process is more amenable to experimentation than the pit initiation. This is briefly discussed here. For a detailed understanding of mechanism readers can refer the review by Strehblow (1995).

Initiation

As far as initiation process is concerned two types of mechanisms are proposed. In the first type, the absorption of aggressive anions, (i) either in preference to oxygen on an active metal surface or (ii) on a passive metal surface, causing localized dissolution of the film is considered to be responsible for pit initiation. In the second type of mechanism the anions are suggested to migrate through the passive film under a high electric field to cause film break down. Each of

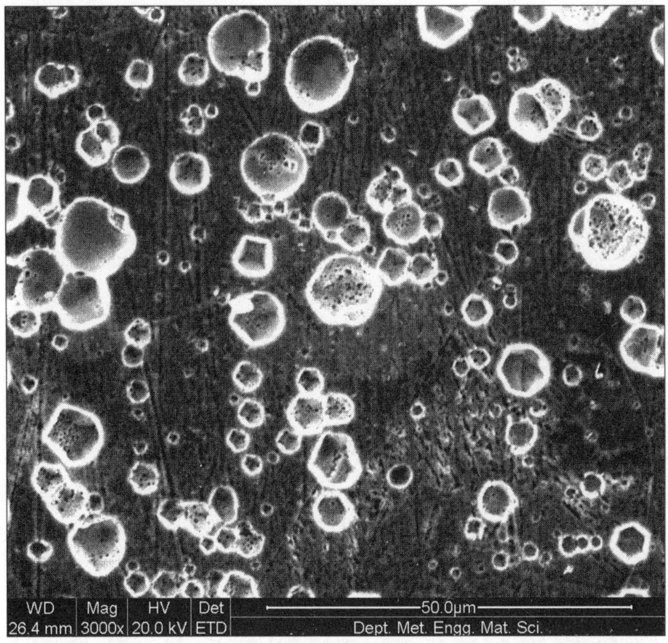

Figure 1 Typical pits at microscopic level on the Fe-18Mn-18Cr-0.45N-0.11C subjected to anodic polarization in 3.5% NaCl solution

these two mechanisms has their merit and demerits in explaining the experimental observation.

Growth

A passive metal dissolves at a rate given by it passive current density which is of the order 10^{-6} A/cm^2, while in the pit the dissolution current density can rise up to 10 A/cm^2. Thus, the dissolution rate of a pit could be a million times higher than that of passive metals. Similarly pits encounter an entirely different environment than what is encountered by passive surface. Stable pits are associated with low pH, high chloride (or halides) concentration and precipitation of salts, which prevent repassivation of pit. Additionally, high potential drop that exists between the pit bottom and the passive surface, keeps the pit growth front in the active region of the anodic polarization curve (will be discussed later). For sustained growth of a pit favorable pit chemistry and potential drop across the pit are essential. If they are not met with, the pits at the first instance remain metastable and cease to become a stable pit. Even the stable pits can cease to grow further, if the above conditions within the pit are not sustained.

Pit Morphology

ASTM Practice G-46 describes different pit morphologies a metal/alloy can exhibit (Fig. 2). Chemistry, microstructure and the nature of inclusions of the alloy influence the pit morphology. Pits in Al-alloys are known to vary from crystallographic to irregular pits depending on the heat treatment.

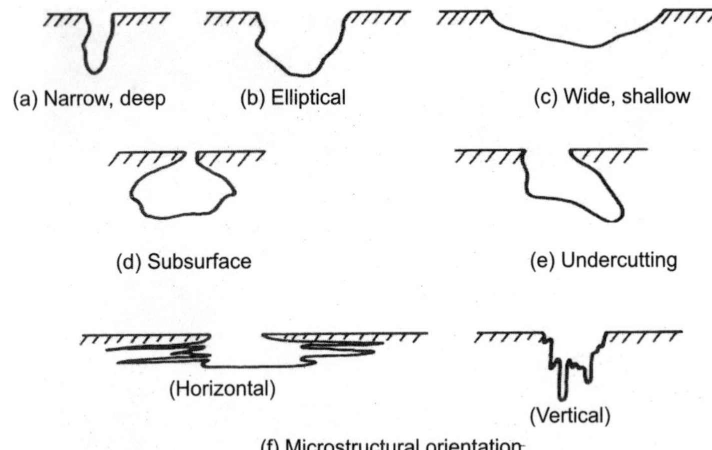

Figure 2 Possible variations in the pit morphology, seen through sample cross-section are shown (ASTM Practice G-46)

Factors Affecting Pitting Corrosion

Pitting tendency of a metal or alloy is governed by many factors. From the perspective of controlling pitting corrosion, it is essential to highlight the factors that influence pitting corrosion. As of now no quantitative relation exists that can predict a pitting failure. Qualitative understanding exists between various parameters that affect pitting tendency of an alloy, which in fact forms the basis for material selection and environment control for preventing pitting failures.

Electrochemical

Anodic polarization curves of passive metals typically exhibit active, passive and transpassive dissolution. A schematic diagram of an anodic polarization depicting these states of a metal is shown in the Figure 3. It is also possible that a metal/alloy do not show active dissolution region, should its corrosion potential lie in the passive region. The high current density at the transpassive state of the passive metal/alloy can be due to passive film break down and or oxygen evolution reaction. Passive films are thin (~100 Å), so even a small potential rise across the film, leads to high electric field of order of 10^6 V/cm that can cause an electrical break down of the passive film initiating pitting, with concurrent increase in anodic current. On the other hand, if the applied potential on the metal exceeds the equilibrium potential (in practice over voltage also needs to be considered) of $2H_2O + 4e + O_2 = 4OH^-$, then O_2 evolution becomes possible. The potential at which film break down occurs is called pitting potential (E_{pit}) or break down potential (E_b). Above this potential pits instantaneously nucleate and grow. Once these pits are formed, they continue to grow even when the potential is reversed. The pits can repassivate only at a much lower potential of the metal called repassivation potential (E_{rp}) or protection potential (E_{prot}). Below this potential the existing pits cease to grow. Accordingly, E_{pit} and E_{prot} values are considered to be indicators of pitting resistance of a metal, though there is some shortcoming in uniquely determining their values. For an effective comparison of these

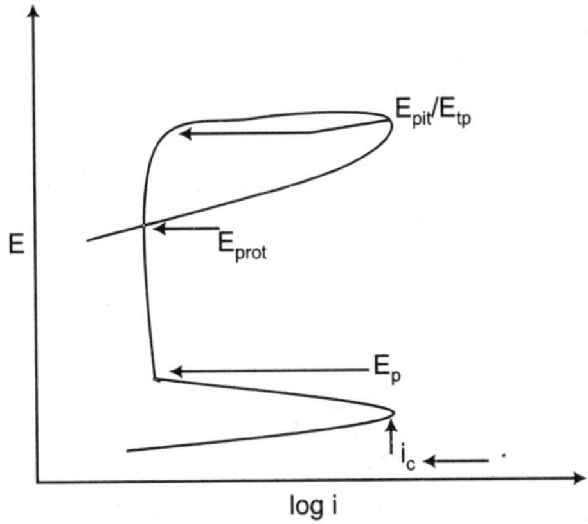

Figure 3 Schematic of anodic polarization curve depicting various electrochemical parameters, namely, pitting potential (E_{pit}) or transpassive potential (E_{tp}), protection potential (E_{prot}), passive potential (E_p) and critical current density (i_c) which are a measure of the pitting tendency of an alloy is shown

values across various metals and alloys, experiments need to be conducted in an identical manner. ASTM G61 standard prescribes a practice to obtain these values.

A metal with high E_{pit} and E_{prot} values and very low difference between E_{pit} and E_{prot} is expected to exhibit high resistance to pitting. When the E_{prot} value of any metal tends to become closer to its corrosion potential (E_{corr}), its ability to resist pitting becomes poor. This is because of the fact that E_{corr} is the steady state electrochemical potential of a metal exhibited in a corrosive environment and so no external agent is required to drive the pit growth. Potentiodynamic polarization curves of AISI type 304L SS and manganese containing austenitic stainless steel (16 Cr-8Mn) shown in Figure 4 illustrate this point. According to the polarization diagrams, the latter should exhibit higher pitting susceptibility than the former. As will be discussed later, Mn in austenitic stainless steels is detrimental to pitting resistance.

Metallurgy

Chemistry of an alloy, the phases present in it and the extent of micro and macro segregation it exhibits have a profound influence on the pitting tendency of an alloy.

In stainless steels, Cr, Mo, W and N play a dominant role, while other alloying elements such as Si, V and Ni seem to play a subtle role in promoting pitting resistance. More over, there seems to be some interdependency between the alloying elements with respect to their ability to exert resistance to pitting tendency of an alloy. For example, Mo is found to be effective, only when Cr is present in the stainless steel and similarly, N seems to be very efficient if the alloy contains both Cr and Mo. Figure 5 summarizes the role of various alloying elements on passivity and pitting tendency of stainless steels. Various attempts have been made to predict the pitting resistance of the alloy based on its composition, without recourse to testing. The following empirical equation relates pitting resistance in terms of an index called pitting resistance equivalent number (PREN) and the alloy content

$$PREN = \% \, Cr + 3.3 \, (Mo + 0.5W) + (16 - 30) \, N.$$

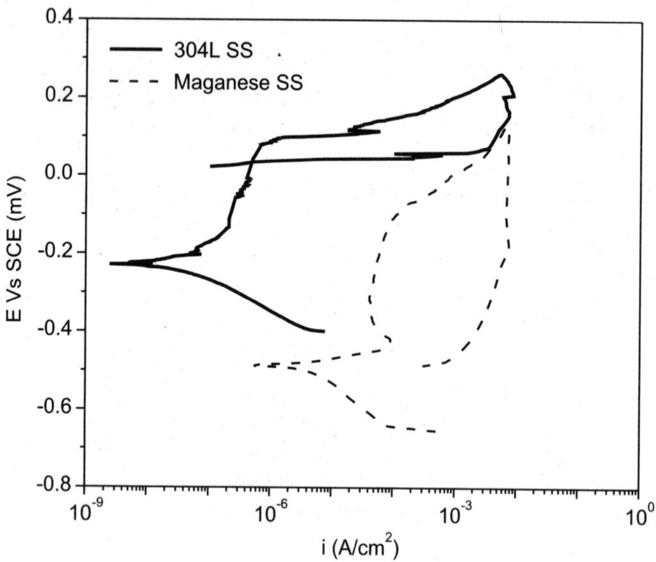

Figure 4 Cyclic polarization curves of type 304L SS and manganese substituted SS obtained in 3.5 wt% NaCl solution. Note the protection potential of the latter alloy is close to its corrosion potential

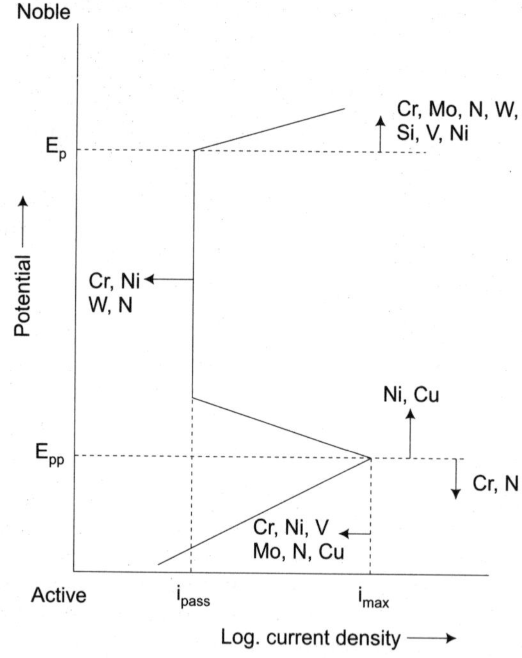

Figure 5 Role of alloying elements on the passivity and pitting tendency of stainless steel is summarized. (Isaacs and Kissel 1972, Heusler and Fisher 1976 and Frankel et al 1987)

The above equation can be used as a rule of thumb to provide preliminary information on the suitability of a stainless steel for a specific application. The increase in PREN indicates a rise in pitting resistance of a stainless steel. Stainless steels with PREN greater than 40 is prefixed with 'super' (such as superaustenitic) to the class of stainless steels, to indicate their high tendency to resist pitting corrosion. Should the stainless suffer from poor microstructures arising out of factors such as improper heat treatment and the presence of inclusions, the application of PREN to assess the pitting resistance of a stainless steel would become redundant. This aspect is discussed below.

Stainless steels become more susceptible to pitting because of (a) inclusions, (b) sensitization, (c) segregation of alloying elements either along the grain boundary in wrought alloy or along interdendritic boundaries / cells in a weld fusion zone and (d) mechanical working.

MnS inclusions are unstable especially in acidic conditions and found to be a source for pitting in stainless steels. The chemical stability of MnS inclusions is known to depend on its S and Cr content. Low Cr and high S levels are detrimental to the chemical stability of MnS inclusions. It should also be noted that S arising out of MnS inclusion can damage the passivity of the alloy.

Sensitization causes a steep drop in chromium content of the grain boundary region (refer the section on intergranular corrosion for details) and so becomes responsible for poor pitting resistance of those stainless steels subjected to improper heat treatment and in some cases welding. Heat affected zones of stainless steels with sufficiently high levels of C can suffer from pitting corrosion. Severe deterioration in the pitting resistance of fusion zone of stainless steel weldments is also a real cause for concern. As shown in Figure 6, not only does the critical pitting temperature[1] (CPT) of the weldment drops far below its wrought counter part, but the more serious concern with the welding process is that it renders Mo that is added to improve pitting resistance of wrought stainless steels, much less effective. It is clear from the Figure 6 that difference between CPT of a wrought alloy to that of its weld counter part increases with Mo content. Micro level segregation of Cr and Mo is responsible for the drop in pitting resistance of the weld-fusion zone. In this respect N is far more beneficial than Mo as has been exemplified by the polarization curves of 904L weld overlay with various N contents (Figure 7). Just 0.25 wt% N addition causes a steep rise in E_{pit} of 904L weld overlay.

Role of metallurgy in relation to Al-alloys is also discussed here, as these alloys find application in several industries and are far more prone to pitting than stainless steels in chloride environment. Holroyd (2001) has reviewed, along with the other forms of corrosion, the pitting corrosion of Al alloys. As a single-phase alloy, Al exhibits high pitting potential (Figure 8) when obtained through sputter deposition along with Zr, Cr, W, Ta and Mo. The ability of these elements to impart pitting resistance to, otherwise pitting prone, Al has been attributed to strong Al-M bond that exists in the case of these elements. In contrast to single phase binary alloys mentioned above, the conventional Al-alloys contain several phases, which influence the pitting resistance of Al-alloy. The age-hardening process, essential for obtaining good mechanical properties, has more often been found to lower the pitting potential of Al. Several studies have been undertaken to understand this aspect. Figure 9 illustrates how the

---

[1]CPT is defined as the lowest temperature at which the alloy shows pitting with in the test duration. This test is done as per ASTM G 48.

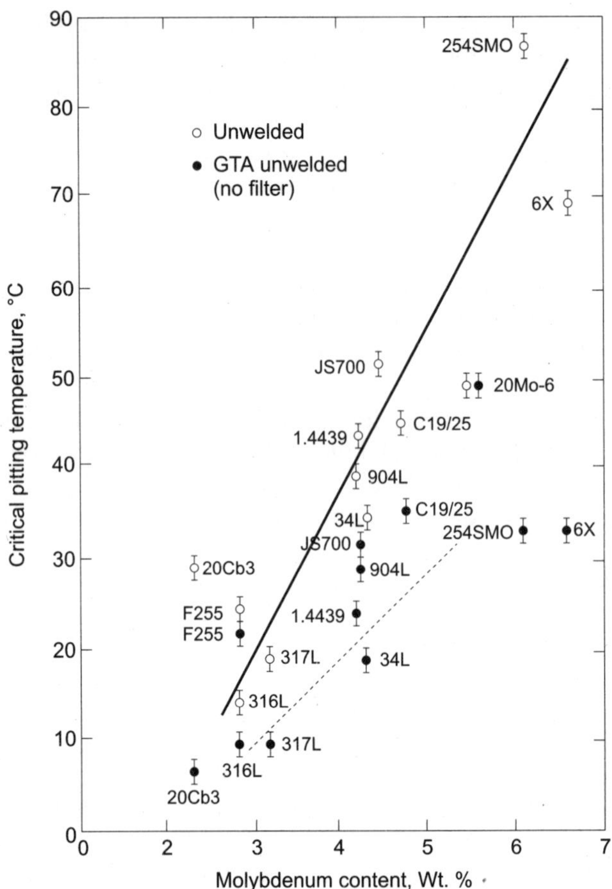

Figure 6 Variation of the critical pitting temperature (determined in ferric chloride solution) with the molybdenum content of various stainless steels (Garner 1985)

pitting potentials at microscopic level, namely grain boundary and matrix of AA 8090 alloy change with the aging time. It is also interesting to note that the morphology of pits that appear in the matrix also change with aging time.

Ti and its alloys offer resistance to most of the aqueous environment barring a few environments, such as Br^- containing solutions and halide containing non-aqueous solutions. Even the pitting corrosion of Ti alloys is reported to be influenced by microstructures. Thus, Raja et al (1993) reported that the Widmanstatten structure of Ti-6Al-2Sn-4Zr-2Mo(0.1Si) alloy brings down the latter's protection potential in 1M NaBr solution.

Environment

Though pitting is a severe form of corrosive attack, the favorable aspect of pitting is that only certain types of anions are responsible for pitting. These ions can be further assisted by other ions - both anions and cations. Chlorides, commonly found in normal water sources, pit

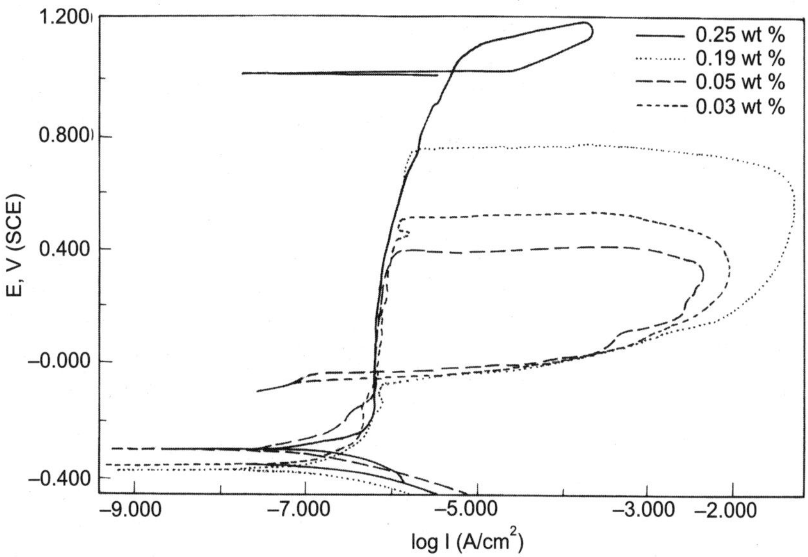

Figure 7 Polarization curves of 904L weld-clad with various nitrogen content in 3.5 wt% NaCl solution show the variation in E_{pit} with N content of the weld-clad (Raja et al 1998)

stainless steels and aluminum alloys. The severity of pitting depends on the concentration of the chloride ions, pH and temperature. Chlorides bring down the pitting potential given by the relation (Leckie and Uhlig 1966)

$$E_{pit} = A - B \log Cl^-$$

Where, A and B are constants. Species such as dissolved O_2, noble metal cations (Cu^{2+}, Hg_2^{2+}, Ag^+) and ions having high equilibrium potentials (Fe^{3+}) can raise E_{corr} of the passive metal above E_{pit} and thereby promote severe pitting in presence of chloride ions. Ions such as SO_4^{2-}, OH^-, and ClO_3^-, CrO_4^{2-} and NO_3^- are found to reduce the pitting tendency of stainless steels.

Environments with high Cl^- and H^+ concentrations demand application of high pitting resistant material. Figure 10 summarises the applicability of different alloys with varying chloride and pH levels. It should also be noted that stagnant conditions of the environment favor pitting, as pit environment is associated with high Cl^- and H^+ levels. Increasing the velocity of the environment on the other hand lowers the pitting tendency of an alloy. For example, the report by International Nickel Inc., (1963) shows that no pits appear on types 316 and 310 SSs weld plates after exposure to seawater flowing at 1.2 m/s, while the alloys were found to suffer severe pitting under stagnant condition.

Metals seem to exhibit strong temperature dependency towards pitting in a given environment. The rise in temperature of the environment is found to accelerate pitting tendency of the alloy. In the case of stainless steels, correlation seems to exist between the pitting resistance of a metal and the critical pitting temperature, (CPT). Similar to PREN, higher CPT values indicate higher pitting resistance of the alloy. So the environments having high pitting tendency require application of stainless steels with high PREN / CPT.

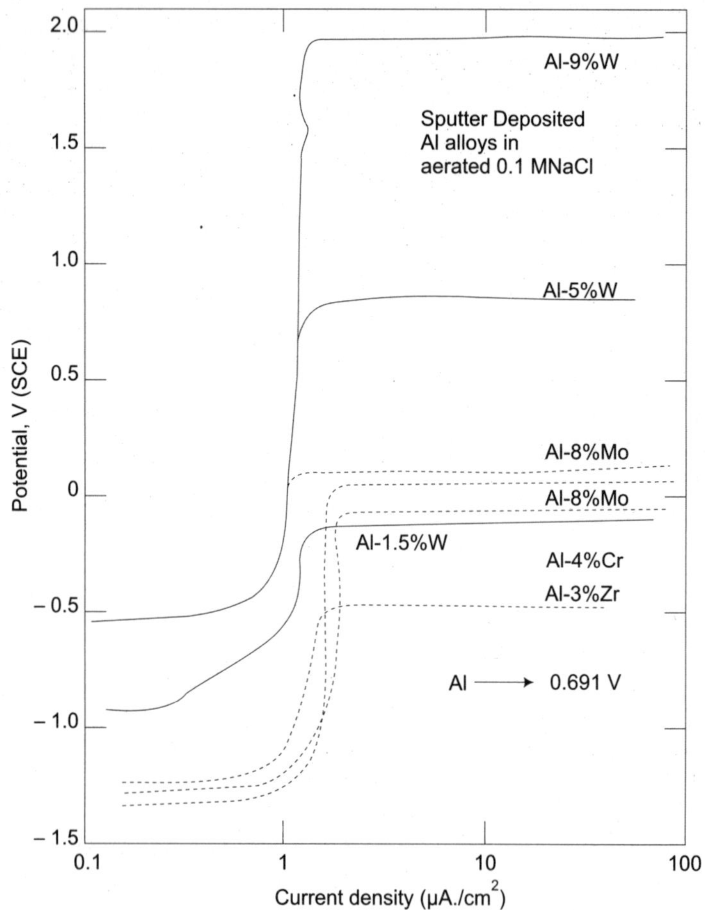

Figure 8 Anodic polarization curves for various sputter-deposited aluminum binary alloys in an aerated aqueous 0.1M NaCl solution. (Davis et al 1991)

Cold working and Surface Condition

Cold work can introduce macro defects such as cracks, roughness and micro defects such as dislocation. All of them can hamper the passivation tendency and hence the pitting resistance of the alloy. Strongly passivating alloys may be less affected than the weakly passivating alloy/environment system. It is worthwhile to point out the fact that cold rolling can induce decohesion between the oxide inclusions and the matrix, when these inclusion have low ductility, and become a cause of enhanced pitting tendency.

Surface condition of stainless steels is an important factor in deciding the pitting tendency of stainless steels. Figure 11 clearly demonstrates that the high pitting resistance of stainless steels can be achieved by a) lowering the surface roughness, b) removing the mill scale and c) passivating the stainless steels (Tuthill and Avery 1992). ASTM designation A380-78 is a recommended practice and is one among several published literature for passivation of stainless steels.

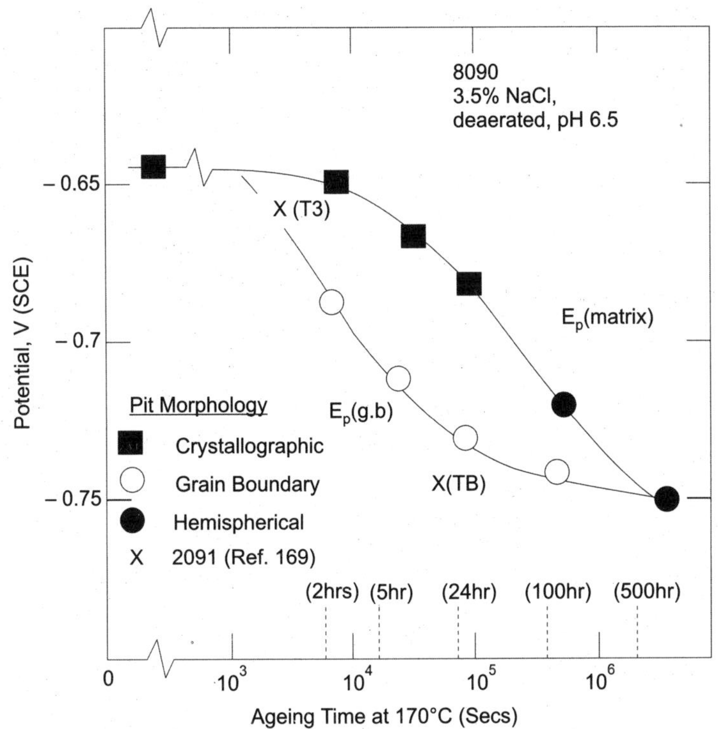

Figure 9 Effect of aging time at 170°C on the grain boundary and matrix pitting potentials for AA 8090 in a de-aerated 3.5 wt% NaCl solution adjusted to 6.5 pH (Ohsaki et al 1990)

Summary

Pitting corrosion is mostly associated with passive alloys and occurs in environments containing aggressive ions. Pitting tendency of an alloy and can be analyzed by means of electrochemical experiments. Metallurgy of the alloy, and environmental conditions greatly influence the severity of pitting. Application of high pitting resistant stainless steels or alloys not susceptible to pitting (such as Cu base alloys,) can prevent the occurrence of pitting corrosion.

DIFFERENTIAL AERATION ASSISTED CORROSION

This form of corrosion, generally, occurs over a larger area of metal surfaces than pitting corrosion and the former is much more severe than the latter. Differential aeration (crevice) corrosion attack on a weld-clad heat exchanger vessel used for seawater cooling system is shown in Figure 12. The 2507 weld-clad was found to be free from pitting corrosion, while crevice corrosion below the gasket occurred within six months of operation. Industrial components due to either their inherent design or operating conditions can encounter regions of restricted access to aqueous environment leading to premature failures. Failures due to such phenomena are described as either crevice corrosion or under deposit corrosion. Typical crevice corrosion is known to occur on flanges, washers, rolled tube ends, threaded joints,

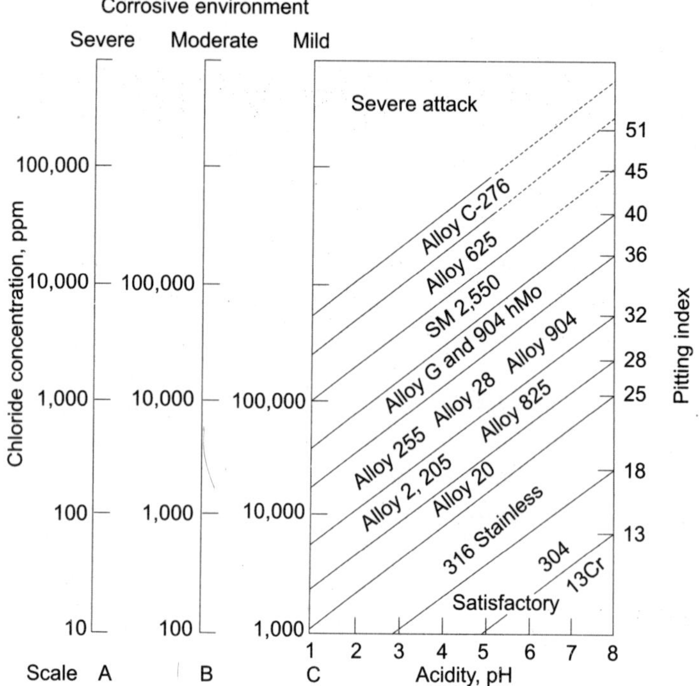

Figure 10 Susceptibility to pitting corrosion with pH and chloride content (Schillmoller and Todd 1987)

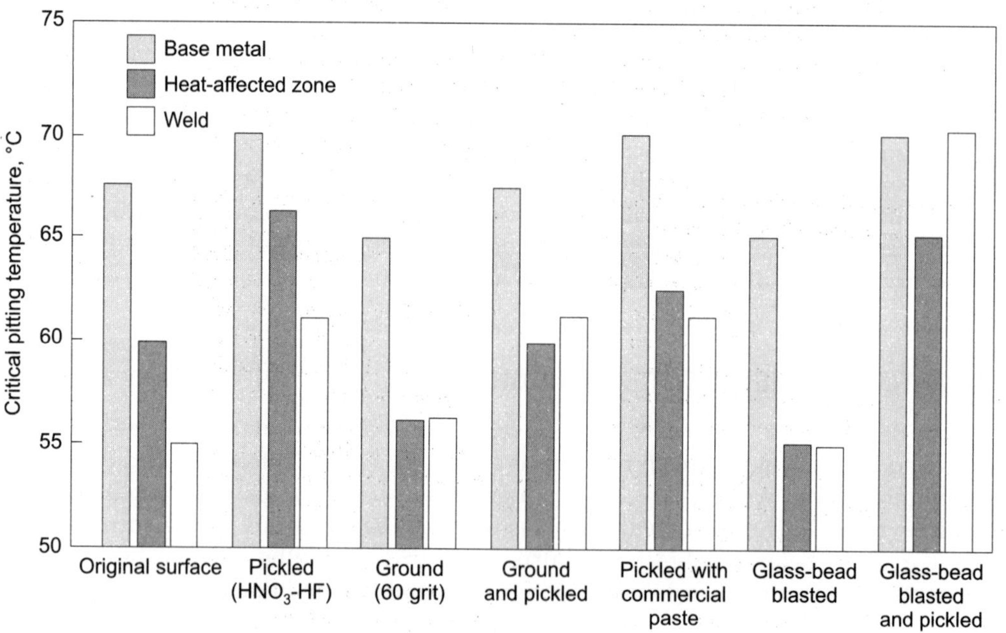

Figure 11 Role of the surface condition on pitting is exemplified in this diagram. Pickling of stainless steel increases the critical pitting temperature in $FeCl_3$ in the base metal, heat-affected zone, and weld areas. Mechanical cleaning treatments that are performed without a subsequent pickling treatment decrease the critical pitting temperature (Tuthill and Avery 1992)

Figure 12 Photograph of a part of heat-exchanger shell affected by crevice corrosion. The carbon steel shell was weld-cladded with SAF 2507 alloy. Note, only the gasket area of the shell suffered from attack; shell side was exposed to seawater

riveted areas, o-rings, gaskets, lap joints, weld fusion zones with inadequate fusion etc. On the other hand, under deposit corrosion is a cause for concern in fouled areas where deposits and sediments are found, as well as within tubercles formed on the metallic surfaces. Though the factors causing these failures are different, they share a common corrosion mechanism of differential aeration. Hence they are treated together.

Mechanism

A metal having an ability to passivate in an environment exhibits a typical anodic polarization curve as shown by the Figure 13. The metal can spontaneously passivate provided the cathodic polarization curve intercepts the anodic curve only with in the passive region and not anywhere else. This enables the metal to exhibit corrosion potential (E_{corr}) in the passive region. In neutral aqueous environments, the predominant cathodic reaction is O_2 reduction process ($O_2 + 4H_2O + 4e^- = 4\ OH^-$, $E° = +0.441V$), in spite of its very low solubility (6 ppm) in water. So, for the oxygen reduction reaction to spontaneously passivate the metal, its limiting current density (i_L) has to be higher than the critical current density (i_{crit}). In various industrial environments, stainless steels barely meet such a requirement. So, when the metal surface has restricted access to oxygen the limiting current becomes so low that metals cannot spontaneously passivate, as explained by the polarization diagram. Given these conditions, a metal can lose its passivity and suffer severe corrosion when it comes in contact with another solid phase, leading to crevice / under deposit corrosion as explained below.

The mechanism of crevice corrosion consists of two stages, namely, initiation and propagation. The electrochemical changes associated with these two processes are explained by means of the Figure 14. Outside the crevice the oxygen can migrate to the metal surface through convection as well as diffusion, where as within the crevice it can migrate only through diffusion. Further the oxygen diffusion (flux) to the metal surface is very low owing to

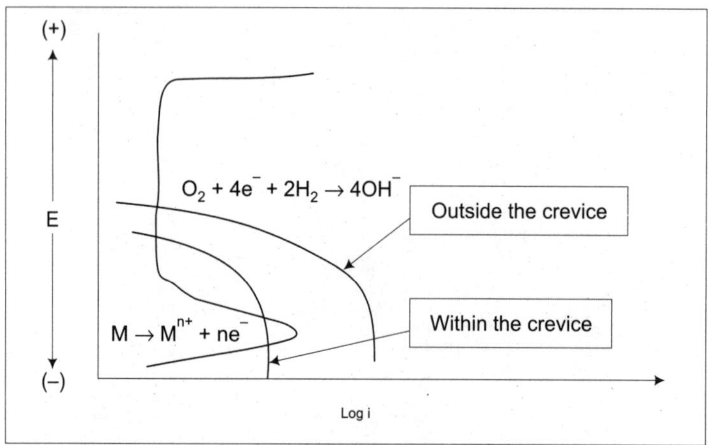

Figure 13 Schematic of anodic and cathodic polarization kinetics that are possible on a passive metal such as stainless steel. Conditions for spontaneous passivity and the lack of it are shown in the figure. See figure 14 for further explanation

very low solubility of oxygen in water (6 ppm). Hence, oxygen within crevice depletes to very low levels within a short span of time. As a consequence, as illustrated by schematic diagram (Figure 14a), the corrosion potential of the passive metal, say –0.2 V, shifts to a more negative potential that lies in the active region of the anodic curve (Figure 13), say –0.8 V. This makes the passive film within the crevice unstable. Under this condition, the metal not only dissolves at a high rate, but also leads to acid generation as given by equation below. The generation of either the metal ion or the H^+ ions upsets the electroneutrality of the occluded environment. Consequently more anions migrate to the occluded area. If halides are present in the environment, they reach occluded area and damage not only the existing passive film, but also prevent the active metal from repassivation. The above processes can be represented by the following equations

$$M \rightarrow M^{n+} + ne^- \quad \text{(accelerated anodic dissolution)}$$

$$M^{n+} + nH_2O \rightarrow M(OH)_n + nH^+ \quad \text{(acid generation)}$$

$$nH^+ + nCl^- \rightarrow nHCl \quad \text{(chloride accumulation)}$$

At this stage, the growth of the crevice resembles that of a growing pit in terms of its electrochemistry and becomes auto catalytic in its growth. The potential drop between the crevice and its mouth keeps the crevice in the active state.

Factors Affecting Differential Aeration Corrosion

Electrochemical

The polarization curves outlined in Figure 13 brings out the fact that crevice corrosion is possible only on those metals which possess a strong negative reaction order with potential (reduction in corrosion rate with rise in potential) as far as their anodic dissolution kinetics are concerned. Passive systems are one such case. In addition, the figure shows that for the metal to

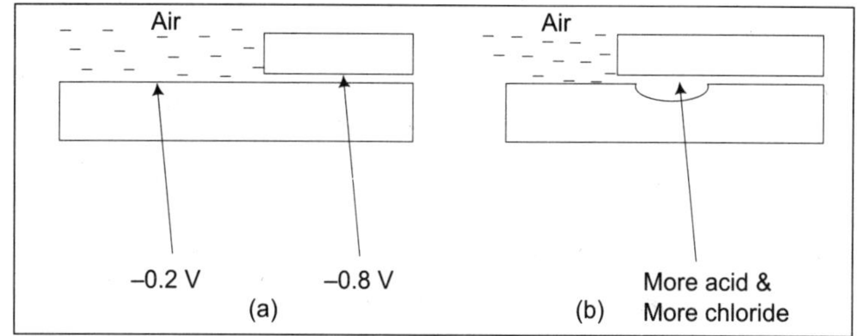

Figure 14 Schematic diagram outlines the chemical and electrochemical changes associated with crevice formation and subsequent corrosion. (a) Active potential, and (b) high chloride and acid levels cause sustained crevice attack

exhibit high resistance to crevice corrosion it must possess very low critical current density, i_{crit}, so that it is able to passivate within the occluded area, even if (a) O_2 content of the occluded cell is very low and/or (b) the occluded cell is slightly polarized anodically. It is well known that Mo addition to stainless steels enhances crevice corrosion resistance of this alloy. This could be possible, due to the fact that Mo lowers i_{crit} of the stainless steels. In addition to this fact, other mechanisms suggest that Mo even modifies crevice chemistry by neutralizing both H^+ and chloride ions – species responsible for crevice growth.

Crevice Gap

As the gap between the mating solid surfaces becomes smaller, convective transport of various species of solution affecting the corrosion, especially O_2, metal ions and chloride ions becomes too small and their migration within the mating surfaces solely depends on diffusion. Furthermore, it helps to prevent dilution in the crevice environment that has high chloride and acid level. A reduction in crevice gap always favors crevice corrosion and more resistant alloys need to be employed under these conditions. Figure 15 illustrating the applicability of various alloys under differing crevice gap conditions aptly brings out the ability of crevice gap to influence crevice corrosion. It is seen that highly alloyed stainless steels are required to prevent crevice corrosion, should the gap between the surfaces become smaller.

With regards to crevice length, unlike the crevice gap, the severity of crevice attack goes through the maxima. That is beyond a point, from the crevice surface, the crevice corrosion tendency gradually decreases.

Environment

Aggressive anions such as chlorides play a significant role in accelerating the crevice attack, without which the crevice corrosion though possible in principle becomes less severe. The importance of chloride content on the choice of alloy for an application is brought out in Figure 16. It should also be emphasized that the reduction in pH of the medium will make the alloy selection further difficult, as it is known to accelerate crevice corrosion.

Temperature exerts a severe influence in promoting crevice corrosion attack. The widely employed AISI type 304 SS suffers from crevice corrosion even at subzero temperature, when

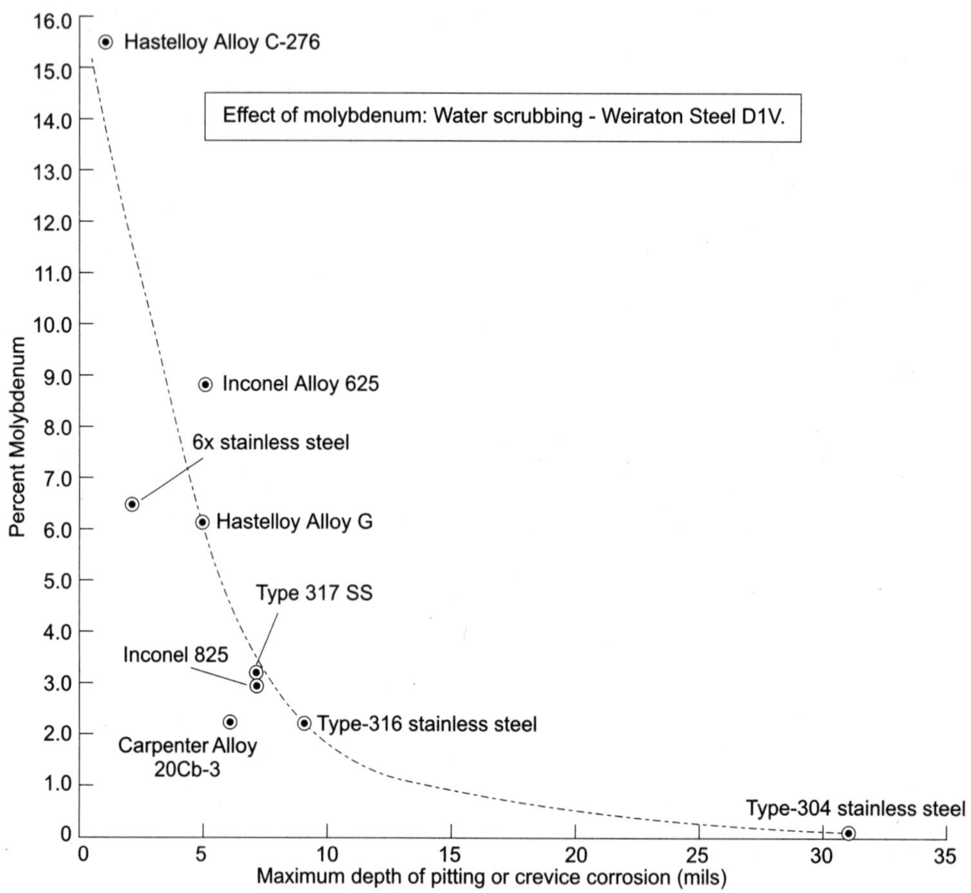

Figure 15 The diagram is a nice illustration of the relation between crevice gap — severity of crevice corrosion and the alloy needed to prevent corrosion under these conditions (Dillon 1995)

exposed to acidified $FeCl_3$ solution. In fact, ASTM-48 test for evaluating crevice corrosion resistance of an alloy involves determining the lowest temperature (called critical crevice temperature, CCT) above which the alloy is found to be susceptible to crevice corrosion. In addition to the chemistry and temperature of the environment, stagnancy and fouling can largely contribute to the crevice corrosion. Corrosion under the fouled areas is termed as under deposit corrosion.

Metallurgy

Differential aeration corrosion is known to occur in passive alloys much the same way pitting corrosion does. Interestingly, the alloys which exhibit high resistance to pitting, also exhibit resistance to differential aeration corrosion. Thus for an alloy to be resistant against crevice corrosion, it can be that the alloy must either exhibit extremely high passivity or no passivity at all. Thus, metals such as Ti, Zr and Ta and their alloys exhibit high crevice corrosion resistance

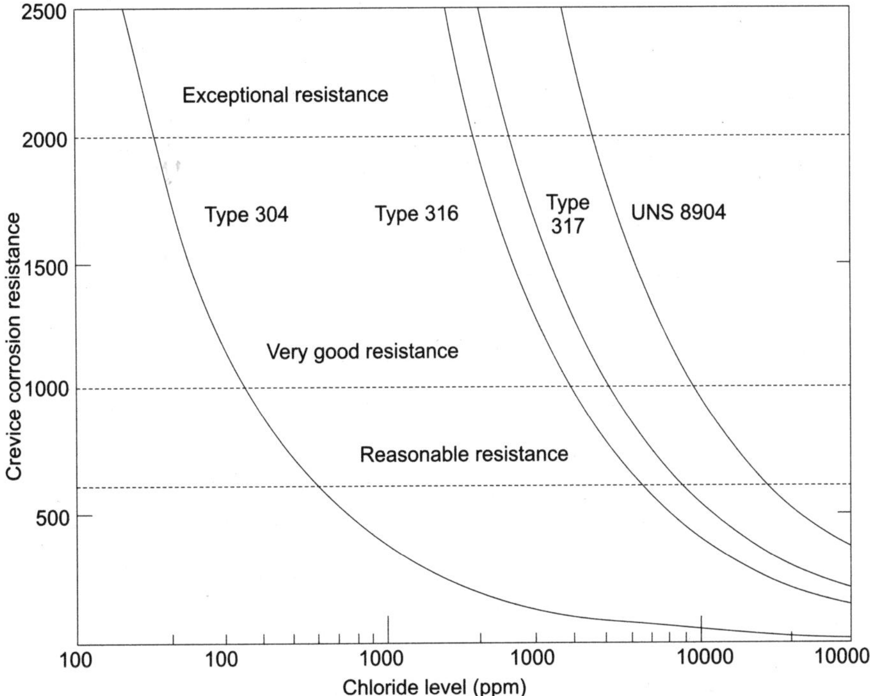

Figure 16 Illustration of the importance of chloride level on crevice corrosion tendency of stainless steels and proper material selection for application (Dillon 1995)

in seawater due to high passivity, while steels do not suffer from such an attack, as they don't passivate. Cu-alloys also are resistant to seawater.

As stainless steels are versatile materials for industrial applications, lot of work has been carried out to develop stainless steels resistant to seawater. A parameter called crevice corrosion index (CCI) is developed to rank the resistance of stainless steels against crevice attack. Steinsmo *et al (1997)* developed an empirical formula for high alloyed stainless steels.

$$CCI = \%\ Cr + 4.1\%\ Mo + 27\%\ N$$

Figure 17 shows the relationship between CCI and critical crevice temperature (CCT) as well as critical pitting index (CPI) and CPT. Such a relation however may not be applicable if alloys contain undesirable inclusions such as MnS inclusions. These inclusions are not only unstable in the corrosive environments but also create a favorable situation for crevice corrosion, as inclusion/metallic interface seems to suffer from crevice attack.

Design

Fouling and stagnancy of an environment in a vessel/reactor cause under-deposit and waterline corrosion respectively. Both of these, to a large extent, depend on the reactor design. Landrum (1989) describes guidelines for preventing these two forms of differential aeration corrosion. Any design that could avoid dead ends and provide good drainage points can minimize crevice corrosion. Wherever possible, joints that encourage crevice formation need to

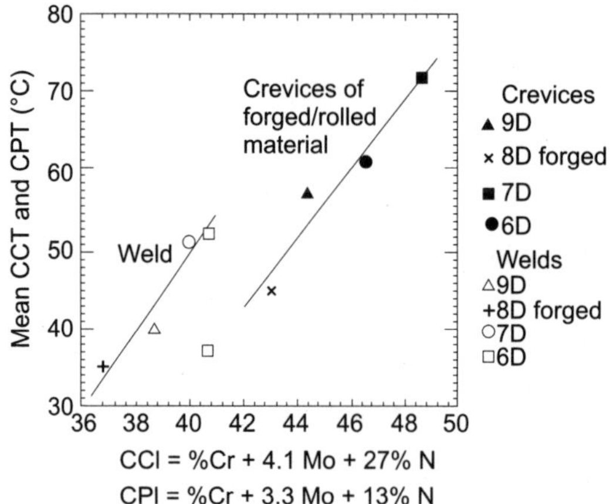

Figure 17 CCT and CPT (welds) Vs critical crevice and critical pitting indices respectively for duplex rolled/forged material (Steinsmo et al 1997)

be replaced with weld joints. It should also be emphasized that weld should be of high quality so that this by itself is not a source of crevice corrosion

Summary

Crevice corrosion, under-deposit corrosion and waterline corrosion share a common mechanism wherein differential aeration triggers the corrosion process. Crevice corrosion is accelerated by chloride ions. Selection of alloys and modification in the reactor design can lower the crevice corrosion tendency of industrial components.

INTERGRANULAR CORROSION

Engineering alloys are predominantly used in polycrystalline form. Grain boundaries between individual grains of the alloys exhibit high crystallographic disorder. They should, as a consequence, exhibit high thermodynamic tendency to corrosion. Further, the high energy associated with the grain boundary has been the cause of chemical segregation and preferential precipitation of phases. The above two factors indeed promote far higher intergranular corrosion than high (thermodynamic) energy associated with the grain boundaries. Hence, intergranular corrosion (IGC) is treated from the latter two aspects namely preferential precipitation and chemical segregation. Among all the engineering alloys, IGC of stainless steels is of utmost importance due to its extensive industrial applications and they are prone to this type of attack and hence more emphasis is given to them than other alloy systems. Cihal (1984) has treated this subject in detail in the monograph on intergranular corrosion of steels and alloys.

Stainless Steels

Stainless steels are known to suffer from intergranular corrosion after they are exposed to high temperatures. The critical temperature over which a stainless steel becomes sensitive to intergranular corrosion depends on the type of stainless steel and its composition. For austenitic stainless steel, this range lies between 450 and 850°C. The reason for the loss in corrosion resistance of stainless steels can be explained as below.

Sensitization Phenomena

The stainless steels, especially, austenitic type comprising 3XX and 2XX series alloys, are produced with carbon levels beyond the latter room temperature solubility limit. So under thermodynamically equilibrium conditions, these stainless steels should exhibit both austenite and chromium carbide ($Cr_{23}C_6$) phases. To prevent the formation of $Cr_{23}C_6$, stainless steels are solution heat treated at 1050°C and quenched to a low temperature. This treatment retains carbon in the soluble form and prevents the formation of $Cr_{23}C_6$. A 450-850°C heat treatment enables chromium carbide formation.

In general any solid-state phase transformation is facilitated by heterogeneous nucleation. So the preferred nucleation sites for the formation $Cr_{23}C_6$, as much as any other phase, is the high-energy grain boundary area. The threat to the grain boundary corrosion resistance arises out of the following fact. In the sensitization temperature range 450-850°C, the diffusivity of Cr in the austenite matrix is much lower than that of C. So, while carbon from the bulk can easily migrate to the grain boundaries the low diffusivity of Cr causes a steep depletion of Cr around the grain boundary (Figure 18). The loss in passivating element, namely, Cr at the microscopic level can affect the corrosion resistance of the grain boundary area. The micrographs of Figures 19 and 20 respectively contrast the unsensitized type 304 SS with that of sensitized type 304 SS. The dark lines -called ditch type attack - of micrograph of Figure 20 is due to intergranular attack. As the other alloy is resistant to IGC, its grain boundary appears very thin and exhibits a step-like structure across the grain boundary.

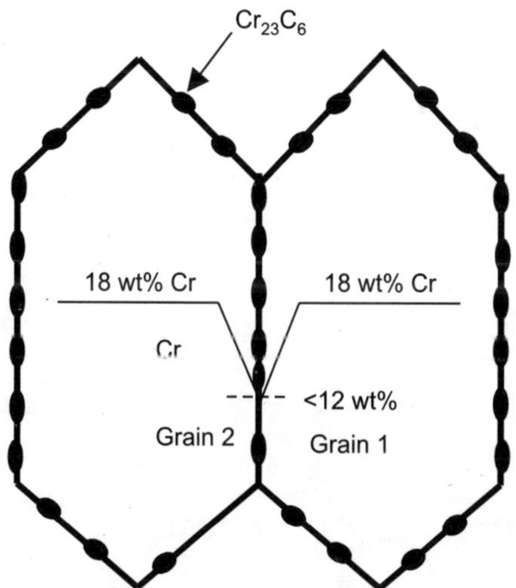

Figure 18 Schematic diagram showing $Cr_{23}C_6$ precipitation and chromium depletion at grain boundaries

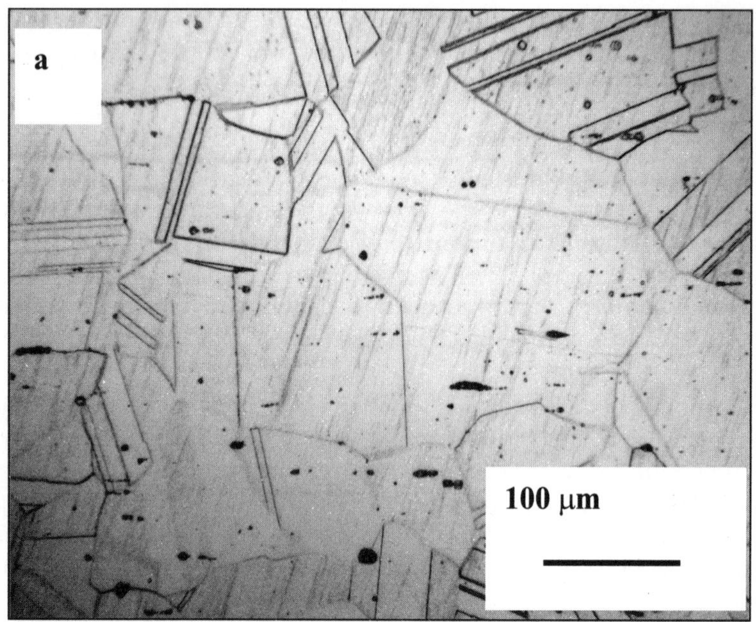

Figure 19 Solutionized type 304 SS subjected to ASTM A262-A test showing step-type grains boundary structure.

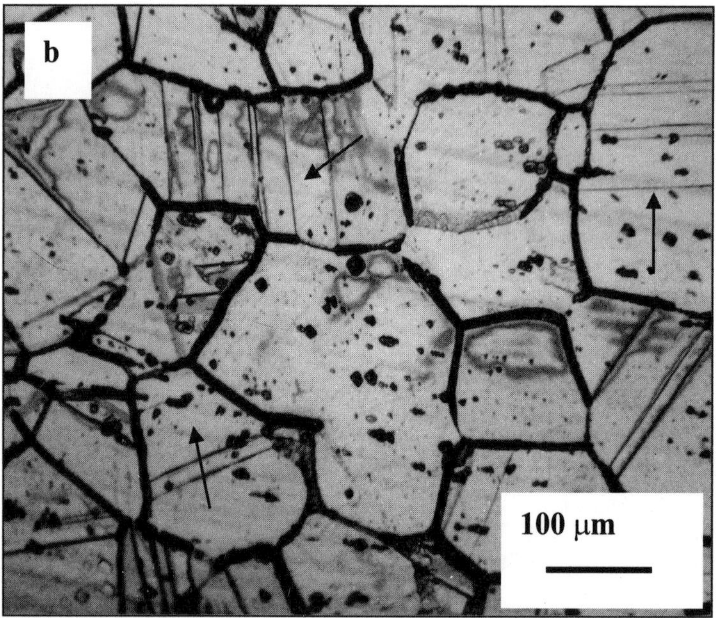

Figure 20 Sensitized type 304 SS subjected to ASTM A262-A test showing ditch-type attack indicative of the IGC susceptibility of the alloy. Arrows highlight some of the twin boundaries

Kinetics of Sensitization

Chromium carbide precipitation, like any other solid state transformation, depends not only on the temperature, but also on the exposure time. Extensive studies have been carried out to determine time-temperature-sensitization (TTS) diagrams. These diagrams are obtained by subjecting stainless steels to various temperatures and time intervals and then examining them for intergranular corrosion attack. Figure 21 exhibits such diagrams showing the relationship between temperature and the time for initiation of IGC. The following characteristics of sensitization can be seen from these diagrams.

1. With temperature, tendency for sensitization goes through a maximum. At low temperatures the growth of $Cr_{23}C_6$ is hampered by sluggish Cr diffusion and at high temperatures the poor nucleation kinetics of $Cr_{23}C_6$ controls the sensitization.
2. The curve moves towards right side, as the carbon content of alloy is lowered. The delay in sensitization kinetics is essentially due to low carbon activity of the alloy.

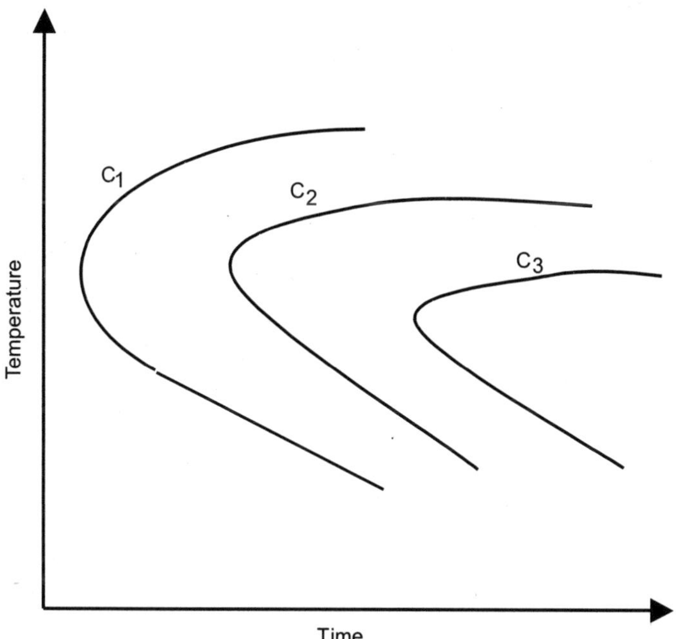

Figure 21 Schematic of time – temperature-sensitization diagram showing the kinetic curves shifting towards right with decreasing carbon content. The carbon content decreases in the order $C_1 > C_2 > C_3$

Factors Controlling Sensitization Kinetics

Alloying elements, cold work, crystal structure and grain boundary orientation have profound influence on sensitization kinetics. The recent review by Dayal et al (2005) brings out these aspects in detail. A brief account of these parameters with a bias on the sensitization related to industrial problem is provided here.

Effect of alloying elements

There are three ways by which sensitization kinetics can be lowered.

1. Lowering the carbon content
2. Adding carbon getters (such as Ti, Nb, Ta) to fix the carbon
3. Lowering carbon activity and controlling the nucleation and growth characteristics of carbides through an alloying element.

Direct relationship between carbon content and sensitization kinetics was discussed in the previous section and the beneficial effect of lowering stainless steel's carbon content was emphasized. Based on this, low carbon (≤ 0.03) and extra low carbon (≤ 0.01) stainless steels are developed which exhibit high resistance to IGC. However, lowering carbon content is met with two disadvantages. Firstly, it raises the cost of stainless steel and secondly, it lowers its tensile strength.

The second approach to lower IGC involves stabilizing the C content of the alloy by adding carbon getters. Ti, Nb and Ta fall under this category of elements. All these elements form carbides at the usual solutionizing temperature of austenitic stainless steels (1050°C), when Cr would be present in solid solution. This treatment is refered to as a stabilizing treatment, when carried out on Ti, Nb and Ta containing stainless steels. On cooling, the carbides, formed during stabilization treatment, remain in the alloy. This class of stainless steels is called stabilized grade stainless steels. Chemical composition of a few of these alloys is given in Table 2. Note that the amount of Ti and Nb and Ta in stainless steel is proportional to carbon content of the alloy. Since, the stabilized stainless steels do not have free carbon; they do not have the ability form carbides on reheating them at 450-850°C. While the main advantage of stabilized stainless steels over low carbon stainless steels arises out of the fact that they exhibit much higher strength than the latter and yet resistant to IGC, this class of stainless steels is prone to knife-line attack. This aspect will be discussed later.

Table 2 Chemical composition (Wt%) of some stainless steels of 300 series

Alloy Type (AISI)	C Max	Si Max	Mn Max	S Max	P Max	Ni Max	Cr Max	N Max	Other elements
304	0.08	1.00	2.00	0.045	0.03	8.5 -10.5	18.0 - 20.0	—	—
304L	0.03	1.00	2.00	0.045	0.03	8.0 -12.0	18.0 - 20.0	0.1 – 0.16	—
316	0.08	1.00	2.00	0.045	0.03	10.0 -14.0	16.0 - 18.0	—	Mo (2.0 – 3.0)
316L	0.03	1.00	2.00	0.045	0.03	10.0 -14.0	16.0 - 18.0	—	Mo (2.0 – 3.0)
321	0.08	1.00	2.00	0.045	0.03	9.0 -12.0	17.0 - 19.0	—	Ti (8 × %C)
347	0.08	1.00	2.00	0.045	0.03	9.0 -13.0	17.0 - 19.0	—	Nb + Ta (10%C)
384	0.08	1.00	2.00	0.045	0.03	15.0 -17.0	17.0 - 19.0	—	—

Extensive research has been directed at understanding the role of various alloying elements on IGC of stainless steels. There are conflicting views on the influence of alloying elements on IGC. What is presented here corresponds to the predominant view expressed by majority of researchers. For more detailed information the recent review on this subject by Dayal et al (2005) can be referred.

Cr, Ni, Mo and N are the major alloying elements of stainless steels, which affect several other forms of corrosion. Hence their effects are presented somewhat in detail. Cr retards the IGC of stainless steels shifting the nose of the TTS diagram towards right. Cr effect towards enhancing IGC resistance of stainless steels is not related to offsetting the loss in overall Cr content of the alloy due to $Cr_{23}C_6$ formation, for Cr content of SS is far in excess of what C can consume to bring down the overall Cr content of the alloy. On the contrary, higher Cr content is believed to enhance its diffusivity to quickly replenish the depleted regions of the grain boundary. In addition, Cr is considered to lower the activity of C in the matrix, which in turn lowers $Cr_{23}C_6$ precipitation kinetics much like low carbon does to stainless steels.

In contrast to Cr, Ni is found to be detrimental to IGC resistance. It decreases the solubility and increases the diffusivity of carbon. In fact, the permissible carbon content of the type 304 SS to avoid IGC increases with Ni content. For example, the permissible C level of 25Cr-25Ni SS is 0.02 wt%, which is more stringent than 0.03 wt% specified for type 304 SS. However, nickel base alloys seem to behave in a different manner. They are less prone to IGC, primarily because of the fact that diffusivity of Cr in nickel base alloys is higher than that in austenitic stainless steels. As a result, Cr levels of grain boundaries of nickel base alloys remain at a much higher level than those found in typical austenitic stainless steels and so are easily passivated.

Mo is found to be beneficial in promoting IGC resistance of stainless steels. The ability of Mo to promote IGC resistance has been attributed to lowered precipitation kinetics of $Cr_{23}C_6$ and enhanced passivation of grain boundary area of Mo containing stainless steels.

N, if present in less than 0.16 wt% significantly retards the sensitization kinetics. It is found to lower the activity of carbon in the matrix, though factors such high passivation tendency of N can be indirectly beneficial to resist the corrosion of grain boundary area of Cr depleted stainless steels. However, if N is added in excess of 0.16 wt%, it becomes detrimental to sensitization resistance. This is because of the fact that Cr forms nitrides such as Cr_2N that cause Cr depletion.

In addition to the above elements, the role of manganese and silicon has also been investigated. Manganese is found to be beneficial in lowering sensitization kinetics of austenitic stainless steels (Raja and Ramkumar 1996) though highly oxidizing environments are found to severely attack grain boundaries of manganese containing stainless steels even in the solutionized condition. Si containing stainless steels seem to suffer IGC in highly oxidizing environment. Si segregates to the grain boundary. Si not only accelerates the $Cr_{23}C_6$ formation but also favours chromium nitride precipitation. Interestingly, silicon is found to lower IGC when it is present above 1 wt% and makes the stainless steels resistant to IGC when its level exceeds 4 wt%. Unlike Si, phosphorus is found to enhance IGC of stainless steels at all concentrations levels, but is similar to Si in the sense that this element also segregates to the grain boundary.

It is instructive to point out that several attempts were made to predict the sensitization kinetics of stainless steels. They were based on relating the efficiency or otherwise of an element vis-à-vis Cr (an element found to resist sensitization) towards sensitization tendency of the alloy. The following empirical relation arrived at by Parvathavarthini and Dayal (2005):

$$Cr^{eff} = Cr + 1.45Mo - 0.19Ni - 100C + 0.13Mn - 0.22Si - 0.51Al - 0.20Co + 0.01Cu + 0.61Ti + 0.34V - 0.22W + 9.2N$$

Taking into consideration only Cr, Ni and C contents Kain et al (1995) showed that type 304 SS and type 304L SS will exhibit IGC resistance to ASTM A262 practice C test if their Cr^{eff} lies greater than 14.0. On the other hand type 304L SS can show resistance to ASTM A262 practice E, even when Cr^{eff} lies greater than 13.5.

Cold Working

Cold working prior to sensitization treatment is found to be detrimental as well as beneficial to IGC resistance of stainless steels depending on the extent of cold working. When stainless steels are subjected to less than 15% cold working sensitization kinetics is advanced, though they suffer less (depth of attack) severe IGC. On the contrary, severe cold working is found to retard the sensitization kinetics. Such opposite effects of cold work on the sensitization kinetics can be understood as follows. Crystalline defects such as dislocations created at low levels of deformation (cold work) enhance the diffusion of both C and Cr, which in turn accelerates the nucleation and growth of $Cr_{23}C_6$ along the grain boundaries. However, because of high Cr diffusivity it can also replenish the depleted grain boundary areas and thus lower the vulnerability of grain boundary to attack. Literature also shows that the low levels of cold working produce favorable size and distribution of grain boundary carbides that promote IGC. At high levels of cold working, the grain interior becomes susceptible to $Cr_{23}C_6$ nucleation, due to high density of dislocation. Hence grain boundary precipitation is minimized.

Role of Crystal Structure: Comparison of IGC of Different Stainless Steels

Ferrite stainless steels by nature contain much lower carbon than other varieties. However, the poor solubility of C in ferrite (bcc) matrix makes them more susceptible to sensitization than the other group of stainless steels. Moreover, unlike austenitic variety, these stainless steels are adversely affected by the presence of N. So sensitization kinetics of ferritic stainless steels are more rapid than austenitic stainless steels. Hence, to prevent sensitization of ferrite stainless steels both C and N content has to be reduced to very low levels ($C + N \leq 0.01$wt%). For high Cr levels, such as those present in super ferritic stainless steels, the acceptable levels of C and N becomes slightly higher. Ferritic stainless steels are alloyed with Ti to stabilize even the low carbon present in the alloy. Types 439 and 444 SS belong to this category.

Duplex stainless steels are resistant to sensitization, though ferritic and austenitic stainless are prone to this problem. The presence of both austenite and ferrite phases in stainless steels has been found to be beneficial due to the following reason. Higher Cr diffusivity in ferrite than in austenite enables preferential growth of carbide within the ferrite phase, with shallow Cr profile ahead of the carbide phase. So the ferrite phase offers more resistance to IGC. On the austenite front, though Cr depletion is severe, IGC is limited to very narrow region, which soon gets eliminated.

As far as **martensitic stainless steels** are concerned, they are prone to carbide formation as they consist of the highest carbon content among all the classes of stainless steels. But the martensitic structure allows a more uniform carbide formation. So the tendency for grain boundary carbide formation, hence IGC, is minimized.

Grain Boundary Orientation

A look at the micrograph of the sensitized 304 SS of Figure 20 shows that the twin boundaries are unaffected, while most of the grain boundaries suffered severe intergranular corrosion. In crystallographic parlance, they are described as special boundaries, having properties markedly different from those of normal grain boundaries. This essentially arises out of the fact that the grain boundary energy of all the grains of a polycrystalline alloy is not the same. An interface boundary becomes special if its Σ values lie below 29. Twin boundaries exhibit Σ3 values. Figure 22 compares the sensitization kinetics of δ-ferrite, high angle grain boundary, incoherent twin boundary and coherent twin boundary. While the reason for high susceptibility of ferrite phase to IGC can be attributed to poor C solubility as has been discussed before, the other three cases are related to the grain boundary energy. The decrease in sensitization kinetics in the order high angle grain boundary > incoherent > twin boundary > coherent twin boundary is related to the decrease in grain boundary energy in the same order.

It is realized that the fraction of special boundaries of an alloy can be engineered through thermomechanical treatment. It is shown that the amount of special boundaries can be raised to above 60% to enhance the IGC resistance. Wasnik (2004) used another concept of the role of grain boundary connectivity on intergranular stress corrosion cracking (IGSCC) tendency of the alloy. In this case, the special boundaries are spatially distributed in a non-random manner in such a way, that a crack progressing along a random boundary gets arrested when it encounters a special boundary. So, should the crack continue to grow, it needs to choose a round-about-path along the random boundary. The path becomes torturous and the alloy becomes resistant to IGSCC. Moreover, when the crack encounters a triple junction of grains, it ceases to grow. While such arrangement of special boundaries can help to lower intergranular stress corrosion cracking, but may not be effective in lowering IGC per se.

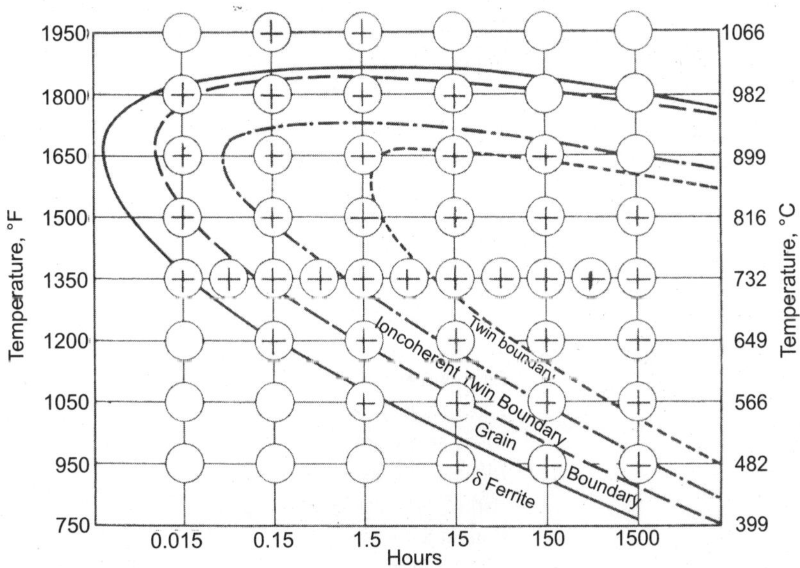

Figure 22 Delineates the difference in sensitization kinetics of different types of grain boundaries (Strickler and Vinckier 1961)

Environmental Effects

Mere sensitization of stainless steel does not cause IGC in all types of environments. Selectivity of a sensitized alloy to an environment to suffer IGC depends on electrochemical characteristics the metal/environment exhibits. The depletion of Cr surrounding the grain boundary can promote accelerated attack provided the corrosion potential lies either in the active or in the transpassive region of the sensitized alloy. Accordingly any environment that exhibits either a strong reducing or strongly oxidizing behavior can cause IGC. In additions, those ions that readily destabilize the passivity of stainless steel are also responsible for IGC. Table 3 provides a few selected environments that cause IGC of sensitized stainless steels.

Weld Decay and Sensitization

Sensitization has been a cause of concern in austenitic stainless steel especially of type 304 SS weldments. Stainless steel weldments suffer intergranular corrosion at heat-affected zone (HAZ) a long strip region of the base alloy located away from the weld fusion zone as well as lying parallel to the latter (Figure 23). HAZ suffers IGC because of sensitization. The fusion zone despite experiencing higher temperature than that experienced by HAZ, does not suffer

Table 3 Some environments that can cause IGC of sensitised stainless steels

Environment
Acetic acid + Silver nitrate
Chromic acid
Sulfite solution
Ferric chloride
Copper sulfate
Nitric acid + $Cr_2O_7^{2-}$
Aluminum sulfate + Water
Sea water
Ammonium chloride
Chromic chloride
Sulfuric acid
Oxalic acid
Acetic acid + Sulfuric acid
Crude oil
Beet Juice
Formic acid
Nitric acid + Chlorides + Fluoride
Nitric acid + Hydrofluoric acid
Hydrofluoric acid + Ferric sulfate
Boiling Sulfurous acid
Water + Starch + SO_2
Hydrogen cyanide
Fatty acids
Sulfurous acid

from IGC. This can be explained with the help of TTS diagram shown in Figure 24. According to the TTS diagram, a stainless steel can be sensitized provided it follows the two conditions namely,

(1) Exposed to sensitization temperature regime.
(2) Held for a time longer than the critical time required for sensitization. This is related to the location of the nose of the curve with respect to the time axis.

Among fusion zone (FZ), HAZ and base metal, only HAZ meets both the criteria and so gets sensitized.

Parameters Controlling Weld Decay

The parameters that can influence weld decay can be classified into those factors affecting cooling rate of HAZ and the other related to alloy. The effect of the latter has already been discussed in relation to sensitization and IGC. In general, weld-cooling rate can be related to weld heat-input, which is governed by welding technique, thickness of the sheet to be welded, welding speed, arc voltage and current and the nature of the weld joints. Among the conventional weld joining techniques arc welding provides the lowest heat input, laser and electron beam welding techniques require even lower heat input to get a sound weld than arc welding. As the heat input becomes smaller, the chances of sensitization of HAZ become minimized. So such welds exhibit high resistance weld decay.

Thick plates require large heat-input. Hence, the sensitivity of the alloy to weld decay can increase with thickness. In order to avoid weld decay in thick plates, multipass welding is practiced. Here the heat is allowed to dissipate after successive weld-passes. Even during multipass welding enough time-gap between each weld passes must be maintained so as to bring the temperature of the weldment to permitted levels.

Weld Design

In fabricating industrial components weld geometry can play a crucial role in either enhancing or stifling heat transfer. It is essential that this factor be taken into account. This happens especially when thick sections are welded to a relatively thinner section of a component. While, thick section dissipates the heat effectively, the thinner section can accumulate heat, due to poor heat transfer, and thereby cause weld decay.

Knifeline Attack

Multipass welding has been suggested as a remedial measure to avoid weld decay of thick stainless steel plates. Such a practice can affect welded components made of stabilized grade stainless steels. As has been pointed out earlier, the temperature of the base metal close to FZ reaches very high value that even carbides of Ti, Nb and Ta tend to dissolve. Because of the fast cooling neither of the carbides, namely, TiC, TaC and NbC reprecipitates. However, if this zone is reheated in the temperature range 450-850°C for sufficiently long time they are prone to $Cr_{23}C_6$ formation. This can occur, when the root of the weld of the thick plate becomes an HAZ for a subsequent weld pass (This happens when multipass welding is adopted) or if a component is subjected to sensitization temperature during its service. Avoiding stabilized stainless steel is only the way to prevent knifeline attack.

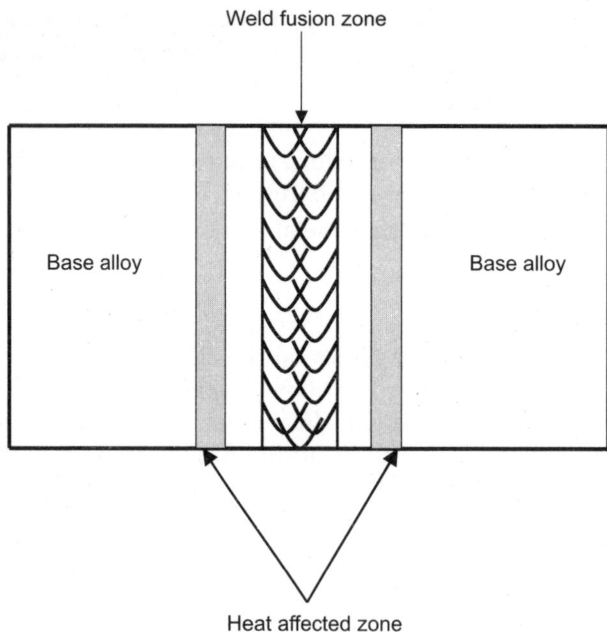

Figure 23 Schematic of weldment showing weld-fusion zone, heat affected zone and base alloy

Intergranular Corrosion of Al-alloys

The difference in the chemistry between the grain boundary region and the adjacent grains provokes microgalvanic cell between these two areas making either of them anodic depending on the alloy composition and heat treatment. Therefore, alloy composition and heat treatment play vital role in selective attack of an alloy. For example, in case of 2xxx series, depletion of copper from the grain boundary areas and the formation of noble $CuAl_2$ precipitates along grain boundary make the grain boundary area vulnerable to selective attack in 2XXX series alloys. On the other hand the precipitation of anodic phases in 5XXX series alloys and Cu-free 7XXX series alloys causes deep attack along the grain boundaries.

What distinguishes Al-alloys from stainless steels in respect to IGC is that the former is prone to exfoliation damage. Exfoliation damage predominantly occurs in mechanically worked Al-alloys, whose grains are elongated with pancake structure. Micrograph of 7010 Al-alloy that suffered a typical exfoliation attack is exemplified in Figure 25 (Bobby 2005). The alloy initially suffers pitting/intergranular attack over a few grains depth. Subsequently, corrosion proceeds parallel to the surface, the attack being confined to the grain boundary area. The corrosion products formed along the grain boundary delaminate the alloy enabling the corrosive environment to have easy access to the subsurface grains for a sustained corrosive attack. Proper thermal treatment can inhibit exfoliation damage in Al-alloys.

Summary

Intergranular corrosion refers to the preferential attack of an environment along the grain boundaries. An alloy can suffer from such an attack if the grain boundary chemistry becomes

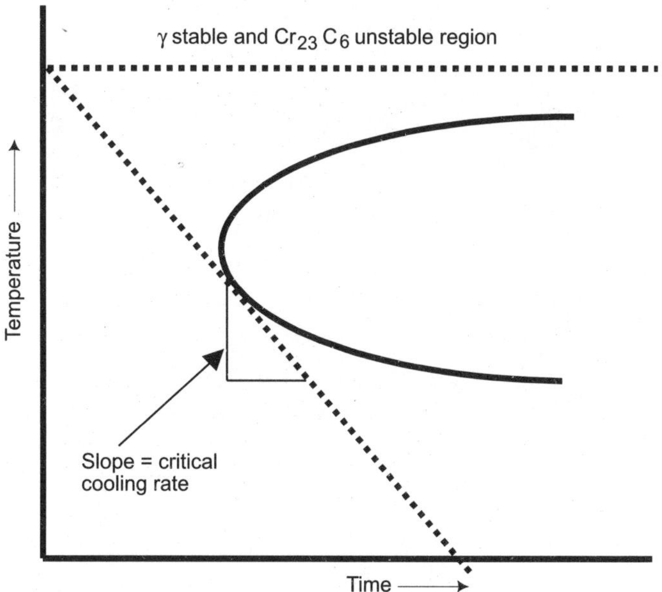

Figure 24 Schematic of TTS diagram illustrates the critical cooling rate and the maximum allowable time for a stainless steel to remain in the sensitization regime

unfavourable for corrosion resistance. In stainless steels (such as type 304 SS), such an effect (called sensitization) is brought out by thermal treatment and consequent Cr depletion along the grain boundary. As welding leads to such an effect along the heat affected zone of the weldment, this zone suffers intergranular attack called weld decay. As the process of sensitization is well understood, it is possible to control sensitization and weld decay. Selection of low carbon/stabilised stainless steels, proper welding techniques help to lower weld decay. The other industrially important alloy, which is significantly affected by IGC, is based on aluminum. The Al- alloys suffer a special type of IGC attack called exfoliation damage, due to pancake grain structure and unfavourable grain boundary chemistry. Proper heat treatment can help to modify grainboundary chemistry and so enhances the exfoliation resistance.

SELECTIVE DISSOLUTION/ SELECTIVE ATTACK

Alloys made of elements highly differing in electrochemical potentials have a tendency to suffer from the less noble element leaching out of the alloy, on exposure to corrosive environments. As a consequence, they lose their mechanical strength leading to premature failure of components made of such alloys. Premature failure of heat exchanger tubes of Cu-Zn and Cu-Ni alloys and underground cast iron pipelines are common examples of this type of failure. In Cu-Zn and Cu–Ni alloys, Zn and Ni are respectively removed, while in cast iron Fe dissolves selectively. Selective leaching also pertains to the process involving selective removal of active elements from phases of metallic alloys. For example, in aluminum alloys, especially alloyed with Cu, elements such as Al and Mg of a phase can preferentially dissolve leaving behind its Cu constituent.

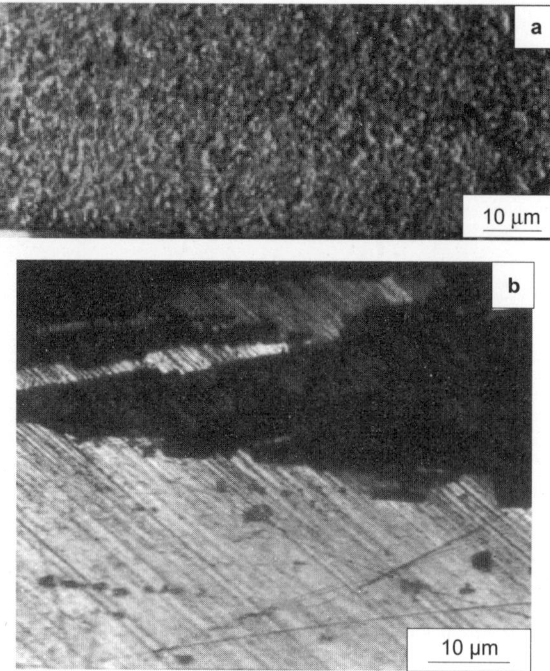

Figure 25 (a) Macro photograph of the exposed surface and (b) micrograph of cross-sectional image of peakaged 7010 Al-alloy subjected to EXCO (ASTM G34) test. The surface roughness seen in (a) and grain boundary attack seen in (b) are due to exfoliation damage (Bobby 2005)

Mechanism

Extensive research has been carried out to understand the mechanism of selective dissolution to possibly explain its characteristics. These mechanisms can be grouped into four types. As a detailed discussion of these mechanisms is beyond the scope of this write-up, they are just highlighted. For a detailed understanding, an exclusive review on this topic by Kaiser (1987) can be referred.

Dissolution and Redeposition

This mechanism involves simultaneous dissolution of both elements of the alloy, followed by redeposition of the noble element through a displacement mechanism. In the case of brass this can be represented as below

$$CuZn \rightarrow Cu^{2+} + Zn^{2+} + 4e \quad \ldots \quad \text{Step I}$$

$$CuZn + Cu^{2+} \rightarrow Cu\ (Cu) + Zn^{2+} \quad \ldots \quad \text{Step II}$$

In the first step, both Cu and Zn dissolve together and in the subsequent step, Cu^{2+} redeposits back on to the surface displacing Zn from the alloy.

This theory does not seem to be applicable to noble alloys such as Cu-Au and Ag-Au, as Au does not undergo dissolution as it only gets oxidized.

Volume Diffusion

According to this mechanism, only the less noble element of the alloy undergoes dissolution. However, this requires high atomic mobility. To account for the high rate of atomic migration to sustain the selective dissolution of the alloy, even at room temperature, divacancies are considered to be injected into the crystal volume below the surface, as a result of selective dissolution. Though the model predicts the occurrence of surface roughness that is noticed in alloys undergoing selective dissolution, it could not explain the high current density associated with selective dissolution of some of the alloys such as Cu-Mn and Cu-Al.

Surface Diffusion

While the less noble element preferentially dissolves, the undissolved noble element on the alloy surface is suggested to migrate to nucleate islands of pure noble metal. As the nuclei coalesce, fresh surfaces of the alloy get exposed forming tunnels and pits. This mechanism also seems to explain the influence of electrolyte, on selective leaching, as they can affect the surface diffusion of the noble element. The inadequacy of the model, however, lies in its suggestion that the dealloyed layer consists only of noble element, which is not true.

Percolation

This model takes into account (i) the presence of active element in the dealloyed layer and (ii) the need for alloying elements beyond a critical value above which selective dissolution is possible (~18% Zn for Cu-Zn alloy and ~14% Al for Cu-Al alloy). While it accepts the surface diffusion concepts of selective dissolution and cluster formation, it further contends that a continuously connected cluster of less noble elements must be present in the atomic lattice of the alloy, in order to provide a continuous active pathway for sustained corrosion to occur.

Characteristics of Selective Dissolution and the Implications

Higher the electrochemical potential difference between the alloying elements, larger is the tendency for selective dissolution. Thus, Cu-Ni alloys exhibit higher resistance to selective dissolution than Cu-Zn alloys. Increase in the content of the noble element decreases the selective dissolution of active element. Thus red brass with 15% Zn does not undergo selective dissolution, while yellow (α) brasses (with about 28% Zn) are prone to selective leading. Furthermore, it is known that $\alpha+\beta$ and β brasses having 40% and above Zn are even more prone to dezincification than yellow brasses.

The alloys susceptible to selective leaching exhibit anodic polarization curves (potential independent current region) similar to that shown by passive alloys (Figure 26). The selective dissolution increases beyond an anodic potential called critical potential (E_c) much the same way the transpassive dissolution of a passive alloy occurs. E_c increases with increase in content of the noble element (Figure 26). On the other hand increasing the oxidizing power of the environment, such as chlorination of the environment, raises E_{corr} beyond E_c and thereby accelerates selective dissolution. In the following sections some of the industrially important alloy systems that suffer selective dissolution are discussed.

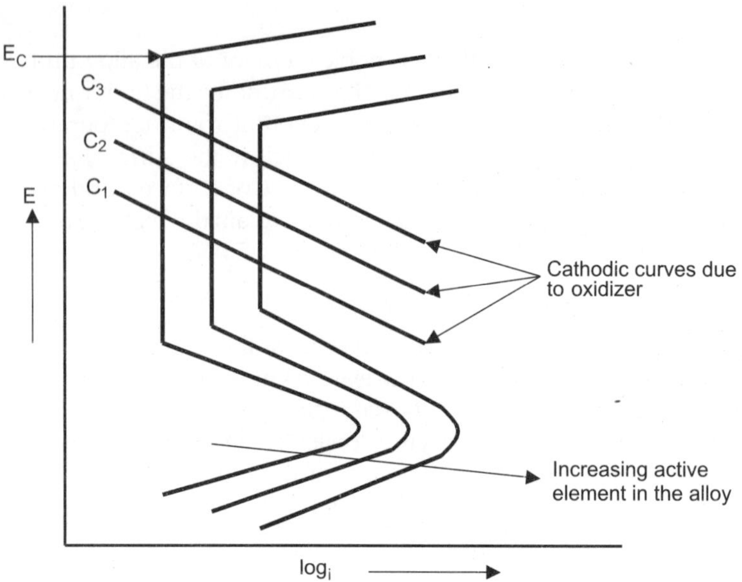

Figure 26 Schematic of polarization curves shown to illustrate the efficiency of (a) increasing active element content of the alloy on critical potential for selective leaching (E_c) and (b) the oxidizer on the selective leaching tendency of the alloy. Note: E_{corr} (intersection of anodic and cathodic Tafel lines) moves towards E_c with the addition of oxidizer, such as chlorine.

Selective Leaching and Industrial Problems

Dezincification

Alpha brass with Cu-30Zn is a commonly used alloy for various industrial applications. The alloy is superior to stainless steels for application in chloride containing environments, as it is not susceptible to pitting corrosion as well as stress corrosion cracking in contrast to the stainless steels. However, alpha brasses suffer from selective leaching of Zn leading to loss of material as well as strength.

Brasses affected by this type of corrosion are characterized by the change in color from yellow of the brass to red, the latter being that of pure Cu. The dezincified surfaces can also exhibit other features, depending on the type of attack the alloy has suffered. The optical micrographs of cross-section of Cu-Zn that suffered dezincification are shown in Figure 27 to illustrate the two types of attack namely (a) uniform or layered type and (b) the plug type inflicting damage to the component. High Zn content of the brass and more acid pH of the environment favour the first type of attack, while, low Zn content and neutral, alkaline and slightly acid environments promote the latter type of attack. In addition chlorine severely affects dezincification of alpha brasses.

Dezincification of alpha brasses (having about 30% Zn) can be controlled by adding 1% Sn. In addition, elements such as arsenic, antimony and phosphorus in the range 0.02-0.06% are added to alpha brasses, as they are very effective in lowering dezincification. Such brasses are called Admiralty brasses. It has been found that these elements are not effective in preventing dezincification of $\alpha + \beta$ and β brasses whose zinc content lies around 40 percent.

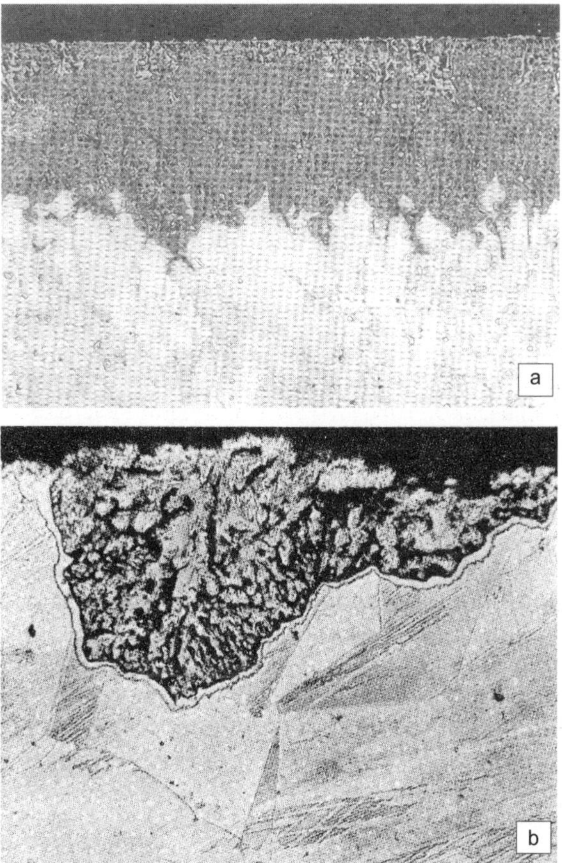

Figure 27 (a) Cross section of leaded brass alloy showing fairly uniform layer-type dezincification after exposure to acidified copper chloride solution for 7 days (Bengough et al 1920) (b) Etched cross section through dezincification plug. (Heidersbach 1982)

Denickelification

When dezincification becomes a severe problem, then cupronickel (Cu-Ni) alloys are chosen for application. Nickel being nobler than zinc, its ability to leach out from the copper alloy is lower than that of zinc. Moreover, Cu-Ni alloys represent better erosion corrosion resistant alloys system than Cu-Zn alloys. Depending on the environmental conditions (high chlorination) even cupronickel alloys can also undergo selective leaching of nickel, the phenomena termed as denickelification. One of the ways to avoid denickelification is to choose low nickel containing cupronickel alloy. For example, alloy with 10% Ni is less prone to denickelification than an alloy that has 30% Ni. However, it should be noted that the latter alloy exhibits better erosion corrosion than the former.

Graphitic Corrosion

This type of corrosion is associated with gray cast iron. Gray cast iron pipelines, due to low cost, were largely in use for transportation of chemicals and water. The underground pipelines were found to suffer from selective dissolution of iron. As shown by the micrograph of Figure 28, gray cast iron consists of graphite flakes embedded in a ferrite matrix. Electrochemically, graphite is more noble than iron and so the former induces the dissolution of the latter. Though, each of the dark graphite flakes, appears to be isolated form the other, they in fact form a graphite network along the three directions of space. In cross-sectional images, they look separated. Such an arrangement of microstructure is very conducive for a sustained galvanic attack within the alloy at microscopic level. Gray cast iron pipelines that have suffered from graphite corrosion lose their strength and ability to withstand impact loading, though they might appear to be intact. Graphitic corrosion is readily caused by soils of low pH and high salt contents.

Graphitic corrosion can be prevented by employing spheroidal graphite iron instead of gray cast iron. These pipelines are popularly known as ductile iron (DI) pipelines. In this type of cast iron graphite is present as isolated spherical particles (Figure 28b). Notably even in such a case graphite particles on the surface exert galvanic corrosion. However, they are detached from the surface, once some amount of iron surrounding the particle dissolves. The galvanic corrosion ceases to operate once the graphite particles are detached from the surface and so DI pipes are resistant to graphitic corrosion.

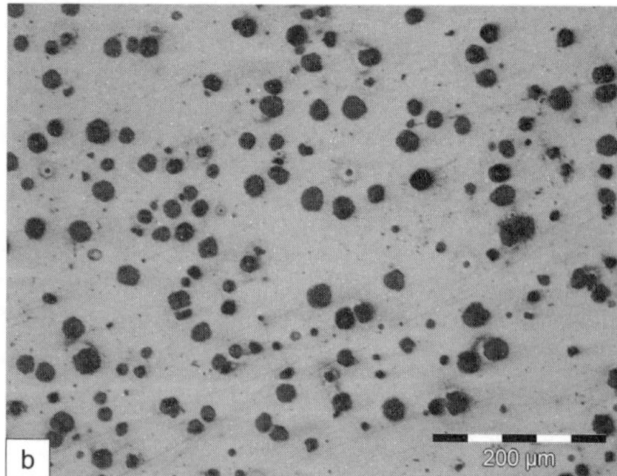

Figure 28 (a) Microstructure of gray cast iron showing the graphite flakes in ferrite matrix, (b) Microstructure of ductile iron showing nodular graphite

Summary

Alloys with both noble and active elements have tendency to undergo selective leaching/ dissolution. Some of the known alloys of this kind are Cu-Zn, Cu-Ni, Au-Cu, Au-Ag, gray cast iron etc. Though the mechanisms of selective leaching is still not well established it is known that the active element leaches out of the alloy. Such a process can bring down the useful mechanical property of the alloy and hence this form of corrosion becomes important. The failure of the alloy by this mechanism is revealed by the change in the color the alloy. Addition of alloying elements wherever possible (Sn addition to alpha brasses), choosing more resistant alloys (Cu-Ni instead of Cu-Zn) and modifying the microstructure of the alloy (flaky to spheroidal graphite in cast iron) are some of the methods to prevent selective leaching of the alloys.

STRESS CORROSION CRACKING

Caustic embrittlement of steel boilers and season cracking of brass cartridges are classic examples of stress corrosion cracking observed as early as 1865 and 1906 respectively. Stress corrosion cracking continues to be an important problem for several industries, as the number of alloy-environment combinations that cause SCC has been steadily increasing. For an historical perspective of this topic the article by Galvele (1999) can be referred.

Stress corrosion cracking (SCC) of metallic components is normally encountered in environments where they hardly suffer uniform corrosion (Figure 29). High resistance to uniform corrosion is usually brought out by the presence of surface films such as passive films and dealloyed/tarnished layers. Film break down has been one of the causes of stress corrosion cracking and hence, this aspect is discussed in this Chapter. Though ceramics and polymers do suffer from such failures, this aspect is not presented here, as it does not fit in to the definition of localized corrosion described earlier in this chapter.

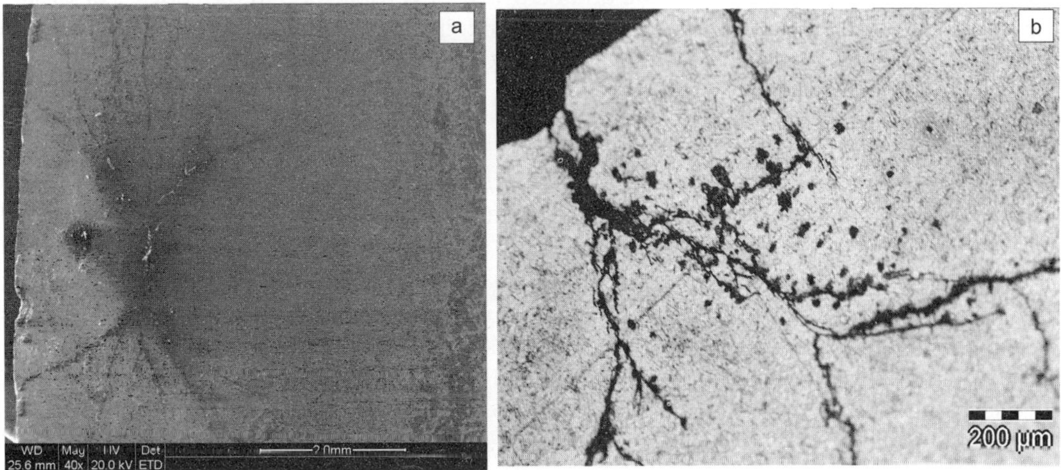

Figure 29 Stress corrosion cracks of type 304 L SS tube of a high-pressure heater of a boiler of a thermal power plant. The cracking occurred a under a deposit (a). Note the SS tube was free from any corrosion except on the cracked region. The cross-sectional image of the cracked area of the tube showed branching cracks, typical of stress corrosion cracking (b).

An alloy is said to have failed by SCC, if the failure is due to conjoint action of tensile stress and environment. The role of environment and stress is one of synergy and must act simultaneously. But what is interesting and also happens to be beneficial from engineering perspective is that only certain alloy environment combinations can induce stress corrosion cracking (Table 4). But the problem associated with SCC is not just the tendency of the alloy to

Table 4 Alloy – environment systems exhibiting SCC

Alloy	Environment
Ordinary steels	NaOH solutions, NaOH – Na_2SiO_2 solutions, calcium, ammonium, sodium nitrate solutions, H_2SO_4-HNO_3 solutions, acidic H_2S solutions and molten Na-Pb alloys
Stainless steels	Acidic chloride solutions, such as $MgCl_2$, NaCl, NaCl-H_2O_2 solutions, H_2S, NaOH-H_2S, sea water and condensing steam from chloride waters
High-nickel alloys	Caustic soda solutions, high purity steam
Aluminum alloys	Aqueous Cl^-, Br^-, and I^- solutions and NaCl-H_2O_2 solutions and sea water
Titanium alloys	Organic liquids, N_2O_4 and aqueous Cl^-, Br^-, and I^- solutions
Magnesium alloys	Aqueous Cl^- solutions, NaCl-$K_2Cr_2O_4$ solutions and distilled water
Zirconium alloys	Aqueous Cl^- solutions, organic liquids and I_2 at 350°C
Gold alloys	$FeCl_3$ solutions and acetic acid-salt solutions
Copper alloys	Ammonia vapors and solutions, amines and water vapors
40 at % Au-Cu	$FeCl_3$, aqua regia

fracture as much as the way it fractures and the stress at which it becomes susceptible to failure. The fracture is almost always brittle in nature and an alloy can fail much below its tensile strength. Hence the failure can be rapid and below the level specified in the codes. The design engineers, therefore, find it difficult to design an engineering component that is expected to suffer SCC. Because of its industrial importance, SCC is widely studied.

Characteristics of SCC

Stress corrosion cracking process constitutes initiation and growth of stress corrosion cracks in an alloy. But the final failure of a component/sample always happens to be due to overload mechanical failure that is not related to SCC.

Stress corrosion cracks are brittle in nature and they exhibit crack branching (Figure 29). Classic cases of caustic embrittlement and season cracking are nothing but failures due to stress corrosion cracking. While SCC surfaces are always brittle in nature, the mechanism of brittle crack growths varies and depends on alloy metallurgy, environment and in some cases the stress levels. Cracks almost travel perpendicular to the stress axis. Typical transgranular stress corrosion cracks (TGSCC) of type 304 SS and intergranular brittle fracture of 7010 Al-alloy observed on the fractured surface lying perpendicular to the stress axis are shown in Figure 30 for illustration. If the cracked samples were to be observed on the surface parallel to the stress axis these cracks will be found to cut across grains and travel along grain boundaries in TGSCC and IGSCC modes of cracking respectively. Fractographic features of stress corrosion cracked alloys are not always restricted to TGSCC or IGSCC. There can be mixed modes of failures.

The type of cracking an alloy might suffer seems to depend on the metallurgy of the alloy and the nature of environment. Carbon steels in several environments are known to undergo intergranular mode of cracking. The solution treated austenitic stainless steels crack transgranularly in chloride medium, but on sensitization the cracking mode changes to intergranular mode. High strength aluminum alloys suffer intergranular mode of cracking in chloride containing environments. Brass can suffer either transgranular or intergranular fracture depending on the pH of the environment.

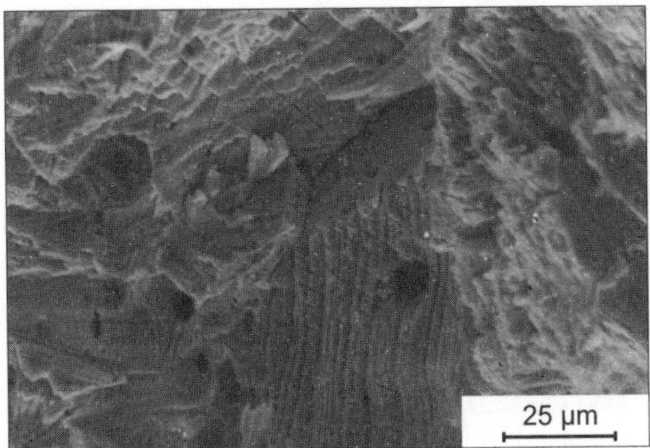

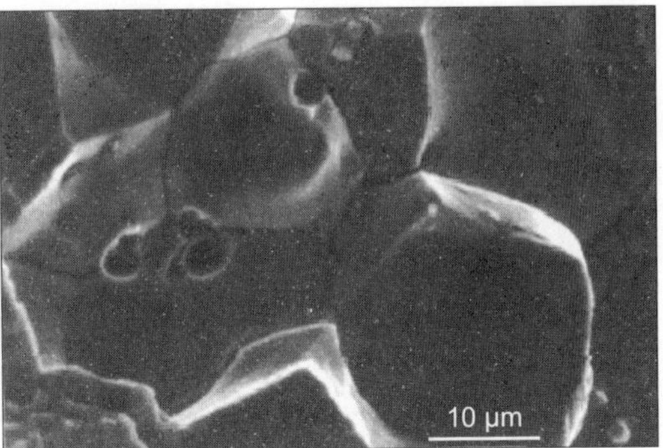

Figure 30 Fractographs of type 316L and peak aged 7010 Al-alloy subjected to SCC tests through U-bend sample showing transgranular and intergranular cracking respectively. Type 316L was subjected to boiling 42%$MgCl_2$, while 7010 Al-alloy was subjected to 3.5% NaCl at room temperature.

Broadly speaking grain boundary chemistry and dislocation structure influence the nature of cracking. Intergranular cracking in carbon steels, sensitized austenitic stainless steels, high strength aluminum alloy are caused due to higher electrochemical reactivity of the grain boundary and/or the grain boundary area than the grains of these alloys. On the other hand, the low stacking fault energy of dislocation and local ordering cause transgranular cracking in austenitic stainless steels and alloys. These aspects will be further discussed under SCC mechanisms.

Effect of SCC on Mechanical Properties

Industrial components fail prematurely, if the conditions viz., environment, alloy metallurgy, temperature and tensile stresses favour stress corrosion cracking. The tensile stress referred here would encompass both applied and residual stresses. The residual stress can arise out of any of the following factors. They are: (a) fabrication processes, such as welding, grinding, machining, rolling etc. and (b) the accumulation of corrosion products within the wedges of a component. The presence of residual stresses lowers the allowable operating stresses.

Stress corrosion cracking can severely affect allowable, as per design, operating stresses and fracture toughness. Notably, failure can occur even as the components experience stress levels much below the yield strength of the alloy. Cases of components failure, even when they were subjected to 20% of yield strength are known. Stress corrosion cracking is a time dependent process. For a given environmental conditions the time to failure depends on the applied stress. The rise in stress lowers the time to failure. One of the characteristics of SCC phenomena is that the alloys undergoing SCC exhibit threshold stress (σ_{th}) below which the alloys become resistant to SCC (Figure 31). Such data is obtained following constant load test. In the same manner growth of stress corrosion cracks depends on the stress intensity factor (K) as has been illustrated by the Figure 32. Examination of published crack growth data shows that the alloys exhibit three crack growth stages. In stage 1, K raises the level of crack growth rate very steeply and in stage 2 the crack growth rate becomes almost independent of K. In stage 3, the rise in K again steeply raises the crack growth rate that eventually leads to failure of the component/specimen. All the alloys exhibit a threshold stress level designated as K_{1SCC}

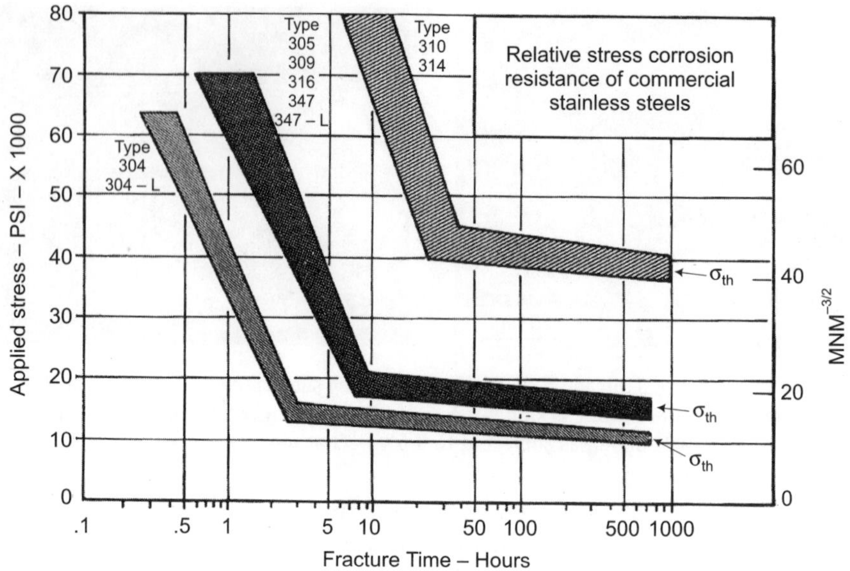

Figure 31 Relative stress corrosion cracking behaviour of major types of austenitic stainless steels in boiling magnesium chloride (Brown 1977)

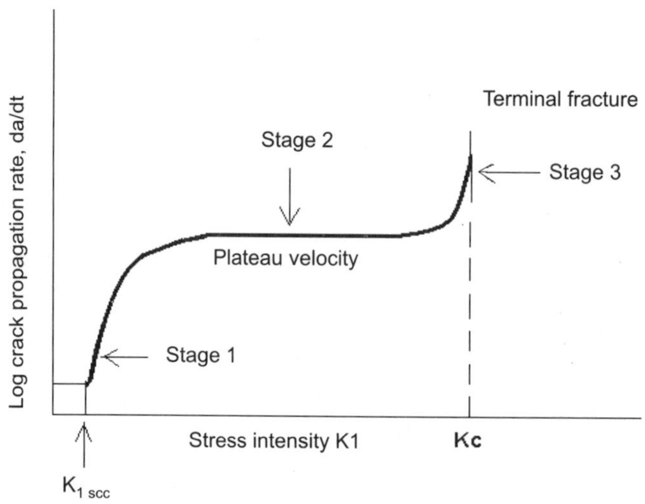

Figure 32 Schematic shows three stages of stress corrosion crack growth of alloys. Details are given in the text

below which crack does not propagate. Stress corrosion cracking resistant alloys are expected to exhibit high K_{1SCC} values. Atlas by McEvily (1990) provides an exhaustive data on both the types of test data for various metals and alloys.

Another indicator of SCC susceptibility of an alloy is the loss in ductility. This value is usually determined from slow strain rate test. Any reduction in ductility (elongation/

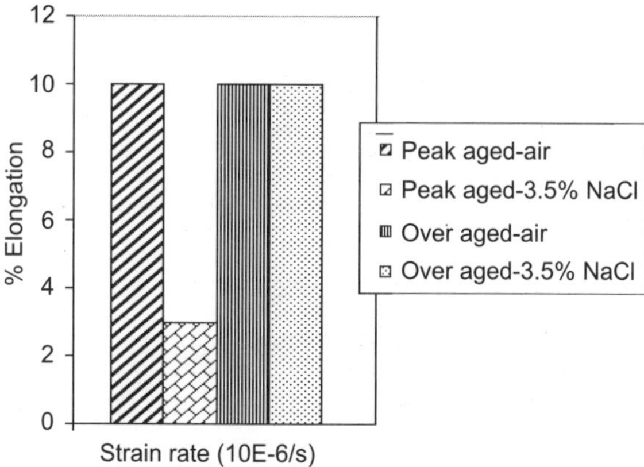

Figure 33 Role of age hardening on the ductility of 7010 Al-alloy in 3.5 wt% NaCl is shown. The results are from slow strain rate test (Bobby et al 2003)

reduction in area) of an alloy in an environment in relation to its ductility obtained in air under the same strain rate conditions is directly related to SCC tendency. As per the Figure 33, the peak aged 7010 alloy suffers severe SCC, as seen by the loss in ductility, when exposed to 3.5 wt% NaCl solution while on overaging it gains SCC resistance.

Factors Affecting SCC

The factors affecting SCC can be broadly classified into two types, those concerning the metal/alloy and the others related to the environment. However, the interaction of the environment with the metal produces another factor, that is electrochemistry. The following issues are the major concern for SCC resistance of an alloy

(a) Nature of environment
(b) Passive film stability
(c) Metallurgy
    (i) Phases
    (ii) Grain boundaries
    (iii) Dislocation
    (iv) Grain size and orientation

These aspects are briefly discussed.

Nature of Environment

It is known that only certain metal/environment combination can cause SCC, though the list of environment causing SCC is expanding. A partial list of metal/environment combinations known to promote SCC has already been given in Table 4. It is also necessary to point out the fact that the definition of environment could mean nature and concentration of the chemical species, conditions of stagnancy or flow and the temperature of the environment. Hence SCC failure avoidance hinges on right choice of material based on proper assessment of industrial environment on one hand and maintaining the environment parameters within in acceptable levels for a chosen material on the other. These aspects are illustrated below.

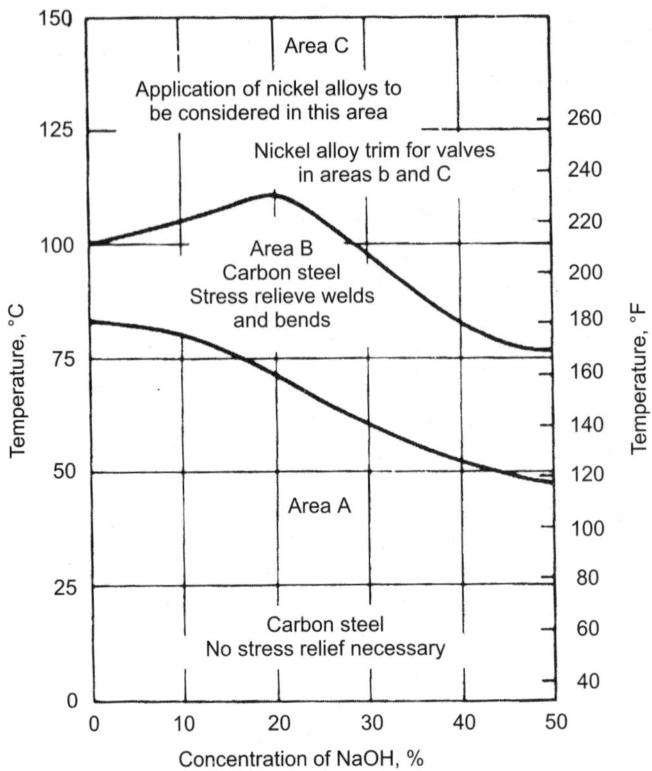

Figure 34 Carbon steels: Temperature and concentration limits for stress-corrosion cracking in NaOH (Corrosion Data Survey 1985).

Carbon steels are known to be susceptible to SCC in caustic environment. But examination of Figure 34 shows that within a certain caustic level and operating temperature the steel can be resistant to SCC even it possesses residual stresses. Should the temperature be raised, the carbon steel component should be stress relieved, without which it will suffer SCC. From the diagram it is also clear that beyond a temperature limit only stainless steels/nickel base alloys can be employed and even stress relieving carbon steel does not help to prevent SCC.

Successful application of stainless steels for various industrial processes again depends on examining the limiting conditions that could cause SCC. Now it is well known that austenitic stainless can be used for cooling water systems only if the cooling water is maintained below 50°C, beyond which they are prone to SCC. Should the temperature exceed either by design or poor operating practice, either Cu-base alloys or duplex stainless steels need to be selected. The successful application of stainless steels to boiler water application, whose operating temperature far exceeds the above limit, relies solely on controlling the O_2 and Cl^- levels of boiler water. Figure 35 brings out the fail-safe environmental regime for type 304 SS.

It is also recognized that, the tendency of cooling water (with the same chemistry) to cause SCC in austenitic stainless tubes depends on the path of cooling water and orientation of the heat exchangers. Heat exchanger tubes exhibit more resistance to SCC when water flows

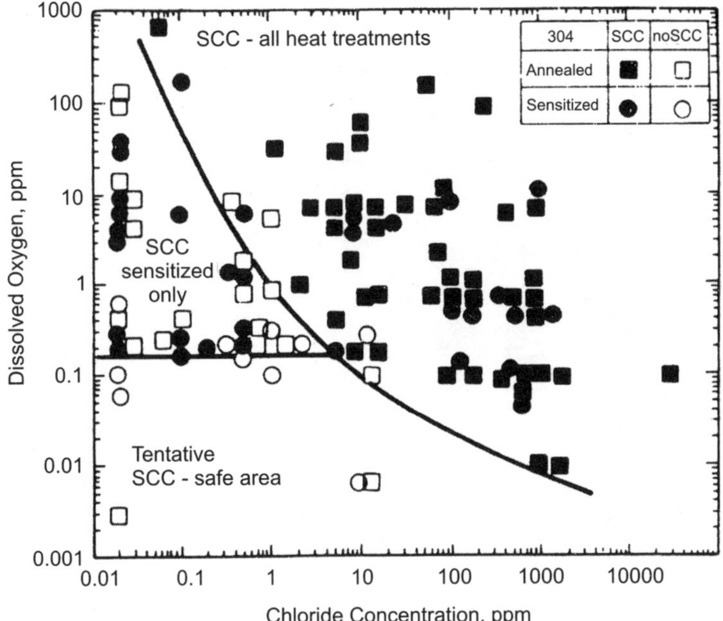

Figure 35 Concentration ranges of dissolved oxygen and chloride that may lead to stress-corrosion cracking of type 304 SS in water at 260 to 300°C. Applied stresses in excess of yield strength and test times in excess of 100 h, or strain rates greater than 10^{-5} /s. (Speidel 1977)

through tube side than through the shell side. High stagnancy in shell side water is responsible for larger incidence of SCC failures of stainless steel tubes of vertical heat exchanger.

Passive Film Stability

Passivity is a thermodynamically unstable state of an alloy, though the corrosion resistance of a metal/alloy is achieved through passive film formation. Potential-pH (Pourbaix) diagrams are means to describe potential-pH regimes of immunity to corrosion, tendency for dissolution and passivity of a metal. SCC is known to occur within the passive regime of an alloy. Hence these diagrams are used to highlight regions of SCC susceptibility of various alloys. Figure 36 illustrates this point. Nitrates, carbonates, caustic and phosphates passivate steel. So they can cause SCC in steel. However, these anions passivate steel in different potential-pH range. Hence, they inflict SCC on steel at different potential pH conditions.

Initiation as well as the propagation of SCC in an alloy very much depend on the passive film stability. This is especially the case, if the SCC mechanism is governed by film rupture process (this will be discussed later). When the potential (corrosion potential/ applied potential) of an alloy lies in the unstable passive region, (such as active-passive) region of the curve (Figure 37), the alloy tends to crack through SCC. Chlorides are known to destabilize the passive film and hence cause SCC. It is also known that pits become active sites for SCC initiation. Recent work is directed towards predicting conditions that transform pits into SCC cracks. Recent data of Turnbull et al (2006.) is shown in Figure 38. The data clearly shows that

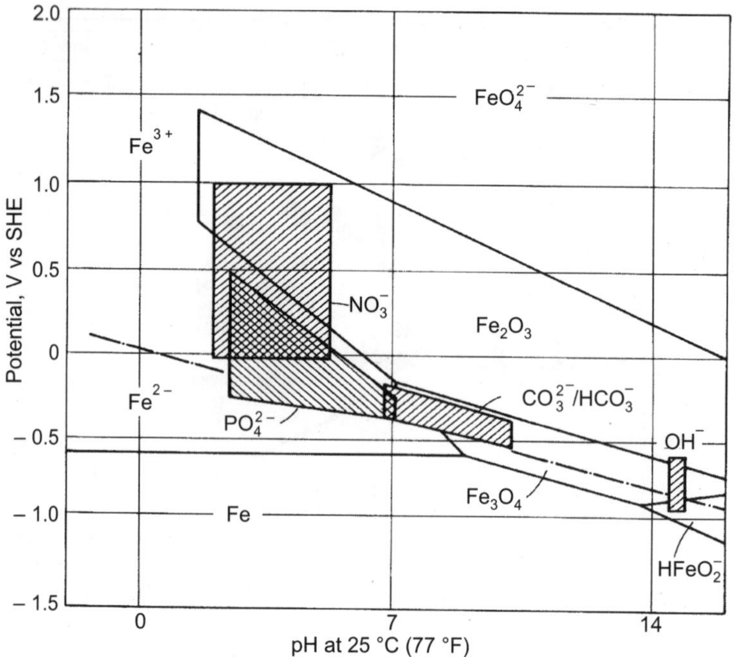

Figure 36 Carbon steels: Severe cracking susceptibility as a function of potential and pH in various environments. (Jones 1987)

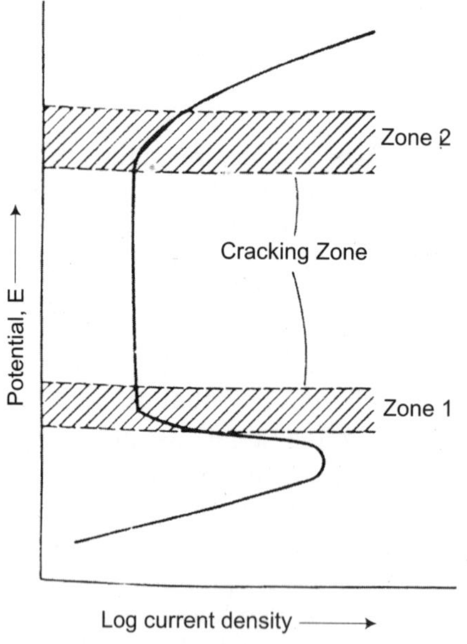

Figure 37 Stress corrosion cracking susceptible areas with respect to anodic polarization curve exhibiting active-passive-transpassive transition

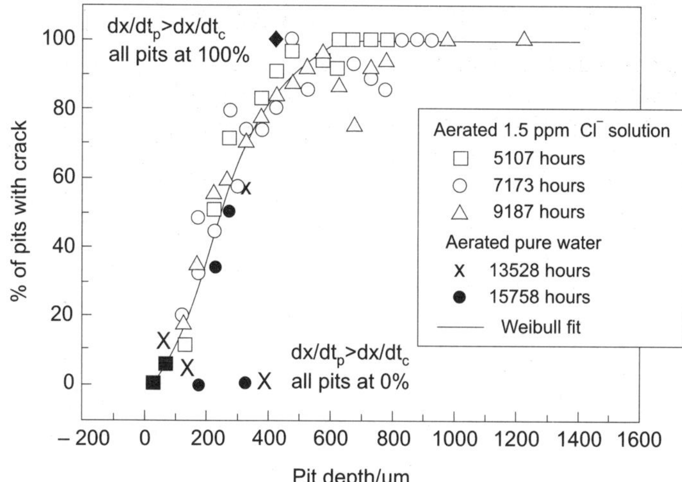

Figure 38 A model to predict the evolution of pitting corrosion and the pit-to-crack transition incorporating statistically distributed input parameters (Turnbull et al. 2006)

the probability of a pit transforming to a crack is directly proportional to the pit depth. So one of the ways to control SCC is to lower the electrode potential to prevent pitting corrosion.

Metallurgy

Alloy chemistry and crystal structure, nature of phases and their distribution, non-equilibrium segregation of alloying elements, type of dislocation, and grain size, morphology and its orientation are some of the most important metallurgical variables which can influence SCC. These are briefly discussed.

Ferritic stainless steels with body centred cubic structure are resistant to SCC, while austenitic stainless steels are susceptible. That crystal structure which favors easy cross-slip for dislocations offers resistance to this type of attack. Addition of Ni brings down the SCC resistance of stainless steels (Figure 39). Similar effect is noticed even with respect to N. Both these elements are known to lower stacking fault energy and resist dislocation to cross-slip. However, at high levels both Ni and N promote cellular dislocations and so become beneficial towards SCC resistance. Notably duplex stainless steels having both ferrite and austenite phases are found to be superior to both austenitic and ferritic stainless steels in chloride environments. SCC of brasses occurs for the similar reason that Zn lowers the stacking fault energy of dislocation and thereby restricts cross-slip. In carbon steel, elements C, P, N influence SCC, by modifying the grain boundary chemistry.

Size, structure and orientation of grains influence SCC tendency of an alloy. There could be several possible reasons for the above influence. For example, lowering grain size can not only raise the yield strength of the alloy but also its SCC resistance. In the case of austenitic stainless steels reduction in grain size can also lower the sensitization, as the level of detrimental impurities along the grain boundary becomes less with decrease in grain size. Similarly low angle grain boundaries lower IGC and thereby promote IGSC resistance (see section on IGC).

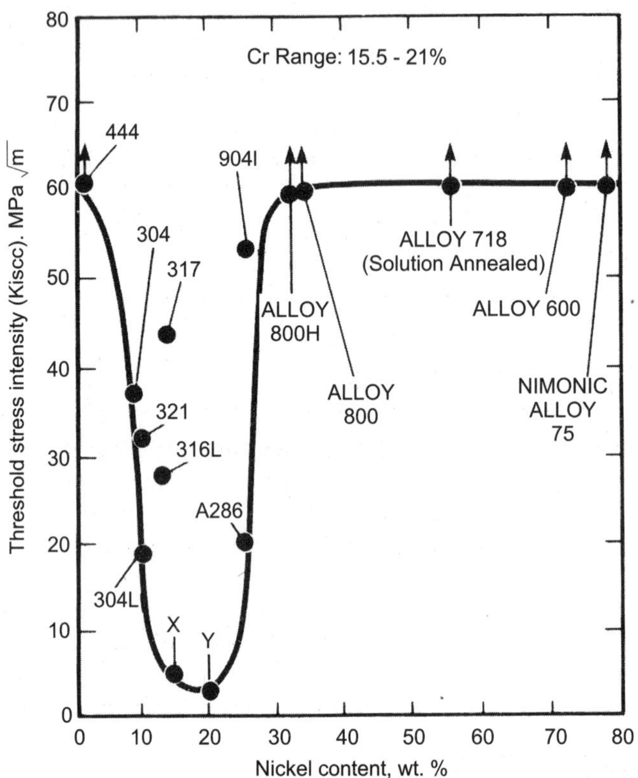

Figure 39 Effect of nickel content on the stress corrosion threshold stress intensity of various alloys in an aerated aqueous 22 wt% NaCl solution at 105°C. (Speidel 1981)

Grain shape and orientation are modified in mechanically worked alloy (say cold/hot rolling / forging). In such cases the SCC resistance of an alloy depends very much on the orientation of the stress axis in relation to forging / rolling direction of the alloy. Such an effect is prominent in Al-alloys.

Heat Treatment

Heat treatment alters the formation and distribution of phases. Alloys or components are heat treated intentionally or otherwise. For example, Al-alloys are intentionally heat treated to obtain different strength levels, where as sensitization of stainless steels due to welding is unintentional. These aspects are illustrated.

Peak aging Al-alloys termed T6 treatment, causes a steep reduction in SCC resistance of high strength Al-alloys of 2XXX and 7XXX series, while overaging recovers the resistance to cracking. Figure 40, exemplifies the above behaviour. Irrespective of the composition, 7XXX series alloys exhibit the lowest K_{1SCC} value and highest crack growth rates on peak aging. Al-alloys suffer IGSCC, as grain boundary areas are vulnerable to preferential dissolution and crack growth. Bobby et al (2003) have found that overage treatment modifies the chemistry of grain boundary precipitates and makes them coarse and less interconnected for crack growth,

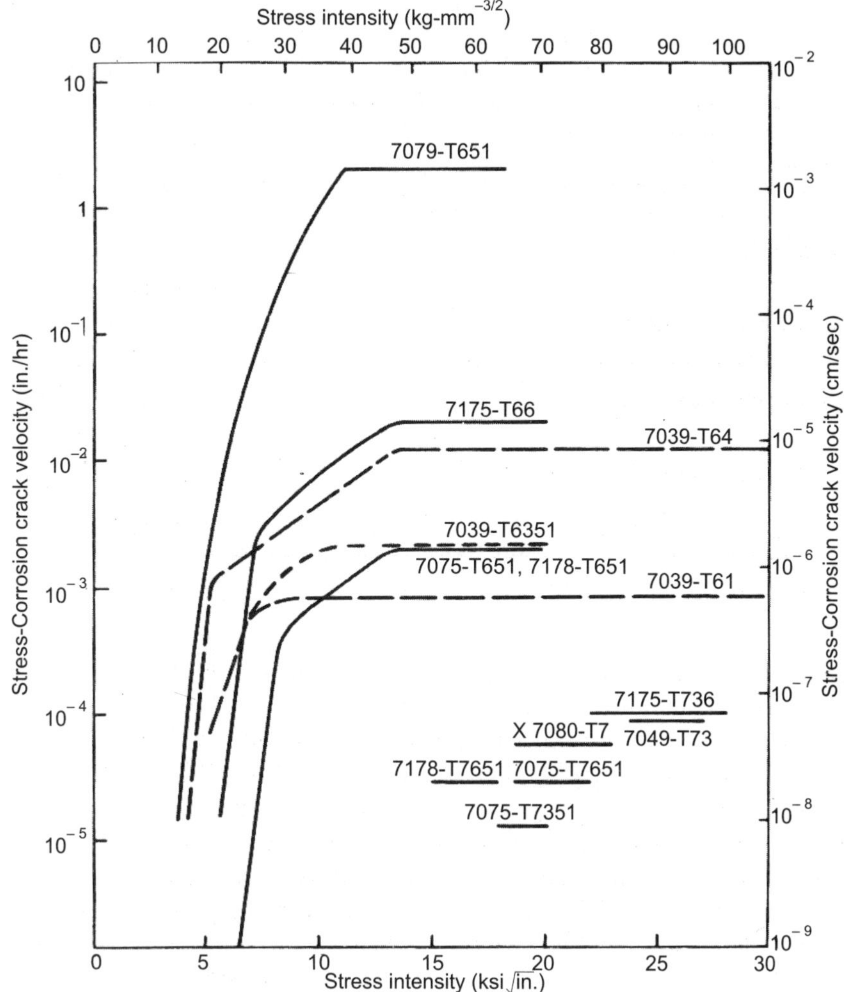

Figure 40 Effect of stress-intensity on stress corrosion cracking velocity for several high strength aluminum alloys immersed in saturated aqueous sodium chloride solutions (Speidel and Hyatt 1972).

(Figure 41) because of which SCC resistance of 7010 Al alloy has significantly increased as illustrated by the Figure 33. Al-alloys are found to suffer SCC when they are subjected to recrystallization. Arresting recrystallization through grain refiners has been found to enhance the SCC resistance of the alloy even in T6 heat treatment condition (Bobby et al 2005).

Stress Corrosion Cracking Mechanisms

Detailed treatment of SCC mechanisms has been made by several authors (Newman 1995, Galvele 1999). A brief account of prominent models is presented here.

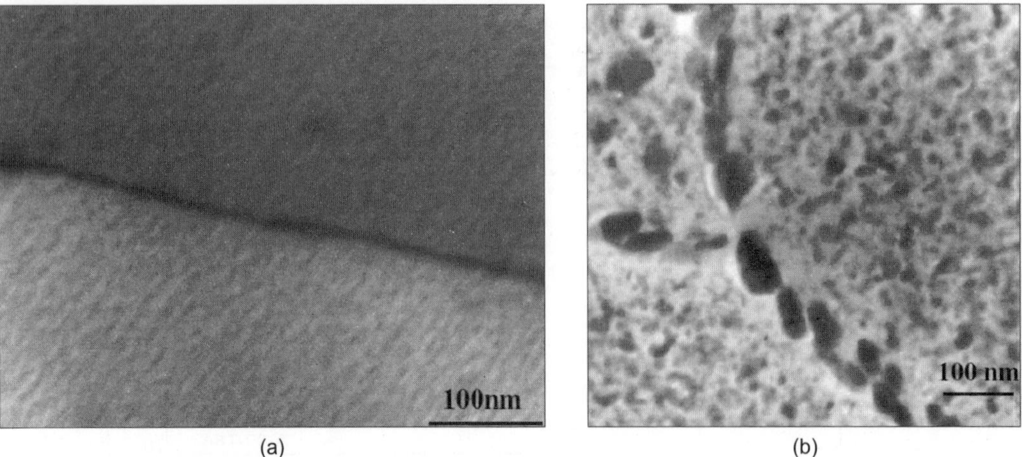

Figure 41 TEM micrographs of over aged 7010 alloy: (a) Peakaged condition: fine precipitates distributed continuously along the grain boundary responsible for IGSCC (b) Overaged coarse precipitates discontinuously located precipitates in the grain boundary becoming obstacle for crack growth (Bobby et al 2003)

Film Rupture or Slip-assisted Dissolution Model

According to this model an emerging dislocation at the crack tip ruptures its passive film creating a bare metal surface. The localized dissolution that follows this event causes crack extension. Subsequently, the crack tip is considered to be repassivated, though some investigators believe that crack tip due to high strain can not be passivated. Crack growth is favored by the passivation of crack-wall and thus localizing the dissolution only at the crack front. The crack growth, according to this model, depends only on metal dissolution, so the crack growth rates are related to the quantity of metal dissolved during slip-assisted dissolution. This mechanism has been found successful in explaining intergranular stress corrosion mechanism observed in carbon steels, Al-alloys and sensitized stainless steels, wherein the grain boundary areas are found to be anodic to the grains. The Figure 42 shows a linear relation between crack growth and the current flow on the strained surfaces of carbon steel and brasses, which crack by IGSCC mode, in different environments. Thus this model seems to be applicable in the case of alloys suffering IGSCC.

Film Induced Cleavage Model

In order to explain discontinuous transgranular cracks of the alloys as well as to account for the observed higher crack growth rates than that can be accounted for by the Faradic current that flows after film rupture, this model has been formulated. This model necessitates that a de-alloyed layer is formed on the surface, due to environment/alloy environment interaction. The high impact caused by the brittle fracture of this layer, due to external stresses, on to the ductile substrate, causes brittle fracture in the latter, to a finite length. Repeated film formation and rupture leads to discontinuous brittle crack growth as observed in many alloy systems such as Cu-Au, Ag-Cu, Cu-Zn, and Cu-Al. This model, however, cannot explain TGSCC of pure Cu in ammonium nitrate solution.

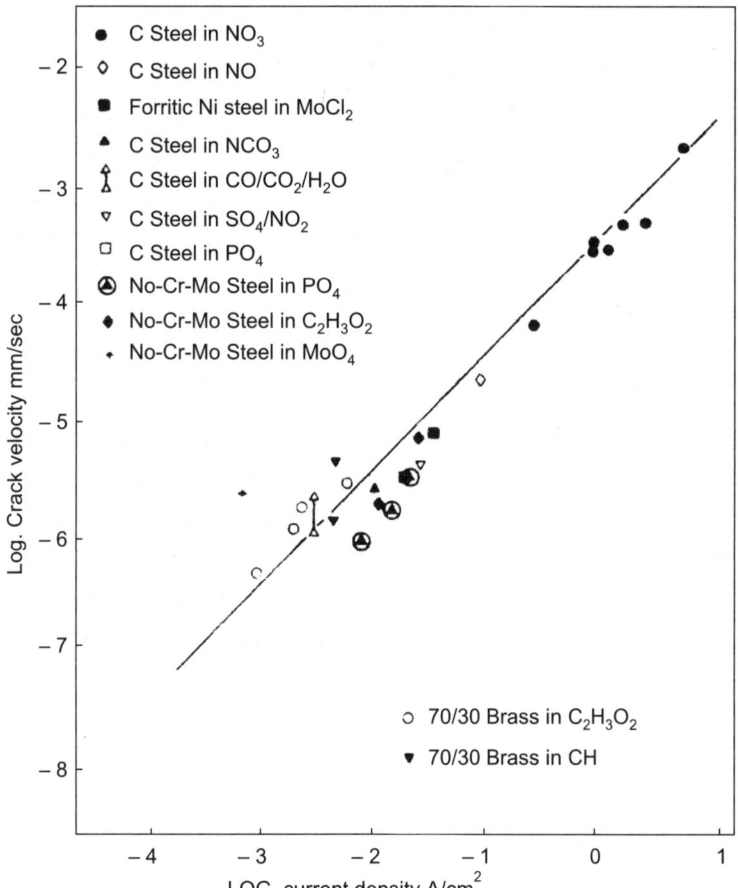

Figure 42 Crack velocities and peak current densities at the same potential for a variety of systems and alloys. (Parkins 1980)

Tarnished Film Rupture Model

Stress corrosion cracks are suggested to grow as a consequence of brittle fracture of surface (oxide) film. The fracture is assisted by tensile stresses. Though the model was originally intended to explain transgranular fracture in alloys, even intergranular stress corrosion cracking has been explained by incorporating the idea that films form along the grain boundary ahead of the crack tip.

Other Models

While the adsorption assisted brittle fracture has been widely used to explain hydrogen embrittlement mechanism, corrosion tunnel mechanism, surface mobility mechanism, adsorption enhanced plasticity mechanism are used to explain special cases of failures.

SUMMARY

Engineering alloys suffer stress corrosion cracking when they meet the criteria of experiencing adequate tensile stress when exposed to specific environments. Under these conditions the alloys always suffer brittle fracture and so SCC becomes an important engineering problem. The brighter aspect of stress corrosion cracking is that a suitable alloy can be found for a given environment that does not cause SCC. Successful development of alloys having high resistance to SCC depends on understanding the interrelation between the environment and the alloy metallurgy.

REFERENCES

Bengough D, Jones R. M, and Pirret R, (1920), J Inst Metals, **23**, p. 65.

Bobby Kannan M., Raja V.S., Raman R., and Mukhopadhyay, (2003), Corrosion, **58**, p. 881.

Bobby Kannan M (2005) "Role of Multi-step Aging and Scandium addition on the Environmental Assisted Cracking and Exfoliation Behavior of 7010 Al Alloy" Doctoral Thesis, Indian Institute of Technology Bombay, India.

Bobby Kannan M., Raja V.S., Mukhopadhyay A.K., and Schmuki, P., (2005), Metall. and Mater. Trans., **36A**, p. 3257.

Bobby Kannan M., Raja V.S., Raman R., and Mukhopadhyay A.K., (2003), Corrosion, **59**, p. 881.

Brown B. F., (1977), Stress Corrosion Cracking Control Measures, National Bureau of Standards Monograph **156**, U. S. Department of Commerce, p. 59.

Bruce C, (2001), "Ferrous Alloys" in Environmental Effects on Engineered Materials, Ed. R. H. Jones, Colorado, p.14.

Cihal V., (1984), Intergranular Corrosion of Steels and Alloys, Materials Science Monographs **18**, Elsevier, New York, NY.

Corrosion Data Survey (1985), Metal Selection, 6th ed. Houston, TX: NACE International, p. 34

Davis G.D., Fitz T.L., Rees B. J, Shaw B.A, and Moshier W.C., (1991), Annual Report MM1- TR90-28c, Martin Marietta Laboratories.

Dayal R.K., Parvathavarthini, N., and Raj B., (2005), International Metals Review, **50**, p. 129.

Dillon C.P., (1995), "Aqueous Corrosion" in Corrosion Resistance of Stainless Steels, Marcel Dekker, New York, p.83.

Falkenberg F., Raja, V.S., and Ahlberg E., (2001), J. Electrochem. Soc., **148**, pp. B132-B137.

Frankel G.S., Stockert L., Hunkeler F., and Boehni M., (1987), Corrosion, **43**, p 429.

Galvele J.R., (1999), Corrosion, **55**, p 723.

Garner A., (1985), Metal Progress, **127**, (No. 5) p. 31.

Heidersbach R. H, (1982), 'Dealloying' in Forms of corrosion: Recognition and prevention, Ed. Dillon C. P, Stillwater, ASTM, 101

Hertzberg R.W., (1983), Deformation and Fracture Mechanics of Engineering Materials, 2nd Ed., John Wiely & Sons, NY, p. 442.

Heusler K. E., and Fischer L., (1976), Wirkstoffe and Korrosion, **27**, p 551.

Holroyd N. J. H., (2001), "Aluminum Alloys" in Environmental Effects on Engineered Materials, Ed. R.H. Jones, Marcel Dekker, New York, p. 173

International Nickel Co., (1963), Corrosion Resistance of the Austenitic Stainless Steels in Marine Environment, NY.

Isaacs H.S., and Kissel G., (1972), J. Electrochemical Soc., **119**, p.1628.

Jones R. H., (1987), ASM Handbook, Vol-13, ASM International, Metals Park, OH, p. 352.

Kain V., Prasad R.C., De P.K., and Gadiyar H.S., (1995), ASTM, J. Test. Eval., **23**, p. 50.

Kaiser H., (1987) "Alloy dissolution" in Corrosion Mechanisms, Ed. F. Mansfeld, Marcel Dekker, New York, p. 85

Landrum R.J., (1989), Fundamentals of designing for corrosion control: A corrosion aid for the designer, NACE, Housten.

Leckie H.P., and Uhlig, H.H., (1966), J. Electrochemical Soc., **113**, p.1262.

McEvily A.J., (1990), Atlas of Stress-Corrosion and Corrosion Fatigue Curves, ASM International, OH.

Newman R.C., (1995), "Stress Corrosion Cracking Mechanism" in Corrosion Mechanisms in Theory and Practice, Ed. P. Marcus and J. Oudar, Marcel Dekker, New York, p. 311.

Ohsaki O, Sato T, Takahashi T, (1990), Aluminium, **66**, p. 565.

Parkins R. N., (1980), Corrosion Science, **20**, p. 147.

Parvathavarthini, N and Dayal, R.K., (2002), J. Nucl. Mater., **305**, p.209

Raja V.S., (2003), Corrosion Reviews, **21**, p.1.

Raja V. S., Angal R. D., and Suresh M., (1993), Corrosion, **49**, p.2-7.

Raja V.S., and Ramkumar A., (1996), Br. Corrosion J., **31**, p.153.

Raja V.S., Varshney S.K., Raman R., and Kulkarni S.D., (1998), Corrosion Science, **40**, p.1627.

Speidel M. O., (1977), Stress Corrosion Cracking of Austenitic Stainless Steels, ARPA Order No. 2616, Ohio State University, Columbus, OH.

Speidel M. O., (1981), Metallurgical Transactions, **12A**, p. 779.

Speidel M. O., and Hyatt M.V., (1972), "Stress corrosion cracking of high strength aluminum alloys", in Advances in Corrosion Science and Technology, p. 115.

Strickler R., and Vinckier A., (1961), Trans. ASM, **54**, p. 362.

Schillmoller C.M., and Todd B., (1987) Opportunities for nickel in the oil and gas market, NIDI Technical Services No. 10013 (January), Nickel Development Institute, Toronto, Canada.

Steinsmo U., Rogne T., Drugli J. M., and Gartland P. O., (1997), Corrosion, **53**, p.26,

Strehblow H. H., (1995), "Mechanism of pitting corrosion" in Corrosion Mechanism in Theory and Practice, Eds. P. Marcus and J. Oudar, Marcel Dekker, New York, p. 201.

Sutula N., and Kurkela M., (1984), "Localized Corrosion Resistance of High-Alloy Stainless Steels and Welds, Stainless Steels and Welds", Stainless Steel, Gothenburg.

Turnbull A., McCartney, L.N., and Zhou, S., (2006), Corrosion Science, **48**, p. 2084.

Tuthill A. H. and Avery R. E., (1992), Advanced Materials & Processes, **142** (No.2), p. 34.

Wasnik D. N., (2004), "Grain boundary nature and localized corrosion in austenitic stainless steels", Doctoral Thesis, Indian Institute of Technology Bombay, India.

CHAPTER 2

High Temperature Corrosion

P.K. Datta, H.L. Du and J.S. Burnell-Gray

Advanced Materials Research Institute, School of Computing, Engineering and Information Sciences, Northumbria University, Newcastle upon Tyne, NE1 8ST, UK

I. INTRODUCTION

Degradation by high temperature corrosion is of major concern in many industrial processes and energy conversion systems - chemical and process plants of various types, furnaces, boilers, engines, incineration plants, gas turbines and in certain types of nuclear plants. Thus high temperature corrosion is a dominant technological problem and its prevention poses a significant challenge.

Five major high temperature corrosion processes may be identified - oxidation, sulphidation, hot corrosion, carburisation/metal dusting and reaction with chlorine and chlorine- bearing gases. Among these five types, the most extensively studied corrosion process is that of **oxidation** and consequently the rationale for the design of materials/alloys/coatings to resist this form of degradation is well developed [1-4]. Sulphidation involving interaction of alloys with sulphur and sulphur-bearing gases in reducing atmosphere is characterised by faster degradation kinetics cf. oxidation [5-12]. Further complication arises from the fact that the use of high temperature alloys with good oxidation resistance does not generally confer good sulphidation resistance and the development of materials/coatings to inhibit this aggressive type of corrosion presents a major challenge. It is worth mentioning that here as well much progress has been achieved in understanding the nature of the degradation processes in sulphur-containing environments thereby allowing a rationale for alloy/coating design to be established [5,13]. Hot corrosion, which embraces accelerated forms of corrosion in oxidising gases contaminated with sulphur and alkali metals, has been well researched [2,14-20]. Understanding mechanisms of hot corrosion - whilst differing in details, say in gas turbine - has been effected by paying careful attention to the alloy/coating compositions, especially the level of alloying additions - Cr, Al and refractory metals such as Mo, Ta, W, V. Two types of degradation processes - carburisation and metal dusting occur due to interaction of alloys with carbon-containing environments. Significant dissolution and diffusion of carbon into the alloy and the formation of carbides within alloys characterise the carburisation degradation and the metal dusting degradation, favoured by low temperature usually characterised by a localised form of attack in the form of pits, the corrosion product consisting of graphite with metal and metal carbides. The environmental conditions and alloy compositions to influence these degradation processes are known and some degree of control is possible [21-25]. Further work to establish the mechanisms of these types of degradation processes is in progress. In recent years, reactions with chlorine and chlorine-containing environments have been studied [26-33] fairly extensively establishing that chloride contamination can cause scale spallation and metal loss by volatile chloride formation. Consequently, the design of chloridation resistant alloys differs in many cases from the principles of design of oxidation resistant alloys.

In general, there are four modes of degradation associated with the processes of high temperature corrosion:

(i) Surface scaling – direct corrosion of metal to corrosion products leading to decreased cross-sectional area and thus load-bearing capacity.

(ii) Internal degradation – formation of particle and/or particulates within the alloy reducing cross-section area and involving mechanical failure mechanisms, e.g. fatigue.

(iii) Surface scale spallation – associated with scale growth stress and stress generated during thermal cycling.

(iv) Corrosion product vaporisation – involves loss of protective scale leading to the depletion of the elements required for the formation of the protective layer by selective oxidation/sulphidation.

This chapter has been devised to provide a review of high temperature corrosion. High temperature corrosion is a vast subject area; this chapter is not meant to be a catalogue of all the published work on high temperature corrosion. We have discussed the corrosion behaviour of selected materials where new information has been generated on the processes of scale formation and scale failure and which have allowed the principles of the design of corrosion resistant alloys and coatings to be established. The chapter is distinguished by its emphasis on generic principles, mechanisms and modelling. A novel aspect of this chapter is the inclusion of in-situ nanoscale studies of the initial stages of high temperature corrosion processes elaborating on the nucleation and growth processes of scale development.

This chapter is structured in eight sections. Section 2 deals with the general principles of high temperature corrosion processes. Section 3 discusses the scaling processes of materials in oxidising environments. As high temperature oxidation is a well-researched topic we have included only the essential information from the literature. Section 4 provides a review of degradation of materials in sulphidising environments with strong emphasis on general principles and mechanisms of scale growth. Brief discussions on hot corrosion and chloridation are included in Sections 5 and 6. Nanoscale studies of the initial stages of oxidation are considered in Section 7. Scale failure and interdiffusion modellings are discussed in Section 8.

2. GENERAL PRINCIPLES

Important aspects of high temperature corrosion involve the processes of scale formation and scale degradation. Two additional modes of degradation which confer susceptibility of materials to high temperature corrosion are intergranular corrosion and scale vaporisation.

The scaling process in high temperature corrosion involves the formation of a thermodynamically stable corrosion product (a scale) which separates the surface of the material from the environment(s) causing corrosion. The formation of a defect (physical)-free, coherent and adherent scale containing lattice defects capable of sustaining only cationic and anionic transport allows progressive scale thickening and diffusion controlled parabolic kinetics. Linear kinetics predominate in the case of an inherently non-protective scale caused by the presence in the scale of inappropriate defect structures, physical defects, or by stress-induced scale spallation. Complex scaling processes characterise alloy corrosion accompanying the formation of multiphase, multilayered scale - each layer growing in a parabolic rate with different rate constants. This steady state scale development is often proceeded by the competitive processes of nucleation and growth of transient corrosion products dictating the mode and nature of subsequent scale growth.

Equilibrium thermodynamics, although not predictive, helps to assess the nature of the possible reaction products and the extent of evaporation or condensation of species and the conditions governing the reaction, the products and the condensed deposits. The standard free energies of formation (ÄG°) of oxides and sulphides as a function of temperatures and the corresponding dissociation pressures of the oxides and sulphides are represented in the well known Ellingham/Richardson diagrams as illustrated in Figures 1 and 2 [34]. Along the ordinates are plotted values of $\Delta G°$ for the oxides and sulphides and of the partial molar free energy of oxygen and sulphur, while the temperature is plotted along the abscissa. The values of $\Delta G°$ refer to the standard free energies of formation of oxides and sulphides per mole of oxygen or sulphur, e.g. $4/3Al + O_2 = 2/3Al_2O_3$.

In an environment containing bi-oxidant, e.g. oxygen and sulphur, the following reactions need to be considered for a divalent metal M

$$M + \frac{1}{2}O_2 = MO \tag{1}$$

$$M + \frac{1}{2}S_2 = MS \tag{2}$$

The following equations define the equilibrium oxygen and sulphur partial pressures:

$$pO_2^{\frac{1}{2}} = \exp\left(\frac{\Delta G°_{MO}}{RT}\right) \tag{3}$$

$$pS_2^{\frac{1}{2}} = \exp\left(\frac{\Delta G°_{MS}}{RT}\right) \tag{4}$$

Equations (3) and (4) allow the establishment of the conditions necessary for oxidation or sulphidation; however, a further reaction must be considered, namely

$$MS + \frac{1}{2}O_2 = MO + \frac{1}{2}S_2 \tag{5}$$

with the equilibrium condition

$$pO_2^{\frac{1}{2}} / pO_2^{\frac{1}{2}} = \exp\left(\frac{\Delta G°_{MS}}{RT} - \frac{\Delta G°_{MO}}{RT}\right) \cdot \frac{a_{MS}}{a_{MO}} \tag{6}$$

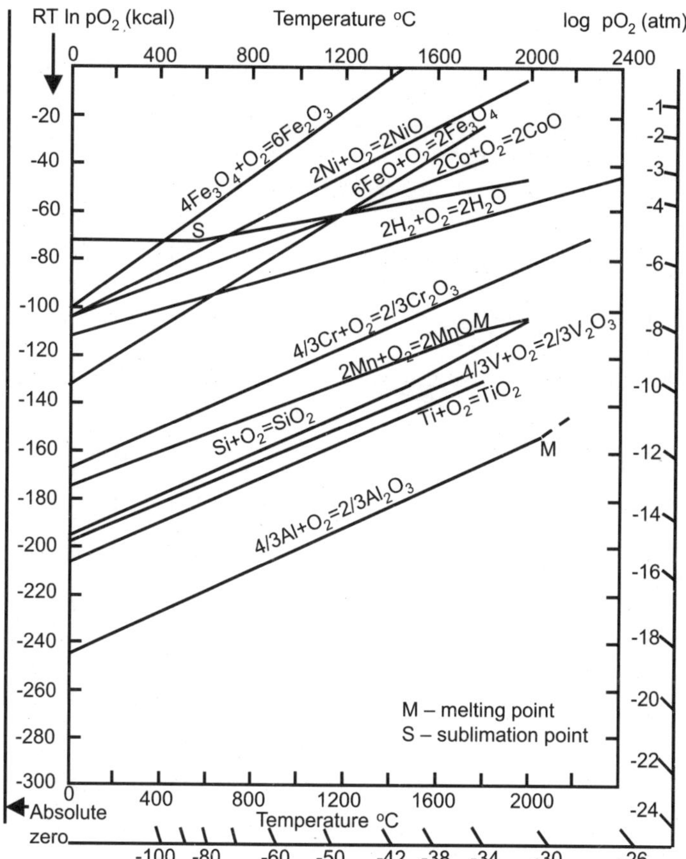

Figure 1 Standard free energies of formation for selected oxides as a function of temperature and oxygen partial pressure

If unit activities are assumed for the phases MS and MO, then expression (6) can be reduced to

$$pO_2^{\frac{1}{2}} \Big/ pO_2^{\frac{1}{2}} = \exp\left(\frac{\Delta G^{\circ}{}_{MS}}{RT} - \frac{\Delta G^{\circ}{}_{MO}}{RT}\right) \qquad (7)$$

Examination of equations (3), (4) and (7) permits the identification of various limiting situations concerning the type of surface corrosion products that may be formed, as follows

- if $(pO_2)_{gas} > (pO_2)_{eq}$ and $(pS_2)_{gas} < (pS_2)_{eq}$ then MO is the only stable surface phase;
- if $(pO_2)_{gas} < (pO_2)_{eq}$ and $(pS_2)_{gas} > (pS_2)_{eq}$ then MS is the only stable surface phase; and
- if $(pO_2)_{gas} > (pO_2)_{eq}$ and $(pS_2)_{gas} > (pS_2)_{eq}$ then both MO and MS should be stable and form as surface products.

Equation (7) indicates the formation of only one phase depending on which of the following conditions prevail:

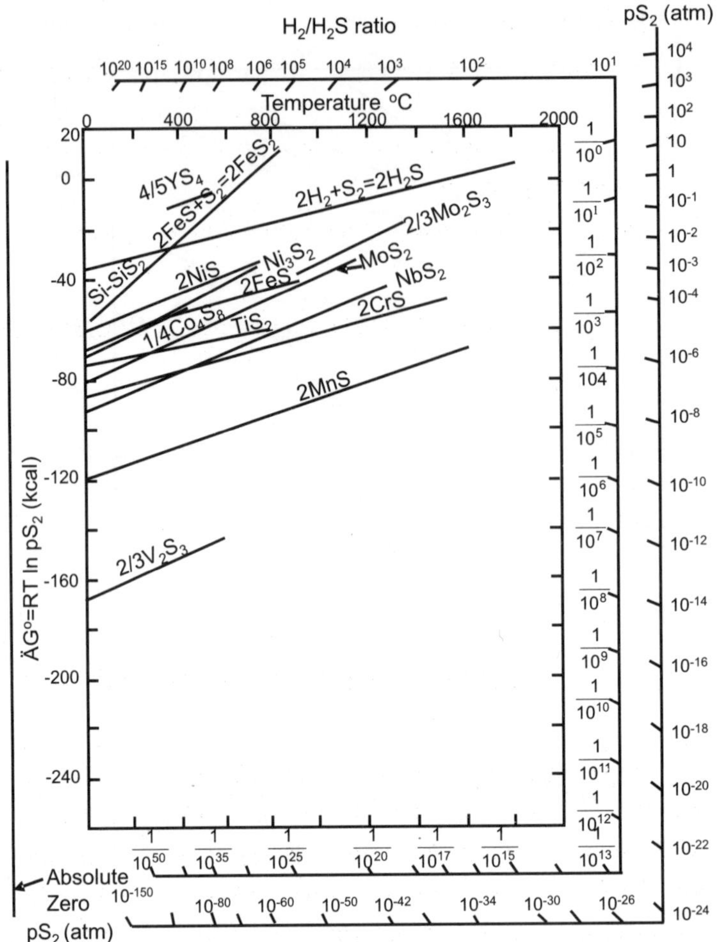

Figure 2 Standard free energies of formation for selected sulphides as a function of temperature and sulphur partial pressure

(a) $(pS_2/pO_2)_{gas} > (pS_2/pO_2)_{eq}$. This condition will favour reaction (5) to move to the left, and MS will be the stable phase where the metal is in contact with the gas;

(b) $(pS_2/pO_2)_{gas} < (pS_2/pO_2)_{eq}$. Here, MO will be the stable phase, and reaction (5) will proceed to the right.

From the known equilibrium partial pressures of the oxidants in the environment, thermodynamic stability diagrams can be constructed for a given temperature as shown in Figure 3 [35]. Such a thermodynamic diagram gives the stability range for all relevant phases, in this case metal, oxide and sulphide at a given temperature. Thermodynamic data for the relevant reactions can be used to calculate the boundaries. The line between the oxide and sulphide remains unchanged by activity changes in the alloy. The corrosion conditions, i.e. the oxygen and sulphur pressures in a given gas atmosphere represent a point on the diagram.

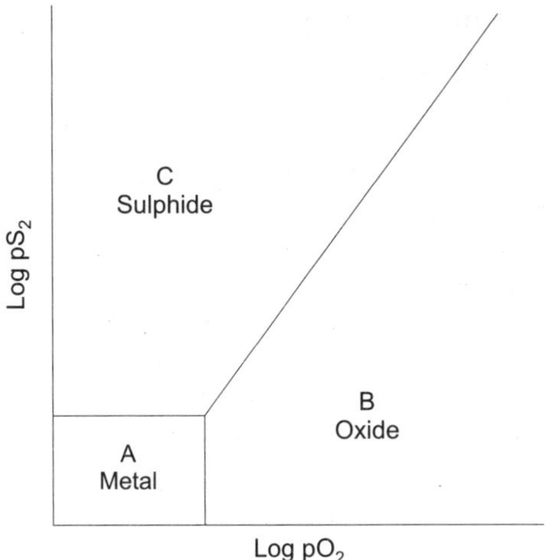

Figure 3 Schematic of a thermodynamic stability diagram showing the stable ranges where oxide or sulphide can be formed at a given temperature

The fields in the schematic thermodynamic stability diagram are indicated:

A - metal is the only stable phase;

B - oxide is the only stable phase;

C - sulphide is the only stable phase;

It should be stated that although the gas equilibrium can be determined for many gas mixtures at a given temperature and pressure, in many cases, even at temperatures as high as 1000°C, gas equilibrium is not established. Additionally, as the surface begins to be either partially or completely covered with corrosion product, thermodynamics cannot be applied to determine the progress of corrosion. As the departure from equilibrium becomes significant the kinetic factors such as diffusivity of the different alloying elements and of different reactive species, the morphologies and phase content of the corrosion products become dominant in determining the scaling processes. The analysis leading to this conclusion is based on the Wagner model [36,37] for the selective oxidation of an active element in a binary alloy to form a continuous external scale in the absence of transient oxidation. Wagner theory predicts the formation of a continuous external layer of oxide on a binary (A-B) alloy when the solute concentration in the alloy exceeds a critical atom fraction (N_B^{crit}), as expressed as:

$$N_B^{crit} = \left\{ \frac{\pi g^*}{3} N_O^{(S)} \frac{D_O V_m}{D_B V_{OX}} \right\}^{\frac{1}{2}} \tag{8}$$

Where $N_O^{(S)}$ is the oxygen solubility in the alloy; D_O and D_B are the diffusivities of oxygen and solute in the alloy, V_m and V_{OX} are the molar volumes of alloy and oxide, respectively; and g^* is the critical volume fraction of oxide scale. It is evident that for high $N_O^{(S)}$ and D_O, N_B^{crit} will also be high. If ($N_O^{(S)}$) and/or (D_O) of oxygen in the alloy decrease then the value of N_B^{crit} can also significantly diminish. Long-term stability of the protective scale is ensued by the high enough flux of solute to alloy/scale interface to prevent oxides of A from becoming stable. Pettit [38], observed two critical concentrations for the formation of alumina scales on Ni-Al alloys: one value required for development of the alumina scale and a larger value required maintaining its stability.

It is obvious that these principles discussed above are very important and useful to understand the formation of oxide/sulphide/chloride scales and high temperature corrosion mechanisms of materials.

3. OXIDATION

Among all the high temperature corrosion processes, oxidation is the best-researched area. Much progress has been achieved in understanding the issues of alloy oxidation and thereby facilitating a rationale to be established within which oxidation resistant alloys and coatings could be developed.

3.1 Cr-, Al- and Si-Containing Alloys

When high temperature alloys (usually based on Ni, Co or Fe) containing a number of elements are exposed to oxygen at elevated temperatures, oxidation may be anticipated in accordance with a design rationale. Thus certain elements (such as Cr, Al and Si), with high affinities for oxygen may be expected to oxidise in preference to those derived from the basis metal(s) with high dissociation pressures. This process of selective oxidation is the concept for developing oxidation-resistant alloys [1,9,11]. In particular, the composition of the alloy is chosen such that the stable oxide is the one that provides the most effective protective barrier. The oxides, Cr_2O_3, Al_2O_3, SiO_2, and possibly BeO are of primary interest because they exhibit low diffusivities for both cations and anions as well as being highly stable. Usually, alumina is an excellent barrier to oxygen at temperature below 1300°C, but at higher temperatures oxygen permeation through silica occurs at a slower rate.

Generally the growth of Cr_2O_3, Al_2O_3 and SiO_2 scales from their respective metals is observed to exhibit parabolic rate constants which are relatively small compared with those for other (Fe, Ni or Co) oxides. Most, if not all, of the current structural and coating alloys rely on the development of at least one of these oxides to impart oxidation resistance [11,39-47]. In this context, silicon is primarily used in coatings on structural alloys since it can result in the formation of low melting point phases when present at concentrations necessary for selective oxidation [11]. One way to overcome this problem is to incorporate silicon in the alloy as a partially dissolved carbide or nitride [48,49]. However, high concentration of Si in the alloy may lead to embrittlement.

Chromia-forming alloys are generally used at lower temperatures (<900°C) to prevent the volatilisation of the protective oxide. The maintenance of a continuous Cr_2O_3 layer or a Cr_2O_3-rich oxide layer is essential to sustain the protectivity of the scale. The effectiveness of a chromia scale in Cr-containing alloys to inhibit oxidation is dependent on the maintenance of a continuous Cr_2O_3 layer or a Cr_2O_3-rich oxide layer requiring a critical level of Cr (15-20wt%) in the alloy allowing maintenance of such a level at the alloy/scale interface. The depletion of Cr at the alloy/scale interface needs to be replenished by the diffusional transport from the depth of the alloy.

The real danger of Cr depletion is the susceptibility of this diluted region to internal oxidation and MO formation. Here it is important to recognise that the interdiffusion coefficient for Fe-Cr alloys is one order of magnitude greater than for Ni-Cr alloys and two orders of magnitude greater than for Co-Cr alloys. Thus the need for higher Cr content increases in the order Fe-Cr > Ni-Cr > Co-Cr alloys. Interdiffusion problems are of particular importance in the design of oxidation-resistant coatings – secondary diffusion between the coatings and substrate may lead to denudation of Cr in the coatings leading to a loss in oxidation resistance.

At high temperatures (>950°C), aluminium is extensively employed to develop oxidation-resistant alloys and, in particular, coatings. Aluminium contents in excess of 5-8 wt% are typically required to form alumina scales on the alloys, while aluminium in excess of 10 wt% is desired for the best resistance to complex atmospheres. The effectiveness of alumina-forming alloys is also dependent on the maintenance of a continuous Al_2O_3 scale requiring a good reservoir of Al to maintain a critical level of Al at the alloy/scale interface in order to replenish the Al-depletion due to scale spallation accompanying progressive oxidation particularly under thermal cycling and by secondary diffusion. Al-depletion may lead to non-protective M-Al spinel formation.

In practical alloys it is usual to have both Cr and Al, the levels chosen to promote the formation of an alumina and chromia scale under the given environmental conditions. It is recognised that the addition of Cr can reduce the critical content of Al required for the formation of protective alumina layer.

The oxidation resistance of the chromia- and alumina-forming alloys or coating alloys can be significantly improved by additions of rare earth/reactive elements [44, 50-60]. Such effects have been observed to be particularly dominant in the case of chromia scales. The improved oxidation resistance, in the main, has been ascribed to: (i) blocking of cation diffusion by the large ionic size of the additive elements; (ii) formation of a barrier layer impeding cation diffusion in the scale; (iii) enhanced chromium diffusion promoting rapid formation of the Cr_2O_3 layer; (iv) promoting inward oxygen transport leading to scale formation at the oxide/metal interface. In the case of alumina-forming alloys these elements act as sulphur scavenger and vacancy sinks and promote "peg" formation which confer improved scale adherence [60]. In addition, precious metals such as platinum or rhodium have been found to be effective in improving oxidation resistance [61,62]. These elements are quite effective when used in aluminide coatings on structural alloys. In particular, it appears that precious metals may improve oxide scale adherence. Recent research has shown that noble metal additions may also produce other beneficial effects influencing the selective oxidation of aluminium [11].

3.2 Ti and TiAl-Based Alloys

Ti, Ti-based and TiAl based-alloys are always of great interest as they can be considered as candidates for applications in advanced aerospace and power generation industries both in monolithic and composite forms [7,63] since these materials possess many desirable properties, such as low density, high melting point and high creep strength. However, their ambient temperature "forgiveness" (ductility, fracture toughness, fatigue crack growth rate) is a concern. Low oxidation resistance of TiAl-based alloys and susceptibility to hydrogen pick-up at elevated temperatures are below the desired levels. Tremendous efforts have been made in order to understand the oxidation mechanisms [64-75].

In general, it is difficult for a protective Al_2O_3 scale to form on Ti-based alloys and TiAl intermetallics; instead multi-layered scales composed of TiO_2 and Al_2O_3 form [64-66]. Such scales grow mostly parabolically with exposure time and consisted of two or even three layers - generally an outer layer of rather pure TiO_2 and an inner layer of a finely dispersed mixture of TiO_2 and Al_2O_3 [64-67], a third intermediate layer enriched in Al_2O_3 may also occur [68]. A study of air oxidation behaviour of Ti-6Al-4V alloy in the temperature range 650-850°C [64] demonstrated that the multilayered oxide scales formed on the oxidised alloy consisted of alternate layers of Al_2O_3 and TiO_2. The number of Al_2O_3 and TiO_2 layers increased with increasing exposure time and temperature. The external gas/oxide interface was always occupied by an Al_2O_3 layer whilst TiO_2 always appeared at the oxide/substrate interface.

The difficulty of forming a continuous, compact and adherent Al_2O_3 scale is attributed to the close affinities of Ti and Al to oxygen so that both compete in developing relevant oxides. A further problem relates to the formation of an oxygen-embrittled zone beneath the oxide scale, which in such alloys would compromise the corrosion protection. Early studies reported this embrittled layer as α_2-Ti_2Al [69], but more recent studies would support the formation of an oxygen-containing cubic intermetallic phase Z-$Ti_{50}Al_{30}O_{20}$ [70]. This oxygen-containing intermetallic degrades the mechanical properties of γ-TiAl in high temperature service.

It is reported [71-74] that oxidation of TiAl intermetallics in air can be significantly faster than in oxygen because air impedes the formation of a protective Al_2O_3 scale. When TiAl alloys oxidise in air it is likely that nitrides will develop. The typical structure of scale formed on a TiAl-based alloy is illustrated in Figure 4 [75]

It is possible to predict the alloying element concentration, the temperature and the oxygen partial pressure of the atmosphere in which preferential oxidation of the alloying element can initially occur provided appropriate thermodynamic data are available. An oxidation mechanism has been proposed by Du, Datta et al [75] for the oxidation of an intermetallic alloy, Ti-46.7Al-1.9W-0.5Si, in air at 750-950°C. The minimum activities of Ti and Al required to form the relevant TiO_2 and Al_2O_3 can be calculated if the oxygen partial pressure (0.21 atm) and experimental temperatures (750, 850 and 950°C) are known. The free energies of formation (J/mole) for TiO_2 and Al_2O_3 [76] are given by:

$$\Delta G^{\circ}_{T,TiO_2} = -910000 + 173\,T \quad (9)$$

$$\Delta G^{\circ}_{T,Al_2O_3} = -1676000 + 320\,T \quad (10)$$

Where T is the experimental temperature in Kelvin. The calculated results of minimum activities of Al and Ti to form Al_2O_3 and TiO_2 are listed in Table 1. Detailed calculations for the Ti-Al binary system [77] show that the activity of Ti is slightly higher than the Al activity in the Ti-46.7Al-1.9W-0.5Si alloy. It appears that the addition of W and Si does not significantly alter the situation concerning the activity of titanium and aluminium. Thus the preferential formation of TiO_2 is predicted during exposure to both environments. The discontinuous nature of the TiO_2 layer at the early stages of oxidation, observed in this study, may be attributed to the two-phase nature of the microstructure of the alloy, which implies that the activities of Ti and Al varied depending on the location of the phases on the alloy surface and it was difficult for TiO_2 to form in certain areas.

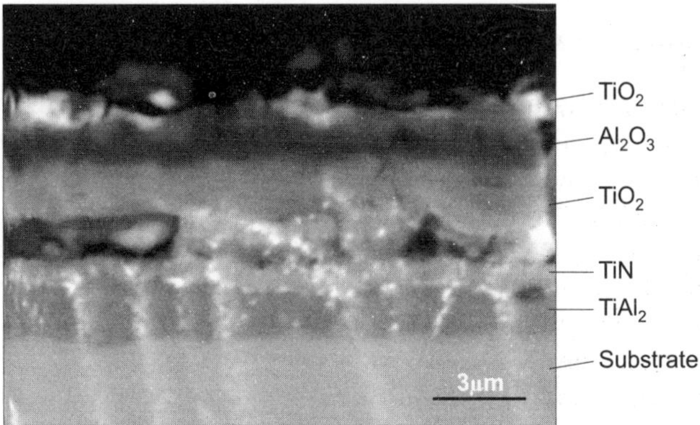

Figure 4 Cross-sectional morphology of Ti-46.7Al-1.9W-0.5Si after 240 hours oxidation at 850°C in air

Table 1 The minimum activities of Ti and Al to form TiO_2 and Al_2O_3 in 0.21 atmosphere of oxygen partial pressure at 750, 850 and 950°C

	750°C	850°C	950°C
a_{Al}	1.19×10^{-34}	7.72×10^{-31}	1.18×10^{-27}
a_{Ti}	1.77×10^{-37}	2.44×10^{-33}	7.05×10^{-30}

During the initial exposure of the Ti-46.7Al-1.9W-0.5Si alloy to air at the experimental temperatures, the high oxygen partial pressure promotes the formation of TiO_2, according to the reaction:

$$Ti\ (s) + O_2\ (g) = TiO_2\ (s) \tag{11}$$

The formation of TiO_2 changes the balance of the activities of Ti and Al and the oxygen partial pressure between the TiO_2 layer and the substrate. The reduction of Ti activity then leads to the development of an Al_2O_3 layer:

$$2Al\ (s) + \frac{3}{2}O_2\ (g) = Al_2O_3\ (s) \tag{12}$$

This development of an Al_2O_3 layer in turn leads to the formation of a Ti-enriched zone beneath the Al_2O_3 layer. At the same time nitrogen migrates through the Al_2O_3 layer to this Ti-enriched zone from the external atmosphere, which creates a favourable circumstance for the following reaction to take place:

$$Ti\ (s) + \frac{1}{2}N_2\ (g) = TiN\ (s) \tag{13}$$

Thus a TiN layer is formed beneath the Al_2O_3 layer. In the mean time oxygen species also diffuse inwards to the interface between the Al_2O_3 layer and the TiN layer allowing a build-up of the oxygen partial pressure. When the oxygen partial pressure reaches a certain level, TiN became unstable favouring the following reaction:

$$TiN\ (s) + O_2\ (g) = TiO_2\ (s) + \frac{1}{2}N_2\ (g) \quad (14)$$

The released nitrogen migrates inwards via the TiN layer. The nitrogen partial pressure gradually increases at the TiN/substrate interface and the nitrogen species encountering titanium from the substrate allowed reaction (13) to take place again. This gives rise to the formation of a titanium-depleted zone demonstrated by the existence of a $TiAl_2$ band beneath the TiN layer. The oxidation mechanisms of Ti-46.7Al-1.9W-0.5Si alloy in air are schematically described in Figure 5. It is apparent that the thickness of the TiO_2 layer increases with exposure time - as does the thickness of the TiN layer as nitrogen migrates inwards from the external environment. However it is not clear why AlN does not develop between the TiN and $TiAl_2$ band, as the affinities of Al and Ti to nitrogen are very close [76]. This can probably be attributed to the faster self-diffusion of Ti in the TiAl substrate than that of Al [78]; TiN became the kinetically favoured product.

4. SULPHIDATION

4.1 General Considerations

The degradation of metallic materials by sulphur at elevated temperatures both in purely sulphidising and in mixed sulphidising environments is an important problem due to the fast

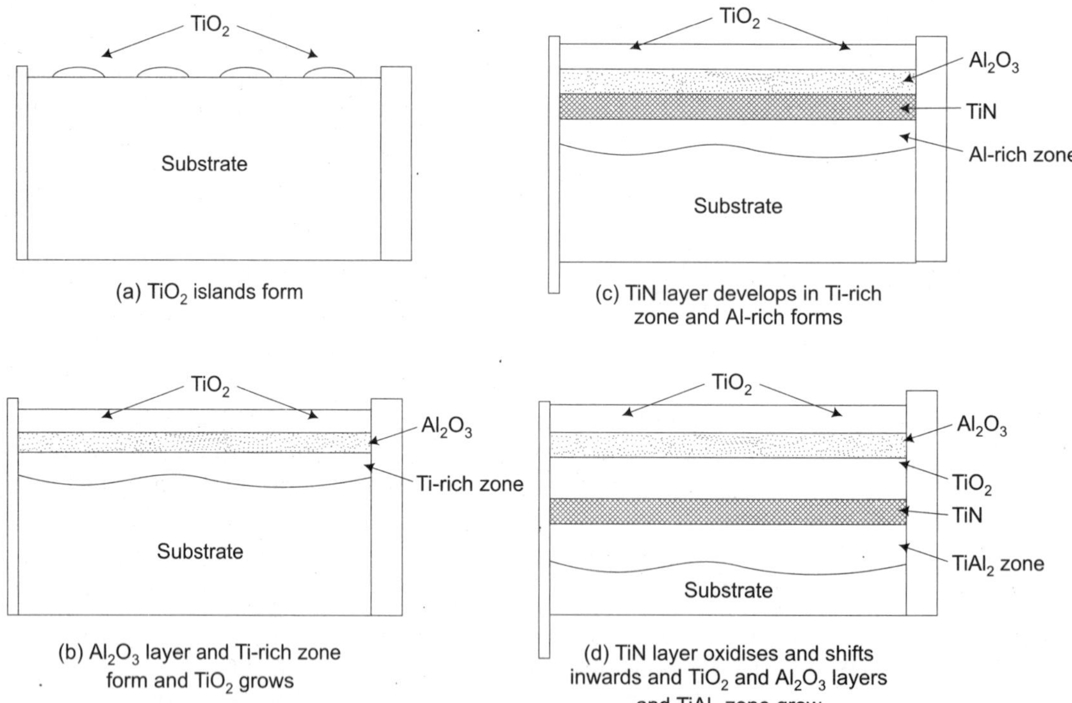

Figure 5 Schematic illustrating the mechanism controlling oxidation of Ti-46.7Al-1.9W-0.5Si in air between 750 and 950°C

rate of corrosion for most common commercial alloys. Strafford and Datta [5] and Datta and Du et al [13] have comprehensively reviewed the process of sulphidation. Compared to oxidation, sulphidation is characterised by faster kinetics and formation of scales with complex defective morphologies. Sulphidation is particularly severe in high pS_2 and low pO_2 environments. Table 2 vividly illustrates the problems of designing sulphidation resistant alloys and coatings. The parabolic rate constants for the sulphidation of three elements Fe, Co and Ni which form the basis of many high temperature oxidation resistant alloys are typically 10^{-6} to 10^{-7} $g^2/cm^4/s$ at 640–800°C and are several orders of magnitude higher than their oxidation rate constants (~10^{-11} to10^{-8} $g^2/cm^4/s$) at 800–1000°C. It is equally important to note that the k_p value for the sulphidation of Cr (~10^{-8} $g^2/cm^4/s$) at 750°C is significantly higher than that (~10^{-13} $g^2/cm^4/s$) at 800°C for the formation of Cr_2O_3, an important component in the development of many oxidation-resistant alloys. In contrast the refractory metals – V, Mo, Nb and W together with Zr and Hf display low rates of sulphidation. These elements when incorporated in sulphidation resistant materials are likely to enhance the resistance to sulphidation particularly under reducing conditions (Table 3).

In environments of low oxygen and high sulphur activities, sulphides formed on certain alloys may offer a moderate measure of protection, in fact analogous to a protective oxide film, but in general, sulphide scales are much more friable and more subject to exfoliation than oxide scales on most high temperature alloys due to larger Pilling-Bedworth ratios. Furthermore, oxide scales formed on the alloys usually melt at relatively high temperatures, above the melting point of the alloys, whereas sulphides have comparatively low melting points and frequently form low melting point eutectics [79]. In addition, the diffusion coefficients of cations in sulphide scales are relatively high because of the greater degree of non-stoichiometry of sulphide structures [79].

Chromium is the principal element in common high temperature alloys conferring a degree of sulphidation resistance to iron-, nickel- and cobalt- based alloys in the presence of

Table 2 Sulphidation and oxidation parabolic rate constants, k_p, of selected metals

	Sulphidation			Oxidation (pO_2 ~10^5 Pa)	
Metal	Temperature (°C)	k_p, ($g^2/cm^4/s$)	pS_2 (Pa)	Temperature (°C)	k_p, ($g^2/cm^4/s$)
Al	500	1.0×10^{-12}	10^5	1000	1.0×10^{-14}
Ti	750	4.5×10^{-12}	10^{-1}	800	3.3×10^{-10}
Zr	750	2.7×10^{-12}	10^{-1}	800	6.6×10^{-11}
Hf	750	7.4×10^{-12}	10^{-1}	800	3.3×10^{-11}
V	750	2.3×10^{-10}	10^{-1}		(linear)
Nb	750	8.2×10^{-13}	10^{-1}		
Cr	750	1.9×10^{-8}	10^{-1}	800	1.0×10^{-13}
Mo	750	2.6×10^{-12}	10^{-1}		(linear)
W	750	2.0×10^{-9}	10^{-1}		(linear)
Fe	800	2.0×10^{-7}	10^5	800	5.5×10^{-8}
Ni	640	1.6×10^{-6}	10^5	1000	9.1×10^{-11}
Co	800	6.7×10^{-6}	10^5	900	2.0×10^{-8}

Table 3 Parabolic rate constants of certain metals and alloys

Metal or alloy	Temperature (°C)	Atmosphere (Pa)		k_p ($g^2/cm^4/s$)
		pS_2	pO_2	
V	750	10^{-1}	10^{-18}	2.3×10^{-10}
Nb	750	10^{-1}	10^{-18}	8.2×10^{-13}
Mo	750	10^{-1}	10^{-18}	2.6×10^{-12}
Fe-20Nb	700	10^{3}		8.4×10^{-8}
Fe-30Nb	700	10^{3}		3.5×10^{-8}
Ni-20Nb	700	10^{3}		7.7×10^{-7}
Ni-30Nb	700	10^{3}		1.3×10^{-7}
Co-20Nb	700	10^{3}		1.6×10^{-7}
Co-30Nb	700	10^{3}		5.9×10^{-8}
Fe-20Mo	700	10^{3}		8.4×10^{-8}
Fe-30Mo	700	10^{3}		3.5×10^{-8}
Ni-20Mo	700	10^{3}		1.0×10^{-7}
Ni-30Mo	700	10^{3}		1.5×10^{-8}
Co-20Mo	700	10^{3}		4.4×10^{-8}
Co-30Mo	700	10^{3}		2.0×10^{-9}
Co-20Cr-3.5Al-1Y-5V	750	10^{-1}	10^{-18}	3.4×10^{-9}
Co-20Cr-3.5Al-1Y-5Nb	750	10^{-1}	10^{-18}	3.1×10^{-9}
Co-20Cr-3.5Al-1Y-5Mo	750	10^{-1}	10^{-18}	4.3×10^{-9}
Co-20Cr-3.5Al-1Y-5W	750	10^{-1}	10^{-18}	8.9×10^{-9}
Co-20Cr-3.5Al-1Y-10V	750	10^{-1}	10^{-18}	7.2×10^{-10}
Co-20Cr-3.5Al-1Y-10Nb	750	10^{-1}	10^{-18}	1.8×10^{-9}
Co-20Cr-3.5Al-1Y-10Mo	750	10^{-1}	10^{-18}	6.2×10^{-9}

oxygen at high enough activity to permit the formation of a protective chromic oxide on the alloy surface. Some alloying elements, such as aluminium and silicon, additionally provide sulphidation resistance. However, low oxygen activity (that is, at activities generally lower than those required to allow formation of oxides of Fe, Ni and Co), coupled with high sulphur activity (for example, pS_2 sufficient to permit the development of sulphides of iron, nickel and cobalt) may have very adverse effects on the corrosion resistance of the alloys since sulphide reaction products rather than oxides might be stable; here the co-formation of sulphides and oxides is particularly deleterious.

One approach to solving such problems is to use conventional oxidation-resistant alloys and subject them to a preoxidation treatment to form a compact, adherent and protective oxide scale that would subsequently act as a barrier to the permeation of sulphur from the working environment [80-82]. In essence, such an oxide scale formed by preoxidation is a type of conversion coating, to be contrasted with a conventional protective extraneous coating that involves the additions of other sulphidation-resistant materials on to the surface. Such an approach is simple, convenient and probably less costly than an extraneous coating. However, preoxidation to form a conversion coating would be a viable alternative only if an adequate self- healing property of the alloys exists: if the preformed oxide scale cracked or spalled, accelerated corrosion attack might occur in complex environments where the oxygen activity was not sufficient to promote self-healing of the cracked oxide scale. In practice physically/ mechanically imperfect scales readily allow inward transport of the second oxidant, e.g. sulphur and breakaway accelerated corrosion is often observed.

Enhanced sulphidation resistance can be achieved employing the same principles used in increasing the oxidation resistance of materials. The incorporation of appropriate elements into the base materials undergoing selective sulphidation leads to the development of a barrier layer capable of sustaining lower ionic transport rates. While selective sulphidation to form a barrier layer sulphide will be governed by the free energy of formation, the kinetics of this selective process and hence the overall rate of sulphidation will be controlled by the defect structure of the sulphides (Table 4), their ability to support fast or slow diffusion rates as measured by self-diffusion coefficients (Table 5) and melting points and mechanical stability indicated by the Pilling-Bedworth ratio.

Strafford, Datta et al [5,83-85] have shown that several metals from Group IV-VI of the periodic table (i.e. Cr, V, Nb, Ta, Mo and W) suffer low or very low rates of degradation in H_2/

Table 4 Physico-chemical properties of sulphides of certain Group III–VI metals

Sulphide	Pilling-Bedworth ratio	Defect structure	Melting point (°C)
Al_2S_3	2.60	n-type	1099
TiS_2	1.11	n-type	1999–2099
ZrS	1.91	n-type	1549
HfS	–	–	2100–2273 (est)
V_2S_3	–	n-type	1799–1999
NbS_2	–	n-type	–
TaS_2	2.42	n-type	999
Cr_2S_3	2.50 (CrS)	n/p-type	1550 (CrS)
MoS_2	3.54	n-type	1457
WS_2	3.47	n-type	>1800
FeS	2.50	p-type	1189
Co_9S_8	2.37	p-type	1080
NiS	2.50	p type	796

Table 5 Self-diffusion coefficients of cations (D_M) in some metal sulphides and oxides

Sulphide	Temperature (°C)	D_M (cm^2/s)	Oxide	Temperature (°C)	D_M (cm^2/s)
$Cu_{2+y}S$	650	5.15×10^{-5}	$Cu_{2-y}O$	1000	1.7×10^{-8}
$Co_{1-y}S$	720	7.0×10^{-7}	$Co_{1-y}O$	1000	1.9×10^{-9}
$Ni_{1-y}S$	800	1.4×10^{-8}	$Ni_{1-y}O$	1000	1.0×10^{-11}
$Fe_{1-y}S$	800	3.5×10^{-7}	$Fe_{1-y}O$	800	1.3×10^{-8}
Cr_2S_3	1000	1.0×10^{-7}	Cr_2O_3	1000	1.0×10^{-12}
Al_2S_3	600	1.0×10^{-13}	Al_2O_3	1000	1.0×10^{-16}

H_2S atmospheres of high pS_2 (~10^{-1} Pa) at elevated temperatures except for Cr. Although the oxidation performances of these refractory elements (with the exception of Cr) are very poor and limiting for their application in most high temperature environments of high oxygen activity, it has been suggested that they might be very useful in environments of low oxygen activity (pO_2~10^{-9} to 10^{-18} Pa) and high sulphur activity (pS_2~10^{-1} to10^{-6} Pa). Here oxidation of the refractory element additions would not be anticipated on thermodynamic grounds.

Jenkinson [86] in this laboratory has carried out sulphidation studies of Cr, V, Nb, Mo and W. It was noted that except V, which at 750°C sulphidised at about only two orders of magnitude slower than Cr (~10^{-8} $g^2/cm^4/s$), all the other metals sulphidised several orders of magnitude slower than Cr. Nb in particular closely paralleled the rate of oxidation of Cr in oxygen (~10^{-13} $g^2/cm^4/s$). This finding is most significant recognising the essential role of Cr in all oxidation resistant M-Cr-type alloys.

4.2 Co-Cr-Al-Y-X-Based Alloys

Some progress has also been made [87] in testing the concept of incorporating refractory metals in Co-Cr-Al-Y alloys to inhibit their degradation due to high temperature sulphidation. Studies of the scaling behaviour of experimental CoCrAlYX-type alloys where X is a refractory metal at 5-10 wt% level after isothermal exposure at 750°C to a predominantly sulphidising environment where pS_2~10^{-1} Pa and pO_2~10^{-18} Pa for exposure periods of up to 240 hours indicate that such alloys have superior sulphidation resistance relative to CoCrAlY-base control alloys without "X" additions. Although some work has been performed on the sulphidation behaviour of Group IV-VI metals, as well as on the alloys containing these elements, much more research effort is required, involving not only basic programmes to provide an understanding of the sulphidation behaviour of individual Group IV-VI metals, but also embracing corrosion evaluation studies on a large number of experimental alloys based on and containing, the refractory and semi-refractory metals.

In details, the sulphidation behaviour of a series of MCrAlYX-type alloys, where M is Co or Co and Fe, and X is V, Nb, Mo or W, added in combination at various levels, has been studied at 750°C in an atmosphere comprising pS_2~10^{-1} Pa and pO_2~10^{-18} Pa [88,89]. For each alloy over prolonged periods of exposure (up to 240 hours) parabolic rate law characterised the sulphidation kinetics. Inclusion of Mo in the alloy containing both V and Nb, and particularly the combined additions of Mo and W led to greatly enhanced sulphidation performance. Also superior sulphidation resistance followed from partial replacement of Co by Fe. The development of the subscale layers - an outer-layer of Co_9S_8 and Co_3S_4, a mid-layer of Cr_3S_4 and an inner-layer of chromium and refractory metal sulphides - characterised the scale formed on the Co-based alloys. The formation of a duplex scale with an outer-layer of Fe_3S_4, Co_9S_8 and Cr_3S_4 and a compact inner-layer consisting of sulphides of chromium and refractory metals was observed to accompany the sulphidation of (Fe,Co)-based alloys. For both categories of alloys the inner-layer of refractory metal sulphides was decorated by an alumina film which inhibited the outward migration of the base elements.

These experimental results and observations have permitted the elucidation of sulphidation mechanisms of these experimental MCrAlYX-type alloys. At the initial stages of sulphidation, the high Co activity favours the formation of Co sulphide nuclei and a thin pellicle of cobalt sulphide (Co_9S_8 and Co_3S_4) rapidly forms on the alloy surface, as pictured

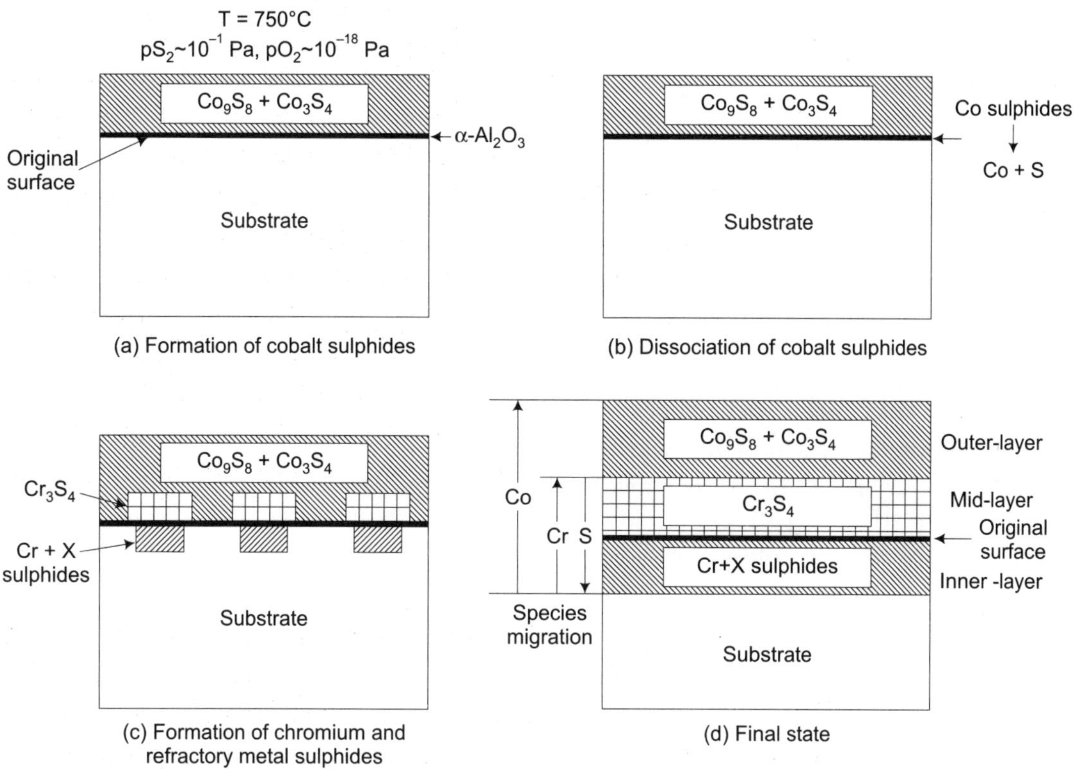

Figure 6 Schematic illustration of the proposed mode for the sulphidised Co-based alloys

schematically in Figure 6. Simultaneously a very thin alumina film forms on the alloy surface, which would be expected to act as a "passivating" layer thereby lowering the overall scaling rates. The presence of such an oxide film, it can be speculated, would be particularly beneficial during these initial stages of sulphidation in retarding the damaging outward diffusion of cobalt ions and assisting in the selective sulphidation of Cr and elements X.

After these initial stages of sulphide scale growth, the sulphur potential between the cobalt sulphide scale and substrate would decrease significantly. Once the local sulphur partial pressure falls below the dissociation pressure of cobalt sulphide, the cobalt sulphide decomposes and releases cobalt cations and sulphur anions – so-called "dissociation mechanism" [90]. The cobalt cations diffuse outward to the scale surface where they react with sulphur to form cobalt sulphides (Co_9S_8 and Co_3S_4). The sulphur anions react with chromium migrated from the substrate and chromium sulphide (Cr_3S_4). Also sulphur anions diffuse inward to the substrate and react with refractory metals as well as chromium to form their corresponding sulphides which constitute the inner-layer. In addition, sulphide growth is probably further facilitated by the migration of sulphur-bearing gas through pores and fissures present in the scaling layers. Significantly the development of the inner-layer by inward diffusion of sulphur anions or sulphur-bearing gas reduces, to some extent, the concentration of vacancies or voids at the interface of the inner-layer and substrate, which may occur due to the outward diffusion of cobalt and chromium, confirmed by the absence of vacancies or voids at the inner-layer/substrate interface. It should be emphasised that the dissociation of cobalt sulphide merely leads to the outward shift of the interface of the outer-/mid-layer and does not contribute to the increasing thickness of the outer-layer. Thickness of the outer-layer increases by the outward migration of cobalt cations from the substrate.

In the case of the (Fe,Co)-based alloys, a thin outer-layer quickly develops, consisting mainly of Fe_3S_4, Co_9S_8 and Cr_3S_4 at the initial stages of sulphidation. This increases the vanadium and niobium activities beneath the outer-layer, the V and Nb sulphides then form initially in a discontinuous manner. After prolonged exposure, the outer-layer thickens and the inner-layer becomes continuous in nature. During the formation of the duplex sulphide scales, Fe, Co and Cr species diffuse outward and form the outer-layer. Likewise, sulphur species would be expected to diffuse inward to the substrate to produce the inner-layer. It is suggested that this compact and adherent inner-layer of V and Nb as well as Cr sulphides effectively improves the sulphidation behaviour of the alloys.

The sulphur species migrate inward by two possible ways:

(1) Sulphur species from the atmosphere diffuse through short-circuit paths, e.g. grain boundaries, microcracks or voids, to reach beneath the outer-layer and react with V and Nb to form V and Nb sulphides.

(2) The formation of an outer-layer on the surface of the alloys reduces the sulphur partial pressure beneath the outer-layer and increases the vanadium and niobium activities. Once the sulphur partial pressure becomes lower than the dissociation pressure of Fe or Co sulphides then these sulphides undergo dissociation and releasing free sulphur which reacts with V, Nb and Cr to form the inner-layer of V, Nb and Cr sulphides.

It should be pointed out that a continuous scale of pure refractory metal sulphides, e.g. V_2S_3, or NbS_2, or MoS_2 or WS_2, does not develop on either the Co-based alloys or the (Fe,Co)-based alloys; this is in contrast to the case of oxidation resistant alloys in which an α-Cr_2O_3 or α-Al_2O_3 scale would form in the purely oxidising environments. Nevertheless, the refractory elements play a significant part in the improvement of sulphidation behaviour, attributable to: (i) the refractory metals provide vacancy sinks so that void formation is greatly reduced; (ii) the selective formation of X_xS_y-rich (where X=refractory metals) pegs at grain boundaries provides a mechanical keying of the scale to the alloy substrate, scale adherence was thereby enhanced.

4.3 Ti- and TiAl-Based Alloys

Degradation of Ti- and TiAl-based alloys in high sulphur and low oxygen environments relates to several important aspects. Pure titanium and Ti-6Al-4V alloy exposed [91,92] at 750°C in $H_2/H_2O/H_2S$ (pO_2~10^{-18} Pa and pS_2~10^{-1} Pa), H_2/H_2O (pO_2~10^{-18} Pa) and air environments for up to 240 hours is characterised by linear sulphidation/oxidation kinetics for Ti and linear-parabolic for Ti-6Al-4V alloy in the $H_2/H_2O/H_2S$ environment and parabolic rate laws in the H_2/H_2O atmosphere (after a transient period) for Ti-6Al-4V; linear-parabolic rate laws characterise the oxidation of Ti and Ti-6Al-4V alloy in air. The exposure of the titanium specimen to the $H_2/H_2O/H_2S$ atmosphere led to the formation of a double scale of TiO_2 with an intervening TiS_2 film between the double layer scale and the substrate and of the Ti-6Al-4V double layer of TiO_2 with a layer consisting of Al_2S_3, TiS_2 and vanadium sulphide at the junction of the inner TiO_2 layer and substrate with some Al_2O_3 precipitation in the external portion of the outer-layer of TiO_2. In the low pO_2 (i.e. H_2/H_2O) the formation of a double layered oxide scale consisting of TiO_2 was identified in both Ti and Ti-6Al-4V alloy; the Ti-6Al-4V alloy also containing an a-Al_2O_3 layer situated between the outer-layer and inner-layer. For both materials, air oxidation accompanies multilayered scale formation. The formation of a multilayered oxide scale of TiO_2 is indicated on the air oxidised Ti whilst a multilayered oxide scale with alternating layers of Al_2O_3/TiO_2 was developed on the air oxidised Ti-6Al-4V alloy.

Work reported on the response of Ti-Al- based intermetallic materials, such as Ti-54Al and Ti-48Al-2Nb-2Mn, exposed to mixed-gas environments with a fixed pO_2 (10^{-14} Pa) and with various pS_2, values (0.0016, 0.11 and 1.6 Pa) at 900°C [93] indicate parabolic kinetics at all pS_2 values for sulphidation of the alloys. After some uncertainties at the early stages of sulphidation, prolonged exposure (168 h at 900°C) indicates that the greatest resistance to sulphidation at the highest pS_2 (Figures 7 and 8). During the initial stages of exposure, the high affinity of oxygen for Ti and Al leads to development of an outer layer of TiO_2, beneath which an Al_2O_3 layer forms. Sulphur diffuses through the TiO_2 and Al_2O_3 layers and reaches the substrate/scale interface where the pO_2 is low enough to promote the formation of TiS_2 and Al_2S_3 (and NbS_2 in the case of Ti-48Al-2Nb-2Mn). Clearly, the presence of high pS_2, in the bulk environment provides a higher driving force for sulphur migration that then supports a higher sulphur flux and favours the formation of sulphides of the alloy constituents. Accordingly, a mixed layer of sulphides and oxides develops in the high-pS_2 atmosphere, thicker than in the low-pS_2 environment, thereby providing higher resistance to sulphidation.

In summary, the superior sulphidation resistance of Ti-54Al and Ti-48Al-2Nb-2Mn can be ascribed to several aspects of the scaling processes. First, the development of an inner layer of sulphides (TiS_2, $A1_2S_3$, NbS_2) provides an effective barrier to cation transport. The presence of refractory metal sulphide, NbS_2, is known to cause significant improvement in the sulphidation resistance of the intermetallic alloy [65,94]. Furthermore, su1phidation resistance of Ti-54Al stems from the formation of a $TiAl_3$ layer at the scale/substrate interface by the diffusion of Ti after development of TiO_2 and TiS_2. Ti released from the dissociation of TiS_2 is likely to promote a thick $TiAl_3$ layer, thus imparting further resistance to sulphidation.

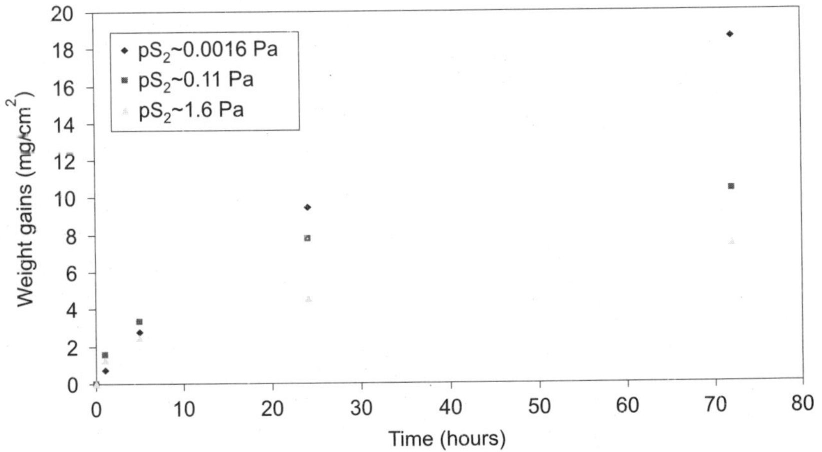

Figure 7 Sulphidation kinetics for Ti-54Al at 900°C

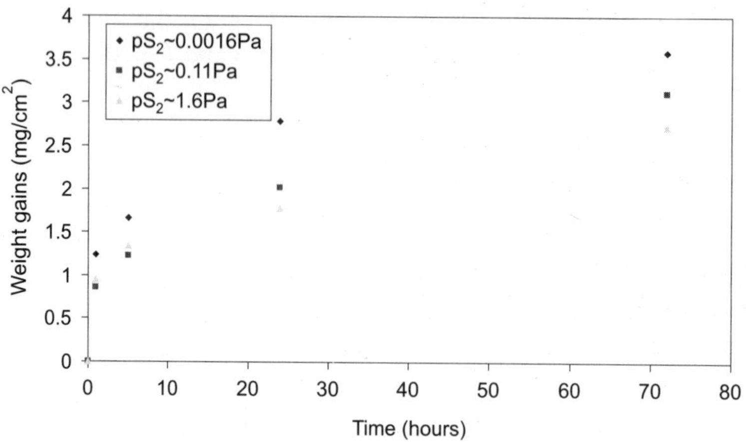

Figure 8 Sulphidation kinetics for Ti-48Al-2Nb-2Mn at 900°C

Investigation by Du, Datta et al [94] of oxidation/sulphidation behaviour of another intermetallic alloy, Ti-56.6Al-1.4Mn-2Mo, with duplex and lamellar microstructure in $H_2/H_2S/H_2O$ (pS_2 ~ 0.1 Pa, pO_2 ~ 10^{-12} Pa) at 750°C and (pS_2 ~ 0.1 Pa, pO_2 ~ 10^{-14} pa) at 900°C reveals parabolic kinetics ($k_p = 10^{-12} g^2/cm^{-4}/s$) at 750°C and cubic kinetics at 900°C and the absence of significant effect on the corrosion behaviour of increased exposure temperature; scaling pattern development resembling that of Ti-Al-Nb-Mn alloy, i.e. with an outermost TiO_2/inner Al_2O_3 layer and base-element sulphides forming between the oxide layers and the substrate [65].

The enhanced corrosion resistance observed at 900°C compared to that at 750°C follows from the differences in the defect structures. Both oxygen vacancies and interstitial Ti ions are important point defects [93], with interstitial Ti ions predominating at low pO_2 and high temperature, whereas oxygen vacancies predominate at high pO_2 and low temperature. At low pO_2 many interstitial Ti ions are expected in TiO_2 and hence an increase in pO_2 would result in a decreased number of interstitial ions. Reducing the interstitial Ti ions would decrease the transport of Ti through TiO_2. The temperature change between 750 and 900°C may not significantly alter the diffusion of Ti and O ions in these alloys. The increased oxygen pressure will increase the oxygen vacancies in TiO_2 and thereby increase the inward diffusion of O and promote $A1_2O_3$, imparting a slow corrosion rate for both materials.

Further work by Du, Datta et al [95] on the oxidation and sulphidation behaviour of Ti-46.7Al-1.9W-0.5Si alloy in an $H_2/H_2S/H_2O$ atmosphere, yielding high sulphur (pS_2~$1.2x10^{-1}$ Pa) and low oxygen (pO_2~$1.2x10^{-15}$ Pa) potentials at 850°C, shows protective kinetics with a parabolic rate constant of $6x10^{-11}$ $g^2/cm^4/s$. Morphological analysis reveals the development of a multi-layered scale consisting of a top rutile (TiO_2) layer, a continuous layer of α-Al_2O_3 beneath the rutile layer and a TiS layer containing scattered pure W particles. It is suggested that fast outward diffusion of Ti within the substrate results in the formation of a zone of high concentration of aluminium ($TiAl_3$ and $TiAl_2$) between the scale and substrate, as illustrated in Figure 9.

Investigation by Du et al [96] on the effects of Y, Hf and Y+Hf on the sulphidation behaviour of Fe_3Al at 900°C in an $H_2S/H_2/H_2O$ environment demonstrates that the additions

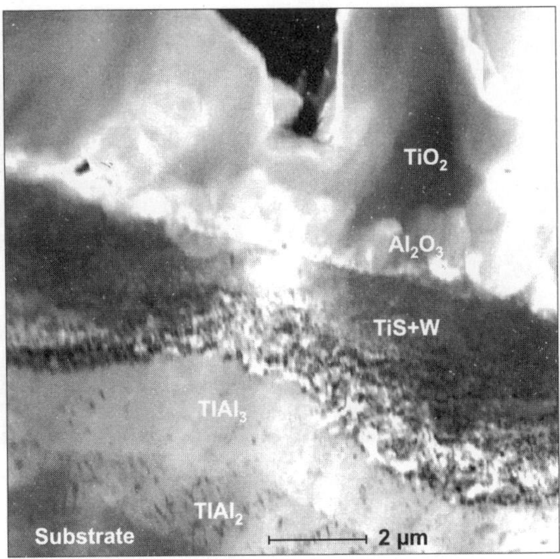

Figure 9 Dark field TEM image of exposedTi-46.7Al-1.9W-0.5Si for 5 hours at 850°C in $H_2/H_2S/H_2O$

of the reactive elements confer significantly increased sulphidation resistance to Fe_3Al. However the combined addition of Y+Hf compromises the benefit to the sulphidation resistance enhancement. It is interesting to note that a superlattice of MnS formed on the sulphidised Fe_3Al with the addition of Y and Hf, as revealed in Figure 10.

4.4 Silicides

Natesan, Datta et al [93] have comprehensively reviewed the sulphidation behaviour of intermetallics. They reported sulphidation performance of the Mo_5Si_3-type intermetallics in a 1.5 vol.% H_2S/H_2 gas mixture and compared the sulphidation resistance with relevant oxidation performance in air. Figure 11 shows the thermogravimetric test data obtained at several temperatures for oxidation and sulphidation of a B-containing Mo_5Si_3. The data infer a protective scaling of the alloy at 500°C under oxidizing conditions, not because of SiO_2 formation, but because the volatilisation rate of Mo oxide is negligible. At temperatures of 800 and 1200°C, the curves demonstrate a sharp drop in specimen weight for ~2 h, after which a plateau is reached and the weight changes little during 50 h of additional exposure, indicating a protective scale. The morphology of the scale after oxidation at 800°C has been observed to consist of a light-coloured MoO_2 phase and a dark grey Si-rich oxide; the scale after oxidation at 1200°C consisting predominantly of Si-rich oxide and almost pure Mo particles. The surface layer shows significant cracking and peeling and appears to remain highly plastic, as indicated by curling rather than spalling of the oxide layer. Such degradation of the oxide layer can expose interior Mo silicide to additional oxidation, and the sequential processes of oxidation and peeling can continue without offering oxidation protection for the alloy over long periods of exposure. The thermogravimetric test data obtained during sulphidation of the material reveal an approximately one order of magnitude smaller decline in specimen weight than in the data obtained during oxidation. It is not clear as to the cause for the initial drop in weight,

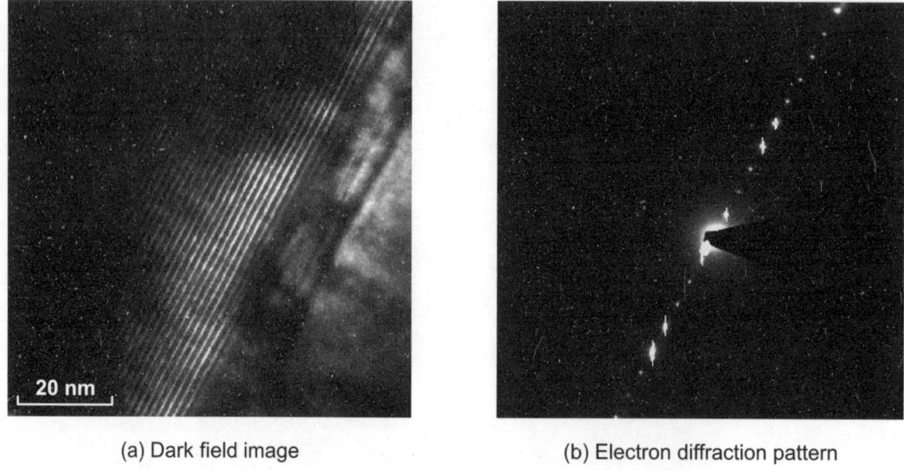

Figure 10 Superlattice layer formed on the sulphidised Fe_3Al+Y-Hf alloy in a H_2-H_2S-H_2O environment at 900°C for 240 hours

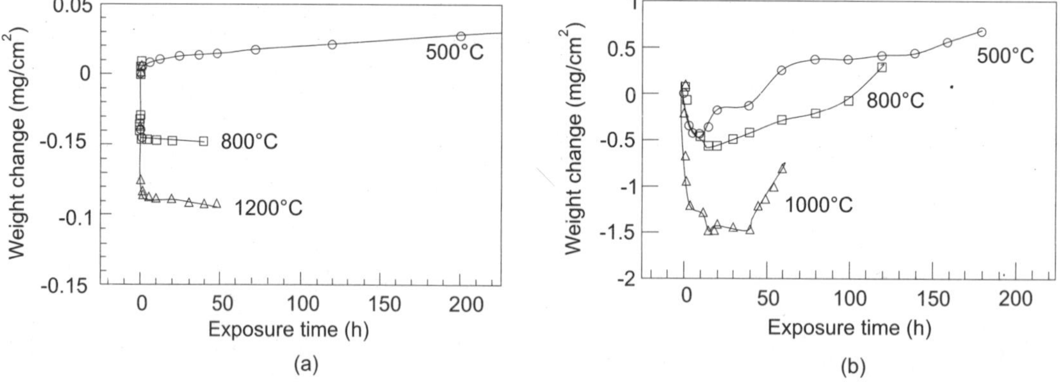

Figure 11 Weight change data during (a) oxidation in air and (b) sulphidation in 1.5vol % H_2S/H_2 gas mixture for B-containing Mo_5Si_3 after exposure at several temperatures

except that the possible presence of residual moisture in the gas mixture can lead, in the early stages of exposure, to formation of volatile oxides such as MoO_3 and/or SiO (especially in the reducing condition used in the experiments), before the development of sulphide reaction products. Figure 12 shows SEM photomicrographs of surfaces of B-containing Mo_5Si_3 alloy after sulphidation at 500, 800 and 1100°C in a 1.5 vol. % H_2S/H_2 gas mixture. After exposure at 500°C, the specimen surface shows isolated regions of Mo sulphide, but no gross oxidation. After 800°C exposure, the specimen exhibits a greater coverage of the surface with Mo sulphide, whereas after exposure at 1100°C, it shows almost complete coverage by Mo sulphide. These preliminary results indicate that the intermetallic material based on Mo silicide can develop protective sulphide scales during service in reducing environments at elevated temperatures.

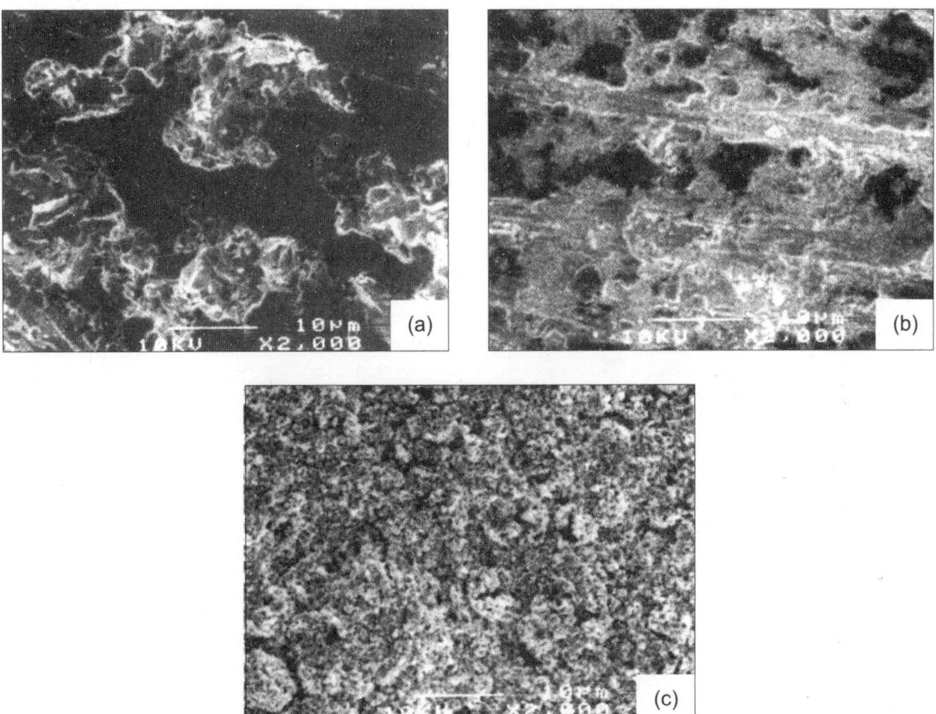

Figure 12 SEM micrographs of surface of B-containing Mo_5Si_3 after sulphidation in 1.5 vol % H_2-H_2S gas mixture at (a) 500°C (b) 800°C and (c) 1100°C

More recently, Du, Datta et al [97] have investigated the sulphidation behaviour of $CrSi_2$, WSi_2, $NbSi_2$, Nb_5Si_3, $MoSi_2$ and Mo_5Si_3 for up to 240 hours in an $H_2/H_2S/H_2O$ environment containing an oxygen potential of $1.2x10^{-15}$ Pa and a sulphur potential of ~$6.8x10^{-1}$ Pa at 850°C. The sulphidation kinetics of all silicides studied, illustrated in Figure 13 indicate extremely good performance in this aggressive atmosphere. The sulphidation resistance shows dependence on the components of the silicides. For example, $CrSi_2$ shows the least sulphidation resistance, which is consistent with poor sulphidation resistance of Cr [5] whilst other silicides reveal superior sulphidation resistance. On the other hand, the Si content in silicides also plays an important role to resist sulphidation attack; the high Si concentration significantly increasing the sulphidation resistance. For example, the sulphidation rates of $MoSi_2$ and $NbSi_2$ are one order of magnitude slower than Mo_5Si_3 and Nb_5Si_3 respectively, as shown in Table 6.

A protective SiO_2 scale observed on all exposed silicides is considered to be responsible for the prevention of further severe environmental attack, as presented in Figure 14. XRD and EDX data confirm also the formation of Cr_2S_3 and NbS_2 on the exposed $CrSi_2$ and Nb_5Si_3. It is interesting to notice that only nodular Cr_2S_3 is formed on exposed $CrSi_2$ and a continuous layer of $CrSi_2$ does not develop even after 240 hours exposure. The virtual destruction of the underlying SiO_2 by the Cr_2S_3 nodules is confirmed by cross-sectioned SEM observation. In contrast, the development of a continuous and uniform NbS_2 layer occurs without affecting the SiO_2 layer which provides the protection for the substrate.

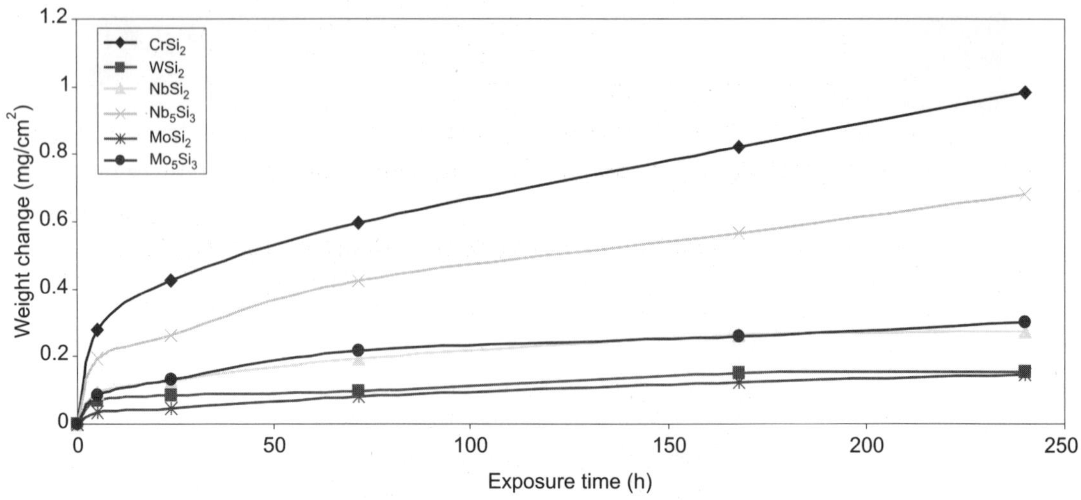

Figure 13 Oxidation and sulphidation of silicides at 850°C

Table 6 Parabolic rate constants (k_p) for sulphidation of $CrSi_2$, WSi_2, $NbSi_2$, Nb_5Si_3, $MoSi_2$ and Mo_5Si_3 at 850°C

Material	$CrSi_2$	WSi_2	$NbSi_2$	Nb_5Si_3	$MoSi_2$	Mo_5Si_3
k_p ($g^2/cm^4/s$)	1.1×10^{-12}	2.7×10^{-14}	8.7×10^{-14}	5.1×10^{-13}	2.4×10^{-14}	1.0×10^{-13}

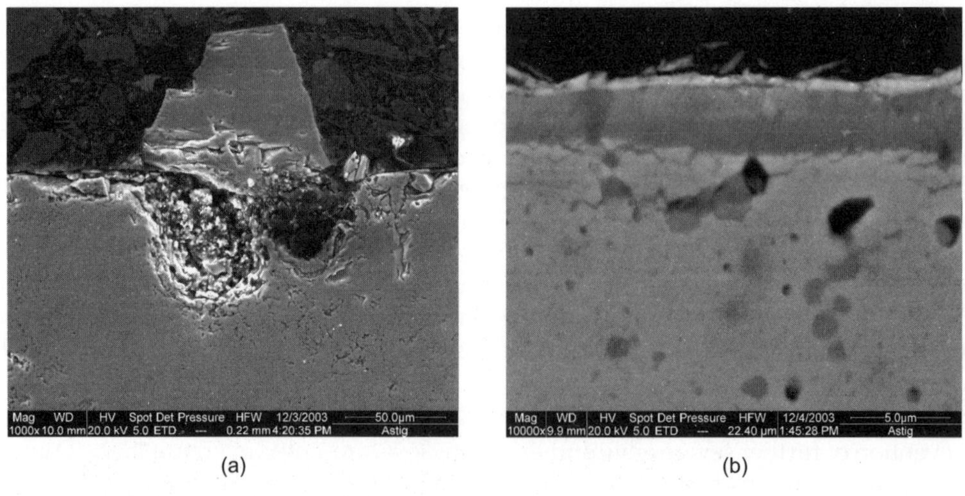

Figure 14 SEM micrographs of exposed $CrSi_2$ and Nb_5Si_3 in $H_2/H_2S/H_2O$ environment at 850°C. (a) $CrSi_2$ and (b) Nb_5Si_3

5. CHLORIDATION – GENERAL CONSIDERATIONS AND REVIEW

There has been an increasing interest in studying the corrosion behaviour of metals and alloys in chlorine-containing environments. The corrosive effects of chlorine at high temperatures are well known as damaging. The corrosion products of metals with chlorine are often low melting point and volatile compounds which are not able to form a protective scale. Chlorine present as a second oxidising species in an oxygen-containing environment, and its role in the acceleration of the reaction of oxygen with engineering materials is more important than the damage done by chloridation alone. Numerous recent publications have demonstrated that in the presence of chlorine, normally protective oxide scales are disrupted and fail to provide protection to the underlying metals.

Studies have indicated that Cr_2O_3-forming alloys (e.g. Inconel 601) exhibit good chloridation resistance at temperature up to 500°C [98]. When exposed at higher temperature (>650°C), however, severe corrosion attack takes place, particularly for Fe-base Cr_2O_3-forming alloys. The corrosion process of Cr_2O_3-forming alloys is characterised mainly by substantial internal attack, probably involving the internal chloridation of chromium. The formation and evaporation of these chlorides cause severe weight loss suffered by such alloys [99,100]. The principal factors that speed up corrosion rates and limit corrosion prevention in chlorine-containing environments are the inherent reactivity of chlorine and the tendency for the metal chlorides formed to exhibit high vapour pressure and low melting points. Table 7 gives a summary of melting points for chlorides, oxides and sulphides of some common metals.

The corrosion resistance of Fe-Cr or Ni-Cr alloys in chlorine-bearing environments has been reported to be essentially dependent upon gas composition, chromium level and exposure temperature [101,102] In bioxidant (oxygen/chlorine) atmospheres, an external Cr_2O_3 scale is likely to be formed on those alloys containing high levels of Cr (> 15wt%) although further corrosion attack of the alloy substrate may not be completely prevented [103]. In HCl or Cl_2 environments, high chromium composition in the alloys often results in the

Table 7 Summary of melting points (°C) for chlorides, oxides and sulphides of some common metals (d=decomposes, s=sublimes)

Metal	Chloride		Oxide		Sulphide	
Fe	$FeCl_2$	677	FeO	1377	FeS	1195
	$FeCl_3$	304	Fe_2O_3	d1462	FeS_2	1171
			Fe_3O_4	1597		
Ni	$NiCl_2$	1030	NiO	1984	NiS	797
		S970			Ni_2S_3	790
					Ni_3S_4	d377
Co	$CoCl_2$	740	CoO	1805	CoS	1100
		s1087	Co_2O_3	d895		
Al	$AlCl_3$	s181	Al_2O_3	2050	AlS	1197
					Al_2S_3	1097
						s1500
Cr	$CrCl_2$	815	Cr_2O_3	2330	CrS	1567
	$CrCl_3$	877			Cr_2S_3	d1350
		s947				

formation of a $CrCl_2$ scale which is protective at temperatures as high as 400-500°C. Above this temperature range, significant evaporation of $CrCl_2$ occurs resulting in the increased corrosion rates of the alloys. The results from studies of Fe-Al alloys clearly indicate the beneficial effect of Al addition in enhancing the corrosion resistance of Fe-Al alloys in oxychlorine environments [104]. The improved corrosion behaviour of Fe-Al alloys is related to the formation of an Al_2O_3 scale which is highly stable in such atmospheres. However, the spallation of the Al_2O_3 scale from the alloy substrates may cause breakaway corrosion.

The corrosion response of a series of HT alloys in various chlorine-containing environments has been reported [105-107] One nickel-based, alumina-forming alloy, HAYNES Alloy 214 in general, has been found to exhibit the best corrosion resistance amongst the commercial alloys tested, including INCONEL 601 and HASTELLOY alloy X. The formation of an external Al_2O_3 scale enhances the environmental durability of HAYNES Alloy 214. Considerable corrosion damage, together with substantial internal attack, and severe weight losses due to the formation and evaporation of chloride species characterise the response of chromia-forming alloys in chlorine-containing environments.

6. HOT CORROSION

Hot corrosion is an accelerated form of corrosion induced by the presence of a thin fused salt film on the surface of a component in the presence of aggressive gases [2,14-21]. The most extensively studied form of hot corrosion is that produced by Na_2SO_4, which is commonly found in deposits on gas turbine hardware. The hot corrosion of nickel-base superalloys and their coatings in these environments has been categorized as Type I (high temperature) and Type II (low temperature) corrosion. Type I is generally observed above the melting point of Na_2SO_4 (884°C) but has been reported to occur as low as 750°C [108]. The Type I morphology is characterized by sulphidation and a depletion zone below the scale and is relatively insensitive to atmosphere composition although it varies from one alloy system to another. Type II hot corrosion is most prevalent at temperatures between 650 and 800°C and requires a significant pressure of SO_3 in the gas phase. The Type II morphology does not involve extensive internal sulphidation or depletion of alloying elements, although both have been observed by cross-sectional transmission electron microscopy [109]. The morphology takes a pitting form in CoCrAl, where it is particularly severe, but is more uniform in other systems such as aluminide coatings.

The kinetics of reaction can usually be characterised by an incubation period where there is little attack followed by a propagation stage during which attack can be catastrophic. The incubation stage is usually evident for Type I corrosion and less so for Type II. The end of the incubation stage results from the penetration of the salt through the protective scale. The path it follows can be cracks caused by thermal cycling, erosion, or the presence of chloride in the deposit or holes produced by chemical attack. The latter likely requires compositional inhomogeneities in the scale as highly perfect scales are very resistant to penetration. Such in-homogeneities have been observed where alloy carbides [110] or yttrides [111] intersected the alloy surface prior to oxidation. The propagation stage varies greatly from one alloy to the next and seems to be different in Type II and in Type I.

The so-called "oxide-scale fluxing" model proposed by Rapp and co-workers [98,100-104] to explain the mechanisms for hot corrosion involves, in essence, the formation of protective oxide scales by high-temperature oxidation are dissolved in or penetrated by, fused salt as anionic species (basic fluxing) or cationic species (acid fluxing) depending on the salt composition. For example, salts equilibrated at high SO_3 partial pressures are said to be acidic. The most important salt-component reactions are sulphidation and oxidation which are similar to those occurring in gaseous atmospheres and reaction with vanadium which forms highly soluble vanadates with most oxides [112]. There is generally agreement that Type I corrosion in vanadium-free environments propagates through some combination of sulphidation or oxidation and basic fluxing [108]. However, the propagation mode for Type II corrosion is still rather uncertain. Most proposed mechanisms involve acid fluxing of oxides since high SO_3 partial pressures are required [19,113,114].

Among all pure metals, hot corrosion of pure nickel in the presence of Na_2SO_4 occurs with a fast linear rate. Under the condition of Type I hot corrosion, the molten Na_2SO_4 dissolves the NiO scale according to either acidic or basic fluxing mechanism depending on the gas partial pressures of SO_3 and O_2. Under the condition of Type II hot corrosion, the reaction products consist of a network of Ni_3S_2, that serves as a rapid diffusion path for the outward diffusion of nickel, thereby resulting in rapid corrosion rate.

Chromia-forming alloys, such as Ni-Cr or Co-Cr based alloys, are more resistant to hot corrosion than pure metals due the their capability to form a protective chromia scale.

7. NANOSCALE STUDIES OF SOME ISSUES IN HIGH TEMPERATURE CORROSION

Nanoscale knowledge of materials permits us to investigate materials phenomena such as corrosion at a most fundamental level. Corrosion issues appear in nanoscience and nanotechnology in two ways: (i) the opportunity to study corrosion phenomena – particularly the early stages of corrosion – in a fundamental way, and (ii) corrosion processes are used in the development of nanotechnology.

7.1 TiAl Oxidation

Nanoscale studies of the early stages of the oxidation behaviour of TiAl allow fundamental information on the scaling processes to be obtained. Such information is necessary to develop structure-based alloy design methodology. Du and Datta et al [115] have carried out nanoscale studies of the early stages of the oxidation behaviour of TiAl using scanning tunnelling microscopy. A two-stage strategy was used. The first stage involved scanning tunnelling microscopy (STM)/scanning tunnelling spectroscopy (STS) investigations of TiO_2(110)-(1x1) surface in order to establish "fingerprints" for the identification of the products formed during oxidation of TiAl. In the second stage a TiAl intermetallic alloy was studied by STM/STS after repeated sputtering and heating (to provide ideal surfaces) both in low and high oxygen potential environments. For TiO_2 it was found that the oxygen vacancies produced create additional defect states in the band gap of stoichiometric TiO_2 (Figure 15b). This state can be taken as fingerprint for the reduced TiO_2 surface indicating the presence of Ti_2O_3. The energy for this state estimated from the STS measurement is in agreement with previous results [115].

Following oxidation at room temperature with 100 Langmuir oxygen, the TiAl surface with well-developed terraces, ledges and kinks, shows islands of oxides of nanometre size and monolayer height (Figure 15a). The tunnelling spectroscopy results recorded on these islands

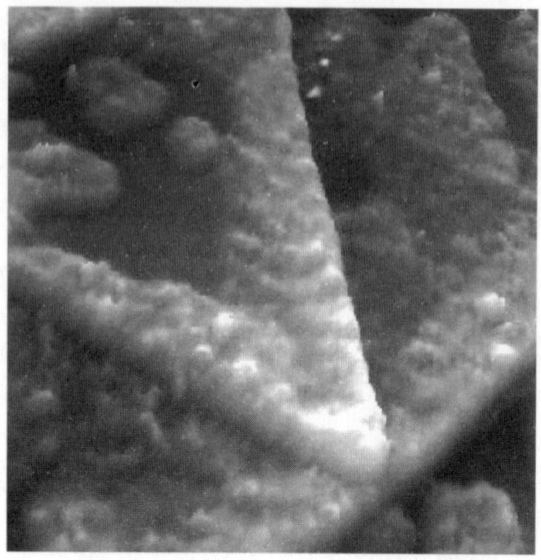

(a) 175 nm × 175 nm STM TiAl topography

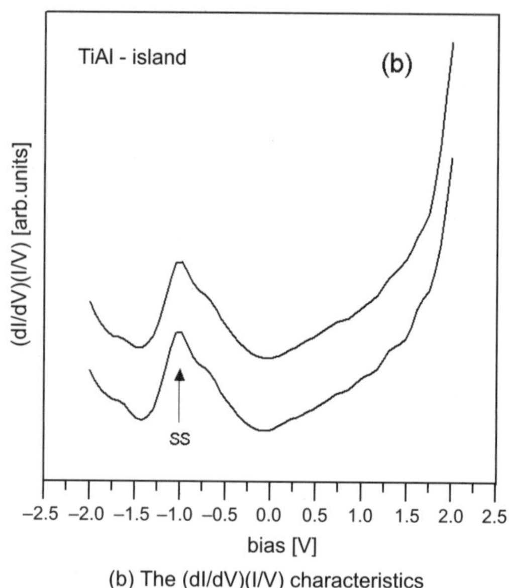

(b) The (dI/dV)(I/V) characteristics

Figure 15 Topography and (dI/dV)(I/V) curves for TiAl after exposure to environment with 100 Langmuir oxygen

show the semiconducting character and peaks at ~1 eV below the Fermi level (SS) on the normalised tunnelling conductance dI/dV curves – Figure 15b. This curve resembles the curve obtained for TiO_2 showing the creation of surface states (SS) were due to the formation of Ti_2O_3. Thus, the curves in Figure 15b can be taken to confirm the nucleation of Ti_2O_3 in a low oxygen pressure environment and clearly demonstrates that oxidation of TiAl begins with the formation of Ti_2O_3. Surface morphology of TiAl alloy after exposure to high (atmospheric) oxygen potential environment is shown in Figure 16a. The I/V curve recorded on this surface (Figure 16b) reveals an asymmetric shape and shows that the tunnelling current is higher for the positive polarization of the sample than for the negative voltage of the same value. Furthermore a well-defined suppression of the tunnelling current, i.e. presence of the I(V)»0 region in -1.5eV to 0.8 eV energy range, is observed. This type of asymmetry and suppression of the tunnelling current can be explained by the presence of TiO_2 material on the surface as amorphous TiO_2 and can be regarded as a 3 eV band-gap binary oxide (suppression of the tunnelling current) with the Fermi level shifted towards the valence band (asymmetry of the tunnelling current) [115]. Such fundamental information is essential to study the initial stages of scale growth and explain the rapid diffusion observed during the onset of oxide nucleation.

7.2 Tribo-Corrosion

Tribo-corrosion is another area where nanoscale studies of the processes involved provide improved insight into mechanisms of high temperature wear. High temperature wear is a serious problem in many situations - power generation, transport, materials processing and nuclear reactors [116-119]. The problem of high temperature wear is accentuated due to faster kinetics of surface oxidation, loss of mechanical strength of the materials constituting the contacting surfaces and change in adhesion between these surfaces caused by the joint action of

(a) 300 nm × 300 nm STM TiAl topography

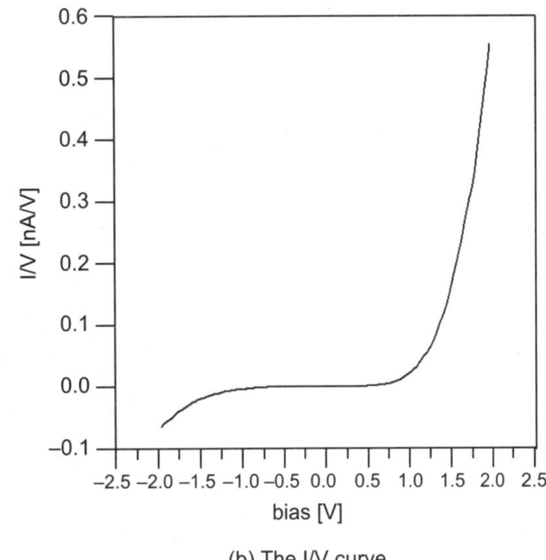

(b) The I/V curve

Figure 16 Topography and I/V curve for the oxidised TiAl after exposure to atmospheric environment

temperature and tribological parameters. The conditions associated with the presence of high temperature severely restricts the choice of coatings and materials that can be used to prevent/ minimise high temperature wear [116-120].

One of the most elegant methods of generating wear resistant surfaces on coated and uncoated materials is to take advantage of the important events - oxidation and debris generation and elemental transfer between the contacting surfaces - which accompany the process of high temperature wear [116-117]. These events under certain conditions of temperature, pressure and speed lead to the formation of surface glazes on the contacting surfaces with immunity to further wear [121]. Fundamental knowledge and research methodology of nanoscience and nanotechnology also allow investigation of the formation and high temperature wear resistance of nanostructured glaze layers and the influence of such layers on the processes of high temperature wear.

These issues are discussed in terms of the analysis of the structures and substructures of nanostructured glaze layers developed during high temperature sliding wear tests of Nimonic 80A (Fe 0.7%, Ni 75.8%, Cr 19.4%, Al 1.4%, Ti 2.5%, Si 0.1%, C 0.08%) against Stellite 6 (Fe 2.5%, Ni 2.5%, Cr 27%, Mn 1%, W 5%, Co 60%, Si 1%, C 1%) were performed as a function of temperature (up to 750°C) under a load of 7N at a speed of 0.314 ms^{-1} [122-124].

Focussing attention on the situation at 20 and 750°C the SEM EDX analysis demonstrates the debris intermixing and oxidation. The dominant phases identified by XRD include $CoCr_2O_4$ and NiCrFe. A cross-sectional composite transmission electron micrograph of the surface formed during wear test at 750°C (Figure 17) shows the presence of the surface layer (glazed surface), the deformed substrate and the glazed layer/substrate interface. The wear-affected region (total thickness ~3 mm) consists of three layers - the top most layer (the glaze layer) showing the presence of grain structure of size 5 to 15 nm with some low dislocation density; some of the grains display contrast.

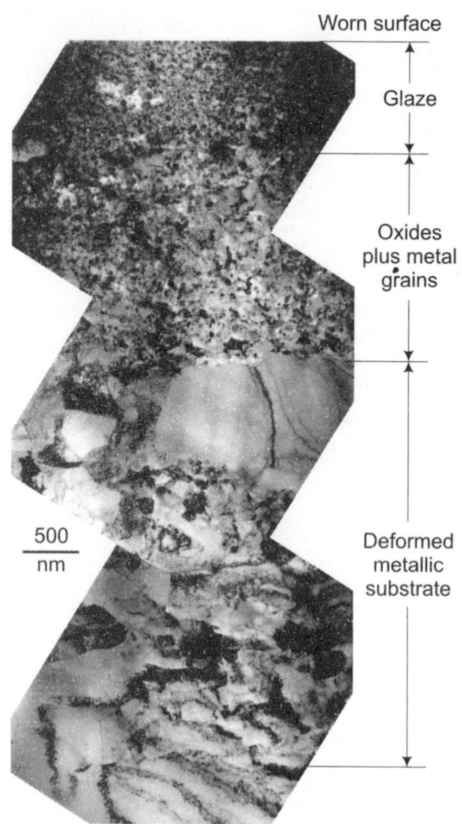

Figure 17 TEM bright field image: wear-induced polycrystalline glaze layer and deformation of substrate

The interfacial layer consisting of grains of size 10–20 nm shows a higher dislocation density. The layer just beneath the interfacial layer shows subsurface deformation and the presence of elongated grains. The structure of the glazed layer and the selected area diffraction (SAD) pattern are separately presented in Figure 18. The SAD pattern consisting of spots arranged in concentric circles indicates the presence of small grains with high angle boundaries, multiple boundaries and large misorientations (formation of misorientated lattice-fragmentation). The poorly defined irregular boundaries indicate non-equilibrium high-energy configuration. The indexed SAD pattern also reveals the presence of oxides of Ni, Cr and Co (indexing not shown here).

The present results clearly demonstrate the formation of a nano-structured glaze oxide layer during high temperature sliding wear of the system Nimonic 80A/Stellite 6 at 750°C using a sliding speed of 0.34 ms^{-1} under a load of 7 N. The creation of nano-structures is also confirmed by the STM topography, as shown in Figure 19. The grain sizes vary from 5 nm to 10 nm with grain thickness ~10 nm. These results together with the wear data demonstrate that such a nano-structure surface is extremely effective in conferring high resistance to wear.

Detailed TEM and STM studies carried out here allow an understanding of the mechanisms of formation of wear resistant nano-structured surfaces. The initial processes

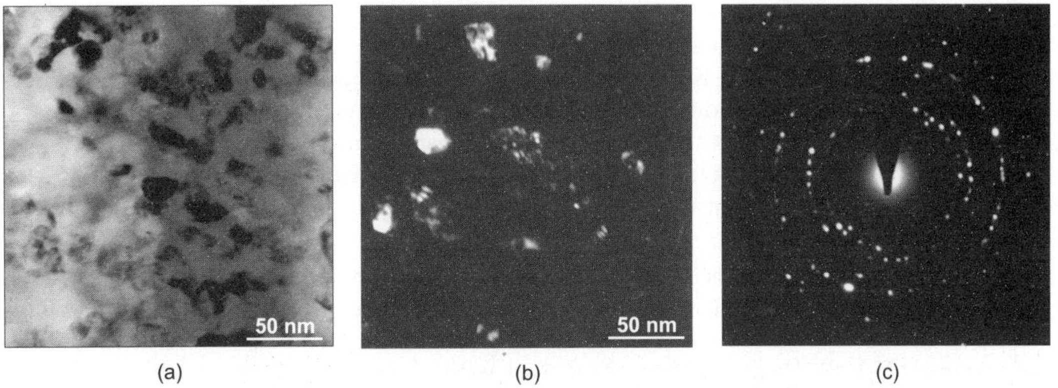

Figure 18 TEM morphological and structural details of glaze layer. (a) Bright field image, (b) dark field image and (c) selected analysis diffraction

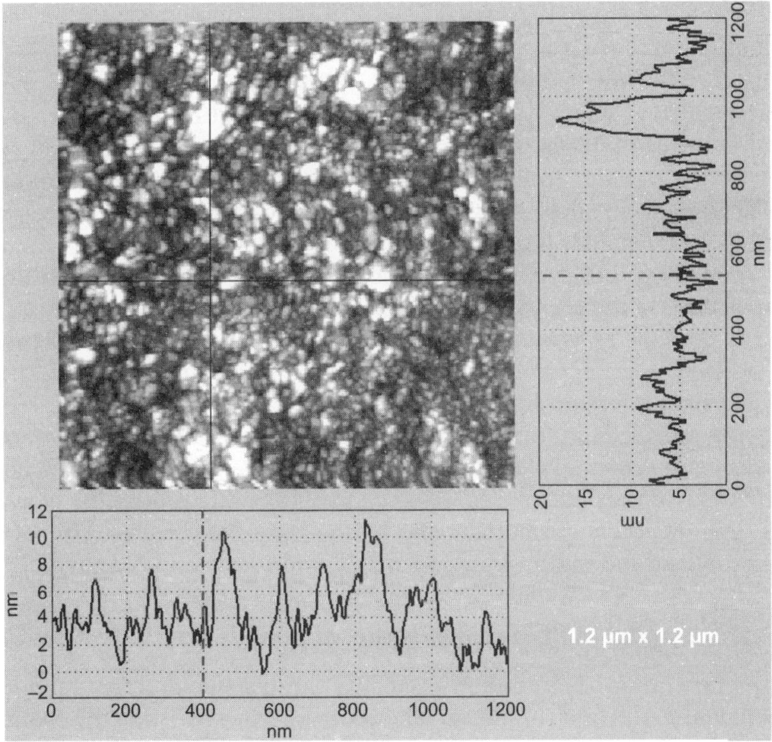

Figure 19 STM surface line profile results on glaze layer formedon Nimonic 80A at 750°C

responsible for generating the glazed layer involve: (i) intermixing of the debris generated from the wearing and the counterface surfaces, (ii) oxidation, (iii) further mixing and (iv) repeated welding and fracture. The next step involves deformation of oxides and generation of dislocations leading to the formation of sub-grains. These sub-grains are then further refined with increasing misorientation giving nano-structure grains with high angle boundaries and non-equilibrium states indicated by poorly defined and irregular grain boundaries. High internal stress is created inside the grains – dislocation density and arrangement depending on the grain size; smaller grains containing fewer dislocations. The process leads to the formation of high-energy grain boundaries with a high defect density. Furthermore, the oxides – TiO_2 and Al_2O_3 – formed beneath the glaze layer play an important role in providing support, additional to that provided by the substrate, to sustain the glaze layer preventing it from collapse as shown in Figure 20.

The superior wear resistance of this nano-structured glaze layer can be ascribed to several factors: the absence of Hall-Petch softening and the occurrence of a significant degree of work-hardening believed to be associated with the difficulties in generating dislocations in nano-sized grains [125,126]. These two factors together with improved fracture toughness, expected to be conferred by the presence of nano-polycrystalline structure, make debris generation an inefficient process.

8. MODELLING

Numeric modelling has broad applications in materials science and engineering including high temperature corrosion. Extensive studies of high temperature phenomena leading to improved understanding of the processes of high temperature corrosion now allow us to model several aspects of high temperature corrosion – scale formation, scale growth, interdiffusion, phase stability (thermodynamics and kinetics), scale deformation and scale fracture. Such activities will allow design of high performance alloys/coatings which produce highly protective scale against high temperature corrosion. The thermodynamic modelling and diffusion modelling will provide scientific principles for designing and optimising the chemical compositions of the alloys/coatings. Such modelling will also allow an assessment of the nature of the possible reaction products in different environments and prediction of the formation of protective scale. On the other hand, the scaling modelling and finite element calculation will help understanding the formation and degradation processes of the scale formed on the alloy surfaces and will allow further optimisation of the alloys/coatings. Based on the preliminary experimental work, life time modelling can be developed which will predict the possible life time for the designed coatings and significantly reduce the workload and save resources. In this chapter, numerical modelling will be limited to consideration of interdiffusion problems and failure mechanisms of oxide scale and coatings.

8.1 Diffusional Transport in Pt-Modified Nickel Aluminide Coatings

For many high-temperature, degradation-resistant coatings, two dominant modes characterise their failure behaviour – (a) and (b) (Figure 21) – both of which denude the coating of elements incorporated to confer resistance to the high-temperature corrosion processes. The development and maintenance of a scale are ensured by the continuous consumption by oxidation of the scale forming elements. Without a sufficient reservoir of such elements the scale regeneration at the interface 1 cannot occur triggering the onset of the base alloy oxidation. The second mode of degradation – Mode (b) – occurs by the outward diffusional transport of damaging substrate elements which residing at the scale/coating interface or

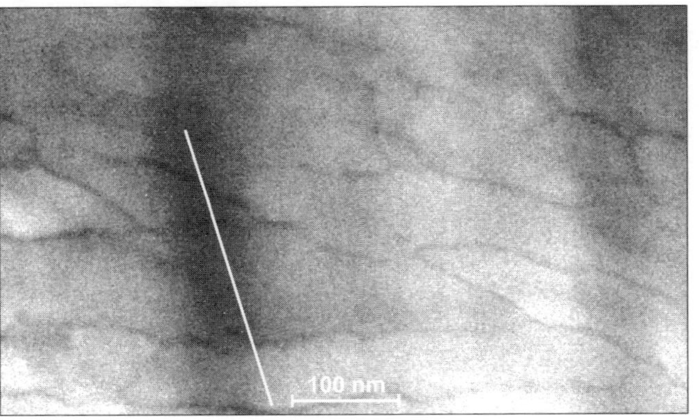

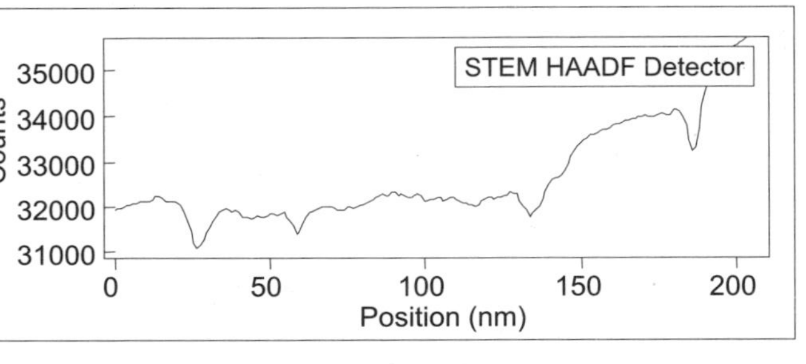

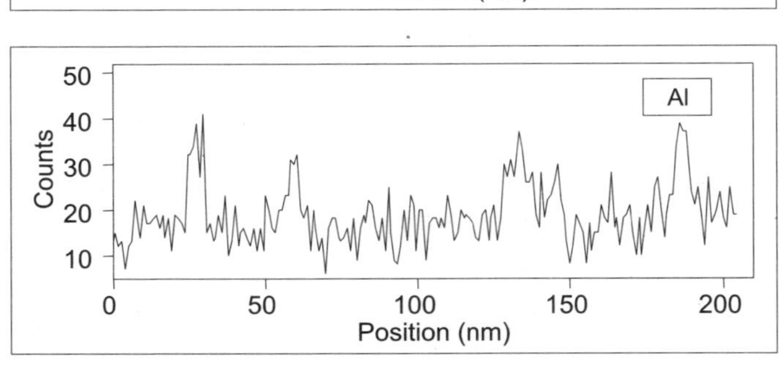

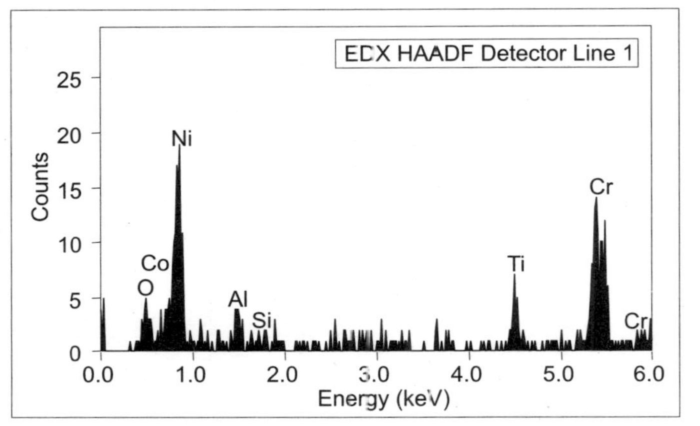

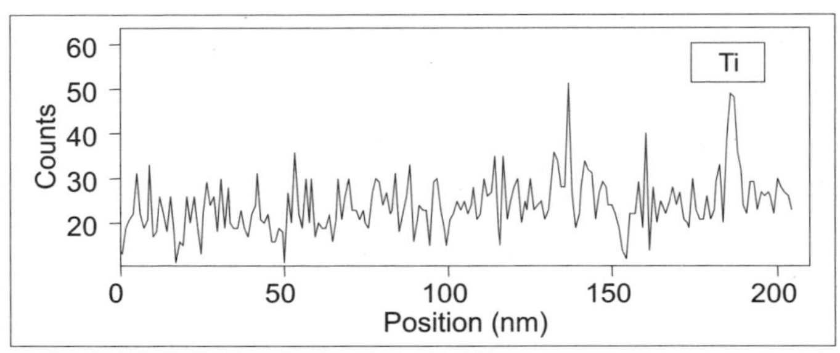

Figure 20 HAADF-STEM (50 mm camera length) overview and EDX line trace revealing the precipitation of light elements (aluminium and titanium) in the Nimonic 80A close to the interface

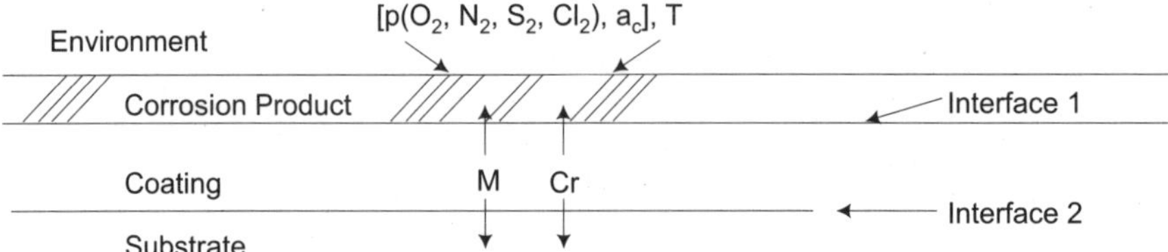

Figure 21 Schematic diagram indicating various high temperature coating degradation mechanisms [p is partial pressure, a is activity]

within the scale undermine the scale integrity. At Interface 2 the inward diffusion of the scale-forming element will also affect the regeneration of the protective scale following spallation.

The problem of interdiffusion is illustrated for an aluminide coating in Figure 22. The coating consists of two layers – a thick outer NiAl layer and a thinner layer containing a mixture of NiAl and Ni_3Al. The substrate elements gathering in the inner-layer inhibits further secondary diffusion. The influence of the substrate type, composition and microstructure on the development of an aluminide coating is illustrated in Figure 23. Increasing Ni increases the diffusional growth rate of the outer NiAl phase, and also suppresses, with benefit, the development of the solid solution or duplex layer. Inhibition or minimization of interdiffusion is critically important. This can be achieved by optimising the base alloy composition and/or through the introduction of a anti diffusion barrier at the coating/substrate interface

In the choice of a suitable anti-diffusion coating, empiricism predominates over understanding and modelling of the solution for interdiffusion and partitioning of elements, even in simple systems, is poorly understood. There is a particular dearth of interdiffusion data. In last few years, interdiffusion modelling work undertaken by Datta et al [127] has made some progress. Such work will significantly facilitate the design of complex coating systems. Interdiffusion problems have been modelled using the Generalized Darken Model [128,129]. The model allows calculation of the concentration profiles in single-phase, multi-component and multi-layer systems [130]. Using this model, interdiffusion in a Pt-modified β-NiAl coating on MAR M002 has been considered [127]. The computed and experimentally measured densities of Al, Pt and Ni at 1073 K after 200 h of diffusional annealing in an argon atmosphere are shown in Figure 24. Other applications of the Generalized Darken Model including calculations of the intrinsic diffusivities and modelling of the interdiffusion in the open system, i.e. selective oxidation of the Pt-modified β-NiAl coating on MAR M002 are presented further in this section.

The Generalized Darken Model [128-130] which facilitates a description of the interdiffusion process in both open and closed systems and when component intrinsic diffusivities vary with composition (the problem is simplified problem by using average values of intrinsic diffusivities), has been used for the interdiffusion modelling in a Pt-modified β-NiAl coating on MAR M002. The average intrinsic diffusivities of Al, Cr, Co, Ni and Pt for the closed system (Pt-modified β-NiAl coating on MAR M002) have been calculated [126] – Table 8 – using the so-called "Inverse Method" [130]. The results of modelling the interdiffusion in the

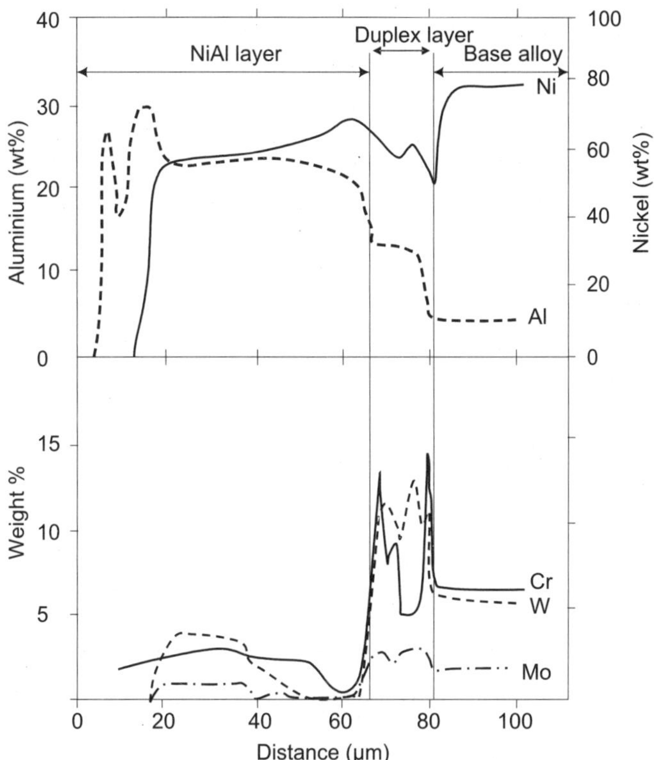

Figure 22 The distribution of elements in an aluminide coating on Ni-based superalloy (reproduced by permission of the Institute of Metals)

Pt-modified â-NiAl/MAR M002 diffusional couple (closed system) at 1073 K using the intrinsic diffusivities (Table 8) are presented in Figure 24.

For the calculations of the concentration profiles in the oxidised coating/alloy system the following data were used:

1. Atomic masses of Al, Cr, Co, Ni and Pt: 26.98, 51.99, 58.83, 58.71 and 195.09 g mol^{-1}.
2. Thickness of the diffusional couple (β-NiAl(Pt)) coating on MAR M002): $2A$ = 80μm.
3. Total concentration: c = 0.114mol cm^{-3}.
4. Annealing time t* = 100 h.
5. Constant (average) intrinsic diffusivities of Al, Cr, Co, Ni and Pt (Table 1).
6. Estimated values of oxygen uptake ($k_p = 10^{-12}$ $g^2cm^{-4}s^{-1}$) at 1173K were used to calculate the flux of oxygen as a function of time ($J_o(t)$) and the equivalent flux of Al through the coating/scale interfaces: J_{Al_L}, $J_{Al_R}(t)$.

The intrinsic diffusivities given in Table 8 have been used in the computer modelling of the interdiffusion in the selectively oxidized, Pt-modified β-NiAl coating on MAR M002 (i.e. Al is the reacting metal forming the Al_2O_3 scale) was undertaken. The computed densities of Al, Ni and Pt in the Pt-modified β-NiAl coating on MAR M002 after oxidation at 1173 K for 100 h are shown in Figure 25. Satisfactory agreement can be seen in the case of Al and Ni distributions,

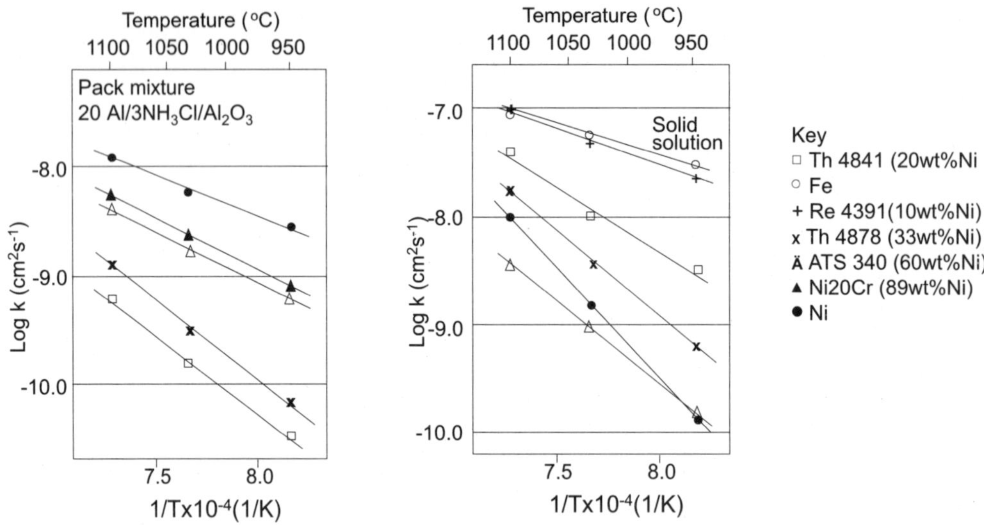

Figure 23 The temperature dependence of the growth of the NiAl phase (left) and the solid solution zone (right) in aluminized substrate alloys with various nickel+chromium contents between 20 and 25wt% (reproduced by permission of Applied Science Publisher)

Table 8 Compound intrinsic diffusivities in the substrate/coating system (Pt-modified β-NiAl coating on MAR M002)

Temperature K	Intrinsic diffusivities, cm^2s^{-1}				
	D_{Al}	D_{Cr}	D_{Co}	D_{Ni}	D_{Pt}
1073	2.73×10^{-12}	8.59×10^{-13}	9.42×10^{-13}	2.50×10^{-13}	8.30×10^{-13}
1173	9.28×10^{-12}	2.78×10^{-12}	4.17×10^{-12}	1.39×10^{-11}	6.28×10^{-13}
1273	2.49×10^{-11}	1.43×10^{-11}	1.05×10^{-11}	1.68×10^{-11}	2.17×10^{-11}
1373	5.08×10^{-11}	2.77×10^{-11}	2.27×10^{-11}	1.06×10^{-10}	4.22×10^{-11}

but the measured Pt densities show marked divergence from the calculated values. This divergence probably stems from underestimation of the values of the intrinsic diffusivities (Table 8). Recent measurements of the concentration profiles of Al, Ni and Pt in the Pt/β-NiAl diffusion couple after 60 minutes of annealing at 1273 K (Figure 26) show the asymmetric nature of the Pt concentration profile suggesting the invalidity of the assumption of constant average diffusivity. Progress in obtaining more precise intrinsic diffusivities in such complex systems may be possible using the Morral-Thompson method of average composition.

Apart from chemical effects of compositional gradients, mechanical stability is affected by the presence of stresses at the coating/substrate or coating/scale interfaces and within the coating itself. The most important sources of stresses are:

(i) the external stresses producing deformation;

(ii) the thermally induced stresses arising from differences in thermal expansion coefficients across the coating/substrate, coating/scale or substrate/scale interfaces; and

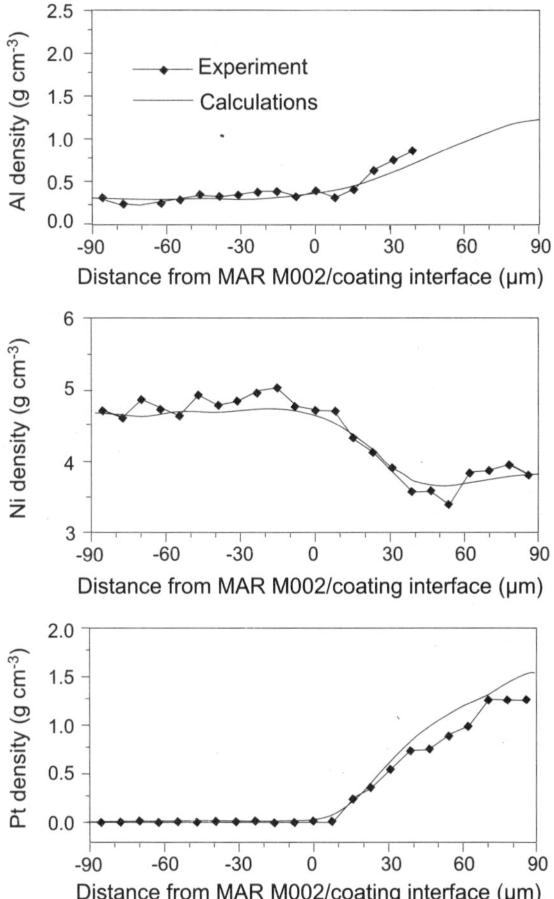

Figure 24 The calculated and measured density profiles of Al, Ni and Pt in the oxidised Pt-modified β-NiAl coating on MAR M002 superalloy at 1073K for 200h

(iii) the stresses in the coating associated with the growth processes, phase separation and precipitation.

In recent years considerable research work at Northumbria University has been undertaken to obtain the quantitative information needed to model the behaviour of the protective scale in the presence of mechanical stresses. Nevertheless uncertainty remains in this area.

8.2 Failure Mechanisms During Oxidation

Recent work at the Northumbria University [131,132] has modelled the deformation and fracture behaviour of Al_2O_3 scales formed on cylinders of Fe_3Al intermetallics with chamfered edges providing improved insight into the scale failure phenomena. The model included bonding between oxide and substrate and the ability of a through-surface crack to develop at a

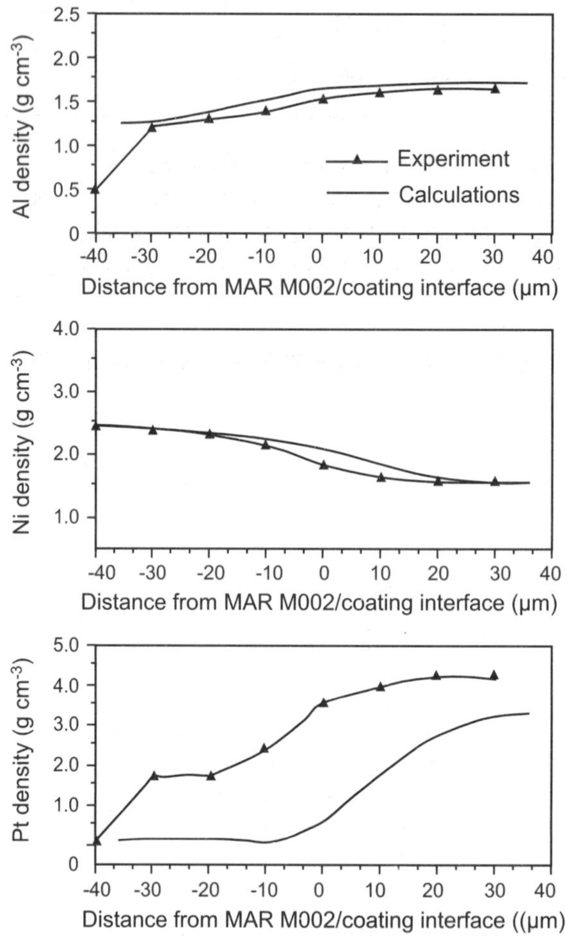

Figure 25 The calculated and measured density profiles of Al, Ni, and Pt in the oxidised Pt-modified β-NiAl coating on MAR M002 superalloy at 1173K for 100h

cylinder edge. A uniform cooling rate of 100 K/s from a temperature of 1273 K to 293 K without creep was assumed together with temperature dependent coefficients of thermal expansion (CTEs).

On running the model the following results were obtained:

1. With embedded defects introduced at both the scale/substrate interface and the oxide edge, the following sequence of events was observed for an elastic analysis:
   (i) the oxide did not fracture;
   (ii) scale spallation commenced at 1091 K, beginning at the oxide/substrate embedded defect;
   (iii) the scale completely spalled, in one piece, at 676K.

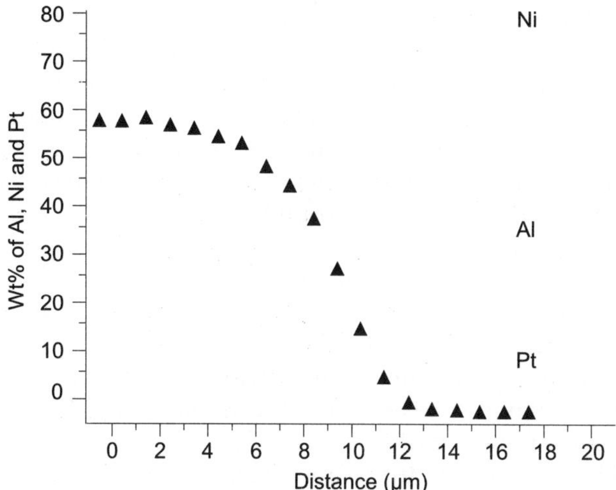

Figure 26 Concentration profiles of Al, Ni and Pt β-NiAl diffusional couple after 60 min of diffusional annealing at 1273K

2. When the same defects as in 1 above were introduced for a perfectly plastic substrate:
   (i) the plastic deformation of the substrate caused the oxide near the edge to behave like an elastic beam in bending and at 1109 K crack growth began in the oxide layer;
   (ii) the oxide started to detach from the substrate at a temperature of 728 K; a much lower propagation temperature than in the linear substrate case, and was due to stress relaxation caused both by plastic deformation of the substrate and by oxide fracture;
   (iii) the oxide itself split at 668 K, and the oxide/substrate interface spalled at 528 K.
3. With only a crack at the interface and no crack at the oxide edge, again for a plastic substrate:
   (i) the onset of delamination occurred at 874 K; and
   (ii) the observed reduced stress relaxation than in the previous scenario was because the oxide remained intact and under tension at the outer edge of the chamfer, as the matrix deformed plastically.

Clearly, corners or edges are the locations of failure, which demonstrates the need to design alloys and coatings which will produce defect-free scales or which are capable of producing crack-healing mechanisms. One approach is to introduce an interlayer. The function of such an interlayer is to promote coating/substrate adhesion by providing a composition/stress gradient interface.

8.3 Modelling of Crack Formation and Degradation of Coating Systems

When ceramics with high melting points and high thermal stability are employed as coating materials to protect the metallic or intermetallic alloys from high temperature corrosion, the formation of micro-cracks, which act as diffusion paths, due to the mismatch of thermal expansion coefficients between the coating and substrate always poses a problem. Attempt has been made [95] to model the formation of such cracks in a TiAlN-coated TiAl system. The development of cracks in the AlTiN coating from pre-existing defects within the coating or at the coating/substrate interface has been explained using numerical modelling in the following way. An AlTiN coating on a Ti-46.7Al-1.9W-0.5Si substrate was modelled using the general finite element (*ABAQUS*). The model used consisted of a 25 × 25ìm substrate on which was a 2.5ìm thick nitride coating. The left and right boundaries of the model were constrained so as to be infinite, thus removing any edge effects, which would otherwise be present. The 25ìm depth of the substrate was chosen to keep the model reasonably sized; although quite thin, this substrate was large in comparison to the 2.5ìm thickness nitride coating. 6840 second order plane-strain elements (determined by mesh-sensitivity analysis) were used throughout the model (Figure 27) and the nitride coating was bonded to the substrate with contact surfaces. An interfacial defect (created by node release method) was placed at a distance of 12.5ìm along the nitride coating/substrate interface, which acted as a stress raiser for both interfacial defects and cracks in nitride coating. Due to the nature of the ceramic materials, a linear elastic analysis was performed which facilitated the use of stress-based failure criteria for the nitride coating and nitride coating/substrate interface. The mechanical properties used included 1.4GPa for the tensile stress of the coating at 600°C (approximated from the hardness and hot hardness data) and the coating/substrate interfacial strength was assumed to be 1.2GPa. The model was subject to a thermal cycle consisting of cooling from the deposition temperature (400°C) to room temperature and then increased to 870°C over a 45-minute period. The modelling results demonstrate that the embedded defect caused rapid mode I failure of the nitride coating at a temperature of 774°C. The through-fracture caused the nitride coating to shrink back leading to delamination around the crack in the nitride coating (Figure 28). It is interesting to notice that the shape of the stress field is very comparable to the observed nodular corrosion products, as illustrated in Figure 29, which may imply that the stress enhanced the diffusion of reactant species. At the end of the analysis each side of the nitride coating had delaminated by 1μm. This through-thickness crack within the coating acted as a rapid diffusion path for the reactant species during the isothermal exposure.

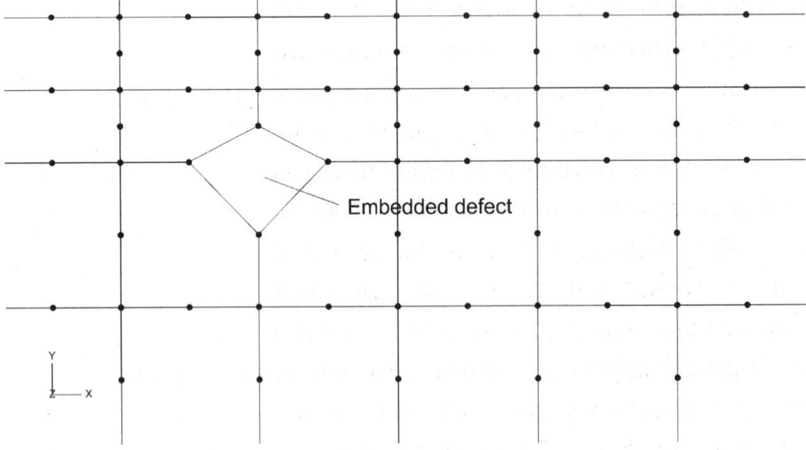

Figure 27 Two dimensional FE model showing details of elements comprising the nitride coating

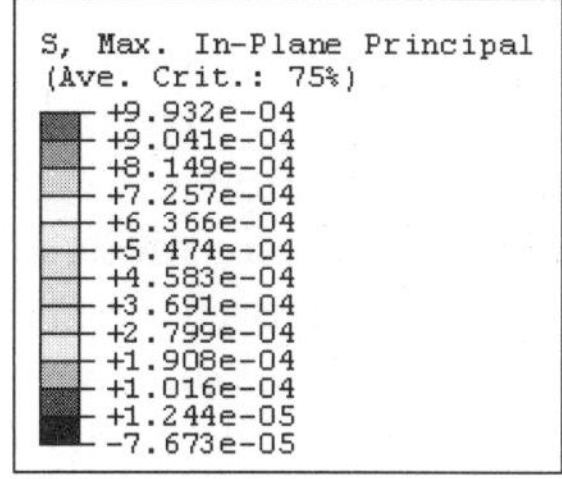

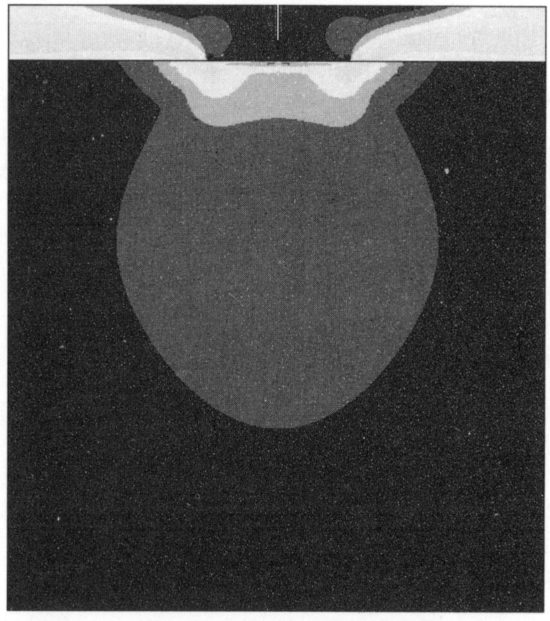

Figure 28 Result of FE analysis showing through-crack in nitride coating

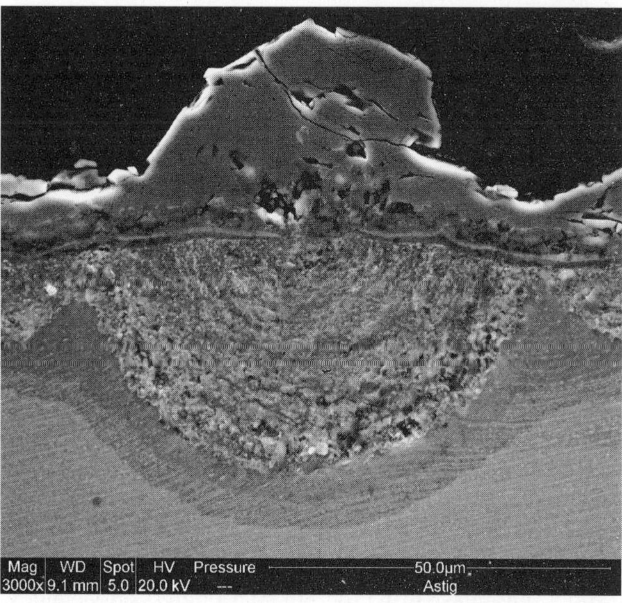

Figure 29 Cross-sectional morphology of AlTiN-coated Ti-46.7Al-1.9W-0.5Si after 240 hours exposure at 850°C in the environment of $H_2/H_2S/H_2O$

9. SUMMARY

In writing this chapter, strong emphasis has been placed on the scientific principles governing the phenomena of high temperature corrosion. The corrosion behaviour of those materials which have provided new information on the mechanisms of scale formation and scale growth has been considered. The development of interdiffusion and scale fracture models has been discussed. Mechanistic information has been used to develop new microstructural models of corrosion. Some of the information presented here originated from the authors' own research. Clearly the lack of space has forced us to be selective in choosing the topics presented here. The novel aspects of this chapter are the inclusion of nanocorrosion and nanoscale elaboration of the processes of wear resistant glaze formation in the area of tribo-corrosion.

ACKNOWLEDGEMENT

Some materials from "P.K. Datta, K. Natesan and J.S. Burnell-Gray, Coatings Technology, *Intermetallic Coatings and Coatings for Intermetallics, Intermetallic Compounds: Principles and Practice Vol. 3 Progress*, J.H. Westbrook and R.L. Fleischer, Ed., 2002" have been reproduced in this chapter with permission from John Wiley & Sons, Ltd.

REFERENCES

1. P. Kofstad, *High Temperature Corrosion, Elsevier Applied Science*, London and New York, 1988.
2. A.S. Khanna, *High temperature oxidation and corrosion*, ASM International, Ohio, 2002
3. F. Gesmundo and B. Gleeson, *Oxidation of Metals*, 44, 211-237 (1995)
4. N. Birks and G.H. Meier, *Introduction to high temperature oxidation of metals*, Edward Arnold, 1983
5. K.N. Strafford and P.K. Datta, *Mater. Sci. Technol.*, 5, 765 (1989)
6. D.J. Young and S. Watson, *Oxidation of Metals*, 44, 239-264 (1995)
7. P.K. Datta, K. Natesan and J.S. Burnell-Gray, Coatings Technology, *Intermetallic Coatings and Coatings for Intermetallics, Intermetallic Compounds: Principles and Practice Vol. 3 Progress*, J.H. Westbrook and R.L. Fleischer, Ed., John Wiley & Son Ltd., 2002, p 561-588
8. K. Natesan, in Proc. Corrosion-Erosion-Wear of Materials in Emerging Fossil Energy Systems, ed., A.V. Levy, 1 (1982)
9. D.L. Douglass, P. Kofstad, A.Rahmel and G.C. Wood, *Oxidation of Metals*, 45, 529-620 (1996)
10. Z.A. Foroulis, *Anti-Corrosion*, 32, 4 (1985)
11. N. Birks, G.H. Meier and F.S. Pettit, *JOM*, Vol. 39, p 28-31 (1987)
12. H.L. Du, PhD thesis, *Studies of high temperature corrosion of some MCrAlYX-type alloys*, Northumbria University, 1991.
13. P.K. Datta, H.L. Du, D. Jenkinson, J.S. Gray, and K.N. Strafford, The design of sulphidation resistant materials: state-of-the-art and future prospects, CORCON-97: Corrosion and Its Control. Vol. I; Mumbai; India; 3-6 Dec. 1998. pp.176-187
14. R.A. Rapp, *Corrosion Science*, 44, 209-221 (2002)
15. J.A. Goebel, F.S. Pettit and G.W. Goward, *Metall. Trans.*, 4, 261-278 (1973)
16. M.G. Hocking and V. Vasantasree, in Proc. 8th Int. Congress on Metallic Corrosion, Frankfurt, 1, 701 (1981)
17. R.L. Jones and C.E. William, *Mater. Sci. Eng.*, 87, 353 (1987)
18. O.H. LeBlanc, Jr., K.L. Luthra and R.W. Haskell, *Oxid. Met.*, 31 293 (1989)

19. K.L. Luthra, *J. Electrochem. Soc.*, 132, 1293 (1985)
20. R.A. Rapp, *Mater. Sci. Eng.*, 87, 303 (1987)
21. H.J. Grabke, *Materials and Technology*, 36, 297-305 (2002)
22. H.J. Grabke, E.M Muller-Lorenz, J. Klower and D.C. Agarawal, *Materials Performance*, 37, 58-63 (1998)
23. H.J. Grabke, *Materials Science Forum*, 369-372 , 101-108 (2001)
24. R.C. Yin, *Materials Science and Engineering A*, 380, 281-289 (2004)
25. J. Zhang, A. Schneider and G. Inden, *Corrosion Science*, 45, 1329-1341 (2003)
26. H.J. Grabke, A Zahs and M. Spiegel, *Corrosion Science*, 42, 1093-1122 (2000)
27. J.W. Kwon, Y.Y. Lee, Y.D. Lee, *Materials at High Temperatures*, 17, 319-326 (2000)
28. X. Zheng and R.A. Rapp, *Oxidation of Metals*, 48, 527-551 (1997)
29. H. Chu, and P.K. Datta, *Oxidation of Metals*, 43, 491-508 (1995)
30. S. Balakrishnaan, J.S. Burnell-Gray, BAUON P.K. Datta, Key Engineering Materials. Vol. 99-100, pp. 279-289. 1995
31. K.N. Strafford, P.K. Datta, G. Forster, *Materials Science and Engineering A*, A120-A121, 61-68 (1989)
32. K.N. Strafford, P.K. Datta, G. Forster, *Corrosion Science*, 29, 703-716 (1989)
33. H. Chu, PhD thesis, *High-Temperature Corrosion of Fe(Ni)CrAlX Coating Type Alloys in Oxygen/Chlorine Environment*, Northumbria University, 1992.
34. F.D. Richardson and J.H. Jeffes, *JISI*, 171, 165 (1952)
35. H.J. Grabke, High Temperature Corrosion in Multi-Reactant Gaseous Environments, *High Temperature Materials Corrosion in Coal Gasification Atmospheres*, Ed., J.F. Norton, Elsevier Applied Science Publishers, 1984, p 59-82
36. C. Wagner, *Z. Elektrochem.*, 63, 772 (1959)
37. D.L. Douglass, *Oxidation of Metals*, 44, 81-111 (1995)
38. F.S. Pettit, *Trans. Metall. Soc. AIME*, 239, 1296-1305 (1967)
39. F.H. Stott, G.C. Wood and J. Stringer, *Oxidation of Metals*, 44, 113-146 (1995)
40. K.N. Strafford and J.M. Harrison, *Oxid. Met.*, 10, 347 (1976)
41. K.N. Strafford and G. Smith, *Oxid. Met.*, 14, 119 (1980)
42. R.G. Miner and V. Nagarajan, *Oxid. Met.*, 16, 313 (1981)
43. C.A. Barrett, R.V. Miner and D.R. Hull, *Oxid. Met.*, 20, 255 (1983)
44. T.A. Ramanarayanan, R. Ayer, R. Petkovic-Luton and D.P. Leta, *Oxid. Met.*, 29, 445 (1988)
45. M. Skeldon, J.M. Calvert and D.G. Lees, *Oxid. Met.*, 28 , 109 (1987)
46. G.C. Wood and F.H. Stott, *Mater. Sci. Technol.*, 3, 519 (1987)
47. J.C. Cannotet, M. Lallemant, J.P. Larpin and G. Bertand, *Mater. Sci. Eng.*, 87, 81 (1987)
48. J. Stringer, D.P. Whittle, V. Nagarajan and M.L. Darsan, in Proc. 8th Int. Congress on Metallic Corrosion, Frankfurt, 1, 655 (1981)
49. S.W. Park and G. Simkovich, in Proc. Symp. on Alternate Alloying for Environmental Resistance, ed., G.P. Smolik and S.K. Banerjee, TMS, Warrendale, PA, 233 (1987)
50. R. Pendse and J. Stringer, *Oxid. Met.*, 23, 1 (1985)
51. D.R. Sigler, *Oxid. Met.*, 32, 337 (1989)
52. M.M. El Gomati, C. Walker, D.C. Peacock and M. Prutton, *Corr. Sci.*, 25, 351 (1985)
53. F.I. Wei and F.H. Stott, *Corr. Sci.*, 29, 839 (1989)
54. A.M. Huntz, *Mater. Sci. Eng.*, 87, 251 (1987)

55. J.G. Smeggil, *Mater. Sci. Eng.*, 87, 261 (1987)
56. K. Wambach, J. Peter and H.J. Grabke, *Mater. Sci. Eng.*, 88, 205 (1987)
57. S. Mrowec, A. Gil and A. Jedlinski, *Werk. Korr.*, 38, 563 (1987)
58. S. Taniguchi and T. Shibata, in Proc. 10th Int. Congress on Metallic Corrosion, IV, 3385 (1987)
59. S. Pandey, S. Parkash and M.L. Mehta, *ibid*, 3507 (1987)
60. F.H. Stott, G.J. Gabriel, F.I. Wei and G.C. Wood, *Werk. Korr.*, 38, 521 (1987)
61. E.J Felten and F.S. Pettit, *Oxid. Met.*, 10, 189 (1976)
62. G. Fisher, W.Y. Chan, P.K. Datta, J.S. Burnell-Gray, *Platinum Metals Review*, 43, 59-61 (1999)
63. T. Noda, *Intermetallics*, 6, 709-713 (1998)
64. H.L. Du, P.K. Datta, D.B. Lewis, J.S. Burnell-Gray, Corrosion Science 36 (1994) 631-642.
65. H.L. Du, P.K. Datta, J. Leggett, J.R. Nicholls, J.C. Bryar, M.H. Jacobs, Advances in Surface Engineering, Vol. I, in: Advances in Coatings and Surface Engineering, in: P.K. Datta and J.S. Burnell-Gray (Eds.), Proceedings, the Royal Society of Chemistry, Cambridge, 1997, pp.53-66.
66. E.U. Lee, H. Waldman, *Scripta Metall.*, 22, 1389-1394 (1988)
67. E. Kasahara, M. Yoshimoto, R. Tanaka, *High Temperature Technology*, 8, 179 (1990)
68. G. Welsh, A.J. Kahveci, Oxidation of High Temperature Intermetallics, The Minerals, Metals and Materials Society, Warrendale, 1988, pp.207
69. A. Gil, H. Hoven, E. Wallura, W. Quadakkers, *Corrosion Science*, 34, 615-630 (1993)
70. H. Copland, B. Gleeson, D.J. Young, *Acta Mater.*, 47, 2937-2949 (1999)
71. S. A. Kekare, P. B. Aswath: *J. of Materials Science*, 32, 2485-2499 (1997)
72. V. Shemet, H. Hoven, W. J. Quadakkers. *Intermetallics*, 5, 311-320 (1997)
73. J. Geng, G. Gantner, P. Oelhafen, P. K. Datta; *Applied Surface Sci.* 158, 64-74 (2000).
74. S. K. Varma, A. Chan, R. N. Mahapatra, *Oxidation of Metals*, 55, 423-435 (2001).
75. H.L. Du, A. Aljarany, P.K. Datta and J.S. Burnell-Gray, *Corrosion Science*, 47, 1706-1723 (2005).
76. D.R. Gaskell, *Introduction to Metallurgical Thermodynamics*, Hemisphere Publishing Corporation, 1981
77. H.L Du, P.K. Datta, A.L. Dowson and M. Jacobs, Effects of Sulphur Partial Pressures on Degradation Behaviour of TiAl in $H_2/H_2S/H_2O$ Environments at 900°C, to be published in Oxidation of Metals
78. Defects and Diffusion in Metals – Annual Retrospective II, edited by D.J. Fisher, 1999
79. S. Mrowec, *Werk. Korr.*, 31, 371 (1980)
80. A. Cooper, PhD Thesis (CNAA), "High temperature corrosion behaviour of iron-based alloys in sulphur and oxygen-containing environments", Newcastle Polytechnic, 1989
81. P.K. Datta, H.L. Du, K.N. Strafford, B. Lewis and J.S. Gray, *"Effect of preoxidation on sulphidation behaviour of Co-based alloys containing refractory metal"*, in Heat-Resistant Materials, Proc. 1st Int. Conf., Fontana, Wisconsin, USA, pp323-332, 1991
82. H.L. Du, B. Lewis, J.S. Gray and P.K. Datta, *"Sulphidation behaviour of preoxidised Inconel 600 and Nimonic PE11 alloys"*, in Proc. 3rd Int. Conf. on Advances in Coatings and Surface Engineering for Corrosion and Wear Resistance and Other Applications, Newcastle upon Tyne, UK, pp27-36, 1992.
83. K.N. Strafford, P.K. Datta and J.S. Gray, in Int. Conf. on Advances in Coatings and Surface Treatments, Newcastle upon Tyne, ed., K.N. Strafford, P.K. Datta and J.S. Gray, 397 (1988)
84. K.N. Strafford and A.F. Hampton, *J. the Less-Common Metals*, 21, 305 (1970)
85. K.N. Strafford and J.R. Bird, *J. the Less-Common Metals*, 223 (1969)

86. D. Jenkinson, "The Sulphidation of Some Refractory Metals", PhD Thesis (CNAA) Newcastle Polytechnic, 1983
87. K.N. Strafford, P.K. Datta, A.F. Hampton, F. Starr, W.Y. Chan, *Corrosion Science*. 29, 775-789 (1989).
88. H.L. Du. P.K. Datta, J.S. Gray and K.N. Strafford, *Oxidation of Metals*, 39, 107-135 (1993).
89. H.L. Du, P.K. Datta, J.S. Gray and K.N. *Strafford, Corrosion Science*, 36, 99-112 (1994)
90. S. Mrowec, *Corrosion Science*, 7, 563 (1967).
91. H.L. Du, P.K. Datta, J.S. Burnell-Gray and D. Jenkinson, *Key Engineering Materials* 99-100, 151-158 (1995).
92. H.L. Du, P.K. Datta, D.B. Lewis and J.S. Burnell-Gray, *Oxidation of Metals*, 45, 507-527 (1996).
93. K. Natesan and P.K. Datta, Sulphidation Behaviour, *Intermetallic Coatings and Coatings for Intermetallics, Intermetallic Compounds: Principles and Practice Vol. 3 Progress*, J.H. Westbrook and R.L. Fleischer, Ed., John Wiley & Son Ltd., 2002, p 707-719
94. H.L. Du, P.K. Datta and S.K. Hwang, *Materials Science Forum*, 251-254, 219-226 (1997).
95. H.L. Du, P.K. Datta, D. Griffin, A. Aljarany and J.S. Burnell-Gray, *Oxidation of Metals*, 60, 29-46 (2003).
96. H.L. Du, P.K. Datta, J.S. Burnell-Gray and J. Bernardi, Microscopy of sulphidised Fe_3Al, to be published in Journal of Materials Science
97. H.L. Du, P.K. Datta, R. Cutting and J.S. Burnell-Gray, Sulphidation Behaviour of Siliconised Mo and Nb by Pack Cementation at 850°C, presented at the International Conference On Metallurgical Coatings And Thin Films ICMCTF 2004, April 19-23, 2004, San Diego, California, USA
98. S. Baranow, G.Y. Lai, M.F. Rothman, J.M. Oh and M.J. McNallan, NACE, Houston, 1984, paper No. 16
99. R. Prescott, F.H. Stott, and P. Elliott, *Oxid. Met.* 31, 145 (1989).
100. P. Elliott and G. Marsh, *Corrosion Science*, 24, 397 (1984
101. A.S. Kim and M.J. McNal1an, *Corrosion*, 46,746 (1990)
102. J. Halfdanason, J and K Hauffe, *Werkstoffe und Korrosion*, 24, 8 (1973)).
103. K.N. Strafford, P.K. Datta and G. Foster, Advances in *Surface Engineering: Processes, Fundamentals and Application in Corrosion and Wear*, Eds Strafford, Datta and Gray, Ellis Horwood, p282, 1990
104. H. Chu, P.K. Datta and K.N. Starfford, Oxidation of Metals, 43(5/6) 491-508 (1995)
105. F.H. Stott, R. Prescott, P. Elliott and M.H.J.H. Al'Atia, *High Temperature Technology*, 6, 115 (1988)
106. P. Elliott and G Marsh, *Corrosion Science*, 24, 397 (1984)
107. J.M. Oh, M.J. McNallan, G.Y. Lai and M.F. Rothman, *Metallurgical Transactions*, 17A, 1087 (1986).
108. J. Stinger, *Mater. Sci. Technol.*, 3, 536 (1987)
109. G.H. Meier, *Mater. Sci. Eng.*, A120, 1-11(1989)
110. J. Litz. A. Rahmel and M. Schorr, *oxidation of Metals*, 30, 105 (1988)
111. S.Y. Hwang, G.H. Meir, G.R. Johnston, V. Provenzano, and J.A. Sprague, in S.C. Singhal (ed), *High Temperature Protective Coatings*, TSM-AIME, Warrendale, PA, 1983, p121.
112. Y.S. Zhang and R.A. Rapp, *Corrosion*, (1987) 348.
113. K.L. Luthra, J. Electrochem. Soc., 127, 2202 (1980).
114. K.L. Luthra, *Metall. Trans. A*, 13, 1643, 1843(1982).
115. H.L. Du, P.K. Datta, Z. Klusek and J.S. Burnell-Gray, *Oxid. Met.*, 62, 175-193 (2004)
116. P.D. Wood, PhD Thesis *"The effect of the counterface on the wear resistance of certain alloys at room temperature and 750°C"*, University of Northumbria, UK, (1997)

117. S. Rose, PhD Thesis "*Studies of the high temperature tribological behaviour of some superalloys*", University of Northumbria, UK, (2000)
118. I.A. Inman, PhD Thesis, "*Compact oxide layer formation of under condition of limited debris retention at the wear interface during high temperature sliding wear of superalloys*", Northumbria University, UK, (2003)
119. F.H. Stott, D.S. Lin, and G.C. Wood, *Corrosion Science*, 13, 449-469 (1973)
120. F.H. Stott, J. Glascott, G.C. Wood, *Wear* 97, 93 (1984)
121. M.G. Gee and N.M. Jennet, *Wear* 193, 133 (1995)
122. I.A. Inman, S. Datta, H.L. Du, J.S. Burnell-Gray and Q. Luo, *Wear*, 254, 461-467 (2003).
123. H.L. Du, P.K. Datta, I. Inman, R. Geurts and C. Kübel, *Materials Science and Engineering*, 357, 412-422 (2003).
124. H.L. Du, P.K. Datta, I. Inman, E. Kuzmann, K. Süvegh, T. Marek and A. Vértes, *Tribology Letters*, 18, 393-402 (2005).
125. R.S. Mishra, A.K. Mukherjee, Superplasticity in nanomaterials, in A.K. Ghosh and T.R. Bieler (Ed.), *Superplasticity and Superplastic Forming*, TMS Warrendale, pp. 109-116, 1998.
126. R.S. Mishra, S.X. McFadden, A.K. Mukherjee, Tensile superplasticity in nanocrystalline materials produced by severe plastic deformation, in: T.C. Lowe and R.Z. Valiev (Ed.), *Investigations and applications of severe plastic deformation*, Kluwer Academic Publications, pp.231-240, 1994.
127. P.K. Datta, M. Danielewski, R. Filipek, R. Bachorczyk and G. Fisher, *Defect and Diffusion Forum, Solid State Phenomena*, 72, 53 (2000).
128. K. Holly and M. Danielewski, *Phys. Rev. B*, 50, 13336(1994).
129. M. Danielewski, R. Filipek, K. Holly and B. Bozek, *Phys. Stat. Sol.*, 145, 339 (1994).
130. M. Danielewski and R. Filipek, *J. Comp. Chem.*, 17, 1497 (1996).
131. D. Griffin, A. Daadbin, P.K. Datta, *Surface and Coatings Technology*, 126, 142-151 (2000).
132. V. Guttmann, F. Hukelmann, D. Griffin, A. Daadbin and S. Datta, *Surface and Coatings Technology*, 166, 72–83 (2003)

CHAPTER 3

Microbially-influenced Corrosion

K.A. Natarajan
Department of Metallurgy, Indian Institute of Science, Bangalore – 560 012, INDIA

INTRODUCTION

Microbially-influenced corrosion (MIC) is extremely harmful to industry and environments such as soil, fresh water and sea water. According to estimation, MIC accounts for more than 30 percent of all corrosion damage of metals, alloys and several building materials. The direct cost of MIC works to be about 30-50 billion US dollars per year. Microoganisms of interest in MIC are mostly bacteria and fungi. Among bacteria, many types such as sulfur-sulfide oxidising, sulfate-reducing, iron oxidising, acid producing, manganese fixing and ammonia and acetate producing are implicated. Among the above, the role of Sulphate Reducing Bacteria (SRB) in MIC has been studied quite extensively. Microbial activities under natural conditions influence many electrochemical reactions through either direct or indirect interference. Microbe-metal interactions of relevance to MIC involves initial adhesion of microbes, biofilm formation, surface colonisation, secretion of polymeric substances and inorganic precipitates and ultimate metallic corrosion and failure of structural components.

In this paper, microbiological as well as physico-chemical and electrochemical aspects of microbially-influenced corrosion are analysed critically. Scientific approach to monitoring, diagnosis and prevention of MIC is illustrated along with suggested remedial measures.

Sea Water as a Corrosive Medium

Since sea water is a typical aggressive corrosive medium for biofouling and microbially-influenced corrosion (MIC), it is essential in the first place to understand the chemical and electrochemical basis of metallic corrosion in sea water.

Seawater contains about 3.4% salt and is a good electrolyte that can cause galvanic corrosion and crevice corrosion. The type and rate of corrosion in seawater is influenced by oxygen content, temperature, velocity and microorganisms. Galvanic series for various metals and alloys in flowing seawater is shown below:

Increasing activity ↓

- Platinum
- Graphite
- Titanium
- Nickel
- Monel alloy
- 90/10 Cu/Ni
- Aluminium brass
- Gun metal
- Admiralty brass
- Copper
- Mild steel
- Zinc

Corrosion by seawater at deeper regions is usually decreased due to lower temperatures. Greater corrosion occurs in the splash zone because of alternate wetting and drying and availability of oxygen. Corrosion rates are also influenced by velocity.

Galvanic series of some commercial metals and alloys in seawater is given below:

↑ Noble	Platinum
	Gold
	Graphite
	Titanium
	Silver
	18-8 stainless steel (passive)
	Nickel (passive)
	Bronzes
	Copper
	Brasses
	Tin
	18-8 stainless steel (active)
	Cast iron
	Aluminium alloy 2024
	Commercial aluminium
	Zinc
↓ Active	Magnesium and its alloys

Table 1 shows the resistance to crevice corrosion and the relative resistance to fouling are given in table 2. At greater depths corrosion by seawater is usually decreased since the temperature is lower (about 40°F).

Table 1 Resistance to crevice corrosion in quiet seawater

Inert	Resistance		
	Less	Medium	Best
Titanium	Nickel-copper alloy Copper	Carbon steel	Bronze Brass
Hastelloy "C"	Incoloy alloy 825	Austenitic Nickel cast Iron Cast iron	90/10 copper nickel 70/30 copper nickel

Table 2 Resistance to biofouling in quiet seawater

Arbitrary rating of biofouling resistance	Materials
Least 0	Low-alloy steels and carbon, Hastelloy "C" , stainless steels
Very slight 10	Nickel-copper alloy
Fair 50	70/30 copper-nickel , zinc, aluminium bronzes
Good 70-90	Bronze and brass
Best 90-100	Copper90/10 copper-nickel

Due to tidal action in rivers and bays near the ocean, chlorides contaminates the brackish water. Although guidelines for materials of construction are approximately the same, corrosion problems are less severe than in full-strength seawater.

If two dissimilar metals having potential difference are placed in contact or electrically connected, there exists an electron flow between them. The more resistant metal becomes cathodic and the less resistant metal anodic. Due to the presence of electric currents and dissimilar metals, this is called as galvanic or two-metal corrosion.

Frequently intensive localized corrosion occurs within crevices and other shielded areas on metal surfaces exposed to corrosives. These are associated with small volumes of stagnant solution caused by lap joints, crevices under bolt and rivet heads, gasket surfaces, surfaces deposits and holes. This form of corrosion is known as crevice or deposit or gasket corrosion.

Definition and Practical Significance

The involvement of microorganisms in the deterioration and destruction of materials can be characterized by three different terms, namely, Biofouling, Biodeterioration and Biocorrosion or Microbiologically-influenced corrosion(MIC). The above terms though different in nature, could be complimentary in their ultimate consequences. Biofouling generally refers to adherence of micro- and macro-organisms onto material surfaces in marine and fresh water systems leading to formation of fouled layers. Deterioration of nonmetallic materials such as glass, concrete, cement, rubber, wood, plastics and some organic and inorganic fluids in the presence of microorganisms is termed biodeterioration. Corrosion of metals and alloys induced by the activities of microorganisms is defined as Microbially-influenced corrosion (MIC).

Microorganisms are omnipresent in nature and their ability to grow and reproduce at amazingly rapid rates accounts for their presence in soil, water and air. Micro- and macro-organisms exhibit extreme tolerance to hostile environments such as acidic and alkaline pH, low and higher temperatures as well as pressure gradients. In microbial corrosion processes, aggressive environments are generated by microorganisms, several interaction factors participating in the mechanisms. The microorganisms often make a contribution without being solely responsible for the failure of materials. In this respect, the microbes can be viewed as living catalysts. As early as in 1891, corrosion of lead sheathed cables was suspected to be caused by bacterial metabolites. Sulphur and iron sulphide accumulation at the interior and exterior portions of water pipes were attributed to the action of iron-sulphur bacteria during 1910. Anaerobic corrosion of bacteria was realised in 1931. Tubercle formation due to microbial growth and reaction products has been reported almost forty years ago. However, a better understanding of biocorrosion processes based on microbiological and electrochemical mechanisms, became available only since the last about 25 years [1-6].

The practical significance of microbial corrosion can be seen from Table 3 where some industrial situations susceptible to microbial corrosion are listed. The universality of microbial corrosion processes is readily evident from the fact that most of the industrially used metals and alloys such as stainless steels, nickel and aluminium-based alloys as well as materials such as concrete, asphalt, bitumen, plastics, and polymers are readily attacked by microorganisms. Even protective coatings, inhibitors, oils and emulsions are degraded under all conditions.

Table 4 lists cases of microbially-influenced corrosion reported more specifically in systems or components in a power plant [7].

Table 3 Industrial situations affected by microbial corrosion

1. Nuclear Power Plants	Cooling water tubes and pipes, sub-sea pipe lines, stainless steel and carbon steel, copper-alloys, aluminium-alloys
2. Underground pipes and machineries	Steels
3. On-shore and off-shore oil and gas exploration	Oil, gas and water handling systems
4. Chemicals	Pipelines, Tanks, Condensers, Joints, etc
5. Constructions industries	Concrete in marine, fresh water and sub-soil conditions
6. Water treatment and Metal working	Heat exchangers and pipes, Breakdown of oils, emulsions and lubricants
7. Aviation	Aluminium fuel tanks
8. Mining	Underground pipelines and machinery
9. Marine atmospheres	Sea-going vessels and marine structures

Table 4 Microbially-influenced corrosion in power plants

Heat exchanger tubing	Aluminium brass, 70:30 Copper-Nickel, 90:10 Copper-Nickel	Pitting Deposits
Water storage tank	316 stainless steel	Nodules of Rust
Service water pipe	316 stainless steel weld	Pitting
Cooling towers	Galvanised steel	Wood rotting, deterioration

Microorganisms and Environmental Parameters

Microorganisms that are known to cause corrosion can be grouped under categories detailed in Table 5.

Table 5 Classification of microorganisms involved in MIC

1.	Bacteria	Sulphate Reducing Bacteria (SRB) Eg: *Desulfovibrio* Sulphur Oxidising Bacteria (SOB) Eg: *Acidithiobacillus* Iron Oxidising Bacteria (IOB) Eg: *Gallionella, Crenothrix, Leptothrix* Miscellaneous Eg: *Pseudomonas* *(Reduction of ferric to ferrous ion)*
2.	Fungi	*Cladosporium resinae* *Aspergillus niger* *Aspergillus fumigatus* *Penicillium cyclospium* *Paecilomyces varioti*
3.	Algae	Blue green algae
4.	Mixed flora	Symbiotic activity among different groups of microorganisms

The sulphur cycle in nature is important to biocorrosion (Figure 1). Sulphur oxidising as well as sulphate reducing bacteria are involved in a number of biogenic oxidation/reduction reactions leading to the generation of various products such as H_2S, metal sulphides and sulphides. All these microbially intermediated processes take part in corrosion reactions in soils and aqueous environments. For example, sulphate reducing bacteria like *Desulfovibrio* are involved in the reduction of sulphate ions to sulphide ions and hydrogen sulphide. These anaerobic heterotrophs thrive under reducing conditions.

$$SO^{(-)}{}_4 + 4H_2 \rightarrow S^{(-)} + 4H_2O \quad ...(1)$$

Sulphur and iron oxidising bacteria such as *Acidithiobacillus thiooxidans* and *Acidithiobacillus ferrooxidans* are essentially acidophilic and aerobic chemolithotrophs bringing out the following reactions:

$$2H_2S + 2O_2 \rightarrow H_2S_2O_3 + H_2O \quad ...(2)$$

$$5Na_2S_2O_3 + 8O_2 + H_2O \rightarrow 5Na_2SO_4 + H_2SO_4 + 4S \quad ...(3)$$

$$4S + 6O_2 + 4H_2O \rightarrow 4H_2SO_4 \quad ...(4)$$

$$Fe^{++} \rightarrow Fe^{+++} + e \quad ...(5)$$

The *Acidithiobacillus* group of bacteria can exist over a range of pH conditions, spanning from highly acidic, to alkaline conditions. *Thiobacillus thioparus* could oxidise sulphur, sulphide and thiosulphate at a pH of 6-10. Pertinent microbiological characteristics of important bacteria

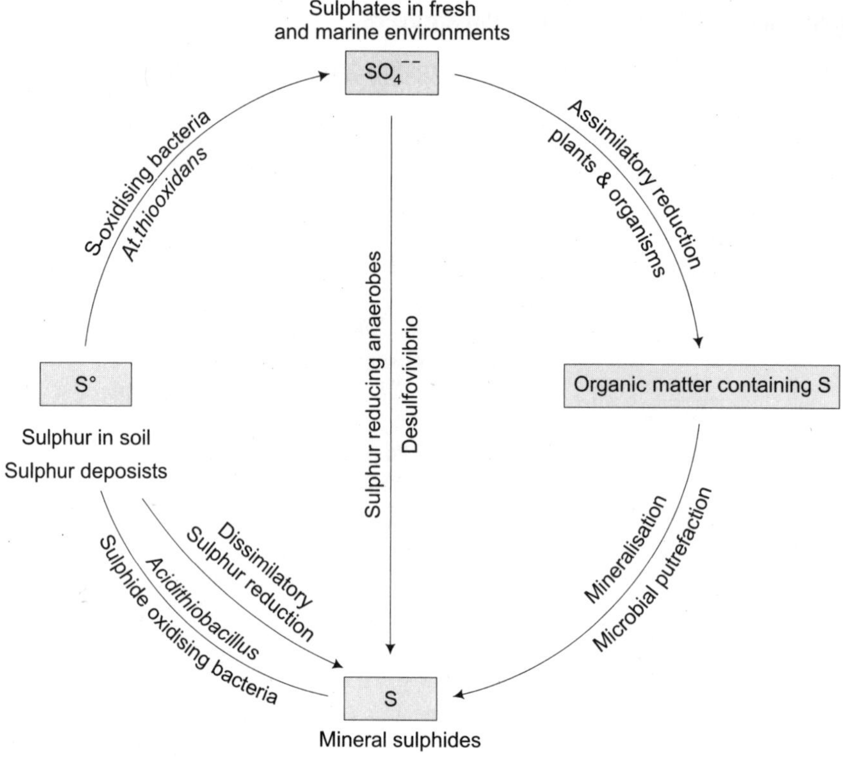

Figure 1 Biological sulphur cycle in nature

involved in MIC are illustrated in Table 6. Morphological features of some bacteria and fungi implicated in MIC are illustrated in Figure 2.

Eh and pH are the most important environmental parameters controlling the growth of various aerobic and anaerobic organisms [8]. Therefore, the stability limits of various types of microorganisms corresponding to optimum activity can be defined through Eh – pH diagrams. Eh-pH diagram for sulphur species wherein the stability and growth regions of various types of microorganisms are represented will be useful in the understanding of MIC. Iron and sulphur-oxidising acidophilic bacteria such as *Acidithiobacillus ferrooxidans and Acidithiobacillus thiooxidans* exist at higher oxidising potentials and acid pH levels. Sulphur and thiosulphate oxidising autotrophs such as *Thiobacillus thioparus* have optimum activity at near neutral pH ranges and relatively higher oxidising potentials. On the other hand, sulphate reducing bacteria (SRB) prefer reducing conditions and mildly alkaline or neutral pH. Iron oxidising heterotrophs are stable and active at neutral pH and higher oxidising conditions.

Such Eh-pH corrosion diagrams can be readily constructed for various metal-water-oxygen systems in the presence of micro-organisms to predict the regions of biocorrosion, immunity and passivation (Figure 3). It should be borne in mind that regular Eh-pH diagram do not strictly represent the corrosion behaviour of metals and alloys in the presence of micro-organisms, since the superimposition of the bacterial stability regions on these diagrams bring about marked changes in the regions of corrosion, immunity and passivation. By a study of

Table 6 Microbiological characteristics of important bacteria involved in MIC

Sl.No.	Types of organism	Habitat	Growth conditions
1.	*Desulfovibrio desulfuricans* (Sulphate reducing)	Mud, sewage oil wells, subsoil conditions	Anerobic, reduces sulphate to sulphides. Optimum pH 6-7.5, Temp. 25-30°C
2.	*Acidithiobacillus thiooxidans* *Acidithiobacillus ferrooxidans*	Sulphur and iron bearing minerals, soils	Aerobic, optimum pH 2-4, Temp. 28-35°C oxidises sulphur, sulphides and iron sulphides producing sulphuric acid
3.	*Thiobacillus Thioparus*	Water, mud, sludge, sulphidic soils	Aerobic, optimum pH 6-8, Temp. 30-35°C, oxidises thiosulphate and sulphur to Sulphate
4.	*Iron Bacteria*	Water containing iron salts	Aerobic, Neutral pH, Temp. 30-40°C, oxidises ferrous iron compounds to ferric form and ferric hydroxide

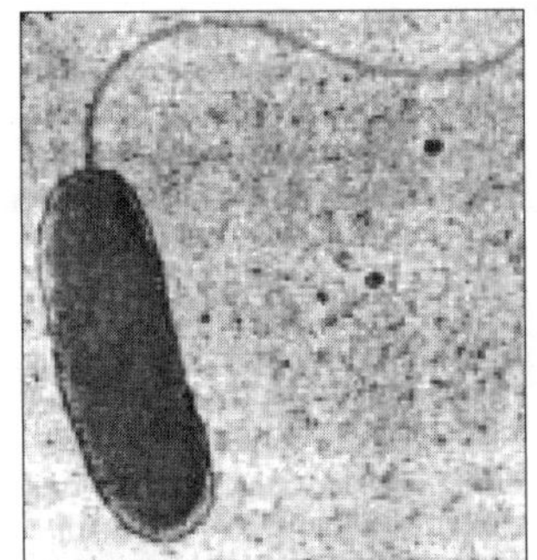

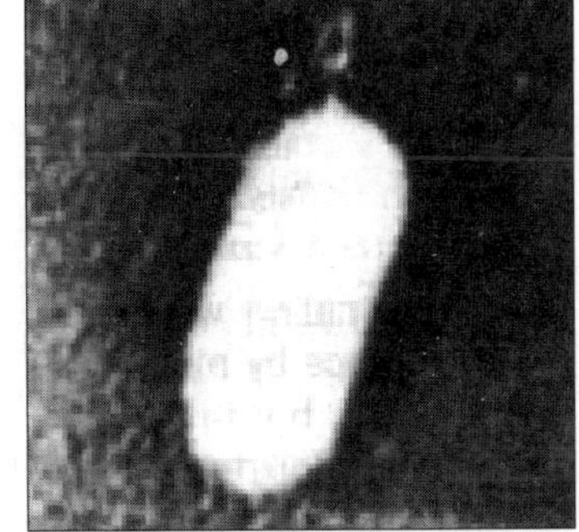

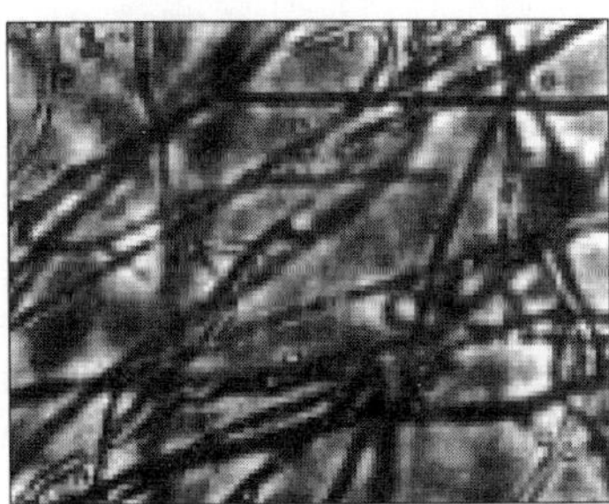

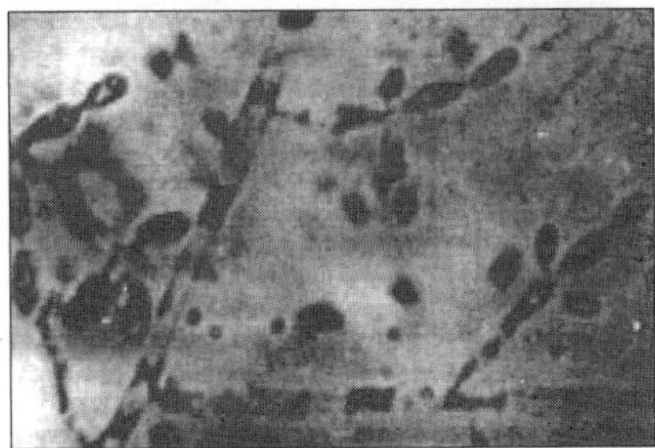

Figure 2 Electron micrographs depicting morphological features of some bacteria and fungi (a) *Desulfovibrio* (b) *Acidithiobacillus* (c) *Aspergillus* and (d) *Cladosporium*

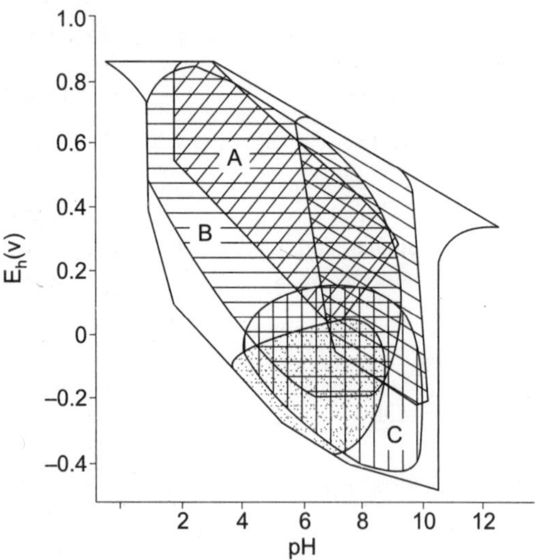

Figure 3 Environmental limits of Eh and pH for some bacteria. A. Iron bacteria, B. Thiobacteria, C. Sulfate-reducing bacteria [8]

such diagrams, one can find examples of metals where, if conditions of corrosion do not exist, bacteria may create them; and conversely, if conditions for corrosion exist, bacteria may at times prevent it as well.

The principal slime forming bacteria include *Bacillus subtilis, Bacillus cereus* and species of *Flavobacterium, Aerobacters and Pseudomonas. Pseudomonas* have been found in systems containing hydrocarbon sources such as oils and emulsions and they use these hydrocarbons as energy source.

Algae range from single cell plants to multicellular species including diverse forms and shapes. They contain coloured pigments, the most important of which is the chlorophyll. Algae generally flourish on wetted surfaces such as cooling towers, screens and distribution systems. Some common algae groups are blue-green algae, the green algae and the diatoms. Due to their ability to generate corrosive organic acids, oxygen and nutrients for other microbes, the algae could play a role in MIC.

Fungi are similar to algae but do not contain chlorophyll. Yeasts and moulds are some major forms of interest. Mould fungi are filamentous in form but most of yeast fungi are unicellular. Some corrosion-causing fungi are *Aspergillus niger, Aspergillus fumigatus, Pencillium cyclospium and Cladosporium resinae.* Production of various types of organic acids such as oxalic acid, citric acid and gluconic acid by fungal metabolism lead to metallic corrosion.

Under actual MIC environments, mixed flora invariably exist, such as anaerobic and aerobic bacteria, fungi, yeasts and algae.

Direct and Indirect Mechanisms

General principles governing microbial corrosion are based on electrochemical reactions, namely,

Anodic: $M \rightarrow M^{++} + 2e$ ...(6)

Cathodic: $O_2 + 4H^+ + 4e \rightarrow 2H_2O$ (aerated, acidic) ...(7)

$O_2 + 2H_2O + 4e \rightarrow 4OH^-$ (aerated, neutral and alkaline) ...(8)

$2H^+ + 2e \rightarrow H_2$ (in the absence of oxygen in acid solutions) ...(9)

Since the microorganisms, very often contribute towards corrosion without being solely responsible for the failure, the mechanisms involved are mostly electrochemical. Both direct and indirect effects could be visualised. In the first case, the microbe interlinks an electrode process with its metabolism, while the indirect contribution may result from formation of differential aeration cells due to coherent microbial growth on metallic substrates and from microbial metabolic products [1,6,7,9].

General mechanisms can be classified as follows:

(a) Oxygen removal by microbial growth leading to formation of concentration cells.
(b) Degradation of additives in lubricants and emulsions.
(c) Liberation of corrosive products such as hydrogen removal.
(d) Microbiological attack resulting in the breakdown or disruption of organic paint coatings, plastic fittings and linings, protective films and inhibitors.
(e) Interaction among the above various factors.

Typical examples of some of the corrosive metabolic products are illustrated below:

MIC can be induced by the following type of microbial acid generation:

(a) Oxidation of inorganic sulphur compounds by *Acidithiobacillus* group of bacteria.
(b) Oxidation of iron sulphides to acidic ferric sulphate by *Acidithiobacillus ferrooxidans*.
(c) Moulds or cellulose bacteria ferment cellulosic material to organic acids. For example, in the presence of organic carbon such as sucrose and glucose, fungi such as *Aspergillus* generate oxalic, citric and gluconic acids.
(d) Exopolysaccharides secreted by *Bacillus* species.

MIC at neutral pH levels could occur due to the following factors:

(a) Corrosion by cathodic depolarization attributable to bacteria which contain the enzyme, hydrogenase.
(b) Corrosion by differential aeration cells attributable to deposits formed by iron bacteria and other microbes.
(c) Corrosive products such as organic sulphides and mercaptides.

Other corrosive metabolic products include volatile phosphorous compounds, sulphur and sulphur compounds, ammonia and amines as well as hydrogen. Organic corrosion inhibitors such as diamines and aliphatics are used as nutrients by certain bacteria. For example, *Nitrosomonas* and *Nitrobacter* oxidise ammonia and amines to nitrite and nitrate, destroying the corrosion inhibition properties of several inhibitors. Ferric oxide coatings are reduced by *Pseudomonas*, exposing the base metal for enhanced corrosion. Iron sulphide films are broken down by *Sulphate Reducing Bacteria*. Protective aluminium oxide layers from aluminium and its alloys could be destroyed by the fungus, *C. resinae*.

Bacterial Adhesion and Biofilm Formation

Bacteria in natural habitat normally range in size from 0.15 to 2.0 μm and could form stable colloidal suspensions in aqueous media. They possess a negative surface charge at natural pH levels. The cell surfaces possess hydrophobic or hydrophilic character depending on the extracellular polymeric substances secreted by the microorganisms. The surface properties of the microorganisms play an important role in their adhesion to solid surfaces leading to the formation of biofilms. It is essential to understand the probable mechanisms in bacterial adhesion to metal substrates before the structure of biofilms could be enumerated. The following forces are involved in the transport of bacteria to surfaces [3-4].

(a) Fluid dynamic forces such as currents in water bodies and eddy diffusion in turbulent flow systems.

(b) Sedimentation, mainly in quiescent waters as flocs.

(c) Chemotactic response towards a substrate under nutrient gradients.

(d) Brownian motion.

Bacterial adhesion to surfaces are also governed by surface charge, surface free-energy and surface roughness. The surface charge and surface free-energy of substrate are modified when immersed in a bacterial habitat and bacterial adsorption is facilitated. The bacterial adhesion could be either reversible or irreversible depending on the forces involved. Electrostatic, hydrophobic and chemical forces are invariably involved.

Biofilm formation on material surfaces is of concern in MIC. Biofilms not only initiate several corrosion processes but also are of nuisance value in several process operations. For example, growth of biofilms could lead to the following problems:

(a) Heat transfer reduction in condenser tubes leading to energy loss.

(b) Increase in fluid friction in pipes and tubing leading to increased pumping power and reduced gravity capacity.

(c) Biofouling of ship's hull and offshore platforms resulting in increased fuel costs and wave loading.

(d) Failure of structures by porosity and general increase in corrosion rates.

A representative scanning electron micrograph of biofilm formed on a metal substrate in sea water is shown in Figure 4.

Biofilms essentially consists of cells immobilised at substrates, along with metabolic and reaction products. The surface layers need not be uniform and could often consist of patches. Several stages are involved in biofilm formation.

(a) Initiation of film formation with a progressive coverage of the surface by multiplication of the primary colonising bacteria and further proliferation by the organisms from the aqueous phase.

(b) A transition stage with multilayers of cells, resulting in increasing population density and competition among growing organisms.

(c) Development of a steady-state biofilm with high population density.

Organisms near the water-biofilm interface are exposed to higher concentrations of oxygen, light, protons as well as nutrients. At intermediate and still lower levels, organisms within biofilms are subjected to greater competition for nutrients and oxygen. Aerobes utilize

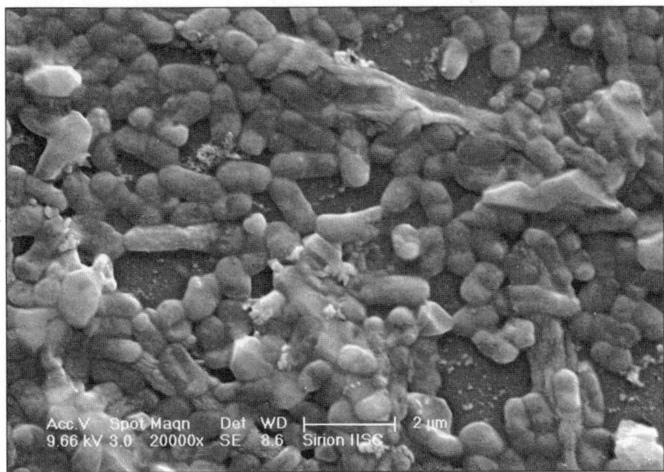

Figure 4 Scanning Electron Micrograph of biofilm formation

the available oxygen creating oxygen-starved conditions favourable for the growth of anaerobic organisms within the film and substrate. Typical case in point is initial growth of aerobes such as *Acidithiobacilli* and iron oxidizers, creating lower oxygen levels for subsequent growth of sulphate reducing bacteria. Such mutualism is favourable for the development of a heterogeneous biofilm on the substrate. Regions, depleted of oxygen, serve as anodes, while those exposed to higher oxygen levels act as cathodes. Differential aeration (oxygen) cells, can thus be initiated, leading to enhanced corrosion at anodic sites. Most reported cases of microbiologically induced corrosion involve localised forms of attack, the prime reason, being the discontinuous nature of biofilms formed by bacterial colonisation (Figures 5 and 6). Common metals and alloys such as steels, aluminium and stainless steel are prone to such corrosive damage. Besides bacteria, algae and fungi are known to be involved in such biocorrosion processes due to their ability to form biofilms.

Profuse growth and colonisation of the rod-shaped bacteria could be seen on all the above substrates under acidic conditions. Extensive corrosion in the form of pits, crevices and general attack could be observed due to such bacterial interaction with metals and alloys. Surface morphology after exposure to bacterial action reveals the presence of exfoliation and crystallisation of reaction products.

Microbiological influences on corrosion are generally associated with microbial colonies, scale formation, debris and exopolymers in the deposits. Initiation of pits under the debris or deposits is common and the pits further propagate resulting in severe structural damage. Surface morphology and surface defects besides electrochemical susceptibility play important roles in pit initiation. The manner in which differential corrosion cells are formed due to bacterial attachment and colonisation is very complex.

Pitting is usually associated with tubercles in isolated area. Influences such as oxygen depletion, chloride iron availability and sulphate reduction enhance probability for corrosion. Mechanical factors such as crevices or faulty fabrication (as in gaskets, rings, welds etc.) promote localised corrosion. Aerobic bacteria such as *Gallionella* and *Leptothrix* aid in the formation of differential aeration cells. Oxygen-depleted regions on the surface of metals and

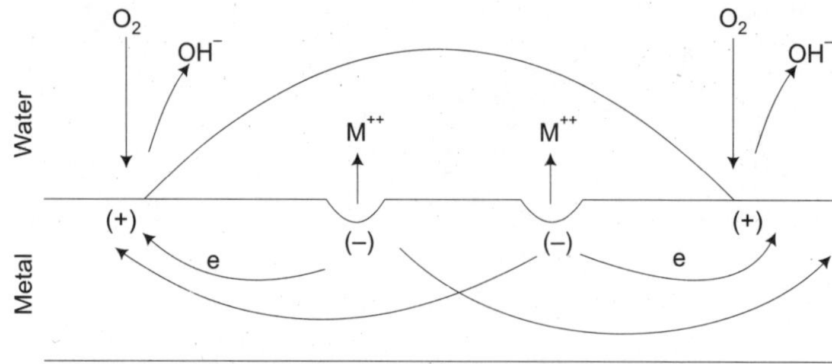

Figure 5 A model depicting coherent mass of microbial growth on a metal surface setting up differential aeration cells

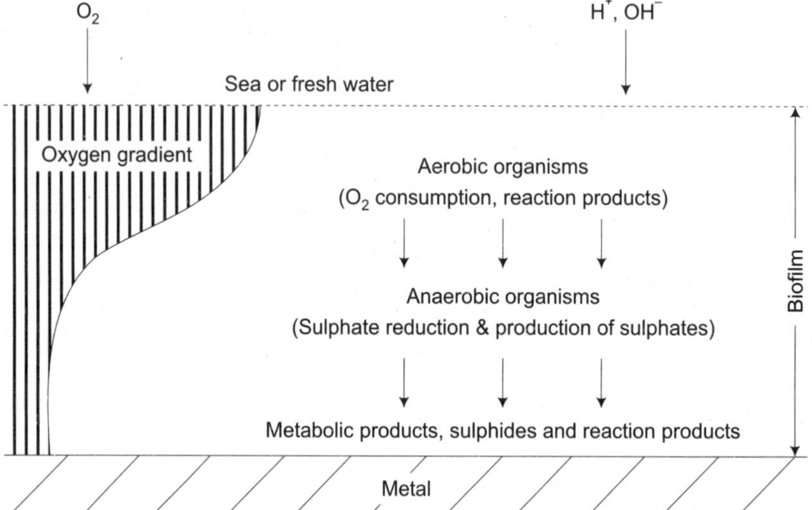

Figure 6 Schematic representation of biofilm formation

alloys are developed. Metallic ions such as ferrous, ferric and manganous and manganic concentrated in the pits and subsequent precipitation of these metal ions to their oxyhydroxides can occur. Consortia of various microorganisms can form. Aerobic bacteria grow on the water side of the tubercle, while anaerobes concentrate on the metal surface beneath deposits.

A schematic representation of cathodic depolarization and generation of concentration cells on a metal surface is illustrated in Figure 7.

Microbially-influenced Corrosion in Anaerobic Environment

Biological corrosion in the absence of oxygen, has been studied quite extensively. Anaerobic heterotrophs, such as *Desulfobibrio desulfuricans* and *D.vulagaris* are good sulphate reducers. Characteristics of some sulphate reducing bacteria are given Table 7. Sulfate-reducing bacteria

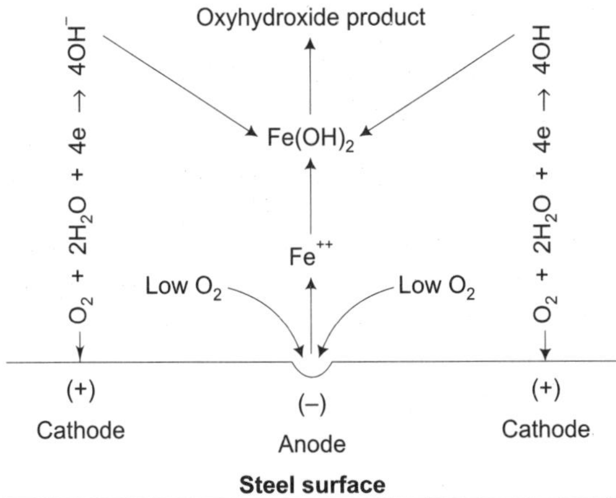

Figure 7 Anodic and cathodic reactions in concentration cells at natural pH

can be found in places such as stagnant points in flow lines, underscale deposits or debris, under sludge or in mud bottom pits, sand/gravel filters, oil-water interface, fresh and sea water environments as well as backfill on the outside of buried pipelines. In natural conditions, they grow in the presence of other organisms and utilise carboxylic acids and fatty acids, which are the common byproducts of other microbes. Sulphate reducing bacteria are abundant in soil, sea water and fresh water and contribute towards the formation of hydrogen sulphide and metal sulphides. Petroleum industries have been plagued by sulphate reducing organisms since they handle large volumes of deaerated water from underground reservoirs. Other environments are natural water-logged soils and water heavily polluted with organic matter. The corrosion of iron-base alloys under anaerobic conditions is often localised and characterised by a black corrosion product and strong smell of hydrogen sulphide. Cast irons are rapidly attacked, while mild steel pipes perforated, especially when buried in clay soils.

Table 7 Characteristics of some Sulphate Reducing Bacteria

1.	*Desulfovibrio desulfuricans*	Monoflagellated, nonspore forming, hydrogenase present
2.	*Desulfovibrio vulgaris*	Curved rods, at times spirilloid, 3-5 μm
3.	*Desulfotomaculum nigrificans*	Thermophilic, 55°C, 3-6 μm variable hydrogenase activity
4.	*Desulfotomaculum orientis*	Fat, curved rods, 3-6 μm hydrogenase absent

Measurements of soil Eh are useful in predicting not only the relative oxidising and reducing conditions existing at different depths of soil and water, but also the predominance of aerobes or anaerobes. Corrosiveness can be thus classified based on the magnitude of Eh, with respect to the activity of sulphate reducing anaerobes.

For example,

<100mV	Severe corrosive conditions
100 to 200 mV	Moderate

200 to 400 mV	Negligible
>400 mV	non-corrosive

An electrochemical mechanism, has been proposed to understand MIC by sulphate reducing bacteria. According to the cathodic depolarization concept, the sulphate reducing organisms utilise cathodically polarised hydrogen from metal surfaces for their metabolic activity leading to depolarisation. Experimental observations, have been recorded using two types of strains, namely hydrogenase-positive *Desulfovibrio vulgaris* and hydrogenase negative *Desulfotomaculum orientis*. While the former organism brought about significant cathodic depolarization, the latter one did not significantly affect the cathodic polarization curves. Free corrosion potentials were made more active, promoting corrosion in the presence of *D.vulgaris*. Correlation between hydrogenase activity of the bacteria and corrosion rates has been shown to prove the utilization of hydrogen by the bacteria. There is strong evidence to suggest that active enzymes are located in the cell cytoplasm and hydrogen utilization actually occurs at the electrode surface. The following reaction sequences have also been suggested.

$$8H_2O \rightarrow 8H^+ + 8OH^- \qquad ...(10)$$

$$4Fe \rightarrow 4Fe^{++} + 8e \text{ (anodic)} \qquad ...(11)$$

$$8H^+ + 8e \rightarrow 8H \text{ (cathodic)} \qquad ...(12)$$

$$SO_4^{-2} + 8H \rightarrow S^{2-} + 4H_2O \text{ (cathodic depolarization)} \qquad ...(13)$$

$$Fe^{++} + S^- \rightarrow FeS \text{ (anodic)} \qquad ...(14)$$

$$3Fe^{++} + 6OH^- \rightarrow 3Fe(OH)_2 \text{ (anodic)} \qquad ...(15)$$

$$4Fe + SO_4^{-2} + 4H_2O \rightarrow FeS + 3Fe(OH)_2 + 2OH^- \qquad ...(16)$$

The indirect role of bacteria in accelerating corrosion need be understood. Cathodic depolarization in SRB cultures may be due to dissolved hydrogen sulfide and a biogenic cathodic reaction such as:

$$H_2S + e \rightarrow HS^- + \tfrac{1}{2} H_2 \qquad ...(17)$$

may be relevant.

The effect of iron sulfide formed due to bacterial precipitation may also be significant. Iron sulfides would serve as cathodes to steel promoting anodic dissolution of steel surfaces. Iron sulfides could also serve as cathodes for hydrogen reduction [5,7,9].

Bio-electrochemical interpretation of microbially-influenced corrosion of carbon steels in anoxic environments could be summarised as follows:

(a) Biogenic sulfides influence localised corrosion

(b) In neutral media sulfide ions lead to formation of a poorly protective film of mackinawite

(c) Cathodic depolarization

(d) Corrosive action of biogenic sulfide enhanced by chlorides.

Suggested mechanisms of microbially-influenced corrosion by SRB include [5,7,9].

(a) Cathodic depolarization by hydrogenase

(b) Anodic depolarization

(c) Sulfide precipitation and formation of iron sulfides
(d) Volatile phosphorous compound
(e) Fe-binding exopolymers
(f) Sulfide-induced stress corrosion cracking and
(g) Hydrogen embrittlement.

MIC of Important Structural Materials

There are no known metals or alloys which can fully resist bacterial film formation and subsequent microbially-influenced corrosion. Behavior of various commonly used metals and alloys in relation to microbially-influenced corrosion is discussed below [3-4].

A. **Copper and copper alloys.** These alloys are commonly used in heat exchangers, pumps, valves and condensers. 90-10 and 70-30 copper-nickel, brasses, aluminium bronzes and admiralty brasses are a few typical examples. SRB's present in marine environments contribute to localised corrosion of the above alloys. All these alloys are also susceptible to microbially-influenced corrosion of different kinds. Extra-cellular polymers secreted by microorganisms can lead to corrosion of copper-base alloys through differential aeration, selective dissolution and cathodic depolarization. Pitting, plug / dealloying and ammonia cracking of brasses and bronzes are also known. Sulphate-reducing bacteria produce tubercles and generate sulphide-rich scales on copper alloys.

   Though the toxicity of copper ions towards microorganisms is well known, copper and copper alloys are not free from biological corrosion. *Acidithiobacillus* group of bacteria can develop higher tolerance to copper ions and leach the metal. Slime forming bacteria along with iron were isolated from the corrosion products of copper-nickel alloy and monel tubes used in a nuclear power plant. Sulphate reducing bacteria can cause corrosion of underground copper tubes and pipes. Biologically generated ammonia can cause stress corrosion cracking of several copper alloys. Corrosion of brass in heat exchanger tubes by ammonia produced ubiquitously by bacteria is known.

B. **Carbon steels.** Generally tubercle formation with pitting underneath is encountered in steel pipes and tubes, resulting in hampered flow and plugging problems. Carbon steels are generally used for water, oil and gas transport under sub-soil or marine environments. Aerobic and anaerobic microorganisms adhere firmly to steel surfaces resulting in complex biofilms. Aerobic bacteria such as *Gallionella, Leptothrix* and *Acidithiobacillus* contribute to corrosion resulting from differential aeration cells. These organisms convert ferrous ions to ferric ions resulting in the precipitation of ferric oxyhydroxides. Anaerobic bacteria such as Sulfate Reducing Bacteria could also inhabit the tubercles formed in mild steel pipe lines.

   Aerobic bacteria can contribute towards corrosion in several ways, namely, formation of slimes, oxidation of iron and sulphides and generation of acidic metabolites. Hydrated slimes cover the metallic surfaces, creating differential aeration cells and provide an environment for subsequent growth of anaerobic organisms. Iron oxidising bacteria listed in Table 8 oxidise ferrous ions to less soluble ferric ions, leading to the

Table 8 Bacteria and fungi involved in Biocorrosion of Steels and Aluminium Alloys

Sl. No.	Genus/species	Temp°C	pH	Nature of Corrosion
1.	*Gallionella*	20-40	7-10	Aerobic, Iron & Steels, Tubercle formation, Oxidation of ferrous
2.	*Sphaerotilus*	20-40	7-10	Aerobic, Iron & Steels, Ferrous oxidation and tubercle formation
3.	*Pseudomonas*	20-40	4-9	Aerobic, Iron & Steels, Some can reduce Fe^{+3} to Fe^{+2}
4.	*P.aeroginosa*	20-40	4-8	Aerobic, Aluminium alloys Corrosion
5.	*Cladosporium resinae*	30-35	3-7	Aluminium allloys, organic acids produced corrode

formation of insoluble tubercles, which consist of hydrated ferric oxides and biological slimes. Steel water pipes are prone to such attack. Massive and tenacious tubercle formation inside steel pipes, not only hinders fluid flow, but also creates severe corrosion problems, such as extensive pitting, fissures and crevices. Areas beneath the deposits become anodes in a differential aeration cell with oxygen being reduced at the surrounding metallic surfaces.

$$Fe \rightarrow Fe^{++} + 2e \quad ...(18)$$

$$Fe^{++} \rightarrow Fe^{+++} + e \quad ...(5)$$

$$O_2 + 2H_2O + 4e \rightarrow 4OH^- \quad ...(8)$$

$$Fe^{+++} \rightarrow Fe(OH)_3;\ \text{hydrated ferric oxides} \quad ...(19)$$

C. **Stainless steels.** Stainless steels are used in many applications in nuclear power plants in high purity and sea water environments. Iron oxidising and iron-depositing bacteria induce corrosion of stainless steels. Microbially-influenced corrosion of austenitic stainless steels is characterized by pitting, usually adjacent to weldments. SRB also attack stainless steels, super stainless steels such as duplex steels and molybdenum steels. Microbial corrosion in under sea water applications is common. Slimes formed by bacteria can create sites for initiation of pits in stainless steels in sea water or fresh water. Microbial corrosion of stainless steels has often been observed at weldments, especially in the heat – affected zones.

D. **Nickel-based alloys.** Alloys such as monels and inconels are susceptible to microbially-influenced corrosion. Nickel-based alloys used in nuclear power plants corrode due to microbial attack under marine environments.

E. **Aluminium and its alloys.** Protective oxide films present on aluminium and its alloys could be disrupted and destroyed through biological attack. Aluminium and 2024, 7075 alloys used in aircraft and fuel storage tanks are susceptible to microbially-influenced corrosion in the presence of hydrocarbons (fuels). The generation of water-soluble organic/inorganic acids by bacteria and fungi can lead to corrosion of aluminium and its alloys. Pitting and intergranular corrosion are common. Aluminium-magnesium (5000 series) alloys used in marine applications are susceptible to pitting, intergranular corrosion, exfoliation and stress corrosion in the presence of microorganisms and biofilms.

It has been well established that aircraft fuel tanks and sea water components of aluminium and its alloys are attacked by organisms such as *Pseudomonas, Leptothrix,* Sulphate Reducing Bacteria and fungi. The fungus, *Cladosporium resinae* can grow on kerosene or paraffins as sole carbon sources, developing pinkish brown colonies. Fuel tanks of especially ground aircrafts are seriously affected by such fungal growth which produces carboxylic acids. The growth can cover large areas of the aluminium alloys and can also result from extracellular enzymatic activity of several microbes. The following microorganisms had been observed in an aircraft tank sludge, namely, *Pseudomonas aeroginosa, Aerobacter aerogenes* and several species of *Clostridium, Bacillus, Desulfovibrio, Fusarium, Aspergillus, Cladosporium* and *Penicillium*. Water associates itself with the fuel as moisture content and through humid air microorganisms can find their way into the tank. Pitting potentials of aluminium alloys could be affected by the organic acids generated by fungi.

F. **Titanium.** Titanium is susceptible to biofouling. SRBs and acid-producing bacteria can generate differential aeration cells leading to pitting of titanium and its alloys [10]. Titanium and its alloys used in marine environments promote biofilm formation involving manganese and iron oxidising bacteria as well as sulfate reducing halophiles. Surface passivity of titanium could be disrupted in the presence of anaerobes.

G. **Concrete.** Concrete is extensively used with reinforcing bars in sub-soil and marine environments. Aggressive soil conditions such as acidity, availability of metal and hydrogen ions, sulphate and sulfide content as well as the presence of different types of microorganisms lead to degradation of concrete structures. Sulphur-oxidising bacteria present in soil induce acid-corrosion of concrete. Anaerobic chemolithotrophs generate sulfides resulting in corrosion of concrete.

Concrete pipes carrying sewage are known to be prone to microbiological corrosion. At moderately alkaline pH levels, a wide variety of *Acidithiobacillus* group of organisms oxidise thiosulphate and polythionates to sulphate, leading to acid formation and decrease in pH. When the pH falls below 5, *Acidithiobacillus thiooxidans* takes over, oxidising sulphides to sulphur and sulphuric acid. Drastic pH drop from about an initial value of 10 to a final value of around 1, could thus occur due to mutualistic role of alkaline and acidophilic autotrophs. Concrete can disintegrate under such conditions. Other examples include collapse of massive concrete cooling towers and corrosion of building stones.

Besides metallurgical factors, microbiological parameters also significantly influence corrosion and failure of materials. The following biological factors are significant:

(a) Microorganisms and their colonies
(b) Scale, debris and biofilm structure and
(c) Exopolymers on the metal surface.

When biofilms form on metal surfaces, a number of factors come into play, such as:

(a) Type and nature of metal/ alloy
(b) Presence of sulfide and metallic inclusions
(c) Surface roughness, surface morphology and defects

(d) Presence of oils, residues and other surface impurities

(e) Microstructure and prior history of working and heat treatment.

All these factors need be considered in the analysis of microbially-influenced corrosion as they influence corrosion mechanisms.

Due to corrosion hard, thin, black and adherent film of ferrous sulphide develops over steel surfaces. The resultant anodic polarization limits corrosion of the metal. In the absence of adherent sulphide films, the corrosion product may be loose and bulky, promoting metallic corrosion. Formation of precipitated ferrous sulphide, as a result of microbial sulphate reduction, plays a significant role. The ferrous sulphide can serve as a cathodic depolarizer. The role of FeS and H_2S in accelerating the corrosion of steels could be seen as a contradiction to the well-known cathodic hydrogen depolarization theory. A possible cathodic role for the bacterially produced FeS would suggest a secondary and indirect role for the bacteria in the overall corrosion process. The presence of a sulphide film in contact with metallic iron could initiate galvanic coupling with FeS serving as cathode and the metal acting as anode, promoting corrosion.

The sulphate-reducers include thermophilic organisms such as *Desulfotomaculum nigrificans* which reduce sulphate at higher temperatures of the order of 60°C - 80°C. Severe corrosion of heat exchangers and cooling systems can result in the presence of the above bacteria.

Biofouling Mechanisms Relevant to MIC

Marine biofouling, involves a complex community of microorganisms which significantly alters and influences the immediate surrounding environment. While this community is highly variable, the major groups of components and their effects are similar. The settlement of living organisms on metallic substrates in sea water confers several advantages to growth nutrition and reproduction. Marine fouling community is dynamic, the organisms growing, dying and reproducing. This intense biological activity results in many changes to the substratum, leading to its degradation.

Biofouling leads to the formation of inorganic and organic deposits on surfaces, which adversely influence equipment performance and reduces its lifetime. Biofouling is the accumulation and growth of living organisms and their associated organic and inorganic products on a surface.

The biological community is extremely sensitive to the environment, especially artificial environments such as the hulls of ships, offshore oil and gas platforms, and cooling-water intakes. The fouling ecosystem can be divided into different components, the presence or absence of which in any particular situation will depend on the physical and chemical parameters of the environment [4].

(a) Bacteria

(b) Fungi

(c) Microalgae - Blue-green algae (Cyanobacteria) and some diatoms

(d) Macroalgae

(e) Protozoa - often associated with bacteria and fungi.

(f) Invertebrates - Essentially marine
(g) Organic molecules - Polymeric extracellular products of living organisms.

Four main stages can be distinguished in the biological fouling of a surface.

(a) Transport of organic molecules from the bulk and its adsorption,
(b) Surface colonization by bacterial species and formation of biofilm,
(c) Incorporation of higher organisms such as protozoa, fungi and microalgae and
(d) Subsequent build-up of biofilms.

The adsorption of organic molecules and the settlement of bacteria, occur in almost any environment where there is an aqueous phase capable of supporting bacterial activity. The degree of initial coverage depends on environmental conditions and the initial surface coverage may be partial. Once the organisms are irreversibly attached to the surfaces, processes of reproduction and alteration to metabolism take place that result in the formation of bacterial colonies. These colonies are cells embedded in voluminous amounts of extracellular polymers.

Fouling organisms mutually interact in an ecosystem, and often establish a dynamic equilibrium.

Biofilms are thus structured assemblages of microorganisms surrounded by extracellular materials and containing several complex subcommunities. The macroscopic organisms provide diverse environments for the microscopic organisms. As the macrofouling community develops, the microscopic film can undergo changes with aerobic and anaerobic regions developing. The creation of sites of anaerobic activity within the biofilm is of particular interest. Juxtaposition of aerobic and anaerobic environments can create considerable dichotomies of chemical and physical conditions at the metal surface.

The conditions at the substratum will be quite different from that in the bulk phase. The interaction of three phases namely, metal, biofilm and the electrolyte results in concentration gradients, influencing processes such as heat transfer, reduction in fluid flow, drag effects and metallic corrosion.

Seawater is a well-known corrosive environment and any biological activity can enhance its aggressiveness further through:

(a) Creation of differential aeration cells
(b) Production of corrosion-promoting metabolites such as acids (both organic and inorganic) an sulfur compounds (especially H_2S).
(c) Promotion of cathodic (sulphate-reducing bacteria (SRB) activity) and anodic reactions and
(d) Damage to protective coatings.

These changes can enhance corrosion by the formation of differential aeration, pH and chemical concentration cells. These cells permanently separate anodic and cathodic sites and thus concentrate metal loss to small areas. The result is severe localized corrosion by 'pitting'. Oxygen concentration cells are the most common.

Microorganisms, through their metabolic activities, are also capable of altering the ionic concentration beneath the biofilm and produce chemical and pH concentration cells that act in a manner similar to that of oxygen concentration cells.

Corrosion by metabolic products of microorganisms and the direct or indirect promotion of anodic and cathodic reactions are similar to other microbially-influenced corrosion situations except that there is likely to be a wider range of environments created around marine fouling. Enhancement of the cathodic oxygen reaction by general aerobic bacterial films has been well known. This effect leads to accelerated pitting and crevice corrosion of stainless steels.

Anodic reactions are promoted by the following:

(a) Production of aggressive metabolites such as acids (e.g. sulfuric in the case of sulfur-oxidizing bacteria and organic acids in the case of fungi).

(b) Production of metabolites enhancing the corrosive action of other anions. (e.g. sulfide anions produced by the reduction of sulfates by SRB, forming corrosive iron sulfide).

Acids attack most metals, with the formation of the metal salt and the production of hydrogen. Acid attack usually requires a pH below 4. There is more buffering capacity and oxygen reduction at the cathode begins to become significant and differential pH cells may become important at higher pH.

Various corrosive inorganic and organic acids are produced by microorganisms. Sulfur and its compounds are oxidized to sulfuric acid by species of the genus *Acidithiobacillus*. Other macrofouling organisms can release acids when damaged, for example species of the alga *Desmerestia*.

Hydrogen sulfide is generated under anaerobic conditions by SRB which cause serious corrosion problems on iron and steel. Macrofouling produces suitable anaerobic environment to provide nutrients for the SRB. Biological production of hydrogen sulfide in combination with the hydrodynamic loading of macrofouling, can produce severe corrosion fatigue problems.

Enhancement of corrosion through the production of H_2S by SRB has been widely noticed in marine environments. The effect of seawater provides a combination of corrosive environment (enhanced by fouling) and stress (wave action, also enhanced by fouling). Corrosion fatigue can be very important because the effect of the major protection method for offshore structures, (cathodic protection) is ambivalent, having been shown to be both detrimental and protective, depending on conditions.

Hydrogen embrittlement involving the entry of atomic hydrogen into a metal, rendering it brittle and stressed is another area of concern. Marine biofouling enhance corrosion fatigue through stress inducement and providing organic sulfur compounds for the production of hydrogen sulfide by the SRB. H_2S is particularly aggressive in corrosion fatigue as the sulfide ions promote anodic dissolution and prevent the combination of hydrogen atoms to form gaseous molecules, thus making more hydrogen available for embrittlement.

A combination of temperature, nutrient availability and suitable areas for H_2S production are probably responsible for the enhancement of corrosion fatigue by SRB activity. Anaerobic environments in the presence of high nutrient levels, as under biofouling conditions will be limited and transitory due to the exhaustion of nutrients and the dynamic nature of the system.

Role of Biofilms in MIC

Within a few hours of immersion in fresh or sea waters the films of microscopic fouling organisms begin to form on structural metals and alloys. The conditioned surface is colonized by a variety of bacteria and other microorganisms followed subsequently by the settlement of macroinvertebrate larvae and the development of a diverse macrofouling layer [3-4].

Electrochemical corrosion reactions take place along with biofilm formation, promoted by differential concentration cells. The biological and electrochemical events are influenced by factors such as pH, temperature and concentrations of various organic and inorganic chemical species at the metal-water interface.

Various types of interaction between microbial films and electrochemical corrosion are known. A growing marine biofilm can influence water chemistry at the metal-film interface. The effects of biofilms on corrosion of passive alloys such as the stainless steels and superalloys, titanium and platinum are indeed significant.

The presence of a microbial film on the metal surface can create chemical conditions at the metal-film interface quite different from ambient environment. Oxygen concentration levels could vary from region to region within the biofilm.

pH changes under marine biofilms, can be much larger and more important than those in the ambient environment. pH values as low as 5 can be expected under aerobic biofilms containing acid producing bacteria. Even more acid pH values, in the range 1 to 2 can be expected under discrete biodeposits. Actual measurements of the pH under biofilms and biodeposits can be carried out using microelectrodes.

The effect of microorganisms on corrosion depends not only on their metabolic reactions, but also on their distribution on the substrate.

Scattered biofilms unlike fully-covered ones do not have much effect on corrosion. When many microorganisms are present, they function in consortia and synergism would prevail. A classic example is the formation in a partial aerated environment, of an anaerobic microenvironment under a developed film or bacterial colony. The aerobic organisms in the outer part of the film have created conditions favourable for the growth of the anaerobic sulfate reducers below them. The SRB may, in turn, produce a conductive environment to the growth of other anaerobes forming a consortia. The SRB may also produce sulfides and other metabolic byproducts that can be utilized by aerobic sulfur oxidizers to produce sulfuric acid.

The electrochemistry of corrosion can also change the chemistry at the metal-solution interface, and this can either promote or inhibit biofilm formation. Corrosion can further promote bacterial attachment and growth by producing chemical nutrient species. It can also produce chemical species that will inhibit biofilm formation. For example, corrosion of copper-based alloys produces toxic copper ions. Cathodic protection can produce enough hydroxyl ions to shift the pH at the metal surface to alkaline values. pH values more basic than about 9.5 are detrimental to most marine bacteria. Another interesting product is hydrogen peroxide, which can be produced by both biological and electrochemical processes on metal surfaces.

Microorganisms can influence the corrosion of passive alloys through an enhancement of the cathodic reaction, with consequent ennoblement of the corrosion potential. Direct initiation of deep pitting under discrete biodeposits could also occur.

The corrosion potentials of passive metals and alloys immersed in natural aqueous environments generally turn more noble with time as the biofilm develops. Ennoblement of the corrosion potential is thus directly attributable to biofilm formation.

Exposure tests on a series of austenitic, ferritic and duplex stainless alloys, superalloys and titanium confirmed ennoblement which was the highest in fresh water and the amount of ennoblement decreased nearly linearly with increasing salinity. Corrosion potential ennoblement due biofilm is variable. Variability can arise from the degree of coverage of the biofilm on the metal surface. No ennoblement was observed under conditions of scant biofilm formation. Complete, coverage, on the otherhand, cannot ensure ennoblement. Mixed potential theory for a passive alloy can possibly explain such ennoblement processes.

Much of the observed ennoblement on stainless steels can be explained by the low pH mechanism. Bacterial enzymes in the biofilm are probably involved both in producing peroxide and in limiting its concentration. The low pH mechanism also provides a basis on which to understand the loss of ennoblement in daylight when photosynthetic algae are present in the film. In the dark, the organisms in the film use oxygen and generate carbon dioxide (a weak acid), thus contributing to acidification at the metal surface.

For passive, corrosion-resistant alloys, biofilms lead to a change in corrosion potential. For less resistant metals, a decrease in the initiation time for localized corrosion is possible. Initiation of pitting occurs when the open circuit corrosion potential exceeds the pitting potential.

The induction period for crevice corrosion is dependent on the time needed for depletion of dissolved oxygen in the crevice. Presence of aerobic bacteria in the crevice could deplete the oxygen. Biological oxygen consumption could influence electrochemical corrosion mechanisms.

In the case of stainless steels, not only does the corrosion potential usually shift in the noble direction due to bacterial biofilm, but the cathodic kinetics are also affected. At the potential range, the limiting diffusion current for oxygen could be also increased. After longer immersion times, the limiting current would decrease and it is uncertain whether the limiting current is really affected over the long term.

There may be yet another consequence related to the increase in cathodic kinetics. Both the open circuit potential and the cathodic kinetics of the stainless steel were observed to be increased in the presence of the biofilm. If copper alloy is connected to stainless steel in a galvanic couple, the corrosion current density for the couple with a biofilm would be higher than that without the film. Presence of biofilm will cause the stainless alloy to be a more severe cathode in galvanic corrosion for anodes such as copper alloys and perhaps steel.

Possible effect of a biofilm on the current density for cathodic protection also needs to be considered. Current requirements in the presence of biofilms are often different from conditions in their absence. Enhanced microbial activity could increase applied current requirements to achieve noble potentials.

Corrosion and Ennoblement through Biogenic Manganese and Iron Oxide Deposition

Microbial colonisation of metals and alloys can influence the mechanisms and rate of electrochemical reactions. For example, ennoblement of passive metals such as stainless steels and titanium can occur due to positive shifts in their open circuit potentials in the presence of microbial colonies. An increase in cathodic current density during cathodic polarization is often associated with enhancement of open circuit potentials. Biofilm formation on stainless steels changed their electrochemical behavior through significant positive shifts in open circuit potentials and increase in cathodic currents. Similar behavior could be observed on titanium as well. Microbial colonisation responsible for such ennoblement results in biogeneration of extracellular substances, organo-metallic complexes and metal-specific enzymes. Microbially-modified acidification at metal-solution interfaces and generation of oxidants such as H_2O_2 can also cause ennoblement. Microbial production of passivating siderophores is yet another possibility. Microbial deposition of iron and manganese oxides / hydroxides can lead to ennoblement of passive metals. Iron and manganese oxidising bacteria inhabiting fresh and sea waters as well as soils can attach and colonise on metal surfaces and precipitate manganic and iron oxyhydroxides.

$$Mn^{+2} + 2H_2O \rightarrow MnO_2 + 4H^+ + 2e \quad ...(20)$$

$$MnOOH + OH^- \rightarrow MnO_2 + H_2O + e \quad ...(21)$$

Depending on the pH of water, the above reactions can influence the measured open circuit potentials to deviate in a nobler direction.

Electrochemical oxidation/reduction of manganese and iron is part of the natural manganese and iron cycles in natural environments. Both manganese and iron oxidising and reducing microorganisms take part in the cycle involving biochemical and electrochemical pathways.

Microbially produced MnO_2 can also corrode active metals such as mild steel.

$$Fe + MnO_2 + 4H^+ \rightarrow Fe^{+2} + Mn^{2+} + 2H_2O \quad ...(22)$$

Manganese-oxidising organisms can corrode stainless steel welds.

Monitoring, Diagnosis, Prevention and Control

Monitoring, detection and diagnosis are important in the study and mitigation of microbially-influenced corrosion. Data collection relevant to microbially-influenced corrosion failures need be undertaken scientifically. Biological sampling, water and sediment sampling and nondestructive examination should be done before clean up and repair of failed structures. Biological corrosion products (organic and inorganic) and exopolymers present in the biofilm of failed structures need be presented, analyzed and appropriately characterized. In water samples, one need to collect information on temperature, Eh, pH, dissolved gases (including oxygen), conductivity, turbidity and suspended solids [3].

Principal stages of a failure analysis include data collection, visual and nondestructive examination, chemical analysis and determination of causes for the failure. Metallographic examination using optical and electron microscopy will reveal microscopic features and data relating to surface morphology. Visual examination itself can be revealing; for example, pitting at low points of pipe, pitting at air-water interfaces, rust-like deposits, small pin holes and weld-decay can be convincingly documented. Nondestructive evaluation methods such as magnetic particle inspection, liquid penetrant inspection, ultrasonic, eddy current and radiographic inspections are very useful besides acoustic emission and stress analysis techniques. On-line monitoring may supplement other sampling and testing systems and could provide early warning of potential microbial-influenced corrosion. Biocide additions and concentrations can be monitored through instrumentation. Performance monitoring of systems such as recording of flow-rates, temperatures and pressures can indicate formation of tubercles or deposits in pipe lines, condenser tubes and heat-exchanger systems. On-line corrosion monitoring could be done using exposure coupons in a process stream, flow-loop, soil or sea water test sites.

Standard on-line corrosion monitoring methods are categorized into three groups [3].

Group 1: Methods for measurement of physical metal loss

a. Corrosion coupons
b. Electrical resistance probes
c. Surface activation and gamma radiography
d. Ultrasonics
e. Direct electrical methods

Group 2: Methods for measurement of electrochemical properties of corroding surfaces

a. Linear polarization resistance probes
b. Galvanic probes (zero resistance ammetry)
c. Potential measurement probes
d. Potentiodynamic polarization probes
e. Electrochemical impedance spectroscopy probes
f. Electrochemical noise probes

Group 3: Other measurements

a. Hydrogen probes
b. Acoustic emission

Environmental factors influencing the aggressiveness of soils as indicated above need be established. Besides measurements of acidity and Eh, electrical resistivity gives a better correlation of corrosivity. The relative oxidising and reducing conditions existing in soil and water (as established by Eh measurements) could provide insight into the presence of aerobes and anaerobes. Soil samples could be tested in the laboratory for the presence of different microorganisms. Based on such tests, efforts should be made for provision of nonaggressive surroundings. Thick backfill of chalk or gravel around the structures to be protected can ensure protection of pipes since this ensures aeration and drainage.

Coatings, which act as a barrier between the metal and the environment, should be resistant to chemical and biodegradation. Coal tar, epoxy and enamels, asphaltic bitumens and several types of resins could be employed. For off-shore applications, teflon, polydimethyl siloxanes and pure silicone rubber have been used as antifouling coatings.

Protective coatings could be barrier, inhibitive or sacrificial depending on whether it acts as a barrier between environment and substrate, control of corrosion or it corrodes preferentially (galvanic effect due to zinc on steel, for example). There are primary, secondary coats and topcoats. Widely used coating materials such as alkyds, acrylics, bituminous, chlorinated rubber, epoxies, ester, latex, phenolics, polyesters, polyurethanes, vinyls as well as inorganic coatings are very effective depending on the circumstances.

Biocides form a part of a larger group of antimicrobial chemicals, consisting of agricultural biocides, industrial biocides as well as health-related ones. Biocides could be introduced either as an adjunct in an industrial process beforehand or incorporated into the finished product. Examples include biocides introduced into machine-cutting fluids and those incorporated with wood, plastics and rubber products. To be effective, the biocides should have the following properties.

(a) Cheaper, stable and long lasting
(b) Acceptable toxicity levels
(c) Safer to handle, non-inflammable and non-explosive and
(d) Slow development of bacterial resistance

Oxidising biocides include chlorine, ozone and other chlorinated compounds while some examples of nonoxidising biocides are organic sulphur compounds, quaternary ammonium compounds and cationic polyamines and amides. Biocides bring about actual destruction of organisms, while biostats maintain them in a state of inactivity and stifled growth.

Difficulties could be encountered in the choice of an appropriate inhibitor at the required dosages. Most of the sulphate-reducing organisms are inhibited by 20 ppm - 50 ppm of cupric ions. However, many *Acidithiobacillus* group of bacteria exhibit very high copper-tolerance, upto 20,000 ppm. Chromates are good inhibitors for sulphate-reducers, but are environmentally very toxic. Field trials are essential to establish the correct dosages of an inhibitor and to assess its environmental impact. It may sometimes be possible to control bacterial growth and activity by removal of an essential metabolite. Elimination of sources of sulphur could prevent the proliferation and deoxygenation could limit the growth of anaerobes and aerobes, respectively. Control of Eh and pH in a bacterial habitat is yet another approach.

Cathodic protection in conjunction with coatings could offer good corrosion resistance. Sacrificial anodes such as aluminium, zinc or magnesium, or impressed current systems could be used. A minimum potential of - 0.95V (vs Cu-$CuSO_4$ electrode) is needed for complete protection of steels.

Use of substitute materials may help in the prevention of MIC. Steel pipes can be replaced by nonmetallic materials such as plastics and polymeric materials. Other preventive measures include steam sterilization of equipment wherever possible at regular intervals. Stagnant solutions and deposits need be removed frequently.

The most common chemical method for controlling biofouling in water systems is the use of oxidising or non-oxidising biocides. Chlorine, ozone and bromine are three typical oxidising biocides. For overall control of fungi algae and bacteria nonoxidizing biocides could be more effective since they are persistent and pH dependant; for example, formaldehyde, glutaraldehyde, isothiazolones and quaternary amines.

New tools for MIC studies have emerged. Use of microsensors for analysis within a biofilm is a recent advance. Equally exciting is the developing of fiber-optic microprobe for studying biofilm structure. DNA probes can be used in biofouling research. Environmental Scanning Electron Microscope (ESEM), Confocal Laser Scanning Microscopy (CLSM) and Atomic Force Microscopy (AFM) allow biofilm observation in real time. Corrosion sensors and electrochemical corrosion assessment technology have received significant attention in recent times. An electrochemical sensor for monitoring biofilms on metallic surfaces have been developed. The system provides an immediate indication of biological activity and can be used to optimise biocide dosage and treatment duration [5].

For natural gas pipelines, guides for identification of internal and external MIC have been recommended based on three types of evidences, namely, metallurgical (corrosion damage appearance), biological (cell counts for aerobes and anaerobes) and chemical (corrosion products and other deposits identification).

Identification and confirmation of MIC can be made based on presence of microorganisms and their metabolic products, morphology of failed substrates, evaluation of corrosion products and assessment of existing environmental conditions. For example, MIC due to SRB leaves behind unique corrosion morphology such as clustered hemispherical pits on stainless steel, aluminium and titanium along with the presence of metal sulfides. On the otherhand, acid producing microbes cause localised corrosion. Slime formers bring about general corrosive attack under slime coatings with appearance of rusted brown surfaces.

Many different methods have been suggested, to combat microbial corrosion. General measures adopted to tackle this problem can be summarised as follows.

Selection and Control of Environment. Establishing the aggressiveness of soil conditions through monitoring parameters of corrosivity such as Eh, pH, resistivity, quality of water, bacterial enumeration and assessment of soil and water composition with respect to sulphides and other mineral and organic matters.

Application of Coatings, Incorporating Preferably Biocidal Components. Coal tar, epoxies, enamels, paints and resins.

Inhibitors. Biocides and biostats; chlorine, ozone, phenolics, heavy metal salts, metal organic compounds, quaternary amines.

Cathodic Protection. Impressed current method and use of sacrificial anodes.

Prevention of microbially-influenced corrosion is always less expensive and thus preferable. One has to follow certain guidelines for establishing and maintaining clean systems.

Design considerations are very significant. Use of wrong materials, improper fabrication and treatment and lack of environmental control can promote microbially-influenced corrosion. In the design of water-containing systems, the following factors need be considered with respect to onset of microbially-influenced corrosion.

- Source of water
- pH, hardness, salinity oxygen content and impurities
- Temperature variations and flow rates

For sub-soil pipe lines adequate soil analysis is required to bring out corrosiveness parameters such as moisture, acidity/alkalinity, oxygen content, stray currents and presence of electrolytes. Soil resistivity and redox potential measurements are very useful and could indicate prevalence of microbial habitats in soils. Manufacturing defects in buried pipes such as improper joints and areas of liquid stagnation could be of concern.

REFERENCES

1. J.D.A. Miller, *Microbial Aspects of Metallurgy,* Medical & Tech. Pub. Co., Lancaster (1971).
2. M.G. Fontana, *Corrosion Engineering,* McGraw-Hill, NY (1987).
3. S.W. Borenstein, *Microbiologically influenced corrosion handbook,* Woodhead Pub. Ltd., Cambridge (1994).
4. C.C. Gaylarde and H.A. Videla, *Bioextraction and biodeterioration of metals,* Cambridge University Press, Cambridge (1995).
5. H.A. Videla and L.K. Herrera, *Microbiologically influenced corrosion: looking to the future,* International Microbiology, 8, pp.169-180 (2005).

6. K.A. Natarajan, *Microbes, Minerals and Environment*, Geological Survey of India, Bangalore (1998).
7. I.B. Beech and C.C. Gaylarde, *Recent Advances in the study of biocorrosion-an overview,* Rev. Microbiol., 30, pp.1-22 (1999).
8. J.E. Zagic, *Microbial Biogeochemistry,* Academic Press, NewYork (1969).
9. R. Javaherdashti, *A review of some characteristics of MIC caused by Sulfate-reducing-bacteria, past, present and future, Anti-corrosion Methods and Materials,* 46, pp. 173-180 (1999).
10. T.S. Rao, A.J. Kora. B. Anup Kumar, S.V. Narasimhan, and R. Feser, *Pitting corrosion of titanium by a fresh water strain of Sulphate Reducing Bacteria,* Corrosion Science, 47, pp.1071-1084 (2005).

CHAPTER 4

Advances in Generation of Design Data for Stress Corrosion Cracking

R.K. Singh Raman
Departments of Chemical and Mechanical Engineering, Bldg 31, Monash University (Melbourne), Vic 3800, Australia

A synergistic action of tensile stress and corrosive medium leading to a premature cracking of materials is called stress corrosion cracking (SCC). The features that make SCC the most dangerous form of corrosion-assisted failure, and a critical consideration in selection of material for equipment operating in corrosive environment include : (a) it can occur at stresses below the design stresses, i.e., those required for failure of the material in the absence of the corrosive environment, (b) environments that are only mildly corrosive may cause severe SCC, and (c) sometimes, the alloy may virtually appear unattacked over most of its surface, and only a few fine and localized cracks may propagate undetected to failure.

SCC continues to account for a great proportion of failures of industrial components operating in corrosive environment. A reliable life assessment and extension of aging plants hinges critically on the reliable, cost-effective and rapid characterization of SCC of in-service as well as failed components exposed to corrosive conditions. A few techniques have been used for monitoring SCC [1-3]. Common techniques, such as the U-bend and C-ring testing, have proved to be too simplistic in approach and the data generated can often represent overly accelerated conditions. Laboratory techniques, such as slow strain rate testing (SSRT), are suitable for providing a broad 'go, no-go' type of assessment[1] but fail to generate data for life prediction of in-service components.

Life extension/prediction of components exposed to in-service corrosive environments, generally involve assessment of susceptibility of existing cracks to environment-assisted propagation. This susceptibility is also a critical consideration for design engineers while selecting construction materials for corrosive environments. Fracture toughness (K_{IC}) is the measure of a given material's resistance to crack growth. The driving force for extension of an existing defect (crack) is the stress-intensity factor, K_I. A crack will propagate when K_I exceeds some threshold value, K_{IC}, i.e., fracture toughness. Determination of the critical value of stress intensity factor (K_I) necessary for propagation of an existing crack (i.e., K_{IC}) is a critical material selection criterion for any mechanical engineering design. For an alloy-environment system conducive to SCC, the threshold stress intensity with the influence of environment at the crack-tip (i.e., K_{ISCC}) is considerably lower than K_{IC}. K_{ISCC} has traditionally been determined using various specimen geometries, as shown in Figure 1.

The prime and most stringent requirement for application of fracture mechanics is the validity of a rigorous plane strain condition (PSC) at the crack-tip, in order to eliminate/minimize the zone of the plastic deformation. In order for specimens of the commonly used

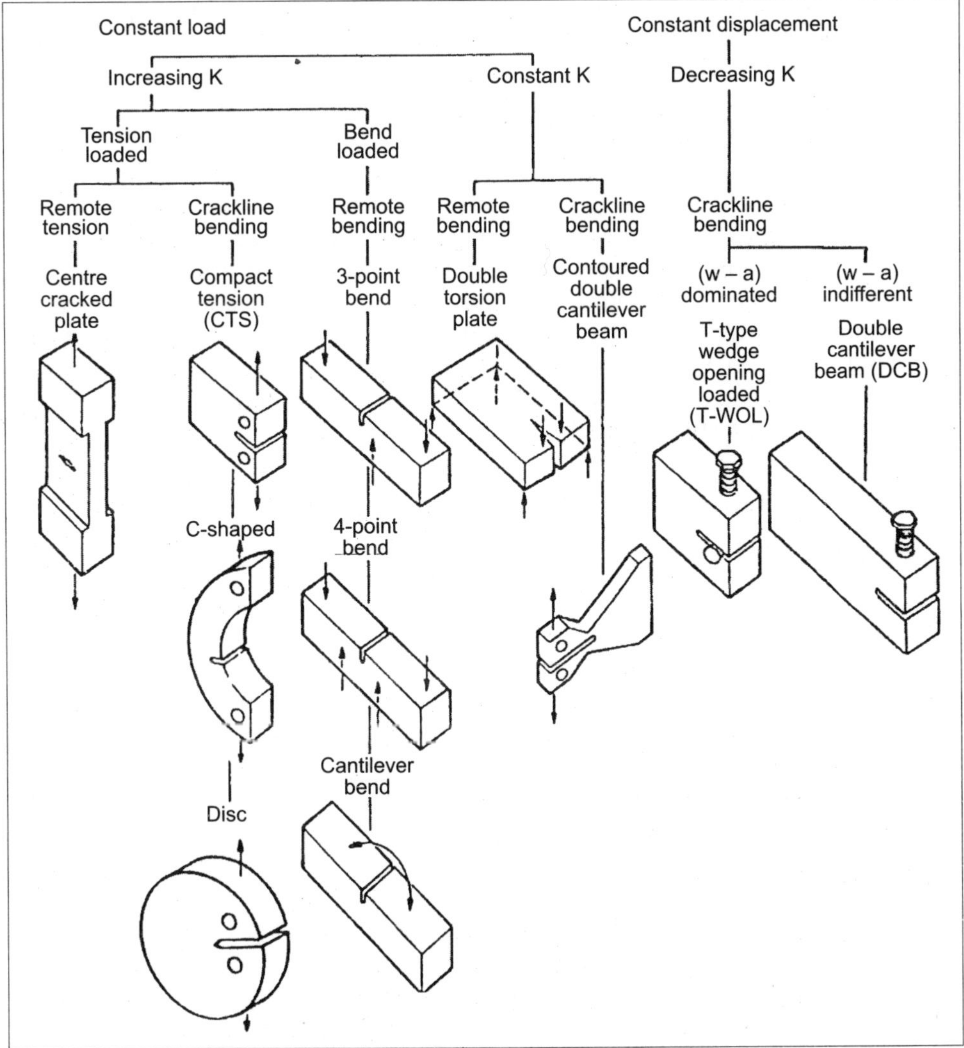

Figure 1 Different specimen geometries for stress corrosion cracking (SCC) testing [2,3]

geometries to comply with the requirements of valid plane strain condition, specimens require to be bulky (viz., compact tension (CTS) and double cantilever beam (DCB) test specimens)[1,2].

Compact-tension (CTS) specimens (shown in Figure 1), one of the most widely used specimen configurations for determination of fracture toughness (K_{ISCC}), is believed not to conform fully to the stringent requirements of PSC, in spite of the bulkiness of the specimens used. It is well known that the theoretical requirements, i.e., uniformly distributed applied stress over the thickness and plane-strain condition, can never be realized as long as the free surfaces exist at both ends of a CTS specimen. The end effects will be further amplified when the thickness decreases to a thin plate. Consequently, specimens of greater thicknesses are used, which can only reduce (never eliminate) the end effects. A large scatter in K_{IC} and K_{ISCC} data generated using CTS specimens are attributed to end effects and to the conservative specimen size, which lead to inability to control the direction of crack propagation, leading to a zigzag or non-uniform crack front. Most importantly, a large scatter in data generated using CTS specimens forces a more conservative allowance in design, such as a large factor for safety. Large scatter in K_{IC} and K_{ISCC} data from CTS testing of the in-service components also leads to a conservative prediction of the remaining life of a given component. Conservative estimates of the design data and remaining life have serious cost implications.

Compact-tension (CTS) and other specimen geometries that require use of bulky test specimens are also fraught with other disadvantages, such as high cost of manufacture/testing of specimens, and requirements of prohibitively long testing time, large number of tests and bulky piece of test materials. Therefore, an accurate, rapid and cost-effective determination of K_{ISCC} is of immense interest to prudent design and maintenance engineers as well as for accuracy in life prediction. With a view to addressing these requisites, this article describes two recent techniques for generation of K_{ISCC} data, viz., circumferential notch tensile (CNT) testing and spiral notch torsion testing (SNTT).

DETERMINATION OF K_{ISCC} BY CNT TESTING

Circumferential Notch Tensile (CNT) Testing

Specimens used for CNT testing, shown in Figure 2, are one of the smallest possible specimens that can produce valid plane strain crack loading conditions [4-6]. Ahead of the notch of the CNT specimen, a uniform pre-crack is developed by subjecting the specimen to a controlled fatigue in a bending-rotating set-up [4-6]. The pre-cracked CNT specimen is then subjected to a constant load, until the specimen fails. Tests are carried out at different loads which translate to different magnitudes of stress intensity (K_I). K_I is correlated with the crack propagation parameters, viz., crack-growth velocity or time-to-failure (T_f) [4-6], in order to determine K_{IC} and K_{ISCC}. However, an accurate determination of K_I is the vital key to the successful application of the technique.

Determination of K_I

The K_I is determined from the fractured specimen, using Equations 1-9[4-6]:

$$K_I = (\sigma_t + \sigma_b)\sqrt{a\pi F_o} \quad ...(1)$$

Where,

$$a = \frac{D-d}{2} \quad ...(2)$$

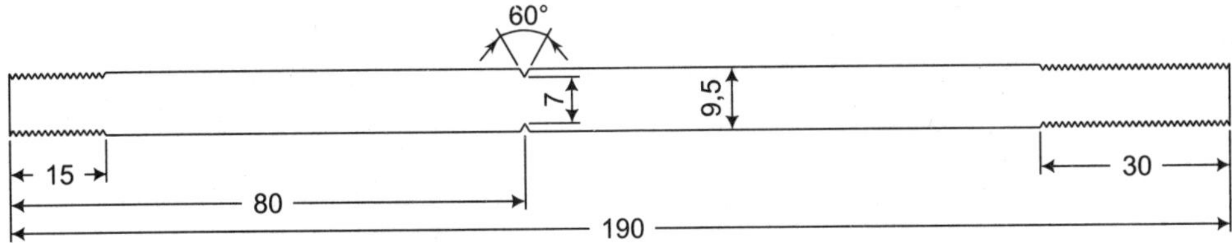

Figure 2 CNT specimen (all dimensions in mm)

$$\sigma_t = \frac{4P}{\pi D^2} \qquad ...(3)$$

$$\sigma_b = \frac{18P\varepsilon}{\pi D^3} \qquad ...(4)$$

$$F_o = Fe^{\alpha(e/D)} \qquad ...(5)$$

$$F = \frac{1.25}{\left[1 - \left(\frac{2a}{D}\right)^{1.47}\right]^{2.4}} \qquad ...(6)$$

$$\alpha = 22.188e^{-4.889(2a/D)} \qquad ...(7)$$

P is the applied load (N); *D*, the specimen's diameter (m); *d*, the equivalent ligament diameter (m); ε, the eccentricity (m); and *a* is the effective crack length (m).

The first estimate of K_I is obtained by using the effective crack length (*a*) in Equation (1). This estimate of K_I is used to get the Irwin correction factor (r_y) from Equation (8) and ($\bar{a}$) from Equation (9) and hence the quoted results for K_I from Equation (1).

$$r_y = \frac{1}{6\pi}\left(\frac{K_I}{\sigma_Y}\right)^2 \qquad ...(8)$$

$$\bar{a} = a + r_y \qquad ...(9)$$

Where σ_y is 0.2% offset tensile yield stress (Pa). The basis of this correction term has been discussed thoroughly in the literature [7-10]. This correction term relates to a certain degree of plasticity at the crack-tip, given that the materials do not behave perfectly elastic and show some plastic deformation at the crack tip. The first estimate of K_I is calculated by using the value of (*a*) in equation (2). Then, K_I is calculated using $\bar{a}$ in place of *a*.

Allowance for eccentricity (ε)

One of the critical issues in determination of K_{IC} by CNT method is the possible effects of the eccentric fatigue crack and the ligaments produced by the rotating bending fatigue crack machine of specimens. As shown in the examples of eccentric pre-cracks in Figure 3, the ligaments are off-centre, which make it difficult to locate the centroid of the ligament. This elliptical fatigue crack or eccentricity occurs due to the insufficient isotropy in its fatigue characteristics. When the ligament is off-centre, the final fracture of the specimen will not result only from tension force only but will also have a contribution from the bending forces.

Figure 3 Fracture surface of CNT specimens showing eccentricity of fatigue cracks produced by the rotating bending [11]

Therefore, the effect of the eccentricity must be taken into account in calculating K_I value. A specific method that is employed to account for the eccentricity is described elsewhere in detail [11,12].

Validity requirements for K_I measurements

Materials in non-ideal situations do not deform completely in elastic brittle manner as the linear elastic fracture mechanics (LEFM) approach assumes. The LEFM approach is only valid if the size of the plastic zone is negligible compared to the specimen geometry. Therefore, specifications for determining the validity of measuring K_I were developed. The validity requirements for measuring K_I are:

$a_f \geq 2r_y$ and $\frac{\sigma_N}{\sigma_Y} \leq 2.5$, where; $a_f = \varepsilon + \frac{(D-2a_m-d)}{2}$ and σ_N is the nominal applied stress in the final ligament (Pa).

Validity limits requires[4-6] the fatigue crack depth to be at least twice the Irwin plastic zone correction in depth, and that the average stress across the ligament after fatigue cracking should not exceed 2.5 times the yield strength. These limits were also applied to those cases where the final ligaments were eccentric to the specimen centreline. With an eccentric ligament, the maximum nominal stress considered was then a combination of tensile and bending stresses, which should not exceed 2.5 times the yield strength. Further, for such eccentric ligament cases the fatigue crack depth considered for the purposes of validity was taken to be the greatest depth of the crack.

CNT Testing for determination of K_{ISCC}

First attempts have recently been made for using CNT testing for generation of K_{ISCC} data [11-14]. Generation of K_{ISCC} data by CNT testing requires relatively small amounts of testing material and small loading devices for achieving high stress intensities, and therefore enables considerable reduction in testing costs and time. Spheroidal graphite (SG) cast iron in hot caustic solution was the first system tested for application of CNT in generating K_{ISCC} dat [15,16]. Choice of cast iron was based on its inherent brittleness and hence likely compliance with the constraints of linear elastic fracture mechanics, viz., plane strain condition (PSC). Cast iron vessels, used for low temperature steps in Bayer process for alumina processing are also known to suffer caustic embrittlement/cracking, as evidenced from the fractograghic and metallographic investigations of ex-service components of SG cast iron (from an alumina processing line) [15,16].

The custom built CNT rig for stress corrosion cracking testing using hot caustic solutions is shown in Figure 4. The major components of the rig include a corrosion cell (made out of Monel 400), facilities for heating of the corrosive solution, temperature control and thermal insulation. The rig also consists of facilities for electrochemical testing, such as the reference and counter electrodes, which are used for determination of K_{ISCC} under imposed electrochemical potentials. The CNT specimens (shown in Figure 2) are machined out of 20 mm diameter rods of the cast iron. The diameter of the specimen is 9.5 mm with a 60° "V" groove in the middle, which gives a minimum diameter of 7 mm. The notched specimens are subjected to fatigue pre-cracking, using a rotating bending fatigue machine. The pre-cracked specimens are cleaned and washed with ethanol and dried, before being installed into the rig. The corrosion cell is filled with 200 g/L NaOH solution and heated to the test temperatures (100 and 120°C) before applying the tensile load to the specimen. This procedure ensures that the applied load did not change as a result of the expansion of the specimen and surrounding metallic fittings. The specimen is held at a given load until it failed. The time-to-failure (T_f) is recorded at the failure of the specimen. The K_I corresponding to a given applied load is determined from the fractured specimen, using a method described earlier.

A plot of the calculated values of K_I against time-to-failure (T_f) of SG cast iron in 200 g/L caustic solution at 120°C is shown in Figure 5. Only those values of K_I that complies with the validity requirements (as described earlier) are used for the plot in Figure 5. Table 1 presents data suggesting compliance of the data with the validity requirements. As shown in Figure 5, K_{ISCC} is determined at 120°C to be 9 $MPa.m^{1/2}$, which is the minimum K_I required for propagation of SCC crack. The specimens in the region below 9 $MPa.m^{1/2}$ should be immune to SCC. Indeed, as shown in Figure 5, a specimen placed in this region did not fail even after nearly 3000 hours of testing.Following the same procedure, K_{ISCC} for SG cast iron at 100°C was determined to be ~ 11 $MPa.m^{1/2}$. Intergranular caustic cracking is evident in metallographic and fractographic micrographs in Figure 6.

Low K_{ISCC} values (i.e., 9 $MPa.m^{1/2}$ at 120°C and 11 $MPa.m^{1/2}$ at 100°C) of the cast iron in hot caustic solutions suggest that once cracks initiate this material will easily suffer caustic cracking. This inference is consistent with the common cases of caustic cracking of cast iron vessels used in alumina processing[15,16]. There are few K_{ISCC} data available for cast iron in caustic solution. However, because of its high brittleness, cast iron was a proper first material for testing the applicability of the modified CNT rig for generating K_{ISCC} data. However, comparison of K_{ISCC} generated by CNT technique with those by other techniques will provide stronger validity of the CNT technique for generating K_{ISCC} data. With a view to investigating this validity, K_{ISCC} of Type 4340 steel in chloride solutions have recently been determined using CNT technique [17].

As shown in Figure 7, K_{ISCC} was determined to be ~ 15 $MPa.m^{1/2}$, which is the minimum K_I required for propagation of SCC crack. In literature, K_{ISCC} data on 4340 steel in 3.5% NaCl are reported [18,19] to lie between 14 and 16 $MPa.m^{1/2}$, as shown by the band in Figure 7. Fracture surface of the failed CNT specimens were examined using scanning electron microscopy (SEM). A fractograph of the area adjacent to the fatigue pre-crack (Figure 8) provides features

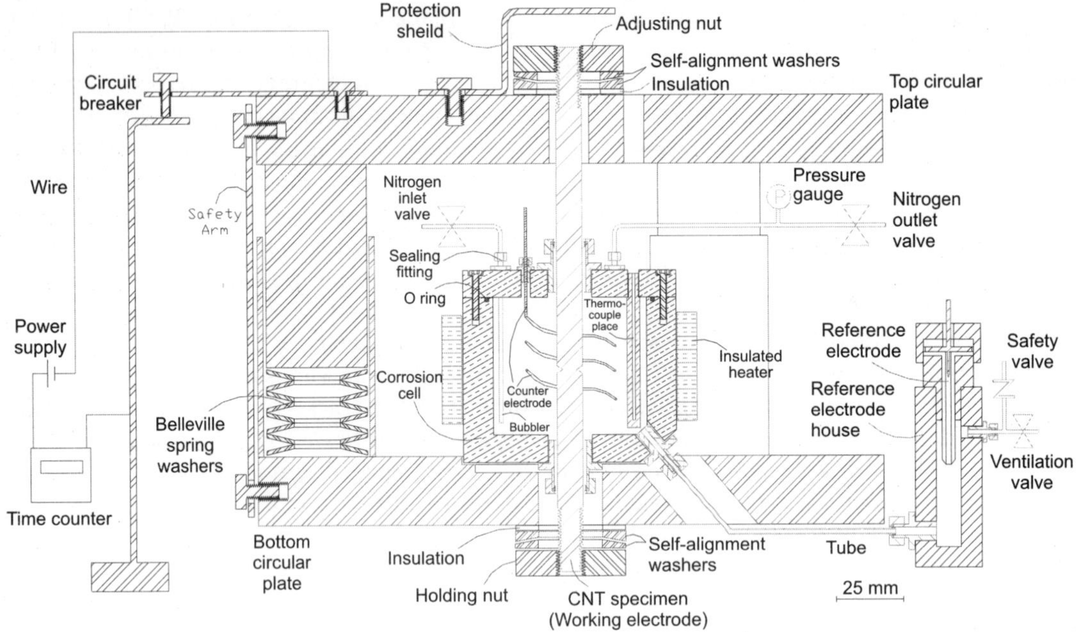

Figure 4 CNT rig for stress corrosion cracking (SCC) testing [12]

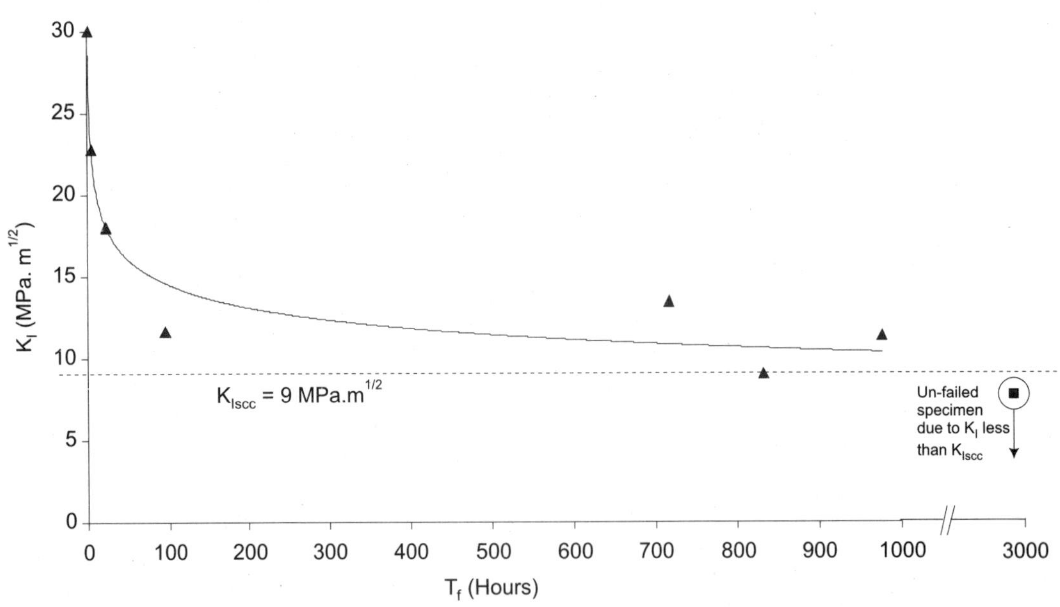

Figure 5 K_I versus T_f plot of SG cast iron in 200g/L NaOH at 120°C [13]

Table 1 Experimentally determined K_I data and proof of K_I data meeting validity requirements of SG cast iron in 5 M NaOH at 120°C [13]

SpecimenNo.	K_I(MPa. $m^{1/2}$)	T_f(Hours)	a_f(mm)	$2r_y$(mm)	$\frac{\sigma_N}{\sigma_Y}$
1	30*	0.2	1.79	1.28	1.71
2	22.8	5.6	1.48	0.62	1.44
3	18.1	22.7	0.77	0.40	1.24
4	18	23	1.52	0.40	1.07
5	13.4	716.3	1.07	0.23	0.87
6	11.6	95.2	0.67	0.17	0.71
7	11.4	976.4	0.45	0.17	0.77
8	9	831.5	0.71	0.11	0.61
9	8.6#	2997.5@	1.05	0.10	0.58

*K_{IC}, # Un-failed specimen, at $K_I < K_{ISCC}$, @ Test duration, no failure.

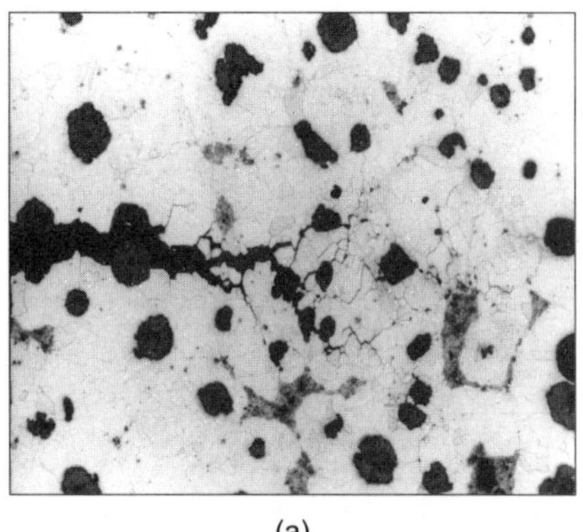

(a)

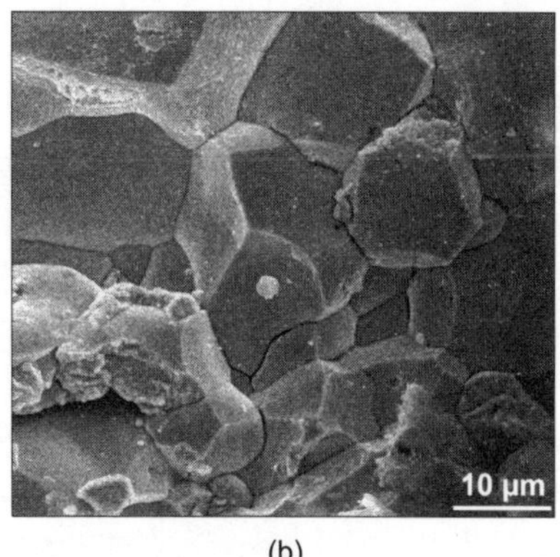

(b)

Figure 6 Intergranular crack propagation and crack branching in the cast iron specimens tested for stress corrosion cracking in hot caustic solution: (a) reflected light micrograph of the longitudinal section of tested specimen [12], (b) scanning electron fractograph [13]

of intergranular cracking (IGC). K_{ISCC} in the previous studies were determined using 'cantilever bend' [18] and 'centre cracked' [19] specimens. It is evident that the K_{ISCC} determined in the present study using CNT specimen (15 MPa.$m^{1/2}$) falls within the band in Figure 7, and is consistent with that determined using three other specimen geometries / techniques [18,19].

The small size of the specimens used in CNT testing has the advantages as listed below:

(a) The small cross-section of the specimen makes it possible to achieve quite high levels of stress intensities by using moderate loads,

(b) CNT testing reduces material requirements, which is a critical requirement when microstructures of limited size such as welds are to be investigated,

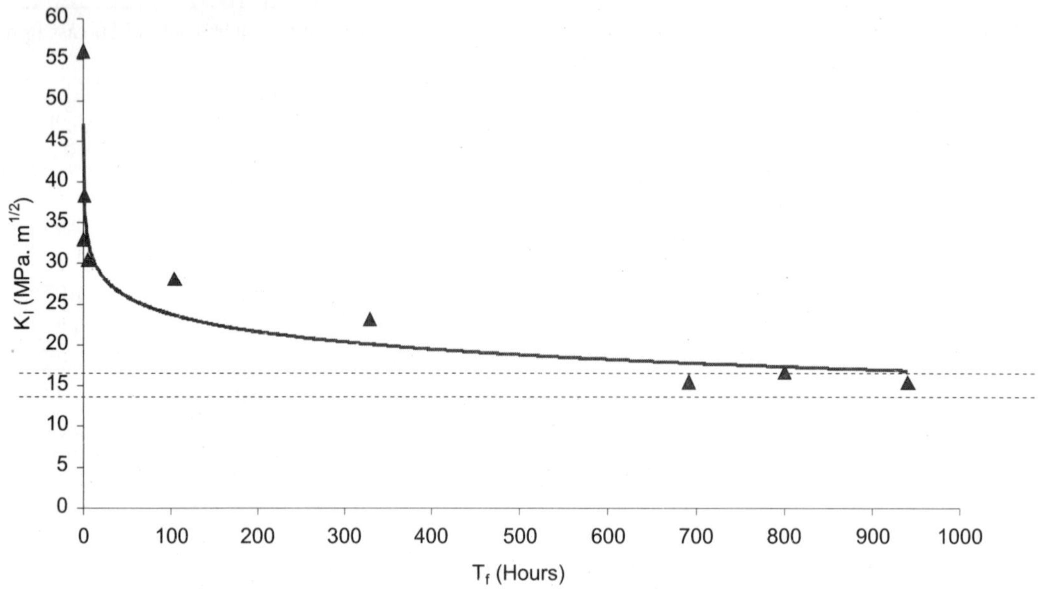

Figure 7 K_I versus T_f plot of 4340 steel in 3.5% chloride solution [17]

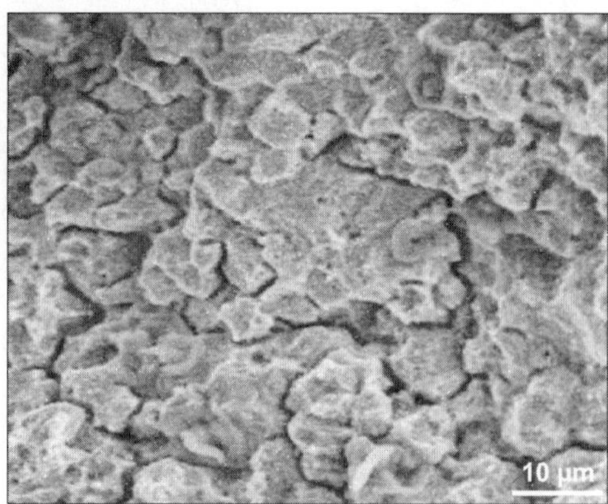

Figure 8 SEM fractographs of the area adjacent to the fatigue pre-crack of the CNT specimen of 4340 steel tested in 3.5% chloride solution, evidencing intergranular crack propagation [17]

(c) In many cases thick section material may not be available. This is characteristic of many pressure vessels constructed out of expensive materials such as stainless steels or duplex stainless steels,

(d) The CNT specimen is cheap to produce because of its small cylindrical shape reducing the cost of each specimen and overall testing.

DETERMINATION OF K_{ISCC} BY SNTT TESTING

Spiral Notch Torsion (SNT) Testing

The SNTT technique, patented in the year 2002 by Wang and Liu [20-25] of Oak Ridge National Laboratory (USA), is a materials testing and analysis system capable of highly accurately determining the intrinsic fracture toughness (K_{IC}) of structural materials. SNTT utilizes an extremely innovative concept of testing round-rod specimens having a V-grooved spiral notch line with a 45° pitch angle, subjected to pure torsion (Figures 9 and 10). Employing a 3-D finite element code, K_{IC} is determined from the fracture torque and final crack length. Unique virtues of the technique include broad latitude in specimen size, controllable crack front propagation, and mixed mode testing.

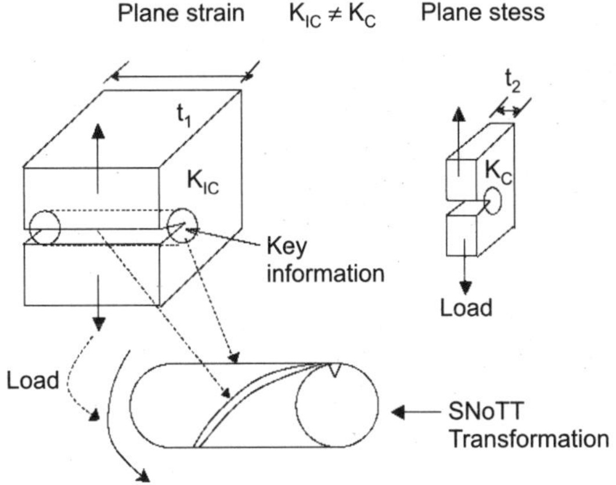

Figure 9 Schematic comparison of CTS and SNTT specimen: size effect [26]

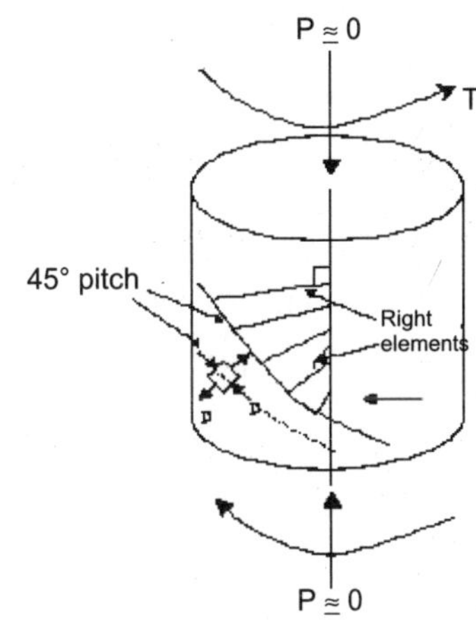

Figure 10 SNTT specimen: pure shear loading [26]

Limitations of the CTS specimens (one of the most widely used specimen configurations for determining K_{IC}) in complying fully with the stringent requirements of plane strain condition (PSC) and the resulting inaccuracies in application of K_{IC} data thus generated to life assessment are discussed earlier in detail. Specimen configuration in SNTT successfully circumvents these limitations of CTS specimens. In conforming to the stress and strain constraint conditions necessary to validate the use of the fracture mechanics theory, a round-rod specimen is subject to pure torsion as shown in Fig. 10. The pure torsion generates an equibiaxial tension / compression stress field on ±45° pitched orthogonal planes along the right conoids, independent of the size of the specimen diameter. In SNTT testing, the plane-strain condition and uniformity of stress are maintained at every point along the spiral notch line of the specimen. Because of the uniformity in the stress and strain fields the crack front is expected to propagate perpendicularly toward the specimen axis along the conoids, as evidenced by the post-testing fractographs.

SNTT Testing for Determination of K_{ISCC}

In order to examine validity of the SNTT technique for generating K_{ISCC} data, first tests were carried out on an alloy-environment system that is well-known to produce SCC. Aluminium alloy, AA7075, which is a relatively recent alloy used in defence and civilian aircrafts, is known to suffer environment-induced cracking, particularly in the T6 temper condition. Specimen design consists of a round rod with a V-groove spiral notch line at 45° pitch angle (Figure 11). Details of fatigue-pre-cracking of SNTT specimens are provided elsewhere [20-25]. Specimens are subjected to pure torsion, resulting in mode I loading and plane strain conditions at the crack tip, using a rising step load torsion device (an adaptation of ASTM F1624) for carrying out SCC tests, which will use the SNTT specimens immersed in chloride solutions. Complete experimental details can be found elsewhere [26].

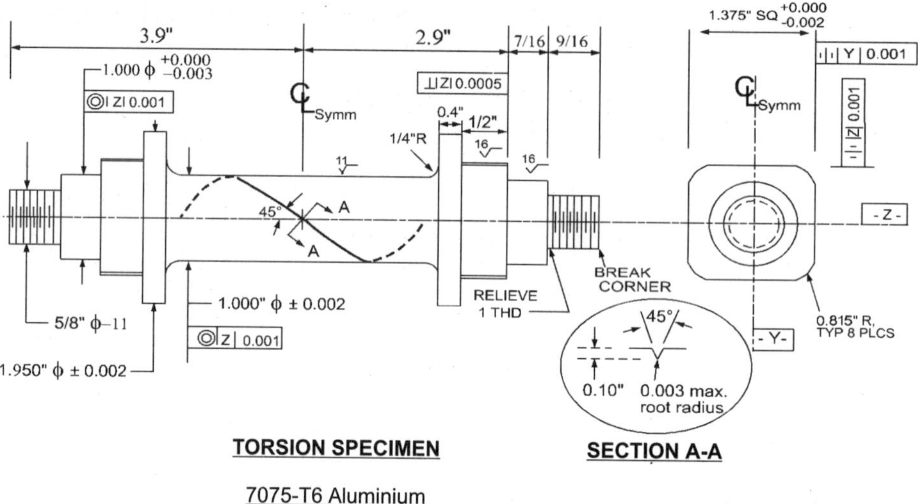

Figure 11 Design of SNTT specimen for SCC testing

Arrangements were made for immersion of the specimen in a corrosive environment during torsion testing in a rig designed specifically for this purpose at the US Naval Research Laboratory (Figure 12). Experiments were carried out by applying static load incrementally to a weight pan attached to a cantilever beam (Figure 12*a*). Through a multiple step-loading technique (ASTM F-1624), torque was increased incrementally while crack mouth opening displacement (CMOD) was measured perpendicular to the notch (using an adapted clip-gauge, as shown in Figure 12*b* and *d*). Fractographic analysis permits the depth of fatigue pre-cracking, extent of crack propagation by fatigue and SCC modes, and final crack length to be measured.

Multiple step-loading (ASTM F-1624) pattern is shown in Figure 13a and a typical plot of CMOD (measured from the clip-gauge) with load in Figure 13b, which is also used to determine the compliance of the specimen. Compliance was in the range of 0.09-0.13 m/MN. Deviations from linearity at low loads occur as a result of pre-loading the specimen and delay in the initial crack opening process, whereas those at high loads are due to sub-critical crack propagation.

Table 2 presents the SNTT test parameters, the test results and the K_{IC} (determined as described later). The failure loads indicate possibility of environment-assisted failure (such as

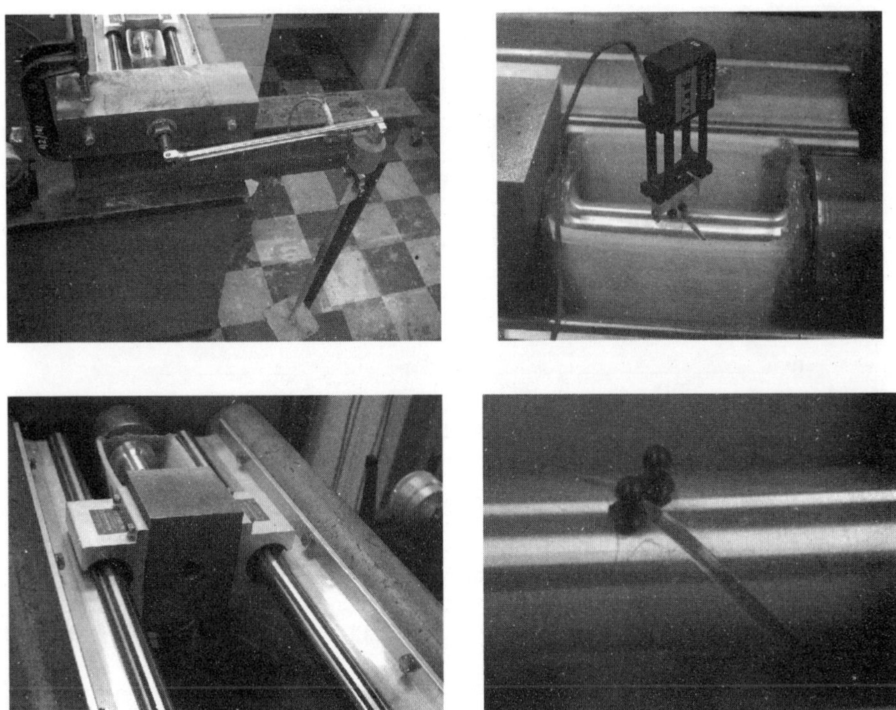

Figure 12 SNTT SCC experimental rig. (a) Cantilever arm in foreground, fitted with load cell and weight pan; (b) SNTT specimen fitted with extensometer and placed in a container of salt-solution; (c) Free-end specimen holder, on roller bearings to ensure zero axial load; (d) Attachments fitted perpendicular to the expanding V-groove for the adapted clip gauge (Ref: 26)

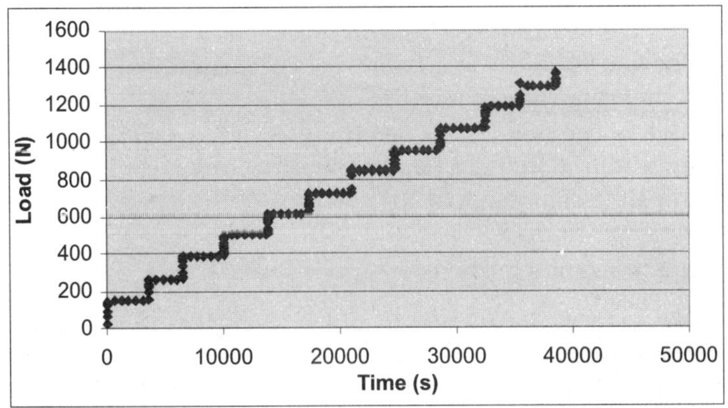

Figure 13a Typical multi-step loading pattern during SNTT experiments [26]

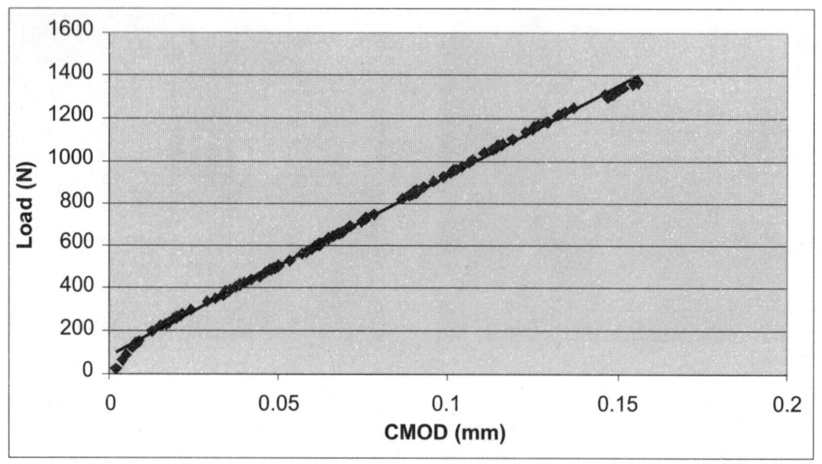

Figure 13b A typical Load-CMOD plot during SNTT experiments[26]

Table 2 Results of SNTT testing of alloy 7075 (T651) in various environments

Test Environment	Pre-crack depth (mm)	Failure load (N)	Torque (N-m)	K_{IC} / K_{ISCC} (MPa$\sqrt{m}$)
Air	6.69	1363	570	27.5*
NaCl-immersion	7.01	1007	421	20.8#
Moist air	*7.24*	*1334*	*559*	*28.1*

* hand book value of K_{IC}, # K_{ISCC} data

embrittlement) in the case of NaCl-immersion specimen. The depth of the fatigue pre-crack in this specimen is relatively close to those measured on specimens tested in dry and moist air, which rules out the possibility that the decrease in failure load was associated with a lower load-bearing area due to a larger fatigue pre-crack.

A simple approach was employed to determine K_{IC}/K_{ISCC} data shown in Table 2. This approach required an assumption that a handbook value for K_{IC} could be applied to the 7075-T6 specimen tested in air presuming that the cracking of this specimen did not involve any environmental effect.

A general fracture mechanics relationship applied was: $K_I = A*T*\sqrt{(\pi a)}$.

Using handbook data for K_{IC} of 7075-T6 in air (i.e., $27.5 MPa\sqrt{m}$), the respective torque (T) from the experimental condition (i.e., 570 N-m), and the pre-crack depth (i.e., 6.69 mm), the value of the geometry constant (A) was determined. This value of A was then used in the above equation to calculate K for the other conditions (Table 2).

A lower failure load in case of specimen tested in the NaCl-immersion has indicated environment-assisted failure. However, K_{IC} of the NaCl-immersion specimen is not drastically lower as compared to the specimen tested in air (Table 2).

Table 3 Comparison of SNTT with other specimen configurations for K_{IC} determination

Compari-son Criteria	**SNTT**	**CT**	**3PB**	**CNT**	**CVN**	**Competitive advantage of SNTT**
Cost of sample	Low	High	Low	Low	Low/	Low/ multiple functions
Size effect	Virtually none	Size dependent	Size dependent	Size dependent	N/A (dynamic test)	Miniature samples on a portable tester
Direct K_{IC} determination	Yes	Yes, with large specimen	Yes, Test adjustt-men	Yes, size dependent	No	Means for direct integrity studies: reduced design/ safety margin
Controlled crack propagation direction	Propagates perpendicularly toward central axis	No	No	No	N/A (dynamic test)	Provides a means to evaluate K_{IC} of interfaces and HAZ in weldments
Statistical aspects	Long/uniform crack front: less data scatter	Large data scatter/ uncertainty	Large uncer-tainty	Large uncer -tainty	Large uncer-tainty	High accuracy, needs less number of test samples to develop valid K_{IC}
Fatigue pre-crack (PC) characteristic /requirement	PC produces uniform crack front for ductile materials	Non-uniform front; PC required for both ductile and brittle materials	Non-uniform front; PC required for all materials	Non-uniform front; PC required	N/A	Uniform crack length produces accurate K_{IC} value
Mixed Mompde testing	Easy to carry out using different pitches for the notch	Needs large/ complicated/special set-up/ specimen	Not available	Not available	Not available	Miniature samples for applications to aging of turbine shaft/steam generating tubing

The K_{ISCC} data presented here are the preliminary results, generated using SNTT, and it will be necessary to carry out tests on variety of other alloy-environment systems, before SNTT could be treated as an established method. However, the preliminary results presented here as well as the previous work [20-25] suggest that SNTT is able to meet the following objectives:

(a) Reducing/eliminating specimen size effect, and enabling specimen miniaturization,
(b) Attainment of a true plane strain condition at the crack tip/front,
(c) Determining intrinsic fracture toughness,
(d) Developing a reliable and cost effective testing method,
(e) Attainment of controllable crack propagation,
(f) Improving statistical aspects, and
(g) Enabling mixed-mode fracture assessment

Other important advantages of SNTT include:

(a) Successful miniaturization of specimen (tests have been conducted using specimens as small as 4 mm in diameter and 40 mm in length, and there is no reason why still smaller cannot be tested),
(b) Minimizing volume change due to zero hydrostatic pressure loading (pure shear),
(c) Sample fatigue pre-cracking requirement can be relaxed for brittle materials,
(d) Possible to test mixed-mode fracture mechanism, and
(e) Uniform crack front and controllable crack propagation results in an accurate K_{IC}.

Table 3 compares advantages of SNTT and CNT with other ASTM standard tests/ specimen configurations for determination of K_{IC} and K_{ISCC}.

ACKNOWLEDGMENTS

CNT data on cast iron presented here constitute part of PhD thesis of Dr. Rihan Rihan (supervised by the author and Assoc Prof R.N. Ibrahim). SNTT data were generated under the author's collaborative project with Dr. R. Bayles (US Naval Research Lab), Dr. J. Wang (Oak Ridge National Lab) and Dr. B. Hinton (Australian Defence Science and Technology Organisation). The author acknowledges contributions of these colleagues in the respective works and the resulting publications.

REFERENCES

1. A. Turnbull, *British Corrosion Journal*, vol 27 (1992) 271.
2. Proc. Environment-sensitive Fracture, Symposium ASTM Committee G-1, ASTM Special Technical Publication 821 (eds: S.W. Dean, E.N. Pugh and G.M. Ugiansky), Pub: ASTM, Philadelphia, 1982.
3. J. Sedriks, Publications of National Association of Corrosion Engineers, US, 1990.
4. R.N. Ibrahim and HL Stark, *International Journal of Fracture* 44 (1990) 179.
5. R.N. Ibrahim and HL Stark, *Engineering Fracture Mechanics*, 25 (1986) 395.
6. R.N. Ibrahim and H. L. Stark, *Engineering Fracture Mechanics*, 28 (1987) 455.
7. G. R. Irwin, "Fracture testing of high-strength sheet materials under conditions appropriate for stress analysis", Report 5486, Naval Research Laboratory, July 1960.

8. G. R. Irwin, "Plastic zone near a crack and Fracture toughness", Sugamore ordinance materials conference, Syracuse University Research Institute, August (1961).
9. R. W. Hertzberg, "Deformation and fracture of engineering materials", Wiley, New York, pp 274-277 (1976).
10. F. A. McClintock and G. R. Irwin, ASTM STP 381, p. 84 (1965).
11. R. Rihan, R.N. Ibrahim and R.K. Singh Raman, *Proc. Structural Integrity and Fracture of Metals Conf. (A. Atrens, J.N. Boland, R. Clegg and J.R. Griffiths, eds.), Brisbane,* Sept 2004, p.317.
12. R.K. Singh Raman, R. Rihan and R.N. Ibrahim, *Metallurgical and Materials Transactions A,* 37A (2006), *in press.*
13. R. Rihan, R.K. Singh Raman and R.N. Ibrahim, *Materials Science and Engineering A,* 407 (2005) 207.
14. R. Rihan, R.K. Singh Raman and R.N. Ibrahim, *Materials Science and Engineering A,* 425 (2006) 272.
15. R.K. Singh Raman and B.C. Muddle, *Engineering Failure Analysis,* 11 (2004) 199.
16. R.K. Singh Raman and B.C. Muddle, *Materials Science and Technology,* 19 (2003) 1746.
17. R.K. Singh Raman, R. Rihan and R.N. Ibrahim, *Materials Science and Engineering A* (communicated).
18. B.F. Brown and C.D. Beacham, Corrosion Science, Vol.5, pp. 745-750, (1965).
19. H.R. Smith and D.E. Piper and F.K. Downey, *Engineering Fracture Mechanics,* Vol. 1, pp. 123-28, (1968).
20. J. A. Wang, K. C. Liu, D. E. McCabe, *Fatigue and Fracture Mechanics: ASTM STP 1417,* W. G. Reuter and R. S. Piascik (Eds.), ASTM, West Conshohocken, PA, 2002.
21. J. A. Wang, K. C. Liu, D. E. McCabe, and S. A. David, "Using Torsion Bar Testing to Determine Fracture Toughness, K_{IC}," *Journal of Fatigue & Fracture for Engineering Materials and Structure,* 23 (2000) 45.
22. J. A. Wang, K. C. Liu, and G. A. Joshi, "Using Torsion Bar Testing to Determine Fracture Toughness of Ceramic Materials," *ASME Engineering Technology Conference on Energy 2002, Composite Materials, Design and Analysis,* American Society of Mechanical Engineering.
23. J. A. Wang and K. C. Liu, "Fracture Toughness Determination Using Spiral-Grooved Cylindrical Specimen and Pure Torsional Loading", *U.S. Patent,* June 2001.
24. K. C. Liu and J. A. Wang, *International Journal of Fatigue.*
25. J. A. Wang and K. C. Liu, "TOR3D-KIC: A 3-D Finite Element Analysis Code for the Determination of Fracture Toughness, K_{IC}, for *Spiral Notch Torsion Test (SNTT), ORNL technical transfer,* December 2001.
26. R.K. Singh Raman, R. Bayles, S. Knight, Jy-An Wang, B.R.W. Hinton, B.C. Muddle, *Proc. Environment-Induced Cracking of Metals Conf.,* Banff, Canada, Sept '04, *(Pub: Elsevier), 2006,* pp. 447-457.

CHAPTER 5

Metallic and Oxide Coatings for Corrosion Protection

S. Ningshen and U. Kamachi Mudali

Corrosion Science and Technology Division, Indira Gandhi Centre for Atomic Research, Kalpakkam 603 102, TN, India

The main function of any coating is to provide an effective barrier in combating corrosion. A major challenge in technological development is to meet requirements for new materials in progressively more severe corrosive conditions and coatings development can provide a protective barrier between metal in such aggressive environment. Usually one or more properties of materials are incompatible with the conditions prevailing in the operating environment. In the material environment configuration, the surface of a component is a vital parameter in determining its optimum performance. This is the basis for the development of the coating technology. In this paper, a review of the present status of the research and technological development in the field of metallic and oxide coatings for corrosion protection is attempted. The ranges of surface coating to modify the properties of a component are too numerous to be covered. Therefore the aim here is to give an insight into most recent development in terms of process fundamentals and applications, and to review the available surface coating and surface modification system.

1.0 INTRODUCTION

The significance of coatings in technology is derived mainly from their anti-corrosion role, for enhancement of tribological properties and for repairs [1]. It could be argued that, in keeping with the saying that 'there is nothing new under the sun', surface coating is basically an old practice or technology. However, the continually increasing engineering demands made of metals and alloys with enhanced properties have provided, especially over the past 50 years, added impetus and drive for the development of newer protective coating technology, as well as the equally significant enhancement of existing practices. It is interesting to note that despite the successful use of many old and newer coatings over many years for corrosion protection, it is a technology which has been continuously improved and adapted. An obvious broadening is noticed of the various areas, i.e., formation, design, investigation and utilization of surface layers, along with their progressing integration. Though the most advanced is the field of methods of manufacture of surface layer by coatings, while the connected field of property testing lags behind. Clearly least advanced is research in the field of design of surface layers of hard and nano-layer coating for corrosion application and much research, however, is needed in this area.

The initiation and propagation of corrosion are major concerns in various technologies. Deterioration of materials by corrosion is an age-long problem that has faced mankind for millennia [2, 3]. From the world's ancient man-made and natural monoliths to today's most modern infrastructures of buildings, bridges and transportation facilities, the longevity of structures are closely controlled by the environments in which they are located. Having little control over these local environments, providing preventive coatings and selecting the materials that are best suited to the conditions to which they are exposed is very important. History shows it is difficult to predict the interaction of structures with their surroundings, and today we see deterioration of historic icons and the loss of cultural relics that mark the development and achievements of humanity due to corrosion. Even after many years of industrial revolution with the development of technologically advanced testing and corrosion monitoring methods, the knowledge of complex interactions of structural materials and with the environment is not sufficient to produce long lasting structures to prevent from corrosion failure.

One feature that is common to the above scenario is mankind's inability to predict the performance of a particular material in a given environment. Although significant effort is generally made to determine the mechanical integrity of structures, very little effort is spent on determining their corrosion properties. Such prediction would typically require the exposure of materials and alloys to different environments, an evaluation of their performance, and the development of a predictive corrosion model. The slow nature of most corrosion processes require testing over long duration and many years. This is often viewed as an unrealistic and expensive undertaking that can delay product marketability, or retard the development. Quite often however, a material's corrosion behavior is investigated after the structure has been built or after the failure has occurred. As such, potentially serious corrosion problems are not identified until after the structure has been placed in service. Predictive corrosion behavior through modeling and on-line corrosion monitoring would greatly reduce the time for evaluating the corrosion initiation process. However, to date this technology has not reached maturity due to the lack of knowledge of the separate and combined contributions each environmental parameter has to the corrosion of a particular material.

Corrosion by simple definition is associated with loss of material due to the electrochemical interaction with the environment and the electrochemical reactions involving electron transfer; therefore, one of the most effective corrosion control techniques is to electrically isolate the anode from the cathode [1]. For example, the formation of passive layer consisting of Cr_2O_3 formed on the surface of stainless steel in oxidizing environments is the main reason for its corrosion resistance and durability. The total cost of corrosion and corrosion related issues in the United States is significant, amounting to about 3% GDP or $300 billion per year [2 - 4]. This figure is divided evenly between direct costs (materials and structures), and indirect costs (loss of productivity). About 90% of the corrosion is associated with iron-based materials. Although corrosion problems cannot be completely remedied, it is estimated that corrosion related costs can be reduced more than 30% by development and use of better corrosion control technologies. A more generic approach to enhance corrosion resistance is to apply protective films or coatings and other corrosion control measures include corrosion inhibitors and cathodic protection and use of alloys resistant to corrosive environment.

Numerous coatings and surface treatments have been developed and applied to mitigate corrosion of engineering components [5 - 10]. The choice of a particular treatment depends on many factors, viz., processing, fabrication temperature, temperature of the working environment, load, relative velocity, material to be used, and the environment itself. The deposition of protective metals and oxide by coating, either chemically or electrochemically, plays an important role in the development of technologies where these metals are used. Protective coatings can be hard or soft, thin or thick, porous or dense, single or multilayer depending on the application. Many physicists, chemists and material scientists focus on novel materials or systems properties that come about by coating to alter the "normal" properties of a material or object to prevent corrosion. There are indications that coating can have a totally revolutionary impact on some areas of science and technology. It works on the simple principle of completely separating the metal from the corrodent or slowing down the reaction that may occur between the metal to be protected and the corrodent. Protective coating applied for various reasons other than corrosion protection (e.g., appearance, abrasion resistance, and electrical properties) will not be discussed in this review. Only the recent developments of coatings for corrosion protection to provide more efficient, reliable and cost effective surface coating will be highlighted in this paper.

2.0 THE OLD AND THE NEW – PROTECTIVE COATING THE KEY TO OPTIMIZED PERFORMANCE

Modern technology has placed increasing stresses on the material used for a variety of technological uses. Both the uses and the stresses have probably grown at a greater rate than the number of materials that can be used to meet them. Although producing new alloys with improved properties has done much, there is a limit to the protection that can be afforded by this means alone. For this reason surface modification and coatings have played an increasing role in protecting the structural metals from corrosion. A protective surface coating provides long term protection under a broad range of corrosive conditions, extending from atmospheric exposure to full immersion in strongly corrosive solutions. In general, coatings in themselves can give little or structural strength, yet they protect other materials so that the strength and integrity of a structure can be maintained. They are the skins, over the skeleton that both protects and beautifies the bone and muscle of the essential structure [7].

The evolution of protective corrosion coating is depicted in Fig. 1. The predominant aspect of coating is the life expectancy of a coated system. System optimization considers the coating composition, structure, porosity, and adhesion together with operating and coating temperatures, substrate/coating compatibility, material availability and costs, coating renewal and in-service repair and maintenance aspects.

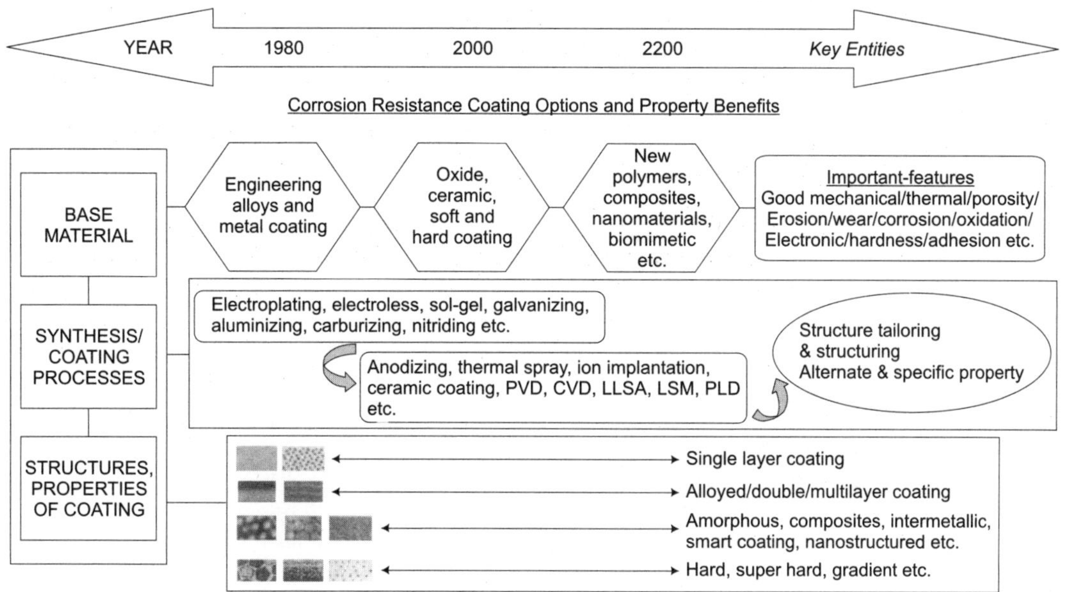

Figure 1 Historical evaluation of corrosion resistance coating.

In the material environment configuration, the component surface is a vital parameter in determining its optimum performance. Metallic thin film or oxide coatings are used in numerous industrial applications. This is the basis for the development of the coating technology for corrosion protection. Definite limits are being approached, or have been reached, in single materials development. A solution is to produce a material, which combines the best properties of different metallic or oxide coatings technology is a challenging field, both for the fundamental and applied research work.

All protective corrosion resistant coatings, however, must fundamentally resist the corrosive environment and prevent it from reaching the basic structure. Thus, there are many variations in the types of coatings depending in forms of corrosion attack. The design of an effective anticorrosive coating is a complex task, which requires an extensive knowledge of not only corrosion principles, but of the science of chemistry and coating formation as well. Without such inclusive information, the development of effective corrosion-resistant coatings would be impossible.

Surface coatings using metallic or oxide is a multidisciplinary activity intended to tailor the properties of the surfaces of engineering components so that their function and serviceability can be improved. The desired properties or characteristic of components include [5, 6].

- Improved corrosion resistance through barrier or sacrificial protection
- Improved oxidation and/or sulfidation resistance
- Improved wear resistance
- Reduced frictional energy losses

- Improved mechanical properties
- Improved electronic or electrical properties
- Improved thermal insulation
- Improved aesthetic

As indicated in Table 1, these properties can be enhanced metallurgically, mechanically, chemically (corrosion), or by adding a suitable coating. The output of literature on coating is varied and voluminous, and it is difficult to do justice to all of this here; a reasonable cross-section is given which are reviews in brief.

Table 1 Coating options and property benefits [5]

Surface treatment/coating type	Primary property benefits
• Changing the surface metallurgy	
- Surface hardening (flame, induction, laser, and electron-beam hardening)	- Improved wear resistance through hard martensitic surface
- Laser melting	- Improved wear resistance through grain refinement
- Laser peening	- Improved fatigue strength due to compressive stresses induced on the surface
• Surface layer or coating	
- Organic coating	- Improved corrosion resistance, wear resistance, and aesthetic appearance
- Ceramic coating	- Improved corrosion
- Slip/sinter ceramic coatings	- Improved wear resistance and heat resistance
- Electroplating	- Depending on the metal or metals being electroplated, improved corrosion resistance
- Thermal spraying	- Primarily used for improved wear resistance, but also used for improved corrosion resistance and oxidation resistance
• Changing the surface chemistry	
- Phosphate/chromium conversion coating	- Used primarily for enhanced corrosion resistance
- Anodizing	- Enhanced corrosion resistance
- Steam treating	- Used on powder metallurgy parts to increase wear resistance
- Carburizing	- To increase resistance to wear, bending fatigue, and rolling-contact fatigue
- Diffusion (pack cementation)	- Improved molten salt hot corrosion
- Boronizing (boriding)	- Improved wear resistance, oxidative wear, and surface fatigue
- Ion implantation	- Improved corrosion, wear and friction resistance for a variety of substrate
- Laser alloying	- Improved corrosion and wear resistance

Fig 2 is an overview of coating-substrate which can be used to control and ensure optimum performance of materials in any corrosive environments. The vast majority of engineering components fail, as a direct consequence of a surfaceinitiated failure such as corrosion or wear. Surface modification by applying a suitable coating can serve to delay the failure significantly or enhance the performance of the component during its lifetime. There are two fundamental approaches to improving performance [3, 4]:

(i) Modification of the existing surface by means of introducing alloying elements or atomic species, e.g. case hardening, ion implantation.

(ii) Application of a coating of a different protective material.

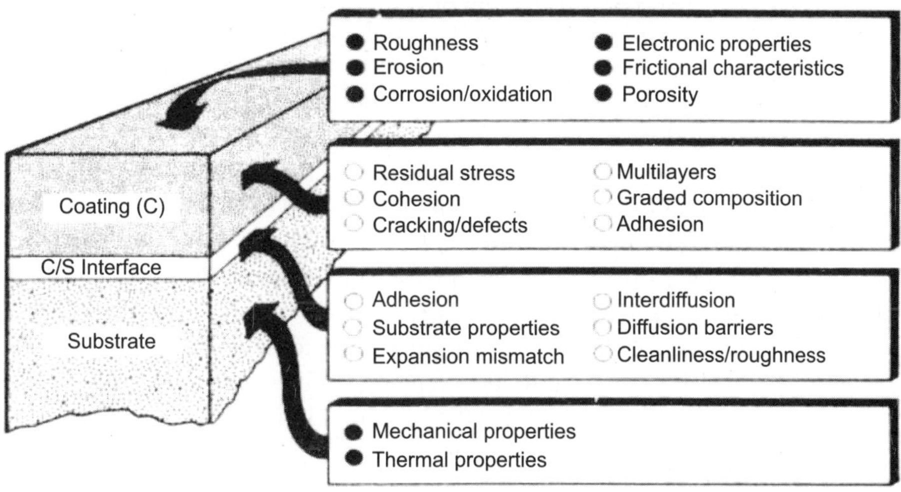

Figure 2 Properties of the coating-substrate which can be controlled to ensure optimum performance in any corrosive environments [6].

Too often a material for a component is specified because of the properties that material exhibit at its surface, even though this may only be a relatively small part of the entire component. Surface modification and coatings allow us to use the material exactly where it is required whilst possibly using a cheaper or less intensive material for the substrate. The potential performance, and consequential environmental benefits in a number of key areas are: (i) Wear resistance, (ii) waste minimization, (iii) thermal performance, (iv) corrosion resistance, (v) manufacturing assistance, and (vi) improved functionality.

3.0 NEW WAYS OF THINKING ABOUT FUNCTIONAL MATERIALS AND COATINGS FOR CORROSION PROTECTION

Coatings technology includes the traditional technologies of painting, electroplating, weld surfacing, plasma and hyper velocity spraying, various thermal and thermochemical treatments such as nitriding and carburising, as well as the new technologies of pulse laser deposition and surfacing, physical and chemical vapour deposition and ion implantation. First, there is a paradigm change or possibly several different paradigm shifts related to the desired structure of a coating. Our ability to design, create and model protective coating systems of many shapes and properties will allow new types of materials with new properties to be created. As one simple example of a different way of thinking, many applications required uniform coatings to achieve reliable predictable performance. Although coatings with a composition gradient in one direction (graded index, for example) are of technological importance, most often a significant degree of uniformity was needed. In a new view, carefully controlled, but highly local nanosized variation in material composition (or structure) can be considered as an important and useful element of system design. Materials and coating of films with previously non-achievable properties can now be produced. Situations that were previously avoided may now be exploited to create new types of coating systems.

At the moment corrosion protections by coating combine to break some of the old paradigms about coating processes and the desired structures of coatings. New approaches can be used to create new types of highly functional systems. Trends and developments in theory, biology, physics, materials and chemistry now enable us to have concepts, synthesize materials or molecules, observe structures and chemistry at the nanometer scale, and measure the properties. In effect, we can imagine, model, grow and test some of our wildest ideas often funded by government agencies, private industry or even venture capitalists. The scientific potential is vast and the technological opportunities large to develop coating to fight against corrosion. Undoubtly, we face the difficult challenge of the rapid transfer of ideas from research and development to application for the benefit of industries.

4.0 COATING TECHNOLOGIES - STATE OF THE ART AND FUNDAMENTAL ASPECTS OF THE PROCESS

The fundamental aspects of coating technologies is to improve durability, reliability and performance of various components; to resist erosion, sliding and fretting wear or to improve surface quality; and to produce corrosion resistant coatings for combating pitting, exfoliation, oxidation and hot corrosion. Protective coatings provide a layer that changes the surface properties of the work piece to those of the metal being applied. The work piece becomes a composite material exhibiting properties generally not achievable by either material if used alone. In any coating, physical integrity of the coating is as important as its chemical barrier properties in many applications. For instance, coatings on impellers that mix abrasive slurries can be abraded quickly; coatings on pipe joints will cold-flow away from a loaded area if the creep rate is not low; and coatings on flanges and support brackets can be chipped or penetrated during assembly if impact strength is inadequate. Selecting the best coating for an application requires evaluating all effects of the specific environment, including thermal and mechanical conditions [5, 6].

There are a number of technologies available for coating alloys. These include electrochemical plating, conversion coatings, anodizing, hydride coatings, organic coatings and vapor-phase processes. One of the major issues related to development of coating is the uniformity that are produced, and in particular the thickness uniformity of the deposit. Some processes in which the coating is produced by a reaction that takes place at the substrate surface give rise to a uniform coating thickness over the entire surface, e.g. electroless plating, CVD etc. In the case of most other coating processes the coating thickness is not uniform. In electroplating there will be a build up of coating at areas of high current density such as corners and a decrease in coating thickness in areas such as holes. This may result in excess material having to be used [5, 6].

In PVD again there can be problems of coating uniformity giving problems with intra batch quality [7]. PVD is essentially a line-of-sight process, but by back filling with gas, gas scattering can take place of the evaporant which can lead some coverage of blind areas. In PAPVD, this effect is enhanced by the sputtering of the surface during deposition that result in a further redistribution of material and a more uniform coating thickness distribution. The effect is not as big as often claimed though, as there is a lack of uniformity in both the distribution of the evaporant and the sputtering of the cathode assembly [6]. Interestingly, it has been shown that better uniformity can be achieved on large workplaces on the surfaces not facing the evaporation source. One way of getting around this problem is to manipulate the samples or the evaporation source within the deposition chamber but this adds to the complexity of the equipment required. However, for many of the processes, line-of-sight processes can be taken. There is an issue related to the production of alloy coatings using EBPVD. Different materials have different vapour pressure profiles, and hence are going to evaporate at different rates [6-8]. However, in the extreme case, such as hafnium, this would have to constitute a significant proportion of the source material even if it is only a minor constituent. This problem is largely solved if sputtering is used as the coating process, because the differential in sputtering rates is rarely more than a factor of 10 [6].

4.1 THE CONCEPT OF COATINGS - BASIC PROCESSES

The concept of coatings most probably originated from the realm of anatomy where it means the cover of animals bodies isolating the organism from the environment and, at the same time, providing contact between the organism and that environment. In animals the skin constitutes the coating. This concept has been imported into the realm of technology [6]. The concept of coatings was given recognition in official documents and specifications in the 1950s. However, to date there is no general terminological specification related to corrosion and corrosion protection, which though assume that the concept of 'coating' is self-evident and defines the various types of coatings.

General definition of coating can be ascribed as a layer of material, formed naturally or synthetically or deposited artificially on the surface of an object made of another material, for protection with the aim of obtaining required technical or decorative properties [6, 7].

The substrate, in other words, the coated object, or in stricter terms, its superficial layer constitutes one phase of the system. The coating constitutes the second phase. Between the coating and the substrate there exists an interface in the form of a layer of certain volume, with intermediate properties, usually facilitating adherence of the coating to the substrate. In some

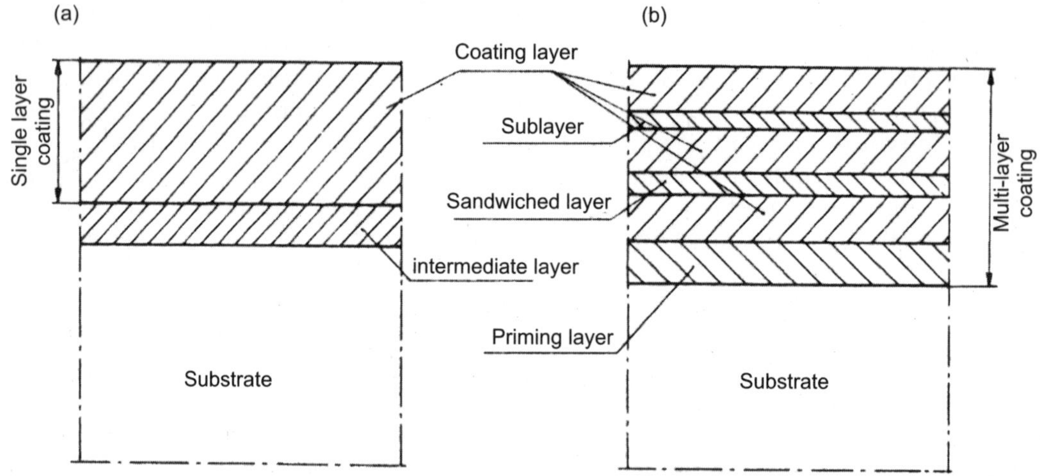

Figure 3 Schematic representation of coating structure: a) single layer; b) multi-layer [7].

cases it is difficult to distinguish between the coating and the superficial layer, particularly in incremental diffusion layer [6, 7]. Factors, which influence the formation of coatings, are:

1. The substrate on which the deposit are formed
2. The temperature of the substrate, and the temperature of the process
3. The medium of deposition
4. The rate of deposition and mass transfer, and,
5. The substrate/coating compatibility

4.2 STRUCTURE OF THE COATING

For obvious reasons, because of the great variety of coatings, both from the point of view of material and technology, which stems from different designated uses; it is difficult to develop one universal model of coating structure. Specific models, pertaining to the particular types of coatings, are given and discussed in literature on the subject, usually with varying degrees of simplification [6 - 10]. The general simplified model of coating structure is shown in Fig 3 using single layer and multi-layer coatings as examples [7].

The single layer coating (single coating, monolaminar) is a coating deposited on an approximately prepared substrate in one process or technological operation, comprising one layer of material. Single layer coatings are divided into:

- Single constituents coatings - consisting of one material (element, compound), e.g. Cr, TiN etc.
- Multi-component coatings - consisting of several material components, e.g. alloying additive, titanium carbonitride Ti(C,N) or titanium aluminonitride Ti(Al,N).

The multi-layer coating is one, which consists of two or more materials. These may be layers of the same material, separated by a sub layer, or they may be of different materials in which case a sublayer may but need not necessary be applied.

The following modifications of multi-layer coatings are known [7, 10].

- **Multiple coating** – comprising two or more layers of the same material, deposited in technological environments that differ only slightly as to physico-chemical properties.
- **Sandwich coating** – comprising several layers of different materials, with at least one of them occurring twice and not directly on top of same material.
- **Self-stratifying paint** - deposited in the form of a liquid or powder mixture, stratifying during drying or melting into a bottom sublayer with strong adhesion to substrate which is usually metallic, and a top sublayer (surface), which has high resistance to environment.

4.3 COATINGS TYPES AND CLASSIFICATIONS

The classification of coatings can be carried out in several ways but any complete system must take into account the type of coating, the process by which it can be applied and some limitations upon the range of choice which necessarily results. In considering only the techniques of application and the types of coating produced coatings may be divided in different ways, depending on the criteria used. The most significant division appears to be by material, by designation and by method of manufacture of the coating. From the point of view of application, coatings can be divided into four group's - protective, decorative, decorative-protective and technical [7, 8].

4.3.1 Division of Coatings by Material

From the point of view of material, coatings may be divided in to two groups: metallic and non-metallic. Very often the name of coating is derived from the coating material used.

4.3.2 Metallic Coatings

Such coatings are made from different metals, metal alloys and metal composites, and deposited on substrates, most often themselves metallic, by different methods. Coatings may be manufactured from all metals and composites. However, not all metals, alloys or composites are applicable in practice on account of their properties and technological difficulties in their manufacture. Most often used are:

- **Coating metals:** Zinc, nickel, chromium, aluminium, tin, cadmium, copper, lead, silver, gold, iron, cobalt, indium, ruthenium, rhodium, palladium, and platinum. Often refractory metals are also used, such as titanium, zirconium, hafnium, vanadium, niobium, tantalum, molybdenum and tungsten.
- **Alloys and coatings composites:** Steels (particularly alloyed, corrosion and heat resistant), brasses as well as alloys of metals:Pb-Sb-Cu, Sn-Ni, W-Co, W-Ni, Ni-Fe, Co-Mo, Zn-Al, Zn-Fe, Zn-Ni, Zn-Mn, Zn-Ce, Zn-Sn, Al-Si, Ni-Cr, Co-Cr, Ni-Al, Pb-Zn, Ni-B-Si, Ni-Cr-B-Si, Ni-Cr-B-Si-C, Co-Mo-Cr-Si, Ni-Cr-Al-Y etc.

4.3.3 Non-metallic Coatings

These coatings, numerous and varied are made from organic materials (paint, rubber, plastic) and inorganic (enamels, ceramics), of natural and synthetic origin and bearing many different trade and chemical names. Most often used coating materials or their basic constituents are: paints, varnishes, resins, varnishing enamels, oils, cements, waxes, asphalts, lubricants, vitreous enamels, metal ceramics, synthetics, rubbers etc.

5.0 DIRECTIONS OF SURFACE COATING AND KEY TO OPTIMIZED PERFORMANCE

In considering the concept of surface layers and coatings, the following distinctions should be made [7, 9]:

- *Technological layers* – are produced as a result of application of various methods, either independently or jointly. Depending on the set of effects utilized to this end, methods of prodution of surface layers may be divided into: mechanical, thermo-mechanical, electro-chemical, chemical and physical. In each group different methods are utilized to produce surface layers of determined thickness and designation.
- *Service-generated layers* – are produced as the result of the utilization of technological layers in conditions either natural or artificial. Utilization causes these surface layers to have properties that differ from those of the initial, technological layers.
- *Designing of surface layers* – This field coating involves such design of surface layers which will allow them to meet service requirements, and is weakly developed. The design of a process, such as to obtain a predetermined structure and properties, and the final decision regarding a manufacturing process which ensures the obtaining of such properties are practiced only in exceptional cases.
- *Investigation of surface layers* – This field of surface engineering involves experimental research of the structure and properties of surface coated layers, relative to various parameters, both technological and connected with service conditions, and acquisition of knowledge about related effects and applicable rules. The results of this research are integrated with the particular manufacturing processes and their parameters and constitute a database of technological know-how serving the design of new surface layers or their composition. The accomplishment of this research requires the implementation of newest methods of investigation, including physical, chemical, biological, corrosion, strength, tribology, etc.
- *Service utilization of surface layers* - This area comprises two problem groups:
- *Service testing* - of behaviour of surface layers in different working conditions. Usually tests cover the change in the behaviour with progressive time of service. Because investigation of layer properties during service encounters numerous difficulties, layers are usually tested after certain predetermined periods or after completed service. Investigation of the structure and properties of surface coated layer require special physico-chemical methods.
- *Production of service-generated* - layers during service, due to interaction with the material, by design, of substrates from the environment, under conditions of forced pressure, temperature, velocity, etc.

6.0 SELECTION OF COATING MATERIALS

The types of materials used for coatings are relatively few metal and a few classes of inorganic (ceramic) compound (oxides, carbides, borides and nitrides) singly or in combination. Note that many useful coating materials, such as paints and plastics, are not mentioned in this chapter. In selecting a coating material it is necessary to consider the uses to which it will be put. Factors to be considered include [6, 7]:

(a) The properties of the coating material itself, its melting point hardness, vapor pressure, density and thermal expansion coefficient
(b) The resistance of the coating material to the attack expected
(c) The compatibility of the coating and substrate over the temperature range of the expected application
(d) The cost.

The following parameters are used to identify the integrity of the coatings; coating method, composition, thickness, hardness, coating, substrate adhesion and friction and wear data.

7.0 CORROSION RESISTANT COATINGS

The classifications of different surface engineering methods involving coatings for corrosion protection are shown in Fig. 4 [10]. The most widely used metallic coating for corrosion protection is galvanizing, which involves the application of metallic zinc to carbon steel for corrosion control purposes. Hot dip galvanizing is the most common process, and as the name implies, it consists of dipping the steel member into a bath of molten zinc. Information released by the U.S. Commerce Department in 1998 stated that about 8.6 million metric tons of hot dip galvanized steel and 2.8 million metric tons of electrolytic galvanized steel were produced in 1997. The total market for metallizing and galvanizing in the United States is estimated at $1.4 billion. This figure is the total material cost of the metal coating and cost of processing, and does not include the cost of the carbon steel being galvanized/metallized [3].

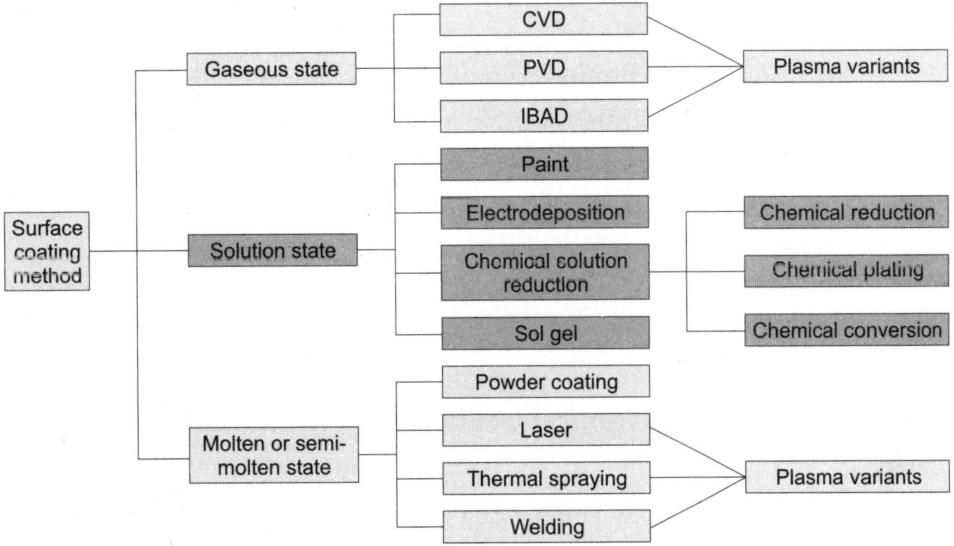

Figure 4 Classification of coating and surface engineering [10].

In industrialized countries the cost of combating corrosion is of the order of 2-4% of GNP, either in the form of corrosion prevention or replacement of corroded parts [2, 3, 11]. Traditionally, the most important coatings for corrosion resistance were electroplated nickel and chromium, and these reduce corrosion by forming a physical barrier between the substrate and the environment. The importance of these coatings lies not only in their corrosion resistance properties but also their wear and erosion resistance, which serve to extend coating lifetime.

7.1 PREPARATION OF SUBSTRATE FOR COATING DEPOSITION

Initial cleaning consists of removing mechanical, organic and chemical contaminants from the surface of the components designated for coating. This initial cleaning may be accomplished by the following means [10 - 13]:

- Mechanical: by removal of scale and permanent discoloration by glass-beading, by rotary finishing and by abrasive technique.
- Chemical: by removal through saponifications, in acidic or alkaline baths, in organic solvents or in alkaline aqueous solutions.
- Physical: by removal of contamination in cleaning baths through their dissolution or emulsification.
- Physico-chemical: by removal through detachment of the more stable contaminants in washing baths (e.g., chemical solvents) and alkaline aqueous solutions, assisted by ultrasonic vibration.

In all coating deposition techniques, good adhesion of the coating to the substrate surface which, in turn, depends on the coated component and its designation. In order for the coating to fulfill its task, the surface of the coated component should feature the following characteristics [6, 7]:

- Hardness: Obtained by heat treatment e.g., hardening and tempering or thermo-chemical treatment e.g., nitriding, chromizing, less often by mechanical treatment
- Smoothness: The surface should be smooth, ground or polished and with deburred edges.
- Cleanliness: The surface should be free of particles of mechanical contaminant (pollutant, dust), organic contaminants (fats, greases, anti-rust protective media) and of products of chemical reactions (corrosion products, e.g., oxides and sulfides).

Not all components are necessarily cleaned by all techniques, as mentioned in the above sequence. In some techniques it is only possible to brush off dust and grease and to degrease chemically.

7.2 INNOVATIVE THIN FILM COATINGS FOR CORROSION PROTECTION - TECHNOLOGY FOCUS

Deposition of thin films with a few microns in thickness is a common technology to improve the performance of tools, dies, and molds for many different applications. Progress in each of these areas depends upon the ability to; (i) selectively and controllably deposit thin films; (ii) thickness ranging from tens of angstroms to micrometers, and (iii) with specified physical properties. This, in turn, requires control often at the atomic level of film microstructure and microchemistry. As mentioned earlier, there are a vast number of deposition methods available and in use today. However, all methods have their specific limitations and involve compromises with respect to process specifics, substrate material limitations, expected film properties, and cost. This makes it difficult to select the best technique for any specific application. The development of thin, adherent hard coatings in the last quarter century has had a major impact on manufacturing, especially in the area of machining. The process involves depositing essentially any thin film metal coating on any metallic substrate using a novel metal coating technology [12, 13].

7.2.1 Proof-of-Concept

Physico-chemical properties of coating differ from those materials from which they are made. Properties of metallic materials in the bulk state, i.e. as obtained in metallurgical processes, are usually different from the properties of coatings manufactured from them. Usually, although not always, the parameters, which describe these properties, assumes higher values for coatings than for bulk materials. Moreover, for the same material, the differences in values of the same parameters depend on the method used in manufacture of the coating.

Corrosion-resistant coatings produced by plasma spraying have exhibited good benefits in certain conditions. It is well known that Al_2O_3, Al_2O_3+TiO_2, Cr_2C_3+NiCr, NiAl, NiCrAl, ZrO_2 and $MgZrO_3$ coatings are resistant to specific corrosion media which are widely employed in industry. With respect to environmental and material factors affecting corrosion processes, the corrosion behaviour of plasma sprayed coatings depend on the following elements [11, 12]:

(1) corrosive media
(2) substrate and their surface state
(3) environmental temperature
(4) composition, structure, porosity, thickness, adhesive strength and other properties of coatings.

Many factors influence the corrosion characteristics of coatings, including their quality and structure. Proof-of-concept selection can include thin films of Nickel (Ni), Zinc (Zn), Cerium (Ce), Indium (In), Bismuth (Bi), Selenium (Se), Molybdenum (Mo), Cobalt (Co), Titanium (Ti), Vanadium (V), Barium (Ba), Tin (Sn), and Copper (Cu) deposited on steel, and copper substrates. Based on the metal coating technique, essentially any single metallic film or two or more component metallic alloys can be deposited on any metallic substrate [13]. However, development and production of thin film for functional coating on special materials is unalterable combined with a permanent quality control and therefore use of sophisticated analysis and measurement techniques. The field of interest involved all steps of the production chain, starting with the characterization of the substrate surface, followed by polishing and cleaning processes up to depth profiling of complex multilayers.

7.2.2 Conventional Thin Film Metallic Coating Techniques

Usually, metallic coatings are deposited using a number of techniques including electroplating, electroless plating, spraying, hot dipping, chemical vapor deposition and ion vapor deposition. For various applications alternative metal deposition methods will replace some of techniques above, and may play a greater role in metal coating in the future. However, low cost metal coating technique stands an excellent chance to compete with traditional metal coating techniques for a large number of metal coating applications, especially in the area of amorphous thin film metallic alloys [10, 11].

Amorphous alloys form a structure very different from crystalline alloys. In the amorphous structure, atoms are randomly placed in a continuous coating, thereby preventing corrosion attack at grain boundaries. The absence of such defects in amorphous leads to the intrinsic properties of amorphous alloys. The absence of dislocations and the lack of boundaries lead to a low friction coefficient, near theoretical strength (and hardness), and outstanding resistance to cavitation. The lack of boundaries improves the resistance to corrosion and resistance to reactions (such as oxidation and sulfidation) at elevated temperatures. A large variety of two or more component metallic alloys can be deposited at comparatively low cost using the metal coating technique for a wide range of potential applications [10, 13].

7.2.3 Potential Applications

Thin film technology has a wide range of potential applications [10, 13]:

- Surface passivation of metallic surfaces, prior to application of paints; finished industrial, and consumer goods metallic products.
- Wear and corrosion resistant coatings for aerospace, oil, gas and chemical engineering.
- Exotic applications such as coatings for dielectric, Piezoelectric and ferroelectric sensors, thin film semiconductors (e.g. CdTe, CuInSe, BiTe, BiSe), and superconductors (e,g, Y-Bi-Cu-O, and Y-Ba-Cu-O) applications.
- Other applications include microelectronics, optics, magnetic, hard and corrosion resistant coatings, micro-mechanics aircraft, automotive, brewing, chemical, defense, food processing, glassware, metallurgy, paper, petrochemical, plastics, power, printing, steel, and textile.

8.0 RECENT PROGRESS AND NEW DIRECTION IN HARD COATINGS

In recent years, research into hard coatings, which previously had concentrated on single-phase films, has entered a new area of composite materials. The search for superhard materials considered so far mainly metastable solids, such as diamond, c-BN and C_3N_4 and single and polycrystalline superlattices. More recently, novel superhard nanocrystalline composites have been developed which appear superior to these materials in terms of stability, resistance against oxidation and relatively simple preparation by means of plasma CVD [14]. More recently, it was shown that similar increase in hardness could be achieved in nanocrystalline composite films [15]. Superhard coatings with Vickers hardness of 40 GPa have attracted large attention because of the scientific curiosity of preparing materials with hardness in the range of diamond (70–90 GPa) and with respect to their industrial applications. Three different approaches towards the preparation of superhard materials can be identified [14 - 16]:

1. Intrinsically superhard materials such as diamond, hydrogen-free "diamond-like carbon" (DLC), and cubic boron nitride (c-BN).
2. Thin coatings where the hardness enhancement is due to a complex, synergistic effect of ion bombardment during their deposition by plasma chemical or physical vapor deposition (PCVD or PVD); and
3. Nanostructured superhard coatings, such as heterostructures and nanocomposites.

A significant challenge is to blend together hardness, low friction, and toughness in a coating such that low wear is realized in diverse environments (e.g., hot, cold, wet, dry, and vacuum). Thin film coating deposition has undergone tremendous advances and it is now possible to grow multilayered, functionally gradient, and nanocomposite coatings that have impressive properties. Nanostructured designs have the flexibility to impart lubricity over many environments, yet maintain hardness, and dramatically increase toughness. Hard coatings prepared by various deposition techniques and conditions exhibit wide variety of microstructures in terms of grain size, crystallographic orientation, lattice defects, texture, and surface morphology as well as phase composition [17]. The technological process of their production and their properties, i.e. hardness, wear and oxidation resistance, however, are continuously being improved. Important milestones in the development of hard coatings are briefly summarized in Table 2 [18]. This table shows a clear effort (i) to decrease the temperature at which hard coatings are formed and (ii) to improve the properties of hard coatings, particularly to increase the hardness and corrosion/oxidation resistance. The oxidation resistance increased up to approximately 1000°C as during high-speed machining the temperature of the tool tip can reach 1000°C and the coating is stable at such high temperatures.

There are a number of good reviews on hard coatings that have been published lately. Monolithic coatings of transition metal carbides/nitrides and oxides dominate the literature [19, 20]; however, attention has recently been focused from monolithic coatings to duplex, functionally gradient, multilayer, and nanocomposite coatings to improve overall friction and

Table 2 Important steps in development of hard coatings [18].

Coating	Material	H (GPa)	Main characteristics
Single layer	TiN, TiC, Al_2O_3	21, 28, 21	CVD at T around 1000°C on cemented carbides
Single layer	TiN, TiC	21, 28	PVD at $T \leq 550$°C on steel substrates
Multilayer	TiC/TiB_2		phase boundaries of TiC/TiB_2
Single layer C	c-BN	50	High chemical affinity of to iron
Single layer	diamond	90	Chemical dissolution of B in iron
Single layer	DLC	65	Amorphous phase
Single layer	CNx	50–60	Substoichiometric ($x = 0.2 - 0.35$) turbostratic structure
Superlattices	TiN/VN, TiN/NbN	~50	Superlattice period 5–10 nm
Single layer	nc-MeN/a-nitride	~50	Superlattice period 5–10 nm
Single layer	Ti0.4Al0.6N	~32	Nanocomposite, oxidation resistance up to 950°C

wear response. So far, less attention has been devoted to carbide multilayer coatings. These coatings can also be superhard, up to 55 GPa, e.g. TiC/VC, 52 GPa [21]; TiC/NbC, 45–55 GPa [22].

8.1 HARD METAL NITRIDE COATINGS

The use of hard metal nitride coatings for the protection of structural components against wear and corrosion is widely recognized. Transition metal nitrides find wide spread technological use as diffusion barriers in microelectronics, hard wear resistant coatings on cutting tools, or as corrosion and abrasion resistant layers on optical and mechanical components. With the increasingly sophisticated microstructural and compositional design of state-of-the-art nitride films, they rely for their performance on a corresponding stability. Examples of microstructurally engineered materials include metastable alloy nitrides such as cubic-phase (Ti, Al)N films, Ti(C, N) films in a state of compressive residual stress, and compositionally modulated nitride films (e.g. nanolaminates; multilayers; superlattices) [23]. Transition nitride coatings like TiN, TiCN, TiBN and TiAlN are well suited for increasing the performance against corrosion. Titanium nitride hard coatings deposited by PVD and CVD have been implemented on an industrial scale for many years. These coatings have distinct chemical and physical properties. As an example, TiN oxidizes steadily at temperatures above 500°C, leading to the formation of poorly TiO_2 adherent rutile phase oxide layer on top of the TiN. The wear protection of TiN coatings is deteriorated accordingly [24]. CrN is more resistant to oxidation than TiN owing to the formation of a dense and passive oxide layer of Cr_2O_3 that restricts further oxidation [25].

Hard coatings based on transition metal nitrides or carbides, both the production process and their properties, i.e. microhardness, wear and oxidation resistance, are continuously being improved. Following this achievement, a variety of TiN based superlattice coatings, such as TiN/WN, TiN/CrN, TiN/TaN, TiN/MoN and TiN/AlN have been studied [26], demonstrating the potential of the superlattice structures. Most of the coating systems were developed under laboratory conditions. Common materials in thin-film coatings are titanium carbonitride, titanium aluminum nitride, titanium nitride, chromium nitride, zirconium nitride and amorphous DLC. Among carbon matrix systems, only hydrogen free DLC coatings were harder than 30-40 Gpa [27].

9.0 SMART COATING FOR CORROSION PROTECTION

Smart coatings could enable components, if corroded or scratched, to detect and heal themselves. There is a large demand for such coatings with new functional properties in almost all industrial branches. Few of these new functionalities, such as easy-to-clean [28], are already available but property such as long time durability is still missing and preventing a high level transfer into the market. Among the known smart coating are electrochemically active organic compounds or polymers that react with metal surfaces, enabling those surfaces to form protective oxides. Coatings that comprise electroactive polymers, or some analogous compounds, can cause corrosion-prone metals behave more "nobly." Some of such chemical compounds includes [28, 29]:

- diphenylamine compounds that protect metals
- polyaniline coatings on aluminum or copper

- polyphenylene ether wash-coats on metal surfaces

In general, the use of "smart materials" for corrosion sensing purposes relies on a material undergoing a transformation through its interaction with the corrosive environment. It is such transformations that can potentially be used for indicating and detecting corrosion damage. Ideally, in principle, the sensing function could be integrated with additional actuation and control functions, designed to control corrosion damage.

Under certain conditions, the driving force of corrosion produces protective oxide films, a process typically referred to as passivity. In cases where passivity is difficult to achieve, for example, for Al alloys in neutral NaCl solutions, it is conceivable and had demonstrated that a 'smart' coating using micro or nano engineered corrosion to generate 'on demand' a corrosion inhibitor for stopping or slowing corrosion at defects [29]. Fig. 5 schematically shows the concept by which a generic machine in a coating, represented by a pump that moves a corrosion inhibitor to a defect, uses the energy of the surface to drive the process. The oxygen reduction half of the corrosion reaction of metal dissolution will initially drive the release of the inhibitor. Several initiatives have been launched on the front of using smart coatings as corrosion sensors, but presently the technologies reside mainly in the experimental laboratory domain. As such, they are "futuristic" in nature.

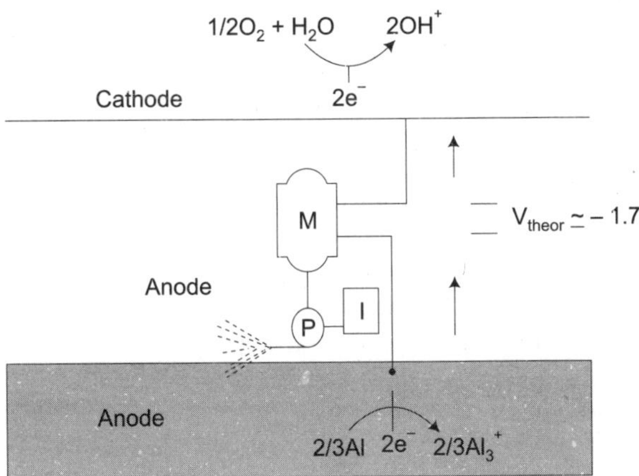

Figure 5 Schematic illustrating an inhibitor delivery system. The metal substrate is the anode. An electrode at the surface of the coating provides the cathode [29].

Some smart coating principles, including those relevant to corrosion sensing, or potentially relevant, include the following [30]:

- Paint systems with color-changing compounds, responding to pH changes that result from corrosion processes.
- Changes of coating compounds from non-fluorescent to fluorescent states, upon oxidation or complexing with metal cations.
- Release of color dyes, on coating damage, from incorporated dye-filled micro-capsules.

- Use of pigments that absorb corrosive chemicals.
- Use of pigments that release corrosion inhibiting chemicals "on demand".
- Piezo-electric thin film applications.
- Fiber optics

9.1 CORROSION PROTECTION BY CONDUCTING POLYMER COATINGS

There has been recently interest on the possible use of conducting polymers as either film forming corrosion inhibitors [31] or as protective coatings [32]. Of the conducting polymers the polyaniline (PANI) families have been the most widely studied, due to their environment stability and ease of synthesis [33]. Polymers were thought of as electrical insulators until the discovery that iodine doped polyacetylene exhibited electrical conductivity many orders of magnitude higher than neutral polyacetylene. This discovery was published by Shirakawa [34], as a result of this pioneering work, they received the 2000 Nobel Prize in Chemistry.

Conducting polymers (CPs) consist of conjugated chains containing electrons delocalized along the polymer backbone. In their neutral form, CPs is semiconductive materials that can be doped and converted into electrically conductive forms. The doping can be either oxidative or reductive, though oxidative doping is more common. There are three states of CPs: non-conducting (uncharged), oxidized (p-doped) where electrons are removed from the backbone, and the reduced (n-doped) (least common), where electrons are added to the backbone. The doping processes are usually reversible, and typical conductivities can be switched between those of insulators ($<10^{-10}$ S/cm) and those of metals (10^{5} S/cm) [35].

Corrosion protection using conductive polymers was first suggested by MacDiarmid [36]. Almost all of the CPs used in corrosion protection fall under the following classes: polyanilines, polyheterocycles and poly(phenylene vinylene). CPs can be synthesized both chemically and electrochemically. It has been observed that most CPs can be electrochemically produced by anodic oxidation [37], enabling one to obtain a conducting film directly on a surface. CPs can be insulating to the conducting state through several doping techniques such as [35, 37]:

- chemical doping by charge transfer
- electrochemical doping
- doping by acid–base chemistry (only polyaniline undergoes this form of doping)
- photodoping and
- charge injection at a metal semiconducting polymer interface.

Fig 6(a) shows the strong influence of deposition conditions for CPs. Choosing different deposition processes (chemically, electrochemically) and solvents (aqueous, non-aqueous) leads to different layer morphology. The deposition of smooth and homogeneous polymer layers is a difficult technical process. Already a small magnification shows in Fig. 6(b) the cauliflower-like surface structure. This structure is the result of electrochemical deposition conditions from diluted, monomer containing solution. In contrast a very homogenous layer from a chemically deposited layer is shown also (formed by spin coating). But a higher magnification level here also reveals some inhomogeneous surface areas. The electrochemically formed layer shows in a higher magnification level a combination of cauliflower and noodle structures [38]. Several known electrochemical characterization

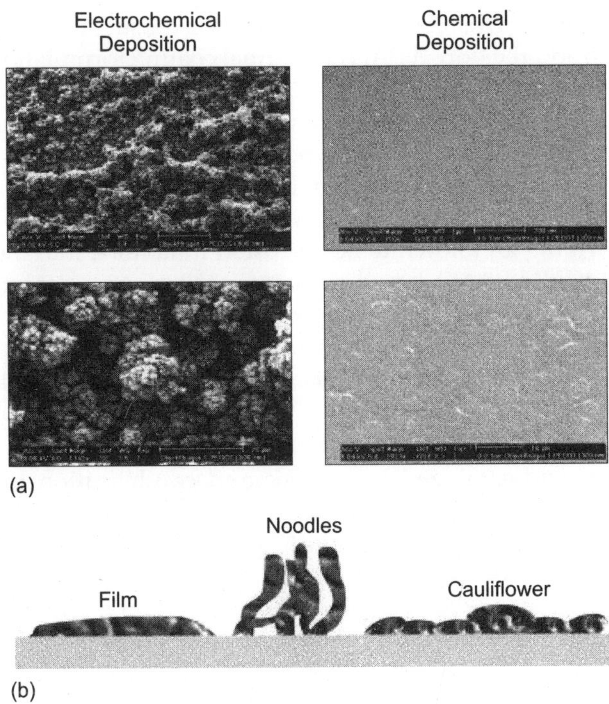

Figure 6 Influence of deposition conditions on surface composition of CPs. (a) Slide with chemically polymerised PEDOT (300 nm); on left side additional electrochemically formed layer (1 mA m^{-2}/100 s/0.05 M in ACN). (b) Surface grow types for CPs [38].

methods are currently used to investigate and quantify the corrosion protecting properties of CPs. Since corrosion is a naturally occurring chemical process in which a metal is destroyed by its environment, both DC and AC electrochemical monitoring techniques can be employed.

The most widely studied CPs for corrosion protection has been polyaniline (PANI). PANI has several advantages over most current CPs. The advantages are [38, 39]:

(1) easy chemical and electrochemical polymerization,

(2) easy doping and de-doping by treatment with standard aqueous acid and base and high resistance to environmental degradation [39].

PANI was first reported in the scientific literature by Letheby [40]. Several hypotheses have been suggested for the mechanism of corrosion protection using CPs, specifically PANI: (a) PANI contributes to the formation of an electric field at the metal surface, restricting the flow of electrons from metal to oxidant; (b) PANI forms a dense, strong adherent, low-porosity film similar to a barrier coating; and (c) PANI causes the formation of protective layers of metal oxides on a metal surface [41, 42].

Of potentially greater interest is the ability to design coatings with special corrosion related properties. The ability to design coatings to replace chromate-based corrosion protection films that are widely used in aircraft and other critical areas is of particular environmental importance. It is now established that conducting polymers, such as doped polyaniline (PANI), are effective at lowering corrosion rates for metals exposed to aqueous solutions [43]. However, the ability to add the doped PANI to normal coating formulations is limited because of low solubility in water and many organic solvents. Viswanathan and coworkers [44] have developed a process using lignosulfonates to enhance the solubility of inherently conducting polymers in aqueous systems. This allows the polymers to be included in water based coating formulations. In addition to aiding dispersion, lignosulfonates are known to complex metals and metal oxides. This makes it possible to incorporate metals or oxide clusters such as oxides of molybdenum or cerium [45] known to be corrosion inhibitors, into the lignosulfonic acid-doped polyaniline (LIGNO-PANI) based nanocomposites. An interesting and potentially important aspect of corrosion coatings based on conducting polymers is the ability of the coating to respond to electrochemical potentials established in the coating if the coating is damaged or scratched. This would allow a healing effect such as thought to operate for the chromate based systems.

Another approach to the formation of a nanocomposite protective film involves a two stage sol–gel process where acid-catalyzed hydrolysis and condensation to form organo-silica nanoparticles with peripheral epoxy functional groups. The nanoparticles are cross-linked through a conventional chemical reaction between amino and epoxy functionalities. The cross-linking and self-assembly occur as the solvent evaporates when a solution of particles, cross-linking species and a solvent are deposited on the surface to be coated. This self-assembled nanophase particle (SNAP) coating process forms a stable continuous and highly adherent protective film on the metal surface and has been used to engineer nanostructured composite, ceramic coatings for aluminum aerospace alloys [46]. The adhesion and structure of these organic/inorganic nanocomposite coatings can be tailored by changing the chemistry of the SNAP process.

10.0 ENHANCING COATING FUNCTIONALITY USING NANOSCIENCE AND NANOTECHNOLOGY

Nanocrystalline materials exhibit various shapes or forms, and possess unique chemical, physical or mechanical properties [47]. In recent years, nanostructured materials have attracted increasingly more attention from the materials community. In the area of coatings, new approaches utilizing nanoscale effects can be used to create coatings with significantly optimized or enhanced properties. The ultimate impact of nanoscience and nanotechnology in the area of coatings, and other potential application areas, will depend on our ability to direct the assembly of hierarchical systems that include nanostructures. Coatings and coating processes will play a major role in the future of nanotechnology regardless of the directions in which they will be taken. Because nanostructured materials, in whatever form they are produced, will be incorporated into larger mesoscopic and macroscopic systems, coating processes will play a critical role in the success of nanotechnology in many different applications [47, 48].

The common definition of a nanosystem is one with at least one characteristic size between 1 and 100 nm. There are indeed characteristics of materials and materials systems that, when extended or scaled to the nanosized region, lead to improved properties as conceptually simple (or scalable) extensions of properties from larger sized systems. A distinction may be drawn between "smaller is better" nanotechnology and a more complex and perhaps richer field of nanoscale functionality which is unobtainable in "bulk" materials or extension of bulk materials properties. In a world where something different is unique, nanomaterials are unique. Highly sophisticated surface related properties, such as optical, magnetic, electronic, catalytic, mechanical, chemical and tribological properties can be obtained by advanced nanostructured coatings, making them attractive for industrial applications in high-speed machining, tooling, optical applications and magnetic storage devices [47, 49]. There are many types of design models for nanostructured coatings, such as nanocomposite coatings, nano-scale multilayer coating, superlattice coating, nano-graded coatings, etc. Designing of nanostructured coatings needs consideration of many factors, e.g. the interface volume, crystallite size, single layer thickness, surface and interfacial energy, texture, epitaxial stress and strain, etc, all of which depend significantly on materials selection, deposition methods and process parameters [49]. Murday [50] states that there are three reasons that nanostructured materials behave somewhat differently from other materials:

(1) large surface to volume or interface to volume ratios,
(2) size effects where cooperative phenomena, such as magnetism, are altered because of the limited number of molecules in the system, and
(3) quantum effects. As already stated, nanostuctured materials are in the size range for which our normal understandings of material properties no longer apply.

The availability of various nanoparticles prepared from vapor, liquid and solid routes have motivated the development of nanocrystalline coatings with superior wear and oxidation resistances as well. There is increasing scientific and commercial interest in the development of nanostructured coatings, particularly those based on low miscibility crystalline/amorphous nanocomposite phase mixtures and 'metallic film'. The ever-growing need for superior materials to withstand severe operating conditions has driven the search for superhard coatings with hardness values above 40 GPa. A nanocomposite coating comprises of at least two phases: a nanocrystalline phase and an amorphous phase, or two nanocrystalline phases. Nanocomposite coatings exhibit hardness significantly exceeding that given by the rule of mixture. Hard materials usually refer to materials with hardness greater than 20 GPa. Materials with hardness above 40 GPa are classified as superhard, and those with hardness

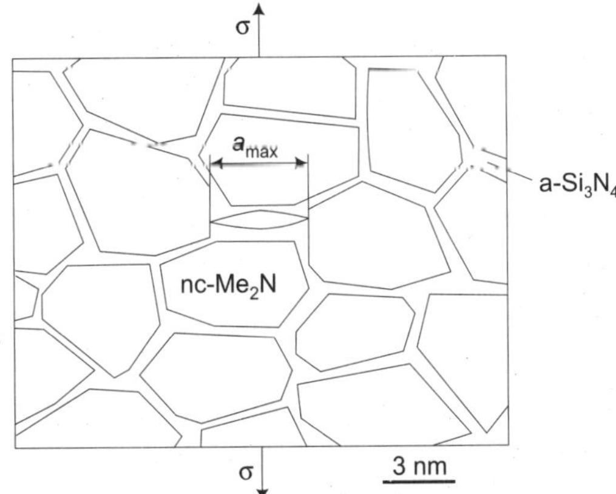

Figure 7 Schematic design of a super-hard composite coating, combining amorphous and nanocrystalline phases, showing restrictions in the initial crack size and crack proagation [58].

above 80 GPa are often called ultra-hard materials. Nanocomposite coatings can be hard, superhard or even ultra-hard, depending on coating design and application [47].

The range of potential applications for nanocomponents in larger systems is immense. Examples of applications where coating processes are likely to be involved include thermoelectric power converters, batteries and fuel cells, corrosion resistant, wear resistant, reflective and other protective coatings, components of sensors and detectors, electronic, optoelectronic and magnetic components, photovoltaic and display systems, and in applications involving catalysis, electrocatalysis and photocatalysis. Because the electronic aspects of nanoscience and properties of individual carbon nanotubes are most widely discussed in the literature, and discussions of the stability and reactivity of nanoparticles are much less common, the following discussion focuses on ideas related to reactivity, corrosion, strength and durability [47 – 51].

10.1 NANOCOATING AND CORROSION

The actual data related to the corrosion of nanostructured materials is limited. Erb et al. [52] noted that the results for nanocrystalline materials made by electrodeposition are mixed in that both detrimental and beneficial effects have been observed. There are, however, other hints of some increase in corrosion resistance for nanosized materials. These include the observations on the slow oxidation of metal clusters exposed to air [53] and an apparent decrease in the reactivity of CdTe quantum dots that combine to form a ring like molecule [54].

There are two reasons why nanostructured materials might have altered corrosion and reactivity relative to other versions of the same materials. First, the change in the energy levels due to quantum effects will alter the chemistry and reactivity of nanosized clusters. They will not react like the bulk material. The tendency of metal clusters to become denser than bulk metals may be consistent with a stronger internal metal-to-metal bond formation in these clusters in comparison to the bulk metals, which might decrease corrosion reactions. The second reason deals with the size of the structures in a nanostructured "bulk" material. Often localized corrosion at grain boundaries or other defects leads to corrosion or material failure. Material for which the grain or feature size is in the nanometer region may not have selected regions where the local chemistry is significantly different than other regions [47]. Thus, there may not be regions for the selective attack that causes the localized corrosion attack. Although nanoscience and nanotechnology offer many possible opportunities, much is unknown on corrosion aspect, detail investigation of such technology will give new insight in combating corrosion.

The initiation of chemical attack of metal surfaces resulting in surface nanostructures with very interesting technological applications have been recently reported in *Nature* [55]. Researchers from Max Planck Institute, the University of Ulm (Germany), and the European Synchrotron Radiation Facility (ESRF) using synchrotron light source, carried out a study to reproduce *in situ* the onset of the corrosive process in a gold-copper alloy. Gold is a very noble metal, which doesn't corrode, whilst copper is less noble and, thus, more prone to chemical attack. At the first moments of corrosion, the copper-gold alloy develops a mechanism to protect itself with an extremely thin gold-rich layer. This layer has an unexpected crystalline and well ordered protective passivation layer structure. When the corrosion process proceeds, this alloy layer transforms into gold nano-islands of 20 to 1.5 nanometres. These islands eventually develop into a porous gold metal layer, which may have many technological applications. These new insights can be applied to a variety of different alloys used in corrosive environments and to materials that can exploit such degradation to form porous metals of technological interest.

10.2 SUPERHARD AND SUPER-TOUGH NANOCRYSTALLINE COATINGS

There is an increasing need in industrial sectors for the development of high performance coatings with better oxidation resistance, higher hardness and longer lifetime than conventional TiN single layer coatings. To meet industrial demands for improved coatings, significant effort has been devoted to the design and synthesis of superhard coatings. Koehler [56] was the first to propose a concept for the design of strong solids or coatings by using two different alternate layers of materials with high and low elastic constants, respectively. The thickness of each layer should be in the nanometer range, and no dislocation source could operate within the layers. If the dislocations could form in the material layer with the lower modulus, they need to overcome a significant repulsive image stress from the higher modulus phase before slip can be transmitted across layers [57], thus, preventing slip from propagating through the whole film. Such multilayer coatings are often termed as "superlattices", and the bilayers of these superlattices can be metal layers, nitrides, or carbides. In designing multilayer coatings, both structural and functional components must be taken into consideration. The former involves the grain size, thickness of individual layers, composition modulation and the number of material interfaces whereas the latter is concerned with the wear, mechanical and physical characteristics. Many different hard materials can be combined in multilayer coatings resulting in optimized material properties. Using similar ideas for restricting dislocation formation and mobility as were used in multilayer approaches to "hardening," nanocomposite coatings can also be superhard [57]. These composites have 3-10 nm crystalline grains embedded in an amorphous matrix and the grains are separated by 1-3 nm. This design leads to ultra-hard (hardness above 100 GPa) coatings reported by Veprek and co-authors most recently [58]. The nanocrystalline phase can be selected from the nitrides, carbides, borides, and oxides, while the amorphous phase include metals and diamond-like carbon (DLC) as shown in Fig. 8 [58, 61]. These suggest that the nanocrystals have strong interaction with the matrix phase to impart super-hardness.

In the course of the development of tough nanocomposite coatings, the following design concepts were formulated:

(1) A graded interface layer is applied between the substrate and crystalline/amorphous composite coating to enhance adhesion strength and relieve stresses (combination of functional gradient and nanocomposite design) [59].

(2) Encapsulation of 3-10 nm sized hard crystalline grains in an amorphous matrix restricts dislocation activity, diverts and arrests macro-crack development, and maintains a high level of hardness similar to super-hard coating designs [57, 58]

(3) A large volume fraction of grain boundaries provides ductility through grain boundary sliding and nano-cracking along grain/matrix interfaces [60].

Fig 8, presents a schematic of a nanocomposite coating design that exhibits "chameleon" behavior. This design was implemented in the fabrication of the YSZ/Au/DLC/MoS_2 and WC/DLC/WS_2 "chameleon" coatings where an amorphous matrix and a hard nanocrystalline

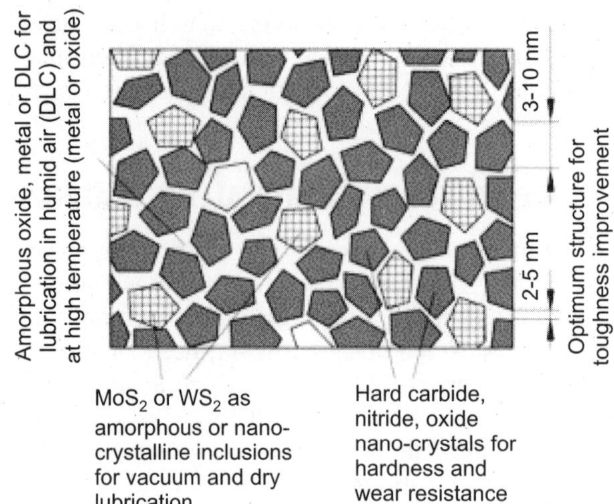

Figure 8 Schematic of a conceptual design for a nanocomposite tribological coating with chameleon-like surface adaptive behavior [61]

phase (e.g., YSZ or WC) were used to produce optimum mechanical performance and load support. Nanocrystalline and amorphous Au, MoS_2, and DLC were added to achieve chemical and structural adjustment of transfer films formed in friction contacts during dry/humid environment and low/high temperature cycling. "Chameleon" behavior is observed during the following sequence. As deposited, all lubricant phases (WS_2 or MoS_2, DLC, soft metals, and oxides) are either amorphous or poorly crystalline and are buried in the coating where they are sealed from the environment [61].

Thus, novel nanocomposite designs for tough tribological coatings are very promising and provide a very attractive alternative to multilayer architectures. Nanocomposite coatings are more easily implemented, since they do not require precise control in the layer thickness and frequent cycling of the deposition parameters, as is required for fabrication of multilayer coatings. They are however relatively recent developments, and suitable scale-up of deposition techniques is currently under intense study.

II. 0 THERMAL BARRIER COATING (TBC)

TBCs are complex, multifunctional thick films (typically 100 μm to 2 mm thick) of a refractory material that protect the metal part from the extreme temperatures (Fig. 9). They comprise thermally insulating materials having sufficient thickness and durability that they can sustain an appreciable temperature difference between the load bearing alloy and the coating surface. The benefit of these coatings results from their ability to sustain high thermal gradients in the presence of adequate back-side cooling. Lowering the temperature of the metal substrate prolongs the life of the component: whether from environmental attack (corrosion), creep rupture, or fatigue. In addition, by reducing the thermal gradients in the metal, the coating diminishes the driving force for thermal fatigue. The current material of choice for TBCs is yttria stabilized zirconia (YSZ) in its metastable tetragonal-prime structure. Since it has proven to be a highly durable TBC material, it is likely to remain the material of choice for turbines

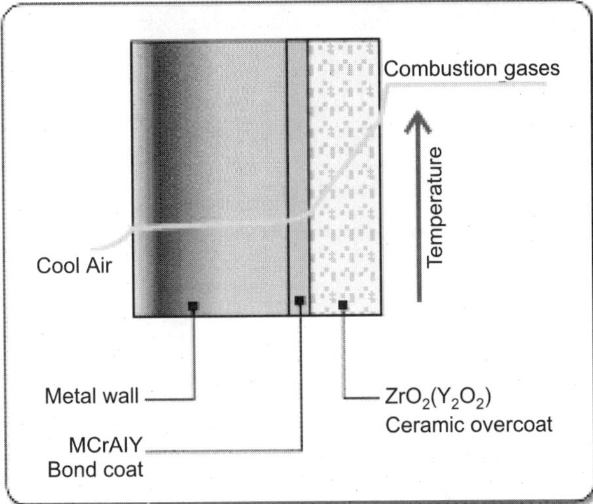

Figure 9 Schematic of the structure of a two-layer thermal barrier coating on a turbine blade surface together with a temperature profile

with current operating temperatures. While the primary function of TBCs is as a thermal barrier, the extremely aggressive thermomechanical environment in which they must function demands that they also meet other severe performance constraints. In particular, to withstand the thermal expansion stresses associated with heating and cooling, either as a result of normal operation or as a consequence of a 'flame-out', the coatings must be able to undergo large strains without failure. Another less stringent but nevertheless rather practical requirement is that the material must not undergo phase transformations on cycling between room temperature and high temperatures. Such phase transformations are usually accompanied by volume changes, which detract from the strain compatibility and reversibility of the coating and, hence, its ability to withstand repeated thermal cycling. Practical TBC materials must also be able to resist erosion, which calls for high resistance to fracture and deformation. Another perhaps less obvious requirement is that the coating material is thermodynamically compatible with the oxide formed by oxidation of the bond-coat.

The search for low thermal conductivity materials for thermal barriers is only just beginning. Producing solid ceramic components is not always the best approach to solving a wear or corrosion problems. In many cases, taking the original metallic part and applying a coating can be the best solution. In the recent years, there has been a growing interest in the use of the high-velocity oxy-fuel (HVOF) thermal spraying process to deposit protective overlay coatings onto the surfaces of engineering components to allow them to function under extreme conditions [62]. With the advent of this process, thermal sprayed coatings, which previously had limited usefulness as corrosion protection coatings due to the presence of interconnected porosity in the structure, have now gained popularity and are being studied extensively for their corrosion resistant properties. In Figure 10, cross-sectional image of a YSZ thermal barrier coating deposited by electron-beam evaporation on a superalloy [63] is shown. Herman et al. [64] reported that HVOF spraying is able to make denser and less oxidized coatings compared with other methods, such as plasma spraying. Furthermore, this spraying system enables

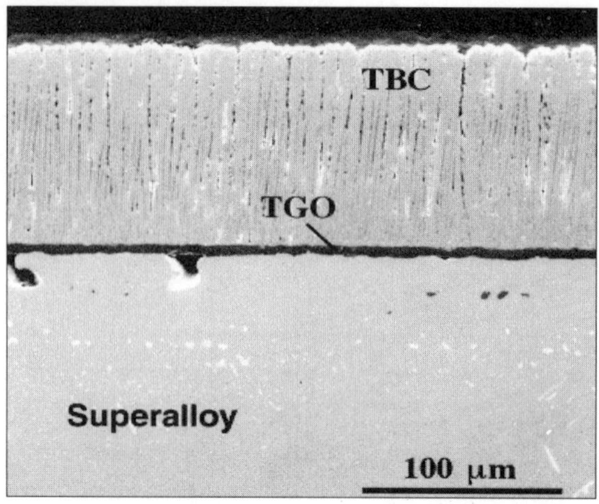

Figure 10 TBC. Cross-sectional image of a YSZ thermal barrier coating deposited by electron-beam evaporation on a superalloy. During use at high temperatures, a thermally grown oxide (TGO) of Al_2O_3 forms on metal beneath the TBC [63]

metals and alloys with high melting point up to about 2000 °C to be deposited on the target surface. These features are suitable for a deposition of corrosion resistant coatings. Helali et al [Fig. 11] indicates the characteristics of HVOF coatings compared with those produced using the standard plasma spraying process [65].

12.0 METALLIC OXIDE COATINGS AND OXIDE ELECTRODE FOR ADVANCED ELECTROCHEMICAL TECHNOLOGY

Metallic oxide coatings are applied to a wide variety of metals from aqueous solutions such as hot alkalis or by controlled thermal oxidation [66-71]. Oxide coatings on the surface of components can produce significant tribological advantages, particularly in preventing adhesive wear and corrosion. Metal oxides, metal oxide mixtures and oxide electrodes are becoming increasingly important in the chemical industry. Metal-supported mixed oxide thin films have recently raised interest in various branches of industrial electrochemistry [66]. Transition metal oxides have shown outstanding activity in a variety of electrochemical processes [66]. Similarly, the discovery of dimensionally stable anodes (DSA-type electrodes) with oxide coatings has led to significant improvements in industrial electrochemistry technology, especially in the chlor-alkali industry [67]. In addition to the traditional DSA electrodes, which are composed of a binary oxide mixture, modified electrodes composed of ternary oxide mixtures have also recently been investigated and found to have enhanced performance [68, 69]. Porosity is a typical feature of oxide electrodes related to the specific procedure of preparation. Porosity leads also to a pH gradient within the electrode [70], which may be detrimental and affect the electrode process. All these aspects open new ways of investigation, which have contributed to raise the interest of researchers for oxide electrodes.

In metal oxide coatings, corrosion can be reduced or minimized by lifting the limit of the redox reaction. This can be done by mixing RuO_2 with other oxides. RuO_2 + IrO_2 are less

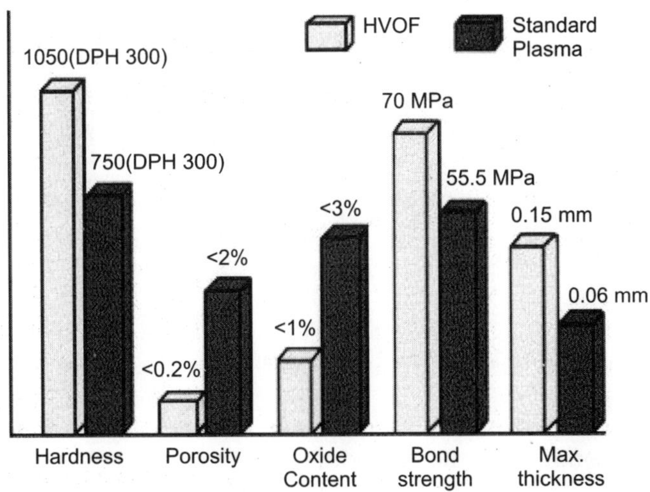

Figure 11 Characteristics of HVOF and standard plasma process coatings [65]

corrodible because the intimate mixing of these two oxides lifts the limiting potential of surface oxidation [71]. IrO_2 corrodes less than RuO_2 by itself; it is interesting that the corrosion rate of IrO_2 is also decreased by the presence of RuO_2 [72]. Thus, the study of these problems offers a wide range of cases where it is possible to identify and test the various parameters, which contribute to the stability of electrodes. In aqueous media and under conditions of simultaneous oxygen evolution, the oxide electrodes found new applications e.g., selective synthesis and destructive oxidation of organic contaminants [73].

Stability is certainly the main reason for industrial success. It is now evident that three factors are to be considered for electrodes [74 -76]:

(a) the formation of surface soluble products

(b) the mechanical strength and

(c) the electronic conductivity

Active oxides were originally those of precious metals, especially RuO_2 and IrO_2. If oxide electrodes are compared with graphite or Pt–Ir alloys, oxides exhibit much higher activity than graphite, comparable with that of the metal alloy [77]. Oxides constituting DSA are not the best catalysts for the given reaction. The aim of modern fundamental electrocatalysis is to establish structure activity correlations and not only activity preparation correlations. In this respect oxide electrodes lend themselves especially to give evidence to true electrocatalytic phenomena. The active layer on the inert (Ti) support consists of crystallites, which play the role of bricks. The size of crystallites increases with the increasing temperature of preparation. As a consequence, the actual surface area decreases, as manifested by the decrease of the voltammetric surface charge, which measures the number of protons exchanged with the solution during a potential cycle [78].

One of the aims of electrocatalysis is to improve the electrode activity by mixing two or more components searching for synergetic effects (Fig. 12). The latter are normally attained as

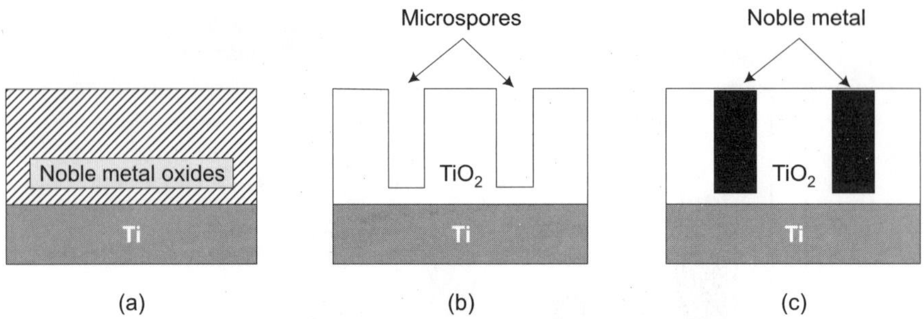

Figure 12 (a) A normal DSA anode, (b) a Ti-microporous TiO_2 system, after anodisation and (c) a Ti/TiO_2/noble metal anode (this is for one type of metal nucleation phenomenon)

the components come into electronic contact. A good example of synergism is found in RuO_2 + IrO_2 mixed oxides. Depending on the procedure of preparation, O_2 evolution changes with composition manifesting the formation either of mixed phases or of solid solutions [77, 78]. Although the surface of the mixed oxide is slightly enriched with IrO_2, the changes cannot be related to surface segregation phenomena but rather to spillover effects, another typical phenomena in electrocatalysis of heterogeneous surfaces. Surface enrichment is normal in mixed oxides as the affinity for oxygen differs among the components. Thus, 'valve metals' prevail in mixture with noble metals [79] but noble metals prevail in mixture with oxides of lower metal–oxygen bond strength such as spinels [80]. Due to the large difference in activity between the two oxides, for instance RuO_2 and Co_3O_4, a maximum in activity is reached at a relatively small concentration of the precious metal, with evident advantage for the consumption of the expensive component.

The success and current efficiency of the electrochemical processes mainly depends on the durability or the corrosion resistance, electrocatalytic activity, and high conductivity of the oxide coated electrodes. Longer life of the electrodes is of paramount importance in order to avoid interruptions in the operations, and to eliminate the risk in frequent replacement or addition of new electrodes in any chemical industry. The anode coatings usually used for oxygen evolution are mixtures of IrO_2, RuO_2 and Ta_2O_5. Alternative coatings to replace noble metals are sought and developed. The purpose of anodes is to supply current to the electrochemical cell. A vast number of materials and their combinations are electrical conductors and may hence be used as anode materials. However, good anode materials must survive in aggressive environments and also be able to pass high current densities. These opposite requirements restrict the amount of possible materials. Good electrical conductivity is important for energy efficiency. Good electrocatalytical properties are important to improve product yield. Long-term stability is also an important property, because electrode wears and corrosion may cause product contamination, increasing energy consumption and material and labor costs due to need of periodical maintenance.

The well-known DSA®, e.g. RuO_2, IrO_2, RuO_2 + TiO_2 and IrO_2 + TiO_2, are frequently prepared by thermal decomposition of the respective chlorides dissolved in acidic solutions [81]. Coatings of pure Ru and Ir oxides, or doped with Ti and/or Sn oxide, have also been intensively investigated. Simpler systems, based on RuO_2 or IrO_2 - SnO_2, are less suitable, either for the poor service-life or for the cost [75, 78]. Thus, the research concerns the development of new anode materials based on $RuO_2 - IrO_2 - SnO_2$ ternary mixtures deposited on Ti supports have been reported [82]. Different techniques may be used to prepare the mixed oxide layer [83, 84]. One most commonly employed is to paint a titanium substrate with a solution containing typically chloride salts of one or more noble metals followed by heat treatment of these painted layers under a controlled atmosphere. This is repeated several times in order to produce a DSA. Such electrodes are characterised by a high content in noble metals, a high cost and a peeling of the noble metal coating followed by passivation under working conditions.

When properties such as high corrosion resistance and physico-chemical stability under high positive potentials are required for an electrode material, dimensionally stable anodes (DSA) are the natural candidates. Discovered by Beer in the 1960s [85] and continuously improved in the 1970s, this designation denotes a class of thermally prepared oxide electrodes where a titanium substrate is covered by metallic oxides. Coatings on titanium include TiO_2, RuO_2, IrO_2 and Ta_2O_5. Combinations such as $(TiO_2) \times (RuO_2)y$ and $(Ta_2O_5) \times (IrO_2)y$ are currently used as electrodes in the chlor-alkali industry. Some DSA type electrodes may receive addition of SnO_2 and Sb_2O_5 in concentrations ranging from minor to main component. Similarly, based on the above DSA concept, mixed oxide coated titanium anodes (MOCTA) were developed [86] for application as electrodes in the electrochemical processes. MOCTA belongs to the category of DSA, which are basically a mixture of oxides of Ru, Ti, Pt, Rh, Ir and Pd coated over the surface of Ti, Ta, W, Cr, or Zr. The MOCTA electrodes developed were Ti substrate coated with $RuO_2 + TiO_2$, with or without an overlay of PtO_2. These electrodes were prepared by thermal decomposition method by which the salt solutions of Ru and Ti were applied over a pretreated Ti surface and thermally oxidised to get an adherent, conductive and electrocatalytic coating.

To increase the current density of operation and service life of these oxide electrodes, a new approach was evolved [87]. An improved process for producing mixed oxide coated titanium anodes having enhanced life in electrochemical applications and the anodes so prepared, namely "Thermochemical Glazing Process". This process consists of noble metal coating over an intermittent MOCTA layer on Ti substrate. Metallic coatings of Pt and Pt–Ir were prepared by thermochemical glazing process on titanium substrates. Titanium substrates were applied first with a single layer of $RuO_2 - TiO_2$. Followed by this, it was applied with Pt and Pt–Ir chemical solutions. Life assessment of MOCTAG (MOCTA-glazed) electrodes in a simulated reprocessing boiling nitric acid (10 N) test solution containing redox ions showed good performance of the Pt–Ir layered electrode, in comparison with MOCTA electrodes. The electrode prepared, by this process, continued to work up to 1840 h at an operating current density of 12.8 mA/cm^2 (Fig. 13).

13.0 SUMMARY

Conventional and well-established surface coatings and modification processes have considerable merit for improving the corrosion and tribological properties of engineering materials. The expanding use of surface coating technologies has generated a need to understand such interrelationship between the mechanical, physical and corrosion properties of the coating substrate system since it is this, which primarily determines coating performance in many engineering application. In several areas, it has been demonstrated that the new

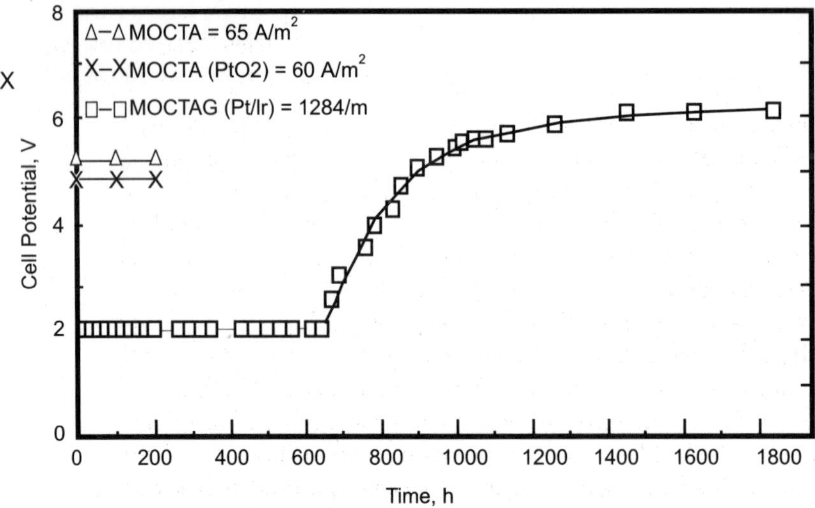

Figure 13 Performance of electrode during electrolysis of simulated reprocessing test solution [87].

surface coatings and treatment technologies arising from use of lasers, electron beams, plasma, chemical vapor depositions etc complement those already in existence. Newer coating technology materials such as nanomaterials, metallic oxide, polymers, ceramics, hard thin film, composites etc enhance the resistance to corrosion and wear and reduce the friction. Corrosion protection coatings can also enhance component operating efficiencies, which are unobtainable by other technique. Thus there is a considerable scope of industry to utilize established and modern surface coatings and surface engineering techniques for corrosion protection to secure improvements in the structural components and economics of industries.

14.0 REFERENCES

1. L.L Shreir, R.A. Jarman and G.T. Burstein, Editors, Corrosion (3rd Ed. ed.), Butterworth-Heinemann, Oxford (1994).
2. G.H. Koch, M.P.H. Brongers, N.G. Thompson, Y.P. Virmani and J.H. Payer, Corrosion cost and preventive strategies in the United States, Federal Highway Administration Technical Report No. FHWA-RD-01-156, March 2002, p. 773.
3. http://www.corrosioncost.com
4. Battelle Memorial Institute, Economic Effects of Metallic Corrosion in the United States, Research Report, 1995.
5. J.R.Davies, Surface Engineering for Corrosion and Wear Resistance, ASM International & IOM Communications, Institute of metals, London (2001), p. 1.
6. D.S.Rickerby and A.Matthews, in "Advanced Surface Coatings: A Handbook of Surface Engineering." Eds. D.S.Rickerby and A.Mathews, Chapman and Hall, New York, (1991).
7. T.Burakowski and T.Wierzchoñ, (Eds), Surface Engineering of Metals - Principles, Equipments, Technologies, CRC Press, New York (1999).
8. T.Biestek and S.Sekowski, Methoden zur Prüfung metalischer Überzüde. Vieweg Verlag, Braunschweig, 1971.

9. T.Burakowski and R.Marczak, The service-generated surface layer and its investigation, Polish Academy of Science - Committee for Machine Building, Book 3 (103), Vol, 30, 1995, p. 327.
10. K. Holmberg and A. Matthews, Coatings Tribology, Elsevier, Amsterdam (1994).
11. Corrosion Prevention by Protective Coatings, Houston (TX): NACE; 1984.
12. C.R.K. Rao and D.C. Trivedi, Coordination Chemistry Reviews, 249, (2005), p. 613.
13. J.Edwards, (Ed), Coating and Surface Treatment Systems for Metals - A Comprehensive Guide to Selection, ASM International, Materials Park, Ohio, USA, (1997).
14. W.D. Sproul. *J Vac Sci Technol. A:* **12**, (1994), p. 1595.
15. S. Veprek and S. Reiprich. *Thin Solid Films,* **268**, (1995), p. 64.
16. S. Veprek, *J Vac Sci Technol. A:* **17**, (1999), p. 2401
17. I. Petrov, P.B. Barna, L. Hultman and J.E. Greene, *J Vac Sci Technol A:* **21,** (2003), p. 117.
18. J. Musil, *Surf Coat Tech,* **125,** (2000), p. 322.
19. B.Bushan and B.K Gupta, Materials, Coatings, and Surface Treatments. In: B.Bushan and B.K. Gupta, (Eds), *Handbook of Tribology,* McGraw-Hill, New York (1991).
20. K Holmberg and A.Matthews, *Coatings Tribology: Properties, Techniques, and Applications in Surface Engineering,* Elsevier, Amsterdam (1994).
21. B.G. Wendler, P. Kula, K. Jakubowski, S. Fauvry, Ph. Kapsa, L. Vincent and D. Heper. In: *Proc. 36th Tagung der Deutschen Gesellschaft für Obrflächen- und Galvanotechnik, October 7–9, Schwabisch Gmünd, Germany* (1998), p. 6–11.
22. B.G. Wendler and M. Molinaro. In: *Proc. 10th Int. Summer School on Modern Plasma Surface Technology, May 10–14, Mielno, Poland* (1998), p. 423–436.
23. L. Hultman, *Vacuum,* **57,** (2000), p. 1.
24. U. Wahlstrom, L. Hultman, J.E. Sundgren, F. Adibi, I. Petrov and J.E. Greene, *Thin Solid Films,* **235,** (1993), p. 62.
25. H. Jensen, U.M. Jensen and G. Sorensen, *Surf Coat Tech,* **74–75,** (1995), p. 297.
26. J. Musil, S. Kadlec, J. Vyskocil and V. Valvoda, *Thin Solid Films,* **167,** (1988), p. 107
27. A.A Voevodin, S.V Prasad and J.S Zabinski, *J Appl Phys,* **82,** (1997), p. 855.
28. M. Thieme, R. Frenzel, S. Schmidt, F. Simon, A. Hennig, H. Worch, K. Lunkwitz and D. Scharnweber, *Adv Eng Mater,* **3,** (2001), p. 691.
29. M.Kendig, M.Hon and L.Warren, *Prog Org Coat,* **47**, (2003), p. 183.
30. http://www.corrosion-club.com/smart.htm
31. S.P. Sitaram, J.O. Stoffer and T.J. O'Keefe, *J Coat Technol,* **69**, (1997), p. 65.
32. T.P. McAndrew, *Trends Polymer Sci,* **5**, (1997), p. 7.
33. S. Sathiyanarayanan, S.K. Dhawan, D.C. Trivedi and K. Balakrishnan, *Corros Sci,* **33**, (1992), p 1831.
34. H. Shirakawa, E.Louis, A. MacDiarmid and A.Heeger, *Chem Commun,* (1977), p. 578.
35. A.J. Epstein, *MRS Bull,* **22**, (1997), p. 16.
36. A.G. MacDiarmid. Short Course on Conductive Polymers, SUNY, New Platz, NY (1985).
37. J. Heinze, *Top Curr Chem,* **152**, (1990), p. 1.
38. J.W. Schultze and H. Karabulut, *Electrochim Acta,* **50**, (2005), p. 1739.
39. J.O. Iroh and R.R. Rajagopalan, *J Appl Polym Sci,* **76**, (2000), p. 1503.
40. H. Letheby, *J Chem Soc,* **15**, (1862), p. 161.
41. T.P. McAndrew, *TRIP,* **5**, (1997), p. 7.
42. D.W. DeBerry, *J Electrochem Soc,* **132**, (1985), p. 1022.

43. D.E. Tallman, G.M. Spinks, A.J. Dominis and G.G. Wallace, *J Solid State Electrochem*, **62**, (2002), p. 73.
44. B.C. Berry, T. Viswanathan, in: T.Q. Hu (Ed.), Chemical Modification, Properties and Usage of Lignins, Kluwer Academic Publishers/Plenum Press, New York, 2002, p. 21.
45. A. Nazari, P.P. Trzaskoma-Paulette and D. Bauer, *J Sol–Gel Sci Technol*, **10**, (1997), p. 317.
46. A.J. Vreugdenhil, V.N. Balbyshev and M.S. Donley, *J Coat Technol*, **73**, (2001), p. 35.
47. C.C. Koch (Ed.), Nanostructured Materials, Processing, Properties and Applications, Noyes Publications, Norwich, USA, 2002.
48. H. Gleiter, *Prog Mater Sci*, **33**, (1989), p. 223,
49. G. Schmidt, *Chem. Rev*, **92,** (1992), p. 1709.
50. J.S. Murday, *Adv Mater Process*, **6,** (2002), p. 5.
51. G.-M. Chowa and L.K. Kurihara, Chemical Synthesis and Processing of Nanostructured Powders and Films, in: C.C. Koch (Ed.), Nanostructured Materials: Processing, Properties and Applications, Noyes Publications, Norwich, USA, 2002, p. 3–50.
52. U. Erb, K. Aust and G. Palumbo, Electrodeposited Nanocrystalline Materials, in: C.C. Koch (Ed.), Nanostructured Materials: Processing, Properties and Applications, Noyes Publications, Norwich, USA, 2002, pp. 179–222.
53. S.A. Nepijko, M. Klimenkow, H. Kuhlenbeck, D. Zemlyanov, D. Herein, R. Schlogl and H.-J. Fruend, *Surf Sci*, **412/413,** (1998), p. 192.
54. M.-H. Na, H. Luo and Y. Liang,, in: D.R. Baer, C.R. Clayton, G.D. Davis, G.P. Halada (Eds.), State-of-the-Art Application of Surface and Interface Analysis Methods to Environmental Material Interactions, vol. 2001-5, The Electrochemical Society, Pennington, NY, 2001, p. 113.
55. F.U.Renner, A.Stierle, H.Dosch, D.M.Kolb, T.-L.Lee and J.Zegenhagen, *Nature*, **439**, (2006), p.707.
56. J.S. Koehler, *Phys. Rev. B*, **2**, (1970), p. 547.
57. S.Veprek and S.Reiprich, *Thin Solid Films*, **268,** (1995), p. 64.
58. S.Veprek, *Surf Coat Technol*, **97,** (1997), p. 15.
59. A.A. Voevodin, M.A. Capano, S.J.P. Laube, M.S. Donley and J.S. Zabinski, *Thin Solid Films*, **298,** (1997), p. 107.
60. A.A. Voevodin, J.G. Jones, J.J. Hu, T.A. Fitz and J.S. Zabinski, *Thin Solid Films*, **401,** (2001), p. 187.
61. A. A. Voevodin, J. S. Zabinski and C. Muratore, *Tsinghua Sci & Tech*, **10**, (2005), p. 665.
62. D.W. Parker and G.L. Kutner, *Adv. Mater. Process*, **4**, (1991), p. 68.
63. D. R. Clarke and S. R. Phillpot, *Materials Today*, **8**, (2005), p. 22.
64. H. Herman, S. Sampath and R. McCune, *MRS Bull*, **25**, (2000), p. 17.
65. M.M. Helali and M.S.J. Hashmi, Proc. of the 10th Conference of the Irish Manufacturing Committee (IMC 10) Galway, Ireland (1992), p. 377.
66. I.M. Kodintsev and S. Trassati, *Electrochim Acta*, **39**, (1994), p. 1803.
67. C. Comninellis and G.P. Vercesi, *J Appl Electrochem*, **21**, (1991), p. 335.
68. D.T. Shieh and B.J. Hwang, *Electrochim Acta*, **38**, (1993), p. 2239.
69. C. Iwakura and K. Sakamoto, *J Electrochem Soc*, **132**, (1985), p. 2420.
70. L.I. Krishtalik, *Elektrokhimiya*, **15**, (1979), p. 462.
71. R. Kötz and S. Stucki, *Electrochim Acta*, **31**, (1986), p. 1311.
72. V.V. Gorodetskii, V.A. Neburchilov and M.M. Pecherskii. *Russian J Electrochem*, **30**, (1994), p. 916.
73. O. Simond, V. Schaller and Ch. Comninellis, *Electrochim Acta*, **42**, (1997), p. 2009.

74. J. Horacek and S. Puschaver, *Chem Eng Prog*, **67**, (1971), p. 71.
75. S. Trasatti, G. Lodi (Eds.), Electrodes of Conductive Metallic Oxides, Part A and Part B, Elsevier, Amsterdam, 1980 and 1981, S. Trasatti. In: J. Lipkowski and P.N. Ross, Editors, The Electrochemistry of Novel Materials, VCH Publishers (1994), p. 207.
76. R. Kötz, S. Stucki, D. Scherson and D.M. Kolb, *J Electroanal Chem*, **172**, (1984), p. 211.
77. G. Faita and G. Fiori, *J Appl Electrochem*, **2**, (1972), p. 31.
78. S. Trasatti, *Electrochim Acta*, **36**, (1991), p. 225.
79. A. De Battisti, G. Lodi, M. Cappadonia, G. Battalin and R. Kötz. *J Electrochem Soc*, **136** (1989), p. 2596.
80. N. Krstajic and S. Trasatti, *J Electrochem Soc*, **142**, (1995), p. 2675.
81. C. Angelinetta, M. Faciola and S. Trasatti, *J Electroanal Chem*, **205**, (1986), p. 347.
82. R. Hutchings, K. Müller, R. Kötz and S. Stucki, *J Mater Sci*, **19**, (1984), p. 3987.
83. C.C. Hu Lee and C.H. Wen, *J Appl Electrochem*, **26**, (1996), p. 72.
84. C.h. Comnimellis and G.P. Vercesi, *J Appl Electrochem*, **21**, (1991), p. 335.
85. H.B. Beer, Br. Patent, 1, 147, (1965), p. 442.
86. U. Kamachi Mudali, R.K. Dayal and J.B. Gnanamoorthy, *Nucl Tech*, **100**, (1992), p. 395.
87. U. Kamachi Mudali, V.R. Raju and R.K. Dayal, *J Nucl Mat*, **277**, (2000), p. 49.

CHAPTER 6

Electro- and Electroless Plated Coatings for Corrosion Protection

T.S.N. Sankara Narayanan[1] and S.K. Seshadri[2]

[1]National Metallurgical Laboratory, Madras Centre, CSIR Complex, Chennai-600 113, India

[2]Department of Metallurgical and Materials Engineering, Indian Institute of Technology Madras, Chennai-600 036, India

ELECTRODEPOSITION

Fundamentals of Electrodeposition

Electrodeposition can be defined as the deposition of an adherent metallic coating for the purpose of securing a surface with properties or dimensions different from those of the base material. Electrodeposits are applied to metal substrates for decoration, corrosion resistance, wear resistance, electrical properties, magnetic properties, solderability, etc. The ability of being decorative as well as protective to substrates distinguishes electrodeposited coatings from metallic coatings applied by hot dipping or spraying. The metallic luster, achieved by bright plating or by polishing after plating, lends a distinctive appearance not provided by other types of coatings. However, corrosion resistance is the primary reason for the use of electrodeposited coatings [1-5].

Metal deposition

In electrodeposition of metals, a direct current is made to flow between two electrodes immersed in a conductive aqueous solution containing the metal salts to be deposited. The flow of current causes one of the electrodes (the anode) to dissolve and the other electrode (the cathode) to become covered with the deposited metal. The metal to be deposited is present as positively charged ions. When the current flows, the positively charged ions react with the electrons and are converted to metallic atoms at the cathode surface. The reverse occurs at the anode where the metal is dissolved to form positively charged metal ions which enter the solution. The schematic of an electrolytic cell employed for electrodeposition of metals is shown in Fig. 1.

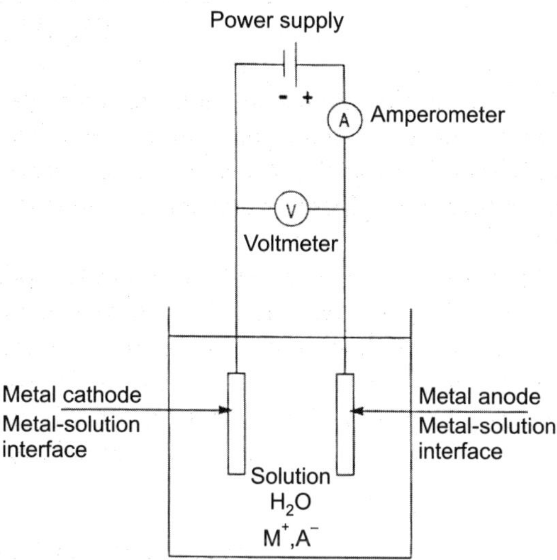

Figure 1 Schematics of an electrolytic cell for plating metal "M" from a solution of the metal salt "MA" (Source: http://electrochem.cwru.edu/ed/encycl - Electrochemistry Encyclopedia)

The Faraday's laws of electrolysis describe the basics of the electrodeposition of a metal as follows:

Faraday's First Law: *The mass, m, of an element discharged at an electrode is directly proportional to the amount of electrical charge, Q, passed through the electrode.*

Faraday's Second Law: *If the same amount of electrical charge, Q, is passed through several electrodes, the mass, m, of an element discharged at each will be directly proportional to both the atomic mass of the element and the number of moles of electrons, z, required to discharge one mole of the element from whatever material is being discharged at the electrode. Another way of stating this law is that the masses of the substances reacting at the electrodes are in direct ratio to their equivalent masses.*

By combining the principles of Faraday with an electrochemical reaction of known stoichiometry, the Faraday's laws of electrolysis that relates the charge density (charge per unit area), q, to the mass loss (per unit area), "m can be written as:

$$\Delta m = \frac{q(AM)}{nF}$$

The amount of metal deposited at the cathode and the amount of metal dissolved at the anode are directly proportional to the product of the current and time:

$$M = 1.095(a)\ (I)\ (t)$$

Where 'm' is the amount of metal deposited at the cathode (or dissolved at the anode) in grams, 'I' is the current that flows through the plating tank in amperes, 't' is the time that the current flows, in hours and 'a' is the current efficiency ratio. The proportionality constant (1.095) grams per ampere.hour equals M/nF where M is the atomic weight of the metal deposited, 'n' is the number of electrons involved in the electrochemical reaction and 'F' is the Faraday's constant (96500 coulombs).

Surface preparation for electrodeposition

Degreasing processes which remove the organic films, greases, etc. and pickling processes which remove dirt, oxide and corrosion products are the pretreatments employed before actual electroplating. Hydrocarbon or mineral oils are removed by solvents; vapour degreasing with chlorinated solvents or emulsification are common alternatives. Greases of animal or vegetable origin are removed by additions of hot aqueous solutions of alkalis. This is called as alkaline degreasing. Electrolytic alkaline degreasing is considerable faster than soak cleaning. Alkaline cleaning solutions are prepared from sodium hydroxide, trisodium phosphate, sodium silicate, sodium carbonate, soaps, detergents, etc. The higher the pH the more effective is the degreasing action, but with non-ferrous metals the greater is the danger of corrosion.

Oxide and corrosion-product films are removed by dissolution in aqueous solutions. Hydrochloric and sulphuric acid are the most common aqueous media used to remove oxides. Concentrations and temperatures are varied according to the substrate. Mixed acids containing wetting agents are supplied as proprietary mixtures. Electrolytic picking is used for special purposes, but hydrochloric acid is unsuitable for this because of its volatility and evolution of chlorine gas. Cathodic pickling of steel in 10-20 wt.% sulphuric acid enables thick rust or scale to be removed without the loss of metal. Anodic pickling in 55 wt.% sulphuric acid is used to remove a thin surface layer from steel. A high current density is used. After 10-20 seconds, the metal becomes passive and dissolution ceases. Disordered and fragmented metal produced by abrasion or machining is removed to leave a surface which favours good adhesion of an electrodeposit. Oxide and corrosion products on copper alloys are removed in hydrochloric or sulphuric acids, less concentrated than used for steel.

Types of plating baths

Table 1 gives the various types of electrolyte solutions used for electrodeposition of metals. Among them, the simple acid solutions are less widely used. Although their polarization characteristics are low, they are less conducive to yield bright electrodeposits than those obtained from the complexant solutions. The complexant solutions offer increased metal solubility and give rise to higher limiting current densities.

The various constituents of the electrolyte solutions used for electrodeposition of metals are listed in Table 2. The metal salts are the source of metal ions. The complexant is vital to stabilize the metal ion in an appropriate form. The plating baths are usually designed to operate with excess of complexant to simplify bath control; depletion of complexant will give rise to greater amounts of uncomplexed metal ion in the solution and this may even arrest the process of electrodeposition. The complexant stabilizer may simply be excess complexant or a pH buffer. In plating baths containing cyanide, it is essential to maintain the pH > 8 with excess alkali to prevent the evolution of the detrimental HCN gas. In some cases a critical pH exists for optimum operation or optimum deposit properties. In Watts nickel plating where solutions have pH between 3.5 and 5, buffering with borate or boric acid is necessary. Depassivators like chloride and to a lesser extent tartarate or citrate are added to maintain smooth dissolution for soluble anodes. In the case of insoluble anodes, such depassivators must be eliminated. To

Table 1 Various types of electrolyte solutions used for electrodeposition of metals

Type	Metal that is getting plated	Electrolyte	Ions in the electrolyte
	Cu	$CuSO_4$	Cu^{2+}
Simple acid	Sn	$Sn(BF_4)_2$	Sn^{2+}
	Ni	$Ni(NH_2SO_2O)_2$	Ni^{2+}
	Cr	H_2CrO_4	CrO_4^{2-}
Complexed acid	Cu	$K_6Cu(P_2O_7)_2$	$Cu(P_2O_7)_2^{6-}$
	Ni	$NiCl_2$	$NiCl_4^{2}$
	Sn	SnF_4	SnF_6^{2-}
Alkaline hydroxide	Sn	K_2SnO_4	SnO_3^{2-}
(amphoteric)	Zn	$Zn(OH)_2$	ZnO_2^{2-}
Complexed alkaline	Cu	CuCN	$Cu(CN)_2^-$

Table 2 Various constituents of the electrolyte solutions used for electrodeposition of metals

Constituent	Functions	Example
Metal salt	To supply metal ions	$CuSO_4$, $NiCl_2$
Complexant	Stabilize metal in soluble form; affect mode of ion discharge	KCN, NaOH
Complexant stabilizer	Excess complexant, acid or alkaline reagent	KCN, NaOH
Buffer salt	Stabilize pH at optimum value	H_3BO_3, Na_2HPO_3
Anode depassivator	Smooth anode dissolution and oxide film disruption	Chloride
Addition agents	Surfactant for hydrogen bubble dispersal, Brightener for bright deposits, Leveler for grain refinement Stress reliever	Detergent, Organic S and N containing compounds

control the properties of the deposit, the special addition agents are most important. Surfactants may be needed to minimize hydrogen absorption or pitting caused by bubbles. Organic additives bring about significant change in the mechanical properties.

The incorporation of byproducts from the additives has only a small effect on mechanical properties of the electrodeposits. However it severely alters the grain size and mode of growth of the electrodeposit resulting in a small-grained, smooth and bright surface. The mode of deposition is affected because the efficacy of an additive may often be indicated by its effect on the activation polarization of the cathode discharge reaction. A more highly polarized reaction (higher activation overpotential) for a given current will yield a brighter deposit. This general rule is valid for complexed-ion solutions also where polarization is high and deposits are brighter. By giving proper attention to such general principles, it is possible to virtually design an electrolyte for depositing any metal.

Throwing power

Metal distribution is influenced by cathodic polarization, the cathode efficiency-current density relationship and the electrical conductivity of the solution. The complex relationship between these factors that influence current distribution and hence metal distribution is called as the throwing power. A plating solution with a high-throwing power is capable of depositing almost equal thickness on both recessed as well as in prominent areas.

Leveling and microthrowing power

The ability of an electroplating solution to fill in the defects and scratches present on the surface preferentially is called leveling. Organic additives present in the plating bath adsorb preferentially at micropeaks, the resulting increase in the local resistance to current flow increases the current density in microgrooves, thereby promoting leveling. Micro-throwing power refers to the ability of an electroplating solution to fill tiny crevices with deposits that follow the contour of a defect or scratch without any leveling action whatsoever.

Electrode reactions

In an ideal case, only metal ions will be discharged at the cathode and they will be completely reduced to atoms which become part of the growing electrode surface. However, many commercial processes are not ideal and many cathodic and anodic reactions do occur. The possible cathodic and anodic reactions that could occur during electroplating are given in Table 3.

Current efficiency

When two or more reactions occur simultaneously at an electrode, the number of coulombs of electricity passed corresponds to the sum of the number of equivalents of each reaction. The current efficiency for the deposition of a metal 'j' is defined as:

$$\text{Current efficiency (CE)} = \frac{w_j}{w_{total}}$$

Table 3 Electrode reactions that occur during electrodeposition

Type of reactions	Actual reaction	Remarks
Cathode reactions	$M^{n+} + ne^- \rightarrow M^0$	Main metal discharge andDeposition reaction
	$2H^+ + 2e^- \rightarrow H_2$	Main secondary reaction causing poor cathodic current density; may cause poor deposit properties
	$M^{n+} + e^- \rightarrow M^{(n-1)+}$	Partial ion reduction and source of inefficiency
Anode reactions	$M^0 \rightarrow M^{n+} + ne^-$	Metal dissolution at soluble anodes; maintains metal ion conc. in solution
	$2OH^- \rightarrow H_2O + ½ O_2 + e^-$	Oxygen evolution is the main reaction on insoluble anodes in basic medium
	$H_2O \rightarrow 2H^+ + ½ O_2 + 2e^-$	Oxygen evolution is the main reaction on insoluble anodes in neutral medium
	$M + H_2O \rightarrow MO + 2H^+ + 2e^-$	Passive oxide formation on anode; raises total cell resistance and inhibits anode dissolution
	$M + nOH^- \rightarrow MOH + ne^-$	Passive hydroxide film formation
	$M^{(n-1)+} \rightarrow M^{n+} + e^-$	Ion oxidation - removal of lower valency metal ions

where w_j is the weight of the metal 'j' actually deposited and w_{total} is that which would have been deposited if all the current had been used for depositing the metal 'j'. For example, during the electrodeposition of copper from a solution of cupric nitrate in dilute nitric acid, three cathodic reactions, namely, the deposition of copper, reduction of nitrate and reduction of hydrogen ions, occur. Hence, the current efficiency of the deposition process will be lesser than 100%, where the remainder of current is used for the side or competing reactions.

Growth mechanisms

The metal ion in plating solution is solvated. As the solvated meal ions approach towards the cathode, they lose some of their solvation sheath and get adsorbed on the cathode. The adion diffuses over the cathode surface until it reaches an atomic step. The adion loses more water of salvation and its freedom is reduced to diffusion along the step. Further desolvation and co-ordination follows when it reaches a kink in the step, at which stage it is immobilized. When other adions following this path eventually join and submerge the first, co-ordination with water in the electrolyte is exchanged fully for co-ordination with metal ions in the metallic lattice.

There are two basic mechanisms for the formation of a coherent electrodeposit [6]: layer growth and three-dimensional crystallites growth. In the layer growth mechanism, a crystal enlarges by spreading of discrete layers, one after another across the surface. In this case, a growth layer or step is the structural component of the coherent deposit. In 3D-crystallites growth mechanism, the structural components are 3D-crystallites built-up as a result of coalescence of these crystallites. The growth sequence of 3D-crystallites growth mechanism consists of four stages: (i) formation of isolated nuclei and their growth to 3D-crystallites; (ii) coalescence of 3D-crystallites; (iii) formation of linked network; and (iv) formation of a continuous deposit.

Role of additives

The use of additives in aqueous electroplating solutions is extremely important as they offer potential benefits which include brightening the deposit, reducing grain size, reducing the tendency to tree, increasing the current density range, promoting leveling, changing mechanical and physical properties, reducing stress and reducing pitting. The additives, when added in small concentrations make striking effects on electrocrystallization processes. Their adsorption on a high energy surface and deposition on growth sites produce a poisoning or inhibiting effect on the most active growth sites [7]. The additives can be organic or metallic, ionic or nonionic, and are adsorbed on the plated surface and often incorporated in the deposit.

Influence of additives on leveling and brightening

Normal electrodeposition accentuates roughness by putting more deposit on the peaks than in the valleys of a plated surface since the current density is highest at the peaks because the electric field strength is greatest in this region. In order to produce a smooth and shiny surface, more metal has to be deposited in the valleys than on the peaks, which is the opposite of the normal effect. The function of certain organic compounds is to produce this leveling in plating solutions. Leveling agents are adsorbed preferentially on the peaks of the substrate and inhibit deposition. An example is coumarin which is used in the deposition of nickel.

It adsorbs on depositing nickel by the formation of two carbon-nickel bonds and inhibits nickel deposition probably by a simple blocking action.

A bright deposit is one that has a high degree of specular reflection in the as-plated condition. Although brightening and leveling are closely related, many solutions capable of producing bright deposits have no leveling ability [1]. If the substrate is bright prior to plating, almost any deposit that is plated on it will be bright, as long as it is thin enough. However, a truly bright deposit will be bright over a matte substrate and it will remain bright even when it is thick enough to hide the substrate completely. Plating solutions without addition agents seldom or never produce bright deposits.

Classification and types of additives

Additives can be classified into four major categories: (i) grain refiners; (ii) dendrite and roughness inhibitors; (iii) leveling agents; and (iv) wetting agents or surfactants [6]. Typical example of grain refiners are cobalt or nickel codeposited in trace amounts in gold deposits. Dendrite and roughness inhibitors adsorb on the surface and cover it with a thin layer which serves to inhibit the growth of dendrite precursors. This category includes both organic and inorganic materials with the latter being more stable. Leveling agents, such as coumarin or butynediol in nickel solutions, improve the throwing power of the plating solution mostly by increasing the slope of the activation potential curve. The prevention of pits or pores in the deposit is the main purpose of wetting agents or surfactants [7].

Additives can act as grain refiners and levelers because of their effects on electrode kinetics and the structure of the electrical double layer at the plating surface [8]. Since additives are typically present in extremely small concentrations, their transport towards the electrode is nearly always under diffusion control and, therefore, quite sensitive to flow variations. The effects of additives are often manifested by changes in the polarization characteristics of the cathode. Many are thought to function by adsorption on the substrate or by forming complexes with the metal. This results in development of a cathodic overpotential which is maintained at a level which allows the production of smooth, non-dendritic plates having the desired grain structure [8].

Mechanisms of additives

Numerous mechanisms have been suggested to explain the behavior of additives: (i) blocking the surface; (ii) changes in Helmholtz potential; (iii) complex formation including induced adsorption and ion bridging; (iv) ion pairing; (v) changes in interfacial tension and filming of the electrode; (vi) hydrogen evolution effects; (vii) hydrogen absorption; (viii) anomalous codeposition; and (ix) the effect on intermediates. These are discussed in detail in a comprehensive review by Franklin [9]. An excellent coverage on the mechanism of levelling and brightening caused by the addition agents during plating is given by Oniciu and Muresan [10].

Properties of electrodeposited coatings

Structure of electrodeposited coatings

The properties of all materials are determined by their structure. Even minor structural differences often have profound effects on the properties of electrodeposited metals [11]. The four typical structures encountered with electrodeposited metals include; (i) columnar; (ii) fibrous; (iii) fine-grained; and (iv) banded [12].

Columnar structure

Columnar structures are characteristic of deposits obtained from simple ion acidic electrolyte solutions that do not contain any addition agents, e.g., copper, zinc, or tin from sulfate or fluoborate solutions, operated at elevated temperature or low current density. Deposits of this type usually exhibit lower strength and hardness than other structures but high ductility.

Fibrous structure

Fibrous (acicular) structures represent a refinement of columnar structure. This type of structure is obtained under conditions that favor the formation of new nuclei rather than growth of existing grains, such as the presence of addition agents, or use of low temperature and high current density. Properties of fibrous deposits are intermediate between columnar and fine-grained deposits.

Fine-grained structure

Fine-grained structures are usually obtained from complex ion solutions such as cyanide or with certain addition agents. These deposits are less pure, less dense and exhibit higher electrical resistivity due to the presence of codeposited foreign material. Deposits from simple ion acidic solutions, such as copper or nickel from sulfate solutions, develop this type of structure if the operating conditions are more extreme, such as, at very high current density or a higher pH [12]. The grain size of this type deposits is of the order of 10^{-5} to 10^{-6} cm. These deposits are usually relatively hard, strong and brittle.

Laminar structure

Laminar (or banded) structures are characteristic of bright deposits in presence of addition agents such as sulfur containing organic compounds in which the grains within the lamellae are extremely small. Gold-copper, cobalt-phosphorus, cobalt-tungsten, and nickel-phosphorus alloy deposits exhibit this type structure. These deposits usually have high strength and hardness but low ductility. Similar laminations can also be found in deposits obtained by pulse plating or periodic reverse current plating.

The crystal structure of electrodeposits strongly depends on the relative rates of formation of crystal nuclei and the growth of existing crystals [13]. Finer-grained deposits result under conditions that favor crystal nuclei formation while larger crystals are obtained under conditions that favor growth of the existing crystals. In general, factors which increase the cathode polarization enable a decrease in crystal size [13]. Fig. 2 shows a pictorial representation of the influence of plating variables on the grain size of electrodeposits [14].

Influence of the substrate

The structure of most electrodeposits is determined by epitaxial and pseudomorphic growth onto a substrate and by the conditions prevailing during deposition. Typically, a deposited metal will try to copy the structure of the substrate and this involves epitaxy, which occurs when definite crystal planes and directions are parallel in the deposit and substrate, respectively [2]. Epitaxy is the orderly relation between the atomic lattices of substrate and deposit at the interface, and is possible if the atomic arrangement in a certain crystal direction of the deposit matches that in the substrate. Another term, pseudomorphism, refers to the continuing of grain boundaries and microgeometrical features of the cathode substrate into the overlying deposit. A deposit stressed to fit on the substrate is said to be pseudomorphic [2]. Pseudomorphism persists longer than epitaxy.

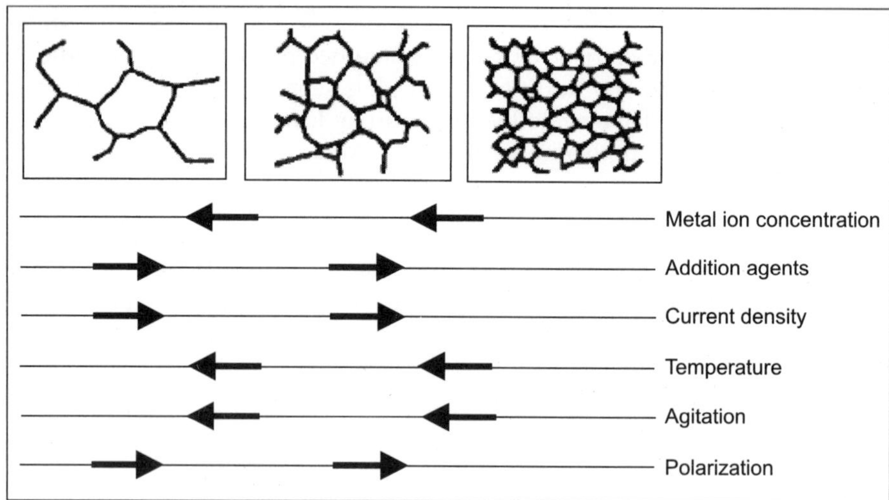

Figure 2 Pictorial representation of the influence of plating variables on the change in grain size of electrodeposits (adapted from Dini [14])

Texture of electrodeposited coatings

Texture, the preferred distribution of grains having a particular crystallographic orientation with respect to a fixed reference frame, is an important structural parameter for bulk materials and coatings [15, 16]. The texture of electroplated coatings can be markedly influenced by solution composition and operating conditions. Depending on pH and current density, five different textures can be obtained from a Watts nickel plating solution [17]. Texture is completely independent of the substrate orientation for thick deposits. Texture can influence a variety of functional properties of electrodeposits.

Mechanical properties

Moduli of elasticity

The moduli of elasticity of electrodeposits are generally smaller than those of the same metal formed in other ways. Irrespective of the method of formation, the moduli should be the same. The smaller moduli of elasticity of electrodeposits could possibly be due to the difficulty of obtaining accurate values and the possibility that the electrodeposits do not behave elastically.

Tensile strength

The tensile strength of electrodeposited metals depends on the operating conditions employed for electrodeposition of the individual metals. In many cases, the electrodeposit is two or three times as strong as the corresponding wrought metal. The primary reason for the higher strength of the electrodeposits when compared with their wrought counterparts is due to the fine grain size of the electrodeposits. The tensile strength of electrodeposited cobalt is more than four times the strength of annealed, wrought cobalt whereas some types of electrodeposited chromium are nearly seven times as strong as cast or sintered chromium.

The type of additives used in the plating bath and codeposition of impurities could significantly influence the tensile strength of electrodeposits. The tensile strength of electrodeposited nickel coating varies from 39 to 250 MPa, depending up on the type additives used in the plating bath. In a copper pyrophosphate plating bath, addition of lower concentrations of dimercaptothidiazole (DTMD) results in copper deposits with low tensile strength. However, when the concentration of DTMD is higher, the formation fine-grained copper deposits enables an increase in the tensile strength. Small amounts of co-deposited carbon in nickel and tin-lead electrodeposits influence the tensile strength. Increasing the carbon content of a sulfamate nickel electrodeposit from 28 to 68 ppm increased the tensile strength from 575 to 900 MPa. Similarly, increasing the carbon content of the tin-lead electrodeposit from 125 to 700 ppm increased the tensile strength from 29 to 41 MPa.

Ductility

Ductility is the amount of plastic deformation that can occur prior to fracture. Fine grained electrodeposits tend to be brittle because the plastic deformation which occurs primarily by dislocation motion is impeded by the grain boundaries. The ductility of electrodeposits, as evidenced by their percent elongation, is higher when they are attached to the substrate as they cannot exhibit highly localized plastic deformation prior to fracture. The lower ductility of electrodeposited free standing foils is probably caused by the more severe local plastic deformation, also called as necking, in the region where fracture occurs subsequently. The free standing foils of electrodeposited nickel and electroless copper have been found to exhibit severe necking. When the deposit adheres well to the substrate, necking is essentially prevented. The additives present in the plating bath have a significant influence of the ductility of the resultant deposits. In a copper pyrophosphate plating bath, addition of lower concentrations of dimercaptothidiazole (DTMD) results in copper deposits with high ductility. However, when the concentration of DTMD is very high, the inclusion of excessive additive in the copper deposit causes the ductility to decrease.

Superplasticity

Superplasticity refers to large tensile elongations, typically 500%, that can be achieved in polycrystalline materials under certain conditions of strain rate and temperature. One of the main requirements for superplasticity is the presence of an ultrafine, equiaxed microstructure, typically 1 to 5 μm diameter that remains stable while being deformed at the superplastic temperature (usually around one-half the melting point). As many electrodeposited metals possess a smaller grain size, electrodeposition offers an exciting prospect for fabricating complex parts by combining a pre-forming by electrodeposition and final forming of the internal structure by superplastic deformation. The two electrodeposited alloys which meet these requirements and, therefore, exhibit superplastic behavior is Cd-Zn and Ni 40-60% Co. Although a number of alloys exhibit superplastic behavior, lead-tin, copper-nickel, and cadmium-tin are of particular interest since these alloy coatings can be deposited from aqueous solution. The growing field of electrodeposition of multilayer coatings by cyclic modulation of the cathodic current or potential during deposition also offers promise for the production of new superplastic alloys. Composition-modulated alloys (CMA) which have been produced by this process include Cu-Ni, Ag-Pd, Ni-Nip, Cu-Zn and Cu-Co.

Residual stress

Residual stress may be defined as a stress within a material which is not subjected to load or temperature gradients yet remains in internal equilibrium [5]. Residual stress in electrodeposits refers to forces created within the deposit as a result of the electrocrystallization process and/or the codeposition of impurities such as hydrogen, sulphur and other elements. The residual stress is either tensile (contractile) or compressive (expansive) in nature. In tensively stressed deposits, the average distance between atoms in the lattice is greater than the equilibrium value, creating a force that tends to drive the atoms closer together. In compressively stressed deposits, the atoms are closer together and the force tends to drive them further apart. Dislocation theory provides a logical explanation of the origin of residual stress in electrodeposits [1].

Residual stresses in electrodeposited coatings can cause adverse effects on its properties. They may be responsible for peeling, tearing, and blistering of the deposits; they may result in warping or cracking of deposits; they may reduce adhesion, particularly when parts are formed after plating and may alter properties of plated sheet. Stressed deposits can be considerably more reactive than the same deposit in an unstressed state. Occasionally stress may serve a useful purpose. For example, in the production of magnetic films for use in high speed computers, stress in electrodeposited iron, nickel, and cobalt electrodeposits will bring about preferred directions of easy magnetization and other related effects [22].

There are a variety of steps that can be taken to minimize stress in deposits: (i) choice of substrate; (ii) choice of plating solution; (iii) use of additives; and (iv) use of higher plating temperatures. Typically, with most deposits, there is a high initial stress associated with lattice misfit and with grain size of the underlying substrate. This is followed by a drop to a steady state value as the deposit increases in thickness. With most deposits this steady state value occurs in the thickness regime of 12.5 to 25 μm. Atomic mismatch between the coating and substrate is a controlling factor with thin deposits. The type of anion in the plating solution can leave a marked influence on residual stress. For example, sulfamate ion provides nickel deposits with the lowest stress, followed by bromide [22]. There are numerous additives, particularly organic, which have a marked effect on the stress produced in deposits. Small quantities (0.01 to 0.1 g/l) of most sulfur bearing compounds rapidly reduce the stress in nickel deposits. When a sulfur compound reduces internal stress compressively, the resulting deposit is brighter. Increasing the plating solution temperature can also help to reduce the stress.

Adhesion

Adhesion of a coating or deposit is defined as the extent to which it sticks to the substrate. Adhesion can be regarded as a measure of the degree to which a bond has been developed from place to place at the coating-substrate interface. Good adhesion is often has no value unless the bond strength is satisfactory. If the electrodeposit separates from the substrate then the adhesion is poor and vice-versa. The adhesion of electrodeposits on metallic substrates is generally better whereas plating on plastics presents a serious adhesion problem. The main factors that are detrimental to adhesion are brittle layers, which can form by diffusion between the deposit and the substrate, especially if annealing is involved.

Porosity

Porosity is one of the main sources of discontinuities in electroplated coatings; the others are cracks arise due to high internal stress. In most cases porosity is considered undesirable as the pores can expose substrate underneath to corrosive agents, reduce mechanical properties and, deleteriously influence density, electrical properties and diffusion characteristics of the deposits. Kutzelnigg [18] suggests that pores may be broken down into two main categories, transverse pores and masked or bridged pores. Transverse pores may be either of the channel type or hemispherical and extend through the coating from the basis metal to the surface of the deposit. They may be oriented perpendicular or oblique to the surface or may have a tortuous shape. Masked or bridged pores do not extend through the coating to reach the surface but either starts at the surface of the basis metal and become bridged or start within the coating and become bridged. Cracks may be regarded as pores much extended in a direction parallel to the surface, but they can also be divided into transverse cracks, enclosed cracks and surface cracks.

The porosity of electrodeposits depends on (i) the nature, composition and history of the substrate surface prior to plating; (ii) composition of the plating solution and its manner of use; and (iii) post plating treatments such as polishing (abrasive or electrochemical) and damage due to wear, deformation, heating and corrosion [19]. At low deposit thickness, porosity of electrodeposited films is largely controlled by the surface condition and characteristics of the underlying substrate. This condition persists up to a limiting thickness, after which the properties of the film itself, primarily crystallographic properties, determine the rate of pore closure [20]. The porosity of electrodeposits drops exponentially with increase in their thickness [21]. A convenient and often very effective way to modify the substrate and reduce porosity is to use an underplate.

Hardness

Hardness is one of the most important properties of electrodeposits, which is also indicative of the strength and ductility as well as the wear resistance of the deposits. The hardness value of electrodeposits is only of limited value unless it is experimentally related to the strength or the ductility of the specific material because for electrodeposits even the qualitative relationship that is commonly observed among hardness, tensile strength and ductility, do not always prevail. For example, it is usually expected that the hardness increases with the tensile strength and decreases with ductility. However, the reverse effect is also common in electrodeposits.

Hall-Petch equation relates the grain size, d, with the hardness, H, of the deposit as follows:

$$H = H_0 + K_H d^{-1/2}$$

where H, and K, are the experimental constants and are different for each metal. The value of H is characteristic of dislocation blocking and is related to the friction stress. K_H takes account of the penetrability of the boundaries to moving dislocations and is related to the number of available slip systems. The Hall-Petch equation has been found applicable for several polycrystalline materials and also for electrodeposited iron, nickel and chromium. Electrodeposited nickel exhibits Hall-Petch relation over a wide range of grain sizes from 12,500 nm down to 12 nm. A microhardness of approximately 700 kg/mm^2 was obtained with the smallest grain size.

Wear resistance

Wear, defined as the damage to the surface of a substrate due to the relative motion between that substrate and a contacting substrate or substances is a complex phenomenon and usually involves a progressive loss of material. The parameters that influence wear include: surface hardness and finish, microstructure and bulk properties, contact area and shape, type of motion, its velocity and duration, temperature, environment, type of lubrication and coefficient of friction. Electrodeposits offer some distinguishing features for wear resistant applications. Chromium plating is more extensively used for wear resistance applications than any other electrodeposited coating. Typical uses include roll surfaces, shaft sleeves, pistons, internal combustion engine components, hydraulic cylinders, landing gear and machine tools. Although the thickness varies with the application, it is usually in the range of 20 to 500 μm. Chromium plating is very effective in reducing the wear of piston rings caused by scuffing and abrasion. The average life of a chromium-plated ring is approximately five times than that of the bare ring made of the same base metal. Hard chromium plating exhibits better resistance to low stress abrasion than hard anodized aluminum and heat treated electroless nickel. Electrodeposited chromium coatings also offer excellent scuffing resistance (evaluated by Falex test) compared to that of electro- and electroless deposited nickel. However, crack-free chromium coating exhibit high wear rate, which is related to its crystal structure. Crack-free chromium has a predominantly hexagonal close-packed crystal structure unlike conventional chromium which is body centered cubic. HCP metals tend to slip on only one family of slip planes, those parallel to the basal plane. This results in larger strains at a given stress level and less dislocation interactions. In addition, the strain-hardening rate is low, leading to rapid localization of deformation, early fracture and an increased wear rate.

Electrodeposited coatings are typically not effective for applications requiring wear resistance at temperatures above 500 to 600°C. Heat treatment of electrodeposited chromium coatings at 800°C for 1 hour drastically affects its hardness from 900 to 450 kg/mm^2, which results in increasingly higher wear rate. However, electrodeposited composite coatings offers promise for higher temperature applications. Co-30Vol. % Cr_2C_3 composite coatings offer better wear resistance of aircraft engines at temperatures up to 800°C. The improved wear resistance is due to the formation of cobalt oxide glaze on the load bearing contact areas during interfacial motion. Electrodeposited composition modulated alloy coatings exhibit novel and interesting wear properties. The layer microstructure of these coatings provides internal barriers to wear damage and increases the wear resistance. Compositionally modulated Ni-Cu coating with alternate layers of nickel and copper having an equal layer spacing of either 10 or 100 nm on AIS1 type 52100 steel is found to be more wear resistant than either pure nickel or copper deposits.

Advantages and disadvantages of electrodeposition

Electrodeposition process is simple, versatile and cost-effective. It is a viable approach for preparing, coatings having a very high electrical conductivity that is commonly used in electronic circuitry, sacrificial anodic coatings that protect the base metal from corrosion, coatings with controllable thicknesses which are amenable to welding and soldering and, coatings with high hardness and wear resistance.

Some substrates like cast iron and some substrate conditions like porosity require special plating procedures. Large metal structures are beyond the capabilities of electrodeposition. The design and geometry of the parts might create problems to obtain uniform coating thickness.

ELECTRODEPOSITION OF ALLOYS

Alloy deposition is subjected to the same principles as single metal plating. The deposition potential of a given metal is determined by the standard electrode potential (E^0) and ionic activity. The ionic activity is proportional to the ionic concentration. The electrodeposition of an alloy requires co-deposition of two or more metals. For achieving the deposition of an alloy, the metal ions must be present in the electrolyte solution that provides a cathode film where their individual deposition potential can be brought close to one another or become identical. If the deposition potential differs a lot then the only way to achieve deposition of the alloy is by controlling the activity of the ions by changing their concentration [23]. Alloy deposition often provides deposits with properties not normally obtained by employing electrodeposition of single metals.

The three main stages of alloy deposition are:

Ionic migration: The hydrated ions in the electrolyte migrate toward the cathode under the influence of the applied potential as well as through diffusion and/or convection.

Electron transfer: At the cathode surface are, the hydrated metal ions enter the diffusion double layer where the water molecules of the hydrated ion are aligned by the field present in this layer. Subsequently the metal ions enter the fixed double layer where because of the higher field present, the hydrated shell is lost. Then on the cathode surface, the individual ion may be neutralized and is adsorbed.

Incorporation: The adsorbed atom wanders to a growth point on the cathode and is incorporated in the growing lattice.

There are five types of alloy plating systems: (i) regular co-deposition; (ii) irregular co-deposition; (iii) equilibrium co-deposition; (iv) anomalous co-deposition; and (v) induced co-deposition.

Regular co-deposition

Regular co-deposition is characterized by deposition under diffusion control. The effects of plating variables on the composition of the deposits are determined by the changes in the concentrations of metal ions in the cathode diffusion layer and are predictable from simple diffusion theory. The percentage of the more noble metal in the deposit is increased by increasing the total metal content of the bath, decrease in current density, elevation of bath temperature and increased agitation of the bath. The effect of current density on the composition of alloys deposited from regular alloy plating systems consistently results in a decrease in the content of the more noble metal in the deposit. At low current density the deposit consists of the unalloyed noble metal; increase in the current density enables an increase in the content of the less noble metal.

Irregular co-deposition

Irregular co-deposition is mostly controlled by chaotic potentials of the metals rather than by the diffusion phenomenon. Irregular codeposition is most likely to occur with solutions of complex ions, particularly with systems in which the static potentials of the parent metals are markedly affected by the concentration of the complexing agents. The relation between current density and alloy composition in irregular co-deposition is complicated and usually unpredictable. The lack of a consistent trend in the content of the more noble metal in the deposit with current density stands in sharp contrast to the consistent behaviour of regular codeposition.

Equilibrium co-deposition

Equilibrium co-deposition is characterized by deposition from a solution which is in chemical equilibrium with both parent metals. The equilibrium plating system is unique in that the ratio of the metals in the deposit (plated at low current density) is the same as their ratio in the bath. However, this equivalence exists only under static conditions (zero current density). Since electrodeposition is always accompanied by a departure of the electrode potential from the static potential, it cannot be expected that the two metals which were initially in static equilibrium in a solution will be in equilibrium with each other during codeposition. Therefore, it cannot be expected to obtain an alloy deposit having the same ratio as that of the bath, unless the deposition is carried out at such a low current density. At higher current densities, the composition of the deposit would be expected to depart to some extent from the equilibrium value.

Anomalous co-deposition

Anomalous co-deposition is characterized by the anomaly that the less noble metal deposits preferentially. With a given plating bath, anomalous codeposition occurs only under certain conditions of concentration and operating variables. Otherwise the codeposition falls under one of the other three types. The relation between alloy composition and current density is also unusual.

Induced co-deposition

Induced co-deposition is characterized by the deposition of alloys containing metals, such as molybdenum, tungsten or germanium which cannot be deposited alone. The relation between the composition of the deposit and the current density in induced codeposition is similar to that observed in irregular codeposition. The effect of current density is not large and there is no consistent trend of the content of the reluctant metal in the deposit with current density.

Some key issues in alloy deposition are as follows:

(i) If an alloy plating bath, which is in continuous operation, is replenished with two metals in a constant ratio, the ratio of metals in the deposit will approach and ultimately take on that value.

(ii) In alloy deposition, the ratio of the concentration of the more readily depositable metal to the other is smaller at the cathode-solution interface than in the bulk of the bath.

(iii) An increase in the metal percentage of a parent metal in an alloy plating bath results in an increase in its percentage in the deposit.

(iv) In the deposition of alloys from the normal alloy plating systems, the most fundamental mechanism is the tendency of the concentration of the metal ions at the cathode-solution interface to approach mutual equilibrium with respect to the two parent metals.

(v) A variation in a plating condition that brings closer together the potentials for the deposition of the parent metals increases the percentage of the less noble metal in the electrodeposited alloy.

Influence of process variables on electrodeposition of alloys

Deposition potentials of metals usually become nobler with increase in temperature, because polarization is decreased. Whether the deposition of the nobler or less noble metal is favoured depends on which deposition undergoes the largest polarization. An increase in temperature increases the concentration of metal in the cathode diffusion layer, because the rates of diffusion and of convection increase with temperature. This is the most important way by which temperature affects the composition of the electrodeposited alloys. Dense solution formed by the dissolution of the anode sinks to the bottom of the cell while the less dense solution from the cathode rises to the surface of the electrolytic cell. As the composition of the electrodeposited alloy is directly affected by the composition of the solution in contact, the dispelling of local variations in bath composition is more important in alloy plating than in deposition of single metal. Agitation of an alloy plating bath or rotation of the cathode can directly affect the composition of the alloy by reducing the thickness of the cathode diffusion layer. This is a purely mechanical action which does not change the electrochemical properties of the solution. Agitation has a more consistent influence on the composition of the deposit than either temperature or current density. Effect of pH on induced co-deposition is more complex than any other type of co-deposition of alloys. In induced co-deposition, the effect of pH exhibit a mixed trend of both an increase as well as a decrease in the alloy deposition.

Addition agents, used in low concentrations, do not affect the static potentials of the metal but usually increase the polarization associated with the deposition process. By using addition agents the dynamic potentials of more noble metals can be brought closer to the less noble metal, thereby making the co-deposition possible. Though the addition agents have some adverse effects like drastic reduction of the mechanical properties of the deposit, they play a very important role of bringing the deposition potentials of the constituent elements close and have a direct influence on the composition of the deposit. The concentration of the addition agent required to produce an appreciable effect is much smaller than that of a complexing agent. Addition agents are effective in baths containing metals as simple ions rather than as complexes.

PULSED CURRENT ELECTRODEPOSITION (PED)

Pulsed current electrodeposition (PED) is a method of depositing metal on a substrate using interrupted direct current [24-26]. The theory behind PED is very simple. The cathode film is kept as rich in metal ions as possible and as low in impurities as possible. During the period when the current is on, the metal ions next to the cathode are depleted and a layer rich in water molecules is left. During the portion of the cycle when the current is off, the metal ions from the bulk of the plating solution diffuse into the layer next to the cathode. Then the process is repeated. Also, during the time the current is off, gas bubbles and impurities that have adsorbed on the cathode have a chance to desorb [27, 28].

PED is accomplished with a series of current pulses of equal amplitude and duration in the same direction, separated by periods of zero current (Fig. 3). These pulses are often employed at a rate of 500 to 10,000 times per second. They favor the initiation of grain nuclei and greatly increase the number of grains per unit area. The intended result is a finer grained deposit with better characteristics and properties than conventionally plated coatings. The pulse rate (frequency) and 'ON' and 'OFF' times (duty cycle) are controllable to meet the needs of a given application. Typical on-time range from 0.1 to 9.9 ms, and typical off-times range from 1 to 99 ms. The physical and chemical properties of deposits can be controlled precisely through the careful selection of pulse-plating parameters.

In PED of metals the pulse on-time should be sufficiently short to stay well below the pulse limiting current density, but sufficiently long to fully charge the double layer. These two requirements determine the available parameter window. The upper limit depends heavily on the metal concentration in the electrolyte, while the lower limit is relatively independent of

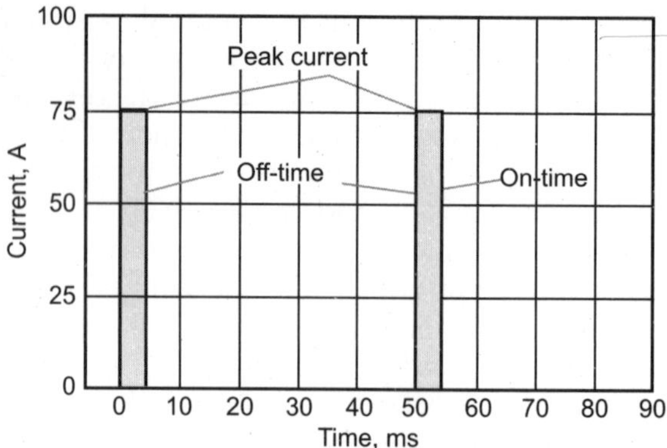

Figure 3 Constant-current pulse plating with an on-time of 5 ms, an off-time of 45 ms, and a peak current of 75 A (adapted from VanHorn [25])

concentration [29]. Electrolyte conductivity must be maintained at a high level to allow the peak pulse current to be completely effective. If the conductivity is not high enough, an excess voltage will be required to attain the desired peak current. Such peaks are power-inefficient and less effective. In periodic reverse plating, the polarity of a constant D.C. output is switched back and forth in a regular pattern. A comparison of the current patterns (ideal pattern) in conventional and periodic-reverse pulsed-current plating is given in Fig. 4. The duration of the current in each direction, called the forward and reverse envelopes, can be individually controlled from 0.1 milliseconds to 99.99 s. Within each envelope a square-wave pulse is generated. The frequency and the duration of the pulses are set independently for the forward and reverse envelopes (Fig. 5). Frequency range is from 10 to 9,999 Hz. Duty cycle settings in percentages determine 'ON' & 'OFF' for each pulse.

The average current density i_m in pulse and pulse reverse plating is given by the pulse current density i_p, the (anodic or cathodic) current that flows during the off-time i_{off} and the duty cycle (γ): $i_m = i_p\gamma + i_{off}(1 - \gamma)$. The pulse period is $t_{pp} = t_p + t_p'$ where $tp = \gamma t_{pp}$ is the pulse on-time and tp' is the pulse off-time. The pulse frequency n often used in the literature is simply the inverse of the pulse period, $n = 1/t_{pp}$.

PED offers increased nucleation rate, reduced diffusion layer thickness, higher diffusion limiting currents, decrease in grain size, reduced fluid dynamic effect on plating uniformity,

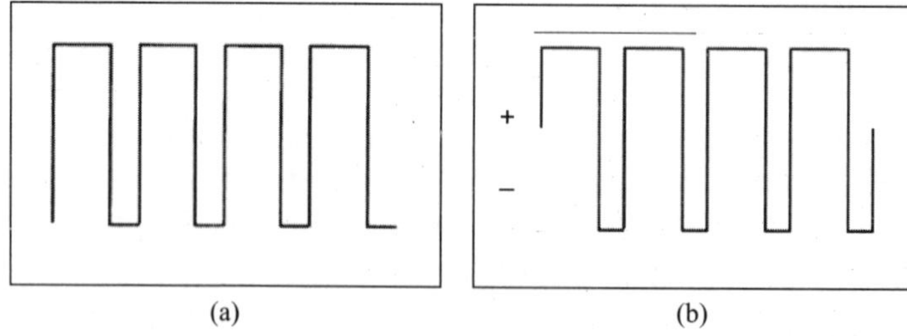

Figure 4 Current patterns in pulsed-current plating (Ideal pattern): (a) conventional plating; and (b) periodic-reverse pulsed-current plating (adapted from VanHorn [25])

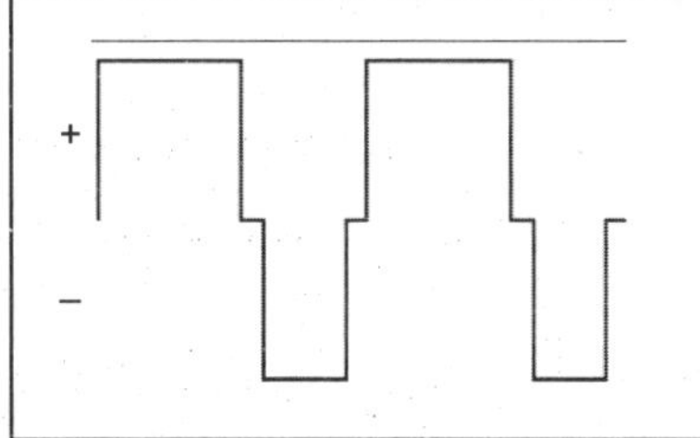

Figure 5 Square-wave current pattern of the forward and reverse envelopes in periodic-reverse pulse plating (adapted from VanHorn [25])

increased current yields, reduced hydrogen embrittlement and improved control in alloy plating. A well-known effect of pulse plating concerns the reduction of internal stresses resulting from codeposition of hydrogen. Using reverse pulses, one can selectively re-oxidize hydrogen dissolved in the deposit and thus reduce the tendency for cracking [30]. PED could dramatically influence the microstructure of the resultant deposits. Table 4 lists conditions that differ from d.c. plating (at same average current density) and how they could influence the physical phenomena that govern the development of the deposit microstructure. PED could also influence the composition of alloy deposits. The choice of applied pulse parameters greatly affects the composition of alloy deposits. To what extent the choice of pulse parameters influences the alloy composition depends on the kinetics of the partial reactions.

The advantages of PED compared to conventional electrodeposition are as follows:

- Deposits are smooth, dense, fine-grained, and almost completely free of pinholes.
- Variation in plate thickness from one part to the next is considerably reduced.

Table 4 Possible effects of pulse current electrodeposition on deposit structure (adapted from Landolt and Marlot [26])

Time interval	Conditions that differ from d.c. plating	Phenomena affected
On-time	Double layer charging	Nucleation rate
	Overvoltage	Growth mechanism (e.g., dendrites)
	Concentration profile near the electrode	Electrode reaction mechanism
	Adsorption (ions, additives, hydrogen)	Codeposition rate (H, alloying elements), additive reactions
Off-time	Double layer discharge	Surface diffusion
	Potential relaxation	Surface recrystallization
	Concentration profile relaxation	Corrosion, displacement reactions
	Desorption (additives, ions, hydrogen)	Passivation, Hydrogen diffusion
Pulse reverse-time	Anodic potential	Selective metal dissolution
	Sign change of double layer charge	Hydrogen re-oxidation
	Concentration profile near electrode	Additive oxidation
	Desorption/adsorption (additives/ions)	Passivation

- Plating speeds can normally be increased.
- Current efficiency generally is improved.
- Use of organic additives, in most cases, can be reduced by 50 to 60%.
- Deposits are free from dendritic growth even if additives are not used.
- For some electrodeposits, such as gold, less metal is required in the plating solution to meet end-use specifications.

The main limitation of PED is that the cost of a pulsed-current rectifier is greater than that of a conventional dc unit.

PED can be used as a means of producing unique structures with improved properties, i.e., coatings with properties not obtainable by d.c. plating. PED can be used to improve the current distribution and to alter the prevailing mass transport conditions. It can also serve to control the microstructure and composition of electrodeposits, for example, nanocrystalline and multilayer deposits. PED permits control of the composition of electrodeposited alloys by varying the electrical parameters. PED has been reported to improve the deposition process and deposit properties such as porosity, ductility, hardness, and surface roughness. The benefits of PED are given in Table 5.

PED can be used to prepare multilayer alloy deposits from a single electrolyte containing a more noble metal in small concentration and a less noble metal in large concentration. The former deposits under mass transport control, the latter under activation control. Hence by periodically varying the potential (or the current) between the value where only the more noble

Table 5 Benefits of Pulsed current electrodeposition

Metallurgical	Denser deposit Finer grain deposit Lower porosity Higher tensile strength Higher elongation Reduced stress Reduce hydrogen embrittlement
Electrical	Higher conductivity Lower contact resistance Better bondability
Physical	Fills submicron trenches Improved adhesion of deposit Reduced stress on photo resists Increased throwing power Control of alloy composition Uniform thickness
General	Reduce additives Reduce plating time 10 - 20%

metal deposits and the value corresponding to deposition of both metals, it is possible to deposit multilayer deposits. In principle, parallel layers of alternating composition are obtained and their thickness, typically a few nanometers, depends on the charge passed during each pulse. At high pulse frequencies the amount of material deposited during one pulse is small, often less than one atomic layer, and alloy deposits of uniform composition are obtained. At low pulse frequencies, with pulse off-times of several seconds, the resulting alloy composition may vary periodically with pulse on-time and off-time leading to formation of multilayer deposits.

BRUSH PLATING

Brush plating, also called as selective plating, differs from conventional electroplating, in that the plating solution is brought to the job (cathode) and applied by a hand held anode instead of immersing both the cathode and anode in the plating bath. This method is usually adopted when the parts are too large to immerse in the plating bath or when a small area of a large part is to be plated or when touch up and repair of components in required. The anodes used in brush plating usually contains some absorbent materials so that the plating solution can be retained in them and can be applied over the cathode. The desired current for plating the metals can be supplied using a d.c. source. The plating solutions used for brush plating have a much higher concentration of metal usually as organometallic salts than do solutions used for tank plating. This permits the use of higher current density which results in faster deposition, better bond strength and less porosity than in tank plating. Continuous movement between the anode and cathode is a key element in obtaining high-quality brush plated deposits. A wide variety of metals and alloys such as, copper, nickel, cobalt, gold, platinum, palladium, silver, rhodium, cobalt-nickel, nickel-tungten, etc. can be deposited by brush plating. Adhesion is good on most of the metals, except Ti, Ta and W, where only a limited adhesion could be achieved. The major advantage of brush plating is portability. Since this method offers precise thickness control, there is no need for subsequent machining. A comparison of the characteristics of conventional electroplating and brush plating is given in Table 6.

Table 6 Comparison of the characteristics of conventional electroplating and brush plating (Source: Brush plating Course Manual, Liquid Development Co.)

Characteristic	Conventional electroplating	Brush plating
Precision buildup capacity	Fair to good	Excellent
Quality of bond	Good	Excellent
Speed of deposit	Slow	Fast
Density of deposits (porosity)	Moderately dense	Very dense
Portability	No	Yes
Requirement for post finishing	Usually required	Not required for thickness up to 0.01 inch on smooth surface
Hydrogen embrittlement	Yes	No

JET ELECTRODEPOSITION

Jet electrodeposition (JED) is a high-speed electroplating technique with special flow characteristics and in this respect it differs from conventional electrodeposition. Conventional electrodeposition is generally composed of three elemental steps: a transfer of metal hydration ions, a transfer of electrical charges and crystallization. The deposition rate is governed by the transfer of hydrated metal ions, which is the slowest of the three steps. The cathodic interface is characterized by a diffusion layer located between the cathode and bulk of the solution. The polarization curve, which exhibits the current-potential relationship, explains the phenomena occurring in the diffusion layer. It is necessary that electroplating is carried out at high current density to obtain a fast plating rate. However, if the current density exceeds the limiting current density, then it could lead to burnt deposits due to the depletion of cations at the cathode interface. At higher current densities diffusion is not fast enough to replenish the cations at the interface. Besides, hydrogen ions are also reduced to hydrogen gas. Replenishment of the interface by metal ions can be enhanced by reducing the thickness of the diffusion layer. In conventional electrodeposition, the diffusion layer thickness can be reduced by cathode movement or by adopting a rotating electrode or by ultrasonic agitation or by modifying the plating cell configuration. In jet electrodeposition, the electrolyte jet forcibly supplies a greater quantity of metal ions and makes the diffusion layer extremely thin. Consequently the limiting current density becomes very high. The increase in limiting current density can be up to several amperes per square centimeter. This enables an increase in the probability of nucleation, which results in the formation of fine grained deposits with increase in microhardness in accordance with the Hall–Petch relation. The grain-size refining effect of the jet electrodeposition is more efficient than conventional plating due to the much higher overpotential and current density.

Qiao et al. [31] have prepared nanocrystalline Ni–Co alloy deposits by JED with much higher current density. According to them, the limiting current density can be over 4.0 A/cm^2 and the highest deposition rate can be up to 47.3 μm/min, which is 90 times faster than that of conventional electrodeposition. The deposits are bright, show no burns, and have fine grain size. JED current density has little effect on the composition of the deposited alloy. Increase in current density from 2.39 to 4.77 A/cm^2 enables refinement of the grain size of the Co–Ni alloy deposits from 19 to 9 nm and an increase in the microhardness from Hv411 to Hv540. Only a few works regarding the use of JED for synthesis of nano-grained materials have been reported [32-36]. Takeuchi et al. [37] have studied the formation of compositionally graded Ni-P and Ni-P-SiC composite coatings by JED. Future development of this high-speed electrodeposition method requires better understanding of the effects of limiting current density and deposition rate on the composition, microstructure, and properties of the deposited alloys.

ELECTRODEPOSITION OF METALS USING IONIC LIQUIDS

The term "ionic liquids", formerly referred to room-temperature molten salts, is generally used for ionic melts with melting points below 100°C. Most of the liquids with which we are familiar (e.g., water, ethanol, acetone, benzene etc.) are molecular. Regardless of whether they are polar or nonpolar, they are basically constituted of molecules. However, the room-temperature ionic liquids are mainly constituted of ions. This gives them the potential to behave very differently from conventional molecular liquids, when they are used as solvents. In contrast to conventional molecular solvents, ionic liquids are usually nonvolatile, in most cases nonflammable, less toxic, good solvents for both organic and inorganic materials, and can be used over a wide temperature range. Moreover, ionic liquids exhibit good electrical conductivity and have a wide electrochemical window whereas the electrochemical window of water is limited to only certain metals that can be deposited from aqueous media. Another advantage of ionic liquids is that problems associated with hydrogen ions in conventional protic solvents can be eliminated, because ionic liquids are normally aprotic. Therefore, ionic liquids have attracted considerable attention as solvents for a wide variety of applications including electrodeposition, batteries, catalysis, separations, and organic synthesis. Room temperature ionic liquids are the promising electrolytes for the ED of various metals because they have the merits of both organic electrolytes and high-temperature molten salts. They can be used in a wide temperature range, so temperature can be elevated to accelerate such phenomena as nucleation, surface diffusion and crystallization associated with the electrodeposition of metals. Since the ionic liquids are nonflammable and volatile, the process can be safely constructed.

A large number of studies on ED of metals from chloroaluminate ionic liquids have been reported whereas only a few studies have explored ED of metals from non-chloroaluminate ionic liquids such as tetrafluoroborate (BF_4^-), hexafluorophosphate (PF_6^-) and chlorozincate ionic liquids. Chloroaluminate ions liquids are highly hygroscopic and upon hydrolysis produce toxic and corrosive hydrogen chloride. Hence they are not suited for practical use. In contrast to chloroaluminate, ionic tetrafluoroborate ionic liquids are more stable against moisture and are applicable for practical use. Moreover, codeposition of metals derived from the ionic liquids does not occur because the tetrafluoroborate anion is not reducible. However, tetrafluoroborate ionic liquids are limited only to a small number of metals for which the anhydrous tetrafluoroborate salts are available. Besides, it is difficult to adjust the Lewis acidity and basicity by the addition of BF_3 and this necessitate the addition of organic chlorides as a Lewis base to dissolve the metal chlorides, which act as Lewis acids to form their chlorocomplex anions. The hexafluorophosphate ionic liquids are similar to the tetrafluoroborate ionic liquids and they are more stable against moisture. However, they undergo hydrolysis when contacted with an acidic aqueous solution. They are not completely miscible with water and sometimes regarded as hydrophobic. Another type of ionic liquid used for electrodeposition of metals is a combination of various quaternary ammonium cations with bis(triflouromethylsulfonyl)imide, $N(CF_3SO_2)_2^-$. These $N(CF_3SO_2)_2^-$ ionic liquids are not only stable against moisture but also immiscible with water. Therefore these $N(CF_3SO_2)_2^-$ ionic liquids are the promising candidates as supporting electrolytes for the practical electrodeposition processes.

A wide variety of metals such as Ti, Cr, Fe, Ni, Co, Cu, Zn, Pd, Ag, Cd, La, Pt, Au, Hg, alkaline and alkaline earth metals (Li, Na, K, Rb and Cs), Ga, In, Tl, Sn, Pb, Bi, Te are deposited using chloroaluminate ionic liquids [38, 39]. Electrodeposition of Cu, Fe, Co, Ag, Cd, Zn-Ni alloy, Te, Ge, Sb, In, In-Sb alloy, etc. has been explored using non-chloroaluminate ionic liquids. Electrodeposition of Ag, Cu, Co, Ni, Zn, Mg. Ga, Ga-As, In-Sb, Nb-Sn has been reported in $N(CF_3SO_2)_2^-$ ionic liquids. The possibility of electrodepositing metals using ionic liquids, which would otherwise difficult to plate using aqueous electrolytes, opens up new avenues in surface modification to achieve desirable characteristics. One notable application is the electrodeposition of Ta on NiTi alloy. Ni-Ti alloys are widely used as orthodontic wires, self-expanding cardiovascular and urological stents, and bone fracture fixation plates and nails. The biocompatibility of NiTi implants depends on their corrosion resistance. The major risk associated with Ni-Ti implants is the breakdown of the passive film, which occurs owing to the aggressiveness of the human body fluids, leading to release of Ni ions that may cause allergic, toxic and carcinogenic effects. Zein El Abedin et al. [40] have shown that ED of Ta using room temperature ionic liquid 1-butyl-1-methylpyrrolidinium bis(tri fluoromethylsulfonyl) imide ([BMP]Tf_2N) containing TaF_5 as a source of tantalum results in the formation of an adherent, dense and uniform layers of Ta of 500 nm thick, which offers better corrosion performance (higher open circuit potential, wider passive region and higher breakdown potential), than the uncoated alloy.

ELECTROLESS PLATING

Fundamentals of electroless plating

Electroless plating was an accidental discovery by Brenner and Riddell in the year 1946, when they tried to electroplate Ni-W alloy on the inner side of a steel tube using a citrate bath. As the Ni-W alloy deposit exhibits a high level of internal stress and cracking due to oxidation of the organic components in the bath, they attempted modification of the plating bath using several reducing agents including sodium hypophosphite. To their surprise, they found that the exterior surface of the steel tube was also coated and this had accounted for the increased current efficiency totaling up to 120% of the theoretical value. After careful analysis they concluded that the coating formed on the exterior of the steel tube might have formed by chemical reduction induced by hypophosphite, which happened to be a good reducing agent (the reduction potential of hypophosphite is 0.499 V vs. SHE at pH 4-6 and 1.57 at pH 7-10). In this process the chemical reducing agent provides electrons necessary to produce a metallic deposit rather than external electric current (as in electroplating). Hence the process was named as chemical nickel plating. Based on its analogy with electroplating process, William Blum coined the term as "Electroless Plating" for this process. Electroless plating process as we know today is an improved version of the process developed by Brenner and Riddell. Among the variety of metals that can be plated using this method, electroless nickel has proved its supremacy for producing coatings with excellent corrosion and wear resistance [41-43]. Compared to electrodeposition, coatings obtained using the electroless plating technique is uniform and they possess a very homogeneous distribution regardless of the substrate geometry [44]. Since no current flow is involved in the electroless deposition process, the rate of deposition on all areas should be equal as long as the solution conditions are maintained properly. This attribute of electroless plated coatings is beneficial when coating complex parts with critical dimensions, such as ball valves or threaded components.

Mechanism of electroless plating

Electroless plating is defined as the autocatalytic deposition (reduction) of metal ions in the presence of a reducing agent (hypophosphite, aminoborane or borohydride. Hypophosphite-reduced plating baths are the most commonly used for commercial electroless nickel plating. The mechanism of the electroless Ni-P deposition reactions taking place in the hypophosphite bath is not fully understood, but it has been postulated that it occurs in microcells of alternating anodic/cathodic polarity on the surface of the substrate. However, the occurrence of the following reactions is most definite as a result of the catalytic dehydrogenation of hypophosphite molecule adsorbed at the surface [43]:

$$3NaHPO_2 + 3H_2O + NiSO_4 \rightarrow 3NaH_2PO_3 + H_2SO_4 + 2H_2 + Ni^0 \quad ...(1)$$

$$2H_2PO_2^- + Ni^{2+} + 2H_2O \rightarrow 2H_2PO_3^- + H_2 + 2H^+ + Ni^0 \quad ...(2)$$

$$Ni^{2+} + H_2PO_2^- + H_2O \rightarrow Ni^0 + H_2PO_3^- + 2H^+ \quad ...(3)$$

$$H_2PO_2^- + H_{ads} \rightarrow H_2O + OH^- + P \quad ...(4)$$

$$3H_2PO_2^- \xrightarrow{\text{catalytic energy input}} H_2PO_3^- + H_2O + 2OH^- + 2P \quad ...(5)$$

Reaction (3), the simplified form of reactions (1) and (2), corresponds to nickel reduction while reaction (5) (a simplified form of reaction 4) corresponds to phosphorus reduction. The aforementioned reactions can be described as the following separate reactions [45]:

$$H_2PO_2^- + H_2O \xrightarrow{\text{catalytic energy input}} H^+ + H_2PO_3^{2-} + 2H_{ads} \quad ...(6)$$

$$Ni^{2+} + 2H_{ads} \rightarrow Ni^0 + 2H^+ \quad ...(7)$$

$$H_2PO_2^- + H_{ads} \rightarrow P + OH^- + H_2O \quad ...(8)$$

$$H_2PO_2^- + H_2O \xrightarrow{\text{catalytic energy input}} H_2PO_3^- + H_2 \quad ...(9)$$

$$Ni^{2+} + H_2PO_3^- \rightarrow NiHPO_3 \quad ...(10)$$

Reactions (6) to (8) are responsible for the formation of the Ni-P alloy coating. In reactions (7) and (8), the H_{ads} produced is consumed and Ni and P are deposited. Not all of the reactions occurring in the bath are favorable to EN deposition. Reactions (9) and (10) are deleterious to the deposition. In reaction (9), hypophosphite forms molecular hydrogen instead of atomic hydrogen, which diminishes the reducing power. As a result of reaction (10), nickel precipitates as nickel orthophosphite. This causes a reduction in the concentration of nickel ions in the bath. Also, the nickel orthophosphite precipitation occurring on the coating surface results in a defective and rough surface [45]. The unwanted occurrence of the reactions (9) and (10) causes a reduction in the efficiency of the electroless nickel coating deposition process.

Paunovic [46] was the first to identify electroless metal deposition in terms of mixed potential theory. He suggested that electroless metal deposition mechanisms could be predicted from the polarization curves of the partial anodic and cathodic processes. In simple terms, mixed potential theory leads to the assumption that electroless nickel plating can be considered as the superposition of anodic and cathodic reaction at the mixed (deposition) potential, E_m. Accordingly, the rates of the anodic reactions are independent of the cathodic reactions occurring simultaneously at the catalytic surface, and the rates of separate partial reactions depend only on the electrode potential, the mixed potential. The partial anodic and cathodic partial reactions are written as:

Anodic reaction

$$H_2PO_2^- + H_2O \rightarrow H_2PO_3^- + H_{ads} + H^+ + e^- \qquad E^0 = 0.50 \text{ V vs. SHE} \quad ...(11)$$

Cathodic reactions

$$Ni^{2+} + 2e^- \rightarrow Ni^0 \qquad E^0 = - 0.25 \text{ v vs. SHE} \quad ...(12)$$

$$2H^+ + 2e^- \rightarrow H_2 \qquad E^0 = 0.00 \text{ V vs. SHE} \quad ...(13)$$

$$H_2PO_2^- + 2H^+ + e^- \rightarrow P + 2H_2O \qquad E^0 = 0.050 \text{ V vs. SHE} \quad ...(14)$$

The oxidation of hypophosphite (11) upon supply of sufficient energy (heat) provides the electrons necessary for reduction of nickel ions. Adsorbed nickel ions consume these electrons (12) and get deposited over the substrate surface with simultaneous hydrogen evolution (13). These electrons are also responsible for the deposition of phosphorus from hypophosphite (14).

Initiation of electroless nickel deposition occurs readily with metals such as iron, nickel, and cobalt. Once the initial nickel layer has deposited on the catalytic substrate, it acts as a catalyst for the process and the deposition continues unaided. This is referred to as autocatalysis of the deposition reaction. This unique property of EN plating makes it possible to coat internal surfaces of pipes, valves, nuts and bolts, and other complex geometries that are very difficult or impossible to be coated by conventional electroplating techniques.

Homma et al. [47] studied the reaction mechanism of electroless deposition using reductants like dimethylamineborane (DMAB), sodium hypophosphite etc., using a molecular orbital approach. It was indicated that the oxidation reaction for the reductants proceed via 5-coordinate intermediates and the calculated value of the heat of reaction could be used for quantitative evaluation of the reducibility of the reductants. Sodium borohydride and DMAB are the most widely used reducing agents for preparing electroless Ni-B coatings. Borohydride is a powerful reducing agent. The redox potential of BH_4 is calculated to be $E^0 = -1.24$ V vs. SHE. In basic solutions, the decomposition of the BH_4 unit yields 8 electrons for the reduction reaction.

$$BH_4^- + 8OH^- \rightarrow B(OH)_4^- + 4H_2O + 8e^- \qquad ...(15)$$

However, it has been found experimentally that one mole of borohydride reduces approximately one mole of nickel. Based on this observation many mechanisms were proposed. The most widely accepted mechanism was proposed by Gorbunova et al. [48].

$$BH_4^- + 4H_2O \rightarrow B(OH)_4^- + 4H_{ads} + 4H^+ + 4e^- \qquad ...(16)$$

$$2Ni^{2+} + 4e^- \rightarrow 2Ni^o \qquad ...(17)$$

Equations (16) and (17) can be combined to give

$$BH_4^- + 2Ni^{2+} + 4H_2O \rightarrow 2Ni^o + B(OH)_4^- + 2H_2 + 4H^+ \qquad ...(18)$$

Reduction of boron

$$BH_4^- + H^+ \rightarrow BH_3 + H_2 \rightarrow B + 5/2H_2 \qquad ...(19)$$

hydrolysis of borohydride

$$BH_4^- + 4H_2O \rightarrow B(OH)_4^- + 4H_{ads} + 4H^+ + 4e^- \rightarrow B(OH)_4^- + 4H_2 \qquad ...(20)$$

This mechanism indicates that the mole ratio of nickel reduced to borohydrate consumed is 1:1, which was supported by experimental evidence [48].

DMAB has three active hydrogens bonded to the boron atom and, therefore, should theoretically reduce three nickel ions for each DMAB molecule consumed (each borohydride will theoretically reduce four nickel ions). The reduction of nickel ions with DMAB is described by the following equations:

$$3Ni^{2+} + (CH_3)_2NHBH_3 + 3H_2O \rightarrow 3Ni^o + (CH_3)_2NH_2^+ + H_3BO_3 + 5H^+ \qquad ...(21)$$

$$2[(CH_3)_2NHBH_3)] + 4Ni^{2+} + 3H_2O$$

$$\rightarrow Ni_2B + 2Ni^o + 2[(CH_3)_2NH_2^+)] + H_3BO_3 + 6H^+ + 1/2H_2 \qquad ...(22)$$

Boron reduction:

$$(CH_3)_2NHBH_3 \xrightarrow{Cat} (CH_3)_2NH + BH_3 + H_2 + H^+$$
$$\rightarrow (CH_3)\ NH_2^+ + B + 5/2H_2 \quad ...(23)$$

In addition to the above useful reactions, DMAB can be consumed by wasteful hydrolysis:

Acid:

$$(CH_3)_2NHBH_3 + 3H_2O + H^+ \rightarrow (CH_3)_2\ NH_2^+ + H_3BO_3 + 3H_2 \quad ...(24)$$

Alkaline:

$$(CH_3)_2NHBH_3 + OH^- \xrightarrow{H_2O} (CH_3)_2NH + BO_2^- + 3H_2 \quad ...(25)$$

In acidic solutions, the first stage of the process is the dissociation of water ($H_2O \rightarrow H^+ + OH^-$) at the catalytic surface. The hydroxyl ions (OH^-) replace the hydrogen in the B-H bond of DMAB and as a result, an electron and a hydrogen atom are produced. The consumption of OH^- ions results in the accumulation of hydrogen ions (H^+) in the solution with a concurrent decrease in pH of the solution. In alkaline solutions, the sources of hydroxyl ions are the basic compounds (NaOH, NH_4OH, etc.) that are added to the plating solution to adjust the pH into the alkaline range of 7.0 to 14.0. As a result of the reaction of OH^- with the B-H bond, the pH also decreases in the alkaline solution. In this case, however, the pH decrease is due to the consumption of OH^- ions rather than the formation and accumulation of H^+ ions.

As can be deduced from reactions (3) to (5), (15) to (19) and (21) to (23), the electroless nickel coating is not pure nickel, unlike for electroplating, but a nickel-phosphorus or nickel-boron alloy coating. Electroless nickel processes are grouped as Ni–P, Ni–B and pure Ni, based, respectively, on the reducing agents used (i.e., hypophosphite, borohydride or dialkyl amino borane and hydrazine) in the plating bath. Hypophosphite reduced electroless nickel plating process has received commercial success because of its low cost, ease of control, and ability to offer good corrosion resistance [41-44]. Borohydride-reduced electroless nickel coatings also received considerable attention in recent years.

The essential components of the electroless plating bath (Table 7) are metal salt, reducing agent, complexing agent, accelerators, buffers, pH regulators, stabilizers and wetting agents. Most applications of the electroless coating are based on their wear and corrosion resistance. Table 8 enumerates the various applications of electroless nickel based coatings [49]. In as-deposited condition electroless nickel is a metastable, supersaturated metal-metalloid alloy. The properties of electroless nickel coatings are directly attributed to their microstructural characteristics. The phosphorus/boron content of electroless nickel coatings defines the physical, mechanical, and corrosion resistance properties of the coating [5].

Electroless Polyalloy Coatings

Development of electroless nickel polyalloy deposits is considered as the most effective method to alter the chemical and physical properties of binary Ni-P or Ni-B alloy deposits. There are many limitations to the production of ternary alloys. It is clear that the main metal (Nickel) must be capable of being catalytically deposited. The reduction of the second metal, which participates in the alloy, is determined by its electrochemical standard potential as well as its catalytic properties in relation to the reduction process. A few metals which do not

Table 7 The various components of electroless plating bath and their functions [43]

Component	Function
Nickel Ion	Source of Metal
Hypophosphite/Borohydride/DMAB	Reducing agent
Complexants	Stabilizes the solution
Accelerators	Activate reducing agent
Buffers	Controlling pH (longer term)
pH regulators	Regulates the pH of the solution
Stabilizer	Prevents solution breakdown
Wetting agents	Increases wettability of surfaces

Table 8 Application of electroless Ni–P coatings (adapted from Agarwala et al. [49])

Application avenue	Components	Coating thickness (μm)
Automotive	Heat sinks, carburettor components, fuel injection, ball studs, differential pinion ballx shafts, disk brake pistons and pad holders transmission thrust washers, synchromesh gears, knuckle pins, exhaust manifolds and pipes, mufflers, shock absorbers, lock components, hose couplings, gear and gear assemblies. Fuel pump motors, aluminum wheels, water pump components, steering column wheel components, air bag hardware, air conditioning compressor components, decorative plastics and slip yokes.	2-38
Air craft/ aerospace	Bearing journals, servo valves, compressor blades, hot zone hardware, pistons heads, engine main shafts and propellers, hydraulic actuator splines, seal snaps and spacers, landing gear components, turbine front bearing cases, engine mount insulator housing, flanges, sun gears, breech caps, shear bolts, engine oil feed tubes, flexible bearing supports, break attach bolts, antirotational plates, wing flap universal joints and titanium thruster tracks.	10-50
Chemical & petroleum	Pressure vessels, reactors, mixer shat, pumps and impellers, heat exchangers, filters and components, turbine blades and rotor assembles, compressor blades and impellers, spray nozzles, valves: ball, gate plug, check and butterfly, stainless steel valves, chokes and control valves, oil field tools, oil well packers and equipment, oil well turbine and pumps, drilling mud pumps, hydraulic systems actuators and blowout preventions.	25-125
Electrical	Motor shafts, rotor blades of stator rings	12-25
Electronics	Head sinks, computer drive mechanisms, chassis memory drums and discs, terminals of lead wires, connectors, diode and transistor cans, interlocks, junction fittings and PCB.	2-25
Food	Pneumatic canning machinery, baking pans, moulds, grills and freezers, mixing louts, bun warmers and feed screw and extruders.	12-25

Table 8 Contd.

Marine	Marine hardware pumps and equipment	25-50
Material	Hydraulic cylinders and shafts, extruders, link drive belts, gears and	12-75
handling	clutches and clutches	
Medical & pharmaceutical	Disposable surgical instruments and equipment, sizering screens, pill sorters and feed screws and extruders.	12-75
Military	Fuse assemblies, tank tarred bearings, radar wave guides, mirrors, motors, detonators and firearms	8-75
Mining	Hydraulic system, jetting pump heads, mine engine components, piping connections, framing hardware.	3-060
Moulds & dies	Zinc dies, cast dies, glass moulds and plastic injection moulds of plastic extrusion dies.	15-50
Printing	Printing rolls and press beds.	~ 38
Rail road	Tank cars, diesel engine shafts and car hardware 20-90.	12-50
Textiles	Feeds and guides, fabric knives, spinnerets, loom ratchets and knitting needles.	~ 30
Wood & paper	Chain saw engine	~ 25
Miscellaneous	Drills and taps Precision tools Shower blades and heads Pen tips	~ 12 ~ 12 ~ 8 ~ 5

possess the relevant catalytic properties in relation to the reduction process and are not catalytic poisons can also be codeposited with nickel. Ni-Re-P is an example of this type. When one of the metals is difficult to reduce, then its share in the alloy composition cannot be high. Ni-Cr-P is an example of this category. Electroless Ni-Cu-P, Ni-W-P, Ni-Mo-P, Ni-Zn-P, Ni-Sn-P, Ni-Re-P, Ni-Cr-P alloy coatings have been explored for a variety of applications. From corrosion protection point of view electroless Ni-Cu-P, Ni-W-P, Ni-Mo-P and Ni-Zn-P coatings assume significance [51-56]. Though alloying of Cu, W, Mo, etc. in the electroless Ni-P or Ni-B coatings enables an improvement in corrosion resistance, alloying of these elements results in a decrease in the P or B content of the coating.

ELECTROLESS COMPOSITE COATINGS

The idea of codepositing various second phase particles in electroless nickel deposit and thereby taking advantage of its inherent uniformity, corrosion resistance and hardenability has led to the development of electroless nickel composite coatings [41-44, 49, 57]. An essential advantage of preparing composite coatings by electroless codeposition compared to electrocodeposition is that the former allows accurate reproduction of the base geometry and eliminates the need for subsequent mechanical finishing. The electroless composite coating is formed by the impingement and settling of particles on the surface of the workpiece, and the subsequent envelopment of these particles by the matrix material as it is deposited. Several factors influence the incorporation of hard and soft particles in an electroless Ni-P matrix including, particle size and shape, relative density of the particle, particle charge, inertness of the particle, the concentration of particles in the plating bath, the method and degree of agitation, the compatibility of the particle with the matrix, and the orientation of the part being plated [57].

ELECTROLESS DUPLEX, GRADED AND MULTILAYER COATINGS

Sankara Narayanan et al. [58] have explored the possibility of preparing electroless Ni-P/Ni-B duplex coatings using dual baths (acidic hypophosphite- and alkaline borohydride-reduced electroless nickel baths) with both Ni-P and Ni-B as inner layers and with varying single layer thickness. The duplex coatings are uniform and the compatibility between the layers is good. Sankara Narayanan et al. [59] have also explored the possibility of preparing electroless Ni-P graded coatings by sequential immersion in three different hypophosphite-reduced electroless plating baths. One important aspect of depositing electroless Ni-P graded coatings is the application of a nickel strike between each layer, so that the deposition proceeds without any hindrance when the substrate is sequentially immersed in electroless plating baths. The graded electroless Ni-P coatings are uniform and the compatibility between the three layers is good. Chen et al. [60] have studied the utility of multi-layered electroless Ni–P coatings to improve the corrosion resistance of powder-sintered Nd–Fe–B permanent magnet. Gu et al. [61, 62] have studied the role of multilayer coatings consisting of different layers of Ni–P and Ni obtained by electro- and electroless deposition processes to improve the corrosion resistance of steel as well as EL Ni-P coatings on AZ 91D Mg alloy.

CORROSION PROTECTION BY COATINGS

Corrosion is affected by a variety of factors including metallurgical, electrochemical, physical chemistry and thermodynamic. When selecting a coating it is important to know its position with respect to its substrate in the galvanic series for the intended application. Coatings are either anodic or cathodic to the substrate. Coatings that are anodic to a substrate in one environment may be cathodic to the same substrate in another environment. The environment itself may change from the surface of the coating to one that exists within any defects like pits, scratches and cracks. These defects are particularly important if the coating is cathodic to the substrate; in this case, if the substrate becomes exposed, rapid corrosion of the substrate may occur as it attempts to protect the coating. On the other hand, if the coating is anodic to the substrate, defects are less important, although they still must be considered.

If the substrate becomes exposed, it will be protected by the corrosion of the coating. Besides galvanic effects, the substrate and the interfacial zone between it and the coating can noticeably affect the growth and corrosion resistance of the subsequent coating since corrosion is affected by structure, grain size, porosity, metallic impurity content, interactions involving metallic underplates and cleanliness or freedom from processing contaminants. The nature of any corrosion products formed also must be taken into account. For example, in industrial environments, zinc is more protective of steel than cadmium is because the zinc sulphate corrosion product formed has a lower solubility than cadmium sulphate. Thus zinc sulphate provides a blocking action to reduce subsequent corrosion. This solution is reversed in marine environments, where the corrosion products are carbonates and chlorides. And, zinc provides more soluble corrosion products than cadmium does in this environment. The most desirable situation is to have the coating anodic to the material beneath it and at the same time to have a very low corrosion rate.

CORROSION PROTECTION BY ELECTROPLATED COATINGS

Electroplated coatings offer corrosion protection of substrate metals in three possible ways: (i) cathodic protection; (ii) barrier action and (iii) environmental modification or control [62, 63]. Cathodic protection is provided by sacrificial corrosion of the coating, e.g., cadmium and zinc coatings on steel. Barrier action involves use of a more corrosion resistant deposit between the environment and the substrate to be protected such as zinc alloy plated automotive parts, copper-nickel-chromium and nickel-chromium coatings over steel. A typical example for environmental modification or control in combination with a non-impervious barrier layer is electrolytic tinplate used in food packaging [1]. The corrosion performance of electroplated coatings is influenced by a variety of factors, which include structure, crystallographic texture, grain size, porosity, impurities and triple junctions, interactions involving metallic underplates and cleanliness or freedom from processing contaminants [64].

Electrodeposits with a dense non-columnar structure offers better corrosion protection compared to deposits with a columnar structure in a water vapor corrosion test. The presence of large voids between the columns in deposits with a columnar structure offers only a minimum protection. Conditions that favor non-epitaxial growth could cause the formation of porosity and voids at the interface between the substrate and deposit. A prolonged acid dip treatment of copper causes pronounced etching and results in the development of the surface having large areas with (111) planes. Such substrates tend to promote non-epitaxial growth of subsequently deposited films with lose adhesion [65].

Texture can noticeably affect the corrosion resistance of electrodeposited coatings. It has been reported that the rate of anodic dissolution of electrodeposited nickel increases in the order (111) < (100) < (110) [66]. The ability of the zinc alloy coating to act either sacrificially or as a protective corrosion barrier is influenced by its texture. Zinc grains with a near (0001) orientation were found to have much lower corrosion currents in sodium hydroxide solutions, than those of other orientations [67]. The corrosion current density in 0.5 N NaOH decreased with increase in packing density of the zinc coating. Raeissi et al. [68] have studied the influence of texture on the corrosion resistance of zinc electrodeposits. Electrodeposition of zinc has resulted in a full epitaxial growth at 10 mA/cm^2, a completely non-epitaxial growth at 200 mA/cm^2 whereas both epitaxial and non-epitaxial components coexist at 100 mA/cm^2. The coatings electrodeposited at 10 and 200 mA/cm^2 have only one texture component and reveal higher corrosion resistance. High percentage of (00.2) basal planes parallel to the substrate provides higher corrosion resistance for the zinc coating obtained at 200 mA/cm^2. However, the full epitaxial growth with fewer grain boundaries offers a higher corrosion resistance for coatings obtained at 10 mA/cm^2. In contrast, the zinc coating formed at 100 mA/cm^2 allows easier hydrogen discharge and exhibit a relatively lower corrosion resistance compared to those obtained at 10 and 200 mA/cm^2.

The small grain size and high-volume fraction of grain boundaries could cause a significant influence on the corrosion behavior of electrodeposits as they would allow easier access of the corrosive species to the coating-substrate interface. Grain boundaries in an electrodeposit tend to corrode preferentially and if there is a range of grain sizes, the fine-grained region tends to corrode faster [69]. The crevices in fine-grained deposits also corrode preferentially. This is due to the fact that the grains in the crevices are even smaller than in the rest of the deposit and they also have a different chemical composition following the incorporation of addition agents [70].

Recently, electrodeposited nanocrystalline (NC) coatings have been identified as promising candidate materials for a variety of applications. It will be of much interest to know how the small grain size and the high-volume fraction of grain boundaries of the nanocrystalline deposits is going to influence their corrosion behavior compared to their polycrystalline counterparts. Studies on the corrosion behaviour of nanocrystalline coatings are rather limited. Only a few studies have addressed the corrosion behavior of nanocrystalline materials [71-78]. Some studies have reported that electrodeposited nanocrystalline coatings offer a better corrosion resistance [72, 74, 75], while some other studies [71–73] report that they are inferior to the conventional polycrystalline coatings.

Rofagha et al. [72, 79] investigated the corrosion behavior of NC nickel (99.99%, 32 nm grain size) and coarse grained nickel (99.99%, 100 μm grain size) in sulphuric acid media. They found that the corrosion potential of NC nickel was shifted about 200 mV positive than that of polycrystalline nickel. Nanocrystalline nickel of grain size 32 nm displayed typical active–passive polarization behavior like pure coarse-grained (100 μm) polycrystalline nickel, but nanocrystalline Ni exhibited higher passive current density [72, 79]. Wang et al. [80] have reported that the corrosion resistance gradually increased with the reduction of grain size from 3 μm to 16 nm. The corrosion resistance of NC Ni is believed to be improved by the rapid formation of continuous Ni hydroxide passive films at surface crystalline defects and the relatively higher integrity of passive films as a result of the smooth and protective nature of the passive films formed on NC Ni coatings. Mishra and Balasubramanian [81] have reported that the passive current densities for nanocrystalline nickel were higher than that for bulk nickel and this has been related to the defective nature of passive film on nanocrystalline nickel. The breakdown potential for fine grain sized nanocrystalline nickel was higher than coarse-grained polycrystalline nickel. The corrosion rate of freshly exposed nanocrystalline Ni was lower compared to bulk Ni, indicating a higher hindrance to anodic dissolution from the nanocrystalline Ni surfaces.

Youssef [78] has reported that NC Zn coatings exhibited improved corrosion resistance compared to electrogalvanized steel in deaerated 0.5 N NaOH solution. The estimated corrosion rate of NC zinc was found to be about 60% lower than that of EG steel, 90 and 229 $\mu A/cm^2$, respectively. The passive film formed on the NC zinc surface seems to be a dominating factor for the observed corrosion behavior. The corroded surface of NC zinc exhibits discrete etch pit morphology, while a uniform corrosion was observed on the EG steel surface.

Vinogradov et al. [82] have shown that the corrosion behavior of NC Cu does not change significantly in comparison with coarse-grained polycrystalline Cu and highly localized corrosion is observed in ordinary polycrystalline Cu with relatively large grains. Yu et al. [83] have reported that the surface of NC copper is more active during initial immersion that leads to the formation of a passive film quickly. The uniform passive film then protects the NC copper from further serious corrosion. Stable passivation does not occur on the surface of conventional polycrystalline copper. Therefore, corrosion pits, formed on grain boundaries, impurities, triple junctions and surface defects are more severe on a polycrystalline surface.

A recent study has shown that the superior corrosion resistance of NC Ti in HCl and H_2SO_4 can be attributed to the rapid passivation of NC Ti and the impurity segregation to grain boundaries in polycrystalline Ti [84]. Aledresse and Alfantazi [85] showed that both NC Co (67 nm grain size) and NC Co–P (50 nm grain size, no bulk P concentration given) did not passivate and the corrosion current densities for NC Co and Co–P alloys markedly increased by a factor of about 2 and 20 times, respectively, when compared to that of polycrystalline Co in 0.25M Na_2SO_4 solution (pH 10.5). NC Co and Co-P specimens show more cathodic potential than conventional polycrystalline Co. Scanning electron micrographs of the NC and polycrystalline Co showed excessive uniform corrosion in nanocrystalline specimen, while less of the same type of corrosion was observed for the polycrystalline Co.

Porosity is one of the main sources of discontinuities in electrodeposits and is an important factor that can influence the corrosion behaviour. The factors that influence porosity in electrodeposits include the substrate, the plating solution and its operating characteristics, and post plating treatments. For sacrificial coatings such as zinc or cadmium on steel, porosity is not usually a problem since these coatings cathodically protect the substrate at the bottom of an adjacent pore. However, with noble coatings such as those used in electronic applications, substrates are subject to corrosion at pore sites. Porosity also permits the formation of tarnish films and corrosion products on surfaces, even at room temperature. Pore corrosion in thin gold plated electrical contacts is a well know example. An effective way to minimize porosity is to use an underplate. Another possibility is to deposit coatings with specific crystallographic orientations which can strongly influence covering power and rate of pore closure. Porosity is related to the thickness of electrodeposits. Generally, thicker coatings are less porous and offer better corrosion protection for the substrate. However, thickness alone is not sufficient enough to offer a better corrosion resistance. For example, the corrosion resistance of a chromium deposit depends not so much on its thickness but on its physical state. If the chromium is crack-free or non-porous, then its corrosion resistance is excellent. However, chromium deposits typically do not remain crack-free in service. A small number of cracks are detrimental; however, the presence of many fine microcracks may be beneficial. This is due to the fact that microcracked chromium deposits cause the galvanic corrosion action to be spread over a very wide area. Therefore, localized corrosion is avoided.

Co-deposited metallic impurities can noticeably influence the corrosion performance. For example, small amounts of sulfur in bright nickel deposits noticeably change the corrosion potential. Bright nickel with 0.04-0.15% sulfur displays a more active dissolution potential than semi-bright nickel containing about 0.005% sulfur [86]. If the two deposits are electrically connected, the rate of corrosion of the bright, nickel is increased, whereas the rate of corrosion of the semibright nickel is decreased [87]. Small amounts of copper in bright nickel plating solutions cause significant reductions in salt spray resistance, e.g., 10 ppm of copper results in a 20% reduction whereas 25 ppm of copper enables a 50% reduction.

Underplates are used for a variety of purposes such as improving adhesion of the plated system, as diffusion barriers, or for improving mechanical properties. They are also important in improving corrosion resistance. One example is the use of a nickel layer between a copper substrate and a final gold deposit to prevent diffusion of the copper to the surface where it would subsequently tarnish [64].

Internal stresses can also be harmful. Bright electrodeposits often possess high tensile internal stress and this leads to stress relaxation by way of cracking, which lowers the protective ability of the deposit. When the combined tensile stresses exceed the tensile strength of the plating, cracks can develop and these expose the base metal to corrosive attack. This is particularly serious in the case of chromium electrodeposits where internal stresses often exceed the yield stress, resulting in cracked deposits. The corrosion of gold plated spectacle frames is yet another example for the effect of stress on the deposit. One source of corrosive attack on spectacle frames is the formation of cracks which exposes the substrate to corrosive attack through perspiration. Substrates with locked-in stresses reduce the corrosion resistance of electroplated coatings relative to the same coating on an annealed substrate. Stressed metal is anodic to annealed or lesser stressed metals and is therefore more prone to corrosion in unfavourable service conditions.

The surface of a substrate must be free from soil and oxides before being plated. Besides assuring good adhesion this also prevents contaminants from being trapped at the interface and subsequently causing corrosion problems. Plating salts on the surface or in the pores of an electrodeposit also need to be adequately removed since they can increase the conductivity of adsorbed water and increase the probability of electrolytic or galvanic corrosion [65].

Electrodeposited Coatings Commonly used for Corrosion Protection

Nickel is widely used as a corrosion-resistant coating and as an undercoat for subsequent coatings. It is not resistant in HNO_3 or in environments containing Cl^- ions. However, it corrodes slowly in environments that do not contain Cl^- ions. Nickel is widely used in automotive industry as an underlayer for microcracked chromium to protect steel. Nickel is usually plated as part of a multilayer coating system in which the electrode potential of each layer is different from those of other layers. The outer layer is more anodic than the layer beneath it. Electrodeposited nickel-phosphorus coatings offer much better corrosion resistance than that of electrodeposited nickel in all environments. If the phosphorus content is above about 10%, the coating is amorphous and therefore lacks grain boundaries or other crystalline defects at which corrosion can be initiated.

Cadmium is generally preferred for the protection of steel in marine environments and aircraft components while zinc is preferred in industrial environments. Both zinc and cadmium could offer sacrificial protection to the steel components under corroding conditions. Cadmium is much more toxic than zinc and tends to embrittle high-strength steel more than zinc does. It is better to avoid cadmium in applications in which its corrosion products may get into the environment. Generally corrosion performance of both zinc and cadmium is enhanced by chromate conversion treatments. Electrodeposited zinc has been widely used for protection of steel from corrosion. A drawback with zinc plating however, is the large amount of corrosion products formed and the very high dissolution rate of zinc when coupled with iron due to the large potentials difference between these two metals. Zn–Ni alloy coatings offer much better corrosion resistance compared to the conventional zinc. The presence of nickel also imparts a good barrier resistance to the coating. The concentration of the alloying metal (Ni) is critical in determining the corrosion resistance properties of the coatings. An enhancement in noble metal composition would lead to a shift in the open circuit potential towards the noble direction, which in turn will reduce the driving force for the galvanic corrosion. Also, the barrier properties associated with such deposits will be superior compared to other coatings. Zn–Ni–X (X = Cd, Cu, P) ternary or quaternary alloy coatings would induce barrier properties to the sacrificial Zn–Ni alloy thereby extending the life of the coating.

Chromium is resistant to atmospheric corrosion, but is soluble in HCl or in alkaline (caustic) solutions. Its resistance to corrosion is because of the formation of an amorphous chromium oxide that acts as a passive film to protect the metal. Chromium is deliberately deposited for decorative and wear applications as a microcracked coating over nickel so that corrosion currents are uniformly distributed over a large area. Tin is widely used as a barrier and as an anodic coating for steel and copper. Since tin is readily solderable and its oxide is conducting, it finds application in electrical conductors and contacts. Its corrosion products are non-toxic. Therefore tinned containers are widely used in the food industry. Lead is stable in most atmospheres and in both chromic acid and sulphuric acid electrolytes. However, it is affected by chlorides. Lead electrodeposits form very inert passive films that tend to reduce the effects of galvanic corrosion.

Copper is not very good is resisting corrosion in the atmosphere and it tarnishes rapidly. When used alone, it should be protected by a suitable corrosion inhibitor. On the other hand, copper is often present in a coating system because, its small grain size reduces the porosity and thus improves the corrosion resistance of subsequent coatings. Gold is often used for decorative purposes and for electrical contact applications. Gold that is plated directly over copper will increase the corrosion rate of the copper through the unavoidable porosity of the gold deposit. Also, gold will diffuse rapidly into copper base metal. Therefore, electroplated nickel or cobalt coatings are normally plated beneath the gold coating.

COMPOSITIONALLY MODULATED MULTILAYER COATINGS

Compositionally modulated multilayers (CMM) are coatings that consist of a number of layers each composed of two sublayers of different metals or alloys. These coatings are produced using either a single bath or dual baths. In dual bath technique, the substrate is successively transferred between two separate plating baths and each layer is deposited alternatively to laminate the sublayer from the relevant bath. Dual bath technique has the disadvantage that it is susceptible to the formation of an oxide layer on the substrate during the transfer between the baths. Such an oxide layer can deteriorate the quality of the multilayers. In the single bath technique, an electrolyte containing two or more metals can be used. Deposition of multilayers from a single bath is normally carried out by periodically varying the current density owing to the difference in the reduction potential of the metal ions present in the electrolyte. Multilayer coatings have better corrosion resistance and physicomechanical properties than those observed in common alloys [88-90].

The use of CMM Zn-Ni alloy coatings for protection of steel substrate from corrosion has been extensively investigated. The cross-sectional morphology of Zn-Ni CMM coating with 12 individual layers is shown in Fig. 7. Kalantary et al. [91] obtained zinc–nickel CMM coatings with an overall thickness of 8 mm by electrodepositing alternate layers of zinc and nickel from zinc sulphate and nickel sulphate electrolytes. Due to the finer structure in each thin layer and the multiple layer effect, a significant improvement in the corrosion resistance is observed for Zn-Ni CMM coatings compared to pure Zn coatings. Chawa et al. [92] have also reported that the corrosion resistance of Zn-Ni CMM coatings obtained from zinc sulphate and nickel

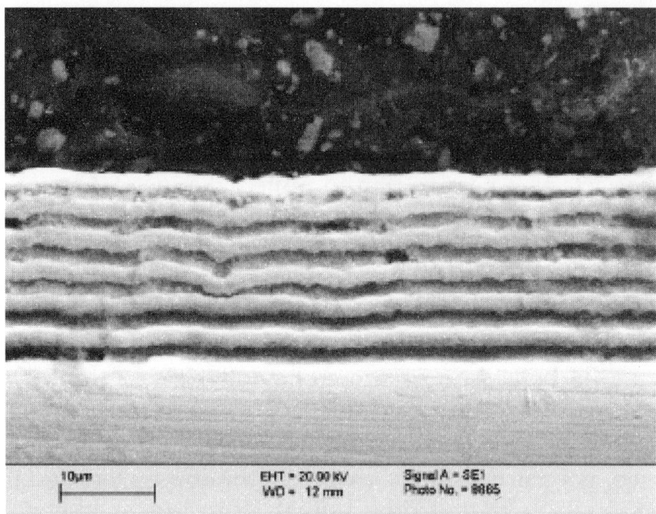

Figure 7 Cross-sectional morphology of Zn/Ni CMM coating with 12 individual layers (reprinted with permission of Elsevier Science, Inc., New York)

sulfamate baths was better than that of zinc or nickel monolithic coatings of a similar thickness. Zhong et al. [93], have studied the corrosion performance of Zn-Fe CMM coatings while Kirilova et al. [94] and Kirilova and Ivanov [95] have reported on the formation and corrosion behaviour of Zn-Co CMM coatings. Ivanov et al. [88, 89] have studied the corrosion performance of zinc–nickel CMM coatings obtained from a single and dual baths. According to them, CMM Zn-Ni coatings consisting of a large number of thin sublayers (11 or 12 of about 0.7 mm thick) are more corrosion resistant than the CMM Zn-Ni coatings consisting of a few thick sublayers (five or six of about 3 mm thick) with the same total thickness. CMM Zn-Ni coatings ending with a Ni oversublayer offer better corrosion resistance than CMM Zn-Ni coatings ending with Zn oversublayer.

Fie and Wilcox [90] have evaluated the corrosion performance of CMM Zn-Ni coatings using a salt spray test. According to them, the corrosion resistance of CMM Zn-Ni coatings was much better than that of a single-layer zinc or nickel deposit, and among the CMM Zn-Ni coating systems, coatings with a nickel sublayer adjacent to the steel substrate and zinc as the top layer (Ni/Zn CMM coating system) have the best protective performance than those with zinc sublayer adjacent to the steel and nickel as top layer (Zn/Ni CMM coating system). For either Zn/Ni or Ni/Zn CMM coating system, the zinc–nickel CMM coatings with the individual sublayer thickness of 2 μm have the longest time to red rust amongst the samples tested. Fie and Wilcox [90] have proposed a probable mechanism to account for the corrosion behaviour of zinc–nickel CMM coatings. Fig. 8 illustrates schematics of the probable corrosion mechanisms for zinc–nickel CMM coatings during corrosion process.

Diagrams A and B show the structures of two Ni/Zn CMM coatings with different individual layer thicknesses. Initially, zinc top layers corrode completely, then the zinc sublayers dissolve through the pores in the nickel sublayers. Diagrams C and D exhibit the structure of two Zn/Ni CMM coatings, with different individual layer thicknesses. The zinc

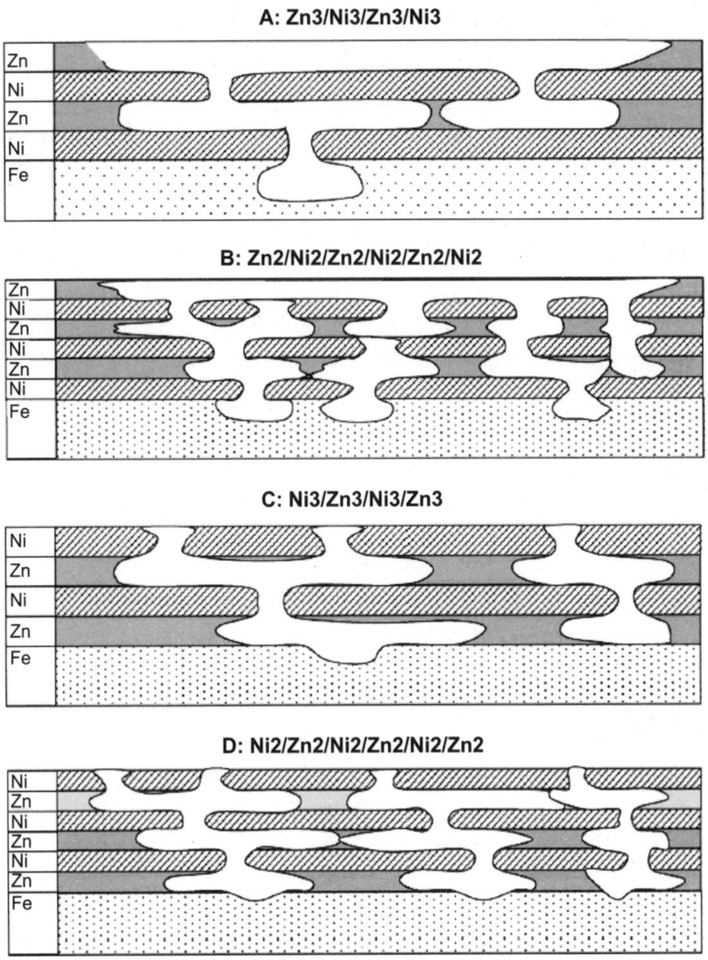
A: Zn3/Ni3/Zn3/Ni3
Zn
Ni
Zn
Ni
Fe
B: Zn2/Ni2/Zn2/Ni2/Zn2/Ni2
Zn
Ni
Zn
Ni
Zn
Ni
Fe
C: Ni3/Zn3/Ni3/Zn3
Ni
Zn
Ni
Zn
Fe
D: Ni2/Zn2/Ni2/Zn2/Ni2/Zn2
Ni
Zn
Ni
Zn
Ni
Zn
Fe

Figure 8 Probable corrosion mechanism diagram for Ni/Zn and Zn/Ni CMM coatings (reprinted with permission of Elsevier Science, Inc., New York)

sublayer beneath the nickel top layer dissolves through the pores existed in the nickel deposits during corrosion process. As a whole, the protection efficacy of zinc–nickel CMM coatings may be thought to depend on the barrier effect of the nickel sublayers and the sacrificial effect of zinc sublayers. Hence it appears that pores and microcracks of the nickel sublayer play an important role in deciding the corrosion resistance of Zn-Ni CMM coatings. Increase in thickness of the nickel sublayer could reduce the porosity but it may lead to the formation of microcracks following tensile stress. It is now observed that the electrodeposition of Zn-Ni CMM coatings with greater than twenty individual layers is quite possible. Attempts on electrodeposition of compositionally modulated alloy multiplayer (CMAM) coatings by depositing alloy sublayers under different plating conditions from single bath has also been made.

CORROSION PROTECTION BY ELECTROLESS PLATED COATINGS

Electroless nickel coatings have excellent corrosion resistance in many industrial environments and is not susceptible to stress corrosion cracking. They are widely used either as protective or decorative coatings in many industries, such as petroleum, chemical, plastic, optics, printing, mining, aerospace, nuclear, automotive, electronics, computer, textile, paper, and food [96, 97]. Recently, the use of EL Ni coatings for corrosion protection of steel reinforcement bar has been suggested by Singh and Ghosh [98]. The corrosion protective ability of electroless Ni-P coatings to improve the life-time of components and assemblies used in a variety of industries is given in Table 13.9. Electroless nickel does not perform as a sacrificial coating in the same way that electrodeposited Zn or Cd performs on steel substrate to provide protection against corrosion. It behaves as a true barrier coating, protecting the substrate by sealing it off from the corrosive environments. Consequently, the thickness of the deposit and the absence of porosity are of great importance. The electroless nickel coating shows superior corrosion resistance compared to electroplated nickel coatings. The most important factors that determine the corrosion resistance are [42]:

- Substrate composition, structure and surface finish
- Pretreatment of the substrate to achieve a clean, uniform surface
- Adequate deposit thickness to meet the severity and time of exposure
- The properties of the deposit (composition, porosity, internal stress etc.) which depends on pH, formulation and prolonged use (turnover) of the plating solution
- Post plating treatments of the coating such as passivation and annealing
- The aggressiveness of the corrosive environment condition

The corrosion resistance of electroless Ni-P coatings varies with the phosphorus content of the coating: relatively high for a high phosphorus but low for a low-phosphorus electroless nickel coating. The better corrosion resistance of electroless Ni-high P coating is a result of its amorphous nature and passivity of the surface film formed on it. Amorphous alloys offer better resistance to corrosion attack than equivalent polycrystalline materials because of the freedom from grain or phase boundaries and because of the glassy film which form on and passivate their surfaces [99].

In general, high phosphorus (10-12 wt.% P) electroless Ni-P coatings are more resistant to acidic environments and products that hydrolyze readily to form acids. Low phosphorus (1-3 wt.% P) electroless nickel coatings are more resistant to strong alkaline environments than medium (6-8 wt.% P) and high phosphorus electroless nickel coatings [100-102]. It is important to recognize here that coating porosity is most significant in acidic environments when ferrous substrates are involved. In alkaline environments porosity is not normally a factor. The assumption that high phosphorus electroless nickel coatings always enhance corrosion resistance is misplaced as low phosphorus electroless nickel coatings are being successfully used in a variety of environments in the caustic soda production industry; corrosion performance of medium phosphorus and high phosphorus electroless nickel coatings in sodium hydroxide environments being relatively inferior.

The better corrosion resistance offered by electroless Ni-high P coatings can be explained as follows: Electroless Ni-P coatings undergo preferential dissolution of nickel even at open circuit potential, leading to the enrichment of phosphorus at the surface layer. The enriched

Table 9 The corrosion protective ability of electroless Ni-P coatings to improve the life-time of components and assemblies used in a variety of industries [97]

Type of industry or application	Component or assembly and its purpose	Type of environment or problem and its severity	Materials used and the normal service life	Recommendations for improvement in service life and the performance of electroless Ni coatings
Chlor-alkali industry	Steel compressors–diaphragm	Build-up of sodium sulphate and ferric chloride resulting in corrosion and erosion.	Steel – Needs repair in less than one year; total life time with repair is 8 years the	High P EL Ni deposits – double compressor life – 15 years – significant cost reduction and production savings
	Control valve used to concentrate NaOH	34% NaOH and 7% NaCl in water; 95°C; 380 L/min at a rate of 1.75 m/sec.	316 SS – 2 weeksN08020 alloy 20 – 3 months coating	A 50 micron low P EL Ni has significantly increased the life time
	Components or assemblies used in transporting brine	Saturated brine; pH :5.6; temperature: 70°C	316 SS – pitting corrosion N02200 – not suitableRubber lining – 8 years	75 microns high P EL Ni. The life time is increased to 15 years.
	Cooling towers used to reduce the temp. of NaOH	Cooling tower water; 35°C; air, Na_2CO_3 and NaCl; pH: 8.5; Flow rate:1.83 m/sec.	Steel pumps – within 2 years – corrosion erosion and cavitation	75 microns high P EL Ni - No corrosion or service problems even after 3 years
Transportation of chemicals by rail cars or tank cars	Nuts and bolts – loading valves on tank cars used for shipping	50% NaOH	641400 (ASTM 193-7) alloy steel – corroded within a short time; risk of failure in transit	25 microns Low P EL Ni – no failure after 3 years; after 5 years – 70% of the bolts are in satisfactory condition
	Safety vents: unloading connections/ barges;valve assemblies	50% NaOH or high purity NaOH	Poor corrosion resistance – discolouration due to iron contamination	50 micron low P EL Ni – Excellent corrosion resistance – prevented discolouration due to iron contamination
Chemical plants	Seawater cooling – Hot taps	Chloride ions - severe crevice corrosion – stagnant seawater	Steel	75 microns high P EL Ni – excellent corrosion resistance – no corrosion or leakage for over 3 years

Table 9 Contd.

Table 9 Contd.

	Flow meters	Aqueous solutions containing 31% KOH plus 19% K_2CO_3; 105°C	316 SS – stress corrosion cracking (SCC) failure	Low P EL Ni - much lower corrosion rate compared to 316SS; Combination of low P EL Ni-316SS -very successful
Distillation of organic compounds	Distillation column	Monochlorotoluene; Severe plugging of the column due build-up of corrosion products	Mild steel – Frequent and plant shutdown	75 microns high P EL Ni – no plugging of corrosion products and no plant shutdown for over 4 years
Food Processing industry	Packaging equipment – bearing, rollers, conveyer systems, hydraulics & gears	NaCl, nitrates, citric acid and acetic acid; humid atmosphere: up to 200°C	Stress corrosion cracking and fatigue failure of process equipment	High P EL Ni – Excellent corrosion resistance - replacement for Cr plating in bakery industry
Oil and gas industry - Surface, subsurface and offshore operations	Ball valve	Crude oil production - associated gas contains 55% H_2S; 80°C; Pressure: 3000 psi	mild steel ball valves –3 months maximum - failure due to corrosion, surface cracking and erosion	75 microns of EL Ni extended the life of the assemblies enormously - No surface deterioration was after two years of continuous service.
	Ball valve used for seawater injection system	Pumping seawater under high pressure	Mild steel	A 75 micron EL Ni coating on the valves – No degradation for more than 4 years of operation.
	Valves, chokes	Khuff gas – 6 mole % CO_2 and 0.1 mole % H_2S; Condensate – Hexane 28 L/Mm^3 & 7 L/Mm^3 water; Pressure: 5000 psi; 90°C; gas velocity: 6 m/sec.	Carbon steel – severe pitting corrosion and corrosion rate: 3-5 mm/year erosion;	25 micron EL Ni for chokes75 micron valve – No corrosion for at least 6 years
	Tubular component	H_2S, CO_2, Cl^-; 260°C	Mild steel – corrosion and erosion; Stainless steel – pitting and galling	50-100 micron high P EL Ni – significantly improves the life-time
	Pumping systems – housing, impellers and discharge barrels	Mixture of clay and water – corrosive and abrasive – a combination of O_2 and H_2S; high pressure	AISI 8630 steel – corrosion fatigue or stress corrosion in seal areas – after 3 months of operation	25-75 micron high P EL Ni – increase the service life of mud pumps for more than 18 months

Table 9 Contd.

Table 9 Contd.

Automotive industry	Pistons, carburetor parts, fuel injectors, engine bearings, gear assemblies	Corrosion of carburetor due to alcohol fuel; corrosion the products plug the narrow channels and orifices	Zinc diecast	5-9 micron high P EL Ni – improve performance of the carburetor
Aerospace industry	Compressor components	Corrosive atmosphere and erosive particles; 425°C; non-uniform thickness; Cd embrittlement	410 SS – Ni/Cd alloy coating by diffusion of Aluminidecoasting	High P EL Ni – excellent corrosion resistance
	high-pressure compressor stator assemblies	Refurbishing	410 SS	8-16 microns High P EL Ni (9-11%) compressively stressed deposit provides excellent gas flow path; greatly enhance the corrosion and erosion resistance for 5800 take-offs and landings
	High-pressure compressor spacer	Overhauling/repair	AMS 6304 alloy	High P EL Ni – uniform coverage and protection that was not possible with other coating systems
	Commercial and military aircraft engine	Compressor case components Corrosion and erosion at 425°C	Martensitic and ferritic iron base alloys and wrought alloy steel	8-25 microns high P EL Ni – improve the performance
	Landing gear of Boeing 727	Corrosion and erosion	High strength steel – ED Cd	8-16 micron high P EL Ni – improve the performance
	Support equipment – catapult covers and tracks used for launching of planes from aircraft carriers	Salt waterSteam at 230°C; force: 4.45 MN Wear and corrosion;	Galvanic corrosion between the track and covers Failure - Within 6-12 months	100 micron EL Ni 12 micron Cd and chromated – 14 to 18 years
Textile industry	Drop wires and needles	Weaving process – various chemicals applied to the fabric; corrosive	Steel – rusting could cause stain in the fabric	High P EL Ni

phosphorus surface reacts with water to form a layer of adsorbed hypophosphite anions ($H_2PO_2^-$). This layer in turn block the supply of water to the electrode surface, thereby preventing the hydration of nickel [103, 104], which is considered to be the first step to form either soluble Ni^{2+} species or a passive nickel film. Such a condition of phosphorus enrichment at the surface would enable hydrolysis of phosphorus and retard hydrolysis of nickel.

Electroless Ni-B coatings are considered to have lesser resistance to corrosion compared to electroless Ni-P coatings. The difference in corrosion resistance between electroless Ni-P and Ni–B coatings is due to the difference in their structure. Since the electroless Ni-B coating is not totally amorphous, the passivation films that form on its surface are not as glassy or protective as those that form on electroless high-phosphorous coatings [105, 106, 58]. The phase boundaries present in these deposits also produce passivation film discontinuities, which are preferred sites for corrosion attack to begin. Also, the inhomogeneous distribution of boron and thallium create areas of different corrosion potential on the surface, leading to the formation of minute active/passive corrosion cells and accelerated the corrosion attack.

Porosity is an important factor that influences the corrosion resistance of electroless nickel coatings [107]. In general, thicker coatings have fewer pores and hence provide a better corrosion resistance [41, 108, 109]. To achieve better corrosion resistance the minimum thickness recommendations are typically 40 μm in mild and moderate environments; but in severe corrosive environments a thickness of 75 μm is preferred. Besides porosity and coating thickness, the corrosion performance of electroless nickel coatings also depends on the surface finish. Generally, smoother substrate surface provide better quality deposits. Mechanical surface treatments and fabrication procedures, such as, grit blasting, casting, rolling, stamping, shearing, lapping, drawing, machining, etc., could create surface imperfections. These surface imperfections could lead to entrapment of solutions which could cause a premature failure. For a given roughness the corrosion potential was found to shift to nobler values and the corrosion current decreased considerably with the increase in thickness of electroless nickel coatings [108]. The tramp constituents co-deposited in electroless nickel matrix influence the corrosion resistance of the resultant coatings. Most of these tramp constituent originate from the stabilizers used in the plating bath. Non-metallic inclusions in the surface of the substrate material such as, carbides, sulphides etc., emanating from alloying elements or impurities can breed pores in the electroless coating, resulting in a decrease in corrosion resistance. Ageing of electroless nickel bath also has a pronounced and detrimental influence on the corrosion resistance of the resultant deposits [110].

For achieving better hardness and improved wear resistance, electroless nickel coatings are usually heat-treated around 400°C for 1 hour. During heat treatment the electroless nickel coatings undergo shrinkage in volume which may cause cracks through the coating, particularly when the microhardness exceeds 900 HV_{100} [96]. This impairs the corrosion resistance of such heat treated electroless Ni-P deposits [41-44]. Low temperature post-plating baking (preferably in vacuum) at temperatures around 200°C provides the optimum corrosion resistance to low and medium carbon steels. When corrosion resistance along with additional hardness is required, a heat treatment temperature of approximately 300°C is recommended to prevent micro-cracking and to limit porosity. At very high temperatures a tenacious oxide film will rapidly form on the coating surface. This film helps to seal the pores and create a passive surface so that electroless Ni-P coatings can withstand the corrosive attack. At temperatures above 650°C the Ni-Fe intermetallic formed at the interface is ductile and dense and reportedly has very good corrosion resistance [102]. Proper and careful post plate treatments can appreciably enhance the corrosion resistance of electroless Ni-P coatings. A 15 minute immersion treatment in hot 1 wt.% CrO_3 solution after plating doubles the salt spray resistance of electroless Ni-P coatings on carbon steel.

Electroless Ni-P and Ni-B based ternary/quaternary alloy coatings, electroless composite coatings, duplex coatings, graded coatings and multilayer coatings are some of the promising developments in electroless plating processes for achieving improved corrosion resistance. The corrosion resistance of the electroless ternary alloy coatings is governed by the phosphorus/boron content as well as the alloying element. The most important factors that determine the corrosion resistance of electroless nickel coatings are also applicable for electroless ternary alloy coatings [111].

The corrosion resistance of as plated electroless Ni-Cu-P deposit is superior to electroless Ni-P deposit or 1Cr18Ni9Ti stainless steel in 50 wt.% NaOH [51]. Liu and Zhao [112] have reported that the anti-corrosion performance of Ni–Cu–P coatings with 17.2 wt.% Cu is superior to the Ni–P coatings and copper in 1 N HCl and 20% NaCl. They have also reported that alloying of Cu in Ni-P and Ni-P-PTFE composite coatings exhibit improved corrosion resistance in HCl and NaCl [113, 114].

Lee and Liang [52] have studied the corrosion resistance of electroless Ni-Mo-P alloy deposits and found that these deposits are more resistant to corrosion attack than are electroless Ni-P deposits. Lu and Zangari [53] have reported that addition of Mo to Ni-P alloys has little or no beneficial effect on corrosion properties.

Aoki and Takano [115] and Bangwei et al. [116] found out that codeposition of tungsten along with nickel and phosphorus in an electroless deposit, increases the protective effect of the film on steel and the extent of protection offered is a function of the tungsten content of the film. The protective effect of electroless Ni-W-P was also addressed by Mallory and Hajdu [42]. Accordingly, deposits containing about 20 wt.% tungsten were resistant to attack by concentrated HNO_3 and 7N HCl. An electroless nickel deposit containing 9.1 wt.% tungsten and 8.2 wt.% phosphorus was considerably more protective to steel than the electroless Ni-P deposit. Tungsten addition increases the corrosion resistance of electroless Ni-P and Ni-B coatings [117].

Tungsten and molybdenum addition are known to improve resistance to uniform, pitting as well as crevice corrosion of Ni in non-oxidizing acid solutions [56]. However, they offer only a limited improvement in corrosion resistance in oxidizing electrolytes. Heat treatment of EL Ni-W-P and Ni-Mo-P coatings also reduces their resistance to corrosion.

Electroless Ni-Zn-P deposit is recommended as a protective coating against corrosion for automobile components [54, 55]. This deposit can be used as a potential alternative for cadmium plating for uniformly coating complex shapes and threaded parts [42]. Veeraraghavan et al. [118] have suggested that EL Ni-Zn-P coatings with 16.2 wt.% Zn could be used as a sacrificial coating for the protection of steel. The high nickel content (74 wt.%) offers better corrosion resistance compared to conventional zinc based coatings.

There is a general belief that the corrosion resistance of electroless Ni-P composite coatings is lesser than that of electroless Ni–P coatings. The codeposited second phase particles present in the electroless nickel matrix are thought to reduce passivity and corrosion resistance. Hence for applications requiring good corrosion resistance, a duplex coating, consisting of an initial electroless Ni–P coating followed by an electroless Ni–P composite coating, is recommended in place of electroless Ni-P composite coatings [119, 120]. However, the corrosion performance of electroless Ni-P composite coatings was found to be satisfactory by Hubbell [121], Hussain and Such [122] and Shoeib et al. [123]. Studies on the corrosion resistance of electroless Ni-P-Si_3N_4 composite coatings in 3.5% sodium chloride solution also revealed a marginal increase in corrosion resistance compared to plain electroless Ni-P deposit of similar thickness [124]. Evaluation of the corrosion resistance of electroless Ni-P-Si_3N_4, Ni-P-CeO_2 and Ni-P-TiO_2 composite coatings by electrochemical impedance spectroscopy suggest that these composite coatings provide better corrosion resistance than plain electroless Ni-P coatings [125].

Sankara Narayanan et al. [58] have studied the corrosion behaviour of electroless Ni-P/Ni-B duplex coatings in comparison with Ni-P and Ni-B coatings of similar thickness (Table 10). Amongst the two sets of duplex coatings, better corrosion resistance is offered by the duplex coating having Ni-P coating as the outer layer and its thickness is 10 μm or higher. The duplex coating having Ni-B as the outer layer also offers better corrosion resistance than EL Ni-B deposits. Though heat-treatment (400°C for 1 hour) tends to reduce the extent of corrosion resistance, the performance of duplex coatings is not drastically reduced. Hence, the electroless Ni-B/Ni-P duplex coating will be a useful replacement for Ni-B and Ni-P coating.

Table 10 Corrosion resistance of electroless Ni-P, Ni-B, Ni-P/Ni-B and Ni-B/Ni-P duplex coatings in as-plated and heat-treated conditions in 3.5% NaCl

System studied	Thickness (μm)	E_{corr}(mV vs.SCE)	i_{corr}(μA/cm^2)	R_{ct}(Ohms.cm^2)	C_{dl}(F)x 10^{-4}
Ni-P as-plated	20	-354	3.62	7960	1.85
Ni-P heat-treated	20	-492	6.89	6498	2.36
Ni-B as-plated	20	-508	9.15	3844	3.53
Ni-B heat-treated	20	-519	14.60	1232	4.22
Ni-P/Ni-B as-plated	10 + 10	-386	3.86	7638	1.89
Ni-P/Ni-B heat-treated	10 + 10	-432	4.93	5339	2.44
Ni-B/Ni-P as-plated	10 + 10	-311	2.46	10130	1.61
Ni-B/Ni-P heat-treated	10 + 10	-356	3.24	7024	1.96

Table 11 Corrosion resistance of non-graded and graded electroless Ni-P coatings in 3.5% sodium chloride solution

System studied	E_{corr}(mV vs. SCE)	i_{corr}(μA/cm^2)	R_{ct}(KΩ.cm^2)	C_{dl}(μF)
EL Ni-Low P coating (L)	-536	4.22	6.90	289
EL Ni-Medium P coating (M)	-434	1.17	24.86	55.60
EL Ni-High P coating (H)	-411	0.60	37.45	49.10
EL Ni-P graded coating (LMH)*	-403	0.41	38.22	36.70
EL Ni-P graded coating (HML)*	-481	1.70	14.77	103

* LMH - Low-Medium-High P; HML – High-Medium-Low P.

Sankara Narayanan et al. [59] have studied the corrosion behaviour of EL Ni-P graded coatings in comparison with EL Ni-P coatings with different P contents (Table 11). The EL Ni-P graded coatings offer a better corrosion resistance compared to EL Ni-P non-graded coatings. Among the two types of EL Ni-P graded coatings, the coating system with EL Ni-high P coating as the outer layer offers better corrosion resistance than its counterpart. However, the corrosion resistance of EL Ni-P graded coating with EL Ni-low P as the outer layer is better than that of electroless Ni-low P coating, due to the barrier properties of the underlying Ni-medium P and Ni-high P layers. Based on their ability to offer corrosion resistance the coatings can be ranked in the following order:

EL graded Ni-P(LMH) > EL Ni-high P > EL Ni-medium P > EL graded Ni-P(HML) > EL Ni-low P

Heat-treatment of electroless Ni-P graded coatings is likely to modify the graded layers so that the gradation is lost. Hence the concept of graded electroless Ni-P coatings is valid only in as-plated condition. Besides, heat-treatment (400°C for 1 hour) drastically reduces the corrosion resistance of these coatings.

Chen et al. [60] have explored the possibility of using multi-layered electroless Ni–P coatings to improve the corrosion resistance of powder-sintered Nd–Fe–B permanent magnet. Since Nd–Fe–B reacts quickly with Cl^- and SO_4^{2} ions in acidic solutions, the Nd-Fe-B magnet was first treated with an alkaline electroless nickel plating bath to deposit a low-P electroless nickel layer. Subsequently, a high-P electroless nickel layer was deposited using an acidic electroless nickel plating bath. Good adhesion of the coating is achieved by the intermediate low-P electroless Ni coating. The electroless Ni-P multilayer coating offers excellent corrosion resistance to Nd-Fe-B magnet. Though there is a loss in magnetic properties due to the multilayer coating, the remaining values are still very attractive for practical applications.

Gu et al. [61] have studied the role of multilayer coatings consisting of different layers of Ni–P and Ni obtained by electro- and electroless deposition processes to improve the corrosion resistance of steel. The difference in the corrosion potential among layers plays a very important role in protecting the steel substrate from rusting. The three-layer coating consists of electroless Ni–low P/electrodeposited Ni/electroless Ni–high P layers, from the surface to substrate, respectively, offers excellent corrosion resistance to the steel substrate. In the salt spray test, the time of the emergence of the first red rust spot for this multilayer coating is about 936 hours, which is 3.5 times of that of Ni–high P coating.

The mechanism of corrosion resistance of this multilayer coating can be explained as follows: when corrosion occurs on the upper electroless Ni–low P layer, some corroded pinholes would lead to localized galvanic corrosion and penetrates vertically from the upper surface layer. When these pinholes meet the electrodeposited Ni layer, the mode of corrosion changes from the original longitudinal corrosion to the extended transverse corrosion since the electrodeposited Ni layer would corrode easily due to its lowest corrosion potential (E_{corr}). This corrosion mode is described in step 1 of Fig. 9, which is similar to that observed in duplex nickel coatings consisting of semi-bright nickel and bright nickel layers [126]. The transverse corrosion of the electrodeposited Ni layer dispersed greatly the corrosion current and provides a sacrificial protection for the upper electroless Ni-low P layer. As a result, the outer surface of coating maintains its original appearance. When the electroplated Ni layer was consumed out by corrosion in some local regions, the upper electroless Ni-low P layer start to corrode and continues to protect the bottom layer from corroding (step 2 in Fig. 9). Thus, the upper and the

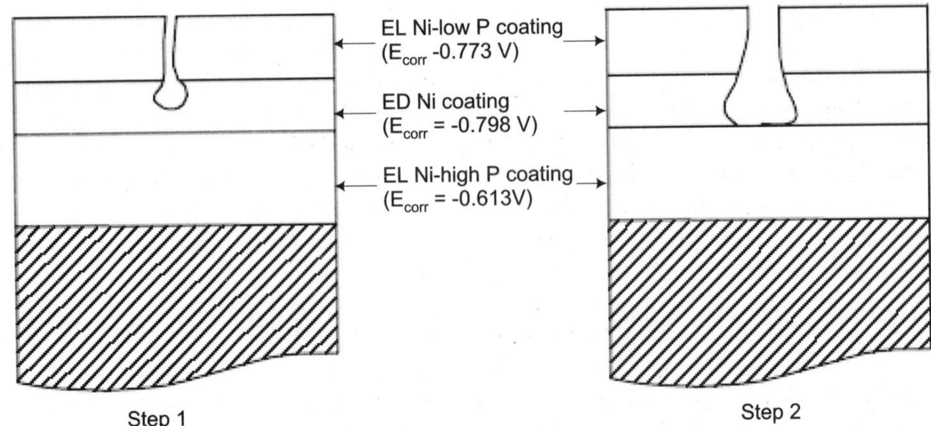

Figure 9 Schematic of the corrosion mechanism of EL Ni–low P/ED Ni/EL Ni–high P coating; Step 1: ED Ni-layer emerged through the pinhole corrosion and corrosion of ED Ni-layer occurs; Step 2: When corrosion reaches the third layer of EL Ni-high P layer, both upper EL Ni-low P and intermediate ED-Ni layers corrode alternately Source: [61] (reprinted with permission of Elsevier Science, Inc., New York)

middle layers corrode alternately in a low corrosion rate and delay the rusting of the steel substrate. Besides, the corrosion products formed during the corrosion process could also block some pinholes and delays the rusting of the steel substrate [127].

Gu et al. [62] have also reported the corrosion protective ability of a 45 μm thick multilayer coating on AZ91D Mg alloy prepared by electroless deposition process. The composition of this multilayer coating from substrate to surface was a protective layer of EL Ni-P deposit, a Ni-P layer with 9.2 wt.% P and another Ni-P layer with 5.4 wt.% P, respectively. Since the corrosion potential of the surface layer (EL Ni-5.4 wt.% P) was lower than that of the middle layer (EL Ni-9.2 wt.% P), when the pitting corrosion penetrates from the surface layer to the middle-layer, the surface layer would corrode preferentially. As a result, the pinning corrosion will not develop vertically to the coatings. This mechanism offers a better corrosion resistance for the AZ91D Mg alloy.

CONCLUDING REMARKS

This chapter presents an overview of the fundamental aspects of electro- and electroless deposition process, the mechanism of deposition, alloy deposition, etc. and the utility of these coatings for corrosion protection. Both electro- and electroless deposition techniques are simple, cost-effective and offer unique advantages for preparing deposits with desirable qualities. In electrodeposition, the plating rate, stability of the bath and the number of turnovers are very high but the resultant coatings lack uniformity on complex shapes and blind holes and they need a post-finishing treatment to achieve the desired performance.

In electroless deposition, the plating rate, bath stability and the number of turnovers are relatively less but the resultant coatings are more uniform and do not require post-finishing treatment. Electroplated coatings offer corrosion protection to the substrate metals in three possible ways: (i) cathodic protection; (ii) barrier action; and (iii) environmental modification or control. The corrosion performance of electroplated coatings is influenced by a variety of factors, which include structure, crystallographic texture, grain size, porosity, impurities and triple junctions, interactions involving metallic underplates and cleanliness or freedom from processing contaminants. Electroless nickel does not perform as a sacrificial coating in the same way that electrodeposited Zn or Cd performs on steel substrate to provide protection against corrosion. It behaves as a true barrier coating, protecting the substrate by sealing it off from the corrosive environments. Consequently, the thickness of the deposit and the absence of porosity are of great importance. The electroless nickel coating shows superior corrosion resistance compared to electroplated nickel coatings. The most important factors that determine the corrosion resistance of electroless plated coatings are: substrate composition, structure and surface finish; pretreatment of the substrate to achieve a clean, uniform surface; adequate deposit thickness to meet the severity and time of exposure to the corrosive environment; the properties of the deposit (composition, porosity, internal stress etc.) which depends on pH, formulation and prolonged use (turnover) of the plating solution; post plating treatments of the coating such as passivation and annealing; and the aggressiveness of the corrosive environment condition. Electro- and electroless deposited ternary/quaternary alloy coatings, composite coatings, duplex coatings, graded coatings and multilayer coatings are some of the promising developments to achieve improved corrosion resistance.

RERERENCES

1. F.A. Lowenheim, *Electroplating,* McGraw-Hill, New York, 1978.
2. L.J. Durney (Ed.), *Electroplating Engineering Handbook,* Fourth Edition, Van Nostrand, Reinhold, 1984.
3. J.W. Dini, *Electrodeposition- The Materials Science of Coatings and substrates,* Noyes Publications, New Jersey, 1993.
4. ASM Handbook, *Surface Engineering,* Vol. 5, ASM International, Materials Park, Ohio, 1994.
5. M. Schlesinger and M. Paunovic (Eds.), *Modern Electroplating,* Fourth Edition, John Wiley & Sons, Inc., New York, 2000.
6. E. Budevski, G. Staikov and W.J. Lorenz, *Electrochemical Phase Formation and Growth,* VCH Publishers, New York, 1996.
7. U. Landau, "Plating-New Prospects for an Old Art" in: *Electrochemistry in Industry: New Directions,* U. Landau, E. Yeager and D. Kortan, (Eds.), Plenum Press, New York, 1982.
8. D. M. Hembree, Jr., Plat. Surf. Finish. 73(11) (1986) 54.
9. T.C. Franklin, Surf. Coat. Technol. 30 (1987) 415.
10. L. Oniciu and L. Muresan, J. Appl. Electrochem. 21 (1991) 565.
11. W.H. Safranek, *Properties of Electrodeposited Metals and Alloys,* Second Edition, American Electroplaters and Surface Finishers Society, Orlando, FL, 1986.
12. V.A. Lamb, "Plating and Coating Methods: Electroplating, Electroforming, and Electroless Deposition", in: *Techniques of Materials Preparation and Handling,* Part 3, R.F. Bunshah (Ed.), Chapter 32, Interscience Publishers, 1968.
13. W.H. Safranek, Plat. Surf. Finish. 75(6) (1988) 10.
14. J.W. Dini, Plat. Surf. Finish. 75(10) (1988) 11.
15. J.P. Celis, J.R. Roos and C. Buelens, "Texture in Metallic, Ceramic and Composite Coatings" in: *Directional Properties of Materials,* H.J. Bunge (Ed.), DGM Informationsgesellschaft mbH, Oberursel, Germany, 1988, p. 189.

16. S.J. Shaffer, J.W. Morris, Jr., and H.-R. Wenk, "Textural Characterization and its Application on Zinc Electrogalvanized Steels" in: *Zinc-Based Steel Coating Systems: Metallurgy and Performance*, G. Krauss and D. K. Matlock (Eds.), The Minerals, Metals & Materials Society, 1990, p. 129.
17. J. Amblard, I. Epelboin, M. Froment and G. Maurin, J. Appl. Electrochem. 9 (1979) 233.
18. A. Kutzelnigg, Plating 48 (1961) 382.
19. M. Clarke, "Porosity and Porosity Tests" in: *Properties of Electrodeposits: Their Measurement and Significance*, R. Sard., H. Leidheiser, Jr., and F. Ogburn (Eds.), The Electrochemical Society, 1975, p. 122.
20. R.J. Morrissey, "Porosity and Galvanic Corrosion in Precious Metal Electrodeposits" in: *Electrochemical Techniques for Corrosion Engineering*, R. Baboian (Ed.), National Association of Corrosion Engineers, Houston, 1985.
21. D.R. Gabe, "Metallic Coatings for Protection", in: *Coatings and Surface Treatment for Corrosion and Wear Resistance*, K.N. Strafford, P.K. Datta and C.G. Googan (Eds.), Chapter 4, Ellis Horwood Limited, 1984.
22. J.B. Kushner, Met. Prog. 81(2) (1962) 88.
23. A. Brenner, *Electrodeposition of Alloys: Principle and Practice*, Academic Press, 1963.
24. J.-Cl. Puippe, F. Leaman (Eds.), *Theory and Practice of Pulse Plating*, AESF, Orlando, FL, 1986.
25. C. VanHorn, "Pulsed-Current Plating" in: *Surface Engineering*, ASM Metals Handbook, Vol. 5, ASM International, Ohio, 1994, pp. 282-284.
26. D. Landolt and A. Marlot, Surf. Coat. Technol. 169-170 (2003) 8.
27. H.Y. Chen, J. Electrochem. Soc. 118 (1971) 551.
28. R.J. Tedeschi, Met. Finish. 69(11) (1971) 49.
29. M. Datta and D. Landolt, Surf. Technol. 25 (1985) 97.
30. R.D. Grimm and D. Landolt, Surf. Coat. Technol. 31 (1987) 151.
31. G. Qiao, T. Jing, N. Wang, Y. Gao, X. Zhao, J. Zhou, and W. Wang, J. Electrochem. Soc. 153(5) (2006) C305.
32. C. Karakus and D. T. Chin, J. Electrochem. Soc., 141, 3 (1994).
33. G. Tzanavaras and U. Cohen, U.S. Patent 5,421,987 (1995).
34. U. Cohen and G. Tzanavaras, Solid State Technol. 2001, 44.
35. U. Cohen and G. Tzanavaras, Solid State Technol. 2001, 61.
36. M. de. Vogelaere, V. Sommer, H. Sprigborn and U. M. Mohammadein, Electrochim. Acta, 47 (2001) 109.
37. H. Takeuchi, Y. Tsunekawa and M. Okumiya, Mater. Trans. JIM, 38(1) (1997) 43.
38. S. Zein El Abedin and F. Endres, Chem. Phys. Chem. 7 (2006) 58.
39. Y. Katayama, "Electrodeposition of metals in ionic liquids" in: *Electrochemical Aspects of Ionic Liquids*, Hirouyuki Ohno (Ed.), Chapter 9, Wiley-Interscience – A John Wiley & Sons, Inc. Publication, 2005, pp. 111-131.
40. S. Zein El Abedin, U. Welz-Biermann and F. Endres, Electrochem. Commun. 7 (2005) 941.
41. G.G. Gawrilov, *Chemical (Electroless) Nickel Plating*, Portcullis Press Ltd., Surrey, 1979.
42. G.O. Mallory and J.B. Hajdu (Eds.), *Electroless Plating: Fundamentals and Applications*, AESF, Orlando, 1991.
43. W. Riedel, *Electroless Plating*, ASM International, Ohio, 1991.
44. Y. Okinaka, T. Osaka, Gerischer and C.W. Tobias (Eds.), *Advances in Electrochemical Science and Engineering*. Chapter 3, VCH Publishers, Weinheim, 1994, p. 55.
45. F.B. Mainier and M.M. Araujo, *IPE Advanced Technology Series*, 2.1 (1994) 63-67.

46. M. Paunovic, Plating, 55(11) (1968) 1161.
47. T. Homma, I. Komatsu, A. Tamaki, H. Nakai and T. Osaka, Electrochimica Acta, 47 (2001) 47.
48. M.K. Gorbunova, M.V. Ivanov, V.P Moiseev, J. Electrochem. Soc. 120(5) (1973) 613.
49. R.C. Agarwala and V. Agarwala, Sadhana, 28(3-4) (2003) 475.
50. S.H. Park and D.N. Lee, J. Mater. Sci. 23 (1988) 1643.
51. Y.W. Wang, C.G. Xiao and Z.G. Deng, Plat. Surf. Finish. 79(3) (1992) 57.
52. S.L. Lee and H.H. Liang, Plat. Surf. Finish. 78(9) (1991) 82.
53. G. Lu and G. Zangari, Electrochim. Acta 47 (2002) 2969.
54. M. Schlesinger, X. Meng and D.D. Snyder, J. Electrochem. Soc. 137(6) (1990) 1858.
55. M. Schlesinger, X. Meng and D.D. Snyder, J. Electrochem. Soc. 138(2) (1991) 406.
56. M. Obradovic J. Stevanovic, A. Despic, R. Stevanovic and J. Stoch, J. Serbian Chem. Soc. 66(11-12) (2001) 899.
57. J.N. Balaraju, T.S.N. Sankara Narayanan and S.K. Seshadri, J. Appl. Electrochem. 33 (2003) 807.
58. T.S.N. Sankara Narayanan, K. Krishnaveni and S.K. Seshadri, Mater. Chem. Phys. 82 (2003) 771.
59. T.S.N. Sankara Narayanan, I. Baskaran, K. Krishnaveni and S. Parthiban, Surf. Coat. Technol. 200 (2006) 3438.
60. Z. Chen, A. Ng, J. Yi and X. Chen, J. Magn. Magn. Mater. 302 (2006) 216.
61. C. Gu, J.S. Lian, G. Li, L. Niu and Z. Jiang, Surf. Coat. Technol. 197 (2005) 61.
62. C. Gu, J.S. Lian and Z. Jiang, Advanced Engineering Mater. 7(11) (2005) 1032.
63. T. Mooney, "Electroplated Coatings" in: *Corrosion: Fundamentals, Testing, and Protection*, Vol. 13A, *ASM Handbook*, ASM International, 2003, pp. 772–785.
64. R.P. Frankenthal, "Corrosion in Electronic Applications", in: *Properties of Electrodeposits, Their Measurement and Significance*, R. Sard, H. Leidheiser, Jr., and F. Ogburn (Eds.), Chapter 9, The Electrochemical Society, 1975.
65. E.C. Felder, S. Nakahara and R. Weil, Thin Solid Films, 84 (1981) 197.
66. J.L. Weininger and M. W. Bretter, J. Electrochem. Soc. 110 (1963) 484.
67. R. F. Ashton and M. T. Hepworth, Corros. 24 (1968) 50.
68. K. Raeissi, M.A. Golozar, A. Saatchi and J.A. Szpunar, Trans. IMF, 83(2) (2005) 99.
69. R. Weil and H. K. Tsowmas, Plating 49 (1962) 624.
70. R. Weil and W. N. Jacobus, Jr., Plating 53 (1966) 102.
71. A. Barbucci, G. Farne, P. Matteazzi, R. Riccieri and G. Cerisola, Corros. Sci. 41 (1999) 463.
72. R. Rofagha, R. Langer, A.M. El-Sherik, U. Erb, G. Palaumbo, K.T. Aust, Scripta Metal. et Mater. 25 (1991) 2867.
73. A.M. El-Sherik, U. Erb, Plat. Surf. Finish. 82 (1995) 85.
74. O. Ekedim, H. S. Cao and D. Guay, J. Mater. Process. Technol. 121 (2002) 383.
75. X.Y. Wang, D.Y. Li, Electrochim. Acta, 47 (2002) 3939.
76. W. Zeiger, M. Schneider, D. Scharnwber, H. Worch, Nanostruct. Mater. 6 (1995) 1013.
77. A. Barbucci, G. Farne, P. Mattaezzi, R. Riccieri and G. Cereisola, Corros. Sci. 41 (1999) 463.
78. Kh.M.S. Yousef, C.C. Koch, P.S. Fedkiw, Corros. Sci. 46 (2004) 51.
79. R. Rofagha, S.J. Splinter, U. Erb, Nanostruct. Mater. 4 (1994) 69.
80. L. Wang, J. Zhang, Y. Gao, Q. Xue, L. Hua and T. Xu, Scripta Materialia 55(7) (2006) 657.
81. R. Mishra and R. Balasubramaniam, Corros. Sci. 46 (2004) 3019.
82. A. Vinogradov, T. Mimaki and S. Hashimoto, Scripta. Mater. 41 (1999) 319.
83. J.K. Yu, E.H. Han, L. Lu, X.J. Wei and M. Leung, J. Mater. Sci. 40 (2005) 1019.

84. A. Balyanov, J. Kutnyakova and N.A. Amirkhanova, Scripta. Mater. 51 (2004) 225.
85. A. Aledresse and A. Alfantazi, J. Mater. Sci. 39 (2004) 1523.
86. INCO Guide to Nickel Plating, International Nickel Co., Saddle Brook, NJ (1989).
87. J. H. Lindsay and D. D. Snyder, "Electrodeposition Technology in the Automotive Industry" in: *Electrodeposition Technology, Theory and Practice*, L. T. Romankiw, and D. R. Turner (Eds.), The Electrochemical Soc. 1987.
88. I. Ivanov, T. Vailkova and I. Kirilova, J. Appl. Electrochem. 32 (2002) 85.
89. I. Ivanov and I. Kirilova, J. Appl. Electrochem. 33 (2003) 239.
90. J. Fei and G.D. Wilcox, Surf. Coat. Technol. 200 (2006) 3533.
91. M.R. Kalantary, G.D. Wilcox and D.R. Gabe, Br. Corros. J. 33 (1998) 197.
92. G. Chawa, G.D. Wilcox and D.R. Gabe, Trans. IMF 76 (1998) 117.
93. L. Yongzhong, J. Jensen, G. Wilcox and D. Gabe, *Proceedings of the Annual Technical Conference SUR/FIN'98*, Session F-Electroless Processes, AESF, 1998, pp. 829–840.
94. I. Kirilova, I. Ivanov and St. Rashkov, J. Appl. Electrochem. 28 (1998) 637.
95. I. Kirilova and I. Ivanov, J. Appl. Electrochem. 29 (1999) 1133.
96. K. Parker, *Proceedings of the 8th International Conference*, Forster-Verlag, Zurich, Switzerland, 1972, pp. 202-207.
97. R. Parkinson, Properties and Applications of Electroless Nickel, NiDI Technical paper Number-10081, Nickel Development Institute, 1997 (www.nickelinstitute.org).
98. D.D.N. singh and R. Ghosh, Surf. Coat. Technol. 201(1-2) (2006) 90.
99. R.N. Duncan, Plat. Surf. Finish. 73(7) (1986) 52.
100. R.P. Tracy, J. Colaruotolo, A. Misercola, B.R. Chuba, Mater. Perform. 25(8) (1986) 21.
101. G. Salvago and G. Fumagalli, Met. Finish. 85(3) (1987) 31.
102. M.K. Totlani, Trans. Met. Finish. Assoc. India, 1(3) (1992) 29.
103. R.B. Diegle, N.R. Sorensen, C.R. Clayton, M.A. Helfand and Y.C. Yu, J. Electrochem. Soc. 135(5) (1988) 1085.
104. J.R. Vilche and A.J. Ariva, *Passivity of Metals*, R.P. Frankenthal and J. Kruger (Eds.), The Electrochemical Society, Princeton, New Jersey, 1978, p. 861.
105. R.N. Duncan and T.L. Arney, Plat. Surf. Finish. 71 (1984) 49.
106. T.S.N. Sankara Narayanan and S.K. Seshadri, J. Alloys Compounds, 365 (2004) 197.
107. C. Kerr, D.Barker and F.C. Walsh, Trans. IMF 74(6) (1996) 214.
108. W.J. Tomlinson and M.W. Carroll, J. Mater. Sci. 25 (1990) 4972.
109. L. Das, D -T. Chin, R.L. Zeller and G.L. Evarts, Plat. Surf. Finish. 82(10) (1995) 56.
110. R.N. Duncan, Plat. Surf. Finish. 83(10) (1996) 64.
111. E. Yang, S.K. Doss and P. Peterson, Plat. Surf. Finish. 75(12) (1988) 60.
112. Y. Liu and Q.Zhao, Appl. Surf. Sci. 228 (2004) 57.
113. Q. Zhao, Y. Liu and E.W. Abel, Mater. Chem. Phys. 87 (2004) 332.
114. Q. Zhao and Y. Liu, Corros. Sci. 47 (2005) 2807.
115. K. Aoki, O. Takano, J. Met. Finish. Soc. Jpn. 39(2) (1988) 81.
116. Z. Bangwei, W.Y. Hu, X.Y. Qu, Q.L. Zhang, H. Zhang and Z.S. Tan, Trans. IMF 94(2) (1996) 69.
117. M. Palaniappa, Ph.D. Thesis, Indian Institute of Technology Madras, Chennai, 2005.
118. B. Veeraraghavan, Bala Haran, S.P. Kumaraguru and B. Popov, J. Electrochem. Soc. 150(4) (2003) B131-B139.
119. P.R. Ebdon, Plat. Surf. Finish. 75(9) (1988) 65.

120. J. Henry, Met. Finish. 88(10) (1990) 15.
121. F.N. Hubbell, Trans. IMF 56 (1978) 65.
122. M.S. Hussain and T.E. Such, Surf. Technol. 13 (1985) 119.
123. M.A. Shoeib, S.M. Mokhtar and M.A. Abd El-Ghaffar, Met. Finish. 96(11) (1998) 58.
124. J.N. Balaraju and S.K. Seshadri, J. Mater. Sci. Lett. 17 (1998) 1297.
125. J.N. Balaraju, T.S.N. Sankara Narayanan and S.K. Seshadri, J. Solid State Electrochem. 5 (2001) 334.
126. E.W. Brooman, Met. Finish. 99 (2001) 100.
127. J.C.A. Batista, A. Matthews and C. Godoy, Surf. Coat. Technol. 142–144 (2001) 1137.

CHAPTER 7

Plasma Methods for Deposition of Cr-based Coatings for Corrosion Control

V.S. Raghunathan and P. Kuppusami
Physical Metallurgy Division, Indira Gandhi Centre for Atomic Research, Kalpakkam-603 102, India

Surface engineering using plasma is the most effective way of exploiting materials for technological applications. In many applications, the properties desired at the surface are different from those at the bulk of the components. Plasma based techniques have become environmentally benign surface structuring process and offer several advantages over other methods. Surface engineering methods can be grouped into two broad categories namely surface modification and overlay coating. This article reviews the application of newly emerging surface engineering methods such as plasma assisted chemical vapour deposition, pulsed plasma nitriding, magnetron sputtering and plasma source ion implantation for the development of Cr- based coatings which can limit corrosion and wear rates. The development of Cr-N coating has been dealt in detail because of its application as a hardfacing material in fast breeder reactors.

I. INTRODUCTION

Coatings have revolutionalized the properties of engineering components not only in terms of improved wear resistance, good corrosion resistance and low friction coefficients but also their aesthetics. Coating of components with thin chromium and Cr-based coatings has been an important engineering practice for many years in the automotive, aerospace, nuclear and decorative industries for various reasons [1-3]. The components operating in aggressive environment like high temperature, corrosive fluids and stresses undergo corrosion and corrosion related degradation. While every attempt is made in the selection of components, proper design and operating environment, anticipated and unanticipated corrosion related degradation of components is inevitable. Therefore, it is important from a technological and economical point of view to protect the components from corrosion degradation. Surface engineering which includes the modification of the substrate surfaces without any external coatings and the deposition of overlay coatings on the surfaces of the substrate has opened up new possibilities to achieve better corrosion and tribological properties.

Chromium and Cr-based nitride coatings have proved to be wear resistant and also resistant to chemically aggressive environments. However, methods for depositing Cr-based coatings by traditional electroplating [4-5] and thermochemical processes [6-8] have inherent pollution problems. Several alternative technologies exist to coat a substrate without using electrolytic solutions or plating baths. These technologies eliminate the use of non-metal toxic components such as cyanide from the plating process. They also can reduce the amount of metal-contaminated waste water and sludge that is generated from plating. For instance, the primary problems with hard chrome plating are that it uses a hexavalent chrome solution and produces large volumes of chrome contaminated toxic wastes. Hexavalent chrome in solution is a known human carcinogen and creates other health problems such as skin and lung cancer [9, 10]. In addition, the films deposited using wet chemical technology do not provide good corrosion protection, as a result of the tendency for cracks to appear in the film, and their hardness decreases with increasing service temperature and therefore their application is limited [11]. Also, surface treatments such as pack chromising and gas nitriding based on chemical vapour deposition (CVD) process require very high process temperatures (1173-1273K), which may deform the substrates. Furthermore, entrapped particles from the pack in the coating may be observed, thus reducing the beneficial properties of such coatings [12].

Among the several methods of deposition, plasma based deposition techniques are more promising and widely used because the chemical reactions that take place normally at very high temperatures can be achieved in a low temperature plasma state. Plasma based techniques have become accepted surface modification techniques over other methods owing to several advantages. The introduction of plasma brings in numerous advantages, such as: (i) low processing temperature (ii) ability to form uniform deposition over large area on metallic as well as ceramic substrates, (iii) deposition on any geometry that is exposed to the plasma, and (iv) low substrate distortion.

Besides Cr, Cr-based coatings such as CrN, TiCrN, CrN/Ni can benefit many industries, which use stainless steel (SS) as the major structural core component [13, 14]. For instance, chromium nitride due to its high hardness, can improve the chemical and mechanical properties like wear, corrosion and fatigue resistance of materials. Ni coatings and its composite coating with CrN can improve the tribological properties of stainless steel substrates. By combining a brittle transition metal nitride like CrN with a soft metal like Ni, the coating is reported to offer better fracture toughness [15]. Mechanical properties including surface hardness, wear resistance and corrosion behaviour can be greatly influenced by Ni as interlayer for CrN coatings [16].

In the present paper, firstly, we investigate the application of plasma in a plasma assisted metalorganic CVD of nanostructured coatings of Cr, CrN and CrN/ Ni using chromium (III) acetylacetonate (Cr (acac)$_3$) and nickel acetylacetonate (Ni [(acac)$_2$ en]) as a chemical vapour source for Cr and Ni, respectively. Secondly, process details and the microstructures obtained in plasma nitrided Cr-plated type 316LN stainless steel are discussed. Finally, a brief outline on the deposition of Cr based coatings by magnetron sputtering and plasma source ion implantation (PSII) is presented. The relation between structural, morphological, mechanical and corrosion properties of Cr-based coatings are the focus of this work.

2. PLASMA DEPOSITION TECHNIQUES

In recent years, application of glow discharge (low pressure) plasma has rapidly expanded owing to a large chemical freedom offered by the non-equilibrium aspects of the plasma. The energetic species in the plasma are free electrons and they gain energy from the electric field much faster than ions. Consequently, the electrons accumulate sufficient kinetic energy to undergo inelastic collisions and to sustain the ionization. In typical low-pressure plasma, the electron temperature is around 3-5 eV and the ion and neutral temperatures are much smaller. With these energies, the electrons will produce a large supply of excited and dissociated molecules and reactive radicals, even for a modest degree of ionization. Thus, plasma will be able to crack relatively stable molecules and encourage deposition at much low temperatures and pressures than would be required in a thermal CVD. In addition, the surfaces exposed to plasma are subjected to bombardment by energetic ions resulting in surface modification of materials. The ionic bombardment of this nature has significant effect on the properties of the deposited film. The bombardment of the surfaces with non-reactive or reactive ions also leads to sputtering.

Among several applications based on plasma, surface modification of materials by pulsed plasma nitriding and deposition of thin films by plasma assisted metal-organic chemical vapour deposition (PAMOCVD) and magnetron sputtering are major deposition methods where direct current (DC) glow discharge plasmas are dominantly used. Recently, plasma source ion implantation (PSII) has been developed as an alternative process to beam line ion implantation because the process is a non- line -of- sight process where entire surface of three dimensional substrate exposed to plasma could be treated.

3. PLASMA ASSISTED CHEMICAL VAPOUR DEPOSITION

Chemical vapour deposition involves the condensation of a compound or compounds from a gas phase onto a heated substrate where reaction occurs to produce a solid deposit. The precursor bearing the deposit material, if not in the vapour state, is formed by volatilization from either a liquid or solid feed and is caused to flow either by a pressure differential or by the action of a carrier gas to the substrate. Though several types of CVD methods are reported, plasma assisted metal-organic chemical vapour deposition (PAMOCVD) has been shown to be the most promising technique for Cr-based coatings. The method consists of transporting metal-organic vapours (precursor) of desired composition to the substrate that is kept above the decomposition temperature of the precursor. In plasma assisted MOCVD, DC glow discharge plasma is superposed in order to effectively decompose the precursor even at low substrate temperatures.

In PAMOCVD technique, vapours of metal-organic compound such as chromium (III) acetyl acetonate (Cr $(acac)_3$) have been used as the precursor for Cr and CrN coatings. Crystals of Cr $(acac)_3$ complexes were prepared from high purity (99.4%) $CrCl_3.H_2O$. Phase pure Cr films were prepared by cracking the vapours of $Cr(acac)_3$ in Ar-H_2 plasma, while CrN films were prepared by decomposing a mixture of $Cr(acac)_3$ added with a small amount of ammonium bi-fluoride (NH_4FHF) in nitrogen-hydrogen plasma. The composite coatings of Ni with CrN was developed using different proportion of the precursors of Ni (Ni [$(acac)_2$en]) and Cr ($Cr(acac)_3$) in N_2 plasma. The schematic diagram of the system is shown in Fig. 1[17]. A rotary vacuum pump was used to evacuate the system. The substrates were placed on the cathode. A laboratory built dc power supply (0-1000V) was used to generate the plasma. The cathode plate was heated using an external heater. The substrate temperature was measured using a K-type (Chromel-Alumel) thermocouple that has been previously calibrated with respect to the actual substrate temperature. Temperature was controlled using a PID temperature controller with an accuracy of ± 2K. Care was taken to electrically isolate the thermocouple from the cathode power supply. Appropriate powders of metal organic precursors were pelletised and placed in the vaporiser chamber, F. Vapours of these precursors were carried into the deposition chamber, A, through a heated gas line H, using nitrogen (99.9% pure) as the carrier gas. Additional N_2 and H_2 (99.9% pure) gases were also fed to the deposition chamber. All the depositions have been carried out at sub-atmospheric pressures, typically 0.2 kPa and at a substrate temperature of 823 K. Typical experimental conditions for Cr deposition are given in Table 1. During the heating cycle of the substrates, the chamber was maintained at a partial pressure of ~0.5 kPa of hydrogen, in order to prevent the formation of a stable passive oxide (Cr_2O_3) film on the hot stainless steel (SS). The deposition time for all the experiments was from 4 to 8 hours. The hydrogen to nitrogen gas flow ratio (**r**) and the amount of the ammonium halide additive were varied during the experiments. Polycrystalline CrN thin films were deposited on well polished 316LN stainless steel(SS) and ceramic substrates such as yttria stabilised zirconia (YSZ), quartz and silicon.

3.1 Cr Coatings

Figure 2(a) shows a typical XRD profile of 4 μm thick Cr coating deposited on SS substrate at a substrate temperature of 823 K. Reflections from (211), (200) and (110) planes of body centered

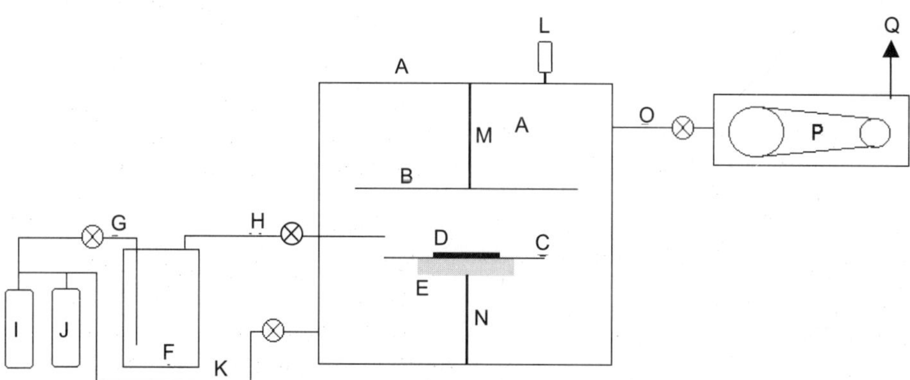

Figure 1 Schematic diagram of the system used for deposition of CrN coating by PAMOCVD technique, where A denotes Water-cooled CVD vacuum chamber. B: Anode Plate, C: Cathode Plate, D: Substrate, E: Substrate Heater, F: Vaporiser chamber (heated), G: Carrier gas line, H: Heated Gas line, I & J: N2 & H2 gas cylinders, K: Gas line, L: High pressure Pirani gauge, M: Insulating stand for anode, N: Insulating stand for cathode, O: Pumping line, P: Rotary pump, Q: Exhaust line and ⊗ Valve.

Table 1 Deposition parameters used to synthesize Cr films

Parameters	Values
Precursor	$Cr(acac)_3$
Substrate temperature	823 K
Deposition pressure	2-4 mBar
Precursor temperature	523 K
Line heater temperature	548 K
H_2 flow rate (bypass)	2-10 lph
Ar flow rate (bypass)	40 lph
Ar flow rate (carrier)	5 lph
Power density	70 mW/cm^2
Duration	4 h

cubic (bcc) Cr are clearly identified signifying the polycrystalline nature of the Cr coating. The crystallite sizes of the Cr coating deposited on SS substrate have been evaluated from the corresponding X-ray diffraction pattern. The angular dispersion corresponding to full width at half maximum (FWHM) was measured for Cr phase [18]. The average grain size of the Cr coatings was calculated using Scherrer's formula [19]. The calculated grain size from the Cr (110), (200) and (211) reflections was about 10 nm. Hence, nanocrystalline Cr coatings were obtained through this technique. Thus, the feasibility of the PAMOCVD technique in the synthesis of Cr coatings has been established.

The surface morphology of the Cr films was examined using scanning electron microscope equipped with an energy dispersive x-ray spectrometer (EDS). Figure. 2 (b) shows the microstructure of the Cr films deposited on SS substrates at 823 K. It is seen that the films are uniform with no porosity and voids throughout the coatings. The coatings exhibit micron sized particles, which are actually agglomerates of several tiny crystallites of Cr (~10 nm). The dense morphology and the absence of cracks in these films are expected to be beneficial especially in corrosion resistance applications by providing an impervious physical barrier to the active corrosive species.

The microhardness measurements, which is an indication of the mechanical strength of the coatings was performed on the Cr films. Hardness values of around 1200 and 830 VHN were obtained for 100 and 200g loads, respectively. This value is higher than the hardness of Cr coatings obtained by either electroplating, pack-chromising or by PVD techniques. Such a high hardness for these Cr films may be attributed to the nano crystalline grains formed in the films.

3.2 CrN Coatings

The XRD patterns of CrN coatings deposited on SS and YSZ substrates are shown in Fig. 3. The hydrogen to nitrogen flow ratio (**r**) during the deposition of these thin films has been maintained at 1.05. $Cr(acac)_3$ with 10 wt% of NH_4FHF has been used as the precursor [17, 20]. Diffraction from (111), (200), (220) and (311) planes of face centred cubic (fcc) CrN are clearly identified in the case of CrN coated on YSZ substrate. Thus the polycrystalline nature of the CrN coating is established. Some of the X-ray reflections from SS substrate overlap with those

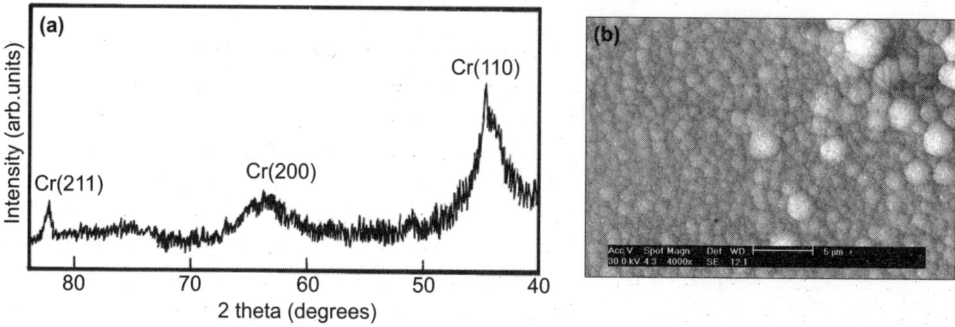

Figure 2 Cr films deposited on SS substrate at 823 K; (a) XRD pattern and (b) SEM Image

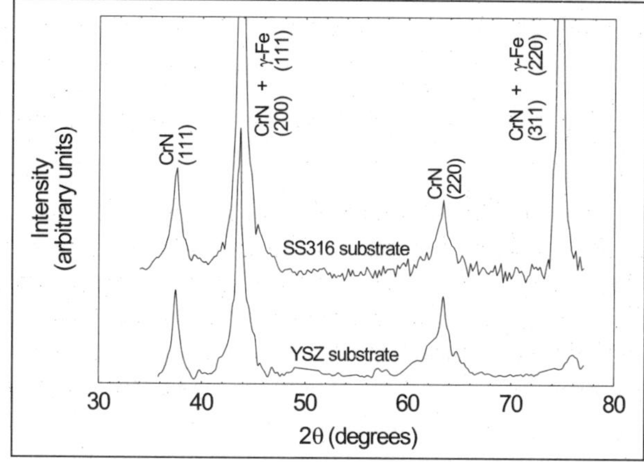

Figure 3 X-ray diffraction patterns of CrN coatings on stainless steel (SS316) and YSZ substrates under identical deposition conditions of hydrogen to nitrogen gas flow ratio(**r**)=1.05 and with 10wt% of NH_4FHF.

of CrN coating. However, the CrN peaks corresponding to the planes (111) and (220) confirm the deposition of polycrystalline CrN films on stainless steel substrates as well. Peaks corresponding to oxides and carbides of Cr or its other nitride such as Cr_2N have not been observed in the XRD spectra. Therefore the feasibility of the PAMOCVD technique in depositing a mono phase polycrystalline CrN coating is demonstrated. Formation of pure CrN has also been reported using PVD techniques [21, 22]. This may be contrasted with the formation of a mixture of CrN and Cr_2N in a diffusion controlled processes, such as, gas nitrided Cr in a NH_3-H_2 environment [23], and plasma nitriding of electroplated Cr [24, 25].

Figure 4 shows the effect of '**r**' on the crystallinity of CrN coatings deposited on YSZ substrates, while using $Cr(acac)_3$ and 10 wt% NH_4FHF as the precursor. It is seen from the figure that when '**r**' is 0, the film is amorphous as exhibited by the broad and diffuse peaks of the corresponding XRD pattern. From the XRD patterns corresponding to '**r**' = 0.01 and 1.05, it is noted that hydrogen addition increases the crystallinity of CrN films significantly. Therefore, at least 1% hydrogen dilution is necessary for the growth of CrN crystallites. It has been suggested [26] that hydrogen is required to promote surface diffusion of the precursors and

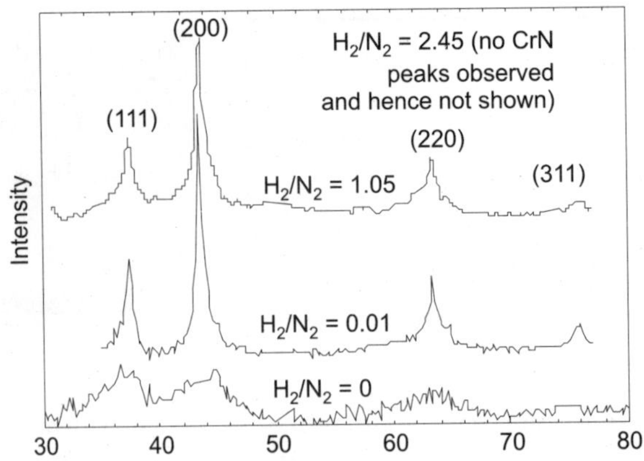

Figure 4 XRD patterns of CrN coatings deposited on YSZ substrates under various hydrogen to nitrogen gas flow ratio ratios. 10wt% of NH_4FHF has been used for these experiments.

prevent random nucleation on the substrate leading to amorphisation of the deposited film. In general, the film nucleates and grows by island formation according to Volmer-Weber mode [27, 28]. Optimum hydrogen coverage of the substrate can increase surface diffusion length of the incoming Cr and N species. This would prevent instantaneous adsorption of the growth species, and instead promote surface diffusion till a suitable growth site for CrN nuclei is encountered. Such a growth mechanism has been established in Si thin films [29,30]. But at a high hydrogen dilution ratio of 2.45, the deposited film did not contain CrN. This is believed to be due to the loss of nitrogen precursor at the growth site. Presence of excess hydrogen may lead to the formation of ammonia gas in the plasma [17]. Therefore, concentration of NH_4FHF in the metal organic precursor and hydrogen dilution of the precursors are two important parameters governing the formation of polycrystalline CrN thin films.

The lattice parameters and crystallite sizes of the CrN coatings deposited on YSZ substrates using different halide precursors and hydrogen dilution ratios have been evaluated from the corresponding X-ray diffraction patterns. Clearly, the measured lattice parameter is in good agreement with the standard value of 0.414 nm for CrN. The crystallite sizes of the CrN films have been calculated using the Scherrer formula. For the same hydrogen dilution ('**r**' = 1.05) CrN crystallite size reduces from ~10 nm to ~5 nm when the halide precursor is changed from NH_4FHF to NH_4I. But for the same NH_4FHF precursor, crystallite size remains almost same when '**r**' is reduced from 1.05 to 0.01. Thus, nanocrystalline CrN coatings were obtained through a right choice of precursor and hydrogen dilution.

3.3 CrN/Ni Composite Films

The composite coatings of Ni/CrN were developed using different proportion of the precursors of Ni and Cr in N_2 plasma at a substrate temperature of 823 K. Prior to deposition, TG analyses of the mixtures of Cr- and Ni- precursors were carried out to understand the thermal behviour of these precursors. TG analysis of the mixtures of $Cr(acac)_3$ and $Ni[(acac)_2en]$ in different proportions (5, 10, 20 and 50% weight of Ni with Cr precursor) were performed in the temperature range 300-573 K. The results revealed that irrespective of the adopted ratios, all are thermally stable at the evaporation temperature and undergo single step weight loss with no decomposition. The pure Cr and Ni vapour sources sublimate in the narrower temperature range 493-533 K. On the other hand, the precursor mixtures are found to have wide temperature for vapourisation. For instance, precursor mixture of 50 wt.% Cr and 50 wt.% Ni has a vapourisation temperature range of 423-558 K [31]. The wide range of the vapourisation temperature can be utilized to control the partial pressure of the precursors. It should be mentioned here that precursor that sublimates over a small range of temperature require precise and expensive temperature flow controllers. The volatile nature of the mixtures of the two precursors can be effectively utilized to engineer composite coatings of desired composition.

A typical XRD profile for the 5 μm films deposited using equal percentage of both the Ni and Cr precursor is shown in Figure 5. It is clear from the figure that the film contains reflections from CrN and Ni phases respectively (JCPDS 76-2494 and 04-0850). No other peaks due to carbides, oxides and nitrides of both the metals could be observed. From the XRD, it is also inferred that the peaks due to CrN are broad (>1°) compared to that of Ni. The grain size determination of the CrN and Ni shows that they are nano crystalline and the grain size of nickel is larger than the CrN particles. The grain size calculation from (111) and (220) reflections of CrN shows that the average grain size to be ~5 nm whereas for nickel, the average grain size is ~13 nm [32,33]. The present study also showed that the process yields a two-phase nanostructured composite coating of Ni/CrN whatever may be the ratio of precursors of Ni and Cr.

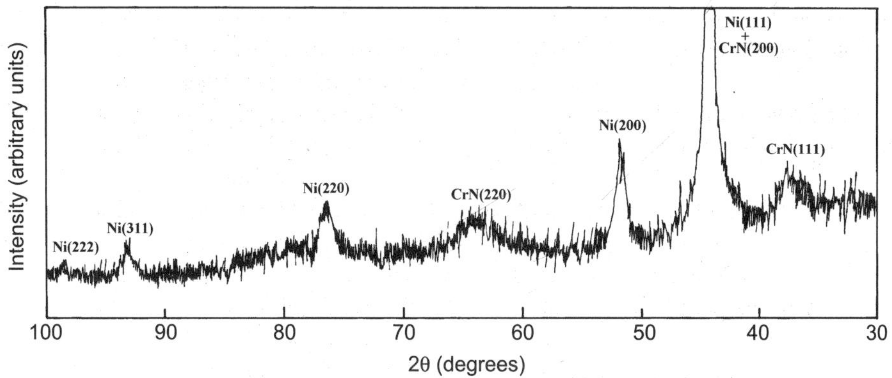

Figure 5 A typical X-ray diffraction profile of the composite coatings of Ni and CrN coated on quartz substrate at 823 K for equal percentage of the Ni and Cr precursor.

The surface morphology of the nano composite films on several substrates (with equal percentage of Ni and Cr precursors) using SEM demonstrated that it forms compact, dense structure, free from pores and cracks. It is also seen that the particle sizes were larger on stainless steel (Fig. 6) substrates than that of the ceramic substrates. This is attributed to the higher surface mobilities and due to the presence of copious amounts of electrically charged species of CrN and Ni on the surface of the conducting substrate. This observation is in agreement with our earlier study on CrN deposition by plasma assisted MOCVD[17].

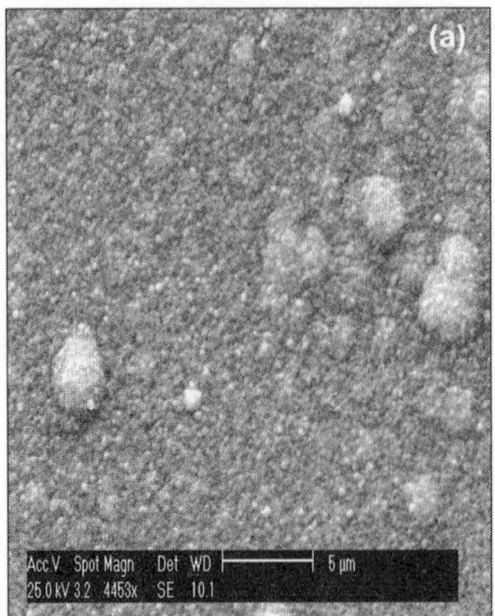

Figure 6 Scanning electron images of the Ni/CrN films deposited at 823 K (a) on quartz substrate and (b) on stainless steel substrate.

The hardness measurements of thin films of Ni, CrN and their composites prepared under identical deposition conditions were carried out. The measurements were performed on the films deposited on the SS substrates under a constant load of 100 g. It has been found that Ni films are much softer than either CrN or CrN/Ni films. A hardness of around 2316 and 440 VHN could be obtained for CrN and Ni films respectively. The hardness reduces to 2277, 1698 and 1134 VHN on adding 25, 50 and 75% of precursor of nickel to chromium precursor (Fig. 7). The addition of nickel makes the CrN films less hard, and provides convenient route to tailor the hardness of the composite coatings.

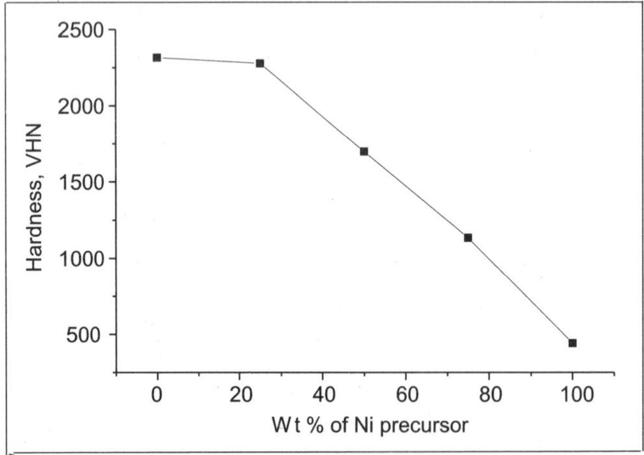

Figure 7 Variation in hardness of CrN-Ni composite coatings by the addition of Ni precursor in different proportions to Cr precursor.

4. PULSED PLASMA NITRIDING OF Cr-PLATED STAINLESS STEEL

Surface modification of materials by plasma nitriding using a glow discharge plasma has become an important environmentally benign surface structuring process to obtain improved hardness, wear resistance and corrosion resistance. Ion bombardment in combination with the plasma induced dissociation promotes chemisorption and surface layer intermixing, with consequence that diffusion coatings can be formed at temperatures lower than in conventional gas nitriding. Plasma nitriding is widely used for improving wear, fatigue and corrosion properties of steel due to its beneficial features such as good reproducibility and low distortion of the treated parts. It is known that depending on the type of metals and corrosion environment, nitriding can either increase or decrease the corrosion resistance [34]. In general, the corrosion resistance of stainless steels is usually adversely affected by nitriding, whereas beneficial effects can be obtained for low alloy steels [35]. It is reported that the improvement in corrosion resistance is related to the presence of γ'-nitride and the predominance of CrN precipitation is responsible for the deterioration in corrosion. The electrochemical corrosion behaviour of AISI 1045, AISI 4340, AISI H13 and AISI 304 without nitriding and after nitriding has been recently reported [36]. The excellent corrosion resistance of stainless steel has been attributed to the presence of native passive layer of Cr_2O_3. The formation of CrN and loss of Cr from SS matrix has led to its poor passivating property. The protection rates were highest in AISI 4340, followed by AISI 1045, AISI H13 and AISI 304 because of the presence of white layer comprising of iron nitrides..

In order to prevent the loss of Cr during nitriding, efforts have been made to electroplate Cr and then convert Cr into CrN by pulsed plasma nitriding [25]. The pulsed plasma nitriding route has been established to develop 70 μm thick Cr-N coating on grid plate sleeve components of PFBR. For this work, type 316 stainless steel samples of dimensions 100 mm × 60 mm × 10 mm were chromium plated and used. The coupons were given heat treatment at about 510 K for 3h to avoid hydrogen embrittlement problem. The coatings in the as deposited condition at room temperature had hardness of about 850-900 HV at a load of 100 g. The chromium plated samples were ultrasonically cleaned in acetone and then loaded onto the cathode plate of the nitriding chamber. Pulsed plasma nitriding was carried out in an indigenously developed pulsed plasma nitriding facility [25]. A schematic of a plasma nitriding system is shown in Fig. 8. The system essentially consists of a vacuum chamber, pumping system, pulsed DC power supply (900 V, 20 A, 1-20 kHz with variable duty cycle), mass flow controllers and temperature controller. The pressure control, voltage control and temperature control were performed through a personal computer interfaced with the nitriding facility. Prior to plasma nitriding, sputter cleaning was conducted in an argon atmosphere at a pressure of about 2 mbar by biasing the samples negatively (500 V pulsed DC) at 523 K for 2 h. Subsequently, plasma nitriding was performed at 0.2-0.5 kPa of nitrogen and hydrogen gas mixture in the ratio 4:1 or 3:1 at a desired temperature for various durations of time. To study the nitriding behaviour of the chromium plated stainless steel samples, the parameters such as nitriding temperature and time were varied. The flow rates of nitrogen and hydrogen were monitored using mass flow controllers and the temperature was controlled by varying the pulse to pause duration and hence the ion current at constant voltage. An auxiliary heater was used to nitride the sample at temperatures above 773 K.

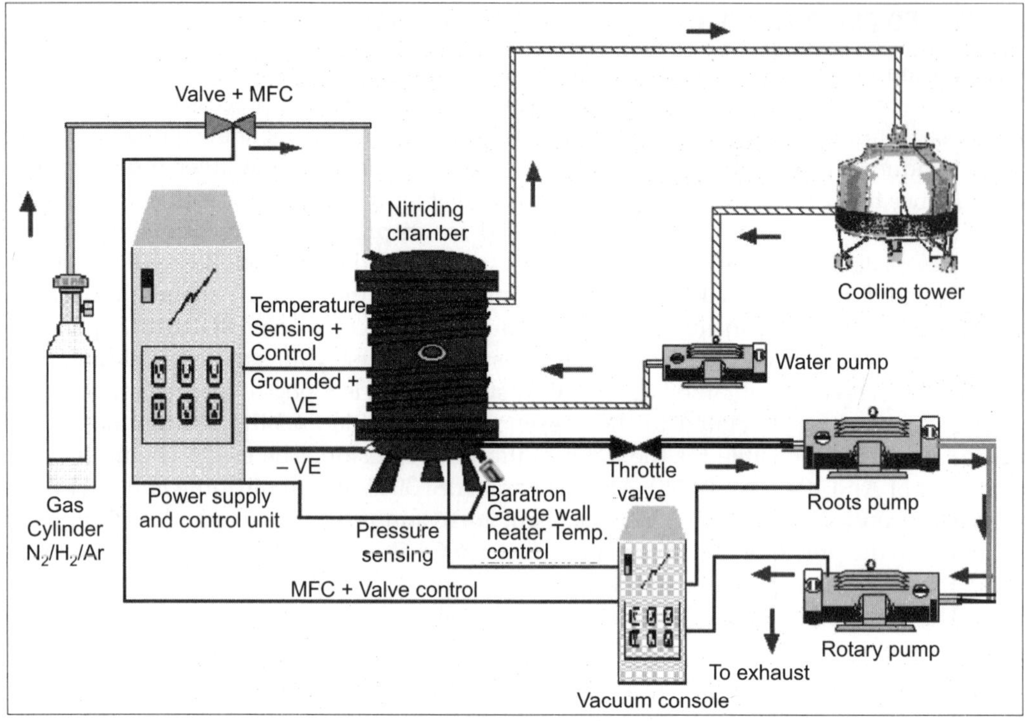

Figure 8 Schematic diagram of pulsed plasma nitriding system

4.1 Microstructure

The plasma nitriding of the samples was carried out for various durations in the temperature range 833-1273K. XRD analysis of the samples indicated the presence of hexagonal Cr_2N phase in the Cr matrix (Fig. 9). The intensity of the Cr_2N diffraction lines was found to increase with increase in the nitriding temperature owing to the increase in thickness of Cr_2N layer with temperature. The presence of untreated chromium was also observed below the surface layer of nitrides. More over, no traces of carbides and oxides were observed within the detectable limits of XRD. It is interesting to note that the nitrides were Cr_2N type and not CrN which were reported in the plasma nitrided type 316 stainless steel [37]. The formation of Cr_2N was found to be more dominant in preference to CrN at an intermediate temperature (~900K) and high temperature ($873K < T < 1273K$) of nitriding of pure chromium. Surface morphology of the nitrided samples showed featureless structure with a random distribution of nitride particles. Also, the surface was found to be etched due to the ionic bombardment of nitrogen and hydrogen ions. A typical cross-sectional SEM microstructure of a sample of plasma nitrided chromium plated stainless steel at 913 K, 45h is shown in Fig. 10. It is seen that the nitrided layer is uniform and the layer boundary is parallel to the sample surface, clearly indicating a planar growth front. Fig. 10(a) also shows the indentation marks due to hardness measurements by a Vickers diamond pyramid indenter. The dimensions of the indentation marks show a gradual decrease in size as a function of distance from the matrix to the nitrided

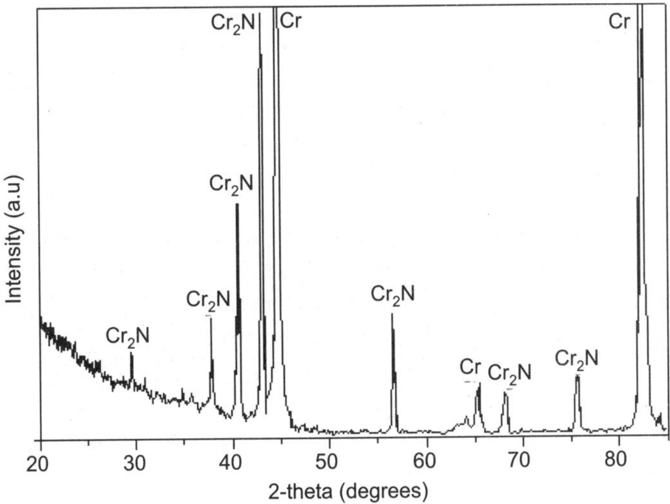

Figure 9 A typical XRD trace of a chromium plated type 316LN stainless steel sample pulsed plasma nitrided at 913 K, 45 h and gas flow ratio of N_2: H_2: 4:1.

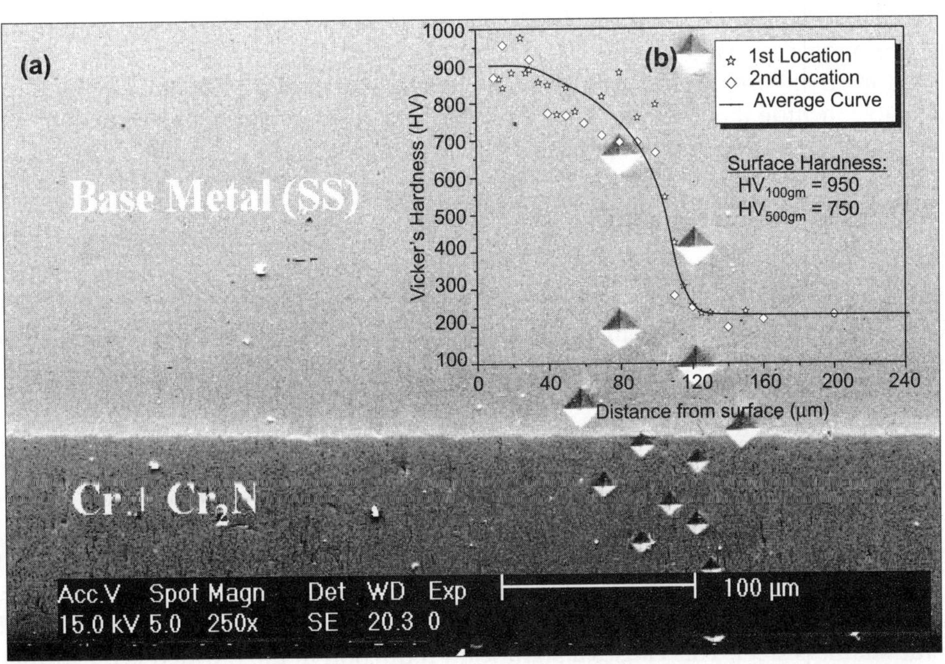

Figure 10 (a) SEM of cross-section of Cr-N coating on SS substrate showing micro-indentation marks; (b) hardness-depth profile across the Cr-N coating/ SS. Cr_2N, Cr and SS316(SS) shown in the figure represent nitrided region containing nitride phase, unconverted Cr and stainless steel substrate respectively.

layer. The hardness-depth profile shown in the inset (Fig. 10(b)) reveals a uniform decrease in hardness as a function of depth for the samples nitrided at the intermediate temperature of 913 K. The sample nitrided at 913 K, 45h has a peak hardness of about 900 VHN for a depth of about 10 μm, hardness plateau of about 700 VHN for a depth of 60 μm and finally followed by a smoothly decreasing hardness profile. When the sample was nitrided at 913 K, 30h, both the hardness and case depth were found to decrease. In both the cases, the decrease in hardness across the chromium-stainless steel interface is not abrupt. In contrast, the hardness-depth profiles have a sharp variation at the interface for the samples nitrided in the temperature range 1173-1273 K [25]. At these temperatures, the nitriding reaction proceeds by the formation of a uniformly hard surface owing to increased reaction rates of chromium with nitrogen which advances progressively into the core. The form of hardness profile developed during the nitriding is known to depend upon the interaction between the nitride forming elements and nitrogen and upon ease of nucleation of a nitride phase [38].

It must also be mentioned about the stability of chromium coating and stainless steel matrix when they were subjected to nitriding at various temperatures. The stability of the these materials was monitored by measuring the surface hardness as a function of nitriding temperature. For example, it was seen that at 823 K, the hardness decreased only very slightly for stainless steel, while it decreased by three times for chromium coating. However, the important advantage of the plasma nitriding even at this temperature was that the surface hardness has increased to about 850 VHN due to the formation of chromium nitride phase. Also, a study on thermal stability of CrN and Cr_2N indicated that there was a decrease in hardness for CrN phase due to decomposition at temperature above 1100K [39]. Therefore, it is plausible to conclude that the high hardness obtained at these high temperatures of nitriding was due to the formation of Cr_2N phase.

4.2 Kinetics of Plasma Nitriding

The observed case depths, hardness and diffusion co-efficient of nitrogen as a function of nitriding temperatures and time are shown in Table 2. It is noticed from the table that when the nitriding temperatures were less than 973 K, the case depths were very small and the hardness obtained were less than 900 HV and at temperatures more that 1073 K, there was a considerable increase in case depth and hardness exceeding 1400 HV. Based on the time of nitriding (t) and case depth (X), the diffusion co-efficients of nitrogen (D) were calculated from the relation, $X \approx 2\sqrt{(Dt)}$. Figure 11 is lnD versus 1/ T plot indicating that the nitrided layer growth in chromium plating is a diffusion controlled process. The activation energy for nitrogen diffusion in chromium determined from the plot was found to be 131.4 kJ/mole [25,40]. Previously, we have reported [41] an activation energy of 69.4 kJ/mole for diffusion of nitrogen in the nitrided layer of type 316 austenitic stainless steel. It is interesting to note that the activation energy for nitrogen diffusion in chromium plated stainless steel is significantly higher than that of austenitic stainless steel. This causes a significant reduction in the nitrided layer formation in the Cr plated stainless steel when compared to austenitic stainless steels especially at low nitriding temperatures. It is suggested that the rate of diffusion of nitrogen has been further restricted due to the formation of hard impervious Cr_2N phase at the surface of the Cr plating.

Table 2 Experimental conditions, case depths, hardness and diffusion coefficients obtained in plasma nitriding of Cr plated Type 316 Stainless Steel

Temperature (K)	Time (h)	Case Depth (μm)	Vickers Hardness (HV)	Diffusion Co-efficient (m^2/s)
833	20	5	550	1.25×10^{-16}
973	3	6	850	3.33×10^{-15}
1073	3	15	1400	20.83×10^{-15}
1173	3	25	1650	57.87×10^{-15}
1273	3	40	1850	148.15×10^{-15}

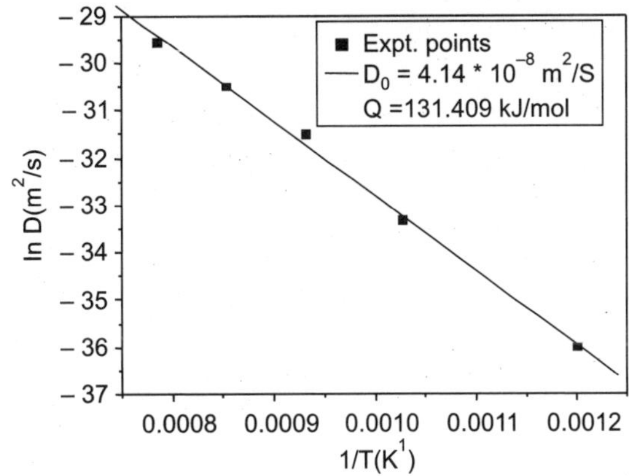

Figure 11 lnD versus 1/T plot for the plasma nitrided samples based on the experimental data given in Table 2.

4.3 Tribolgical Properties

In the case of as deposited chrome plated stainless steel, the thickness of the chromium plating was ~100 μm, surface hardness was ~780 VHN and it was confirmed to be free from macro pores. After plasma nitriding, the surface hardness of the coating was 950 VHN and 750VHN for a load of 100g and 500g, respectively. The case depth was nearly the same as that of the Cr-plating (~100 mm). The coating passed 'U-bend' test and did not form flakes. There was no noticeable dimensional distortion on sleeve due to nitriding. The abrasive wear test carried out in air, at room temperature using Calowear equipment had indicated that the chrome nitride coating exhibited very good wear resistance in comparison with chromium plated stainless steel (Fig. 12).

5. MAGNETRON SPUTTERING

Plasma can be very effectively used in a sputtering process where ejection of atoms from the surface of a material takes place by bombardment with energetic ions. The ejected or sputtered atoms can be condensed on a substrate to form a thin film. One of the important variants of sputtering techniques is magnetron sputtering, where magnetic fields are usually applied by

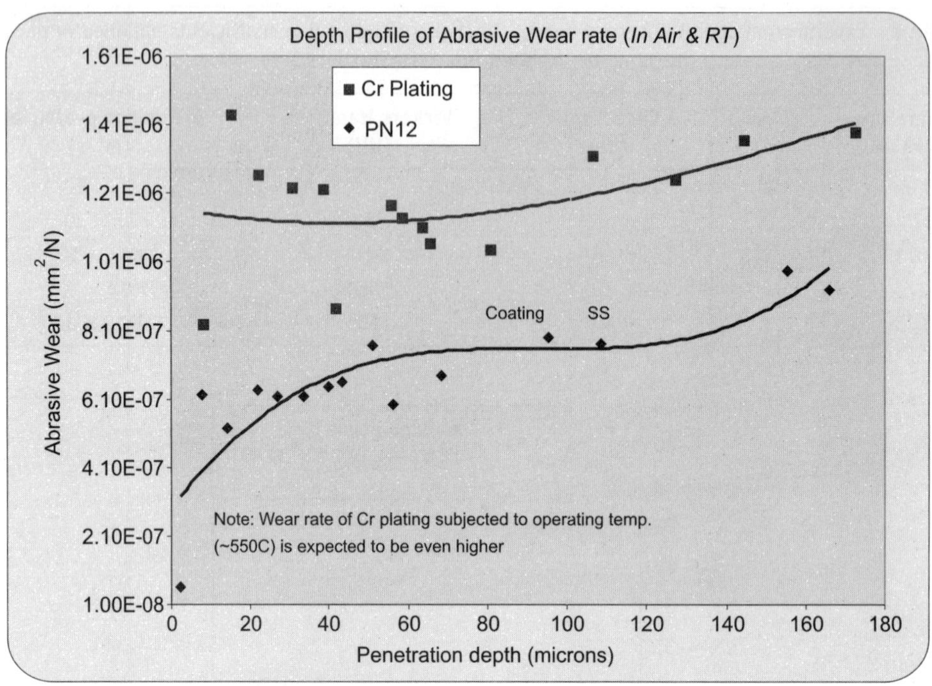

Figure 12 Calowear test conducted on plasma nitrided chromium plated stainless steel and chromium plating.

fixed permanent magnet pole pieces in close vicinity to the target. The fundamental difference between magnetron sputtering as a plasma process and thermally excited thin film preparation methods (evaporation, chemical deposition methods) is the high energy input into the growing film, that can be achieved by magnetron sputtering. In order to tailor the film properties one has to adjust the energy input into the substrate. Major advantages of the technique are (i) low substrate temperature (down to room temperature), (ii) good adhesion of films on substrates, and (iii) alloys and compounds of materials with very different vapour pressures can be sputtered easily by reactive sputtering in rare /reactive gas mixtures and many compounds can be deposited from elemental (metallic) targets.

A schematic diagram of a DC planar magnetron sputtering is shown in Fig. 13. The magnetic field confines the electrons in front of the target in a torus-like plasma region, which causes a non-uniform erosion of the target. The potential distribution in the discharge, i.e. between the target and the substrate, is essential for the sputtering and the deposition of the film. This potential distribution determines the energies of the ions and neutral species, which contribute to the deposition process. The external discharge parameters such as working pressure, discharge power, design of the magnetic field (i.e. balanced or unbalanced magnetrons) and the excitation mode (DC or RF) influence the potential distribution and hence the particle energies. During the process, the positive ions are accelerated towards the cathode, leading to the sputtering of the target. On the other hand, electrons and negative ions move from the target to the substrate. Together with reflected neutral argon atoms, energetic negative ions can reach the substrate and will influence the layer growth.

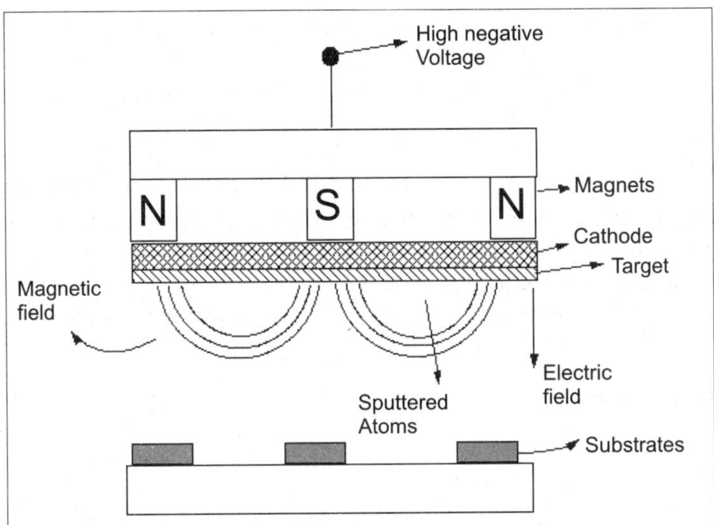

Figure 13 Schematic cross-sectional view of a typical DC planar magnetron sputtering configuration

In DC planar magnetron configuration due to the crossed electric and radial magnetic fields, the path of an electron between collisions will be a series of small cycloids in a large closed loop above the area of erosion. Here the electric and magnetic fields are assumed to be locally uniform. The motion of the electrons along this large closed loop causes a large drift (rotation) current [42, 43]. Some of them make ionizing collisions with gas particles. The ions produced are collected straight by the cathode and result in discharge currents.

The main advantage of the DC sputtering over RF is the continuous and effective magnetic confinement of the electrons near the cathode. In the case of RF sputtering, during the positive half cycle, the target acts as an anode. The deposition rate is reduced under the condition of constant discharge power. The DC process yields deposition rates that are a factor of 1.5 – 2.0 higher than that for RF sputtering. For DC sputtering the deposition starts at relatively low threshold and the deposition rate is proportional to the discharge power [44]. In case of RF excitation, an higher threshold power is required due to low discharge voltages. Additionally the deposition rate is determined by the energy of the ions at the sputtering target and the energy of reflected argon atoms is higher for DC than for RF excitation.

Unlike the thermo-chemical treatment of CrN, most of the synthesis using sputtering techniques has been carried out at substrate temperatures less than 873K. In general CrN coating is characterized by its fine grained and low stress structure, which permits the deposition of much larger thicknesses than conventional PVD coatings of a few μm. Adhesion to steel substrates as well as corrosion resistance are usually improved by the deposition of 0.1-0.5 μm thick Cr intermediate layer [45]. Suzuki et al. [46] have reported the synthesis of CrN by DC reactive sputtering, while nanostructured CrN/Cr multilayers have been prepared by RF magnetron sputtering with varying argon to nitrogen flow ratio [47]. From the powder x-ray diffraction experiments, the crystal structure of CrN was found to be of NaCl type with a lattice parameter of 0.4155 nm. The compound is stable up to 1028K, above which a part of CrN compound begins to decompose and forms Cr_2N phase and at temperatures > 1293 K, Cr_2N

and Cr coexist. It has been found that the hardness of the CrN/Cr multilayers increases monotonically from 15 to 27 GPa with layer thickness decreasing from 120 nm down to 11 nm [47].

The corrosion and oxidation resistance of selected Cr-based hard coatings were systematically studied. Compared to uncoated AISI 304 substrate as well as TiN coatings, CrN coatings show better corrosion resistance in the lower potential range (E < 1.0V) in 0.5M H_2SO4 solution [48]. Among several Cr based coatings exposed to high temperature oxidation, only TiCrN showed lower oxidation rates than CrN coatings as determined by weight gain method [49]. Moulds for plastic processing are made of hot or cold working tool steels, high speed steels and stainless steels. Often these surfaces are protected with 10-20 μm hard chrome. Improvement of these moulds with a hard TiN coating has brought a large step forward in processing of all plastic materials, where the moulds were subjected to abrasive, adhesive and chemical wear. CrN coating was more resistant to all types of corrosion than TiN and TiCN coatings[50]. Therefore, Cr-based coatings are expected to replace TiN coatings used in plastic processing industries in near future.

6. PLASMA SOURCE ION IMPLANTATION

Plasma source ion implantation (PSII) is a new cost–effective process developed for surface modification of materials. In PSII, substrates to be modified are placed directly in a plasma source and then pulse biased to a high negative potential. A plasma sheath forms around the substrate and the entire surface of the substrate is bombarded by ions. Fig. 14 shows schematically the experimental equipment [51]. The vacuum chamber is made up of type 316 stainless steel chamber and a base pressure of 7×10^{-5} Pa in the PSII chamber was obtained by a turbo molecular pumping system. It uses a pulsed power supply with a maximum voltage of 25kV and a current of 10A with varying frequency from 10-5kHz. In order to obtain the full ion energy at the substrate surface, the pressure must be kept sufficiently low (<0.5Pa) in order to avoid ion-neutral collisions in the sheath. Thus, low pressure plasmas have to be employed. Mainly in order to avoid excessive power loads to the substrate, the high voltage has to be applied in pulses of typically some 10 ms duration. As a further advantage of PSII, it can be combined with the thin film deposition system and the method is popularly known as plasma immersion ion implantation assisted deposition (PIID). Like ion beam assisted deposition, PIID may be used with non-reactive or reactive gases, enabling the formation of compound films.

At present, PSII techniques have not yet been established in broad fields of industrial production although several applications appear to be very promising. PSII has to compete with a number of techniques which are well established and have been developed through many years. A comparison of plasma nitriding and PSII treated stainless steels shows the superiority of PSII over plasma nitriding, however at higher capital cost. The main advantage of PSII is the availability of higher ion energies, with a number of potential beneficial consequences for practical applications. For example, PSII of nitriding of austenitic stainless steel is of particular interest. It is possible to limit surface nitride formation by PSII due to relatively large depth of implantation. In such cases the so called expanded austenite is formed, which has the austenite structure with lattice expansion of about 7%. By diffusion through this layer, nitrogen is continuously added to the interface with the underlying bulk, thus extending the thickness of the nitrided layer without having the precipitates of CrN [52]. No or merely a

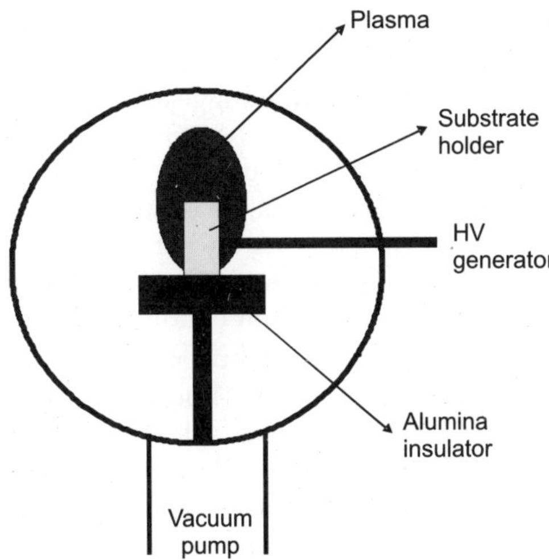

Figure 14 Schematic diagram of plasma source ion implantation system

very small fraction of Cr-N can be detected. Compared to the untreated surfaces, the wear rates are drastically reduced, by about three orders of magnitude and the corrosion resistance is not deteriorated in the case of stainless steel. The potential use of ion implantation in tribology has been recognized long. Nevertheless, cost of the process has impeded a major breakthrough, although ion implantation is routinely applied to the hardening of surgical prostheses [53]. Therefore, with its prospectives for cost reduction, PSII is a promising new technique in the development of biomaterials.

7. CONCLUSIONS

In the present work, we have illustrated a few emerging plasma methods of deposition of Cr-based coatings for corrosion control. They are: (i) plasma assisted metal-organic CVD, (ii) magnetron sputtering (iii) pulsed plasma nitriding, and (iv) plasma source ion implantation. While the first two methods are used to deposit Cr- based coatings as a overlay coating on the substrate, the latter two are used to modify the surfaces using plasma as an effective medium for surface engineering. Some important conclusions of the present study are given below:

(i) Using chromium acetylacetonate and an optimum concentration of ammonium bifluoride precursors, deposition of Cr and CrN thin films by a plasma assisted metal organic chemical vapor deposition technique has been developed successfully using appropriate flow rates of argon/nitrogen and hydrogen during deposition. Nanocrystalline (~10 nm) and adherent CrN films could be deposited both on metallic (SS) and ceramic substrates.

(ii) The composite coatings of CrN/Ni were developed using different proportion of the precursors of Ni (Ni[(acac)$_2$en]) and Cr ((Cr(acac)$_3$) in H_2/N_2 plasma at a substrate temperature of 823 K. Hardness measurements on the Ni, CrN and their composites deposited under identical conditions revealed that Ni films are softer and addition of Ni to CrN reduces its hardness.

(iii) The plasma nitriding behaviour of Cr-plated stainless steel was investigated as a function of nitriding temperature and time. An activation energy of 131.4 kJ/mole for the diffusion of nitrogen in chromium suggested that the rate of nitriding in pure chromium was low and the formation of hard Cr_2N phase was found to retard the diffusion rates of nitrogen in Cr- plated austenitic stainless steel. Plasma nitriding at an intermediate temperature of about 900K for about 45 h produced a peak hardness of about 900 HV with a uniform decrease in hardness as a function of depth. SEM, microhardness and calowear tests revealed that Cr-N coatings are significantly very hard, wear resistant and have an excellent bonding with the stainless steel.

(iv) Application of magnetron sputtering and plasma source ion implantation to deposit Cr-based coatings has been highlighted. Though, magnetron sputtering is a line of sight process, it has been developed into a well accepted industrial process because of its inherent advantages. On the other hand, aspects such as safety, duty time of the cycle, up-scaling and in particular the processing cost must be established in the case PSII. The PSII has the potential to accommodate a few PVD methods and to offer solutions to some of the challenges in the area of surface engineering.

ACKNOWLEDGEMENTS

The authors gratefully acknowledge Dr. Arup Dasgupta, Dr. Antony Premkumar, Mr. E. Mohandas and Dr. M. Vijayalakshmi and for many a fruitful discussion during the course of the investigation. They also thank Dr. Baldev Raj, Director, IGCAR for the support and encouragement.

REFERENCES

1. J.W. Seok, N.M. Jadeed and R.Y. Lin, *Surf. Coat. Technol.*, 138 (2001) 14.
2. K.C. Walter, J.T. Scheuer, P.C. McIntyre, P. Kodali, N. Yu and M. Natasi, *Surf. Coat. Technol.*, 85 (1996) 1.
3. J.P. Ge, *Plating and Surface Finishing*, (1996) 146.
4. G.B. Hoflund, A.L. Grogan, J. Douglas, A. Asbury, H.A. Laitinen, S. Hoshino, *Appl. Surf. Sci.*, 28 (1987) 224.
5. G. Bolelli, V. Cannillo, L. Lusvarghi and S. Ricco, *Surf. Coat. Technol.*, (2005), in Press.
6. E.J. Perez, F. Pedraza, M.P. Hierro, M.C. Carpintero and C. Gomez, *Surf. Coat. Technol.*, 184 (2004) 47.
7. H.M.J. Mazille, *Thin Solid Films*, 65 (1980) 67.
8. W. Hanni and H.E. Hintermann, *Thin Solid Films*, 40 (1977) 107.
9. K.O. Legg, M. Graham, P. Chang, F. Rastagar, A. Gonzales and B. Sartwell, *Surf. Coat. Technol.*, 81 (1996) 99.
10. R.K. Guffie (ed) Handbook of Hard Chrome Plating, Gardner, 1986.
11. E. Lunarska, K. Nikiforow, T. Wierzchon, I.U. Pokorska, *Surf. Coat. Technol.*, 145 (2001) 139.
12. B. Eastwood, S. Harmer, J. Smith, A. Kempster, *Mater. Sci. Forum*, 102-104 (1992) 543.
13. F. Schuster, F. Maury, J.F. Nowak, C. Bernard, *Surf. Coat. Technol.*, 46 (1991) 275.
14. F. Maury, L. Gueroudji, C. Vahlas, *Surf. Coat. Tech*nol., 86/8 *(1996)* 316.
15. P. Karvankova, H.D. Mannling, C. Eggs, S. Veprek, *Surf. Coat. Technol.*, 146-147 *(2001)*, 280.

16. F.B. Wu, J.J. Li, J.G. Duh, *Thin Solid Films,* 377-378 (2000) 354.
17. Arup Dasgupta, P. Kuppusami, Falix Lawrence, V.S. Raghunathan, P. Antony Premkumar, K.S. Nagaraja, *Mater. Sci. & Engg.* A, *374 (2004)* 362
18. P.Antony Premkumar, Arup Dasgupta, P. Kuppusami, K.S. Nagarajan and V.S.Raghunathan, *Materials Letters* 61 (2007) 50-53.
19. A. Guiner, X-ray diffraction (San Francisco: Freeman), 1963, p. 143.
20. P. Kuppusami, Arup Dasgupta, O.M. Sreedharan, Falix Lawrence, V.S. Raghunathan, Baldev Raj, K.S. Nagaraja, Antony Premkumar, Indian Patent Application Number 637/MUM/2001 (2001).
21. A Kimura, H Hasegawa, K Yamada and T Suzuki, J. *Mater. Sci. Lett.,* 19 (2000)601.
22. S K Wu, H C Lin and P L Liu, *Surf. & Coat. Technol.,* 124 (2000) 97.
23. J G Buijnsters, P Shankar, J Sietsma and J J ter Meulen, *Mater. Sc. & Engg.* A 341 (2003) 289.
24. E Manthe and K T Rie, *Surf. & Coat. Technol.,* 112 (1999) 217.
25. P Kuppusami, A Dasgupta and V S Raghunathan, *J. of The Iron and Steel Institute of Japan International,* 42, (2002) 1457.
26. S Ghosh, A Dasgupta and S Ray, *J. of Appl. Phys.,* 78 (1995) 3200.
27. M Volmer and A Weber, *Zeit Phys Chem.,* 119, 123 (1926).
28. V S Raghunathan and P Kuppusami, *Trans. Ind. Inst. Met.* 52 (1999) 251.
29. J J Boland and G N Parsons, *Science* 256 (1992) 1304.
30. S. Veprek, *Mat. Res. Soc. Symp. Proc.,* 164 (1990) 39.
31. P. Antony Premkumar, K.S. Nagaraja, P. Kuppusami, Arup Dasgupta, C. Mallika, V.S. Raghunathan, Baldev Raj, Indian Patent Application Number 272/CHE/2005 (2005).
32. P. Antony Premkumar, Arup Dasgupta, P. Kuppusami, K.S. Nagaraja and V.S. Raghunathan (Eds) D.S. Patil in *Proceedings of DAE-BRNS Workshop on Plasma Surface Engineering 2004,* Mumbai, India, pp 286.
33. P. Antony Premkumar, Arup Dasgupta, P. Kuppusami, P. Parameswaran, C. Mallika, K.S. Nagaraja, and V.S. Raghunathan, *Chem. Vap. Deposition,* 12 (2006) 39.
34. Z.L. Zhang and T. bell, *Surf. Eng.,* 1 (1985) 131.
35. C.K. Lee and H.C. Shih, *Corrosion,* 50 (11) (1995) 848.
36. *Plasma Processing Update,* (Ed) K. S. Ganesh Prasad, Issue 46 (2004) 6.
37. D.Sundararaman, P. Kuppusami and V.S. Raghunathan, Surf.& Coat. Technol., 30 (1987)343.
38. K H Jack, in Heat Treatment "73, (The Metals Society, London, 1973) p. 39.
39. C Heau, R Y Fillit, F Vaux and F Pascaretti, Surf. & Coat. Technol., 120-121 (1999) 200.
40. V.S. Raghunathan and P. Kuppusami (Eds) D.S. Patil et al. in *Proceedings of DAE-BRNS Workshop on Plasma Surface Engineering, 2004,* Mumbai, India, (Allied Publishers Ltd:Mumbai) pp 165-180.
41. P Kuppusami, A L E Terrance, D Sundararaman and V S Raghunathan, *Surface Engineering,* 9 (1993) 142.
42. W. Knauer, *J. Appl. Phys.,* 33 (1962) 2093.
43. S.M. Rossnagel and H.R. Kaufman, *J. Vac. Sci. Technol.,* A, 5 (1987) 88.
44. R. Kukla, T. Krug, R. Ludwig and K. Wilmes, *Vacuum,* 41 (1990) 1968.
45. B. Navinsek, P. Panjan, and I. Milosev, *Surf. &Coat. Technol.,* 97 (1997) 182.
46. K. Suzuki, T. Kaneko, H. Yoshida, Y. Obi and H. Fujimori, *J. Alloys and Compounds,* 280 (1998) 294.
47. E. Martinez, J. Romero, A. Lousa and J. Esteve, *J. Phys. D: Appl. Phys.,* 35 (2002) 1880.

48. I. Milosev, H.H. Strehblow, and B. Navinsek, *Surf. Coat. Technol*. 63 (1994) 173.
49. P. Panjan, B. Navinsek, A. Cvelbar, A. Zalar, and I. Milosev, *Thin Solid Films*, 281-282 (1996) 298.
50. *Product information*: Plastic processing: Fewer Problems, lower costs with BALINIT hard coatings, BD 802 008 AE (9302) Balzers AG, Liechtenstein, 1993.
51. K. Baba and R. Hatada, (Eds) D.S. Patil et al.in *Proceedings of DAE-BRNS Workshop on Plasma Surface Engineering, 2004*, Mumbai, India,(Allied Publishers Ltd:Mumbai) pp 28-44.
52. S. Mukherjee, (Eds) D.S. Patil et al. in *Proceedings of DAE-BRNS Workshop on Plasma Surface Engineering, 2004*, Mumbai, India, (Allied Publishers Ltd: Mumbai) pp 181-199.
53. P. Sioshansi and E.J. Tobin, Surf. Coat. Technol., 83 (1996) 175.

CHAPTER 8

Coatings for Corrosion Resistance

G. Sundararajan, L. Rama Krishna, Nitin P. Wasekar, G. Sivakumar, and A. Jyothirmayi
International Advanced Research Centre for Powder Metallurgy & New Materials (ARCI), Balapur (P.O), R.R. District, Hyderabad –500 005, India

I. INTRODUCTION

Corrosion and its mitigation has direct relevance in very diverse areas such as automobiles, trucks and two wheelers, furniture and home appliances, buildings, bridges and structures made using reinforced steel bars and most importantly in industrial plants. The total cost of metallic corrosion in U.S.A in the year 1995 was around US $ 300 billions of which around US $ 100 billion was an avoidable cost [1].

Coatings protect the substrate materials from corrosion by three basic mechanisms namely, barrier protection, chemical inhibition and galvanic (sacrificial) protection [1]. As the term implies, barrier protection is achieved by complete isolation of the substrate from the environment. In this case it is important that the coating is free of pores, cracks and impervious to corroding species. Chemical inhibition is achieved by addition of required amounts of the inhibitor in the paint or coating. Finally, galvanic protection is achieved by coating the component with a more active metal (in that particular environment), thereby ensuring that the substrate becomes the cathode in the corrosion cell. Galvanized zinc wherein Zn coating is provided over the steel substrate is the most well-known example.

The overall objective of this chapter is to review the scientific literature in the broad area of corrosion behaviour of various types of coatings, specifically designed for enhancing the corrosion resistance of the substrate/component on which they have been deposited. However, before defining more precisely the scope of the review, it should be worthwhile to analyze the available literature in the area. Such an exercise is carried out in Section 2.

2. AN ANALYSIS OF LITERATURE AND SCOPE OF REVIEW

A review of literature from 1995 to present (2006) using the "Web of Science"and "Science Direct" indicates that the number of published scientific papers pertaining to all forms of corrosion of all types of coatings is extremely large. Thus, as a first step, it has been decided to restrict the review to only uniform and galvanic corrosion and the role of coatings in reducing these forms of corrosion. With such a restriction, the number of relevant papers published in SCI journals (Period: 1995 to 2006) amounts to 963 and the distribution of these papers on the basis of their year of publication is presented in Fig. 1. From this figure it is clear that the number of publications is continuously increasing with year, thereby attesting to the ever growing popularity of the field of coatings for corrosion protection among the world-wide scientific and engineering community. In the year 2005, around 170 papers were published in the area and it is likely that the corresponding number of papers for the year 2006 will exceed 200. Further analysis of the 963 papers published during the period 1995 to 2006, presented in Table 1, indicates that Surface and Coatings Technology is the most popular journal (22% of the 963 publications) for publishing scientific work in the area of uniform and galvanic corrosion of coatings with Corrosion Science (8.7% of the publications) and Corrosion (7.6%) being the 2nd and 3rd most popular journal.

A perusal of the 963 papers under analysis also reveals that more than 80% of them are related to corrosion behaviour of coatings deposited on steel substrates. This is not surprising since many of the engineering structures, industrial plant components and civil structures like buildings and bridges prone to uniform and galvanic corrosion utilize steel as the basic material of construction. Therefore, the cost savings achieved by protecting such steel structures by coatings is also substantial. Apart from steel, the other materials that have been protected against corrosion through coatings include magnesium, aluminium and titanium alloys. However, the present review will be restricted to protection of steels against uniform and galvanic corrosion through the use of coatings.

Finally, even with the above restrictions, an analysis of the literature indicates that a large variety of coatings have been considered for conferring resistance to corrosion (see Table 2). In all 11 types of coatings have been evaluated for improvement in corrosion resistance and of these, ion implantation, electro deposition, physical vapor deposition (PVD), conversion

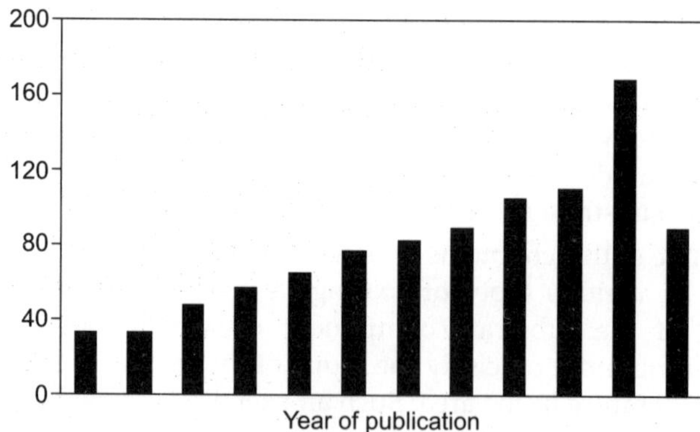

Figure 1 Number of published articles pertaining to uniform and galvanic corrosion protection by coatings from 1995 to 2006 (* till May 2006)

Table 1 The distribution of papers published in the area of uniform and galvanic corrosion of coatings among SCI Journals (Period: 1995-2006)

S.No.	Name of Journal	Number of papers (1995-2006)	Percentage of total paper
1	Surface & Coatings Technology	213	22.1
2	Corrosion Science	84	8.7
3	Corrosion	73	7.6
4	Electrochimica Acta	36	3.7
5	Journal of the Electro Chemical Society	31	3.2
6	Wear	31	3.2
7	Thin Solid Films	30	3.2
8	Materials and corrosion-werkstoffe und korrosion	24	2.5
9	Progress in Organic Coatings	22	2.3
10	Protection of Metals	22	2.3
11	Applied Surface Science	21	2.2
12	Anti Corrosion Methods and Materials	19	2.0
13	Journal of Thermal Spray Technology	19	2.0
14	Transactions of the Institute of Metal Finishing	18	1.9
15	British Corrosion Journal	17	1.8
16	Electrochemical Methods in Corrosion Research	17	1.8
17	Synthetic Metals	17	1.8

Table 2 Number of papers published in the areas of uniform and galvanic corrosion in SCI journals categorized on the basis of type of coatings (Period: 1995-2006)

S.No.	Type of coatings	Number of papers (1995-2006)	Percentage of total papers
1	Ion implantation	217	22.5
2	Electro deposition	178	18.5
3	Physical vapor deposition	119	12.4
4	Conversion coatings	99	10.3
5	Sol-gel coatings	89	9.2
6	Diffusion coatings	83	8.6
7	Thermal spray coatings	74	7.7
8	Anodising / Micro arc oxidation	44	4.6
9	Chemical vapor deposition	31	3.2
10	Laser surface modification	19	2.0
11	Galvanizing & dipping	10	1.0
	TOTAL	963	100.0

coatings, sol-gel coatings, diffusion coatings and thermal spray coatings appear to be the most popular in that order (Table 2).

Taking into account the need to provide an in-depth review of selected areas rather than a broad, superficial review, considering the expertise of the authors writing this review and also keeping in view the emerging developments in the area of coatings for corrosion resistance, it has been decided that this review will concentrate on the uniform and galvanic corrosion of electrodeposited, PVD and thermal spray coatings deposited essentially on steel substrates.

3. ELECTRODEPOSITED COATINGS

Electrodeposition (ED) involves the deposition of an adherent metallic or composite coating on a component surface which acts as the cathode. ED is a time-tested technique for obtaining coatings which enhance the aesthetics, corrosion resistance, chemical inertness and wear resistance of the components upon which it is applied. However, protecting metallic components against corrosion has been the most common reason for utilizing ED coatings. Some of the well-accepted applications for ED coatings include [1]

(a) Zn plating of steel strips, sheets, stampings, wires, forgings and machine parts for a variety of applications in automobile and other sectors.

(b) Sn or Cr plating of steel strips for food packaging and other container uses.

(c) Zn and Cd plating of fasteners and other hardware items.

(d) Bright Ni-Cr plating for household appliances, interior auto hardware.

(e) Hard chromium plating of gun barrels, aerospace components.

(f) Electroless Ni-P and Ni-B coatings especially for applications requiring a combination of corrosion and wear resistance.

The ED coatings can protect the steel component surface either by ensuring that it is electro negative to steel and thus corrode sacrificially (as in the case of Zn and Cd coatings in a marine environment) or by acting as a barrier between the environment and the component and thus reducing or eliminating corrosion. In the latter case of barrier coatings, it is important that the coating is not only defect free (free of pores, cracks etc.) but also sufficiently thick to make the coating impervious to the environment.

A perusal of the literature pertaining to corrosion resistance of electrodeposited coatings reveals that 178 articles have been published in SCI journals pertaining to the above area from 1995 till date. A more detailed inspection of the data reveals that the number of papers published per year in the area has in general increased with increasing years, as illustrated in Fig. 2. In the year 2005, 38 papers pertaining to corrosion behaviour of electrodeposited coatings were published in SCI journals and the number for the year 2006 is expected to be equally high or even higher based on the fact that 17 papers have already been published in the area in the first three months of 2006. The distribution of the papers related to corrosion behaviour of electrodeposited coatings among the various journals is presented in Table 3.

The purpose of this review is not to chronicle the development of various types of ED coatings over the years; rather it aims to describe some of the recent and innovative developments in the field. In line with the above objective, some of the recent work in the area of novel electrodeposited composite coatings, nano zinc coatings obtained by pulsed electro deposition, newer developments in the field of electroless Ni-P type coatings, Zn-Ni coatings

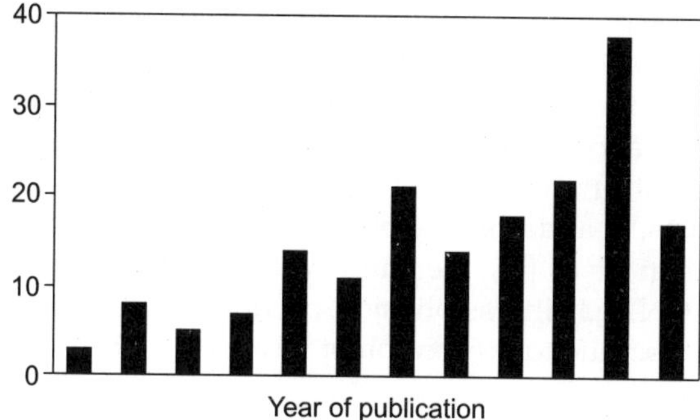

Figure 2 Number of publications pertaining to corrosion behaviour of electrodeposited coatings from 1995 to 2006 (till may 2006)

Table 3 Distribution of papers in the area of corrosion of electrodeposited coatings among SCI Journals (Period: 1995 till date)

Sl.No.	Journal name	Number of papers	Percentage of total (%)
1	Electrochimica Acta	20	11.2
2	Corrosion	14	7.8
3	Corrosion Science	13	7.3
4	Surface & Coatings Technology	11	6.1
5	J of Electrochemical Society	11	6.1
6	Protection of Metals	10	5.6
7	Trans Inst. Metal	9	5.0
8	British Corrosion Journal	5	2.8
9	Russian J Electrochemical	5	2.8
10	Applied Surface Science	3	1.6
11	Electro chemistry communication	3	1.6
12	Bulletin of Electrochemistry	3	1.6
13	Synthetic Metals	3	1.6

as replacement to Cd coatings and ceramic coatings by electro deposition will be highlighted in this section.

3.1 Electrodeposited Composite Coatings

Electrodeposition, as a technique, is ideally suited for depositing composites of metallic and non-metallic constituents. ED coatings containing carbides [2,3], oxides [4] and diamond [5] with improved wear and corrosion resistance have been demonstrated. Recently, carbon nanotubes (CNTs) have been used as the reinforcing phase in a Nickel matrix utilizing the electrodeposition process [6,7]. It is well known that CNTs exhibit exceptional strength and stiffness values and therefore it is expected that composites of CNTs will have exceptional properties as well. Chen et al [7] employed a sediment electrodeposition technique with a nickel sulfate bath to obtain a Ni-CNT composite coating of thickness 10 to 15 μm on steel substrates. They evaluated the corrosion resistance of this composite coating in a 3.5 wt % NaCl solution of PH 6. The corrosion rate and the corrosion potential (E_{corr}) of the Ni-CNT composite coating is compared with electrodeposited Ni and uncoated steel in Fig. 3 and Fig. 4. It is clear from Fig. 3 that the corrosion resistance of Ni -CNT composite coating is a factor of 7 higher than that of bare steel and about a factor of 2 higher than electro deposited Ni coatings. The corrosion potential is also strongly modified by CNT addition with the E_{corr} values increasing towards more noble values. In addition, the pitting potential also increased with CNT addition to Ni. According to Chen et al [7], the improvement in corrosion resistance with CNT reinforcement is due to CNTs acting as physical barriers to the corrosion process by filling in crevices, cracks etc, and also due to prevention of localized corrosion thereby leading to lower general corrosion.

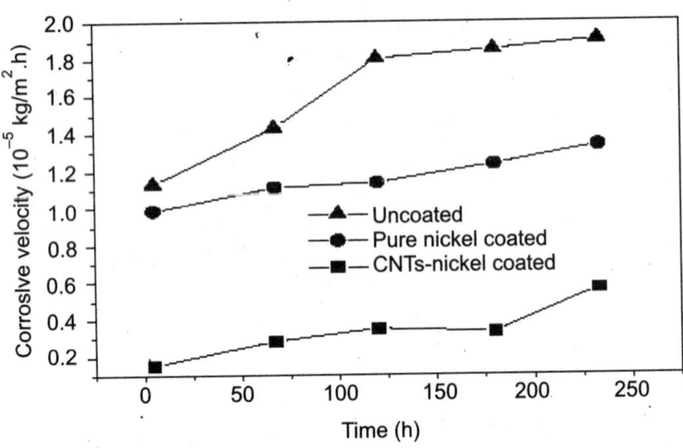

Figure 3 Variation of the corrosion rate (mass loss) with immersion time for uncoated, pure nickel coated and CNTs – nickel coated samples in 3.5 wt.% NaCl solution [reproduced from ref: 7]

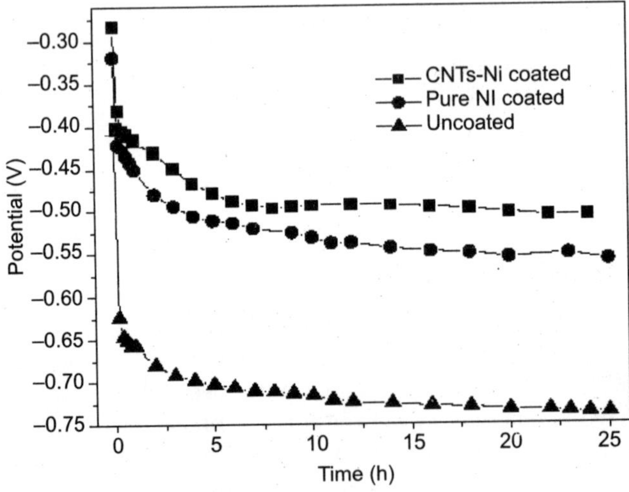

Figure 4 Variation of the corrosion potential (E_{corr}) with the immersion time in in 3.5 wt.% NaCl solution [reproduced from ref: 7]

Automotive industry uses a wide range of metallic and polymeric coatings with a view to improve tribological performance and paintability of a number of automobile parts. However, these coatings also require adequate corrosion resistance. In a recent study, Rossi, Chini, Straffelini, Bonora, Moschini and Stam pali [8] have compared the corrosion resistance of Ni-PTFE composite coating of 20 μm thickness obtained by electroless process on a carbon steel substrate with bronze reinforced PTFE, Zinc phosphate coatings and also with a 2 -layer coating composed of phosphated coating on the steel followed by a layer consisting of MoS 2 particles in a resin. Their corrosion results obtained from salt-spray tests indicated that Ni-PTFE coatings out performed the other coatings as illustrated in Table 4. Similar behaviour was observed with respect to the variation of corrosion potential (E corr) with immersion time in 3.5 wt % NaCl solution as illustrated in Fig. 5. In the case of Ni-PTFE coatings, E_{corr} values remained constant (-285 ± 20 mV Ag/AgCl) up to an immersion time of 2 days and the corrosion potential of the Ni-PTFE coating was nobler than the E corr value of steel ($\simeq$-560 mV Ag/AgCl). Further, AC impedance analysis of the coatings by Rossi et al [8] as a function of immersion time in 3.5% Nacl solution also demonstrated the superiority of Ni+PTFE coatings. This aspect is illustrated in Fig. 6 wherein the variation of total impedance value (R_{tot}) as a function of immersion time is presented for all the coatings.

3.2 Nanostructured Zn Coatings

Pulsed electrodeposition is an excellent technique capable of depositing nanostructured metallic coatings with outstanding properties in terms of hardness, wear resistance etc; [9,10].

Table 4 Results of salt spray tests [ref: 8]

S.No.	Coatings	Time (days) for substrate attack
1	Ni+PTFE	25
2	PTFE + bronze particles	3
3	Zinc Phosphate	1
4	Zinc Phosphate + MoS_2 layer	1 to 2

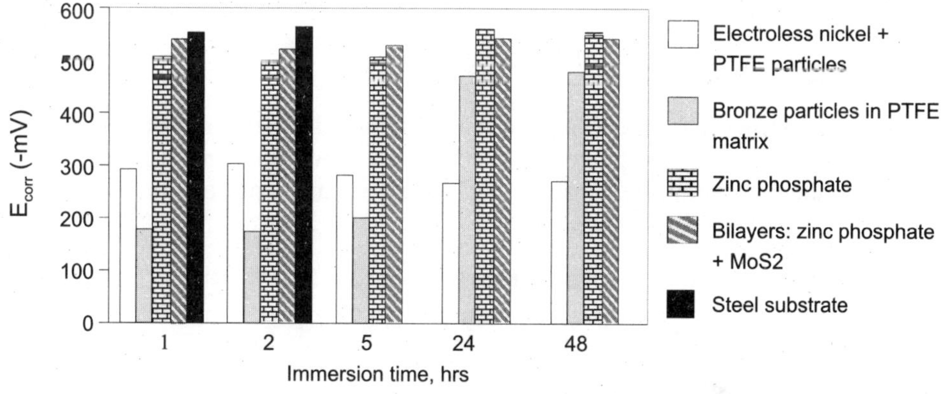

Figure 5 E_{corr} (-mV Ag/AgCl) of different samples during immersion in 3.5 wt.% NaCl solution [ref. 8]

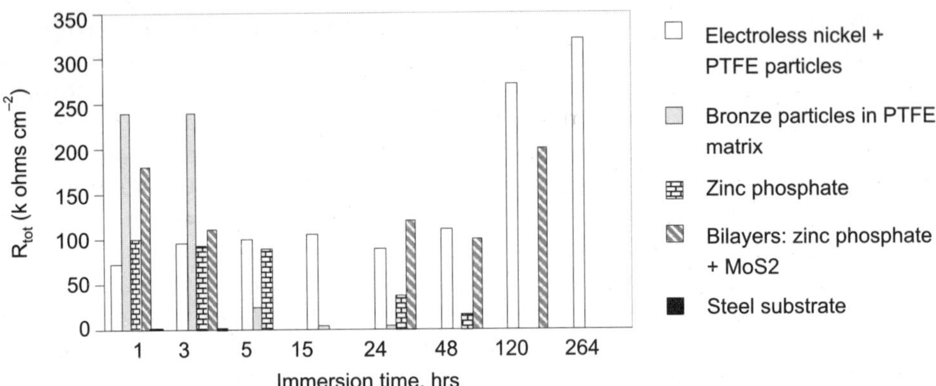

Figure 6 R_{tot} (k ohms cm^{-2}) with immersion in 3.5 wt.% NaCl solution [ref. 8]

Pulse deposited nanocrystalline Ni coating has been the most studied system and the corrosion behaviour of nanocrystalline Ni coatings (grain size 32 nm) is only marginally better than coarse grained Ni coating obtained by conventional electro deposition [11]. However, evaluation of nanocrystalline Zn coating for corrosion resistance is more relevant, given the fact that Zn is most widely used to confer corrosion resistance to numerous steel products. The corrosion behaviour of coarse grained Zn coatings on steel substrate have been extensively studied [12-14]. However, similar data on nanocrystalline Zn coatings obtained using pulse deposition technique is not available except for the work of Youssef, Koch and Fedkiw [15]. These authors deposited nanocyrstalline Zinc (nc Zn) coating on steel (56 nm grain size) and compared its corrosion behaviour with electroglavanized (EG) steel (grain size:10-20 μms) in a 0.5N NaOH solution. Potentio dynamic anodic polarization resistance (R_p) and corrosion current density (i_{corr}) are significantly lower for nc Zn as compared to EG steel even though the E corr value for nc Zn was slightly more negative than the E_{corr} value for EG steel (see Table 5). Such a result has been rationalized on the basis that a nanocrystalline structure enhances both the kinetics of passivation and stability of the passive film as confirmed by the lower values of passive current density (i_p; Table 5). The above postulate was also confirmed by the AC impedance technique which revealed a lower average capacitance value (C_e) for nc Zn as compared to EG steel (see Table 5).

3.3 Ni-P Monolithic Coatings

Electroless Ni-P coatings have been applied successfully in many industries due to their excellent corrosion and wear resistance [16, 17]. The trend towards nanostructured electroless Ni-P coatings while increasing the hardness and wear resistance, more often compromises on

Table 5 Electrochemical parameters of nc Zn and EG steel [ref: 15]

Sample	E_{corr} (mV)	R_p (Ohms/cm^2)	i_{corr} (μA/cm^2)	C_e (μF/cm^2)	i_p (μA/cm^2)
nc Zn	−1470	140	90	69	210
EG steel	−1455	77	229	227	828

corrosion resistance. Therefore, there has been an effort to improve the corrosion resistance of nanocrystalline Ni-P coatings by incorporation of elements like Cu and W. For example, Gao et al [18] have compared the corrosion resistance of Ni -W-P alloy coatings with that of binary Ni-P coatings (both on steel substrate) in a 0.5 M H_2SO_4 solution. These authors obtained as plated deposits of Ni-P and Ni-W-P (30 ± 5 μm thick), with nanocrystalline, mix and amorphous structures using appropriate bath compositions. All the above 6 coatings were examined for corrosion resistance using anodic potentiodynamic polarization technique. In addition, the Ni-P and Ni-P-W deposits having mix structures were annealed to obtain single phase nanocrystalline Ni for evaluation of their corrosion behaviour. The results are presented in Table 6. It can be observed from Table 6 that the binary Ni-P deposits with nanocrystalline or mix structure (as plated and annealed) exhibit much higher i_{corr} and lower R_p and E_{corr} values as compared to N-P deposit with amorphous structure. Such a behaviour can be attributed to the acceleration of corrosion caused by the formation many more micro electrochemical cells between the high volume fraction of grain boundaries and the matrix in nanocrystalline and mix structures as opposed to deposits with amorphous structure. In contrast, Ni-W-P deposits especially in the annealed nanocrystalline condition, exhibit a lower i_{corr} and higher E corr values than even the Ni -W-P deposits having amorphous structure. Such a behaviour is most probably related to the passivation element W which diffused more rapidly through the nanocrystalline deposit to from a dense WO_3 film during the annealing process [19].

The influence of Cu addition to Ni-P deposit is known to improve its corrosion resistance [20,21]. In a recent study, Liu et al. [22] have investigated the corrosion behaviour of Ni-Cu- P deposits (around 16 μm thick) on copper substrates by exposing them to HCl and NaCl solutions and measuring the weight loss as a function of time. As can be observed from Figs.7 and 8, the addition of Cu to Ni-P improves the corrosion resistance (as indicated by lower weight loss) in a HCl environment, with the Ni-P deposit containing 17.2 wt % Cu providing the best results. Similar results were obtained in NaCl solution as well.

3.4 Ni-P Layered Coatings

In recent years, multi layer and graded Ni-P type electrodeposited coatings have gained importance since they are able to confer excellent corrosion resistance even at low or medium levels of P in the coating while retaining the crystalline structure. Gu et al [23] have recently evaluated the corrosion resistance of two -layered Ni-P coatings with monolith Ni -P coatings. In this study, 6 types of coatings, on steel substrates as shown in Table 7, were evaluated using

Table 6 The corrosion data on Ni-P and Ni-W-P deposits [ref: 18]

	Ni-P deposits			Ni-W-P deposits		
Deposit details	E_{corr} (mV)	i_{corr} (μA/cm²)	R_p (ohm/cm²)	E_{corr} (mV)	i_{corr} (μA/cm²)	R_p (ohm/cm²)
as plated amorphous	-154	2.88	328.90	-216	7.50	4,196.0
as plated mix structure	-437	195.70	58.7	-344	12.50	383.3
as plated nanocrystalline	-430	1198.00	23.1	-378	181.20	110.3
Mix structure annealed for 100 min	-396	67.51	159.2	-208	1.45	28,920.0

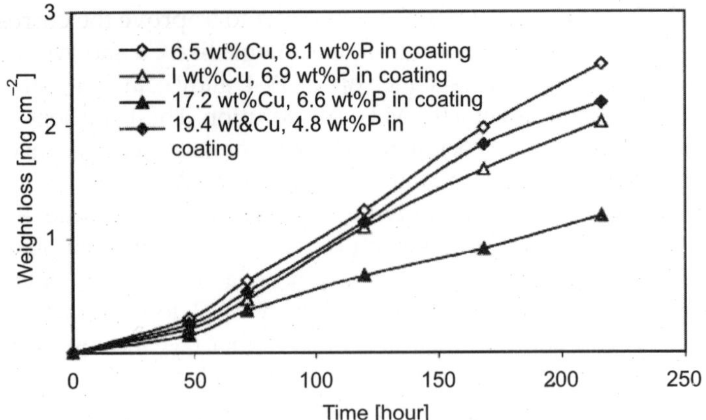

Figure 7 Effect of copper content on corrosion rate of the coating in 0.1N HCl [reproduced from ref: 22]

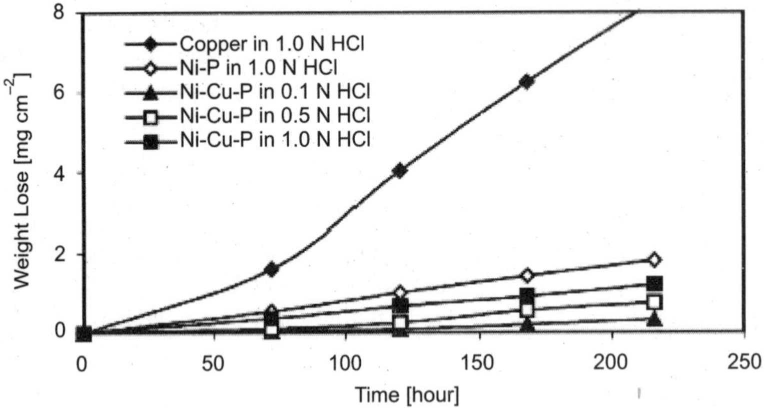

Figure 8 Comparison of Ni-Cu-P with Ni-P and copper in inhibiting HCl corrosion [reproduced from ref: 22]

salt spray tests. The results of the salt spray tests are presented in Fig. 9. It can be seen that the time for the emergence of first red rust spot, defined as beginning of substrate corrosion, is the highest for sample 6 and the least for sample 1. Among the single-layer coatings (sample 1 to 3); the coating with the highest P content (sample 3) and having an amorphous structure exhibited the best resistance under salt spray conditions (see Fig. 9). It is also clear from Fig. 9 that the 3 layer coatings (samples 5 and 6) gave the best performance. The above result has been rationalized by Gu et al [23] on the basis that the intermediate Ni layer with lowest corrosion potential allowed the spread of corrosion in the transverse direction (across the Ni layer) from the original longitudinal direction through the pinholes in the top Ni-P layer

Sankara Narayananan et al. [24] have obtained electroless graded Ni -P coatings on mild steel substrates by sequential immersion in three different plating baths. The corrosion resistance of these graded coatings has been evaluated in a 3.5% NaCl selection. The results are presented in Table 8. It is clear that the 3 layer coating with Ni-high P coating on top gave the best performance, even better than Ni-high P (single layer).

Table 7 Various types of Ni-P coatings evaluated for corrosion resistance [ref: 23]

Sample No.	Coating architecture (from surface to substrate)
1	20 μm (low P, Ni-P deposit)
2	20 μm (medium P, Ni-P deposit)
3	20 μm (high P, Ni-P deposit)
4	6.5 μm (low P) + 13.5 μm (high P)
5	9.5 μm (med. P) + 1 μm (Ni) + 9.5 μm (high p)
6	9.5 μm (low P) + 1 μm (Ni) + 9.5 μm (high p)

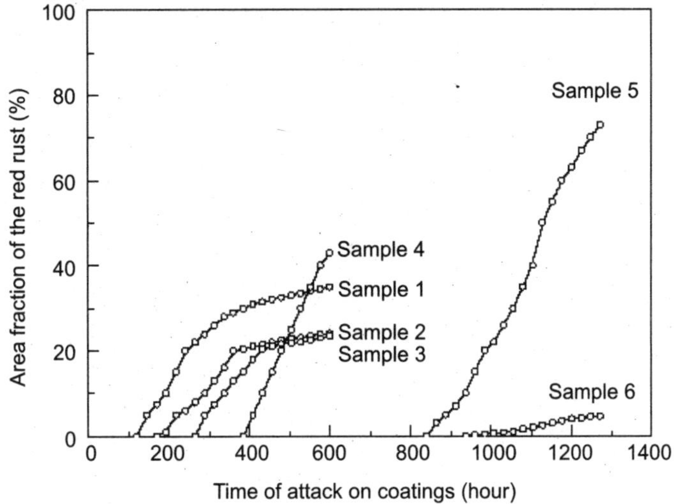

Figure 9 The variation in area fractions of the red rust on the coating surface with time [reproduced from ref: 23]

Table 8 Corrosion resistance of graded electroless Ni-P Coatings in 3.5% NaCl solution [ref: 24]

Sl.No.	Coating architecture (from surface)	E_{corr} (mV)	i_{corr} (μA/cm^2)
1	Ni-low P coating (single layer)	–536	4.22
2	Ni-med. P coating (single layer)	–434	1.17
3	Ni-high P coating (single layer)	–411	0.60
4	Ni-low P + Ni-med P + Ni-hi P (3 layers)	–403	0.41
5	Ni-high P + Ni-med P + Ni-low P (3 layers)	–481	1.70

* Thickness of each layer was around 10 μms.

3.5 Zn-Ni Coatings

Electrodeposited Zinc alloy coatings, especially Zinc-Nickel alloys, have the potential to replace electrodeposited Cd coatings (which are now strictly regulated with respect to its use) for anticorrosion applications [25]. In order that the Zn alloy coating is sacrificial in relation to the steel substrate, the choice of substitutional metals to be added to Zn is most critical [26]. It has been observed that an addition of 10 to 15 wt% of Ni to Zn results in the best corrosion resistance and that beyond 15%, the Zn-Ni deposits become softer than the substrate [27,28].

In a recent study Kin et al [29] have compared the corrosion behaviour of electrodeposited Cd, Zn-Ni and Zn-Ni-Cd and Zn-Ni-Cd coating has the highest polarization resistance among the coatings. The corrosion data obtained from the polarization plots (Fig. 10) is also provided in Table 9. It is obvious from Table 9 that the addition of Ni to Zn does decrease the corrosion rate by a factor of 2.5 while maintaining the corrosion potential. However, the Zn-Ni-Cd coating provides the best performance which is even better than that of pure Cd deposit.

Another approach to obtain corrosion resistance better than that of Zn is through the use of compositionally modulated multilayer (CMM) coatings based on alternate Zn and Ni layers [30-33]. Fei and Wilcox [33], in a very recent study, compared the corrosion resistance of CMM Zn-Ni coating with monolithic Zn and Ni coatings of same thickness by means of neutral salt spray test. The cross -sectional morphology of a typical CMM Zn/Ni coating with 12 individual layers obtained by these authors is illustrated in Fig. 11 while the results from salt spray test are presented in Fig. 12. While the bare steel plates developed red rust corrosion product within 4 hrs of salt spray test, all the coatings performed better and exhibited the first red rust after longer periods (24 hrs and more). Of the coatings, the Zn-Ni CMM coatings exhibited a longer time of red rust than monolithic Zn and Ni coatings (Fig. 12). Among the

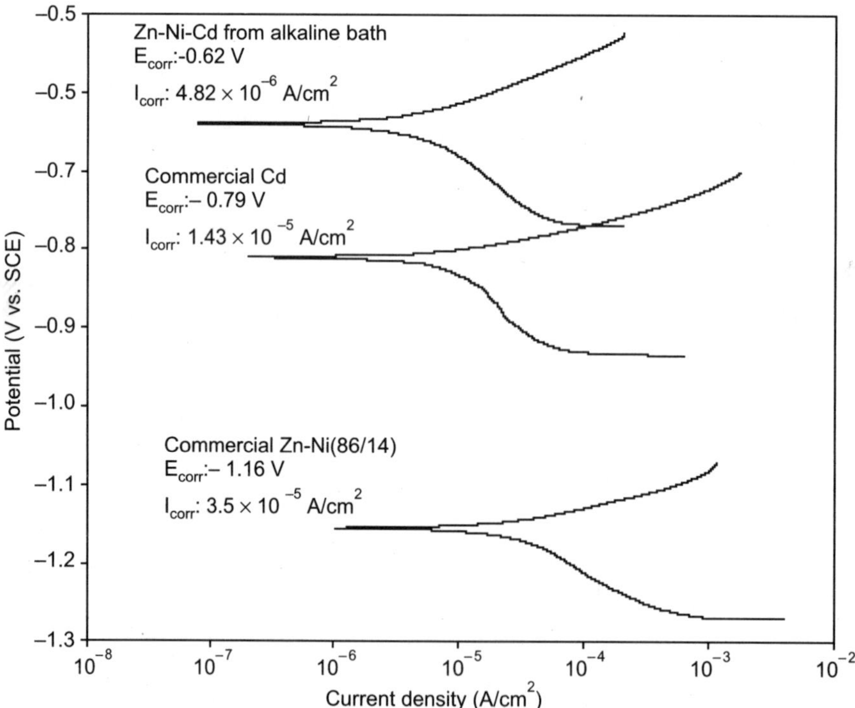

Figure 10 Tafel polarization plots of Zn-Ni-Cd deposited from the alkaline sulfate solution and Zn- Ni, Cd deposits obtained from the commercial baths [reproduced from ref: 29]

Table 9 Corrosion rates of Zn, Zn-Ni, Cd and Zn-Ni-Cd coatings [ref: 29]

Deposit	Composition wt%			E_{corr} (mV)	Corrosion rate (mm/year)
	Zn	Ni	Cd		
Zn	100	0	0	-1.14	1.24
Zn-Ni	86	14	0	-1.16	0.50
Cd	0	0	100	-0.79	0.31
Zn-Ni-Cd	50	28	22	-0.62	0.07

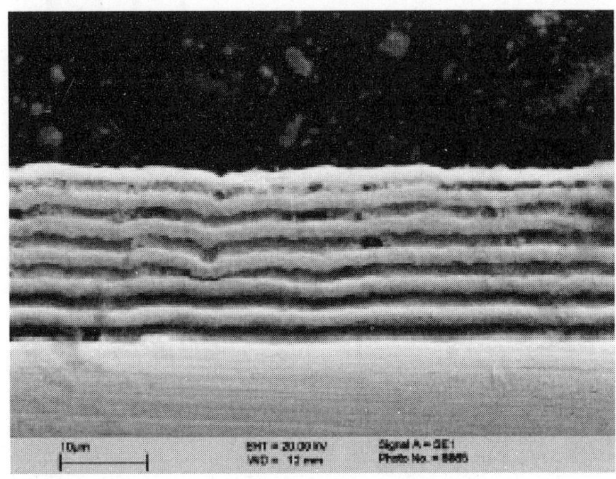

Figure 11 Cross-sectional morphology of Zn/Ni CMM coating with 12 individual layers [reproduced from ref: 33]

CMM coatings, those coatings with Ni as the top layer gave better performance than those with Zn as the top layer. This aspect is brought out in Fig.12 to be noted that though the Zn- Ni CMM coatings are nobler than pure Zn coatings, they are still cathodic to steel (see Table 10). The results have been rationalized by the authors on the basis that while corrosion proceeds strictly through pin holes and cracks in Ni layer, it occurs rapidly in the Zn layer. Thus, having Ni layer on top provides better performance.

4. THERMAL SPRAY COATINGS

Among the family of surface engineering techniques, thermal spraying offers wider options to deposit relatively thick coatings ranging from 50 μm to few mm thickness encompassing the deposition of wide variety of materials to combat corrosion and/or wear. Thermal spray coatings are advantageously placed due to their economical and versatile features. Fig. 13 illustrates the continuously increasing interest in terms of ever-growing number of world-wide research articles published in different journals from 1995 to 2006 evaluating the corrosion protection obtainable by the thermal spray coatings. Table 11 illustrates an analysis of the journals in which the papers pertaining to corrosion of thermal spray coatings have been published. In Table 11, all the journals in which atleast 2 papers pertaining to corrosion of thermal spray coatings have been published during the period 1995 to 2006, are provided. It is

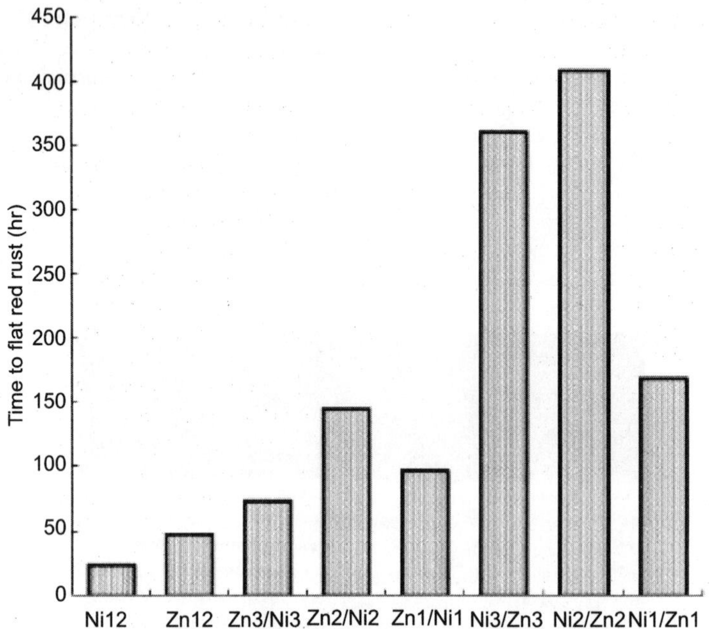

Figure 12 Dependence of corrosion resistance on the zinc-nickel CMM coating configurations [reproduced from ref: 33]

Table 10 Dependence of corrosion potentials (E_{corr}) on coating architecture [ref: 33]

Sl. No.	Coating system (from substrate)	Number of layers	Top layer	E_{corr} (mV)
1	Fe	-	Fe	-566
2	Nickel	1	Ni	-396
3	Zinc	1	Zn	-1045
4	Ni/Zn/Ni/Zn	4	Zn	-1024
5	Ni / Zn (repeated 3 times)	6	Zn	-1033
6	Ni/Zn (repeated 6 times)	12	Zn	-1049
7	Zn/Ni/Zn/Ni	4	Ni	-894
8	Zn/Ni (repeated 3 times)	6	Ni	-924
9	Zn/Ni (repeated 6 times)	12	Ni	--990

* All coatings had a total thickness of 12 μm

clear that the Surface and Coatings Technology is the most popular journal (18% of the publications in the area) followed by Journal of Thermal Spray Technology and Corrosion (9.5% and 8.1% of publications respectively).

Most materials ranging from plastics-to-metals-to-ceramics, that can withstand melting without decomposing, can be thermally sprayed to form a coating. Thermal spray process involves feeding a solid coating material, usually in powder, wire, or rod form, is fed into the

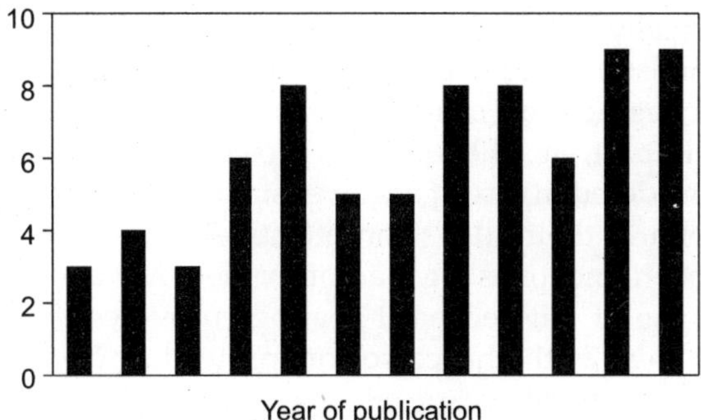

Figure 13 Published articles pertaining to corrosion behaviour of thermal spray coatings as against the year of publication from 1995 to 2006 (till may 2006)

Table 11 Number of articles pertaining to evaluation of corrosion protection of thermal spray coatings published in different journals from 1995 to 2006.

Sl. No.	Name of Journal	Number of papers (1995-2006)	Total papers (%)
1	Surface & Coatings Technology	13	17.6
2	Journal of Thermal Spray Technology	7	9.5
3	Corrosion	6	8.1
4	Wear	5	6.8
5	Electrochemical Methods in Corr. Res. VI	4	5.4
6	Electrochemica Acta	2	2.7
7	Journal of Materials Engg. & Performance	2	2.7
8	Materials Performance	2	2.7
9	Materials Science and Engineering A	2	2.7
10	Materials transactions	2	2.7
11	Surface Engineering	2	2.7
12	Tappi Journal	2	2.7
13	Thin Solid Films	2	2.7
	Total	74	

high enthalpy region, which gets converted into molten or semi-molten droplets. These droplets are accelerated toward the substrate surface through a gas supply stream where the impacted melt rapidly solidifies to form the coating. The optimum in-flight particle temperature and velocity combination aids better adherence and good coating characteristics. During repeated particle buildup, the coated structure exhibits certain porosity levels which significantly depend on the thermal spray variant employed. Usually, the substrate temperature is maintained around 150°C. The use of sealants over thermally sprayed layer has become a common practice to seal the pores and avoid penetration of the corrosion media. However, recent advancements in thermal spray technologies enable tailored coatings with minimal porosity, increased adherence and enhanced properties.

In general, the thermally sprayed coatings exhibit complex corrosion behaviour due to their heterogeneous, non-isotropic, lamellar microstructural features. Moreover, corrosion is related to the chemical degradation caused by the environment rather than other physical degradations like wear, impact, etc. The corrosion protection not necessarily requires very hard coatings as normally considered in case of wear resistant applications; even softer coatings like zinc, aluminium, nickel and their alloys can effectively limit the corrosive attack [34-35]. Usually, the corrosion phenomenon is often accompanied by wear in actual applications, and in such situations, the use of tailored hard coatings provides the optimum solution. The following sub-sections deal with the corrosion protection of thermal sprayed coatings evaluated in different environments.

4.1 Metallic Coatings

Many drawbacks of traditional organic or inorganic coatings like thermal degradation, low thermal conductivity and cost ineffectiveness etc. can be addressed by depositing metallic coatings. Conventionally, most thermally sprayed metallic coatings are related to Zn or Al and their alloy coatings [34-35]. Since Zn or Al alloys are sacrificial, the coatings can protect the substrates from further corrosion even if some areas of substrate are exposed to electrolyte directly. However, they cannot provide the corrosion protection for long duration because they corrode sacrificially. Instead, if noble materials are used to deposit coatings, they can provide long duration protection provided the coatings are pore and crack-free.

Zhao and coworkers [36, 37] have evaluated the corrosion resistance of NiCrBSi coatings deposited on carbon steel substrate utilizing the thermal spray technique. The thickness of the NiCrBSi coating was around 500 mm. The DC polarization tests of these coatings were carried out in a variety of solutions and the resulting Tafel plot was utilized to determine the corrosion rate (i_{corr}). The results are presented in Table 12. It is clear from Table 12 that the High Velocity Oxy Fuel (HVOF) sprayed NiCrBSi exhibits excellent corrosion resistance in alkali solution and also in NaCl (neutral) solution. However Cl^- and SO_4^{2-} ions in the electrolyte cause severe corrosion of the coatings. It was also noted by these authors [36, 37] that presence of porosity and inclusions in the NiCrBSi coatings are responsible for their poor performance especially in acidic environment.

A variant of the NiCrBSi coating is the NiWCrBSi HVOF coatings containing 17%W addition. Gil et al [38] have investigated the influence of process parameters like powder feed rate, spraying distance and fuel to oxygen (F_i) ratio on both the porosity and corrosion rate (in 3.5% NaCl solution) of a HVOF NiWCrBSi coating deposited on AISI 1020 steel substrate. The

Table 12 The corrosion behaviour of NiCrBSi HVOF coatings on a variety of electrolyte solutions [ref: 36]

Sl. No.	Solution	I_{corr} (mA/cm^2)
1	1.0 N H_2SO_4	84.2
2	1.0 N HCl	165.8
3	3.5% NaCl	7.0
4	3.5% NaCl (acidified)	53.3
5	1.0 N NaOH	0.8

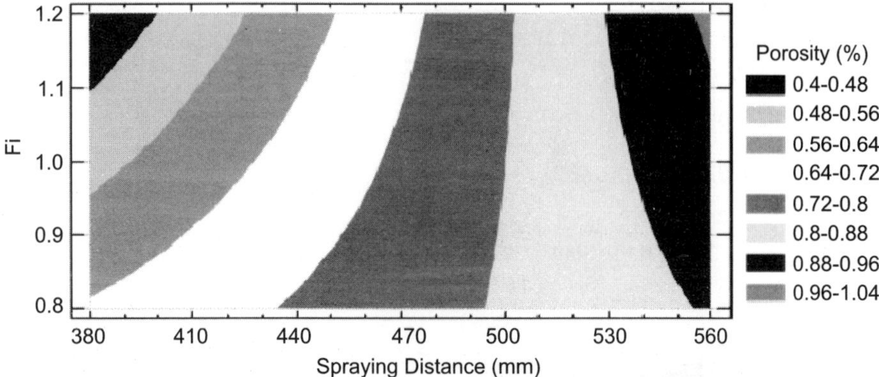

Figure 14 Contours of constant porosity as a function of the spraying distance and Fi ratio. Powder feed rate is maintained constant at 60 g/min [reproduced from ref: 38]

results from such a parametric study are presented in Figs. 14 and 15. From Fig. 14 it is obvious that the lowest porosity levels (0.4 to 0.48%) are obtained at the lowest spray distance and powder feed rate and at the highest fuel to oxygen ratio. Interestingly, as can be observed from Fig. 15, the same process conditions also result in the lowest corrosion rate (5 to 10 mA/cm^2). Though these authors concluded that the lowest porosity in the HVOF NiWCrBSi coating is essential for adequate corrosion resistance, they also point out that the coatings may require post deposition treatment for improving their resistance to corrosion further.

Boudi et al [39] have evaluated the efficacy of HVOF Inconel 625 coatings deposited on carbon and stainless steel substrates against corrosion (immersion in 0.1 N H_2SO_4 + 0.05 N NaCl solution for various periods) on the basis of tensile bond strength of the Inconel 625 coating with the substrate before and after immersion in the corrosion medium. The results are presented in Fig. 16 and Fig. 17 for Inconel 625 coatings deposited on stainless steel and carbon steel substrates respectively. Both these figures indicate that the tensile bond strength of the Inc onel 625 coatings with the substrate decreases dramatically with increasing periods of corrosion exposure. Such a poor performance of the HVOF Inconel 625 coatings has been attributed to their porosity (1.5% to 4.5%) being sufficiently high to be interconnected thereby allowing easy penetration of the electrolyte into the coating.

In another study, Zhang et al [40] have investigated the influence of two different variants of HVOF process on the properties and corrosion performance of the resulting Inconel 625 coatings. The two variants of HVOF systems utilized gas (propane) and liquid (kerosene) as fuels respectively. Of these, the liquid fuel based HVOF resulted in higher particle velocities but lower particle temperatures as compared to gas fuel based HVOF. Salt spray tests conducted on the Inconel 625 coatings indicated that the time for first appearance of rust is of the order of 350 hrs for liquid fuel based HVOF as opposed to only 18 hrs for gas fuel based HVOF. The improved performance of the Inconel 625 coatings obtained by liquid fuel HVOF has been attributed to denser coatings (low porosity) and lower oxide content (Cr_2O_3).

Shrestha et al [41] investigated the influence of thermal spray coating technique on the corrosion behaviour of 316L stainless steel and Inconel 625 coatings in a 3.5% NaCl solution. The four coating systems evaluated are listed in Table 13 along with the porosity and oxide

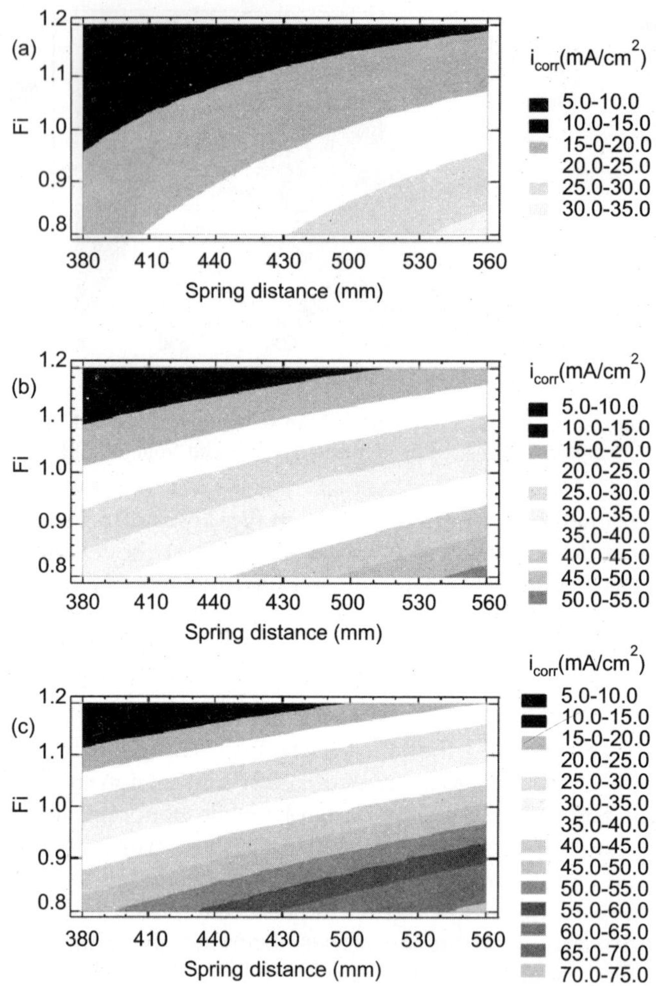

Figure 15 Contours of constant current density (i_{corr}) as a function of the spraying distance and the F_i ratio. Powder feed rate is maintained constant at (a) 60 g/min; (b) 90 g/min; (c) 120 g/min. [reproduced from ref: 38]

content of the stainless steel (SS 316 L) and Inconel 625 coatings obtained with each coating system. The corrosion parameters, i.e., corrosion potential (E_{corr}) and corrosion current (i_{corr}) are also given in Table 13. In general, Inconel 625 coating exhibits superior corrosion resistance to SS 316 L coating, irrespective of the coating system used. However, the coating system which gives the combination of lowest porosity and oxide content provides the coating with the best corrosion resistance. It is also clear from Table 13 that though the optimized SS 316 L and Inconel 625 coatings are superior to the carbon steel substrate in terms of corrosion resistance, they are still not as good as bulk material of the same composition. This is not surprising given the fact that even the best coating still has substantial porosity (1.5% to 2%) and oxide levels (0.8 to 3.0%). Similar results were obtained by Kawakita et al [42] since they also observed that the

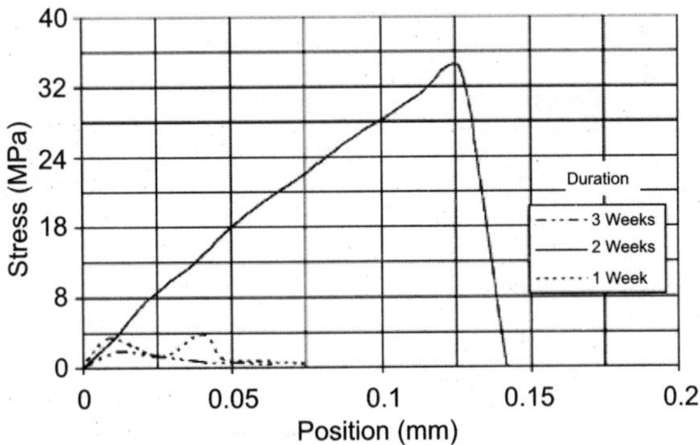

Figure 16 Tensile bond strength of stainless steel [reproduced from ref: 39]

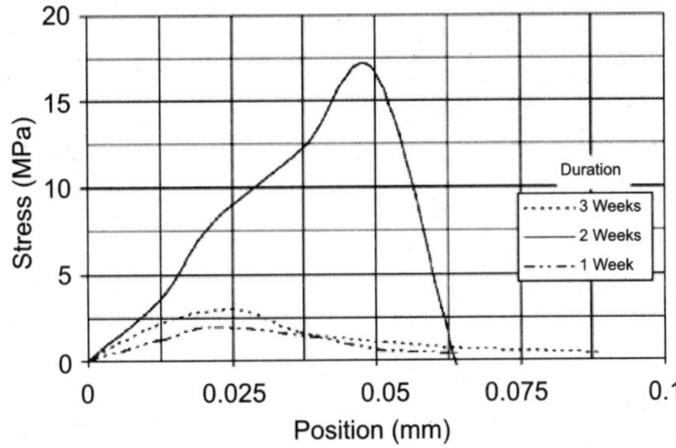

Figure 17 Tensile bond strength of carbon steel [reproduced from ref: 39]

SS 316 L coating deposited by HVOF spray technique was inferior to bulk SS 316 L in terms of corrosion behaviour in seawater.

It is generally accepted that addition of Mo to stainless steel improves its corrosion resistance especially in a saline environment. In a recent study, Wu et al [43] have investigated the influence of Mo addition to HVOF stainless steel coatings on their resistance to naphthenic acid corrosion (NAC) predominantly encountered by components of the petroleum refineries. In this study the original stainless steel HVOF coating was modified to include 2 to 3 wt% Mo. The corrosion rate of HVOF coated Mo bearing stainless steel is compared with carbon steel in Fig. 18. It is clear that the corrosion rate of Mo bearing stainless steel HVOF coating is considerably lower than carbon steel both at room temperature (Fig. 18a) and at higher temperatures up to 280°C (Fig. 18b).

4.2 Ceramic Coatings

Oxide ceramic coatings like Al_2O_3 are expected to have excellent combination of corrosion and wear resistance due to their chemical inertness and high hardness. Rosso et al [44] have evaluated the corrosion resistance of Al_2O_3 and Al_2O_3-TiO_2 air plasma sprayed coatings on

Table 13 Corrosion performance of 316 L and Inconel 625 HVOF coatings [ref:41]

Coating system	Porosity, %	Oxide level, %	E_{corr}, (mV)	i_{corr}, (mA/cm^2)
316 L stainless steel coating				
1. Top gun HVOF	0.40	11.8	-461	6.500
2. JP 5000 HVOF	1.50	0.8	-514	4.300
3. Diamond Jet HVOF	2.40	3.8	-478	57.800
4. High Velocity Flame Spray	0.30	26.0	-446	High
Inconel 625 coating				
1. Top gun HVOF	2.10	7.3	-474	0.625
2. JP 5000 HVOF	2.20	3.3	-125	0.006
3. Diamond Jet HVOF	2.00	3.1	-140	0.004
4. High Velocity Flame Spray	0.30	20.0	-98	0.078
Bulk materials				
1. Inconel 625	—	—	-55	0.001
2. 316 L stainless steel	—	—	-120	0.002
3. Carbon steel substrate	—	—	-715	>10.000

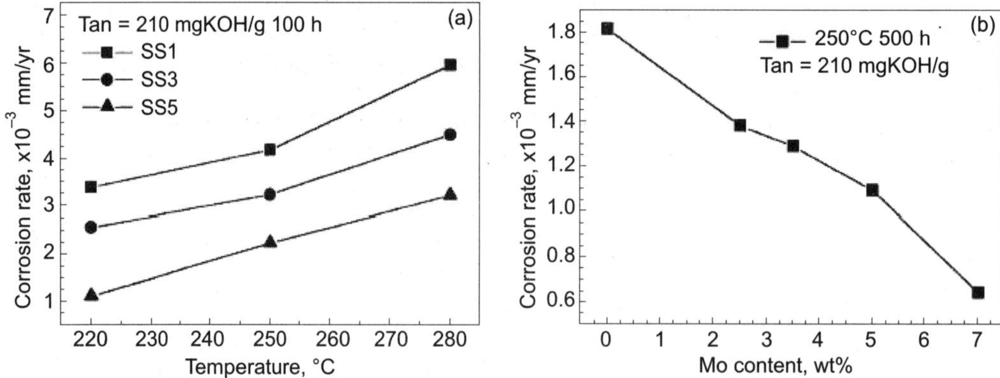

Figure 18 Dependence of NAC rate of Mo-bearing stainless steel on (a) temperature and (b) Mo content [reproduced from ref: 43]

steel substrates with and without sealant applied on coating top surface. The corrosion resistance was determined by immersing the coated samples in a weak acidic solution (pH:2.5) for 192 h and measuring the mass loss suffered. The results from the above work are presented in Fig. 19. This figure indicates that among all coating compositions, the sealed Al_2O_3-TiO_2 (97%-3%) coatings exhibited the highest corrosion resistance while the unsealed Al_2O_3-TiO_2 (87%-13%) coatings were the worst. Further, in all the coatings, sealing the top surface helped in improving the corrosion resistance.

Celik et al [45] have evaluated the corrosion resistance (1N H_2SO_4 solution) of air plasma sprayed Al_2O_3, Al_2O_3-TiO_2 and Cr_2C_3+NiCr coatings as a function of their thickness. Their

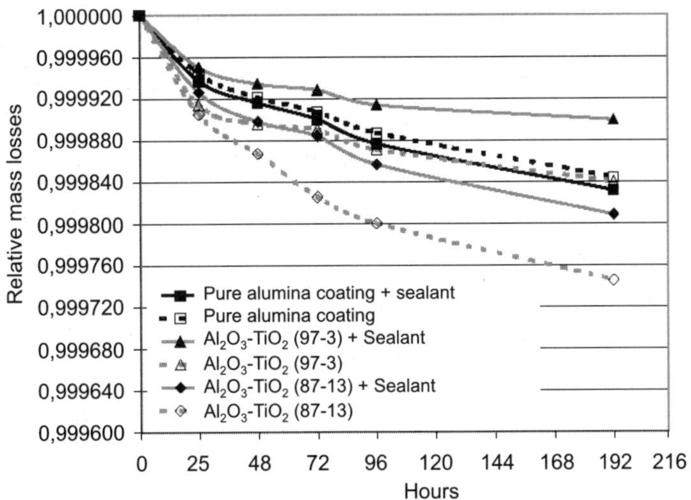

Figure 19 Mass losses curves concerning the different coated pistons subjected to static immersion corrosion test [reproduced from 44].

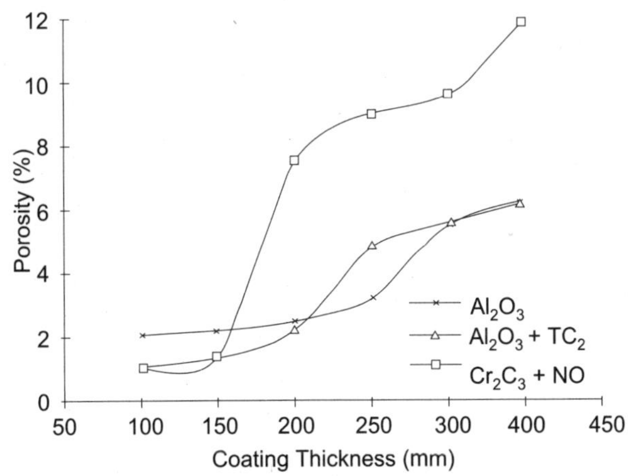

Figure 20 Porosity as a function of coating thickness for Al_2O_3, Al_2O_3+TiO_2 and Cr_2C_3+NiCr coatings [reproduced from ref: 45]

results are presented in Fig. 20 and Fig. 21. It is clear that the corrosion rate increases with increasing coating thickness (Fig. 21) for all the coatings studied and such a behaviour is due to increased porosity in the coatings with increasing thickness (Fig. 20). The increase in the tensile residual stress within the coating with increasing thickness could be yet another reason for observed behaviour.

Celik and coworkers [46] have also investigated the corrosion behaviour of a range of air plasma sprayed thick ceramic coatings on AISI 304 L stainless steel substrates in a 1N H_2SO_4 solution environment. The results from the above study are presented in Table 14. A perusal of Table 14 once again reveals the importance of porosity level with regard to corrosion resistance.

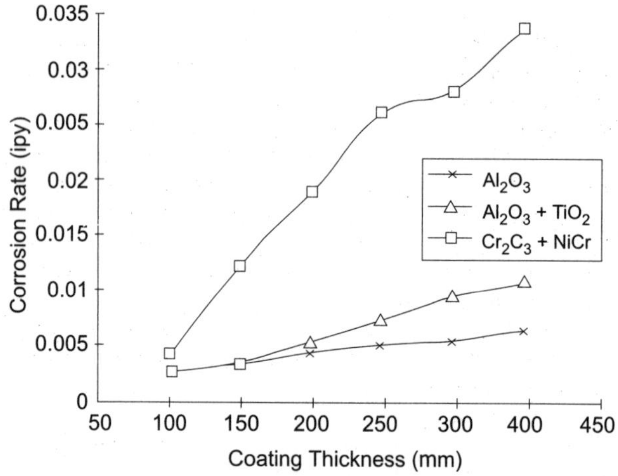

Figure 21 Corrosion rate versus coating thickness for Al_2O_3, Al_2O_3+TiO_2 and Cr_2C_3+NiCr coatings [reproduced from ref: 45]

Table 14 Electrochemical measurements on plasma sprayed coatings in 1N H_2SO_4 [ref: 46]

Coating	Coating thickness (mm)	E_{corr} (mV)	I_{corr} (A/cm^2)	Porosity (%)
AISI 304 L	—	-431	17.26	—
Al_2O_3	200	+273	15.11	0.49
Al_2O_3 + TiO_2	220	-553	267.42	7.70
Cr_2C_3 + NiCr	220	-473	263.35	13.68
ZrO_2 + NiAl	500	-337	5.13	0.38
$MgZrO_3$ + NiAl	300	-355	0.94	0.13
NiCrAl	150	-403	105.05	8.50

4.3 Cermet Coatings

Among the several available thermal spray coating materials, WC based cermet coatings are most widely used because they provide an excellent combination of hardness and wear resistance. However, in actual applications, such coatings have to exhibit adequate corrosion resistance as well. The corrosion resistance of cermet coatings is primarily determined by the corrosion behaviour of metallic binders.

In a recent study, Cho et al [47] have compared the open circuit potential of the constituents of WC based cermet coatings with the substrate (Fe) in aerated 5 wt% H_2SO_4 solution (see Fig. 22). The open circuit potential of WC and W_2C is nobler than the conventional binder materials (Co, Ni) and also the steel substrate (Fe). Thus, microgalvanic corrosion is expected between the WC and the anodic binder phase. Further, since the anodic binder phase occupies only 10 to 20% of the coating volume, the above microgalvanic corrosion is expected to be severe due to the area ratio effect. It is also obvious from Fig. 22 that Cr as the binder material should result in reduced microgalvanic corrosion.

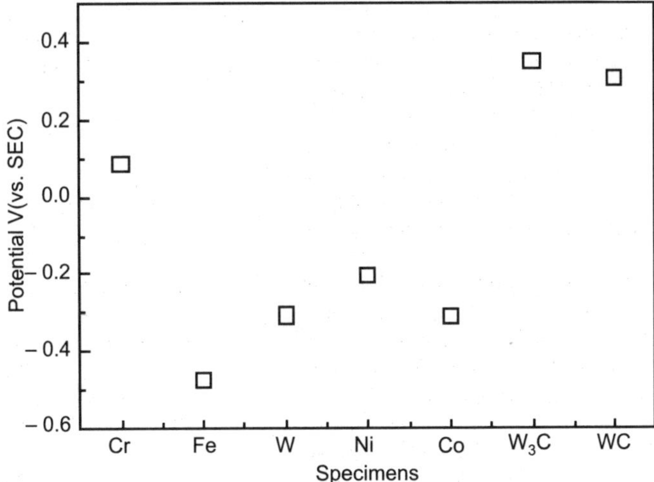

Figure 22 Comparison of the open circuit potential of the constituents of coating materials and substrate in the aerated 5 wt% H_2SO_4 solution [reproduced from ref: 47]

Consistent with the above observation, Voorwald et al [48] observed that WC-10Co-4Cr HVOF coatings performed substantially better than WC -17Co HVOF coatings in a salt spray test conducted as per ASTM B117 in 5 wt% NaCl solution. The WC-17Co HVOF coating showed corrosion products on the surface after the salt spray test as can be seen in Fig. 23. In contrast, the WC-10Co-4Cr coating did not show any surface damage (see Fig. 24).

Figure 23 Salt spray test results for WC-17Co HVOF Coatings of 1,250 μm thickness [reproduced from ref: 48]

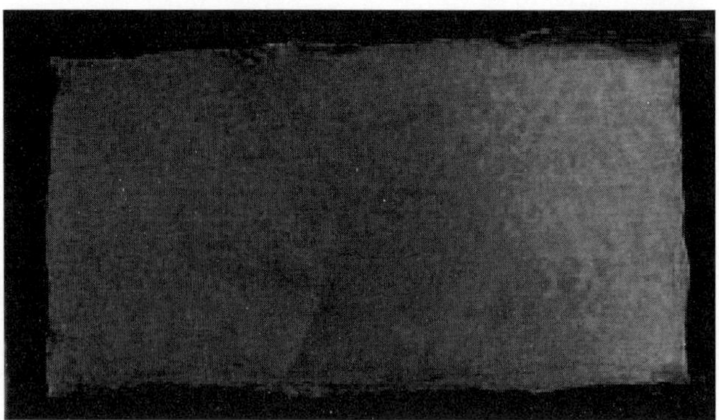

Figure 24 Salt spray test results for WC-10Co-4 Cr,150 μm thickness [reproduced from ref: 48]

In an interesting study, Natishan et al [49] have compared the salt fog corrosion behaviour of HVOF T400 (Tribaloy 400) and HVOF WC-Co coatings with electrodeposited hard chromium (EHC) coatings and more importantly on different substrates, i.e., 7075 aluminium alloy, 4340 steel and 13-8 PH stainless steel. The results from the above study, in the form of protection rating as defined in ASTM B 537-70 standards, are indicated in Table 15. The main conclusions from Table 15 are that EHC coating provides the best protection on 7075 Al substrate while WC-Co coating is the best on 13-8 PH stainless steel and 4340 steel substrates.

Monticelli et al [50] have compared the protection afforded to carbon steel by WC -12Co and WC-17Co HVOF coatings in a 3.5% NaCl solution. Their results presented in Fig. 25 indicate that WC -17Co coating is more protective than WC -12Co coating especially at high coating thickness. The increase in the binder phase in WC-17Co coating and the resulting change in pore morphology from interconnected to isolated pores is most probably responsible for the observed behaviour.

Table 15 Average value of protection ratings after salt fog test [ref: 49]

Coating	Substrate	Protection rating	
		Face	Edge
T 400	7075AL	9.0	3.0
WC-Co	7075 Al	10.0	2.0
EHC	7075 Al	10.0	10.0
T 400	4340 steel	1.6	1.0
WC-Co	4340 steel	3.4	3.2
EHC	4340 steel	3.2	2.0
T 400	PH 13-8 SS	8.6	6.8
WC-Co	PH 13-8 SS	10.0	10.0
EHC	PH 13-8 SS	10.0	—

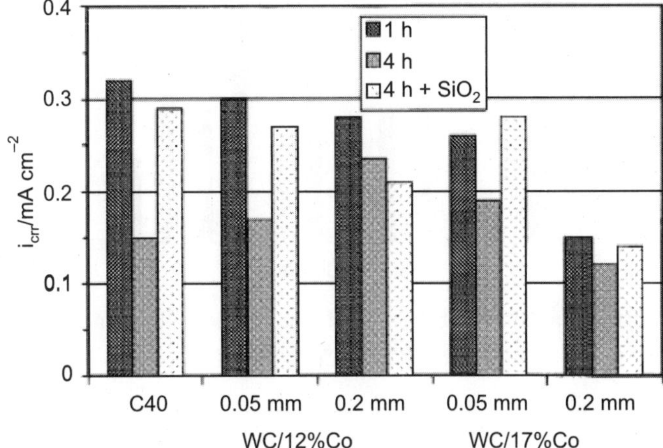

Figure 25 Corrosion rates of uncoated and 0.05 mm and 0.2 mm WC-12Co, WC-17Co coated steel after 1, 4 h immersion in 3.5% NaCl solution, 4 h immersion in 3.5% NaCl solution+0.25% suspended silica, 3000 rpm [reproduced from ref: 50]

5.0 PHYSICAL VAPOR DEPOSITION

Physical Vapor Deposition (PVD) coatings are already in extensive use in cutting tool industry and for decorative applications. PVD coatings like TiN, TiC and ZrN are the most commonly used on cutting tool inserts while TiN based coatings are particularly suited for decorative applications such as watch cases, eye-glass frames, door plates etc; [51-53]. In addition, PVD hard coatings exhibit good biological compatibility thereby making them popular as coatings for medical implants and instruments [54, 55]. Many of the above applications often require adequate corrosion resistance on the part of the PVD coatings.

Invariably PVD hard coatings are nobler than the steel substrates and therefore have to provide protection against corrosion by acting as an impervious barrier. Otherwise the substrate will be subjected to galvanic corrosion [56]. It is therefore important that PVD hard coatings are sufficiently thick and more importantly pore, crack and defect free. PVD has also been utilized to deposit soft coatings which are anodic to the substrate and thereby sacrificial protection of the substrate is possible. In these cases, the PVD coating has to compete with the well established electrodeposited and hot dip coating techniques.

In such a scenario, there has been a significant increase in research aimed at understanding the corrosion behaviour of PVD coatings. Fig. 26 illustrates the number of research articles published in SCI journals from the year 1995 to till date. The continuously increasing interest among the global scientific community in examining the corrosion resistance of different coating systems is evident from Fig. 26.

Further analysis of literature with respect to the number of articles published in different journals during the said period, as shown in Table 16, clearly illustrates the fact that the subject concerned is published in large spectrum of journals. However, the "Surface and Coatings Technology"has retained the premier position as the largest source for published articles in this area.

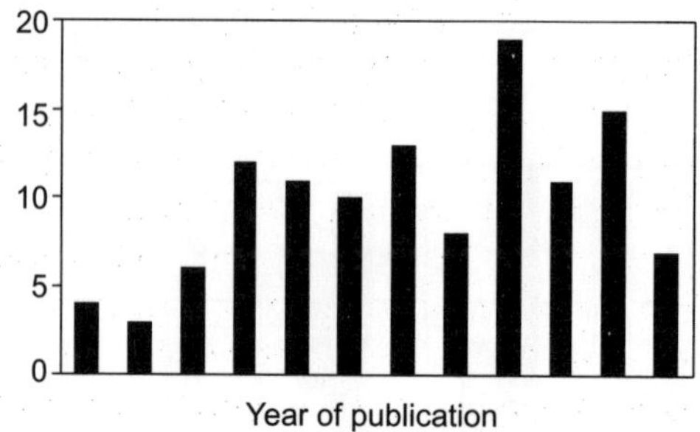

Figure 26 Published articles on corrosion behaviour of PVD coatings with year of publication from 1995 to 2006 (till may 2006)

Table 16 Articles published in different journals related to corrosion behavior of PVD coatings

S.No.	Name of Journal	Number of papers (1995-2006)	Percentage of total papers
1	Surface & Coatings Technology	46	38.65
2	Corrosion Science	8	6.72
3	Thin Solid Films	5	4.20
4	Wear	5	4.20
5	Vacuum	4	3.36
6	Corrosion	3	2.52
7	Materials & Corrosion	3	2.52
8	Materials Science and Engineering A	3	2.52
9	Electrochemical Methods in Corrosion Res.	2	1.68
10	Electrochemica Acta	2	1.68
11	High. Temp. Corr. & Protection of Metals	2	1.68
12	J. of Materials Processing Technology	2	1.68
13	J. of Vacuum Science & Technology A	2	1.68
14	Metal Finishing	2	1.68
15	Surface Engineering	2	1.68

From the reasonably large volume of literature available, articles have been carefully chosen so that the overview brings out the recent and emerging work in the area of corrosion behaviour of PVD coatings.

5.1 Nitride Coatings

PVD nitride coatings have been the workhorse of both the cutting tool and decorative application industries for the past 20 years. However, the corrosion behaviour of PVD nitride coatings has not been studied extensively. In this section, the corrosion behaviour of both monolithic and multi-layered nitride coatings will be reviewed.

5.1.1 Monolithic (single layer) nitride coatings

The excellent properties of thin films of TiN such as high hardness, good wear and corrosion resistance, high electrical conductivity, chemical stability and good adhesion have led to many useful applications. As is well known, bulk TiN has such an excellent corrosion resistance in most acidic and alkaline solutions [57-59]. However, thin TiN coatings are not corrosion resistant to aqueous or gaseous media due to the defects like pinholes existing in the coating [60-63]. The pitting corrosion often occurs when the coating thickness is less than 6 mm. However, preparation of thick TiN coatings is difficult by conventional PVD methods since the thick TiN coating is prone to delamination from the substrate when the coating thickness exceeds 6 μm [64-68].

By utilizing the higher ionization offered by arc ion plating process (AIP) as compared to the conventional ion plating methods, 6-30 μm thick TiN coatings were successfully deposited on steel substrate at a relatively lower temperature of 523K by Lang et al [69]. Fig. 27 shows the polarization curves obtained by these authors on TiN coatings deposited on A3 steel (~0.2%C) with a coating thickness of 6,12,18 and 30 μm and also the bare substrate in 1N H_2SO_4 solution. It is evident from Fig. 27 that a significant improvement of corrosion resistance is obtained in the case of TiN coatings of thickness 12 μm and above. In particular, the critical current density of the thicker TiN coatings is about 0.02 $\mu A/cm^2$ as compared to 200 $\mu A/cm^2$ for steel substrate. Further, passivation current density for the thicker TiN coatings is only 2-3 $\mu A/cm^2$ as opposed to $10^5 A/cm^2$ for the steel substrate.

Fig. 28 (a-b) shows the surface morphology of 6 μm and 30 μm thick TiN coatings after corrosion testing [69]. The TiN coating of 6 μm thickness is too thin to inhibit the penetration of

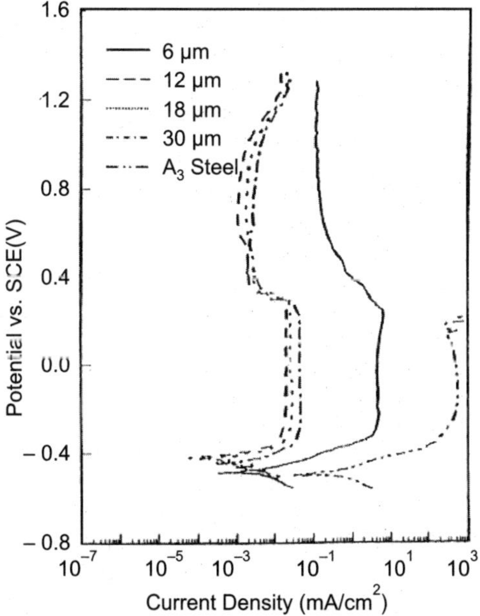

Figure 27 Polarization curves of A3 steel without and with TiN coatings of 6,12,18 and 30 μm in 1N H_2SO_4 solution [reproduced from ref:69].

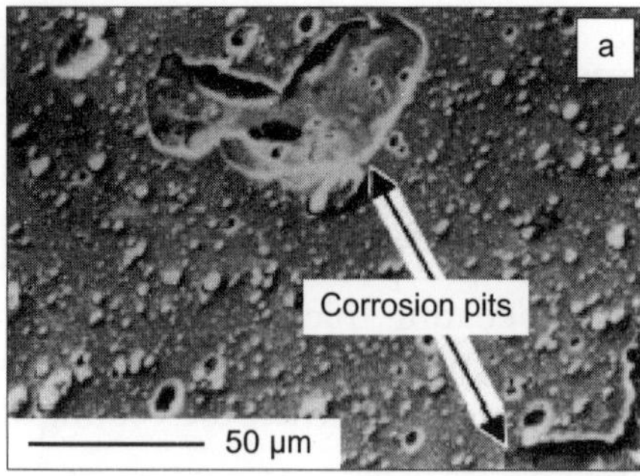

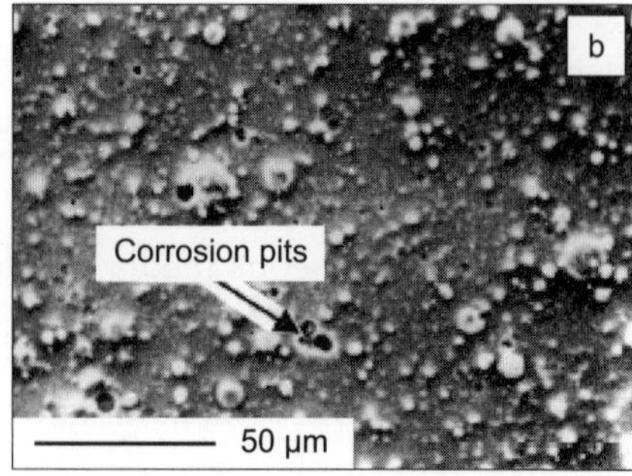

Figure 28 SEM surface micrographs of TiN coatings after corrosion testing (a) 6 μm and (b) 30 μm [reproduced from ref: 69].

corrosion media (1N H_2SO_4) because of the physical defects present and thus exhibits a large number of corrosion pits. However, no clear corrosion pit is found on the surface of the TiN coating of 30 μm thickness.

Chen et al [70] have carried out a detailed study on the corrosion behaviour of TiN coated tool steel in 5% NaCl and 20% H_3PO_4 solution and also through a 500h salt spray test. These authors have defined the packing factor (i.e., normalized density or inverse of porosity percent) and thickness of the coating as the two important parameters influencing the corrosion behaviour. In Fig. 29 the variation of corrosion current density in 5%NaCl solution with the parameter "packing factor (PF) × thickness (T)" is illustrated. It is clear that, irrespective of TiN coating thickness or the presence/absence of Ti interlayer, corrosion current density decreases linearly with increasing value of parameter "PF × T". Similarly, salt spray test results also indicate that beyond a critical value of "PF × T", the TiN coating is immune to salt spray attack (see Fig. 30).

The other approach that has been followed is the use of duplex coating by combining the conventional low temperature ion-nitriding as a pretreatment before depositing the TiN coating on steel substrate [71]. The TiN coating of 2 mm thickness deposited after the nitriding treatment by means of ion plating technique exhibited superior adhesion in combination with improved corrosion resistance. Fig. 31 shows the relation between the natural and pitting potential. Potentials of the steel tempered at 530°C and coated with TiN at 450°C are included in the Fig. 31. It is evident that the best corrosion resistance of the steel is obtained by ion nitriding at 410°C followed by PVD coating of TiN at 450°C. Similar results have been reported by Dong et al. [72]. These authors observed that plasma nitriding of the steel substrate prior to depositing TiN, CrN and (TiAl)N enhanced the corrosion resistance as compared to PVD coatings deposited on untreated steel substrate. Other examples of duplex coatings resulting in improved corrosion performance of the TiN include electroless plating of Ni/P followed by PVD TiN coating [73-75].

Apart from TiN coatings, the CrN is the most widely recognized PVD coating among the family of nitride coatings for tribo-chemical purposes [76-77]. The PVD techniques like cathodic arc deposition and magnetron sputtering techniques are in general employed for depositing CrN based coatings. The deposition of Cr-N coatings at different nitrogen partial pressures leads to the formation of different phases, such as α-Cr (cubic), CrN (cubic) and Cr_2N

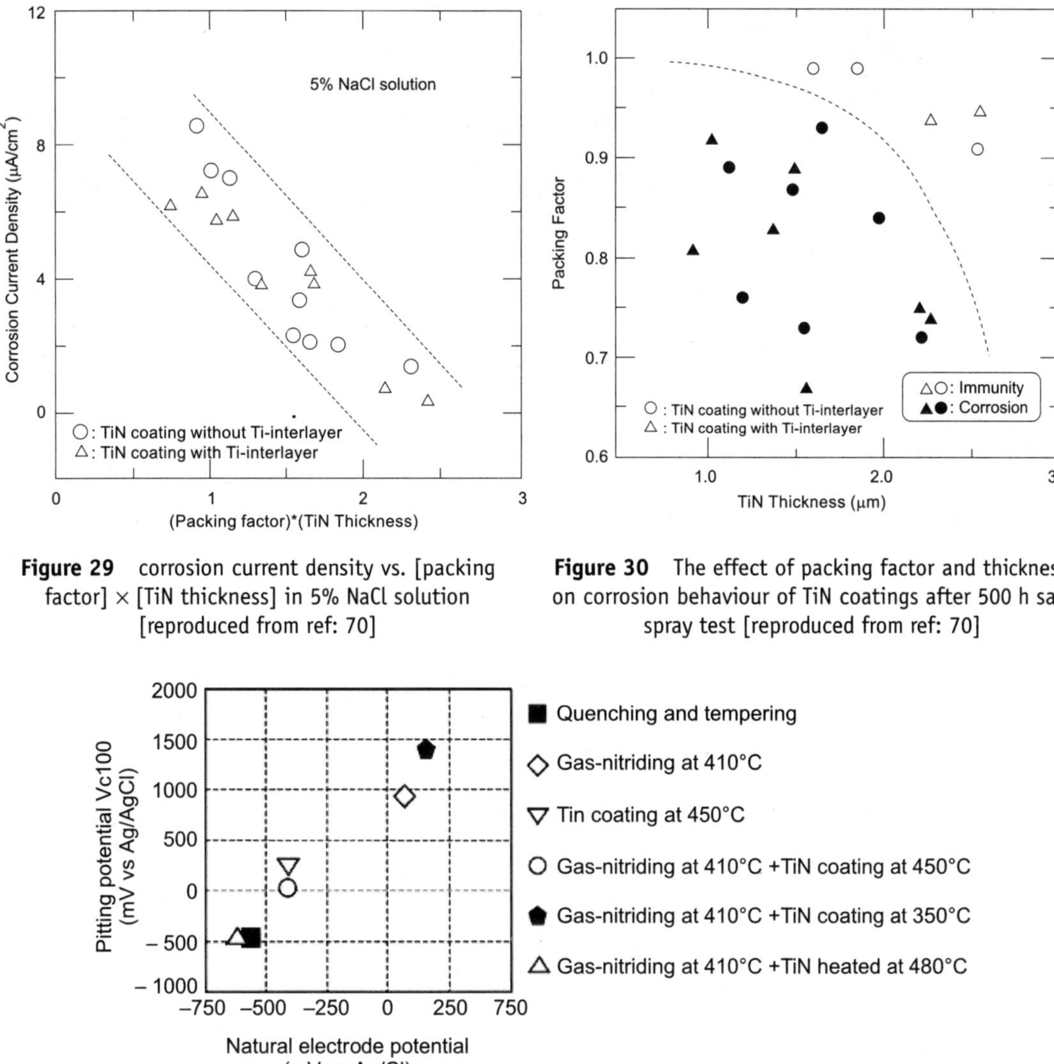

Figure 29 corrosion current density vs. [packing factor] × [TiN thickness] in 5% NaCl solution [reproduced from ref: 70]

Figure 30 The effect of packing factor and thickness on corrosion behaviour of TiN coatings after 500 h salt spray test [reproduced from ref: 70]

Figure 31 Relation between the natural electrode potential and pitting potential [reproduced from 71]

(hexagonal) as demonstrated by Ahn et al [78]. Fig. 32 (a-c) illustrates the planar and cross-sectional morphologies of α-Cr, CrN and Cr_2N coatings respectively deposited on die steel by cathodic arc deposition [78].

The CrN coatings exhibit a dense, columnar grain sized microstructure with minor intergranular porosity and pinholes (see Fig. 32b). However, of all the coatings illustrated in Fig. 32 (a-c), the α-Cr exhibited the densest structure. Of the three types of coatings, CrN exhibited the best overall performance in NaCl environment as indicated in Table 17, even though all the coatings were noble compared to steel (E_{corr} = -726 mV).

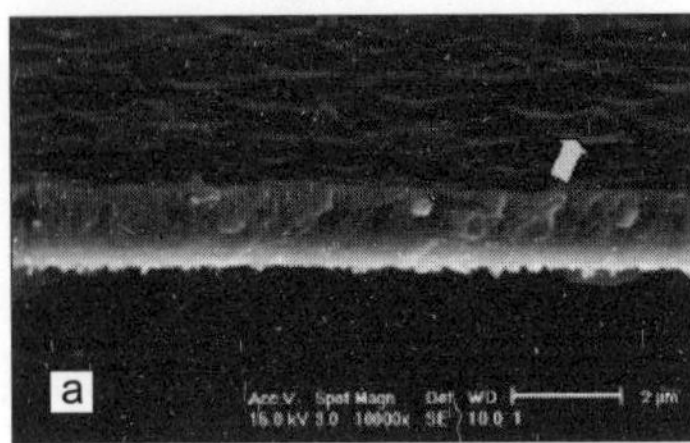

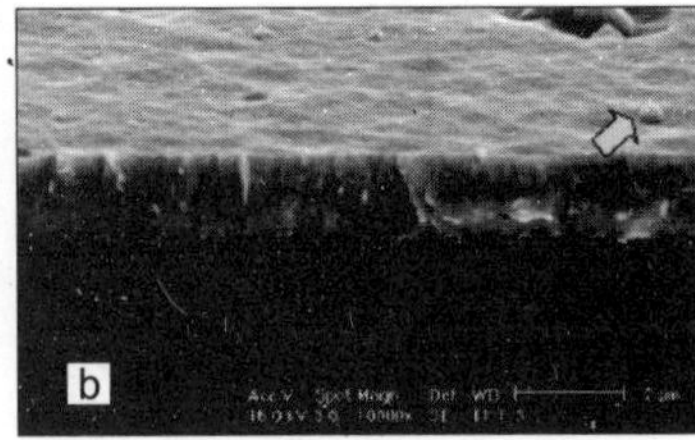

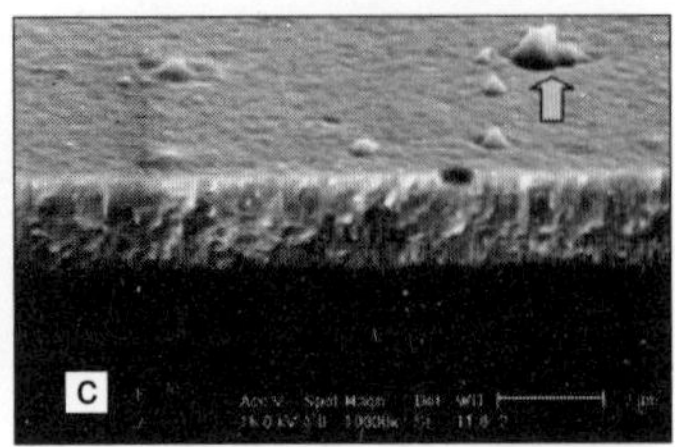

Figure 32 Scanning electron micrographs of cross-section of (a) a-Cr, (b) CrN and (c) Cr_2N coating (0: droplet) [reproduced from 78]

Table 17 Impedance data obtained for different coatings [ref: 78]

Sample	E_{corr} (mV)	i_{corr} (mA/cm^2)	Impedance data (72 h exposure)			
			C_{coat} (10^{-4} F/cm^2)	R_{coat} Ohms cm^2	C_{pore} ($\times10^{-4}$ F/cm^2)	R_{coat} Ohms cm^2
α-Cr	-581.5	0.203	1.09	27.9	12.120	3081
CrN	-556.7	0.039	1.11	32.4	14.340	5353
Cr2N	-558.2	0.536	1.40	12.9	0.003	138

Duplex coatings, combining PVD CrN coating with another type of coating, has also proved effective in combating corrosion. For example, Vencovsky et al [79] coated the tool steel first with electroless Ni-P coating of various thicknesses (12 and 22 μms) followed by PVD of CrN coating (of thickness around 4 μm) at two different temperatures (250°C (LT) and 400°C (HT). The corrosion resistance of these duplex coatings was compared with that of stand alone NiP and CrN coatings and also with the bare steel substrate in 1N H_2SO_4 solution. The results obtained by these authors [79] are presented in Table 18.

It is clear from the above Table that a duplex coating of NiP and CrN has a lower corrosion rate than the standalone CrN or Ni-P coatings. Liu and Coworkers [80, 81] have undertaken a detailed study of the comparative performance of TiN and CrN PVD coatings. SEM micrograph of the cross-section of the two coatings (Fig. 33) indicates that the TiN coatings are

Table 18 Corrosion current and corrosion potential values of different materials [ref:79]

Coating	Thickness (μm)	Corrosion current (μA/cm^2)	E_{corr} (mV)
Steel Substrate	—	2100.0	-458
NiP	12.0	10.0	-330
NiP	22.0	5.5	-317
CrN HT	3.3	3.4	-443
CrN LT	2.9	2.8	-461
NiP + CrN HT	12.0 + 3.3	1.4	-434
NiP + CrN HT	22.0 + 3.3	3.7	-430
NiP + CrN LT	12.0 + 2.9	0.6	-325
NiP + CrN LT	22.0 + 2.9	0.2	-368

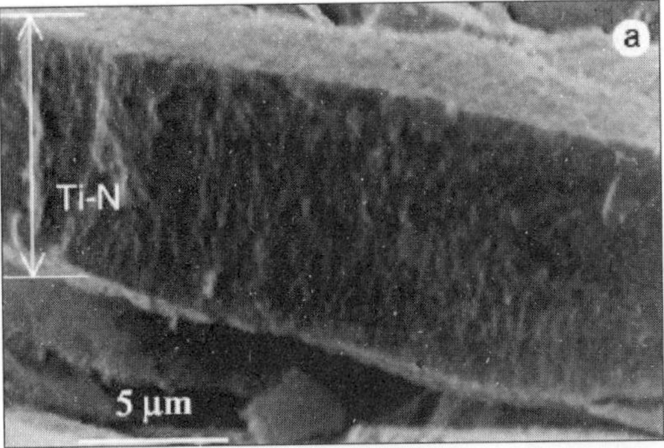

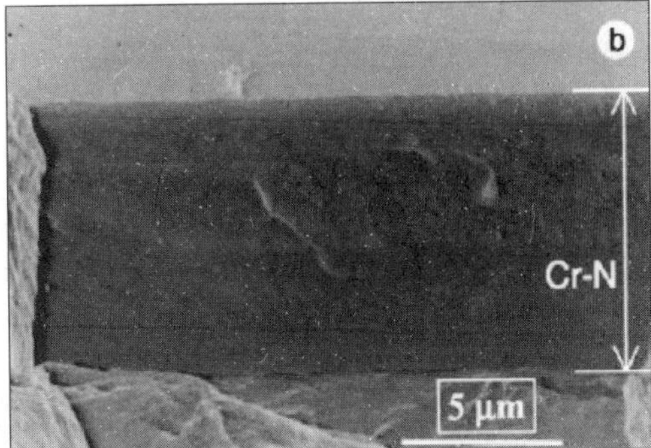

Figure 33 SEM micrographs of fractured cross-section of (a) TiN coating with columnar crystallite structure and (b) CrN with equiaxed crystallite structure [reproduced from81]

composed of fine columnar crystallites while the CrN coatings exhibit a finer and denser equiaxed crystalline structure. CrN PVD coating provided a superior resistance to corrosion than TiN PVD coating largely due to lower porosity / pinholes density and lower oxygen diffusion rate in equiaxed CrN as against columnar TiN [81].

Mendibide et al [82] in a recent study have compared the corrosion resistance of magnetron sputtered CrN and TiN coatings in saline environment. They found that the CrN coating imparts better corrosion protection than TiN. Although both these coatings formed a thin oxide passive film spontaneously in solution, the difference was in that the CrN formed a favourable p-type semiconductive film while the TiN formed an n-type semiconductive film. The p -type semiconductive film formed on the surface of CrN coatings slowed down the reduction kinetics and thus also the corrosion rate [82].

5.1.2 Multilayer nitride coatings

From corrosion point of view, nitrides based coatings are known to be often cathodic when deposited onto steel and there is therefore a risk of galvanic localized attack through the open porosities as discussed earlier. A key-parameter is thus the size of microporosities [83-85]. The size of most pinholes in PVD coatings are below 5 μm in diameter, a few are above 20 μm. Fig. 34 (a-c) illustrates the typical surface morphology depicting an inclusion of 8 μm and a pin hole of 3 μm in size respectively [80]. Based on these typical sizes of defects, it was opined that the coatings should be substantially thicker than the size of the inclusions within it [86]. In order to overcome the harmful effect of microporosities and inclusions, several strategies are adopted. One such an approach is to deposit the layers of TiN with an ultra thin Ti metallic interlayer [70,87]. Such a multilayered coating is expected to improve the corrosion resistance due to a number of reasons [88] as described below:

- The increase in coating thickness, which statistically reduces the possibility of through-coating defects (e.g. pores);
- Discontinuous crystallite boundaries in the columnar structure which decrease the opportunity of pore formation; and
- Alternating interlayers of different compositions (and therefore electrical behaviour) which can redirect the current flow between coating and substrate

A multilayered coating will have lower porosity than a single layer coating, since the open structure, reaching from the surface to the substrate, will be interrupted by repeated nucleation at the interfaces between sublayers [89]. An improved protection of the substrate was observed by Herranen et al. [90] using multilayer coatings with thick sublayers. In the case of ion plated TiN coatings deposited on tool steel, which typically exhibit the columnar structure, by introduction of Ti sublayers with a high degree of perfection was found to hinder the columnar growth and thereby improve the corrosion protection in 0.1 M H_2SO_4 medium. Further, the introduction of few thick sublayers was found to be more efficient than many thin sublayers.

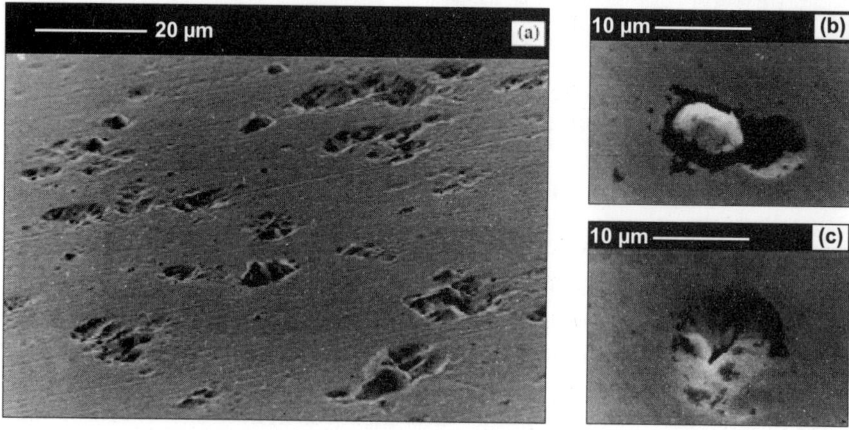

Figure 34 SEM surface micrographs of TiN coating (a) overall surface morphology, (b) an inclusion and (c) a pinhole [reproduced from ref: 80]

The Ti interlayer was found to be preventing pitting corrosion by forming a protective titanium oxide atop the Ti interlayer at coating defects. Such graded interfaces were found to reduce the interfacial stresses, decrease the porosity and increase the corrosion protection [87]. The overall corrosion protection offered by the TiN coatings with Ti interlayers is superior to that of single layer TiN coatings of identical thickness as observed by Liu et al. [88]. Fig. 35 illustrates the potentiodynamic polarization measurements made in 0.5 N aqueous sodium chloride solution of single layer and multilayer TiN and CrN coatings deposited on mild steel substrate in comparison with the substrate itself [88]. Clearly, in comparison with the corrosion potential (E corr) of the bare mild steel substrate (-0.65V, SCE), E_{corr} was increased by applying the TiN or CrN coatings to a value of -0.4V and -0.45V respectively for four-layer and single layer coatings. This increase represents a nobler electrode potential being achieved by Liu et al. [88], thus resulted in improvement of corrosion resistance of mild steel. A small, but significant E_{corr} increase (~50 mV) of four- layer coatings compared to single -layer coatings suggests that the four -layer system is nobler than the single-layer treatment.

The intermediate etching –or in other words, ion bombardment of the grown film surface – results in a high number of defects in the near surface area and is thought to create new nucleation sites [91] for the subsequently deposited nitride film in the magnetron sputter deposition process resulting in less through-pores and pinholes. The intermediate etching is performed by Fenker et al. [92] by using higher substrate bias voltages in the order of 1000 V.

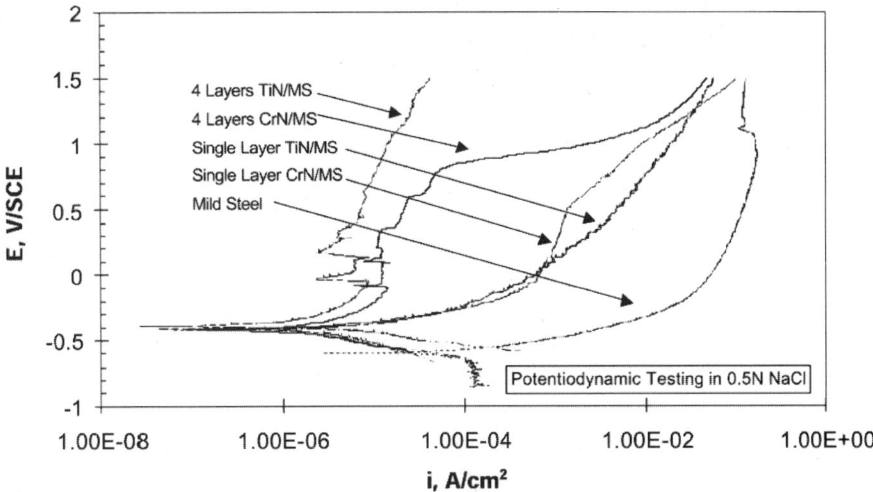

Figure 35 Potentiodynamic polarization curves for the bare mild steel substrate and single-layer or four-layer TiN and CrN coatings in 0.5N NaCl solution [reproduced from ref:88]

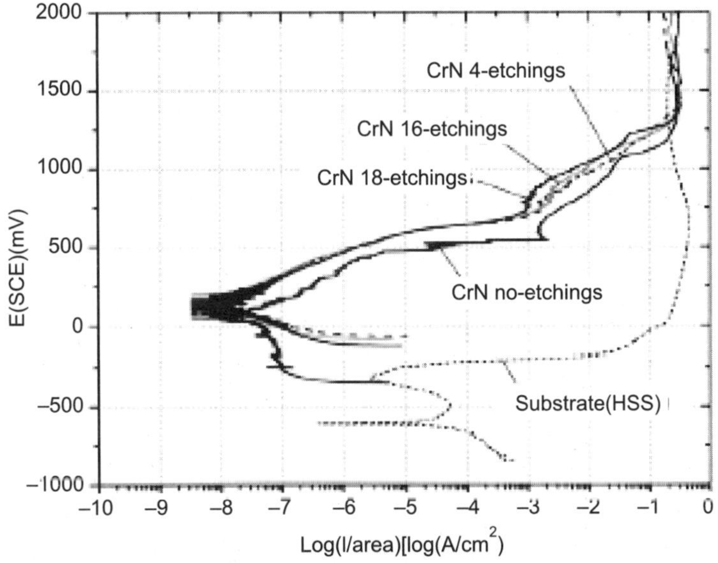

Figure 36 Potentiodynamic corrosion test of intermediate etched CrN thin films (0.8 M NaCl solution) [reproduced from ref: 92]

Fig. 36 shows the potential-current density curves of the high speed steel as well as the non-etched and the intermediate etched CrN films with four, eight and 16 intermediate etching steps. Comparison of the curves reveals that the higher the number of intermediate etching steps, the higher the corrosion resistance of the system CrN/steel, i.e., the corrosion resistance is approximately proportional to the number of generated interfaces. In contrast, the intermediate etching employed in the case of NbN/steel deposited by the same technique could not alter the corrosion behaviour.

Superlattice coatings are nanometer-scale multilayers composed of two different alternating layers with a superlattice period, i.e., the layer thickness of two materials, ranging from 18 to 20 nm [93]. The compositional modulation resulting in superlattice formation substantially enhances coating barrier properties. The innovative arc bond-sputtering (ABS™) process represents one of the recent major advances in PVD coating technology and is successful in producing highly adherent and wear-resistant coatings with both monolithic and layered structures. The ABS technique is capable of producing superlattice coatings of CrN/NbN coatings which exhibit superior corrosion protection than the monolithic CrN coatings [94]. However, given a well-recognized drawback of an arc process, namely the production of macroparticles within the coatings [95], it is of no surprise that macroparticles are also present in coatings prepared by the ABS process. Some of these macroparticles creates through-thickness defects which are deleterious to the overall corrosion protection. To counteract this problem, arc filtering technique is recently being used.

Table 19 presents typical superlattice period, lattice parameter, coating thickness and microhardness of CrN/NbN coatings deposited at high (400°C) and low (200°C) temperatures obtained by Tomlinson et al. [77] by utilizing the ABS process. It is to be noted that all coatings had microhardness values in excess of 2750 kg mm^{-2} and adhesion failure loads in excess of 25N. Figure 37 shows typical polarization data for 304L, CrN and CN22 and CN45 coatings [77]. It is clear from Fig. 37 that the monolithic CrN coating has only a marginal effect on the corrosion performance of the 304L substrate. Thus, the corrosion potential is raised slightly however the pitting potential and pitting rate are essentially identical as evident from EIS studies conducted by the authors. The CrN/NbN superlattice coatings have significantly raised pitting potentials and reduced passive currents and, after corrosion testing, considerably less surface discoloration than the monolithic CrN coating [77].

The novel superlattice coatings composed of WC-$(Ti_{1-x}Al_x)N$ with stepwise changing Al concentrations (WC-$Ti_{0.86}Al_{0.14}N$, WC-$Ti_{0.72}Al_{0.28}N$, WC-$Ti_{0.58}Al_{0.42}N$) were deposited on steel substrate by Ahn et al. [96]. The Al concentration was controlled by using evaporation source for Al and fixing the evaporation rate of the metals (WC alloy and Ti). The measured galvanic corrosion currents between coating and substrate in deaerated 3.5% NaCl solution indicated that WC-$Ti_{0.72}Al_{0.28}N$ coating exhibited the best resistance. The potentiodynamic polarization tests showed that this coating had passivation tendency and lower porosity indicating the improved corrosion resistance.

Table 19 Composition and properties of CrN/NbN superlattice coatings deposited on 304 L [ref:77]

Sample and deposition temperature	Nitrogen:metal ratio (GDOES)	Superlattice period (nm)	Lattice parameter (nm)	Thickness (μm)	Micro-hardness (kg mm^{-2})
CN45:400°C	1.05	4.20	0.4283	4.23	3400
CN46:400°C	0.74	3.38	0.4352	5.41	2750
CN22:200°C	1.29	2.25	0.4251	2.75	3320
CN24:200°C	1.12	2.88	0.4282	3.00	3200

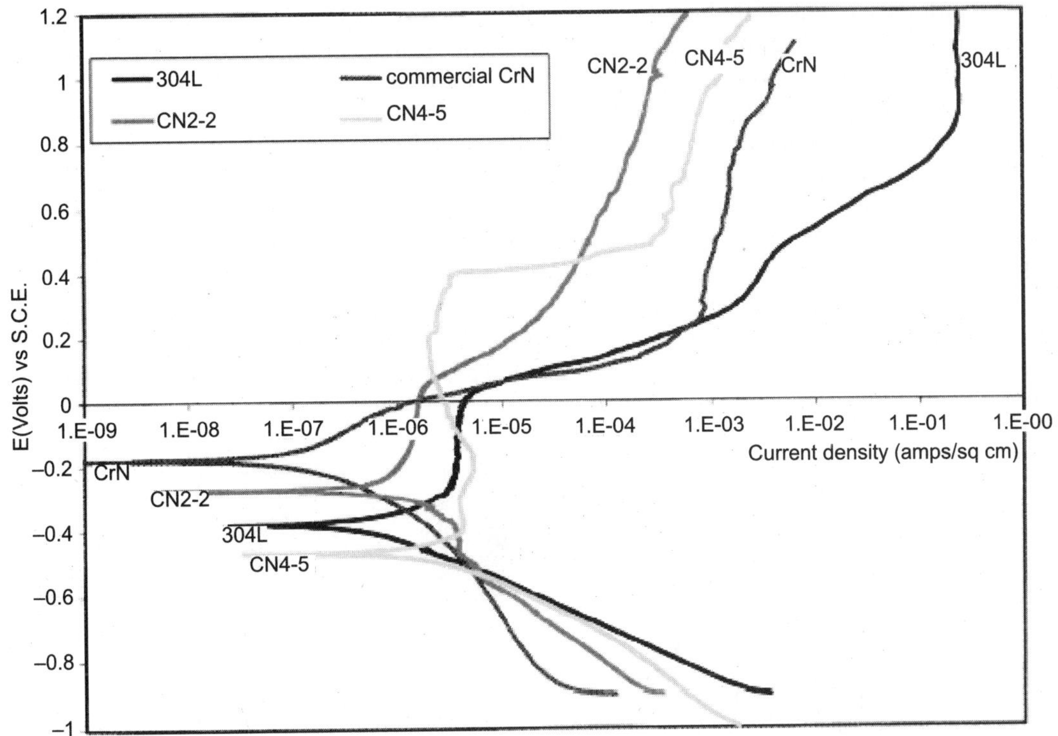

Figure 37 Superimposed potentiodynamic polarization data for 304L, CrN and typical results from low (CN22) and high (CN45) temperatures CrN/NbN superlattice coatings [reproduced from ref: 77].

5.2 Metallic Coatings

For many years zinc and its alloy coatings have been used to improve the corrosion resistance of steel sheets. Conventionally, these coatings are deposited by hot dip and electroplating processes. A possible alternative coating process is PVD which includes the techniques of evaporation, ion beam, arc and sputter deposition [97-99]. However, the disadvantage of the PVD process is the low deposition rate compared to electrochemical or hot dip methods. Despite of this, these processes are environmentally friendly and have the following additional advantages:

(a) They allow the deposition of any type of Zn alloy without the restrictions inherent in electroplating processes due to different electrochemical potentials of the alloying materials;
(b) There is no danger of steel embrittlement as a consequence of hydrogen evolution during the electrochemical process; and
(c) PVD coatings can be produced with very compact microstructures which have been assumed to further increase the corrosion resistance, and might also advantageously influence steel sheet processing properties such as formability, weldability and paintability.

The influence of substrate bias voltage on the structure and barrier protection of PVD Zn - Al, Zn and Zn -Ni coatings has been studied by Li and Mowak et al [100], Musil et al [101] and Bowden et al [102] respectively. These studies have indicated that the application of a negative bias voltage results in compact and fine grained coatings with superior barrier properties as compared to less compact electrodeposited coatings of similar composition.

Maeng et al [103] have compared the Cr coatings produced by cylindrical magnetron sputtering with Cr coating formed by electrodeposition. The polarization curves for the electrodeposited and PVD Cr coatings are shown along with the steel substrate in Fig. 38. The PVD Cr coating exhibits a more noble corrosion potential value of -75 mV (vs. SCE) than that of electrodeposited Cr coating. In addition, the corrosion current density of the former is four orders of magnitude smaller than the latter. This result demonstrates that the PVD Cr is more corrosion resistant than that of the electrodeposited coating, potentially due to the formation of passive oxide film.

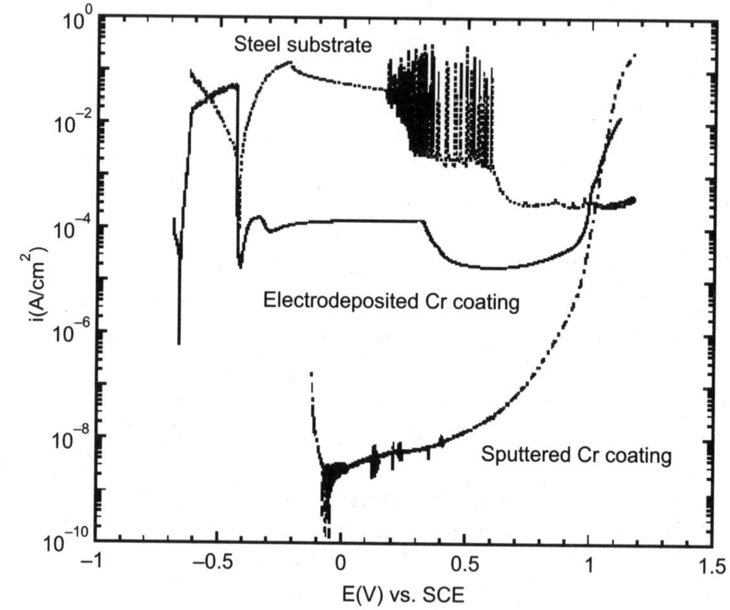

Figure 38 Anodic polarization curves of electrodeposited Cr and sputtered Cr coatings and the steel substrate (ASTM A723) after 1 h immersion in 0.5 M H2SO4 deaerated with N2 at room temperature [reproduced from ref: 103].

Zykova et al [104] have evaluated the corrosion resistance of Ti and Zr PVD coatings deposited on stainless steel (SS) and NiTi (Ti-48.5% & Ni-48.5%) substrates when exposed to artificial isotonic physiological solution for 6 h. The results are presented in Fig. 39 in terms of corrosion current vs. corrosion potential. It is clear from Fig. 39 that PVD Ti coating on SS and NiTi substrates exhibits the lowest corrosion rate which is even lower than that exhibited by PVD ZrN and TiN coatings. It was also noted, on the basis of inculcation of implants coated with various coatings as above in distal part of rat femoral bone and leaving it there for 4

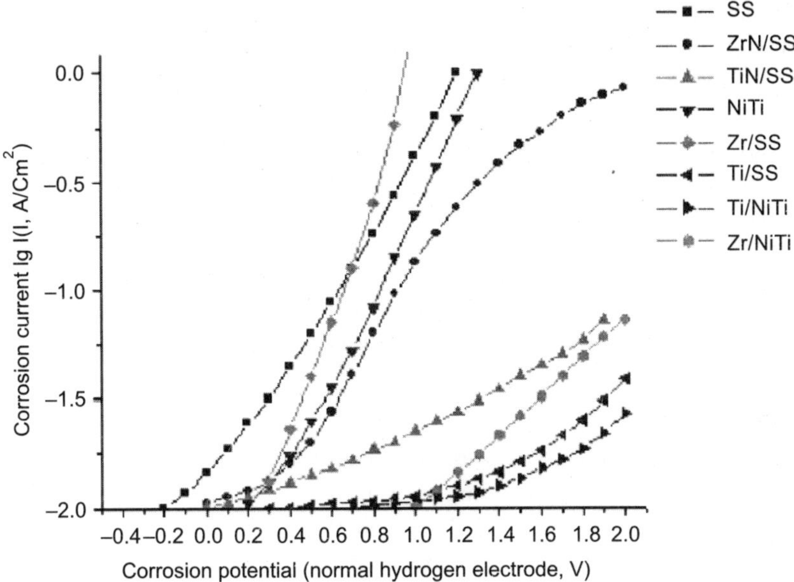

Figure 39 Dependency of corrosion current on corrosion potential for various types of coatings on stainless steel samples, Ti, Zr and titanium nickelide (NiTi) in 0.9% NaCl isotonic solution (pH = 6.0) [reproduced from ref: 54]

months, that the migration of the toxic microelements contained in the implant coatings to the kidney, was minimal in the coatings having the best corrosion resistance.

6.0 CONCLUSIONS

The corrosion behaviour of coatings on steel substrates have been reviewed. Further, among the variety of coatings that have been utilized to enhance the corrosion resistance of steels, this review has been restricted to recent advances inelectrodeposition, thermal spray and physical vapour deposition (PVD) coatings.

Electrodeposition (ED) is a well -established technique and ED zinc, Cd and Cr coatings have found extensive applications in the industry. As a result, the review has concentrated in the recent developments in the field of ED and electroless coatings. In particular, recent work on the corrosion behaviour of carbon nanotubes and PTFE reinforced ED composite coatings, nanostructured Zn obtained by pulsed deposition, Zn-Ni ED coatings as replacement for environmental regulated Cd coatings, electroless Ni-W-P and Ni-Cu-P coatings in comparison to electroless Ni-P coatings and layered Ni -P coatings as compared to monolithic Ni-P coatings have been reviewed in this chapter.

Thermal spray coatings (metallic, ceramic and cermet) are extensively used in applications requiring wear resistance. However, in many applications, the thermal spray coatings should also in addition possess adequate corrosion resistance. Recent drive to replace electrodeposited hard chromium coatings with more environmentally friendly thermal spray coatings has also enhanced interest in the study of corrosion behaviour of thermal spray coatings. In this chapter, recent work related to corrosion behaviour of steel substrates coated with NiCrBSi and

NiWCrBSi HVOF coatings, HVOF Inconel 625 coatings, HVOF 316L stainless steel coatings and Mo bearing stainless steel coatings has been reviewed. In addition, the corrosion performance of Al_2O_3 based ceramic coatings and WC-Co cermet coatings have also been described.

PVD coatings are increasingly considered for corrosion applications in view of the recent developments in PVD coating technology which allows the deposition of thick (>10 μm) PVD coatings on one hand and multilayered PVD coatings on the other. Some of the recent works on the development of corrosion resistant nitride (TiN, CrN) coatings taking advantage of the above developments are reviewed in this chapter. The use of the concept of duplex coatings to enhance the corrosion performance of PVD coatings and also the corrosion behaviour of metallic PVD coatings (Zn-Ni, Zn-Al, Cr, Ti, Zr) have also been described.

RERERENCES

1. Corrosion: Understanding the Basics, J.R. Davis (ed.), ASM International, U.S.A. (2003).
2. J. Zahavi, H. Koebel, Proce. Inter finish (1980) 208.
3. J. Zahavi, H. Koebel, Plat, Surf. Finish, 69 (1982) 76.
4. M. Pushpavanam and B.A. Shenoi, Met. Finish, 75 (1977) 38.
5. J. Zahavi, J. Hazan, Plat. Surf. Finish 72 (1983) 57.
6. X.H. Chen, J.C. Pong, X.Q.Li, J.X.Wang, W.Z. Li, J. Mater. Sci. Lett. 20 (2001) 2057.
7. X.H. Chen, C.S. Chem, H.N. Xia, F.Q. Cheng, G. Zhomg and G.J. Yi, Surface & Coatings Technology 191 (2005) 351.
8. S. Rossi, F. Chini, G. Straffelini, P.L. Bonora, R. Moschini and A. Stampali, Surface and Coatings Technology, 173 (2003) 235.
9. D.H. Jeong, U. Erb, K.T. Aust, G. Palumbo, Scripta Materialia 48 (8) (2003) 1067.
10. I. Brooks, U. Erb, Scripta materialia 44 (5) (2001) 853.
11. R. Rofagha, R. Langer, A.M. El-Sherik, U-Erb, G. Palambo, A.K. Aust, Scripta Metallurgica et Materialia, 25 (1991) 2867
12. L.M. Baugh, Electro Chimica Acta, 24 (1979) 657
13. D.A. Jones and N.R. Nair, Corrosion, 41 (1985) 357
14. G. Barcelo, M. Sarnet, C. Muller, J. Pregonas, Electro Chimica Acta, 43 (1998) 13.
15. Kh.M.S. Youssef, C.C. Roch and P.S. Fedkiw, Corrosion Science, 46 (2004) 51.
16. M. Bin-Sudin, A. Leyland, A.S. James, A. Matthews, J. Housden, B. Garade, Surf. Coat. Technol. 81 (1996) 215.
17. R.E. Miller, Plating Surf. Finish 74 (1987) 52.
18. Y. Gao, Z.J. Zheng, M. Zhu and C.P. Luo, Mat. Science and Engg. A 381 (2004) 98.
19. A. Warren, A. Nyland, I. Olefjord, Int. J. Refract. Hard. Mater. 14 (1996) 345.
20. H.A. Soruhabi, H. Doalti, J. Manzorri, Appl. Surf. Sci. 185 (2002) 155.
21. Y.W. Wang, C.G.Xiao, Z.G.Deng, Plat. Surf. Finish, 79 (1992) 57.
22. Y. Liu, Q. Zhao, Appl. Surf. Sci. 228 (2004) 57.
23. C.Gu, J. Lian, G. Li, L. Niu and Z. Jiang, Surf. Coatings Tech., 197 (2005) 61.
24. T.S.N. Sankara Narayanan, I. Baskaran, K. Krishnaveni, S. Parthiban, Surf. Coatings Tech; 200 (2006) 3438.

25. M. Gavrila, J.P. Millet, H. Mazille, D. Marchandise, M.M. Cuntz, Surf. Coatings Technology, 123 (2000) 164.
26. D. Marchandise, J.M. Curtz, Surfaces 232 (1992) 24.
27. G.D. Wilcox, D.R. Gabe, Corrosion Science 35 (1993) 1251.
28. N. Gage, D.A. Wright, Proc. Asia Pacific Interfinish, Singapore (1990) 65.
29. H. Kim, B.N. Popov and K.S. Chen, Corrosion Science 45 (2003) 1509.
30. M.R. Kalantory, G.D.Wilcox, D.R. Gabe, British Corrosion Journal 33 (1998) 197.
31. G. Chawa, G.D. Wilcox, D.R. Gabe, Trans. Ins. Of Metal Flireshing 76 (1998) 117.
32. I. Jvanov, T. Valkova, I. Kirilova, J. Appl. Electrochemistry 32 (2002) 85.
33. J. Fei and G.D.Wilcox, Surf. Coating Technology 200 (2006) 3533.
34. N. Parkansky, R.L. Boxman, S. Goldsmith, Yu. Rosenberg, Surf. Coat. Technol. 76-77 (1-3) (1995) 352.
35. P.R. Sere, M. Zapponi, C.I. Elsner, A.R. Sarli, Corros. Sci. 40 (10) (1998) 1711.
36. W.M. Zhao, Y. Wang, T. Han, K.Y. Wu, J. Xue, Surf. Coat. Technol. 183 (2004) 118.
37. W.M. Zhao, Y. Wang, L. X. Dong, K.Y. Wu, J. Xue, Surf. Coat. Technol. 190 (2005) 293.
38. L. Gil, M.H. Staia, Thin Solid Films 420-421 (2002) 446.
39. A.A. Boudi, M.S.J. Hashmi, B.S. Yilbas, Journal of Materials Processing Technology 155-156 (2004) 2051.
40. D. Zhang, S.J. Harris, D.G. McCartney, Materials Science and Engineering A344 (2003) 45.
41. S. Shrestha, A.J. Sturgeon, Surface Engineering 20 (4) (2004) 237.
42. J. Kawakita, T. Fukushima, S. Kuroda, T. Kodama, Corrosion Science 44 (2002) 2561.
43. X.Q. Wu, H.M. Jing, Y.G. Zheng, Z.M. Yao, W. Ke, Corrosion Science 46 (2004) 1013.
44. M. Rasso, A. Scrivani, D. Ugues, S. Bertini, International Journal of Refractory Metals & Hard Materials 19 (2001) 45.
45. E. Celik, I.A. Sengil, E. Avci, Surf. Coat. Technol. 97 (1997) 355.
46. E. Celik, I. Ozdemir, E. Avci, Y. Tsunekawa, Surf. Coat. Technol. 193 (2005) 297.
47. J. E. Cho, S.Y. Hwang, K.Y. Kim, Surf. Coat. Technol. 200 (2006) 2653.
48. H.J.C. Voorwald, R.C. Souza, W.L. Pigatin, M.O.H. Cioffi, Surf. Coat. Technol. 190 (2005) 155.
49. P.M. Natishan, S.H. Lawrence, R.L. Foster, J. Lewis, B.D. Sartwell, Surf. Coat. Technol. 130 (2000) 218.
50. C. Monticelli, A. Frignani, F. Zucchi, Corrosion Science 46 (2004) 1225.
51. P. Seserko, U. Kopacz, S.Schulz, Galvanotechnology 80 (1989) 4272.
52. R. Riedl, Galvanotechnology 80 (1989) 3391.
53. S. Bastian, Galvanotechnology 81 (1990) 2706.
54. A. Wisbey, P.J. Gregson, M. Tuke, Biomaterials 8 (1987) 477.
55. H. Brauner, Surf. Coat. Technol. 62 (1993) 618.
56. H.A. Jehn, Surf. Coat. Technol. 125 (2000) 212.
57. C. Pfohl, K.T. Rie, M.K. Hirschfeld, J.W. Schultze, Surf. Coat. Technol. 112 (1999) 114.
58. H. Kusamachi, Titanium metal & its applications, Nikkan, K. Shibun-sha, Tokyo,1983, p.42.
59. A. Kobayashi, Surf. Coat. Technol. 132 (2000) 152.
60. H. Dong, Y. Sun, T. Bell, Surf. Coat. Technol. 106 (1997) 91.
61. A. Kobayashi, Weld. Int. 4 (4) (1990) 276.
62. L. Swadzba, A. Maciejny, B. Formanek, Surf. Coat. Technol. 78 (1996) 137.

63. J.S. Huang, A.S. Oates, J. Zhao, Thin Solid Films 365 (2000) 110.
64. Z.M. Z.M. Yu, Z. Jin, C.Q. Liu, L. Yu, S. Dai, J. Vac. Sci. Technol. A13 (1995) 2303.
65. M. Nordin, M. Herranen, S. Hogmark, Thin Solid Films 348 (1999) 202.
66. L. Yu, F.Q. Lang, Z.M. Yu, Z. Jin, Corros. Sci. Protect. Technol. 11 (1999) 84.
67. L. Cunha, M. Andritschky, L. Rebouta, R. Silva, Thin Solid Films 317 (1998) 351.
68. C. Rebholz, J.M. Schneider, H. Ziegele, B. Rhale, A. Leyland, A. Matthews, Vacuum 49 (1998) 265.
69. F. Lang, Z. Yu, Surf. Coat. Technol. 145 (2001) 80.
70. B.F. Chen, W.L. Pan, G.P. Yu, J. Hwang, J.H. Huang, Surf. Coat. Technol. 111 (1999) 16.
71. A. Kagiyama, K. Terakado, R. Urao, Surf. Coat. Technol. 169-170 (2003) 397.
72. H. Dong, Y. Sun, T. Bell, Surf. Coat. Technol. 90 (1997) 91.
73. M.J. Park, A. Leyland, A. Matthews, Surf. Coat. Technol., 43-44 (1990) 481.
74. R.B. Diegle, N.R. Sorensen, C.R. Clayton, M.A. Helfand, Y.C.Yu, J. Electrochem. Soc. 135 (5) (1988) 1085.
75. Y.I. Chen, J.G. Duh, Surf. Coat. Technol., 48 (1991) 163.
76. H.W. Wang, M.M. Stack, S.B. Lyon, P. Hovsepian, W.D. Munz, Surf. Coat. Technol. 126 (2000) 279.
77. M. Tomlinson, S.B. Lyon, P. Hovsepian, W.D. Munz, Vacuum 53 (1999) 117.
78. S.H. Ahn, Y.S. Choi, J.G. Kim, J.G. Han, Surf. Coat. Technol. 150 (2002) 319.
79. P.K. Vencovsky, R. Sanchez, J.R.T. Branco, M. Galvano, Surf. Coat. Technol. 108-109 (1998) 599.
80. C. Liu, Q. Bi, A. Leyland, A. Matthews, Corros. Sci. 45 (2003) 1257.
81. C. Liu, Q. Bi, A. Matthews, Corros. Sci. 43 (2001) 1953.
82. C. Mendibide, P. Steyer, J.P. Millet, Surf. Coat. Technol. 200 (2005) 109.
83. M.A.M. Ibrahim, S.F. Korablov, M. Yoshimura, Corros. Sci. 44 (2002) 815.
84. C. Liu, Q. Bi, A. Leyland, A. Matthews, Corros. Sci. 45 (2003) 1243.
85. K.H. Lee, C.H. Park, Y.S. Yoon, H.A. Jehn, J.J. Lee, Surf. Coat. Technol.142-144(2001) 971.
86. S. Bababeygy, Corrosion protection by vacuum arc coatings, Proc. Interfinish 88, Paris, October 1988.
87. L.A.S. Ries, D.S. Azambuja, I.J.R. Baumvol, Surf. Coat. Technol. 89 (1997) 114.
88. C. Liu, A. Leyland, Q. Bi, A. Matthews, Surf. Coat. Technol. 141 (2001) 164.
89. M. Larsson, P. Bromark, P. Hedenqvist, S. Hogmark, Surf. Coat. Technol. 76-77 (1995) 202.
90. M. Herranen, U. Wiklund, J.O. Carlsson, S. Hogmark, Surf. Coat. Technol. 99 (1998) 191.
91. R.F. Bunshah, R.S. Juntz, J. Vac. Sci. Technol. 9 (6) (1972) 1404.
92. M. Fenker, M. Balzer, H.A. Jehn, H. Kappl, J.J. Lee, K.H. Lee, H.S. Park, Surf. Coat. Technol. 150 (2002) 101.
93. J.S. Yoon, H.S. Myung, J.G. Han, J. Musil, Surf. Coat. Technol. 131 (2000) 372.
94. C. Liu, A. Leyland, S.B. Lyon, A. Matthews, Surf. Coat. Technol. 76 (1995) 615.
95. H.D. Steffens, M. Mack, K. Moehwald, K. reichel, Surf. Coat. Technol. 46 (1991) 65.
96. S.H. Ahn, J.H. Yoo, Y.S. Choi, J.G. Kim, J.G. Han, Surf. Coat. Technol. 162 (2003) 212.
97. C. Bowden, A. Matthews, Surf. Coat. Technol. 76-77 (1995) 508.
98. L. Li, W.B. Nowak, J. Vac. Sci. Technol. A 12 (1993) 1587.
99. J. Musil, J. Mattous, V. Valvoda, Vacuum 46 (1995) 203.
100. L.Li, W.B Nowak, J. Vac. Sci. Technol. A12 (1993) 1587.

101. J. Musil, J. Mattous, V. Valvoda, Vacuum, 46 (1995) 203.
102. C. Bowden, A. Matthews, Surf. Coat. Technol. 76/77 (1995) 508.
103. S. Maeng, L. Axe, T.A. Tyson, D. Cote, Surf. Coat. Technol. 200 (2006) 5676.
104. A.V. Zykova, V.V. Luuyanchenko and V.Z. Sofonov, Surf. Coat. Technol. 200 (2005) 90.

CHAPTER 9

Volatile Corrosion Inhibitors

K. L. Vasanth
Naval Surface Warfare Center Carderock Division, 9500 MacArthur Blvd., West Bethesda, MD 20817, USA

A literature review of the past several years of work on Volatile Corrosion Inhibitors also called Vapor Phase Corrosion Inihibitors (VCIs) has been summarized. Definitions, classification, structural features, corrosion inhibition mechanism and methods of evaluation of Volatile Corrosion Inhibitors are also discussed. Practical methods of using VCIs for corrosion protection of ferrous and non-ferrous alloys are explained with some specific examples.

INTRODUCTION

According to a recent report, [1] metallic corrosion costs the United States about $300 billion a year. The report, released by Battelle, Columbus, Ohio and Speciality Steel Industry of North America, Washington DC, claims that about one-third of the cost of corrosion ($100 billion) is avoidable and could be saved by using corrosion-resistant materials and the application of state-of-the art corrosion control technologies. The more we know about corrosion, the various causes and the mechanism of attack of metals and alloys, the better equipped we will be to combat it. Atmospheric corrosion of metals and alloys is responsible for a large percentage of total corrosion cost. Apart from direct cost of corrosion, indirect consequences are critical such as depletion of the natural resources, structural damages leading to failure of industrial and military systems.

DEFINITION AND CLASSIFICATION

The use of inhibitors is one of the corrosion control technologies directed towards combating corrosion. The definition of an inhibitor favored by the National Association of Corrosion Engineers International (NACE) is: a substance that retards corrosion when added to an environment in small concentrations [2]. It is well established that inhibitors function in one or more ways to control corrosion: by adsorption of a thin film onto the surface of a corroding material, by inducing the formation of a thick corrosion product, or by changing the characteristics of the environment resulting in reduced aggressiveness [3]. One of the recent classifications of inhibitors given by Uhlig and Revie [4] in their book is that inhibitors are designated as a) passivators, b) organic inhibitors and c) vapor-phase inhibitors.

Passivators are usually inorganic oxidizing chemicals such as chromates, nitrites and molybdates that passivate the metal and shift the corrosion potential several tenths of a volt in the noble direction. In general, the passivating-type inhibitors are very efficient and reduce corrosion rates to very low values. Primers containing hexavalent chromium have been the work- horse for passivating aluminum alloys and steel. Because it is known to be a carcinogen, several research programs are on-going to replace the hexavalent chromium. There is a critical concentration for passivators at which they show optimum inhibition. Below the critical concentration the passivators behave as active depolrizers and will lead to localized corrosion. Some alkaline compounds such as NaOH, Na_3PO_4, and $Na_2B_4O_7$ are known to facilitate passivation of iron by favoring adsorption of dissolved oxygen.

Organic inhibitors are also called adsorption-type inhibitors since the organic compounds adsorb on the metal surface and suppress metal dissolution and reduction reactions. These affect both anodic and cathodic processes [5]. Organic amines and carboxylates are typical examples of this class. Typical effective organic pickling inhibitors for steel are quinolinethiodide, *o*-and *p*-tolylthiourea, propyl sulfide, formaldehyde and *p*-thiocresol. These are used to remove mill scale, save steel, save acid and reduce acid fumes caused by hydrogen evolution [4].

Volatile corrosion inhibitors or vapor phase corrosion inhibitors are similar to organic adsorption-type inhibitors. These possess moderately high vapor pressure and consequently can be used to inhibit atmospheric corrosion of metals without placing VCIs directly on the metal surface. VCIs are usually effective if used in enclosed spaces such as closed packages or the interior of machinery during shipment [6]. The advantage of VCIs is that the volatilized molecules can reach hard-to-reach spaces commonly found in electrical and electronic enclosures, between two metal flanges, void spaces and similar other systems. In the case of precise scientific instruments, electrical and electronic equipment, VCIs offer a definite advantage over classical methods of corrosion protection. A VCI can be defined as a single chemical or combination of chemicals (mostly organic) having high vapor pressure that can prevent atmospheric corrosion of metallic materials. According to Miksic and Miller, "volatile corrosion inhibitors are secondary electrolyte layer inhibitors that possess appreciable saturated vapor pressure under atmospheric conditions, thus allowing vapor-phase transport of the inhibitive substance." [7]

Fodor [8] has documented an extensive literature review on vapor-phase corrosion inhibition. His report discusses theoretical and practical considerations including definitions, classifications, corrosion inhibition mechanism and evaluation of VCIs. Protection of ferrous and non-ferrous alloys is discussed with some specific examples. Also included in Fodor's report are the structural identification of several VCI organic compounds and a summary of commercial preservation procedures.

A review by Miksic [9] discusses the atmospheric corrosion inhibition of both ferrous and non-ferrous metals using VCIs. To be effective VCIs, the chemical compounds need to possess high passivating properties, strong tendencies towards surface adsorption and the ability to form a comparatively strong and stable bond with metal surfaces. The delicate and yet complex nature of atmospheric corrosion processes occurring under thin films of electrolyte preordains the transport mechanism of the corrosion inhibitor that, in order to become effective, must diffuse through the electrolyte film and cover a substantial portion of the surface. Certain physical and chemical properties such as vapor pressure, molecular structure, availability of functional groups for surface physical and chemical bonding, the polarity, contamination and the conductivity of the electrolyte are critical. Miksic[9] also states that there is considerable controversy over the importance of and the relationship between the saturated vapor pressure and its influence on the effectiveness of the specific compound.

In a relatively comprehensive recent review by Singh and Banerjee [10], the authors categorized VCIs based on their application to specific groups of metals, e.g., for ferrous metals and alloys, for copper and copper alloys, for silver and for aluminum.

VCI ACTION MECHANISM

It is widely accepted that a VCI must be a volatile compound or a mixture of such compounds. It must be capable of forming a relatively stable bond at the metal interface, thus producing a protective layer that limits the penetration of corroding species [11, 12]. According to Balezin [11] every corrosion inhibitor including volatile ones, should:

(a) be capable of establishing a stable bond with the metal surface in a given environment of certain range of acidity and pressure and
(b) create an impenetrable layer for corroding ions.

The mechanism of inhibition as illustrated by Balezin is shown in Figure 1. Two functional groups are attached to the nucleus R_0: R_1 is responsible for adsorption of the inhibitor on the metal's surface, R_2 gives the thickness and impenetrable nature to the protective inhibitive layer.

Vapor pressure is a critical parameter in determining the effectiveness of a VCI. A VCI reaches the metal surface that it must protect through the vapor phase. This transport mechanism requires the VCI to possess an optimum vapor pressure. Too low a vapor pressure of the order of 10^{-6} Torr at room temperature, leads to slow establishment of a protective layer that may result in insufficient corrosion protection. Furthermore, if the space that houses the equipment and the VCI is not sealed, sufficient inhibitor concentration may not be reached. On the other hand, if the vapor pressure is too high (approximately 0.1 Torr at ambient conditions), VCI effectiveness will be limited to a short time period, as its consumption rate will be high. Therefore, a VCI must not have too low or too high vapor pressure, but some optimum vapor pressure. Both corrosion rate of a metal and the volatilization of a VCI are a function of temperature. Because of this similar temperature dependent nature, the available vapor phase concentration of a VCI may adjust to aggressiveness of the environment [12, 13]

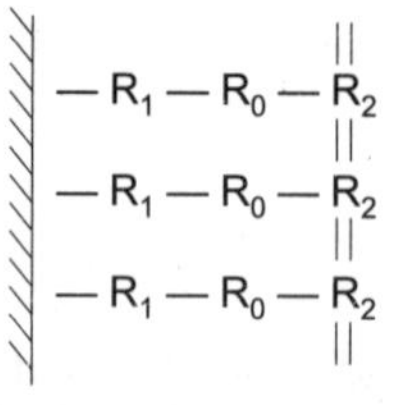

$$\begin{array}{c} -R_1-R_0-R_2 \\ -R_1-R_0-R_2 \\ -R_1-R_0-R_2 \end{array}$$

Figure 1 The mechanism of corrosion inhibition, Balezin[11]

Kuznetsov [14] studied the role of irreversible adsorption in the protective action of simple amines and heteroalkylated amines such as diamines, aminoalcohols or aminoketones on steel, copper and zinc. He concluded that VCIs of the amine type are capable of forming the protective layers up to 3 months and exhibit post-treatment effect on the metal surfaces due to irreversible adsorption of them. The tendency of irreversible adsorption of a VCI depends on its chemical structure. As a rule, low-molecular amines are inclined to desorption and do not exhibit post–treatment effect on steel, or zinc and initiate copper corrosion. However, some higher molecular weight amines have hydrophobicity and can exhibit post-treatment effect. Yet, low volatility of these amines, limits their application. The best results were obtained with certain heteroalkylated amines that have more stable adsorption due to cyclic structure and presence of two reactive groups. According to Kuznetsov, the most effective VCIs, as a rule, form on metal surfaces protective polymolecular films that consist of thin chemisorbed and physically adsorbed layers.

Andreev and Kuznetsov [15] have provided some basics of the theory of VCIs. They discussed the role of volatility, solubility, basicity and other features of VCIs towards their efficiency against CO_2 corrosion. According to them, the effect of chemical structure on these parameters can be quantitatively estimated using the linear free energy relationship.

STRUCTURE AND SYNTHESIS OF VCIs

The efficiency of a VCI depends on its adsorption ability on the metal's surface, which in turn is related to its vapor pressure. Both adsorption and vapor pressure depends on the chemical structure of the inhibitor compound. Rosenfeld et al [16] studied the effect of structure and the influence of substituents such as OH, NO_2, NH_2 on the effectiveness of inhibitors. It was shown that for ferrous metals, aromatic nitrogen bases were ineffective corrosion inhibitors, even when additional hydroxyl or nitro groups were introduced. The majority of strong bases of aliphatic and alicyclic compounds were effective VCIs. He found that cyclohexylamine, hexamethyleneamine, piperidine, morpholine and benzylamine were the most effective bases. It was claimed that the effectiveness of benzoates could be increased by the introduction of a highly electrophilic nitro group into the ortho position or two nitro groups into the meta positions with respect to the carboxyl group. These groups reduce the electron density in the aromatic ring. Miksic [17] claims that the nitro group-induced polarization shifts the steady-state potential of the metal in the positive direction, inhibiting the anodic reaction. The introduction of carboxylic group into the compound promotes the reduction of the nitro group and thus increases the oxidizing properties of the inhibitor. Complex ethers and weak aromatic amines were unsuitable inhibitors. Rosenfeld et al. also evaluated several inorganic salts and found that ammonium carbonate and copper ammonium carbonate to be effective inhibitors [16].

Trabanelli listed the inhibiting efficiency values of various aliphatic, alicyclic and aromatic amines deduced from accelerated corrosion tests on Armco iron in presence of 10 ppm of sulfur dioxide [18] and are given in Table 1. He concluded that the aliphatic and alicyclic compounds are superior to those of the aromatic substances having similar vapor pressures. Atmospheric corrosion described by some as vapor phase corrosion results from the individual and combined action of oxygen, moisture and atmospheric pollutants [19]. Trabanelli also listed several references that dealt with organic compounds for VCI purposes and have been documented by Fodor [8]. In a recent paper, Andreev et al. [20] have reported the study of

Table 1 Inhibiting efficiencies of different organic substances, deduced from accelerated atmospheric corrosion test on armco iron in 10 ppm SO_2 polluted atmosphere[19]

Substances	% Inhibiting Efficiencies
None	—
n-hexylamine	92
n-dihexylamine	57
n-octylamine	94
n-decylamine	84
n-dodecylamine	50
cyclohexylamine	80
dicyclohexylamine	95
N,N-dimethylcyclohexylamine	93
N,N-diethylcyclohexylamine	94
Aniline	40
Diphenylamine	16
Triphenylamine	–17
N,N-dimethylaniline	87
N,N-diethylaniline	23
N,N-dimethyl-meta-toluidine	52
N,N-dimethyl-para-toluidine	80
N,N-diethyl-ortho-toluidine	15
N,N-diethyl-meta-toluidine	12
N,N-diethyl-para-toluidine	44

adsorption on iron and the protective after-effect of some ethanolamines on ferrous and non-ferrous metals. They report that N, N-diethylaminoethanol, the most volatile and effective ethanolamine forms a polymolecular film. The effect of the chemical structure of VCIs of this type on desorption kinetics and protective after-effect of adsorption films on the metal surface in air has been analyzed. They report that IFKhAN-111, a new VCI composition based on N, N-diethylaminoethanol effectively protects both ferrous and non ferrous metals.

Kuznetsov et al.[21] have reported the protective action of beta-aminoketones, $R(1)R(2)N(CH2)_nCH(R(3))C(O)(CH2)_mCH3$, as volatile corrosion inhibitors in systems supersaturated with water, and the effect of preliminary exposure to inhibitor vapor of the passive state of metals. The protective action of VCI under a thin electrolyte was determined by electrochemical measurements and ellipsometric data on the kinetics of VCI adsorption and desorption has been reported.

Subramanian et al.[22] synthesized morpholine and its derivatives such as carbonate, borate and phosphate and evaluated their performance on protecting mild steel using weight loss and visual methods under continuous condensation test. A comparative assessment of their performances is given in the paper.

Cole, Dixon and Keohan[23] have reported their work on the synthesis and evaluation of new low toxic, multi-metal VCIs. Their approach consisted of reacting potential corrosion inhibitors having active hydrogens with appropriate silylating agents. The most versatile silylating agents they used are trimethylsilylacetamide, bis (trimethylsilyl)acetamide, hexamethyldisilazane and trimethylchlorosilane. Most silylation reactions were conducted neat by simply stirring the candidate organic base compound with the desired stoichiometric ratio of silylating agent. The reactions were run under a nitrogen atmosphere and at temperatures not exceeding 100°C. The crude derivatives were either filtered under reduced pressure to remove solid byproducts or vacuum distilled to remove volatile contaminants. They used Skinner's method [36] and slow scan potentiodynamic method to evaluate the effectiveness of these VCIs on cold rolled steel, brass, Al 7075-T6 and 70-30 Cu-Ni alloy. They reported that several combinations of new compounds, showed significant vapor phase corrosion inhibition, compared to any single-component, on all four alloys.

VAPOR PRESSURE OF VCIs

It is well known that a chemical compound used as a volatile inhibitor must possess neither too high nor very low vapor pressure, but an optimum vapor pressure. Two methods, the Rosenfeld-Martin-Knutsen effusion method [24] and the dynamic flow method [25] have been used for determining the saturated vapor pressure of VCIs and its dependence on temperature. There are certain inconsistencies in vapor pressure values published by different authors. It is believed that the reproducibility of data depends on experimental parameters that are inherent properties of the test method used. It was shown that measured vaporization rates in vacuum and in the atmosphere are not equal, and that differences for the same compound could be relatively large [26]. According to Rosenfeld [26], the following effects are possible:

(1) the change in the total pressure modified the Gibb's free energy for the components of the condensed phase,

(2) an extraneous gas slows down the vaporization of the condensed phase components (kinetic effect) and ,

(3) part of the vapor, for example, water vapor, dissolves in the condensed phase and modifies the Gibb's free energy of the latter.

Andreev and Kuznetsov [27] have documented the capabilities of correlation methods based on linearization of free energy and xi (R) constants of volatility for estimation of the vapor pressure of organic VCIs. Ways to optimally choose a VCI and control modification of the structures of compounds belonging to various classes for achieving optimum volatility of inhibitors are given. Andreev [28] has shown how the volatility of organic substances can be predicted from their boiling points and given examples of calculating the vapor pressure of several inhibitors.

Most of the effective VCIs are the products of a weak volatile base and a weak volatile acid. Such substances, although ionized in aqueous solutions, undergo substantial hydrolysis, the extent of which is almost independent of concentration [9]. In the case of amine nitrites and amine carboxylates, the net result of those reactions may be written as:

$$RNH_2NO_2 + H_2O \rightarrow (RNH_2)^+ : OH^- + H^+ : (NO_2)^-$$

In further support of VCI transport mechanism, it can be noted that amine salts, such as dicyclohexylammonium nitrate or diisobutylammonium sulfate, which are not extensively dissociated by water, do not give significant vapor-phase inhibition. The same is true for slightly hydrolyzed alkali metal salts such as sodium nitrite or sodium benzoate, although the latter is an excellent rust inhibitor when their solutions are in direct contact with metal surface [9].

Baker and Zisman [29] stated that the three nitrite compounds, diisobutyl, diisopropyl and dicyclohexylammonium nitrite, sublime readily at room temperature and inhibit rusting by adsorbing onto exposed metal surface to form a protective film. A hydrophobic monolayer is formed, on top of which there may deposit a loosely adhering layer of crystalline inhibitor. The hydrophobic film had a contact angle of 55° for the diisopropyl and 50° for the dicylohexylammonium nitrite at 20°C. The hydrophobic property is attributed to the polar-nonpolar structure of the substituted ammonium nitrite molecule. Because of the low molecular weight of the hydrocarbon portion of the organic nitrite molecule, prolonged contact with bulk water will result in dissolution or desorption of the protective film and allow corrosion of the substrate. Film desorption and solubility of the inhibitor in water lead to limited vapor-phase inhibition under conditions of extreme humidity. Recent developments suggest the use of a mixture of two or more volatile corrosion inhibitors to provide protection in extremely humid conditions.

VCI EVALUATION METHODS

According to Romanov and Khanovich [30] the effectiveness of VCIs can be evaluated in terms of several parameters:

- time of inhibitor loss,
- relative loss of inhibitor,
- time of appearance of first corrosion products,
- relative residual quantity of inhibitor when corrosion starts,
- rate of corrosion, and
- a factor that allows for the nature of corrosion.

Various accelerated corrosion tests are in use. Before using any of these tests, one must exercise caution [9]; experimental results of accelerated tests in which artificially increased concentrations of corrosive agents, e.g., SO_2, H_2S, or NaCl are used should be compared to actual field performance only after proper calibrations. A similar warning was also expressed by Rosenfeld: "reproducibility of data depends upon experimental parameters that are inherent properties of the test method employed." [26]

A widely used evaluating method to describe the relative abilities of inhibitors to prevent rusting of steel is described under ASTM D1748 [31]. In this method, steel panels are prepared to a prescribed surface finish, dipped in the test oil, allowed to drain and then suspended in a humidity cabinet at 48.9 +/−1.1°C (120°F +/− 2°F) for a specified number of hours. The oil fails or passes the test according to the size and number of rust dots on the test surfaces of the panels.

An essentially identical test method is described under Federal Test Method Standard No. 791B, Method 5329.1, "Corrosion Protection (Humidity Cabinet)." Federal Test Method Standard No. 101C, Method 4031 [32] consists of two procedures: Procedure A is for testing VCI materials in crystalline or liquid form; Procedure B is for testing VCI – coated or VCI-treated materials. Determinations may be made in the "as received" condition or on samples exposed to accelerated conditions.

The German Vapor Inhibiting Ability (VIA) Test [33] method is used to determine the corrosion inhibiting effect of VCI papers or foils. A test object of non-alloyed construction steel with high corrosion sensitivity in regard to condensation, together with the VCI packing aid to be tested, is placed inside a tightly sealed Erlenmeyer flask. The flask is stored at 23+/−2°C for 20 hours. At the end of storage time, a freshly prepared glycerol/water mixture is poured into the flask and it is closed immediately. During handling of the flask, care is taken not to submerge the samples in glycerol/water mixture. The test flask is stored in a thermost at maintained at 40 ± 1°C for 2 hr ± 10 minutes. At the completion of the test, the test sample of steel is examined visually for signs of corrosion. A control test is run using the test object of steel without a VCI paper/foil. A number grading system of 0-3 is used to rate the extent of inhibiting effect being no corrosion inhibiting effect and 3 means good corrosion inhibiting effect.

Christoph Kraemer [34] reported a VCI foil testing procedure that can be used for testing five different metals simultaneously. The basic principle of the test is similar to the German VIA test. The author claims that this is a quick and easy method to implement it as a VCI testing procedure, which demonstrates various approaches as to how deficiencies in the existing test procedures can be avoided with regards to practice orientation, lack of alignment to meet the demands of new VCI recipes and the usability of the test results for marketing purposes. Another test similar in principle to the German VIA test is the Japanese Industrial Standard[35] for VCI treated paper, JIS Z 153535.

William Skinner [36] published a paper that describes a new method for quantitative evaluation of VCIs. The experimental assembly used by him was a preserving jar with a lid that had drilled openings. Any type of inhibitor including a VCI is placed at the bottom of the jar. Prepared metal samples are mounted on the openings of the lid using a rubber seal ring. Metal weights are used to keep the metal samples in place. The whole assembly is then placed in a heated water bath. After a suitable inhibitor film-forming period, electrolyte is added. Condensation takes place on the inner surfaces of the jar and the metal samples. After the test duration, metal samples are removed for visual inspection and mass loss determinations. Based on corrosion rates with and without an inhibitor, inhibitor effectiveness is calculated. Skinner writes that the assembly enables a fairly accurate simulation of operational conditions and that reliable quantitative results are obtained by using a new set of jars, lids, glasses and seal rings for each test.

Recently, Paul Jaeger, Luis Garfias-Mesias and Markus Buchler [37] have described the use of a four-point probe method based on resistance measurements. Other test methods used to evaluate VCIs include electrochemical, [38-40] quartz micro-balance, [41] capacitance measurements of the electrical double layer, [42] radiochemical, [43] mass spectroscopy [44] and IR spectroscopy [45]. Typical slow potentiodynamic scans of Al 6061-T6 in 3.5% NaCl with and without a proprietary inhibitor spray, VCI F is shown in Figure 2, to illustrate the application of an electrochemical method in the evaluation of VCIs [39].

VCI APPLICATIONS

Just as catalysts are unique in catalyzing a reaction, VCIs are unique in their corrosion protection capabilities. In other words, compounds that protect ferrous alloys may not protect non-ferrous alloys. Instead, they may cause corrosion and vice versa. It is therefore, important that the user

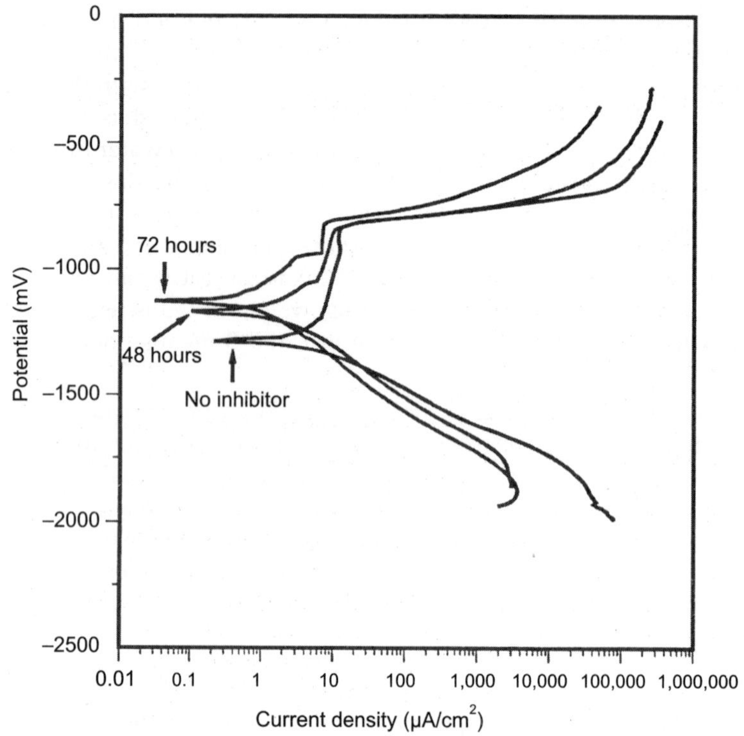

Figure 2 Slow potentiodynamic scans of aluminium 6061-T6 in 3.5% NaCl with and without VCI F

must take utmost precaution in the selection of VCIs for a given application. In the absence of supporting data, VCIs must be tested using sample materials that need to be protected, in a simulated environment that mimics actual conditions.

The applications of different VCIs used for protecting ferrous metals, copper and its alloys, and silver are summarized in Tables 2 through 4 along with respective original references [46-65].

Protection of Aluminum

Aluminum and its alloys are very susceptible to attack in marine environments. Bun Sung Lee [66,67] et al. prepared VCI papers, powders and tablets from a mixture of hexamethylenetetramine and sodium nitrite in a molar ratio of 1 to 4. They used 6% milk casein or 7% gum Arabic as binders. They showed that aluminum samples exposed to 90% RH at 40°C could be protected using these inhibitors. These researchers claim that the rust-inhibiting mechanism of this system is due partially to the fact that both the ingredients are hygroscopic; as they reduce moisture, corrosion is controlled. More importantly, both ingredients decompose as they absorb water forming formaldehyde and nitrous acid that prevent surface corrosion of aluminum due to their reducing action. Another decomposition product, ammonia, neutralizes the acidic substances in the sealed atmosphere [8].

Table 2 VCIs for ferrous metals protection

VCI Compound	Details of Protection & Conditions	Reference(s)
Di(cyclohexyl)amineDicyclohexylammonium nitrite(DICHAN)	Protection of ferrous metals and alloys.	46, 47
Amino nitrobenzoate Dintroresorcinate	Effective passivator	48
Hexamethylenetetramine	Protects iron in humid environment.	49
Tall oil plus industrial oils	Considerable protection of steel in sulfurdioxide atmosphere.	50
Ricinoleic acid hydroxide or its derivatives with ethylene oxide	General formula:CH3(CH2)5C(OH) HCH2CH=CH(CH2)CONHNH2 Protects iron, steel & cast iron	51
Naphthalene vapors at 250-350°C	Effective for steel in 1 N HCl vapors.	52
Hexamethylene tetramine with various binders, fillers & reducing agents	$C_6H_{12}N_4 + 4NaNO_2 + 10H_2O = 6HCHO + 4NH_4NO_2 + 4NaOH$ $4NH_4NO_2 = 4NH_3 + 4HNO_2$ Reduce iron corrosion	53
NaNO2, triethanolamine & a low molecular weight surfactant e.g., naphthanic or aliphatic acids	Good for protecting ferrous alloys in field tests.	54
Acylation product of lanolin or multibasic alcohols as erythritol, glycerol, sorbitol in oil/organic solvents	Good for protection of iron & steel in humid atmosphere.	55
Patented formulation of aliphatic amines, R.CH=NR where R=alkylGroup	Good for protection of steel in acid & NaCl atmosphere.	56
Alkylation product of 3- & -5 methylpyrazoles	Protection of steel in H_2S environment.	57

Table 3 VCIs for Cu and Cu alloys protection

VCI (s)	Details of Protection & Environment	Reference
Thiourea derivatives, e.g.Phenylene Ihiourea	Protection of brass	58
Acetylenic alcohols	Protection of atmospheric corrosion of Cu	59
Triazole compounds, e.g., tolyltriazole, Benzotriazole (C6H5N3)	Excellent protection of Cu & Cu alloys.	49, 60
Aq.solution of 5-50% complex phenol Carboxylic acid or its esters (tannic, caffeic, Gallic etc) and .1-20% water soluble Heterocyclic compounds such as ethylene Thiourea, mercaptobenzothiozole etc.	Protection of Cu in aggressive environments during storage.	61
Sodium mercaptobenzothiozole	Protection of copper in presence of soaps	62

Table 4 VCIs for protection of silver

VCIs	Details of Protection & Environment	Reference
Benzotriazole	Protection of Ag in H_2S atmosphere	49
Packaging paper impregnated with Alcohol/ketone solution of Chlorobenzotriazole	Forms a protective layer of silver complex on silver metal in H_2S atmosphere	63
Stearic acid or ethyleneglycol monostearate (3-5%) in a non-polar solvent, CCl4	Protection of Ag in industrial atmosphere.	64
Solution of KOH (5-100 g/L), K2Cr207 (5-100g/L), 2-mercapto-benzothiozole at 35-45°C for 4-10 minutes	Improves resistance to tarnishing & keeps the luster	65
Organosilicates e.g., 1- aminohexamethylene amino methylene triethoxisilanes & diethylamino methylene triethoxisilane in alcohol.	Inhibits silver plated on steel	

Protection of Ferrous and Non-ferrous Metals

In practice, there are generally multi metals involved and it is important to know the combination of VCIs that can be used to protect them from corrosion. Singh and Banerjee [10] have given an excellent collection of such information in their review article and it is summarized in the following paragraphs.

Benzimidazole, 2-benzimidazolethiol, benzotriazole and 2-mercaptobenzotriazole are reported to be effective for Cu, Al, Zn and their alloys. [68] Reaction products of methylbenzotriazole with water soluble guanidine class of compounds such as guanidine carbonate or nitrate or methyl, dimethyl, trimethyl, ethyl or diethyl guanidine in a molar ratio of 1:0.5 – 1:2 are effective atmospheric corrosion inhibitors for Cu, brass, Sn, tin plated steel and Al [68]. Products obtained by reaction of benzotriazole, ethylbenzotriazole and methylbenzotriazole with soluble urea compounds e.g., urea, methyl, ethyl, propyl or dimethyl urea in a molar ratio of 1:0.5 –1:2 at 10 – 60°C are effective for Cu, steel, Al and Zn. [69]

Reaction products of the reaction of 1M benzotriazole, ethyl or methylbenzotriazole and 1-5 M amino compounds such as methylmorpholine, tridodecylamine, didodecylamine, dicyclohexylamine and dimethylamine are effective for Fe, Cu, Al, Zn and other metals. [70] The resultant inhibitor is soluble in machine, transformer, engine lubricating, and other mineral and vegetable oils.

For protecting metals during storage, a compound formed by an incomplete reaction at 110 – 170°C for 0.5 – 2 hours of glycerine or ethyleneglycol with nitric acid and an alkanolamine with a primary or secondary alkylamine having more than six carbon atoms, is used. The product is dissolved in a mineral oil. Mineral oil solution of salts of Zn, Mg or Pb in naphathenic acid, together with 1-aminoethylimidazoline, is found suitable for temporary protection against atmospheric corrosion. [71]

The reaction product of benzotriazole with polyamine, e.g., ethylene diamine and carboxylic acid and dissolved in spindle oil, effectively resists atmospheric corrosion of Fe, Cu, Al and Zn. [72]

Table 5 VCIs for common metals[86-89]

Name of Compound	Vapor Pr. (mmHg)	% Efficiency for protection of					Ref#
		MildSteel	Brass	Cu	Al	Zinc	
6-Methoxy-aminobenzothizole	1.4×10^{-7}	97.9	96.6	95.8	NT	NT	86
6-Methoxy-aminobenzothizole cinnamate	4.0×10^{-7}	99.7	98.3	98.9	NT	NT	86
6-Methoxy-aminobenzothizole nitrobenzoate	4.5×10^{-7}	99.2	98.3	96.9	NT	NT	86
6-Methoxy-aminobenzothizole succinate	3.4×10^{-7}	98.1	97.5	94.8	NT	NT	86
6-Methoxy-aminobenzothizole	5.5×10^{-7}	98.7	96.6	95.8	NT	NT	86
Diaminohexane-cinnamate	5.0×10^{-8}	97.3	NT	NT	91.0	99.2	87
Diaminohexanenitrobenzoate	3.0×10^{-9}	60.1	NT	NT	85.3	94.1	87
Diaminohexanephthalate	6.0×10^{-9}	90.8	NT	NT	92.5	95.5	87
Diaminohexaneorthophosphate	5.0×10^{-9}	76.4	NT	NT	82.2	97.3	87
Diaminohexanemaleate	8.0×10^{-9}	93.3	NT	NT	78.9	95.5	87
1-(2-aminoethyl-2-dec-9-eny-2-imidazoline salicylate	91×10^{-7}	94.8	73.1	58.6	81.1	NT	88
1-(2-aminoethyl-2-dec-9-eny-2-imidazoline nitrobenzoate	69×10^{-7}	95.7	78.7	58.9	81.3	NT	88
1-(2-aminoethyl-2-dec-9-eny-2-imidazoline phthalate	16×10^{-6}	96.2	86.0	74.1	81.8	NT	88
1-(2-aminoethyl-2-dec-9-eny-2-imidazoline cinnamate	14×10^{-6}	96.6	88.2	79.3	82.5	NT	88
4-Amino-5-mercaptopropyl- triazole	$1.7 \times^{-6}$	98.4	98.3	96.9	NT	NT	89
4-Amino-5-mercaptopropyl- triazole cinnamate	2.7×10^{-7}	98.9	98.3	96.9	NT	NT	89
4-Amino-5-mercaptopropyl- triazole nitrobenzoate	3.9×10^{-7}	99.2	98.0	97.9	NT	NT	89
4-Amino-5-mercaptopropyl- triazole maleate	2.7×10^{-6}	98.9	97.5	97.9	NT	NT	89
4-Amino-5-mercaptopropyl- triazole succinate	5.0×10^{-7}	99.5	98.3	96.9	NT	NT	89

Baligim et al. [73] have reported that the mixtures of benzotriazole with benzoates of ammonia, guanidine, hexamethylene-diamine and monoethanolamine showed a synergistic protective action to steel, Zn and Cd. Mixture of benzotriazole with guanidine benzoate and ammonium benzoate afforded the best protection. Solidified products of cyclohexylamine, dicyclohexylamine, propylamine and butylamines with a concentrated solution of benzylidene sorbite in organic acid are reported to have good protecting properties towards ferrous and non-ferrous metals. Also 3, 5-diphenyl or 3, 5-bis-(R-phenyl)-2, 2, 4-triazole where R is hydrogen, alkyl or hydroxyl, is reported to protect gaseous or aqueous attack on Cu, Ag, Fe, Al, Ni and their alloys. [74]

Quarternary ammonium salts namely tetra-butyl, tetra-methyl of valeric, caprioic, caprilic, butyric, lauric, myristic trydecylic, palmitic, stearic, benzoic and carbamic acids and also 0-, and p-nitrophenalates are effective for steel, Cu and brass. [75] In another study, tetrabutyl ammonium 0-nitro phenate is most effective for steel and has no effect on Cu and brass. [76]

In atmospheres polluted with SO_2 and H_2S, 4-H-1, 2, 4 triazole or its derivatives in combination with known inhibitors such as dicyclo-oxylamine nitrite, urea, $NaNO_2$ etc. is effective for Fe, Al, steel, Cu, Ag, Zn, Ni, Sn, Cd, Mg, Pb and their alloys. [77] A Dutch patent [78] describes the solution of an aromatic hydrocarbon, an alkyl metal nitrate and alkyl substituted benzotriazole impregnated in paper to be effective for metal articles made of Al, Zn, Cu, steel and other metals and alloys.

Substituted pyrazoles dissolved in water, methanol or ethanol and impregnated in paper offer more protection than dicyclohexylamine nitrite towards atmospheric attack on Al, Cu, Zn, Sn and other non-ferrous metals. [79] Anan et al. [80] have tested a number of wood industry products as vapor phase inhibitors and have reported that 2, 5-dimetheylfuran is an excellent inhibitor for non-ferrous metals. Cast iron, steel, copper, aluminum, solder and other metals may be protected by a reaction product of 1 mole of cinnamic acid with 1-1.05 mole amine (diethyl amine, triethylamine, diethanolamine, cyclohexylamine etc.) with or without a solvent like water. [81] A product of 4- or 5-halohydroxy and/or nitro substituted benzotriazole with a primary, secondary or tertiary amine is reported to be highly effective for Al, cast iron, steel, brass solder, copper and other metals and alloys. [82]

A mixture of benzotriazole with oxidative compound e.g., nitromethane, nitronaphthalene, dinitrophenol, m-nitobenzoic acid and dissolved in acetone or dimethylphthalane provided good protection to steel, gray cast iron, Cu, brass, galvanized iron and Pb in humid atmosphere. [83] A ternary mixture of benzotriazole, sodium carbonate and urotropine is highly effective towards steel, and Cu. [84] A composition containing by 3-30 parts of benzotriazole and higher aliphatic amine (Octadecylamine) ensures protection of ferrous metals, Cu and Cu alloys, Al and Al alloys, Zn, Cd, Pb, Ni and Cr and phosphate and oxide coatings. [84]

In a long exposure test conducted for six years in humid atmosphere by Komrova [85] on steel, Cu, brass and galvanized iron using dibutylphthalate, isoamylcinnamate, triamylborate or the nitrobenzoate and other esters, it was concluded that these esters (apart from dibutylphthalate) provide complete corrosion protection towards steel, Cu and brass at relative humidity values up to 98% and of galvanized iron at relative humidity values up to 85%.

Quraishi et al [86-89] have synthesized several VCIs, measured their vapor pressure and tested their corrosion protection efficiencies on common metals such as mild steel, brass, copper, aluminum and zinc. These VCIs are listed in Table 5.

In addition to the above formulations, there are a number of other patents in the literature but provide very little information about their exact nature. A supplement to the recent Materials Performance [90] has a number of articles on VCI applications such as preservation of Air Force vehicles and equipment, Trans Alaska pipeline protection, VCI coatings, electronics protection, packaging and storage tank protection. Just before concluding the current review article, the author found yet another review article on the subject matter and it has been listed under the reference [91].

CONCLUSIONS

Papers and reports that were available in the open literature have been reviewed. Although it is not an exhaustive survey of papers on volatile corrosion inhibitors, it is hoped that this review provides information about the basic structural features of volatile corrosion inhibitors, corrosion inhibition mechanism and methods of evaluation. Wherever possible practical methods of using volatile corrosion inhibitors for protecting ferrous and non-ferrous alloys are discussed.

REFERENCES

1. Materials Performance, 5, No.6, 1995, p 5.
2. NACE Glossary of Corrosion Terms, Mat. Proc., 4, No.1, (1965) p 79.
3. N. Hackerman in "Fundamentals of Inhibitors," NACE Basic Corrosion Course (Houston,TX : NACE International, 1965)
4. Herbert H. Uhlig and R. Winston Revie., "Corrosion and Corrosion Control,"John-Wiley &Sons., N.Y., 1985.
5. Fontana and Greene., "Corrosion Engineering," McGraw-Hill Book Co., N.Y., 1978.
6. Riggs, Olen L. Jr., "Corrosion Inhibitors," (Ed:Nathan, C.C); (Houston, TX: NACE International, 1973).
7. B. A. Miksic and R. H. Miller., European Symposium on Corrosion Inhibitors, 5th proceedings, 107th Manifestation of the European Federation on Corrosion, September 1980, p 217-236.
8. G.E. Fodor, "Inhibition of Vapor-Phase Corrosion-A Review," Report No. BFLRF-209, Southwest Res. Institute, San Antonio, TX, October 1985; CORROSION/89 Symposium Proc., "Reviews on Corrosion Inhibitor Science and Technology," eds. A. Raman and P. Labine, p 11-17-1 (Houston, TX: NACE International, 1993).
9. B. A. Miksic, "Use of Vapor Phase Inhibitors for Corrosion Protection of Metal Products," in CORROSION/89 Symposium Proc., "Reviews on Inhibitor Science and Technology," eds. A Raman and P. Labine, p 11-16-1, (Houston, TX: NACE International, 1993).
10. D.D.N. Singh and M. K. Banerjee, "Vapour Phase Corrosion Inhibitors- a review," Anti-Corrosion, 31,6 (1984) p 4-8.
11. S. A. Balezin, Comptes Rendus de 2eme European Symposium Sur le Inhibiteurs de Corrosion, Ann.Univ., Ferrara, Italy, N.S., 1966, p277.
12. I. L. Rosenfeld, V.P. Persiantseva, M. N. Polteva., Inhibitors (Houtson, Tx : NACE International, 1972) p 606-609.
13. B. A. Miksic, Chem. Engg., September 1977, p115-118
14. Yu. I. Kuznetsov, " The Role of Irreversible Adsorption in the Protective Action of Volatile Corrosion Inhibitors," CORROSION/98, Paper # 98242 (Houston, TX: NACE International, 1998).
15. N. N. Andreev and Yu. I. Kuznetsov., "The Volatile Inhibitors for CO2 Corrosion," CORROSION/98, Paper # 98241 (Houston, TX: NACE International, 1998).
16. I. L. Rosenfeld, V. P. Persiantseva, M. A. Novitskaya, Z.F. Shustova, B.P.Terent'ev, M.N.Polteva., Tr.Vses. Mezhvuz. Nauchn. Knof. Po.Vopr. Bor'bys. Korroziei, Azerb. inst. Neft i Khim (1962), p 352-365; Chem Abstracts 60, (1964) p8958g.
17. B. A. Miksic, Anti-Corrosion Methods Materials, 22,3 (1975) p 5-8.
18. G. Trabanelli, A. Fiegna, V. Carassiti., La Tribune du Cebedeau 288 (1967).

19. G. Trabanelli, F. Zucch, 74th Manifestation of the European Federation on Corrosion (October 1974) p 289-301.
20. N.N.Andreev, N. P. Andreva, R.S Vartapetyan, YI Kuznetsov and T.V edotova., Protection of Metals, 33: (5) 470-475 SEP-OCT 1997.
21. Y.I. Kuznetsov, N.N. Andreev, N.P. Andreeva, D.V. Tolkachev and T.V. Fedotova., Protection of Metals 32: (5) 483-487 SEP-OCT 1996.
22. A. Subramanian, R. Gopalakrishnan, C. Bhoopathi, K. Balakrishnan, T.Vasudevan, M. Natesan and N. S. Rengswamy., Bull. Electrochem., 14: (10) 289-290 Oct 1998.
23. D. E. Cole, B.G. Dixon and F. L. Keohan, "New Low Toxicity, Multi-metal Active Vapor Phase Corrosion Inhibitors," CORROSION/99, Paper # 495 (Houston, TX: NACE International, 1999).
24. T. Martin Jr., J. Chem. Engr. Data 10, 3 (1965).
25. V. P. Persiantseva et al., Zaschita Metallow 7, 4 (1971) p 392.
26. I. L. Rosenfeld et al., Zaschita Metallow 10, 4, (1974) p 339.
27. N.N. Andreev and Y.I. Kuznetsov, Protection of Metals 32: (2) 148-154, MAR-APR 1996.
28. N.N. Andreev, Protection of Metals 34: (2) 101-111, MAR-APR 1998.
29. H. R. Baker and Zisman., "Liquid and Vapor Corrosion Inhibitors," U.S.Naval Res. Lab. Report, NRL-3824, May 1951.
30. V. A. Romanov, R. P. Khanovich., Energetika 4 (1975): p132.
31. ASTM D 1748, "Rust Protection by Metal Preservatives in the Humidity Cabinet," American Society for Testing and Materials, Philadelphia, PA.
32. Federal Test Method Standard No. 101C, March 1980; A. Furman and C.Chandler, 9th European Symposium on Corrosion Inhibitors, Italy, 2000.
33. German VIA test Method, TL 8135-0002, issued 3/12/98, W. Kohlhammer Gmbh, Post box 800306, 70508 Stuttgart.
34. Christoph Kraemer, "A Procedure for Testing the Effect of Vapor Phase Corrosion Inhibitors on Combined Metals," CORROSION/97, Paper # 97178 (Houston, TX: NACE International, 1997).
35. Japanese Industrial Standard, JIS Z 1535-1994.
36. W. Skinner, Corr. Sci. Vol 35, Nos. 5-8, PP 1491-1494, 1993.
37. Paul Jaegar, Luis F. Garfias and Markus Buchler, "A Novel Technique to Studying the Effect of a VCI," CORROSION/99, Paper# 99132 (Houston, TX: NACE International, 1999).
38. M.N. Desai and G.H. Tanki, Australian Corr. J., 13,19 (1969)
39. K.L. Vasanth and C.M. Dacres, "Vapor Phase Corrosion Inhibitors for Navy Applications," CORROSION/97, Paper # 179 (Houston, TX: NACE International, 1997).
40. J.M. Bastidas and E.M. Mora, "A Laboratory Study of mild Steel Vapour Phase Corrosion and it's Inhibition by Dicyclohexylaminenitrite," Canadian Metallurgical Quarterly 37: (1) 57-65 JAN 1998.
41. Paul Jaeger, "Characerization of VCIs using QCM and Supporting Techniques," CORROSION/ 97, Paper# 97180 (Houston, TX: NACE International, 1997).
42. L. G. Kar et al, Korrozia izashchita met. 82, 4, (1978)
43. K. Schwava, DECHEMA Monographien, 45, Verlag chemie-GMBH Weinheim/ Bergstr. 273 (1962).
44. Y. F. Yu Yao, J. Phys. Chem., 101, 68 (1964).
45. T. L. Rosenfeld, F. I. Rubinstein, V. P. Persiantseva and S. V. Yakubovich, IInd. European Symposium on Inhibitors Annali VVI, Ferrara N.S., Sez. V. Supply n 4, 7 (1966).
46. L. Cavallaro, G. Mantovanu, Werkstoffe und Korrosion 10 (1959): p 422.

47. K. Schwabe, Z fur Phys. Chem., Leipzig 1 (1964): p 226.
48. O. I. Golyanitskii, Proc. Chelyabinsk Inst. Mechaniz and Electrifik S. Kh. 24 (1977): p 126.
49. R. J. Lee, R.E. Karll, U.S. Patent No. 1,696,147.
50. M. S. Chernov, S.P. Gusev and I. L. Lomanov, "Nanch. Tr. Mosr-in-t nar. Kh.Va. 131, 3 (1974).
51. Nakumnra Kazuhiko, Japanese Patent C1.12 A 82 (c. 23f 11/14) No. 49-32416.
52. N. G. Klyuchnikov and M. Ya, Rutten Coll. Abst. Of papers Perun, Conference on Protection of Metals against Corrosion, Pru 104 (1974).
53. Lee Bum Sung, Seno Manabu and Asahara Teruzo, "Kinzoku Hyomen Jijutsu," J. Metal Finsih, Soc. Japan 392, 25 (1974).
54. O. I. Golyanitskiy, Tr. Chelyabinsk. In-ta mekhaniz. I.elektnifik. S. Kh. 24, 126 (1977).
55. U. Hiro and N. Shoechi, Japanese patent appl.class 12 A 8 (C 23 F 11/00) No. 53-95842.
56. N. A. Savirina et al, USSR Patent 318,315 class 23 F 11.
57. A. I. Bocharrikov, A. A. Sorochenko, V. Ya, Mudrakova, V. I. Seneya and V.A. Dubroovskay, Zashchita Metallov, 705, 15 (1979).
58. V. S. Agarwala, K. C. Tripathi, Werkstoffe und Korrosion 18 (1967): p 15.
59. J. Barbior, C. Fiand, Metause 49, (1974): p 271.
60. M. Ryokichi, Japanese Patent, Class 12 A 82 (C 23 f 11/00), No. 49-19508.
61. N. Hideo, M. Seishi and N. Takao, Japanese Patent class 12A 41, (c 23f 7/100) No.52-10164.
62. T. Czarnecki, Z. Zaczyriska, Polish patent, Class 22 g 5/08 (co 9 df/08), No. 72609.
63. Yosino Goro et al, Japanese Patent Class 12 A 82; Corr. Abst. 48, 9 (1977).
64. Y. M. Polukarov, Au th.cert class C 23 F 11/02, No. 539092; Corr. Abst. 48, 9 (1977)
65. G. G. Georgiyer, V. A. Kanazirska and D. P. Spasova Bulgarian Patent Class 23 b 6/44, c23 C 17/00 No. 16882.
66. B. S. Lee, M. Seno and T. Asahara, J. Soc. Metal Finish of Japan, 23, 10, p493-499 (1974).
67. Lee Bun Sung, Seno Manabu and Asahara Terguzo, J. Soc. Metal Finish of Japan, 25, 398 (1974).
68. Morita Ryokichi, Japanese Patent, Class 12 B 82, (C 23 f 11/00) No. 49- 20862.
69. Morita Ryokichi, Japanese Patent, Class 12 A 82, (C 23 f 11/00) No. 49- 19508.
70. Morita Ryokichi, Japanese Patent, Class 12 A 82 (C23 f 11/14) No. 50-13751.
71. Czarnecki Tadensz and Zaczyriska Zofia, Polish Patent Clas 22 g 5/08, (co 9df/08) Nc. 72609.
72. Morita Rekiti, Japanese Patent, Class 12 A 82 (C 23 f 11/14), No. 50.
73. S.A. Baligim et al Izv. VUZ Khimiya i. Khim teknol, Ivanovao 9 (1977).
74. No Kotachi, J. Hori and M. Yasuda, Japanese Patent, Class 12 A 82 (C 23 f 11/14) No. 51-10186.
75. R.V. Berentseva, Jr., Proc. Chelyabinsk Inst. Mekhaniz and Elektrif,. R. S. Kh 51, 126 (1977).
76. R.V. Berentseva Jr., and A. P. Gugnina, Trans. Chelyabinsk
77. K. Neaki, Y. Mashiro and H. Takeshi, Japanese Patent, Class 12 A 8 (C 23 f 11/14) No. 52-1377.
78. C. M. R. Davidson, Dutch Patent, C z 3 f 11/14 No.152605.
79. T. Tadasi, T. Kiiti, Y Seizo and K. Takashi, Japanese Patent, Class 12 A, (C 23 f 11/00) No. 53-47775.
80. G. F. Anan, Yeva and B. F. Nikandrov, Coll Chemical and Mechanical processing of wood and wood wastes, Leningrad 53, 4 (1978).
81. Macda Akiro, Japanese Applications, Class 12 A 82 (C 23 f 11/14) N0. 52-155148.
82. A. P. Gugina, Proc. Chelyabinsk Inst. Mechaniz and Electrif Agric. 61, 146 (1978).
83. E. G Z ak, G. B. Rotmistrova and L. Z. Zavitayeva, Metal Corrosion Inhibitors, Moscow, 154 (1979).

84. P. Mityska and V. Novak, Czechoslovak Inventors Certificate, Class (C 23 F 11/02), No. 177381
85. L. G. Komrova, Proc. Chelyabinsk Inst. Mechaniz and Electrif Agric., 67, 146 (1978).
86. M. A. Quraishi and J. Rawat, "Influence of some 6-methoxy- aminobenzothiazole derivatives on corrosion of ferrous and non-ferrous metals under vapor phase conditions," Corrosion (accepted for publication).
87. M. A. Quraishi and D. Jamal, "Development and testing of all organic volatile corrosion inhibitors, Corrosion 58, 387-391 (2002).
88. M. A. Quraishi and D. Jamal, "Development of testing of newvolatile corrosion inhibitors for multimetal systems," Materials Performance, January (2003).
89. M. A. Quraishi and J Rawat, "Triazole derivatives: New class of vapor phase corrosion inhibitors," in Corrosion causes and Mitigation, ed., A Kumbhavedekar, Quest Publication vol. 2, 115-122 (2000).
90. A Supplement to Materials Performance, January 2001. (Houston, TX: NACE International, 2001).
91. D. M. Bastidas, E. Cano and E. M. Mora, "Volatile corrosion inhibitors: a review," Anti-Corrosion Methods and Matreials, vol.52, No.2, p71-77, 2005.

CHAPTER 10

Aqueous Corrosion Behaviour of Intermetallics

V. Shankar Rao
Jindal Stainless Limited, Hisar 125 005, India

I. INTRODUCTION

Intermetallic compounds are widely recognized as a potential material for high temperature structural applications. This is mainly due to their high melting point and good oxidation resistance, which are essential requirements for a high temperature material [1]. Apart from their use as a structural material, coatings of intermetallics, like aluminides, are very frequently used to protect parts of turbo engines, combustion chamber van and blade against oxidation. The NiAl compound is formed in the case of nickel base superalloy whereas CoAl is formed on the cobalt base superalloy [2].

Though intermetallics are mainly developed for structural applications at elevated temperature, understanding of aqueous corrosion behaviour of these materials is needed for its durability. Aqueous corrosion behaviour can be a significant problem even without immersion in electrolyte. For example, materials used in gas turbines (both aircraft and industrial turbines) and nuclear reactors must be able to withstand against corrosion resulting from continuous or periodic exposure to humid environments. Further this structural material will not always be at the operating temperature, and corrosion damage during maintenance could lead to catastrophic failure during service. Electrochemical properties of intermetallics are also of interest due to emerging of new advance areas for their applications such as in magnetic applications, catalyst for the hydrogen production in fuel cells, energy storage, electronic devices and orthopedic implants [3-6]. In a recent development, formation of nanotubes on Re containing nickel aluminides surface by electroselective dissolution of alloying element was reported [7]. Reliability of the developed materials profoundly depends on the corrosion resistance of the alloy in a specific environment. These considerations and renewed interests led to an increase in number of research works related to the aqueous corrosion behaviour of intermetallics in the last ten years or so. Nonetheless, less attention has been paid on the aqueous corrosion behaviour of intermetallics in related textbooks and against stainless steels.

Therefore, this paper outlines the work done on the electrochemical corrosion behaviour of the intermetallics in past two decades. A full review of the state of the art would exceed the time and space limitation, thus the scope of the review on the corrosion behaviour of intermetallics is limited to the most widely used intermetallics based on aluminide, namely iron aluminides, titanium aluminides and nickel aluminides. During the reviewing of the subject emphasis will put on:

- the electrochemical corrosion behaviour of aluminides and their constituent elements
- chemical structure and chemistry of the passive film
- effect of alloying elements on the corrosion resistance of the aluminides and
- a possible comparison with stainless steels or other alloys, in each section.

There are some excellent textbooks and review papers related to intermetallic compounds [1, 8, 9], reader is directed for the alloy compositions, crystal and phase structure, mechanical and metallurgical properties, and further details prior to understanding on their aqueous corrosion behaviour.

2. ELECTROCHEMICAL CORROSION BEHAVIOUR OF IRON ALUMINIDES

Much of the aqueous corrosion behaviour of aluminides to date has been performed on iron aluminides despite a fact that commercial use of these compounds in high-temperature applications is relatively low compared to its counterpart nickel aluminides and titanium aluminides. This is mainly because iron aluminides are also considered for room temperature applications being less costly than conventional stainless steels such as pipes and tubes for heating elements, and ecological useful alternative to stainless steels due to avoiding use of Cr and Ni [3].

Considering a wide range of data on the corrosion behaviour of iron aluminide, this study is divided into two sections. First section is considered for the corrosion and passivation behaviour of iron aluminide and its constituent elements Fe and Al, while second section consider for the effect of alloying elements on the corrosion behaviour of iron aluminide.

2.1 Corrosion and Passivation Behaviour of Iron Aluminides

In this review, let us begin at the beginning, the role of constituent elements Fe and Al on the electrochemical corrosion behaviour of iron aluminides. During the discussion we will also report the corrosion behaviour of Fe-Al alloys along with iron aluminide to get a coherent view of the subject.

Accordingly, Fig. 1 compares the polarization curves of Fe_3Al based iron aluminide, pure Al (99.99%) and Fe (Armco iron) obtained in 0.25 M sulfuric acid (H_2SO_4) [10]. Notably, during anodic polarization, the current density in the passive region of Fe is lower than that of Al, whereas the latter has a wide passive potential range compared to the former. The combined effect of Fe and Al seems to benefit iron aluminide over a wide potential range. At low potential, close to the corrosion potential of the aluminide, the spontaneous passivation

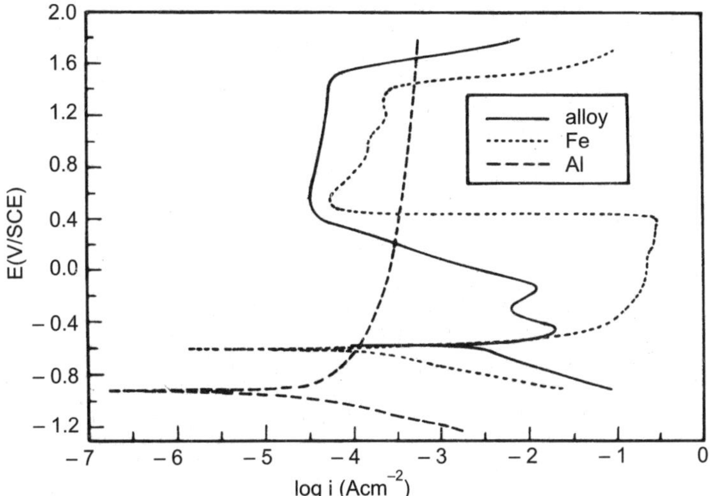

Figure 1 Potentiodynamic polarization curves of iron aluminide, pure Al and Fe obtained in 0.25 M H_2SO_4.

tendency of Al helps to reduce active dissolution of iron aluminide by formation of an Al_2O_3 film, while at higher potential the presence of iron oxide provides an additional resistance to the passage of current with Al_2O_3. Thus the current density of iron aluminide in the passive region is lower than that of pure iron. Despite these beneficial facts of iron aluminide in passivation behaviour over Fe and Al, the difference between corrosion potential (E_{corr})-passive potential (E_{pass}) is relatively high compared to the other passivating alloys such as stainless steels. Plus, the anodic critical current density (i_{corr}) of aluminide is more than its limiting current density for oxygen reduction (0.1 mA cm^{-2}), which makes the aluminide difficult to self passivate in aerated conditions. A more detailed discussion on this aspect has appeared with Choe et al. [11], Frangini and co workers [12, 13], and Shankar Rao et al. [10].

Frangini [14] and his group [12-13, 15] made extensive studies on the electrochemical corrosion behaviour of iron aluminides. In one of their studies, they compared the potentiostatic polarisation behaviour of B2 FeAl (24.4%Al) with Fe in 0.5 M H_2SO_4 solution [13]. After setting a potential in the passive region, the change in current density with time during initial transient period follows a similar trend in both cases, with a slope log(*i*) *vs* log(*t*) of approximately –1. However, FeAl (4 μA/cm^2 after 1 h) attains the steady state more rapidly compared to Fe (7 μA/cm^2 after 20 h). They suggested that the linear current transient during the passivation process follow the high-field ion conduction and the place exchange mechanism for the oxide growth, though not much discussion is available on that.

To avoid the mechanism of preformed oxide film the repassivation behaviour of Fe_3Al along with pure Al and Fe were investigated in 0.25 M H_2SO_4, using the rapid scratched electrode technique [16]. Typical current transient curves from the scratched surface of the passive film, which were formed at 1 V_{SCE} (see Fig. 1), are shown in Fig. 2. It is well accepted that the corrosion current measured during repassivation is primarily associated with the formation of a passive film provided pitting corrosion did not occur. Therefore, the measured current corresponds to the repassivation process which follows in the order of $Fe_3Al < Al < Fe$. The peak current density of iron aluminide is about two times lower than that of Al while

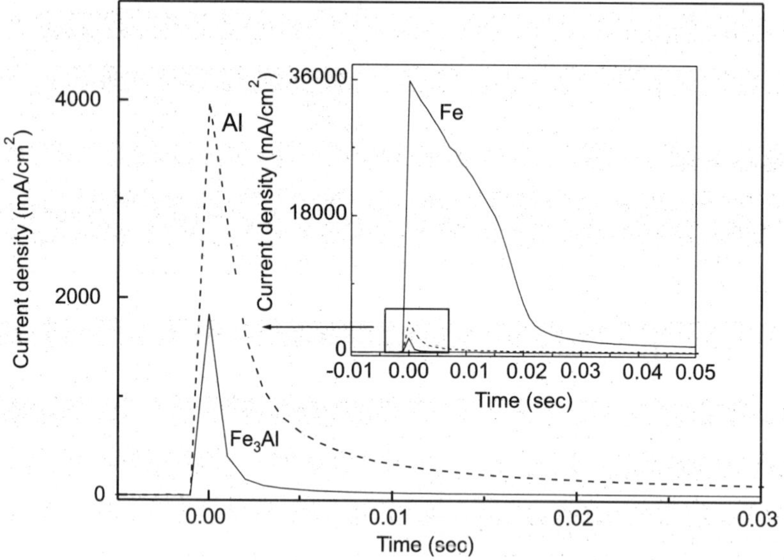

Figure 2 Repassivation kinetics of iron aluminide, Fe and Al obtained in 0.25 M H_2SO_4.

compared to Fe it is much lower, about seventeen times. Notably, the rate of anodic current decay of Fe is much slower than that of Al and iron aluminide. Not only that, Al and iron aluminide display similar kinetics during repassivation, whereas Fe seems to differ from them. A slower repassivation rate of iron despite its low passive current density (over Al) casts doubt on the stability and protectiveness of the passive film. Importantly, the faster repassivation rate of iron aluminide substantiates its superiority on the passivation behaviour over Fe and Al. More studies related to the repassivation behaviour of Fe-Al alloys with different (5-25) % Al *(unless otherwise noted, all composition will be given in weight percent)* contents can be found in the work of Madsen and Adler [17].

Meanwhile, keeping the kinetics of Fe and Al in mind we will continue our discussion by reviewing the works related to corrosion behaviour of Fe-Al alloys. Aqueous corrosion behaviour of Fe-Al alloys was first published with Nachman and Duffy in 1974 [18] but it was not an electrochemical study. These authors evaluated corrosion behaviour of Fe-Al alloys for different Al contents (6-16%) in sea water solution by weight loss measurement and then most promising alloy was tested for erosion corrosion in a flow of sea water. Their results show a minimum 10% Al should be there in Fe-Al alloy for a good corrosion resistance and to maintain a protective oxide film.

Electrochemical corrosion behaviour of Fe-Al was first reported by Defrancq, as early as 1977 [19] and it was reported that the addition of 25% Al in cast iron markedly changes its polarization behaviour. His results are shown in Fig. 3 where he compares potentiokinetic polarization behaviour of the cast Fe-25.74Al alloy with that of Fe-2.25Al alloy in sulphuric acid media. In the low-Al content alloy, two anodic maxima peaks appeared at +250 mV_{SHE} and +650 mV_{SHE} which are associated with iron dissolution, while in the case of Al rich cast iron these peaks disappear. This was attributed to the formation of a protective film of aluminium oxide on the electrode surface during polarisation, which restricts the anodic iron dissolution even at higher potential.

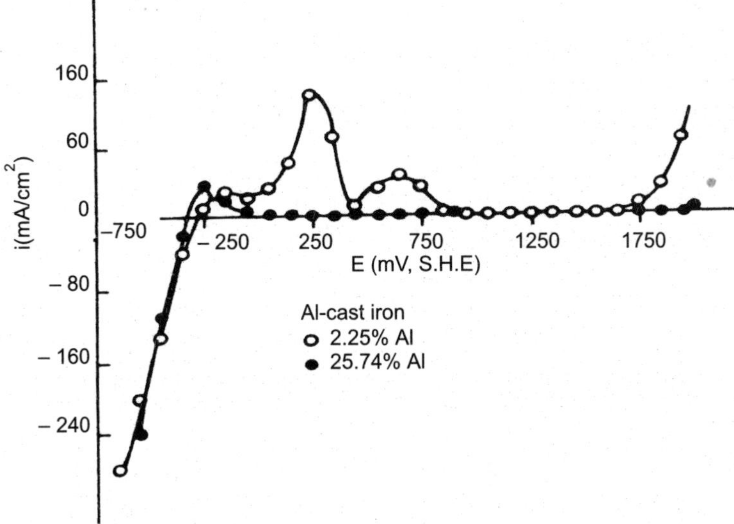

Figure 3 Potentiodynamic polarization curves of cast Fe-2Al and Fe-25Al alloys obtained in 0.5 M H_2SO_4 [19].

Electrochemical behaviour of iron aluminide is reported to be pH dependent. Schaepers and Strehblow [20] have investigated the passivation behaviour of Fe-Al alloys with Al content of 4, 8 and 12% in acidic and alkaline electrolytes. In the borate buffer solution there is no significant difference in the polarisation behaviour of Fe-Al alloys with variation of Al content. Owing to lower currents in this electrolyte support for passivation by Al was not detectable in the polarisation curves. Whereas, the polarisation curves in H_2SO_4/Na_2SO_4 solution display higher current density than those obtained in alkaline solution and pronounced changes are observed between low and high Al content alloys. The alloy with 4% Al shows a large dissolution peak, which grows due to surface roughening with increasing number of cycles, whereas the alloy with 12% Al shows a lower anodic current. The decrease in anodic current at -0.1V_{SHE} with increasing number of scans suggests that a part of the oxide is not reduced and that it remains at the surface and supports repassivation of the alloy in the next cycle.

Cristofaro et al. [21] have studied the passivity and its breakdown of FeAl (24% Al) in sulphates and chlorides containing borate buffer solution, in which it was reported that the voltammogram response of iron aluminides is similar to Fe. In agreement with Frangini et al [13] the peaks in cyclic voltammogram were suggested to be iron oxidation/reduction processes. Sulphate addition to the buffer solution makes the passive film more stable. According to them a mixed iron (II-III) oxide is responsible for passivation of FeAl in the sulphate-borate solution in contrast to only iron (III) rich oxide in the borate solution without sulphate. Moreover, iron suffers from pitting corrosion in sulphate-borate solution, while aluminide remains un-attacked. This implies that the presence of Al in iron aluminide enables the formation of a more protective passive layer. However, this does not apply with the addition of Cl^- in buffer solution as well. In which case, the pitting potential of iron aluminide increases linearly with log Cl^- with a slope of 0.40V/decade and produces nucleation of pits on iron aluminide with addition of 6×10^{-4} M NaCl in buffer solution. This shows the choice of electrolyte and presence of ions in it has a critical influence on the corrosion behaviour of iron aluminides.

Kim and Buchanan [22] tried to evaluate the effect of change in structure (DO3 *Vs* B2) on corrosion behaviour of an Fe_3Al alloy in aerated 200 ppm Cl^- at pH 4. In this study the alloy with the DO3 structure possesses low passive current density and high breakdown potential compared to the B2 structure. A study similar to that by Kim and Buhman [22] was conducted by Garcia-Alonso et al [23] where corrosion behaviour of Fe_3Al with different crystal structures was evaluated in Hank's solution. Their studies show that corrosion rate of the aluminide is independent of the crystal structure but the alloy with order states B2 as well as DO3 possesses more pitting resistance than the disordered states.

So far we have seen the beneficial effect of Al on the passivation behaviour of iron aluminide above its open circuit potential (OCP) and in the passive region. However, the story seems to be the other way around at the potential below open circuit and in the cathodic region. Frangini [14] revealed the detrimental effect of Al on the corrosion behaviour of iron aluminide who used linear polarization resistance method as an alternative method for determining the corrosion current density at the OCP. According to them the corrosion rate of FeAl aluminide remains equal to pure Fe during initial stage of corrosion in mild sulphuric acid media but with increasing exposure time the corrosion rate of iron aluminide increases and becomes higher than that of iron. Later Shankar Rao et al [10] explained the high corrosion rate of iron aluminide over that of iron by taking into account their respective hydrogen exchange current density ($i_{oH^+/H2}$) values. They found the $i_{oH^+/H2}$ value in 0.25 M H_2SO_4 of the alloy Fe-16Al-0.05C (2.5×10^{-2} mA cm^{-2}) is higher than that of iron (2.0×10^{-3} mA cm^{-2}). It is well known that the higher the exchange current density of an alloy higher is its corrosion rate. It is worth mentioning here that the diffusivity of librated nascent hydrogen in the lattice of iron aluminide is very low and generally decreases with increasing Al content in intermetallics [24].

Frangini and Lascovich [25] employed a.c. impedance technique to study the passive film properties grown on iron aluminide. They measured impedance behaviour of FeAl alloy in 0.5 M H_2SO_4 as a function of passive film thickness and passive potential between 0.5 to 1.5 V_{SCE}. Their results show two overlapping time constants at lower frequency, which were attributed, respectively to the charge transfer processes at the film/solution interface and ion conduction within the passive film. There was a decrease in the capacitance value and the film thickness with increase in potential above 1.0 V_{SCE}. This was attributed to the transformation of the initially formed compact passive layer to a more porous passive structure. By evaluating iR values as a function of the potential drop inside the film, it was suggested that the transport of ions in the passive film on FeAl occurs predominantly through low electric field strength regions.

Generally, a thermally generated alumina layer (Al_2O_3) on the iron aluminide surface provides excellent oxidation resistance. Taking this into account, Escudero et al. [26] reported that by prior oxidation of Fe-40Al alloy its corrosion resistance can be improved when subjected to the Hank's solution (8g NaCl, 0.14g $CaCl_2$, 0.4g KCl, 1g glucose, 0.1g $MgCl_2$, 0.35g $NaHCO_3$, 0.06g $Na_2HPO_4.2H_2O$, 0.06g KH_2PO_4 and 0.04g $MgSO_4.7H_2O$ in one liter distilled water). By impedance studies on a sample oxidized in air at 1000°C and on an unoxidized sample they found a higher impedance value in the entire range of frequencies of a Bode plot in the unoxidized sample compared to the well-formed alumina sample. They noticed that the protection given to the intermetallic was highly variable from sample to sample and suggested that only a continuous and defect free thermal oxide layer can ensure the good resistance for iron aluminide.

Several studies were directed on the structure and chemistry of the passive film formed on iron alumindes with different Al contents in the alloy, different electrolytes concentration and pH, and different formation potentials [12, 13, 20]. Despite these variations among the existing studies some consensuses can be found from them, which are as follow.

- An enrichment of Al in the form of Al (III) oxide/hydroxide on the surface during passivation. This enrichment is even more than the Al bulk concentration of the alloy.
- Al content in the passive film increases with increasing Al content in the alloy.
- Coexistence of iron and aluminium oxides on the surface during passivation.
- Addition of alloying elements such as Cr and Mo facilitate the formation of their respective oxides (Cr_2O_3 and MoO_3) in the passive film along with Al_2O_3.

Some investigators have intended to compare the chemistry of passive films with that of the air-oxidized films in order to differentiate film properties in wet and dry conditions [12, 27]. Such a study is illustrated in Figs. 4a,b where X-ray photoelectron spectroscopy (XPS) spectra of the iron aluminide exposed to the 0.25 M H_2SO_4 at 1 V_{SCE} and iron aluminide oxidized in O_2 atmosphere at 800°C for 10 min are compared. In Fig. 4a appearance of the peak in the oxidized sample at 74.2 eV is assigned to the binding energy (B.E) of Al^{3+} in Al_2O_3 [28]. Notably, in the case of the passivated sample there is a shift in the Al^{3+} peak towards high B.E, on deconvolution of the spectra, it revealed coexistence of Al^{3+} as Al_2O_3 and Al-OH at 74.49 eV and 75.29 eV, respectively [28, 29]. In Fig. 4b the peaks of iron at 711 ± 2 eV correspond to Fe^{3+} [30]. After deconvolution of the passive spectra it reveals multiple peaks assigned to oxides and sulfates [30, 31]. Unlike in the case of Al, there is no difference in peak position of Fe^{3+} between the oxidized and passivated samples. This shows that Al exists in the form of an oxide as well as a hydroxide whereas the presence of Fe in the passive film is only as an oxide. In a study similar to this, Frangini et al. [12] reported that the Fe^{2+} concentration is more in an anodically formed film than in an air formed film, though after sputtering it disappears from both the films. From depth profile analysis they found that the Fe/Al ratio is high in the case of the anodically formed film compared to the air formed film, which indicates an enrichment of iron in the outer layer of the anodic film with respect to the air formed film. Their work further shows that the outer part of the passive film predominantly consists of mixed Al-Fe oxy-hydroxide whereas the inner part is of mostly an Al-rich oxide phase.

In another study, investigation of the passive layer by elastic ion scattering and XPS depth profiling shows maximum enrichment of Al (III) within the centre of the passive film [20]. The initially formed Al (III)-rich film contains a large amount of hydroxide. However, like any other passivating alloy during longer passivation time and at more positive potential the layer composition changes into a mixed oxide/hydroxide film with a larger amount of additionally formed iron oxide.

To get an idea about the mechanism behind the passivation behaviour of iron aluminides we take a look at the precise chemical reactions which take place after its immersion in electrolyte. For this, anodic polarization behaviour of the iron aluminide in acid media will be discussed.

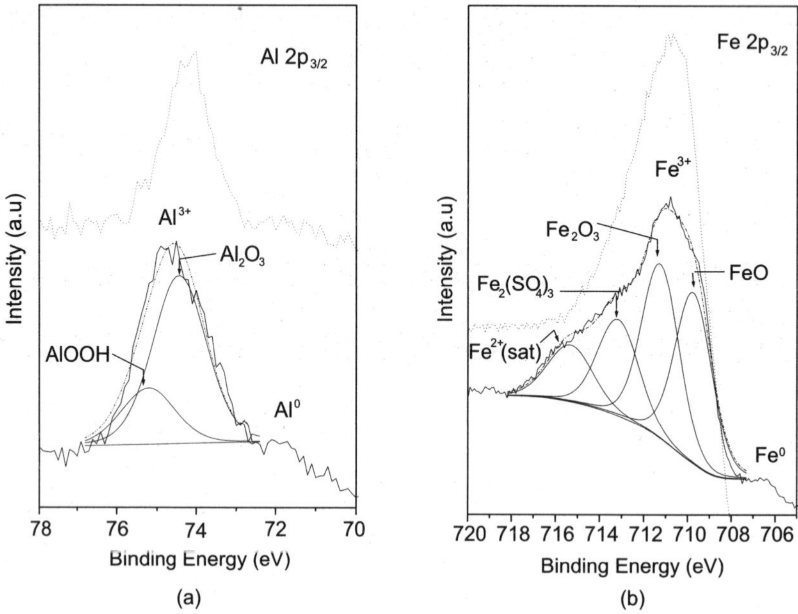

Figure 4 XPS spectra for (a) Al2p and (b) Fe2p; for thermally oxidized Fe_3Al at 800°C (dotted lines) and passivated Fe_3Al in 0.25 M H_2SO_4 at 1 V_{SCE}.

During active dissolution of the alloy the following reactions are possible depending upon alloy compositions whether it is FeAl or Fe_3Al.

$$FeAl + 6H^+ = Fe^{3+} + Al^{3+} + 3H_2$$

$$Fe_3Al + 6H^+ = 3Fe^{3+} + Al^{3+} + 3H_2 + 6e^-$$

$$Fe_3Al + 6H^+ = 3Fe^{2+} + Al^{3+} + 3H_2 + 3e^-$$

These reactions are balanced by the following reduction reaction which consumes electrons

$$2H_2O + 2e^- = H_2 + 2OH^-$$

Thermodynamically Al_2O_3 has more negative free energy compared to iron oxide [20] and kinetically, the mobility of Al^{3+} is also low compared with iron within the barrier film [13]. Therefore, after active dissolution from iron aluminide, aluminum quickly converts into Al_2O_3 on the surface by following reaction, whereas iron ions remain in the solution.

$$2Al^{3+} + 3H_2O = Al_2O_3 + 6H^+$$

This Al_2O_3 layer is expected to act as a diffusion barrier similar to the case of pure Al for dissolution of iron from the surface. However, due to porosity in nature it allows some iron to dissolve.

As the potential is moved from E_{crit} towards the pre-passive potential iron oxides start forming on the surface by the following reactions.

$$2Fe^{3+} + 3H_2O = Fe_2O_3 + 6H^+$$

$$Fe^{2+} + H_2O = FeO + 2H^+$$

It is postulated that the formation of these iron oxides on the surface of the passive film blocks the porosity of the Al_2O_3 passive film. Plus the presence of iron oxides on the surface provides additional resistance to protectiveness of the passive film. Thus a combined resistance of Al_2O_3 and Fe_2O_3 oxides drop the passive current density of iron aluminide to its minimum value in the passive region.

As mentioned above with the shift of passive potential towards the noble direction more Al-OH starts forming. This can be explained by the following reaction:

$$Al_2O_3 + 3H_2O \rightarrow 2Al(OH)_3$$

since the above-mentioned oxides and hydroxide remains stable from their passive potential to the oxygen evolution potential for a given pH. Therefore, the transpassive region corresponds to the following reaction:

$$2H_2O \rightarrow O_2 + 4H^+ + 4e^-$$

This corrosion and passivation behaviour of iron aluminides have been covered at length in a review paper [32]

2.2 Effect of Alloying Elements on Corrosion Behaviour of Iron Aluminides

Addition of third alloying elements in iron aluminides has been an intelligent approach to improve their mechanical properties and oxidation resistance. But at the same time this can affect aqueous corrosion behaviour of iron aluminide as well. In that respect, several researchers contributed to this subject.

When it was proven that the addition of Cr in iron aluminides improves their ductility as well as their oxidation resistance, a number of studies were directed to test the effect of Cr addition on aqueous corrosion behaviour of iron aluminides [12, 22, 33]. Kim and Buchanan [22] studied the role of Cr on pitting and crevice corrosion of iron aluminides (28 at.% Al) in mild acid chloride solution. Their studies indicate beneficial effects of Cr in terms of an increase in breakdown potential of the passive film with increasing Cr content in the alloy. In another study, Balasubramaniam [33] reported that the corrosion rate of Fe-28Al-5Cr alloy is about two orders less than that of Fe-25Al in (pH 4) H_2SO_4 solution containing 200 ppm Cl^-. They tried to visualize this difference in corrosion behaviour of both the alloys by applying mixed-potential theory. Frangini et al. [12] conducted some experiments to determine the role of Cr in the passive film by incorporation from the solution rather than from the alloy. For this, Cr ions were incorporated into the passive film of FeAl from alkaline chromate solution through repetitive oxidation-reduction cycles and its corrosion behaviour was compared with Fe-(12-24)% Cr stainless steels in 0.5 M NaCl containing borate solutions. Their results show that the pitting potential of this alloy after treatment is comparable to those of stainless steels despite the fact that the Cr enrichment in the passive film of FeAl was not as high as in the passive film of the compared stainless steel. Based on this finding the protective property of the passive film was suggested to be the synergistic action of Cr and Al.

Addition of Mo in stainless steel is well known for improving its pitting resistance and in iron aluminide it is added generally to enhance the adhesion resistance of the Al_2O_3 on the surface. Choe et al. [11] investigated the effects of 1 at% Mo along with 2-6 at% Cr, in order to check its effect on corrosion behaviour of Fe_3Al (28 at% Al). Their results show the addition of Mo along with Cr in iron aluminide increases the corrosion potential, pitting potential and repassivation potential in 0.25 M H_2SO_4, whereas the critical current density and passive current density are decreased. In addition, Mo impeded Cl^- adsorption on the surface. An advantage of Mo addition over Cr addition in iron aluminide was seen during their cyclic polarisation tests in 0.1 M HCl solution, in which pitting and repassivation potentials of the aluminide increased with addition of Mo as well as Cr, however, Mo has a much greater effect in this regard than Cr. These authors also investigated the effect of B on corrosion behaviour of iron aluminide and found that unlike Mo and Cr it was not very beneficial in improving passivation behaviour, but it was more beneficial in increasing the activities at grain boundaries in aluminide compared to the other two. Their electrochemical potentiokinetic reactivation (EPR) studies in 0.5 M H_2SO_4 + 0.01 M KCN solutions show that the degree of intergranular attack on these alloys increases in the order of Mo > Cr > B. In another work Choi and Kim [34] investigated the passive film behaviour of 28Al-4Cr-xMo (at%) aluminide in thiosulfate-chloride solution (chloride 50 ppm and thiosulfate 10 ppm) as a function of Mo 0.5-2 at% using the a.c. impedance technique, at three different states namely breakdown, passivation and repassivation. Their studies show that the Mo bearing alloy possesses higher charge transfer (R_t) values than that of the Mo-free alloy in all three states. Furthermore, they suggested a synergistic effect of Mo and Cr in improving the pitting resistance of iron aluminides.

Babu et al. [35] experimented to see the effects of mischmetal (Mm: 43Ce, 23La, 18Nd, 5Pr, 3Sm and 8Fe in %) addition on the electrochemical corrosion behaviour of the Cr-alloyed iron aluminide, as mischmetal are known for inducing passivity of an alloy like Cr. By comparing potentiodynamic polarization behaviour of Fe-28Al-2Cr and Fe-28Cr-0.15Mm in 0.05 mol l^{-1} H_2SO_4 solution they found that the addition of mischmetal is beneficial on the passivation behaviour only in terms of reducing the passive current density, whereas there is no significant affect on other electrochemical parameters such as i_{crit}, passive range, zero current potential and breakdown potential [35].

It was not until the middle of the nineties that Fe_3Al-Fe_3AlC dual-phase iron aluminide reported to possess better hydrogen embrittlement resistance than single-phase Fe_3Al [36]; since then, investigations on the electrochemical corrosion behaviour of these dual phase iron aluminide have increased [10, 27, 37, 38]. Plus, carbon a common alloy impurities is believed to be present in iron aluminides which led to affect their corrosion resistance. With such a motivation the effect of C (0.14-1%) on the corrosion behaviour of Fe_3Al (16%Al) in 0.25 M H_2SO_4 solution is illustrated in Fig. 5. The value of electrochemical parameters such as $i_{OH+/H2}$, i_{crit}, and current density in the passive region, which were calculated by their polarization curves found to increas with increasing C content. Microstructural and phase analysis of these dual-phase alloys reveal an increase in the volume fraction of the carbide with increasing C content. The nature of attack on an Fe_3Al-Fe_3AlC alloy after immersion in acid media reveals the preferential dissolution of the carbide phase. Hence, deterioration in corrosion resistance of iron aluminide with the increasing addition of C is related with the increase in the volume fraction of the carbide. At the same time in a complementary study the effect of Al on the

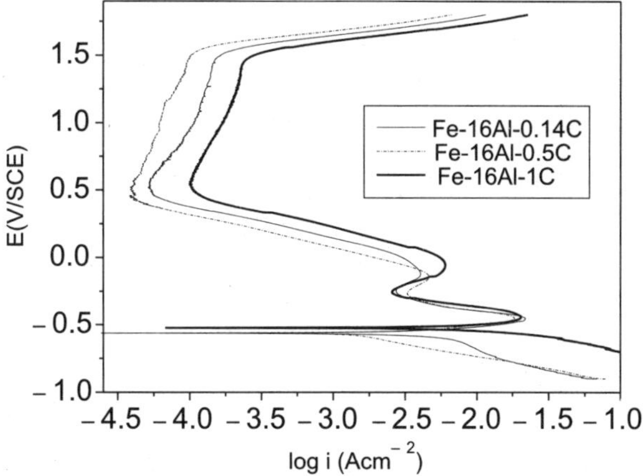

Figure 5 Potentiodynamic polarization curves of Fe-16Al-(0.14-1)C alloys obtained in 0.25 H_2SO_4.

corrosion behaviour of Fe-Al-C was investigated in 0.25 M H_2SO_4 [37]. Like binary Fe-Al alloys, lowering the Al content of the Fe-Al-C alloys resulted deterioration of the passivation behaviour in terms of increase in i_{pass} and i_{crit} values though the corrosion rate of the alloy is not much affected. An important finding in these investigations is that in high Al content alloys the carbide phase preferentially attacked whereas in low Al content alloys the both matrix and carbides dissolve at similar rate. This has been linked to the compositional variation in the phases with variation of Al and C in alloys.

Sriram et al. [38] investigated the corrosion behaviour of Fe-20Al-2C, Fe-18.5Al-3.6C and Fe-19Al-3.3C-0.07Ce alloys in 0.25 m/l H_2SO_4 by polarization as well as immersion tests to reveal the effect of Ce on the corrosion resistance of Fe-Al-C alloys. They found that the addition of Ce destroy the passivity of Fe-Al-C alloys by decreasing its breakdown potential. However, the immersion test revealed that the Ce containing alloy has superior corrosion resistance over the alloy without Ce content. This has been attributed to the modifications in surface film with Ce addition.

Before concluding this section we compare the aqueous corrosion behaviour of iron aluminides with stainless steels as mentioned during the introduction. The first comparative study between 304L stainless steel and iron aluminide was reported by Kim and Buchanan [22] where the pitting and crevice corrosion behaviour of Fe_3Al (28 at %Al) is reported to be the less resistance compared to 304 L stainless steel in mild chloride solution (200 ppm Cl^- + 6.25×10^{-5} M H_2SO_4) with pH adjusted to 4. At about the same time in an another communication Buchanan et al. [9] reported a relatively high corrosion rate of an Fe_3Al (28 at %Al) compared to 304L stainless steel in acid and sulfur-bearing environments, but in an alkali environment (1 M per litter NaOH) both the alloys were reported to have equivalent corrosion resistance.

To provide one to one comparison between the corrosion behaviour of iron aluminide and stainless steel potentiodynamic polarization curves of iron aluminide (Fe-16Al) and Fe-18Cr alloy (430 SS) obtained in 0.25 M H_2SO_4 are compared in Fig. 6. There is no significant difference in the general corrosion behaviour of iron aluminide to that of 430 SS, as difference in the i_{corr} values as well as in the E_{corr} values is very minimal.

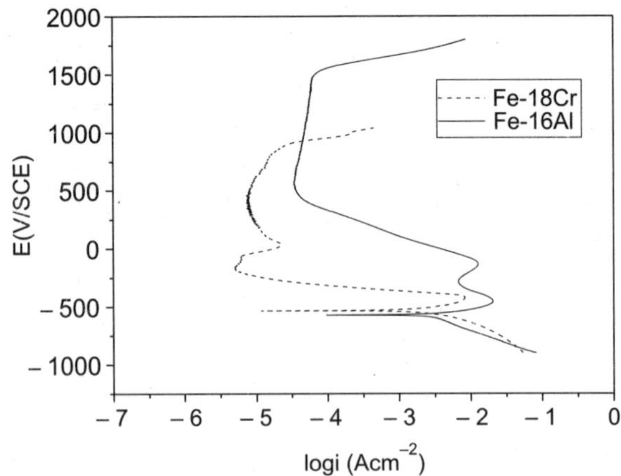

Figure 6 Potentiodynamic polarization curves of Fe-16Al and Fe-18Cr alloys obtained in 0.25 M H_2SO_4.

The i_{corr}, which is an important parameter to assess the dissolution tendency of an alloy, though appears to be low in the case of stainless steel compared to the iron aluminide but this difference can be ignored while considering relative high Cr content in 430 SS (Fe-18Cr) compared to Al content in Fe-16Al alloy. A pronounced difference in their anodic polarization behaviour appears in the pre-passive region, where Cr-based alloy possesses clear advantage over Al-based alloy. As the transition period from the active to passive region in case of stainless steel is about 10 min whereas for iron aluminide it is 30 min, for a sweep rate of 0.5 mV/sec. This lends credence to the porous nature of the Al_2O_3 oxide film. It should be pointed out that a remarkable enrichment of Cr ions occurs in the passive film of Fe-Cr, which is mainly attributed to its passivation [39], whereas, as brought out in this section, Fe and Al oxides coexist in the passive film formed on iron aluminide. While comparing both alloys in their passive regions we can find that the iron aluminide has noble passive potential and at the same time the passive region appears to be much wider compared to the alloy 430 SS but its passive current density is about one order higher than that of stainless steel. Coming to the transpassive region the high transpassive potential of iron aluminide leads this region to the oxygen evolution process whereas in stainless steel it generally corresponds to Cr dissolution from the passive film.

3. AQUEOUS CORROSION BEHAVIOUR OF TITANIUM ALUMINIDE

The advantage of titanium aluminide over other aluminides is low density, which makes it to receive much attention as potential aerospace and biomaterials. There are three intermetallic phases in titanium aluminides, namely Ti_3Al, TiAl and $TiAl_3$. Among three, Ti_3Al and TiAl are interest for structural applications, whereas $TiAl_3$ is only for academic interest. Hence, the aqueous corrosion behaviour of these intermetallics has been studied in detail.

We start reviewing this subject with the role of constituent elements Ti and Al on the steady-state corrosion behaviour of titanium aluminides. Ziomek-Moroz et al. [40] investigated corrosion behaviour of Ti_3Al (16 % Al) and TiAl (39 % Al) in alkaline and acid solutions. They provided a comparative study on steady-state passive current density of Ti, Ti_3Al, TiAl and Al, in 2N NaOH and in 2N H_2SO_4, which is shown in Fig. 7.

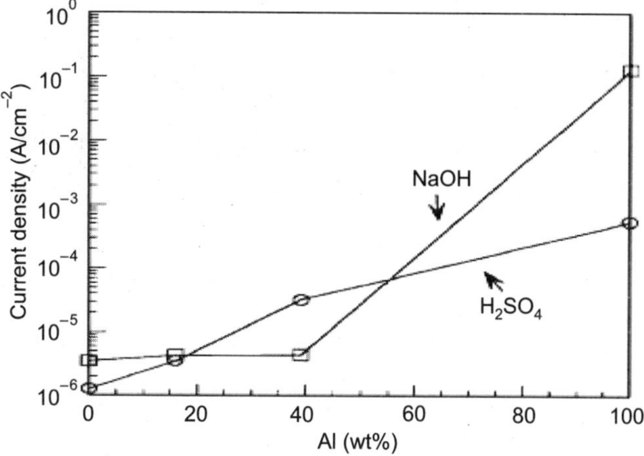

Figure 7 Quasi-state passive current density of Ti, Ti_3Al, TiAl and Al in 2N NaOH and 2N H_2SO_4 [40].

In both the alkali and acid solutions, the steady-state current density of pure Ti is lower than that of pure Al. This difference in steady-state current density between Ti and Al is much more in alkaline solution compared to acidic solution. In 2N NaOH, Ti, Ti_3Al and TiAl have the same steady-state current density, despite a fact that Al possesses much higher steady-state current density than the Ti. In other words, the effect of Al on steady-state current density of titanium aluminides is less differentiable in alkali media. While in 2N H_2SO_4, steady-state current density of the titanium aluminide increases with increase of the Al content. From Fig. 7 it is evident that Ti_3Al behaves equally well in 2N NaOH and 2N H_2SO_4 solutions, and the current density in both the solutions is the same, while TiAl has higher current density in 2N H_2SO_4 than in 2N NaOH.

Having seen steady-state current behavior of titanium aluminides we now explore their polarization behaviour. Accordingly, polarization curves of TiAl (Ti-36.6Al), Ti (99.97%) and pure Al (99.99 %) obtained from 0.5 N H_2SO_4 are compared in Fig. 8. The i_{pass} value increases in the order of Ti, Ti_3Al, TiAl and Al. Cathodic polarization curve of TiAl resembles to that of Ti and its cathodic current density for a given potential is higher than that of Al. During anodic polarization there is a typical active to passive transition in TiAl, while Ti and Al start passivating from their OCP itself. E_{corr} value of TiAl is between Al and Ti but closer to Al. From Fig. 8 it is very much clear that the corrosion and passivation behaviour of TiAl in 0.5 N H_2SO_4 lies between Ti and Al. We extend this discussion to the work of Ziomek-Moroz et al. [40] who studied anodic polarization behaviour of Ti_3Al (16 % Al) and TiAl (39 % Al) in much more strong acid and alkali solutions. According to their work, in 2N H_2SO_4 Ti_3Al behaves similar to pure Ti and with increase of Al content in titanium aluminide the shape of polarization curve shifts towards Al behaviour. In addition, there was appearance of a secondary anodic peak at 1.8 to 2 V_{SCE} in Ti_3Al and TiAl during polarization. Intensity of this peak was weaker in TiAl compared to Ti_3Al [40]. Such a peak with Ti can be found in Fig. 8, however, this peak was not observed with TiAl, which could be due to the high Al content in the alloy and comparatively low acid concentration of solution. In 2N NaOH, titanium aluminides display better passivation compared to in 2N H_2SO_4 with respect to passive potential range and low passive current density. However, there was no perceptible difference in the polarization behaviour among pure Ti, Ti_3Al and TiAl, despite a fact that passive current density of Al is about four orders higher than that of Ti in 2N NaOH.

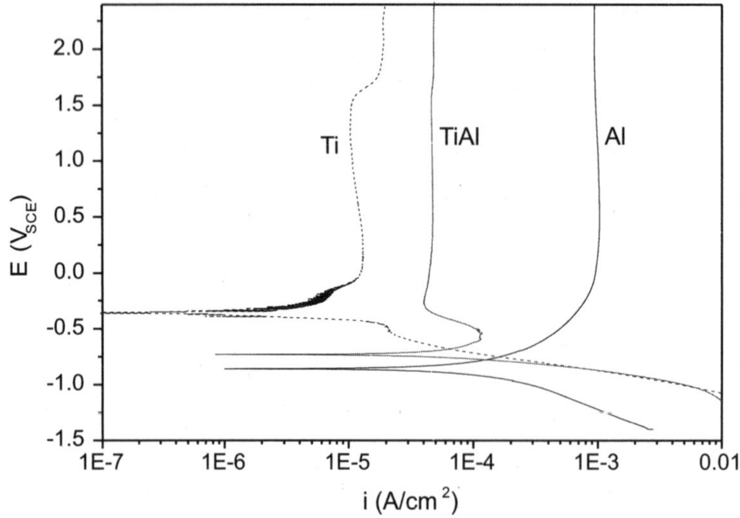

Figure 8 Polarization behaviour of TiAl, Ti and Al obtained from 0.5 N H_2SO_4.

So far, it has been demonstrated that Ti has better passivity over Al. However, loss of passivity results in pitting corrosion; hence, it is important to consider this subject. Since Ti is also known as the most pitting resistant material, we will try to analyze if this advantage appears in titanium aluminide. We refer the work of Saffarian et al. [41] who investigated corrosion behaviour of all types of titanium aluminides, along with constituent elements in neutral aggressive solution, and mild acid solution. Fig. 9 reproduced from their work illustrates anodic polarization behaviour of Ti_3Al, TiAl and $TiAl_3$ along with pure Ti and pure Al in 0.5 M NaCl.

Following points can be drawn from this study.

- E_{corr} and pitting potential (E_{pit}) of Ti are nobler than those of Al.
- With increasing the amount of Al in Ti, E_{pit} value of the titanium aluminides decreases.
- E_{corr} of the titanium aluminides remains closer to that of pure Ti irrespective to their Al content.
- There is a decrease in passive range of titanium aluminides with increase of Al content.

Further, examination of the alloy surface after polarization tests showed all titanium aluminides are susceptible to pit formation but pure Ti is not. These facts reflect pitting corrosion resistance of titanium aluminide is between Ti and Al in neutral pH solution containing Cl^- ions. However, in absence of Cl^- ions at neutral pH such as 0.25 M Na_2SO_4 solution there was not much difference in electrochemical corrosion behaviour between Ti and Al. So, no substantial difference in corrosion behaviour among titanium aluminides with variation of Al content was observed. More precisely, in 0.25M Na_2SO_4 solution anodic

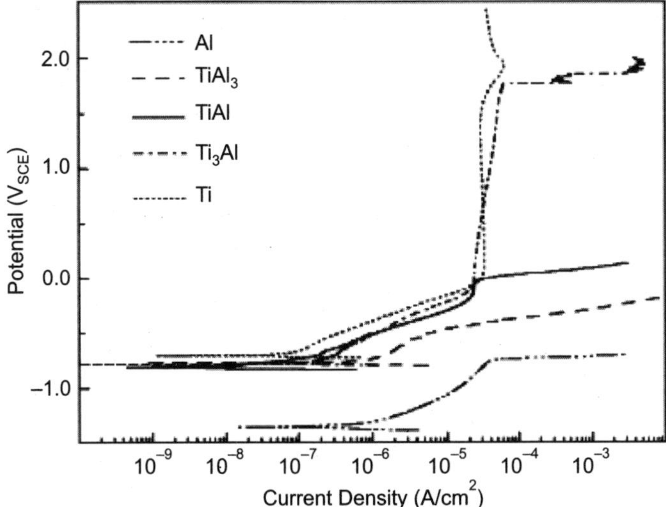

Figure 9 Comparison of anodic polarization behaviour for titanium aluminides, pure Ti and Al in 0.5 NaCl solutions at pH 6 [41].

polarization behaviour of Ti_3Al and TiAl are similar to that of pure Ti. The i_{pass} value of these three was nearly the same (25 ± 3 μAcm^{-2}), while for $TiAl_3$ it was about three times higher than Ti_3Al and TiAl. In a recent work, it was shown that the corrosion rate of TiAl (48%Al) in seawater is about three times higher than in 3.5% NaCl solution [42]. However, this alloy does not display stable passive region in both the solutions. This was attributed to the breakdown of the passive film at potentials close to the E_{corr}.

TiAl most often encountered misleading interpretation on susceptibility to hydrogen induced cracking due to the cathodic reaction on the alloy surface during hydrogen charging. Gao et al. [43] investigated cathodic corrosion behavior of TiAl (49.5 at.%Al) by hydrogen charging in various aqueous solutions e.g. pH = 2 (0.5 M H_2SO_4), pH = 2.5 (0.5 M Na_2SO_4 + H_2SO_4), pH = 13 (0.5 M NaOH), pH = 6 (0.5 M Na_2SO_4) and molten salt at 60°C. TiAl undergoes cathodic corrosion in all the aqueous solutions. Cathodic corrosion of TiAl in acid solution was ten times higher than in salt solution. Fig. 10 illustrates the severity of the cathodic corrosion in acidic solution, where the cathodic corrosion rate is much higher than the anodic corrosion rate for a same range of potential shift from OCP. This investigation was further extended to know the effect of hydrogen poisonous compound, As_2O_3, on the cathodic corrosion rate of TiAl. As_2O_3 addition resulted in an increase in hydrogen concentration on the surface and in bulk of the alloy, which led to soaring of cathodic corrosion rate. X-ray diffraction analysis of the cathodically charge specimen in 0.5 M H_2SO_4 revealed the formation of a hydride phase as $(TiAl)H_{0.5}$. Disruption of the surface film by formation of hydride was considered to be responsible for the occurrence of cathodic corrosion during cathodic polarization.

The corrosion behaviour of Ti-13.4Al-29Nb alloy was evaluated in marine electrolytes, industrial environment and acid environment at 25°C and 50°C [44]. Performance of this alloy was found to be good in 3.5 % NaCl and 0.5 N Na_2SO_4. However, in 0.5 N H_2SO_4 its corrosion

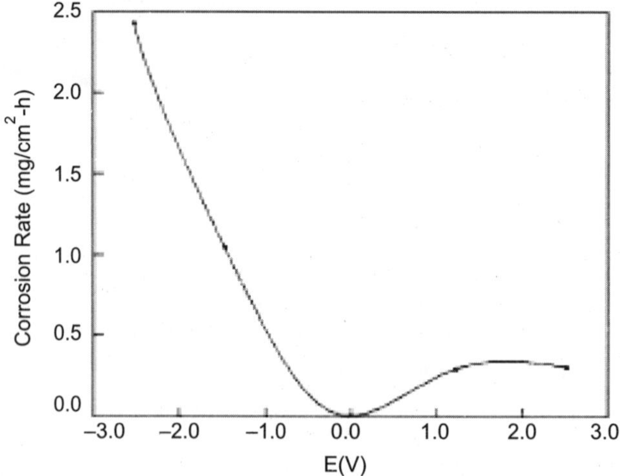

Figure 10 Variation of anodic dissolution and cathodic corrosion of TiAl with applied potential in 0.5 M (Na_2SO_4 + H_2SO_4), pH – 2 [43].

resistance was inferior. This was speculated to the dissolution of the protective oxide layer in low pH solution. The author claimed about change in the degradation mode of the alloy in marine and industrial environments with increase of temperature from 25°C to 50°C. At low temperature the alloy suffered from pitting corrosion and at high temperature uniform corrosion. Nonetheless, it needs to be confirmed.

Ti based alloys have long been favorable materials for orthopedics implantation due to their high corrosion resistance. In this field of biomaterial, research activities have been focused on the development of titanium aluminides with additions of various alloying elements to improve corrosion resistance for a possible replacement of existing Ti alloys. This is because titanium aluminides possesses more strength over Ti based alloy, a prerequisite for biomaterials. One such example can be found in the work of Choubey et al. [45] who investigated electrochemical corrosion behaviour of Ti_3Al based intermetallic without and with Nb, in simulated body environment (Hank's solution) at 37°C and compared with Ti-6Al-4V. The chemical compositions of the alloys and their respective corrosion data are tabulated in Table 1. Addition of Nb enhances the corrosion resistance as well as passivity of the TiAl, which is evident from the shift in OCP towards positive side and decrease in i_{pass} value. Simultaneously, there is much higher shift in E_b value towards the positive side. Though there is marginal increase in the corrosion rate with addition of Nb but it is still less than of Ti-6Al-

Table 1 Chemical compositions of TiAl alloys and their electrochemical corrosion parameters [45].

Sample (wt.%)	OCP (mV_{SCE})	E_b (mV_{SCE})	i_{pass} ($\mu A/cm^2$)	icorr ($\mu A/cm^2$)	Corrosion rate (mils per year)
Ti-15Al	-534	750	1.6	0.08	0.028
Ti-13.4Al-29Nb	-447	1097	1.0	0.15	0.054
Ti-6Al-4V	-231	1277	3.0	0.16	0.055

4V. Notably, the passive range (E_b-OCP), of TiAl with addition of Nb is almost equivalent to that of Ti-6Al-4V. Nevertheless, the passive current density of all the three alloys is in same order of magnitude and their corrosion rates are comparable.

In a parallel study, Escudero et al. [46] evaluated the performance of g-TiAl (Ti-45Al-2W-0.6Si-0.7B at.%) for the endoprothesic applications such as hip and knee prosthesis, dental implants. For *in vitro* corrosion resistance evaluation, the electrochemical corrosion behaviour of the TiAl in as-received condition and after surface modification by oxidation at two different temperatures 700°C and 1000°C was investigated in simulated body solution (8 NaCl + 0.2 $CaCl_2$ + 0.2 KCl + 1 $NaHCO_3$) g/l at pH 7.4 known as Ringer's solution. The purpose of choosing two ranges of temperature was to form different oxide layers. At 700°C, the oxidized layer mainly consists of Al_2O_3, whereas at 1000°C it is TiO_2. Fig. 11 compares the cyclic polarization behaviour in Ringer's solution of the TiAl as-received and pre-oxidized at 1000°C for 1 h.

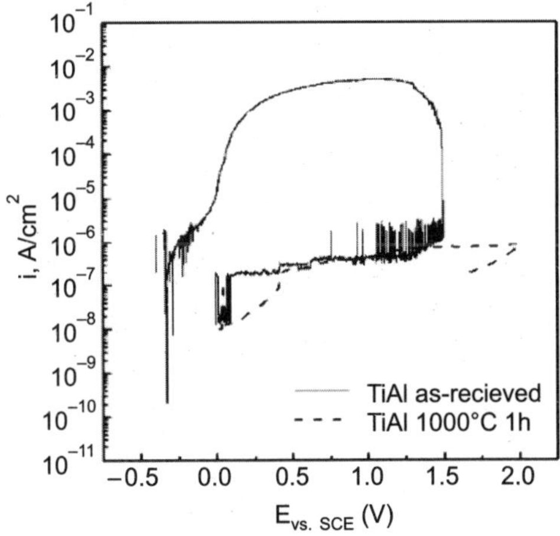

Figure 11 Anodic polarization curve for the TiAl as-received condition and after pre-oxidation at 1000°C for 1h, in Ringer's solution [46].

The superiority of pre-oxidized specimen is evident as it displays low passive current density and high breakdown potential over unoxidized specimen. These authors further report that the corrosion resistance of TiAl alloy pre-oxidized at 700°C is even superior to the pre-oxidized at 1000°C. This was attributed to the formation of lower concentration of Ti in oxide film formed at 700°C. In another wok, Delgado-Alvarado and Sundaram [47] evaluated the corrosion resistance of the alloy Ti-48Al-2Cr-2Nb (at.%) in Ringer's solution without and with pre-oxidation at different temperatures. In agreement with Escudero et al. [46] they reported that the pre-oxidation process improves corrosion resistance of the alloy. The alloy oxidized at 500°C showed higher polarization resistance than that of the alloy oxidized at 800°C. Lower corrosion resistance of the alloy oxidized at 800°C was attributed to the formation of porous oxide layer.

We have discussed earlier that the steady state current density of titanium aluminides is independent on the Al content in NaOH but not in H_2SO_4. To credence this feature, the chemical compositions of the passive films formed in respective electrolytes during steady-state current measurement were analyzed using XPS [40]. The passive film formed in NaOH consists of mainly Ti-oxides and there was no sign of Al-oxides in outer layer of the passive film. Whereas the passive film formed in H_2SO_4 solution was found to be enriched with Al ions. In a similar study, high-resolution XPS spectra of the Ti taken from the passive film formed on Ti_3Al in NaOH solution, before and after sputtering by Ar^+ ions are illustrated in Fig. 12a and Fig. 12b, respectively [41]. They reflect the structural change in the passive film with the film thickness. In the outermost layer of the passive film Ti is present in the form of Ti^{+4}, whereas the inner layers consist of Ti^{+4}, Ti^{+3} and Ti^{+2}. An analysis of potential-pH diagrams of Ti-H_2O and Al-H_2O systems support the results obtained from anodic polarization and surface analysis. According to these diagrams, Ti dissolves only at low pH in the potential range between –1.8 to 0.1 V_{SHE} and stays passive above this potential value over the whole pH range. On the other hand, Al corrodes at low and high pHs, while at pH close to neutral it passivates.

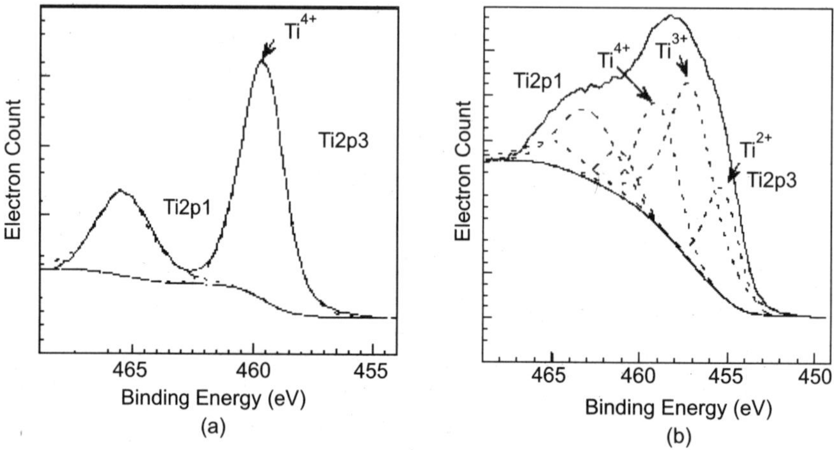

Figure 12 Ti_{2p} XPS spectra (a) of the topmost surface, (b) after sputtering, show different chemical states of Ti, for Ti_3Al passivated in NaOH [41].

4. AQUEOUS CORROSION BEHAVIOUR OF NICKEL ALUMINIDE

A typical Ni-Al phase diagram represents the presence of four types of intermetallics namely $NiAl_3$, Ni_2Al_3, NiAl and Ni_3Al [48]. However, out of these only NiAl and Ni_3Al are of interest for structural material at high temperature due to required physical properties and high melting point. The reported works on the corrosion behaviour of nickel aluminide is limited primarily for two reasons. Firstly, nickel aluminides have never been considered as a material for room temperature applications due to high cost of Ni. Secondly, the development of Ni based alloys and superalloys, which possess better corrosion resistance in aggressive environments than nickel aluminides.

In section 2.1 we have described the role of alloying elements in electrochemical corrosion behaviour of iron aluminides. A similar line can be drawn for understanding the corrosion behaviour of nickel aluminide with its alloying elements. Ni is more corrosion resistance compared to Fe and Al. This advantage appears in corrosion resistance of nickel aluminide over iron aluminide, which we will find while discussing the subject. Ever since nickel aluminides came in the practice for use in structural applications a detailed study on the corrosion behaviour of these alloys was first reported by Bertocci et al. [49]. Taken from their work, Fig. 13 illustrates OCP value of Ni_3Al based nickel aluminide (Ni-22.2Al-0.42Hf-0.087B at.%) without and with annealing process, and its constituent elements Ni and Al, in acidic, neutral and alkaline solutions. Irrespective to degree of variation, OCP value of Ni_3Al is closer to Ni than Al in all solutions. The difference in OCP between cold worked Ni_3Al and annealed Ni_3Al is less distinguishable. Moreover, unlike in case of Al, there is less influence of electrolyte (ions and pH) on OCP of nickel aluminide.

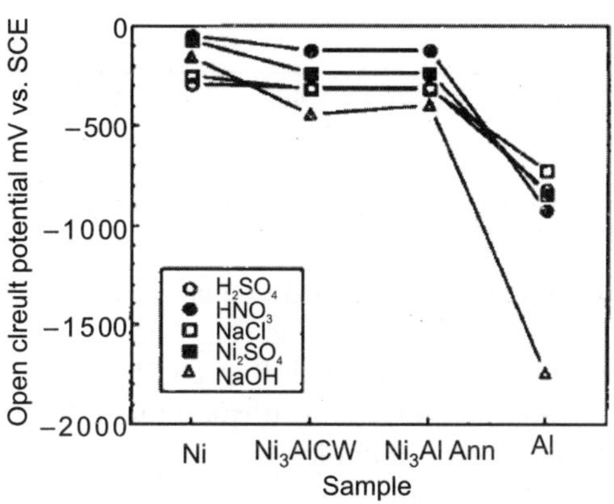

Figure 13 Open circuit potential of Ni_3Al, Ni and Al in various solutions: acidic, neutral and alkali solutions [49].

These authors further investigated passivity and breakdown of passivity of Ni_3Al. Figure 14 shows the potentiodynamic polarization curve of Ni_3Al in 0.5 M H_2SO_4. Typical active-passive transition behaviour is evident from the polarization curve. The curve displays two anodic peaks, and intensity of these peaks depends upon the scan rate. At low sweep rate, intensity of the peak closed to OCP is high, while at high sweep rate the other peak. The polarization behaviour of this nickel aluminide was reported to be similar to that of pure Ni in 0.5 M H_2SO_4. Even in pure Ni there is appearance of two anodic peaks. Further, examination of the nickel aluminide after polarization test for unstable passive potential region i.e. between two anodic peaks (-125 mV_{SCE} and +50 mV_{SCE}) revealed formation of pits on the surface. Notably, no such report exists in case of pure Ni [49]. An explanation provided for the root of such unusual pit formation in acid media was the drop of ohmic resistance inside the pit, which drives the potential into the active region. These observations suggest that the passive film formed on the Ni_3Al surface is marginally less stable than that of pure Ni in acid media. On the other hand in 0.5 M NaOH the polarization curves of Ni_3Al and Ni solution appear similar in nature [49]. While quantitatively the current density of Ni_3Al at the same potential was about one order less than that of pure Ni, which indicates Ni_3Al is less active than Ni with respect to the hydrogen evaluation reaction.

In another study, Lillard and Scully [50] evaluated electrochemical corrosion behaviour of order NiAl and its constituent elements in buffered solutions of pH 2, 7 and 12.5. Figs. 15a and b show anodic polarization behaviour of NiAl, Ni and Al in pH 2 (0.5 tartaric acid) and pH 12.5

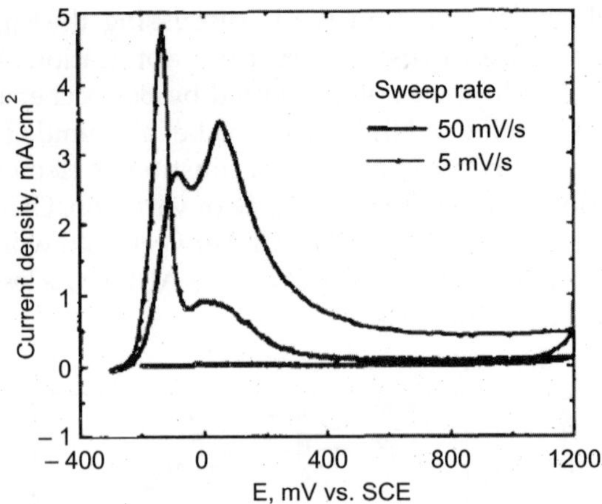

Figure 14 Potentiodynamic polarization behaviour of Ni_3Al in 0.5 M H_2SO_4 at different scam rate [49].

(adjusted with NaOH) solutions. In pH 2 solution Al spontaneously passivates, while Ni and NiAl exhibit active-passive-transpassive transition (Fig. 15a). The addition of Al in Ni resulted decrease in critical anodic current density and shift in OCP marginally towards negative side. In high pH electrolyte Al corrodes actively, as confirmed by an OCP of -2.24V and high anodic current density (Fig. 15b). Ni is a metal of choice for many alkali media, which is also evident from its polarization curve in Fig. 15b. This is because formation of a protective nickel oxide on the surface. However, Al has little effect on the corrosion behaviour of Ni, as nature of polarization curve of NiAl remained similar to pure Ni. On the whole, this analysis shows that passivation behaviour of Ni_3Al follows to Ni irrespective to pH of the electrolyte.

Here some contradiction appears on the role of Al on the electrochemical corrosion behaviour of nickel aluminide. Bertocci et al. [49] reported that the presence of Al does not

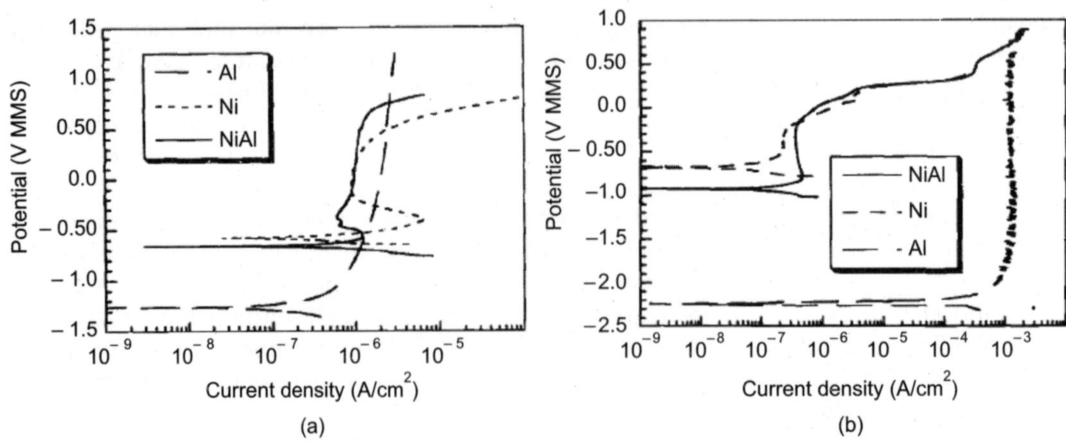

Figure 15 Polarization behaviour of NiAl, Al and Ni: (a) pH 2, 0.5 tartaric acid, (b) pH 12.5 adjusted with NaOH [MMS = mercury/mercury sulfate electrode] [50].

confer additional corrosion resistance, whereas according to Lillard and Scully [50] Ni and Al function cooperatively in passivation of NiAl, Ni improves passivity of NiAl in high pH solution while Al improves in low pH solution.

Ricker [51] performed scratched test to study the repassivation kinetics of Ni_3Al (Ni-11.53Al-0.03Zr-0.02B), pure Ni and pure Al in 0.5 M H_2SO_4 and 0.5 M NaCl. In both the solutions, repassivation behaviour of Ni_3Al was similar to that of pure Ni and its repassivation potential does not drop below the hydrogen evaluation potential. Based on thermodynamic data they calculated Al^{3+}/Al equilibrium potential, which was much active with respect to hydrogen evaluation potential. This finding is significant as Al makes 25% of this compound and yet has little influence on the potential transients that follow from the rupture of the passive film. This revealed that the potential drop during the repassivation process does not cause significant hydrogen evaluation and absorption.

This review so far shows that the electrochemical corrosion behaviour of nickel aluminide closely follows to Ni despite presence of Al. The next point for the discussion is chemistry and chemical structure of the passive film formed on nickel aluminides. The work of Lillard and Scully [50] provides a detail study on this issue. They analyzed the passive film formed on NiAl in various pH solutions using XPS. In pH 2 (0.5 tartaric acid) the passive film formed at -0.1 V vs Hg/Hg_2SO_4 consists of Al_2O_3, NiO and $Ni(OH)_2$. A mixed oxide of $NiAl_2O_4$ was also envisaged in the passive film by reaction between NiO and Al_2O_3 in the passive film. The concentration of Al^{+3} and Ni^{+2} in a passive film formed at different pH is presented in Fig. 16 in terms of atomic fraction. Notably, Al concentration is more than 70% in pH 2 solution hence it is reasonable to conclude that the presence of Al_2O_3 lower the i_{crit} value of NiAl. The presence of 10-30% Ni ions throughout the oxide film alter the oxide conductivity or catalytic properties of NiAl sufficiently to enable the oxygen evaluation reaction which is not observed on Al. The passive film formed in pH 7 and 12 solution contains the same compounds those formed at pH 2 but differ in ratio. At high pH had nearly equal amounts of Ni^{+2} and Al^{3+} in surface contrary to domination of Al_2O_3 in low pH. Notably, at high pH concentration of Al ions is low, which could be possibly due to dissolution of the Al_2O_3 oxide in alkali media.

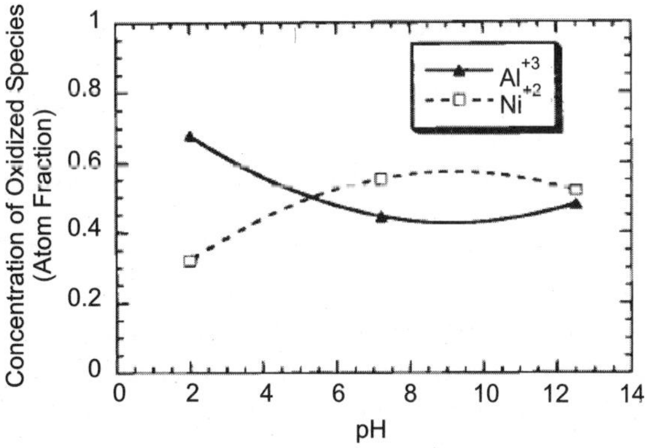

Figure 16 Atomic fraction of Al^{3+} and Ni^{2+} as a function of pH obtained from the XPS data of the passive film formed on nickel aluminide [50].

Table 2 Summary of the E_{corr} and i_{corr} values obtained from polarization curves for the NiAl-based alloys in 0.5 M HNO_3 and 0.5 M H_2SO_4 [52]

Alloy nominal composition (at. %)	0.5 M HNO_3		0.5 M H_2SO_4	
	E_{corr} (mV_{SCE})	i_{corr} (mA/cm^2)	E_{corr} (mV_{SCE})	i_{corr} (mA/cm^2)
Al-44Ni	−212.2	0.940	−287.4	0.003
Al-42Ni-6Fe	−224.1	0.402	−378.7	0.015
Al-41Ni-6Mo	−200.7	0.041	−387.1	0.029
Al-41Ni-6Ga	−185.7	0.347	−288.8	0.003
Al-40Ni-6Fe-2Ga	−241.3	0.126	−283.9	0.005
Al-40Ni-6Fe-2Mo	−205.2	0.336	−368.2	0.010
Al-39Ni-6Fe-6Mo	−303.5	0.004	−303.8	0.004
Al-42Ni-2Mo-2Ga	−181.1	0.308	−278.7	0.002

In a recent study, Albiter et al. [52] reported the effect of alloying elements addition namely Mo, Ga and Fe on the electrochemical corrosion behaviour of nano crystalline NiAl in two different acidic solutions 0.5 M H_2SO_4 and 0.5 M HNO_3. The electrochemical data obtained from their work are reproduced in Table 2, where corrosion rate of unalloyed nickel aluminide in 0.5 M HNO_3 is about 300 times higher than in 0.5 M H_2SO_4. Though, their work reflects little on corrosion behaviour of nano crystalline NiAl, but the data shown in Table 2 have practical implication and are useful for comparative analysis. The effect of alloying elements on the corrosion parameters appears selective to electrolyte. For example, addition of Mo in nickel aluminide is the most effective for decreasing the corrosion rate of NiAl in HNO_3 whereas the same alloy shows the least corrosion resistance among all in H_2SO_4. Likewise, Ga is very effective in HNO_3 both in shifting E_{corr} towards noble direction as well as decreasing the i_{corr} value, whereas in H_2SO_4 it shows little change in these parameters. The addition of 6% Fe in nickel aluminide causes shift in E_{corr} of the alloy towards negative side in both the acid electrolytes. At the same time, it makes decrease in the corrosion rate to half in HNO_3 electrolyte, while in H_2SO_4 electrolyte markedly enhances corrosion rate to about five times. A synergetic effect on the electrochemical values is visible from Table 2 by addition of two elements in nickel aluminides. 6Fe+6Mo is the most effective alloying addition among all under investigation as this cause decrease in i_{corr} of the alloy in both the electrolytes HNO_3 as well as H_2SO_4. Though this makes E_{corr} more negative. In support of these findings same authors investigated the change in corrosion rate of the nickel aluminides with immersion method. In agreement with the electrochemical studies it was found that the NiAl-6Mo alloy has highest corrosion rate among all in H_2SO_4

Ni is important not only in nickel aluminide but it is main element of nickel based superalloys and one of the major constituent elements in austenitic stainless steels. It would be interesting to know how Ni behaves in these three different types of alloy, Songbo et al. [6] evaluated the corrosion behaviour of NiAl intermetallics [Ni-29.3Al], Ni based superalloy GH217 [Ni-23.2Cr-10Co-2.2Ti-1.2Mo] and 18-8 stainless steel [Fe-18.1Cr-9.7Ni] in a molten carbonate fuel cell environment (62 mol% Li_2CO_3 + 38 mol% K_2CO_3) using cyclic voltammogram. According to their report, the pitting potential and the repassivation potential of NiAl are the most positive among three alloys. The high corrosion resistance of the NiAl was attributed to the formation of a protective oxide layer as $LiAlO_2$ along with $Li_2Ni_8O_{10}$ on the surface.

As mentioned in Section 1, electrochemical studies in aluminides are not limited to analysis the corrosion behaviour. With the invention and development of technology this technique is also now in use of some creative works. For example, Hassel et al. [7] utilized the electrochemical corrosion method for formation of nanopore arrays on the surface of Ni-Al-Re alloys. A directional solidified eutectic (Ni-29.9Al-5.5Re) alloy was chemically treated in an electrolyte containing 3.3 % HCl + 3% H_2O_2 + distilled water, for selective etching of Re. This process led to the formation of Re nanofibers on Ni-aluminide surface. An SEM image of the alloy obtained from such process is shown in Fig. 17 from the work of Hassel et al. [7]. In yet another work, on the surface of the same alloy nanopores was formed through electrochemical etching in 1 M acetate buffer solution (pH 6). In this process, Re is selectively dissolved from passivated NiAl surface.

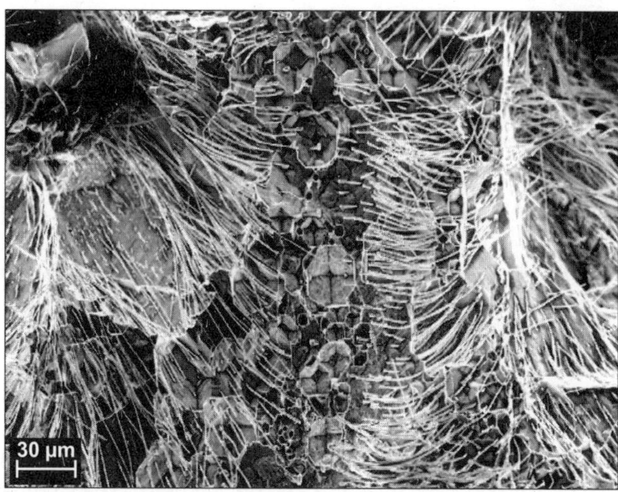

Figure 17 SEM micrograph of Ni-29.9Al-5.5Re after etching in 3.3% HCl + 3% H_2O_2 + distilled water [7].

5. SUMMARY

The aqueous corrosion behaviour of iron aluminides, titanium aluminides and nickel aluminides were reviewed for high temperature structural, biomaterial and marine applications. The degree of corrosion resistance as well as passivation behaviour of aluminides depends on ratio of their respective constituent elements and exposed environment. These intermetallics overall show good corrosion resistance and pronounced passivity. In order to make corrosion resistance of aluminides equivalent to stainless steels, much work needs to be done. The concern with titanium aluminides is their higher cathodic current density compared to anodic current density. Pre-oxidation enhance the corrosion resistance of the aluminide in chloride containing solution and body fluids. However, the degree of corrosion resistance depends upon the physical as well as chemical nature of the oxide formed during oxidation process

In future studies, aqueous corrosion data at high temperature need to be generated. Theoretical analysis on the corrosion behaviour of aluminides is required in understanding and enhancing their corrosion resistance.

ACKNOWLEDGEMENT

This review was accomplished during Author's stay in the Department of Materials Science Engineering of Korea Advanced Institute of Science and Technology, South Korea. The author wishes to thank Prof. R. Balasubramaniam of Indian Institute of Technology, Kanpur, India and Dr. L. K. Singhal, Director, Jindal Stainless Limited, Hisar for motivating to complete this task.

REFERENCES

1. C.T. Liu, J.O. Stielger, F.H. Fores "Ordered Intermetallics" In Metals Handbook ASM Vol.2 Metals Park, OH, (1990) p.913.
2. Robert Pichoir, "Aluminide coating on nickel or cobalt-base superalloys: Principal parameters determining their morphology and composition" in High temperature Alloys for Gas Turbines, eds. D. Coutsouradis, P. Felix, H. Fischmeister, L. Habraken, Y. Lindblom, M.O. Speidel, Applied Science Publishers Ltd, London, 1978, p.191
3. N.S. Stoloff, C.T. Liu, S.C Deevi, Intermetallics 8 (2000) 1313.
4. Ya Xu, Satoshi Kameoka, Kyosuke Kishida, Masahiko Demura, An-pang Tsai, Toshiyuki Hirano, Intermetallics 13 (2005) 151.
5. M.L. Escudero, M.A. Munoz-Morris, M.C. Garcia-Alonso. E. Fernandes-escalante, Intermetallics 12 (2004) 253.
6. Xu Songbo, Zhu Yongda, Huang Xing, Zhou Bangna, Yu Zhongxing, J. Power Sources 103 (2002) 230.
7. A. W. Hassel, Belen Bello-Rodriguez, Srdjan Milenkovic. André Schneider, Electrochim. Acta 50 (2005) 3033.
8. C.G. McKamey, C.G. DeVan, J.H. Tortorelli, V.K. Sikka, J. Mater. Res. 6 (1991) 1779.
9. R.S. Buchanan, J.G. Kim, R.E. Ricker, L.A. Heldt, in Physical Metallurgy and Processing of Intermetallic Compounds, eds. N.S. Stoloff, V.K. Sikka, Chapman and Hall, (1996) p.517.
10. V. Shankar Rao, R.G. Baligidad, V.S. Raja, Corros. Sci. 44 (2002) 521.
11. H.C. Choe, H.S. Kim, D.C. Choi, K.H. Kim, J. Mater. Sci. 32 (1997) 1221.
12. S. Frangini, J. Lascovich, N.De Cristofaro, Mater. Sci. Forum 185-188 (1995) 1041.
13. S. Frangini, N.De Cristofaro, A. Mignone, J. Lascovich, R. Giorgi, Corros. Sci. 39 (1997) 1431.
14. S. Frangini, Corrosion 55 (1999) 89.
15. S. Frangini, N.De Cristofaro, J. Lascovich, A. Mignone, Corros. Sci. 35 (1993) 153.
16. V. Shankar Rao, Corros. Sci. 47 (2005) 183.
17. B.W. Madsen, T.A. Adler, Wear 171 (1994) 215.
18. J.F. Nachman, R. Duffy, Corrosion 30 (1974) 357.
19. J.N. Derfancq, Werkstoffe und Korrosion 28 (1977) 480.
20. D. Schaepers, H.H. Strehblow, Corros. Sci. 39 (1997) 2193.
21. N.De Cristofaro, S. Frangini, A. Mignone, Corros. Sci. 38 (1996) 307.
22. J.G. Kim, R.A. Buchanan, Corrosion 50 (1994) 658.

23. M.C. Garcia Alonso, M.F Lopez, M.L. Escudero, J.L. Gonzalez-Carrasco, D.G. Morris, Intermetallics 7 (1999) 185.
24. W.C. Luu, J.K. Luu, Corros. Sci. 43 (2001) 2325.
25. S. Frangini, J. Lascovich, Electrochim. Acta 40 (1995) 637.
26. M.L. Escudero, M.C Garacia-Alonso, J.L. Gonzalez-Carrasco, M.A. Munoz-Morris, M.A. Montealegre, C. Garcia Oca, D.G. Morris, S.C. Deevi, Scripta Mater. 48 (2003) 1549.
27. V. Shankar Rao, V.S. Raja, Corrosion 59 (2003) 575.
28. C.D. Wagner, W.M. Riggs, L.E Davis, J.F. Moulder, Handbook of X-ray Photoelectron Spectroscopy, (G.E. Mouilenberg Ed.) Perkin-Elmer Corp. (1978) p.50.
29. T.L. Barr, J. Phys. Chem. 82 (1978) 1801.
30. N.S. Mcintyre, D.G. Zetaruk, Anal. Chem. 49 (1977) 1521.
31. D. Brion, Appl. Surf. Sci. 5 (1980)133.
32. V. Shankar Rao Electrochim. Acta 49 (2004) 4533.
33. R Balasubramaniam, Scripta Matell. 34 (1996) 127.
34. Yoon-seok Choi, Jung-Gu Kim, Mater. Sci. Eng. A 333 (2002) 336.
35. N. Babu, R Balasubramaniam, A.V.R. Kumar, J. Mater. Sc. Lett. 19 (2000) 1061.
36. R.G. Baligidad, U. Prakash, V. Ramakrishna Rao, P.K. Rao, N.B. Ballal, ISIJ Int. 36 (1996) 1453.
37. V. Shankar Rao, V.S. Raja, R.G. Baligidad, R. Raman, Corros. Eng. Sci. Tech. 38 (2003) 235.
38. S. Sriram, R. Balasubramaniam, M.N. Mungole, S. Bharagava, R.G. Baligidad, Corros. Sci. 48 (2006) 1059.
39. I. Olefjord, Mater. Sci. Eng. 42 (1980) 161.
40. M. Ziomek-Moroz, W. Su, B.S. Covino Jr., Corrosion, 55 (1999) 635.
41. H.M. Saffarian, Q. Gan, R. Hadkar, G.W. Warren, Corrosion 52 (1996) 626.
42. C. Delgado-Alvarado, P.A. Sundaram, Corros. Sci. 49 (2007) 3732.
43. K.W. Gao. J-W. Jin, J. Qiao, W.T. Chu, C-M Hsiao, Corrosion 52 (1999) 3.
44. I. Gurrapa, Intermetallics 11 (2003) 867.
45. A. Choubey, B. Basu, R. Balasubramaniam, Intermetallics 12 (2004) 679.
46. M.L. Escudero, M.A. Munoz-Morris, M.C. Garcia-Alonso, E. Fernandes-escalante, Intermetallics 12 (2004) 253.
47. C. Delgado-Alvarado, P.A. Sundaram, Acta Biomaterials 2 (2006) 701.
48. A. Brown, J.H. Westbrook "Formation Techniques" in: Intermetallic Compounds Ed. J.H. Westbrook, Robert E.KriegerPublishing company, New York, (1977) p.304.
49. U. Bertocci, J. L. Fink, D. E. Hall, P. V. Madsen and R. E. Ricker, Corros. Sci. 31 (1990) 471.
50. R.S. Lillard, J.R. Scully, J. Electrochem. Soc. 45 (1998) 2024
51. R.E. Ricker, Mater. Sci. Eng. A 198 (1995) 231.
52. A. Albiter, M.A. Espinosa-Medina, J.G. Gonzalez-Rodriguez, R. Perez, Int. J. Hydrogen Energy 30 (2005) 1311.

CHAPTER 11

High Nitrogen Steels as Specific Corrosion Resistant Alloys

Jacques FOCT
Laboratoire de Métallurgie Physique et Génie des Matériaux – UMR CNRS 8517 – Université de Lille I – 59655 Villeneuve d'Ascq Cedex, France

Results of research and development, technological and industrial achievements and economics of raw materials are steering new upgraded alloys among which high Nitrogen stainless steels (H.N.S.) are very important. The reasons and the mechanisms which induce many improvements such as mechanical properties and corrosion resistance by Nitrogen alloying are presented. The aim of this chapter is to explore the state of the art of H.N.S. science and technology with a special attention to the role of Nitrogen alloying on mechanical properties and corrosion resistance as well as the resulting impact on fabrication and applications.

I. INTRODUCTION AND PRELIMINARY REMARKS

The largest part of metallic alloys are based on the first transition series elements, Ti, V, Cr, Mn, Fe, Co, Ni, Cu and Zn. Steel and Nickel alloys usually contain appreciable amount of these alloying elements as well as other transition elements of the second and third series such as Nb, Zr, Mo and W. Besides the transition elements, alloys may contain non transition atoms such as Al and Si which originate strong interatomic interaction with the host matrix atoms. All above quoted elements replace each other, they occupy substitutional sites. When light and small non metallic atoms such as H, C, N and to a lesser extent B and O are present in the alloy they occupy interstitial sites which were initially empty. The fact to replace nothing by "something" originates large modifications of the alloys, therefore interstitial alloys containing H, C and or N constitute specific alloys which demand a special attention because the interactions with lattice, with point defects, with dislocations, and with two dimensional imperfections such as stacking faults, twins and grain boundaries, phase boundaries, surfaces and interfaces are important.

In addition with these strong interactions, the atomic diffusivity of H, C and N is several order of magnitude larger than that of metallic alloys, therefore interstitial alloys show specific types of phase transformations obeying to high kinetics rates. The large atomic mass difference between transition metals and hetero-interstitial atoms may induce a biased view on the importance of the presence of H, C or N in the alloy, roughly the mass percentage must be multiplied by 50 for H and by 5 for C and N to evaluate the corresponding atom fraction.

The previous very basic remarks concerning interstitial atoms enlighten the fact that the intentional addition of Nitrogen in stainless steel results in a specific class of materials called "H.N.S.", meanwhile this family is likely to extend to H.N.A. (High Nitrogen Alloys) and even to H.I.A. (High Interstitial alloys).

Since the early eighties of the twentieth century the specificity of H.N.S. became more explicit and several international meetings and books [1-17] have been devoted to this subject. There is neither evident nor unanimous definition of "high" concentration, higher than 0.4 wt% in austenitic steel [7], more than 0.08 % in martensitic steel. It is also proposed to consider that "high" means intentionally raised N content by appropriate alloying, by pressure, by powder metallurgy, etc.[14]. In our view, a steel can be considered as high nitrogen alloy when the nitrogen content induces an unambiguous (more than 10%) change of the property into consideration. Concentration such as 0.01, 0.05, 0.12 and 0.06 wt% N, respectively for ferritic, martensitic, austenitic and duplex steels can be considered as significant because in all these cases either driving force for precipitation or grain boundary segregation or interaction with dislocation or corrosion properties, are appreciably affected. Furthermore, because interactions between N and other interstitial atoms such as H or and C are important the definition of high interstitial alloyed steel could be "flexible". Although some documents refer steel by the symbol "N_2" this should be considered as a lapsus (slip of pen) except when bubbles are formed.

This article is intending to emphasize the corrosion properties of H.N.S. in relation with the specific role of Nitrogen. Meanwhile, because corrosion resistance usually does not constitute the unique criterium for an alloy to be utilized in a given application, other properties have to be satisfied such as mechanical behaviour for materials which have to resist to mechanical loading under aggressive chemical conditions. The interaction of mechanical constraints with chemically active environment leads to specific phenomena such as stress corrosion, corrosion fatigue, corrosion-erosion, etc. which can be considered to refer to corrosion-deformations interactions (C.D.I.). Although other properties such as thermal behaviour, thermal conductivity, electrical conductivity, expansion coefficient, magnetic properties may also interfere with corrosion damage, only mechanical and corrosion properties with their link with the microstructure resulting from fabrication and steel making, will be discussed.

II. THERMODYNAMICS OF ALLOYING AND STEEL MAKING

In contrast with carbon, in normal conditions Nitrogen exists as a diatomic gas N_2, according to the Sievert's law the solubility of N in a liquid alloy obeys a square root relation of the N_2 gas pressure :

$$[X_N]_A = K(T)\sqrt{P_{N^2}}\ \gamma_N^{-1}$$

$[X_N]_A$: concentration in the alloy A. γ_N : activity coefficient of N in A.

P_{N^2} : pressure of the N_2 gas phase. K(T) : constant of equilibrium of the reaction $\frac{1}{2} N_2 \leftrightarrows [N]_A$

The chemical potential μ_N of the alloy is generally high in comparison with its value in the gas phase, therefore in order to increase the solubility of nitrogen in the alloy two routes are explored:

- decrease μ_N (and γ_N) in the alloy by suitable elements such as Cr and Mn.
- increase of the N_2 gas pressure.

The variation γ_N as a function of the j alloying element concentrations is given by [18] :

$$L_n \gamma_N = \sum_j \varepsilon_N^j x_j + \text{second order terms}$$

ε_N^j are interaction parameters of j elements on N. At 1600°C the following order is observed :

$$\varepsilon_N^B \rangle \varepsilon_N^{Si} \rangle \varepsilon_N^{Co} \rangle \varepsilon_N^{Ni} \rangle \varepsilon_N^{Mo} \rangle \varepsilon_N^{Mn} \rangle \varepsilon_N^{Ta} \rangle \varepsilon_N^{Cr} \rangle \varepsilon_N^{Nb} \rangle \varepsilon_N^{Y} \rangle \varepsilon_N^{Ti}$$

As shown by the former series the stability of transition metal nitrides and the solubility of nitrogen are decreasing when the density of 3d and 4d electron increases. This observation is consistent with the studies devoted to nitrides, carbides and interstitial compounds [19].

When elements such as Ti, Zr or V, Nb which present a strong affinity for nitrogen are added to the steel the dramatic increase of solubility they induce at low concentration stops as soon as the corresponding very stable nitrides appear *(Fig. 1)*. Therefore in order to increase the solubility of N in steel the main alloying elements are Cr, Mn and Mo. In contrast with these elements, Ni, Co and Cu are reducing solubility as well as to a larger extent Al, Si, C and P [10, 18]. As shown by *figure 2*, under atmospheric pressure (0.1 MPa) the solubility of N dramatically varies with the composition (more than 1 wt - % in FCC γ phase of 13% Cr – Mn steel and less than 0.05 wt - % in F.C.C. γ iron), with the liquid or solid state, with the crystallographic structure C.C. α, δ or C.F.C. γ, and with the temperature.

The influence of Cr, Mn, Mo, Ni and Co on solubility of nitrogen lead to the development of many new stainless steels as well as Chromium, Nickel and Cobalt high alloys [19]. Because of the important variation of the nitrogen solubility with X_i, T and P thermodynamics calculations

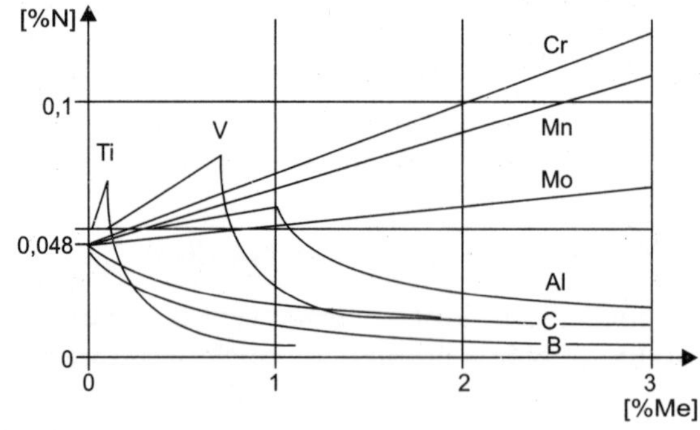

Figure 1 Solubility of nitrogen (mass %) versus alloying element concentration (mass %) in iron based alloy at 1600°C. Cr, Mn, Mo increase the solubility, once the formation of Ti and V nitrides starts the increase of solubility induced by Ti and V ceases.

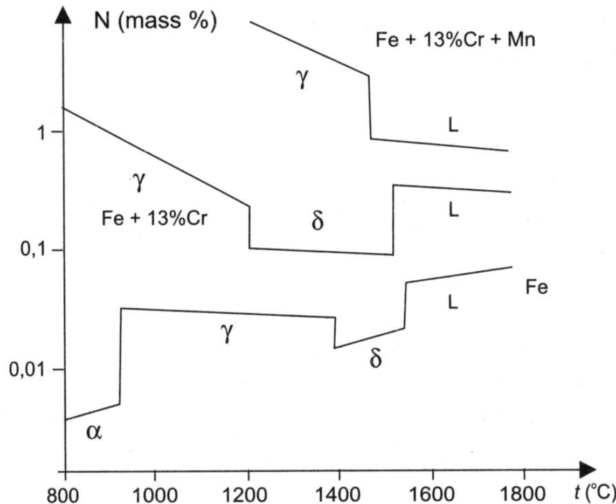

Figure 2 Nitrogen solubility under 0.1 MPa Pressure versus temperature for pure iron, Fe-Cr and Fr, Cr, Mn alloys – α, δ ferrite – γ austenite – L liquid.

of the stability domains and phase diagrams are necessary to optimize the composition of alloys.

In order to introduce N in the chemical potential of x in the source μ_N^S has to be higher than that in the alloy target μ_N^A:

$$\mu_N^S > \mu_N^A$$

Once the aimed content obtained, the ulterior treatments: solidification, cooling, thermal treatments, welding, etc. do not have to originate any solubility well likely to reduce the concentration of the alloy or to induce concentration gradient, segregation and in the worst-case N_2 bubbles and porosity. This is specially the risk when the B.C.C. α or δ phases appear during peritectic solidification or during high temperature formation of ferrite.

From the principles briefly quoted above, different industrial and experimental processes are resulting, of which the main features are summarized on *table 1*.

Clearly the most widely exploited industrial process is the "N.O.D." which is based on classical steel making with a suitable alloying composition. Pressure Electro Slag Remelting (P.E.S.R) is an elegant and efficient process which succeeds to produce high performance H.N.S. such as P900 utilised to manufacture retaining rings for electromechanical rotors [21], meanwhile P.E.S.R. demands special technological care and in many cases does not seem cost effective. High Pressure N_2 gas steel making may offer an alternative to P.E.S.R. from the point of view of industrial investments [12].

As shown by *figure 2* the solubility of N in γ austenite is increasing when T is decreasing, this property and the high diffusivity of interstitial N constitute the main grounds for the success of high temperature solid steel nitriding [14] and of the powder metallurgy process [22, 76].

Table 1 Overview of the different routes of H.N.S. making

Process	Pressure MPa	Reaction	Nitrogen source	Remarks
Modified AOD "N.O.D."	~ 0.1	N_2 + Liquid	N_2	N % ~ 0.1 – 0.4 %
High Pressure	1-6	N_2 + Liquid	N_2	Risk of Bubble counter pressure
P.S.E.R.	2-4	Si_3N_4 + Liquid	Si_3N_4 + CrN.. Mn_4N	Pressure Electroslag remelting volume of liquid alloy
Plasma	< 0.1	N Plasma + Liquid	Nitrogen plasma	Homogeneity of the plasma
H.T. gas-nitriding	~ 0.1	N_2 + Solid alloy	N_2	Gradient materials high temperature surface treatment
NH_3 gas	< 0.1	N + Solid alloy	$NH_3 \leftrightarrows N + {}^3/_2\, H_2$	Diffusion controlled reaction T < 600°C – Depth of nitriding
Powder Metallurgy and HIP	0.1 -	CrN, Mn_4N + Stainless steel diffusion	Nitrides in metastable conditions	Possibility to design very different new alloys
Mechanical Alloying + PM	~ 0.1 MPa + $\sigma_{ii} \sim Y_S$	Driven diffusion	Nitride	Small amount of new product, new alloys

Because the corrosion properties resulting from the presence of nitrogen in steel do not only concern H.N.S. but also the nitrided coating of low alloyed steel some remarks concerning the Fe-N diagram are necessary *(Figure 3)*. This very unusual phase diagram does not correspond to the equilibrium of a given amount of iron with a given quantity of nitrogen but to the Fe-N solid solutions and nitrides obtained under given chemical potentials of μ_N. The variation of μ_N in a large domain can be obtained by increasing dramatically the gas pressure in a reaction $Fe + X/2\, N_2 \rightarrow FeN_x$ till some thousands of atmospheres, but in the zone of interest (T < 900 K, x < 0.5) the variation of μ_N is easier to control through the dissociation of NH_3, for which μ_N is determined by $(P_{NH3})\,(P_{H2})^{-3/2}$. This reaction is exploited in industry for gas surface nitriding and with $NH_3 - CH_4$ or NH_3 – CO mixtures for carbonitriding. These C V D processes are competing with P V D surface treatments such as plasma and or ion nitriding and carbonitriding.

For the production of large amounts of high nitrogen alloys the increase of the chemical potential of N in the plasma has been explored carefully [23] meanwhile it appeared that some fluctuations of the plasma and of T may create some heterogeneities and segregations. Compared to powder metallurgy (PM) techniques, mechanical alloying (MA) is likely to offer new possibilities to achieve alloys inaccessible by other routes. It has been shown that under impact conservative equations such as $(1 - x)\, Fe + x\, Fe\, N_y \rightarrow Fe\, N_{xy}$ are obeyed [24] and that during milling, diffusion presents features leading to a specific atomic mechanism of "drift diffusion" for which extended defects dislocations, grain and phase boundaries play a major

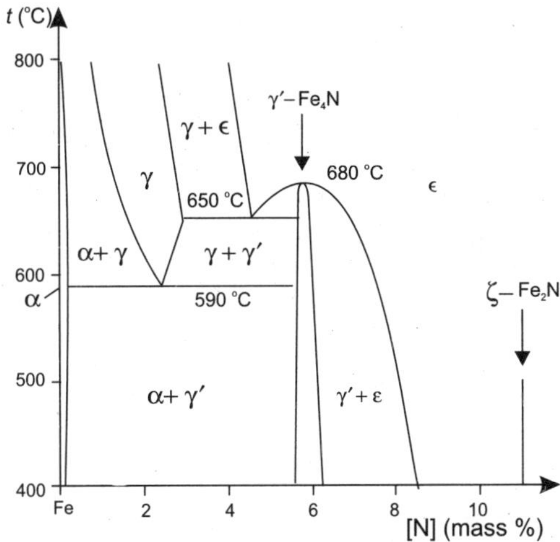

Figure 3 Low temperature part of the Fe-N Phase diagram obtained under controlled chemical potential of Nitrogen.

role [25]. In addition to their technological interest samples obtained by a "MA + PM" route can be utilized to precise the role of N in the corrosion phenomena.

All processing techniques which have been discussed above in the frame of steel and specially of stainless steel are likely to be transferred to alloys in which the major element is not iron. This is specially the case of Cr, Mn, Co, Ni and Cu based alloys as shown by Ni based super alloy. The aim is then to take advantage of all the improvements resulting from interstitial alloying in the mechanical and corrosion properties. Although carbon alloying does not always appear beneficial, it has been shown that the association C + N can suppress some specific drawbacks resulting from carbon alloying.

III. MICROSTRUCTURE OF H.N.S.

3.1 Impact of Microstructure on Properties

Microstructure of metallic alloys (crystallographic structure + phase distribution + grain size + texture + etc.) plays a major role on mechanical and corrosion properties. Microstructure does not only results from the chemical composition of the alloy but also from the complete thermomechanical history which includes all the thermal, thermochemical, thermomechanical steps of alloy making and fabrication. The variability of possible microstructures of H.N.S. is boundless, therefore this paragraph only aims to identify the main features of microstructure evolution related with Nitrogen alloying. Although many different alloys and steels are likely to be developed, the major industrial achievements concern stainless steels and mainly wrought rather than cast ones. Meanwhile in the case of cast stainless steel it is generally considered that a reasonable criterion lies in the carbon content [17]. The carbon content of corrosion resistant casting steel is usually aimed to be lower than that of heat-resisting steels of similar composition. Cast H.N.S. obtained under atmospheric pressure, for which the high Cr + Mn content leads to about 0.25 % N and avoids any loss, are not likely to suffer from carbon induced sensitization, Nitrogen alloying is beneficial.

Because Stainless Steels (S.S.) contain more than 12% Cr the solubility of N is appreciable and most of H.N.S. are developed from classical S.S. grade quoted in textbooks [16-17]. In case of grades such as the 304 (18-20 Cr, 8-12 Ni, 2 Mn, 0.08 C) and 316 (16-18 Cr, 10-14 Ni, 2 Mn, 2-3 Mo, 0.08 C) they become 304 N and 316 N with content between 0.1 – 0.16. A decrease of the nickel concentration of about twenty times the introduced N wt - % is then possible without changing the nature of the steel. The dramatic difference between the cost of N and Ni constitutes a very convincing argument for stainless steel industry to replace totally or partially Ni by nitrogen or by a combination N + Mn. High Manganese, High Nitrogen Stainless Steel have been developed in different countries such as India and U.S.A. Fabrication of these grades may appear more delicate than classical nickel grades but continuous R&D progress should lead to an increasing replacement of Nickel S.S. by Nitrogen – Manganese S.S.

3.2 Microstructure Inherited from High Temperature Solution Treatment

Depending on the cooling rate from the liquid state to the temperature of solution treatment and of the concentration X of γ former (Ni, Mn, C, N, etc.) and α-former (Cr, Mo, Si, etc.) elements the high temperature structure of H.N.S. can be either γ austenitic (F.C.C. lattice) or α ferritic (B.C.C.) or austeno-ferritic duplex ($\alpha + \gamma$). After cooling or quenching, when the concentration of γ former elements is not high enough to stabilize the γ phase, α' martensite forms and gives rise to martensitic steels when the steel was purely austenitic at high temperature (about 1050°C) or to dual steel when the steel was duplex before quenching. Although both steels of martensitic and dual type contain martensite and retained austenite, they strongly differ by the presence of ferrite in dual phase and by their morphology.

The M_S temperature which corresponds to the temperature at which martensitic transformation starts, is adjusted to a linear function of the composition X, is of the following form:

$$M_S\,(X)\ °C = (500;\,1300) - (30;\,60)\ \text{wt}\ \%\ \text{Ni} - (12;\,75)\ \text{wt\%}\ \text{Cr}.$$

in which the bracketed interval results from the heterogeneity of the different regression analyses already published and which refer to different steel grades. The large interval of variation of the coefficients in equation $M_S(X)$ may induce a too negative and therefore erroneous image of this type of formula which results from empirical regression analyses valid for well defined and limited domains of chemical composition. In contrast with the $T_0^{\gamma-\alpha'}$ temperature which corresponds to the intersection of the $G^{\gamma}(T)$ and $G^{\alpha'}(T)$ free enthalpies of the γ and α' phases and which constitutes the necessary condition: $T < T_0^{\gamma-\alpha'}$ for α' martensite to appear, the $M_S(X)$ temperature is not only determined by thermodynamics but also by mechanical and kinetical parameters. This makes the modelling of $M_S(X)$ uneasy. In the first approach, the main features of the microstructure can be predicted from the knowledge of the phase equilibrium i.e. of the phase diagram at high temperature corresponding to solution treatment and homogeneization. This is obtained due to computational thermodynamics codes (such as thermocalc) which have proven their efficiency for the design of new alloys [26] and in the frame of a more global and qualitative approach with the Schaeffler diagram.

The Schaeffler diagram, *figure 4*, was established to provide an approximative estimate of the structure resulting from welding it also appears to be a very useful tool to predict the

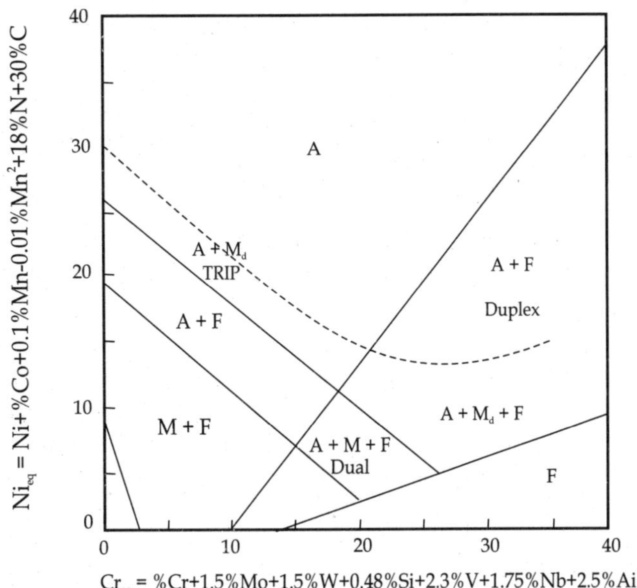

Figure 4 Schaeffler diagram for estimating the structure of steel from Ni_{eq} and Cr_{eq}. A : austenite – MC : martensite from cooling – M_d Martensite from deformation – F : ferrite.

structure of stainless steel as a function of the Chromium and Nickel equivalent given by semi empirical expression such as [9, 16]:

$$\text{Ni equi.} = \%\,\text{Ni} + \%\,\text{Co} + 0.1\,\%\,\text{Mn} + 18\,\%\,\text{N} + 30\,\%\,\text{C}$$

and

$$\text{Cr equi.} = \%\,\text{Cr} + 1.5\,(\%\,\text{Mo} + \%\,\text{W}) + 2.3\,\text{V} + 0.48\,\text{Si} + 2.3\,\%\,\text{V} + 2.5\,\%\,\text{Al}$$

Despite the large uncertainties on the borders of the different domains, *figure 4* presents the principal types of H.N.S., *1)* Austenite; *2)* Martensite; *3)* Ferrite; *4)* Duplex (Austenite + Ferrite). Meanwhile the different domains show very different features according to the composition (Ni or Mn rich) to the stability of the austenite (T.R.I.P. steel) to the thermal treatments (strain ageing, marageing, etc.) and are likely to be "splitted" into several "sub" grades.

T.R.I.P. steels are characterized by an additional contribution to plastic deformation provided by the martensitic transformation which is induced by the combined influence of stress and strain resulting from mechanical loading below M_d, additions of carbon and nitrogen lowers M_s and M_d of approximatively :

$$\Delta M_d{}^\circ C \sim (450 \pm 50)\,(C + N)$$

Classical T.R.I.P. steel are obtained with Al or Si additions. In order to underline the possibility to design T.R.I.P. and Dual alloyed steels, a qualitative modification of the Schaeffler diagram showing the corresponding domains is given (broken lines on *figure 4*).

3.3 Atomic Redistribution and Precipitation Resulting from Thermal Treatment

From the end of the solidification to the final stage of fabrication, the steel is submitted to different thermal and thermomechanical treatments : homogenisation, solutionizing, metal forming, rolling, forging, continuous cooling, quenching, ageing etc. The aim is to optimize the microstructure. Meanwhile the resulting microstructure rather corresponds to a compromise between different properties than to an absolute optimum for all properties. Besides changes of the morphology (texture, grain size) of the single or multiphase H.N.S. a redistribution of atomic species (ordering, clustering, preprecipitation, precipitation) is taking place during all above quoted processes. Let us examine how does Nitrogen modify the atomic redistributions phenomena in the different types of alloys.

3.3.1 The case of low interstitial concentration steels.

When C and N concentrations are absent from the stainless steels matrices thermal treatments lead to redistributions of substitutional atoms and mainly to the precipitation of three intermetallic phases : σ, χ and η [16, 17].

The composition of the σ, χ and η phase may vary in a very important range depending on the composition of the steel, the crystallographic structure and the thermal treatment, therefore σ, χ and η correspond to families of intermetallic compounds which are very widely documented [28]. In all cases, the concentration of Cr column transition elements Cr, Mo, W is high. These elements as well as those with low d electron densities (Ti, Zr, V, Nb) are referred as "A" elements. At the right of Mn, the d electrons rich transition elements, Fe, Co, Ni are "B" elements, Mn is considered to belong, either to A or to B elements. The composition of the intermetallic compounds can be described by the generic formula $A_a B_b$, meanwhile additional complications may result from the role of minor alloying elements which may accelerate (Si) or delay (Al) the precipitation of σ and from the large number of non equivalent crystallographic positions in the unit cell.

In order to keep the precipitation of σ and χ under control, isothermal and continuous cooling transformation diagrams (T.T.T. and C.C.T.) have been established. Meanwhile the corresponding diagrams devoted to mechanical or corrosion properties resulting from the corresponding thermal treatment may be shifted from the T.T.T. and C.C.T. curve because of the influence of the parameters of the microstructure which are not taken into account by the transformation diagram.

Despite the large difference of diffusion rate in ferritic and austenitic matrices the temperature domain of formation of σ and χ do not greatly differ (550– 900°C), meanwhile kinetics are much faster in ferritic and martensitic steels than in austenitic ones. In most cases, in alloys containing more than 12% Cr σ formation is promoted by Mo, Si, V, W, Ti, Nb and χ specially by Mo and W. In contrast with these elements, C and N as well as Ni at high concentration retard σ formation. The necessary condition for σ or χ to form during thermal treatment is the existence of a driving force ΔG for precipitation. The determination of this driving force is based on thermodynamical calculation and computer codes such as Thermocalc which have been used successfully [29, 30].

Although the thermodynamical calculations and the corresponding computer codes demonstrated their power, the Achilles heel must not be hidden: the success of the approach results on experimental results and data which are sometimes missing or unprecise, the complexity and variability of the intermetallic phases render the determination of corresponding stability and energy difficult. In addition, data related with nucleation, growth, kinetics are even more difficult to determine. For all these reasons the necessary computerized thermodynamical approach demand complementary experiments.

An additional phenomenon which mainly affects the mechanical properties of ferritic stainless steel is the 475°C embrittlement which results from the spinodal decomposition of chromium alloyed steels, the influence of nitrogen on this phenomenon is not yet unambiguously established.

To a large extent the understanding of the role of intermetallic compounds in alloys results from the studies on chemical bonding, crystallography and properties of these phases.

The σ phase belongs to topologically close packed structures [28], its structure is complex tetragonal with 30 atoms per cell. This phase may form by a peritectic reaction (Co-V) or by precipitation (Fe-Cr, Fe-V) depending on the alloy system. It can be stable at room temperature or undergo eutectoid decomposition to α B.C.C. and χ intermetallic compound (Nb-Re). An important feature of σ is its strong resemblance to a close packed hexagonal one, meanwhile the displacements of atoms from the H.C.P. positions originate five different sites corresponding to the crystallographic positions referred as a, f, ic, id and j of the space group $P4_{2}$, mnm $- D_{4h}^{14}$. The corresponding number of atoms occupying a, f, ic, id and j are 2, 4, 8, 8 and 8 [28].

Atomic size factor and electronic factors determine preferential occupation of sites 'a' and 'id' by B-like elements (Fe, Co, Ni) and of 'f' sites by A elements (V, Cr, Mo) ; 'ic' as well as 'j' shows a less clear type of occupation which becomes "mixed". A model of electronic structure predicted a high electron state density at the Fermi level when the 'd' band is filled up to more than one-third [28]. Among so called topologically close packed structures, the crystallography of σ does not appear as the most complex and other intermetallic such as α-Mn type, P, R, μ, Δ etc. are closely related to sigma phase [28].

The α-Mn type χ phase is frequently observed in thermally treated transition metal alloys, depending on temperature and on alloy composition, χ forms before, after or simultaneously with σ. The structure of χ is cubic ($a_x \sim 0.9$ nm), the unit cell contains 58 atoms distributed on four sublattices a(2), c(8), g_3(24), g_4(24). [19], typical compositions observed are $Fe_{28}Cr_{23}W_7$ and $Fe_{36}Cr_{12}Mo_{10}$.

The covalent and ionic contributions of chemical bonding which characterize the intermetallic compounds frequently lead to well defined stoicheometry. Many compounds at the A_2B composition present the structure in $MgCu_2$ (Cubic) or $MgZn_2$ (hexagonal) or $NiZn_2$ (hexagonal) and are referred as Laves phase [28]. Mo_2Fe which plays a key role in the properties of Marageing steel belongs to the family of Laves phases, its structure is that of $MgZn_2$ [16].

The other A_2B type intermetallic compound with a Ti_2Ni cubic structure, which precipitates during thermal treatment of stainless steel, is referred as η phase sometimes as Laves phase and more frequently as η carbide, because C and N greatly modify the stability of its structure to be discussed in connection with the role of C and N on precipitation.

From the point of view of applications, the relation between properties and heat treatment is essential, meanwhile properties are largely determined by microstructure which itself is a consequence of heat treatments. In Interstitial Free (I.F.) stainless steel the presence of intermetallic phases mainly σ and χ deteriorate mechanical properties (loss of ductility, reduced toughness, embrittlement, increase of ductile brittle transition temperature, etc.) as well as corrosion resistance (pitting corrosion, sensitization, corrosion rate, intergranular corrosion, high temperature oxidation, etc.). On the one hand, the deleterious effect results from the intrinsic properties of σ and χ (brittle, hard, low corrosion properties), and on the other hand, from the morphological characteristics of the precipitates (amount, distribution, size, factor of form) which originate gradients in the stress field (stress intensity factor) and in the electrochemical potential (fluctuation of the Cr concentration). Because the crystallographic structure of σ and χ largely differs from the simple B.C.C. and F.C.C. ones the phase boundaries are incoherent and the interface energy is high. The resulting energy barrier acts against homogeneous nucleation, when the driving force for precipitation is moderate, interphase, intergrain, twin nucleation prevails.

3.4 Role of Interstitial Alloying (C, N) on Atom Distribution and Precipitation phenomena

In order to understand the impact of C and N interstitials on the ordering phenomena, pre-precipitation and precipitation in steel, it is necessary to comment on the atomic distribution taking place in binary Fe-C and Fe-N phase and to examine points such as: – the importance of C-N interaction – some feature related to the stability of binary carbides and nitrides versus the d electrons density – the solubility of C and N in the intermetallic compounds.

Altough binary Fe-C and Fe-N ferrite, austenite and martensite are isomorphous, the ordering phenomena resulting from ageing and tempering strongly differ :

- in $FeC_{0.09}$ α' martensite ageing (200°C) creates clustering of C atoms, then formation of intermediate FeC_x ε-like carbide ($x \sim 0.5$) and cementite Fe_3C,
- in $FeN_{0.1}$ α' martensite ageing (160°C) is characterized by N interstitial ordering resulting in the formation of α'' $Fe_{16}N_2$ nitride which can be considered as a double step ordering of N in a Fe B.C.C. lattice : the final nitride is Fe_4N γ' [31, 32],
- in Fe-C as well as Fe-N α ferrite, α' martensite, γ austenite, interstitials C or N occupy octahedral interstitial sites,
- in Iron, Manganese and Chromium carbides of the typical stoichiometries M_3C, M_7C_3, $M_{23}C_6$, etc. C is inside a trigonal prism, the structure of these carbides is described by stackings of trigonal prism sheets. The trigonal prism appears to be specific of these carbides and does not seem to be observed in nitrides. Attempts were made to introduce sensible amount of N in Fe_3C cementite failed [33],
- in the nitrides observed in the Fe-N system under controlled chemical potential of N: γ', ε and ξ. $\gamma' Fe_4N$ shows a perowskite – like cubic structure, ε FeN_x an hexagonal structure and a large domain of solubility $0.17 < x \leq 0.5$ and the orthorhombic structure of Fe_2N ξ results from the ordering of the occupied sites. For all these nitrides the interstitial sites are the octahedral ones,
- interesting to observe that the stable ε nitride and the metastable ε carbide are isomorphous, ε $Fe(NC)_x$ is obtained by carbonitriding. This suggests that solubility of carbon in nitrides is usually not vanishing although very low in γ Fe_4N in contrast with the vanishing solubility of Nitrogen in carbides such as cementite which is characterized by the occupation of trigonal prism interstitial sites,
- in the interaction coefficients of N and C which lead to a reciprocal increase of chemical potentials μ_C and μ_N giving rise to changes of the electron density at the Fermi level which produce optimized mechanical properties of some C + N interstitial alloyed steel grades [14],
- as a general rule the stability of carbides and nitrides diminishes with the filling of the 3d, 4d and 5d layers, large for Ti, Zr, Hf, moderate for Mn and Fe very low for Co and Ni [19].

Experimental results on phenomena which take place during ageing and thermal treatment of stainless steel show that Nitrogen alloying modifies the nature and the kinetics of formation of precipitates. Schematically three cases : *i)* precipitation of nitrides is dominant ; *ii)* Nitrogen inhibits the formation of precipitates ; *iii)* Nitrogen and carbon interact or interfere.

- At high temperature, above 600°C, nitrogen alloying in austenitic stainless steel promotes the precipitation of the hexagonal Cr_2N nitride. This process is faster with higher nitrogen contents, typically at 800°C it starts after about 1 s for 0.6 % wt N, and 400 s for 0.15 % wt N. The nucleation is made easy by the supersaturation and therefore by the thermodynamical driving force and very likely by a moderate interface energy between nitrides and the matrix in consistency with the corresponding structural coherency. Although reliable data about interface energies are still missing it is reasonable to suggest that this factor is responsible of the moderate coalescence and ripening of nitrides precipitates in comparison with intermetallic compounds and carbides. Meanwhile at high temperature (about 800°C) the driving force for homogeneous nucleation of Cr_2 N is too low and precipitation starts heterogeneously at grain boundaries. The growth develops according to a cellular, also called discontinuous, process which leads to large domains of "eutectoid-like" morphology (layers of Cr_2 N and d) which deteriorate the mechanical resistance [34].
- In contrast with carbon alloying, nitrogen alloying of ferrite and martensitic steels leads to much finer and homogeneous distribution of precipitates. Mixed Fe-Me enriched zone Me = (Ti, Nb, V, Cr) zones transform into small and coherent precipitates of hexagonal and cubic structure coherent or semi coherent with the matrix. This process is consistent with the ordering phenomena leading to the formation of $Fe_{16}N_2$ α'' during tempering of α' Fe-Nx martensite [31, 32, 33, 35].
- When the solubility of nitrogen in an intermetallic phase is low, because of structural "incompatibilities", alloying is likely to inhibit or even to suppress the precipitation of this intermetallic. This case is experimentally observed with the σ phase and to a lesser extent with χ.

The difference between the σ intermetallic compound which dissolves no carbon and χ arises from the χ–lattice interstices likely to accommodate carbon atoms. The (stoichiometry) of χ carbide may be written as $Me_{18}C$ ($Fe_8Cr_{6.3}W_{3.7}C$) [19], and the possibility for χ to dissolve also nitrogen leads to form a carbonitride according to Evans and Jack [36].

The role of nitrogen on the precipitation of σ and χ in steel can not be completely attributed to its solubility in σ and χ, the necessary condition for the precipitation to occur is a positive driving force : $\Delta G = G_i - G_f$ large enough to overcome the nucleation barrier, G_i and G_f being the free enthalpies before and after the precipitation (i = initial ; f = final). Therefore, the determination of the variation of ΔG versus the concentration has been achieved by Hertzman [30] for Cr20 Ni18 Mo6 Nx and by Jargelius-Patterson in Cr20 Ni22 Mo5 Mn5 Nx type steels [29]. Both the authors demonstrated that $\frac{\partial \Delta G}{\partial X_N}$ is negative and therefore nitrogen reduces the precipitation of σ and χ, meanwhile because the decrease of the driving force for the precipitation of σ is larger than for χ, the induced reduction of precipitation is more favourable on σ than on χ.

Depending on the precipitation sequences the α-Mn type χ phase can be replaced by the β-Mn type Π phase for which the stabilization by addition of carbon is necessary, the β-Mn cubic unit cell is built from 20 metal atoms, the lattice parameter is about 0.64 nm and the stoichiometry $M_{20}C$ [19].

Tempering of Fe-C and Fe-N α martensites demonstrated that C and N behave differently in the steel matrix, promoting either clustering or ordering [32-14]. This difference resulting from the different 2p electron number neither excludes nor forbids the substitution of one interstitial by the other depending of the phase which is considered, therefore the simultaneous presence of N and C in steel affects the precipitation sequence. Depending on the ageing temperature 500-700°C and on the alloying elements fine precipitation of hexagonal Me_2(N-C) and cubic Me(N-C) have been observed in martensitic steels [37, 38]. The comparison of secondary hardening in martensitic steels containing similar total interstitial concentrations 0.6 wt % C; 0.6 wt % N and 0.3 wt % C + 0.3 wt % N has shown that maximum hardening observed near 50°C in Fe-C martensite and near 500°C in Fe-N martensite transforms into a quasi-plateau from 100°C till 500°C and therefore that the mixed interstitial hardening stabilizes the fine distribution of interstitial precipitates [14](*Fig. 5*).

The role of nitrogen on the sequence of precipitation is determined by its influence on : - thermodynamical factors such as the G functions of the different phases and therefore the driving forces, interface energies, gradients of chemical potential, etc. - kinetical factors like diffusion, nucleation rates – and structural characteristics. All these factors are interdependant. The calculation of free enthalpies of quaternary phases is reliable as long as the data concerning these phases are correct, the knowledge of factors controlling kinetics are most often obtained due to indirect methods based on the analysis of experimental data likely to be implemented by atom-scale computer simulations [39, 40]. The crystal chemistry of phases likely to precipitate and whose structure is a center of competition between C and N provides the most concrete

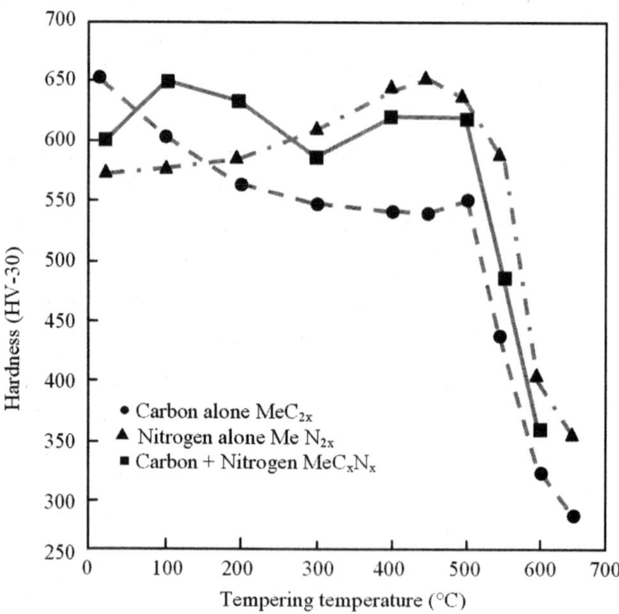

Figure 5 Tempering treatment of martensitic stainless steel containing the same concentration 2 x of interstitial atoms either C or N or C + N. Schematic view of the hardness evolution (after 14).

and comprehensive explanations. Beside the (FeMe) $(NC)_x$ α, γ, α′ and ε phases the case of η offers some enlightenment.

The η cubic structure, lattice parameter of the unit cell of about 1.1 nm, 96 metals atoms and 16 interstitial sites in the cell, can be described on the basis of the distribution of octahedrons and tetrahedrons on two cubic sub-units (a) and (b) as shown on *figure 6*. In order to satisfy the cubic symmetry the positions of (a) and (b) sub-units is analogous to the distribution of α Fe and β FeAl sub cells in ordered Fe_3Al *(Figure 6)*. As shown by *figure 6* the interstitial site in the η structure is the middle of the segment joining the centres of the closest metal octaedra. The Ni_2 Ti type η intermetallic is stabilized by C or N interstitials and is therefore called η carbide, resp. nitride and carbonitride. The generic formula is $(Me)_6$ (CN) and is built from two metals belonging to different transition lines *(Figure 6)* for which the d electron number differs. The interaction between the different metal atoms leads to ordering phenomena characterized by two variants η_1 W_3Co_3 C and $\eta_2W_4Co_2$ C. The reference to the mixed tungsten-cobalt carbides results from the importance of these phases in the cutting tool industry.

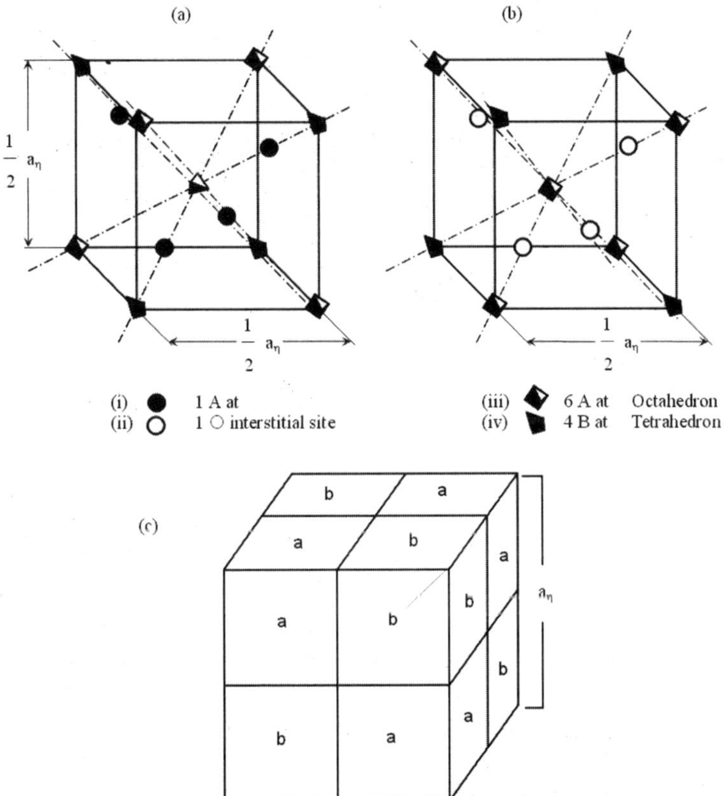

Figure 6 η carbide, A atoms such as Ti, Cr, Mo, W form octahedrons which are centred on 4 of the 8 summits of two subcells (a) and (b). B atoms such as Ni, Co, Fe form tetrahedrons which are centred in the "reciprocal" summits of the sub-cells. In sub-cell (a) the center is occupied by a tetrahedron whereas, in (b) by an octahedron. The figure (c) shows the ordered packing of (a) and (b) in the Ni Ti_2-like structure of η carbide after Mueller and Knott, (Transactions of the Metallurgical Society, Vol. 227, (1963), p. 674-8.

It is worth to remark that the octahedral interstitial site of the η structure are built from the metal atoms Ti, V, Nb, Mo, Ta, W for which stable carbides and nitrides exist. Isothermal treatments between 600 and 1050°C of $Cr_{17}Ni_{13}Mo_5$ austenitic steels containing Carbon or Nitrogen have shown that χ and $M_{23}C_6$ phases precede the formation of η. Without carbon (and without surprise) $M_{23}C_6$ does not form nor any isomorphous nitride, the first to precipitate is the α-Mn type χ phase. In consistency with the low structural compatibility of N with the χ structure the influence of nitrogen is a monotoneous retardation of χ precipitation *(figure 7)*. In contrast with χ but in agreement with the stabilizing role of nitrogen on η, the high temperature precipitation of η is eased by high nitrogen concentration. Meanwhile at about 600°C the influence of nitrogen appeared to be related with kinetics and consistent with a nitrogen induced lowering of Cr diffusion. This role of nitrogen on the diffusivity of chromium is suggested by the analysis of creep behaviour by Mathew and Srinivasan [41].

Nitrogen also influences the formation and the mobility of lattice defects [40] (vacancies, self interstitials dislocations, grain and phase boundaries, etc.) therefore the morphology (size, shape, texture, etc.) of heavily deformed alloys as well as the recovery, recrystallisation, phenomena [42] which originate important modifications of the final properties of the steel. Up to now a generic and complete explanation of the involved mechanisms is not achieved even when the observed changes clearly appear as shown by the transformation of "bamboo" into "pearl" morphologies observed in heavily deformed and recrystallized duplex H.N.S.. [56]

In classical carbon steel, intermediate thermal treatments between the temperatures of perlitic and martensitic transformations lead to the bainitic structure for which the role of interstitial carbon is essential. Because the bainitic structure produces excellent properties the possibility to develop new bainitic nitrogen alloyed steels should be speculated.

IV. PROPERTIES OF H.N.S. AND H.N.A.- INFLUENCE OF NITROGEN ON MECHANICAL AND CORROSION RESISTANCES

Interactions of atoms, lattice, defects, etc. with interstitials are generally important, therefore interstitial alloys, solid solutions, carbides, nitrides, hydrides present a broad spectrum of interesting properties concerning: electric conductivity, superconductivity, thermal conductivity, thermoelectric power, magnetism, high temperature resistance, thermal expansion, energy storage, electrochemical properties, mechanical resistance. An exhaustive analysis of the role of nitrogen on these properties is beyond the scope of this section which only aims to examine some points concerning the mechanical and corrosion properties of high alloyed steels.

Two extreme and correct descriptions of the influence of nitrogen on properties could be considered. According to the first very concise one : "there is no mechanism involved in mechanical and corrosion properties of H.N.S. likely to be independent of N alloying"; neither contradiction nor demonstration have been opposed to this postulate. The second description, based on the numerous experimental results, aims to analyse, quantify, discuss, interpret and compare the numerous experimental results, it corresponds to extended reviews such as those published in well documented references and monographs [1-15]. Only some results are discussed below, those which may exemplify the cases for which the role of nitrogen is the most dramatic.

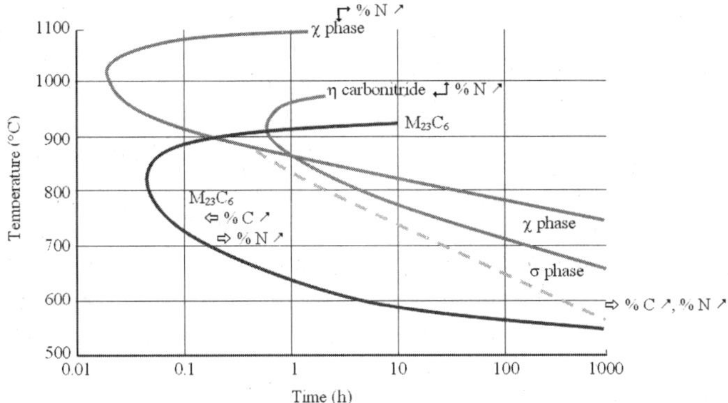

Figure 7 Schematic influence (arrows indicate the sense of displacement when % C or % N increases) of N and C alloying on the precipitation of 1) σ phase, 2) η carbonitride, 3) χ phase, 4) $Cr_{23}C_6$ – like $M_{23}C_6$ Carbide in which C interstitial occupies square antiprism. The precipitation of Cr_2N for which the maximum precipitation rate occurs around 800°C is not shown and increases with the Nitrogen content. The data refers to results obtained by Thier et al. (1967, 1969) on a $Cr_{17}Ni_{13}Mo_5$ steel base [14].

4.1 Mechanical Properties of H.N.S.

Generally, when the yield stress and the hardness increase, the ductility and toughness decrease, but in most cases Nitrogen alloying of stainless steel results in an improvement of the mechanical resistance, the yield stress and ultimate tensile stress increase, meanwhile ductility and toughness are maintained important. This beneficial effect is particularly interesting for austenitic stainless steel grades for which nitrogen alloying became usual as long as it concerns classical compositions and steel making processes. High nitrogen concentrations appeared beneficial for austenitic as well as for martensitic and duplex steels. That a nitrogen weight concentration which could be considered as moderate such as 0.2 – 0.4 % dramatically modifies the mechanical behaviour of steel is not astonishing : these concentrations correspond to interstitial site occupation rate over 1 % and to a proportion of metal atoms first nearest neighbour of N of about 10 % and sometimes more.

A striking example of the efficiency of the impact of nitrogen on mechanical properties is given by the steel P900 designed by G. Stein and co-workers [21, 43] Cr18Mn18Mo2N0.8 for which the yield strength is over 1500 MPa and the critical stress intensity factor likely to reach 200 MPa$\sqrt{m}$. This steel grade was dedicated to build non magnetic retaining rings of generators shafts submitted to very high centrifugal acceleration. Due to this P900 grade the disastrous crashes which were observed with previous materials have been avoided and unambiguous evidence is provided that H.N.S. and H.N.A. constitute a strategic family of "specialty metals" [44]. In case of H.N.S. such as P900 or similar for which the nitrogen concentration is about 0.6 – 0.8 wt % it is worth to underline that the mechanical resistance is not far from that of some maraging steel with the additional advantage of an excellent corrosion resistance of H.N.S. to be compared with the usual low one for maraging steels. Due to the large diversity of stainless steels and H.N.S., and to the relative lack of systematic data bank devoted to the mechanical resistance, the parameters and results concerning a well defined H.N.S. and well defined thermomechanical treatment and fabrication conditions have to be found in journals, papers reports documents of which only a part is referenced hereafter. Although the present chapter is not suitable to provide the numerous answers necessary to predict all the characteristics of the mechanical behaviour of a given H.N.S. it should help to identify which mechanisms are hit by nitrogen and how it could interfere with corrosion resistance.

Solid solution hardening resulting from Nitrogen has been demonstrated to be the highest among all alloying elements. For ferritic steel the slope of the variation of the yield stress σ_y versus the Nitrogen at % is about 600 MPa roughly 10 times more than with Si and 100 times more than with Mo [14, 16]. Although the hardening rate induced in austenite is smaller than in ferrite the interest for applications is higher because austenite is frequently too soft for many usages.

A shown by the *table* 2 the hardening of austenite induced by Nitrogen measured by the variation of the yield stress $\Delta\sigma_y$ is correlated with a "strengthening" effect i.e. an increase of the ultimate tensile stress $\Delta\sigma_M$.

Table 2 Hardening $\Delta\sigma_y$ and strengthening $\Delta\sigma_M$ induced by 1 at % of two interstitial N and C and two substitutional Cr and Nb alloying element in austenite and corresponding variation of the lattice parameter

Solute	$\Delta\sigma_y$ per at %	$\Delta\sigma_M$ per at %	$10^3 \frac{\Delta a}{a}$ per at %
N	124	213	2.4
C	77	116	2.2
Cr	3	116	0.7
Nb	67	132	-

In comparison with C the influence of N on $\Delta\sigma_y$ and $\Delta\sigma_M$ is much more important and can not be simply the result of a simple elastic interaction between dislocation and interstitials because the lattice expansion induced by C and N are not different enough. More over in the F.C.C. structure of austenite the deformation tensor around C and N is cubic and does not strongly interact with the stress field of screw dislocations, $\Delta\sigma_y$ and $\Delta\sigma_M$ result from many phenomena.

In the case of solute Nitrogen in austenite it has been shown that :

- The electronic structure of iron atoms nearest neighbours of foreign interstitial atoms noticeably depends on the nature of the interstitial C and N [44, 40]. In principle ab-initio calculations, optimization of interatomic potentials and molecular dynamics lead to determine the interaction energy of dislocation with interstitial as well as the dislocation structure. The contribution of the elastic misfit parameters $\eta_{ij} = \frac{1}{C_{ij}} \frac{dC_{ij}}{\partial x}$ (C_{ij} elastic constant) and size misfit $\delta = \frac{1}{a} \frac{da}{dx}$ results from interatomic energy and is therefore taken into account by the calculations. So far, the calculations brought partials answers about the influence of N on mechanical properties which are consistent with experimental evidences given by spectroscopic analyses: - the N-Fe bond appears less covalent than the Fe-C bond – this is agreeable with the "compatibility" of Nitrogen with the host metallic matrix already discussed above in relation with to the coherency of nitride precipitates.

- The interstitial interaction is clearly repulsive for Nitrogen in f.c.c. as well as b.c.c. matrix.
- Solute Nitrogen in austenite influences the stacking fault energy S.F.E.. Meanwhile neither the sign of the derivative of S.F.E. vs. the Nitrogen concentration x $\left(\frac{\partial SFE}{\partial x}\right)_T$ nor of the derivative vs. the temperature $\left(\frac{\partial SFE}{\partial T}\right)_x$ are constant. A possible correlation of S.F.E. with the electron state density at the Fermi level has been explored by SHANINA and GAVRILJUK [14].

Additional arguments concerning the influence of Nitrogen on S.F.E. in austenite are suggested by the formation of the hexagonal ε phase which is promoted by interstitials specially N and H. The stacking fault formation is favoured when the free energy difference of the ε and γ phase $|\Delta G_\gamma(x, T) - \Delta G_\varepsilon(x, T)|$ is reduced. Most often the S.F.E. in Nitrogen alloyed austenite is moderate about 40 mJ/m^2 or less [46]. Dislocation appears therefore dissociated, cross slip is impeded, and collective dislocations structure is planar. The planar morphology of dislocations structure was clearly evidenced by T.E.M. on 316 type alloys containing 0.08 – 0.25 wt % N submitted to Low Cycle Fatigue L.C.F. under total deformation amplitude form 0.6 % till 2.5 % [46-47]. Because the corresponding life time of the specimens submitted to L.C.F. tests increased when the Nitrogen concentration grows it was concluded that Nitrogen promoted the reversibility of the L.C.F. dislocation structure likely to delete the cell formation, consequently the vein – channel structure, the formation of persistant slip bands and intrusions-extrusions. Similar beneficial influence of alloying has been observed for high cycle fatigue tests [48]. The possibility for short range order to originate planar dislocations structure in nitrogen alloyed austenite has been explored by simulation [49], because the kinetics of destruction – reconstruction of interstitial short range order SRO by the passing through of successive dislocations is difficult to analyse this question should still be a matter of study. A possible link between SRO and hardening or softening of samples submitted to constant amplitude vibrations either in fatigue or internal friction has not been completely deciphered.

The modelling of solid solution hardening received numerous different expressions, from which a common generic form can be extracted : $\Delta\sigma_y = f^\alpha C^\beta \theta^{-\gamma}$.

$\Delta\sigma_y$ is the increase of the stress induced by the solute atoms for concentration C, which create pinning forces f on the dislocation line of which the tension is θ, α β and γ are positive exponents which, depending on the model chosen, slightly differ from 1. The previous relation makes sense for primary solid solution therefore only a power expansion limited to the linear term is utilised to plot the experimental results on a straight line of which the slope represent the solute hardening rate which increases with f and decreases with θ. As shown by the data on previous table the slope of $\Delta\sigma_y$ is about 120 MPa per 1 N at %, meanwhile depending on the Cr and Mo concentration the slope may show very large variations the higher the Cr and Mo concentration, the higher the hardening rate. This effect very likely originates from the formation of Cr-N or Mo-N or Cr-Mo-N clusters which constitute stronger obstacles "f " to the dislocation displacement than the isolated solute atoms.

The mechanical behaviour of mixed interstitial alloys such as stainless steel containing important and similar concentrations of C and N also show that the resulting hardening is not a simple additive phenomena and that the interactions between differents solutes originate synergistic effects [14].

In steel the non linear influence of mixed interstitial alloying N + C, N + H, C + H received some enlightment from the electronic structure either studied by calculation or experimentally tanks to spectroscopy and microscopy. The beneficial influence of mixed N + C alloying on strength and toughness appears to result from a more homogeneous deformation. Concerning the role of N on hydrogen embrittlement discussed above the most likely reason of its favourable action also seems to originate from the promotion of a less localized deformation modes and from an effect of nitrogen atoms on the diffusivity of H.

The previous paragraph devoted to precipitation underlined the difference between nitrides precipitates and others such as carbides and intermetallic compounds, very often small nitrides are coherent and form efficient and shearable obstacles to the displacement of dislocations. In contrast with large incoherent precipitates which are bypassed by dislocations after the formations of a belt of loops and important stress concentration, the coherent nitrides increase simultaneously the yield stress without decreasing so much the ductility and therefore the toughness. An additional advantage of H.N.S. results from the care necessary to the processing of alloying which leads to the elaboration of "clean" alloys with very low inclusion content.

Since Hall and Petch established well known relations between grain size and mechanical behaviour, the optimisation of the grains morphology and texture in order to optimize properties of metals and alloys constitutes a permanent concern for materials science and grain boundary engineering. Experimental results obtained on H.N.S. proves that nitrogen increases the efficiency of grain boundaries [50, 51] and also contributes to slowdown coarsening of the microstructure [52, 53].

The yield stress of polycrystalline H.N.S. versus the concentration x and the grain size d(mm) obeys the following relation :

$$\sigma_y(x, d) = \sigma_{yo} + \Delta\sigma_{yo}(x) + k_y(x)\, d^{-1/2}$$

σ_{yo} : yield stress of large grains alloy without N.

$\Delta\sigma_{yo}(x)$: solid solution hardening.

$k_y(x)$: Hall-Petch coefficient as a function of x at %.

From heterogeneous origin data it is possible to propose the following approximation:

$$k_y(x)\ \mathrm{MPa}\sqrt{\mathrm{m}} = 10 + 30\,x \text{ at \% N.}$$

Although the Hall-Petch – type relations are very well accepted and utilized the exact description of the mechanisms involved is still in question: role of dislocations pile-up, nucleation of dislocations from the boundary, passing through of dislocations, role of the boundary structure, role of grain boundary segregation? A subject of fascinating and fruitful controversies concerns the role of interstitial segregation studied by comparison of the effects of C and N. Experimental results and interpretations differ, the support of an interpretation based on dislocation ejection from the grain boundary proposed by BATA and PERELOMA resulted from similar segregation of C and N on the grain boundary [54], when GAVRILJUK demonstrated that affinity of N for grain boundaries is weaker than that of Carbon and that the binding energy of N with dislocations is clearly higher than that of C [55]. Therefore it was concluded that the stress transferred from one grain to an other likely to activate or unpin dislocation displacement was determined by the bonding energy of dislocation greater with N than with C.

The arguments in favour of the role of the strong - dislocation interaction are also consistent with the induced increase of the work hardening rate leading to very high σ_M and with a lowering of the recovery and recrystallisation kinetics [52-53]. In order to take advantage of a "kinetically" stable refinement of the microstructure high duplex steel was studied before and after rolling as well as the corresponding influence of thermal treatment [56]. The variation of the concentration offers means to modify the relative role of austenite versus ferrite i.e. matrix – inclusion, relative harness of both phases, nucleation of deformation in one or the other phase, ... From the point of view of fatigue, toughness, hardness, a broad optimum of concentration 0.15 – 0.3 wt %, of duplex steel with similar phase partition fraction was determined and lead to fine annealed microstructure showing specific bamboo and pearl microstructure [42, 56].

The hardening phenomena discussed above mainly concern the mechanical behaviour in the broad domain of intermediate temperatures. The stress barrier to overcome (back stress σ_B) results from the long range stress field originated from the "trees of the dislocations forest" and the long range distribution of obstacles, it corresponds to the so called athermal component of the strength. At low temperature short range interactions have to be overcome by thermal activation and originate σ_T a strongly temperature dependant contribution to the strengthening : the thermal component. Therefore the flow stress can be written :

$$\sigma = \sigma_B + \sigma_T$$

The contribution of applied stress to overcome the enthalpy barrier ΔH is assumed to be the product of the effective stress $\sigma - \sigma_B$ by an activation volume "b.a." of the order of the Burgers vector "b" multiplied by the increment of area swept by the dislocation jump "a".

According to SEEGER [57] the deformation rate can be written :

$$\dot{\varepsilon} = \dot{\varepsilon}_0 \exp - \frac{\Delta H - (\sigma - \sigma_B)ba}{kT}$$

$\dot{\varepsilon}_0$ is the maximum deformation rate observed above a critical temperature T_c.

Below T_c thermal activation is necessary for dislocation to pass obstacles and $\sigma(T)$ is given by the following expression :

$$\sigma = \sigma_B + \frac{1}{ba}\left(\Delta H + kT \operatorname{Ln} \frac{\dot{\varepsilon}}{\dot{\varepsilon}_0}\right)$$

Above T_c $\quad \exp- \dfrac{\Delta H}{kT} \rightarrow 1$

and $\quad \sigma = \sigma_B$

Two reasons command to discuss the influence of N alloying below T_c : - from the point of view of applications stable austenitic steel is a strategic alloy for cryogenic applications (tankers, alternators, particules accelerators, ...) which needs non magnetic alloys, low expansion coefficient, high mechanical resistance; - from the point of view of interpretation of ductile – brittle temperature transition which is induced in FCC metals by too high interstitial concentration.

Careful experiments achieved by NYILAS and OBST [58] at low temperature have shown that several pining mechanisms have to be overcome at low temperature with different and rather important activation energies: jogs on screw, jogs and kinks on edge dislocations, constriction necessary for extended dislocations to cross slip. These mechanisms far to be completely understood demonstrate that nitrogen intervenes through the bonding with dislocation, through the S.F.E., and possibily through the core structure of the dislocation. In addition the strong interactions which are determined between N and point defects [55] are also likely to be involved in relation with the point defects resulting from the displacement of slipping dislocations across the dislocations forest.

In any case, for very high concentration, when nitrogen concentration in austenite is exceeding 4 – 5 at % the ultimate tensile stress is no more much higher than the yield stress and cleavage is observed below room temperature for austenitic [59] as well as for duplex [60] H.N.S. which suffer from a ductile-brittle temperature transition.

The embrittlement mechanisms do not only result from the excessive interstitial concentration but also from the dislocation structure which as a result of an intense localization originates high stress concentration factor specially for coarse grain structure. In contrast with the embrittlement examined above, usual concentration associated with the suitable grain size is proven to improve mechanical resistance even at low temperature, L.C.F. life time at 77 K is clearly increased by Nitrogen alloying of austerite [61].

At low temperature the free enthalpy of austenite is increasing and the driving force for martensitic transformation also, under low temperature test the crossing of planar slip bands originate ε and later α' martensite [62], this phenomenon contributes to provide a supplementary deformation mode by transformation induced plasticity T.R.I.P. effect [63] which increases the mechanical resistance of the steel. Beside Al and Si alloyed T.R.I.P. steels high nitrogen T.R.I.P. steel could be an interesting alternative which does not seem to have been explored deeply.

At the other extremity of the temperature domain of utilization of steels, above 500°C Nitrogen alloying under the limit of solubility determined by the Cr, Mn, Mo alloying for concentration of 0.15 to 0.25 N wt % lead to a significant improvement to the creep resistance of austenitic steel [41, 64] which mainly results from the fine intergranular precipitation of nitrides and carbonitrides. A complete discussion of the different precipitates involved and of the creep data is beyond the scope of this chapter, meanwhile it is worth to underline that the trend should be to design creep resistant H.N.S. thanks to a fine and stable dispersion of nitrides and therefore to launch N.D.S. Nitrides Dispersed Steels in analogy with O.D.S. Oxides Dispersed Steels.

History of sciences shows that during last century the discovery of physical concepts likely to provide interpretation and modelling of relativity, electronic structure, magnetism, nuclear reactions, ... often arrived much earlier than those necessary to understand properties which may seem much more trivial such as the mechanical behaviour of solids. This delay which mainly results from the large number of mechanisms and parameters staged on the scene and from the complexity of their interplay is still valid in the present field, subjects such as the determination of dislocation velocity versus stress and temperature, modelling of the collective behaviour of dislocations sets, etc. are still uncompletely known.

Therefore further diagnostic experiments, round Robin tests, development of new concepts, virtual simulations and calculations on H.N.S. are necessary. Parameters such as the variation of surface energy of H.N.S. and elastic moduli of which the product is directly involved in the resistance to fracture are not impossible to measure or to calculate. Some observations resulting from mechanical alloying reactions between nitrides and iron and showing conservation of the total content in the solid phase [25] and crack healing of H.N.S. under severe cold rolling [65] suggest that nitrogen increases the void-alloy interface energy in H.N.S. and may be an important source of toughness. This point demands verification.

4.2 The Role of Nitrogen Alloying on Corrosion Properties

Despite the numerous excellent analyses on the role of on corrosion resistance of H.N.S. which have been published [66-69] no definitive interpretation of the corrosion mechanisms involving nitrogen seems to satisfy completely the corrosion community therefore the sentence written by H. J. GRABKE in 1996 [67] "Since more than 10 years the corrosion of H.N.S. has become a favourite topic of corrosion research especially in connection with surface science, ..." should still be up to date at the present time as well as in a near future. Although a general consensus, mainly based on experimental and empirical results, agrees with a quasi-systematic beneficial role of N on corrosion resistance specially concerning localized corrosion, contradictory observations and interpretations are found. At least four types of interpretations of the role of N are proposed [69] :

- Preferential enrichment of by anodic segregation in the passive film.
- Formation of ammonium ions reducing the acidity and therefore improving passivation.
- Formation of a dense protective nitride or oxynitride layer.
- Formation of nitrates and nitrites acting as corrosion inhibitor.

This frustrating difficulty to identify the mechanisms results from the intrinsic complexity of corrosion phenomena which depends on :

- The processing and fabrication conditions.
- The microstructure of the alloy (grain size, precipitation, ...).
- The chemical composition and the interactions between alloying elements.
- The interactions between plastic deformation and the electrochemical reactions, ...
- The nature and composition of corrosion products, ...

In addition, the characterization of the local conditions which govern the corrosion phenomena and of the exact nature of the involved electrochemical reactions is usually difficult and often unprecise. Understanding the corrosion properties is far from being only based on corrosion tests as underlined by KAMACHI MUDALI [69], surface analysis by X-ray photoelectron spectroscopy (X.P.S.), Auger electron spectroscopy (A.E.S.), secondary ion mass spectroscopy (S.I.M.S.), scanning tunnelling microscopy (S.T.M.) and atomic force microscopy are providing data on the chemical state and concentration of the alloying elements at or near the reactive zone.

Reviews on corrosion of H.N.S. in acidic and anodic dissolution [67, 69] underlined that in austenitic solid solution nitrogen can either enhance the anodic dissolution and shift the Flade potential to more positive values leading to a broadening of the active range, whereas in contrast other experiments achieved with different alloying composition the potentiodynamic polarization curves showed much lower active and passive current when the concentration was increased. For the binary FeN alloy containing about 0.8 wt% Nitrogen induced enhanced acid and anodic dissolutions but the comparison of results obtained with a FeN alloy containing about 1 wt % N with the behaviour of austenitic stainless steel could be misleading if the role of the martensitic structure and of the possible self tempering of the binary alloys is not taken into consideration.

The beneficial role of nitrogen on localized corrosion led OSAZAWA and OKATO [70] to suggest a reaction of N with electrolyte leading to the formation of NH_4^+ ions, this was confirmed by chemical analysis by JARGELIUS [71] and WEW-TA TSAI *et al.* [72]. The presence of other ions such as NO_2^- and NO_3^- has also been detected. The reaction leading to these ions is still a matter of debate therefore an equation which spares the future such as following one can be proposed :

$$Me + [N] + 2H_2O + [?] + \ldots \rightarrow [NH_4^+] + [?] + \ldots$$

The formation of NH_4^+ ions increases the pH of the solution which is less acidic and favours the repassivation of corrosion pits. The formation of nitrate discussed by KAMACHI MUDALI *et al.* [69] corresponds to a mechanism of corrosion-inhibition.

When the potential-current conditions coincide with a low corrosion rate and therefore to a low current, a protective film forms at the interface between the alloy and the electrolyte. The thickness of the passive film, some nanometers, increases with the passivation potential and corresponds to a steady state resulting from the equality of the mass flux crossing the metal-film interface and of the flux across film-electrolyte. Clearly the structure, the composition and the thickness of the passivation film determine its efficiency, therefore many different characterization techniques are exploited.

In ferritic Fe-Cr stainless steel XPS Analysis of the passivation film shows a high Cr^{3+} concentration of about 60% of metallic cations which is about five times higher than the chromium concentration in the steel [67]. Similar analyses achieved on austenitic steel containing Nickel and Molybdenum show a significant increase of the amount of Ni and Mo cations [69].

Since the beginning of the development of H.N.S. the presence of nitrogen in the passive film suggested by many authors among whom Truman [66] was confirmed by further experimental measurements [69] of which important points are commented below:

- The segregation of N at the interface of steel-passive film coincides with an interfacial Chromium enrichment. Meanwhile the determination of the electron binding energy is different enough from that of bulk nitride to conclude that this part of the film corresponds to chemisorption of nitrogen [67, 69].
- Concerning the structure of the passive film, several models have been proposed quoted in corrosion conference proceedings as well as in articles devoted to H.N.S.. These models agree rather well with the following description : starting from the steel an oxide-like layer, then an hydroxide layer containing different cations and anions as well as water molecules. Because the surface analyses are usually achieved on dry

surface the results do not allow to discriminate between all the possible structures of the passive layer and to characterize completely the corresponding properties. In the case of H.N.S. the results of surface analysis showing the presence of N in the passive film and chemical analyses of the electrolyte revealing NH_4^+ ions suggested that the following reaction could take place :

$$N^{3-} + 4\,H^+ \rightarrow NH_4^+$$

To our knowledge no direct evidence of N^{3-} has been given, it is usually proposed that N would bear a negative charge $\delta\, e^-$ corresponding to $N^{-\delta}$, meanwhile δ has never been precisely determined. The presence of the negative $N^{-\delta}$ in the passivation film at the interface steel/oxide is decreasing the potential in the passive layer and originates a decrease of the passive current and of the penetration of corrosive anions such as Cl^-.

Even if some details concerning this mechanism may be a subject of discussion (for example : the influence of the hydratation of the film, the role of the polarization of ions and of the resulting weak chemical bonds), the influence of Nitrogen on the passive film is considered as beneficial.

The influence of nitrogen concentration on the formation of passive film may originate some controversial results. Experiments achieved on Fe-20 Cr-20 Ni-6 Mo with (0.1 – 0.19 wt %) and without nitrogen [69] have shown that the thickness of the passive layer versus potential is only very slightly modified by nitrogen content and that the evolution of the metal composition during the polarization under different values of the potential is nearly independent of the potential. It is characterized by a continuous decrease of the Chromium concentration from 20 to about 12 wt %, and of the Iron from 54 to about 33 wt %, by a stable concentration of Molybdenum and by a continuous increase of Nickel till about 60 – 70 wt %.

When the surface analysis is associated with ionic erosion the variation of the composition versus time leads to the concentration profiles of the different alloying elements in the passive film. In contrast with the results quoted above SIMS measurements [69] obtained on 316 Low (0.015 wt %) and High (0.56 %) nitrogen showed a dramatic influence of the content on the concentration profiles. The profiles which are observed can be considered to be of five different shapes, starting from the zone which was the closest of the electrolyte to the base metal : stable (S), continuous increase (C.I.), continuous decrease (C.D.), increasing to a maximum (I.Max) decreasing to a minimum (D.Min). This method of characterization leads to distinguish the cases for which an enrichment is observed at the extreme surface of the passivation layer and which corresponds to a nitrogen concentration higher than in the base metal from those where the concentration at the surface is lower than in the bulk.

The comparison of the concentration profiles obtained by different authors on three different steels, low alloyed Ferritic Fe-N, austenitic 316-like 0.015 N wt % and 316 0.56 N wt % [69] shows that in most cases a enrichment is observed at the surface of the passive layer and that the concentration in the layer does not vary monotonously. In case of alloying with Molybdenum an apparent correlation between N and Mo concentrations suggests [69] a synergistic effect likely to involve also Ni which according to Clayton et *al.* [73] could correspond to a ternary nitride Ni_2Mo_3N.

In contrast with these interpretations, an investigation of the influence of nitrogen on the pitting corrosion of austenitic Cr-Ni-Mo steels containing different amounts of Mo and concentrations of nitrogen between 0.032 and 0.051 wt % does not conclude to any synergistic effect [74]. This conclusion results from the linear correlation which is established [74] between the pitting resistance equivalent PRE and the concentrations which is linear and without any product such as % Mo x % N :

$$PRE = Cr + 3.3\ Mo + 2.5\ N$$

At the present time, facing the limits of the utilized techniques of characterization, the restricted domain of concentration which is usually explored, the complexity of the phenomena involved in passivation, pit initiation and growth, repassivation ... as well as the distribution of in the passive layer the description of the mechanisms likely to describe completely the role of nitrogen is not achieved. Meanwhile the effect of nitrogen alloying is rather well identified thanks to pragmatic studies which showed that nitrogen alloying is beneficial for the resistance to pitting, crevice and stress corrosion as commented below.

The resistance to pitting corrosion is measured either by the critical pitting corrosion temperature C_{PCT} or by the critical pitting corrosion potential C_{PCP} or by the current density : the higher the temperature or the potential the higher the resistance and the lower the current. Three mechanisms are usually proposed to understand the local break down of the passive film and therefore the initiation of pits [67, 69, 77] :

- penetration of corrosive anions such as Cl^- or SO_4^{2-} through the film,
- mechanical stress and strain induced damage of the film,
- adsorption of anions on the passive film.

The adsorption mechanism is considered to be the most probable. In complement with the usual arguments in favour of this interpretation, the role of the additional degree of freedom resulting from the introduction of chemical species containing nitrogen in the film could, by analogy with the theory of diffusion path in multiple component systems, explain local enrichments of aggressive anions.

When the potential increases the breakdown of the film shows a transition zone which is exacerbated by nitrogen. In the interval $E_r - E_b$, successive and sudden burst phenomena correspond to initiation repassivation phenomena *(Fig. 8)*. Above the potential E_b pitting is irreversible and the growth takes place. In presence of nitrogen the formation of NH_4^+ ions in the pit lead to an increase of the pH which reduces the growth rate.

Similar phenomena are considered to take place during crevice corrosion therefore the effects of nitrogen on pitting and crevice corrosion have been compared by Marcus SPEIDEL [75] : the critical pitting corrosion temperature T_{CPC} and critical crevice corrosion temperature T_{CCC} follow parallel linear variations versus the same composition parameter called MARC :

$$MARC = Cr + 3.3\ Mo + 20\ (C + N) - 0.5\ Mn - 0.25\ Ni$$

$$T_{CPC} = 3.5\ (MARC) - 65\ (°C)$$

$$T_{CCC} = 3.5\ (MARC) - 112\ (°C)$$

As shown by these relations T_{CCC} is about 50°C below T_{CPC}.

Because the experimental conditions may differ (pH, composition of the electyrolyte, microstructure and composition of the samples, etc) the pitting corrosion equivalent PCE and correlations between T_{CPC}, T_{CCC} and concentrations may be characterized by different numerical coefficients, meanwhile the trends are similar. According to published results

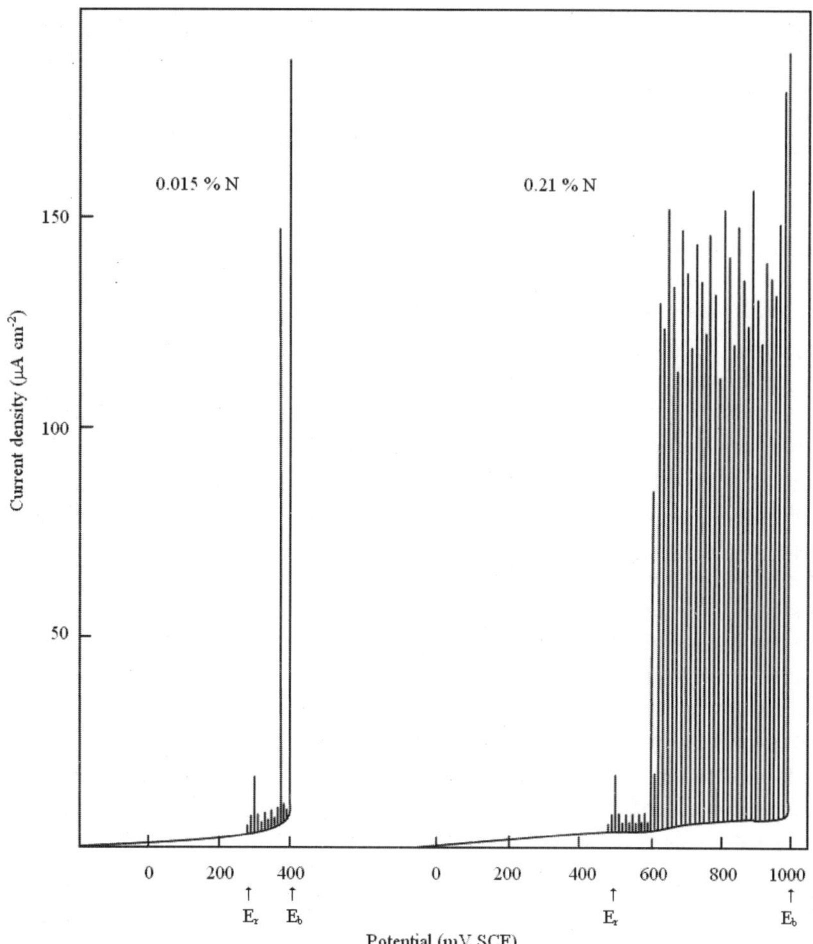

Figure 8 Potentiodynamic curves for 20 %Cr-25%Ni-4.5%Mo-steel in 1 M NaCl at 75°C. When Nitrogen concentration increases from 0.015 mass % to 0.21 mass % the passive domain is becoming broader, as well as the domain of repassivation which starts at E_r and stops at the break down potential E_b. Schematic diagram from data R. J. Argelius (1986) quoted by H. J. Grabke [67].

[75-78] the coefficient which characterizes the efficiency of nitrogen content is dramatically modified when N and C are present in the alloy. Schematically it seems that N is becoming less efficient when the Carbon concentration increases and that the role of carbon is becoming less damaging when C and N are both forming an interstitial solid solution.

As underlined by the paragraph devoted to microstructure the precipitation phenomena involving either C or N greatly differ, in agreement with the induced retardation of the precipitation of $M_{23}C_6$ and σ phases it is observed that even when some Cr_2N precipitation replaces the precipitation of carbides the sensitisation is reduced by nitrogen alloying [69].

When mechanical loading superimposes to corrosive environment, it appears that alloying of stainless steel is beneficial for the resistance to stress corrosion and corrosion fatigue. Two main reasons are identified :

- A shift of the potentiodynamic polarization potential E_{SCC} which characterizes the transition from the negligible corrosion cracking SCC to the domain where SCC becomes important. This shift toward more positive potential is about 50 mV when the concentration increases up to 1 At % [67]. It seems that this shift of the transition "N0 SCC - SCC" is correlated with the shift of E_B versus nitrogen concentration.
- The second reason is related with the influence of nitrogen on the yield stress and on the plasticity. Stress corrosion and corrosion fatigue originates from corrosion - deformation interactions C.D.I.. In most cases alloying increases the yield stress, reduces the plastic deformation under constant external stress loading and therefore the mechanical induced damage of the passivation layer.

The response of steel under combined mechanical loading and chemically active environment, is involving different mechanisms of interactions between chemical reactions and plastic deformation which are revealed through numerous experiments among which "corrosion creep" achieved by HÄNNINEN et *al.* [78]. These authors showed that *i)* the steady state deformation rate $\dot{\varepsilon}_{SS}$ is dramatically increased by the presence of corrosive medium, *ii)* that a possible mechanism could result from the formation of vacancies, *iii)* that the effect of alloying is not monotonously beneficial and that the optimum has to be determined. The coupling of corrosion behaviour with mechanical properties is exemplified by many other cases such as corrosion fatigue and stress corrosion of duplex stainless steel [79] which also demonstrate the importance of optimal nitrogen concentration. In many cases the optimal concentration results from a transition of the nature of the crack propagation which is impeded by nitrogen when it starts intergranular and which becomes transgranular at high concentration [9, 67, 69].

During the corrosion of steel by electrolytic solution the reduction of H^+ ions originate atomic hydrogen which diffuse into the steel and induce embrittlement. Among the principal models which are proposed to interprete hydrogen induced embrittlement: *1)* formation of brittle hydrides - *2)* decrease of the Young modulus and of surface energy through the influence of H on atomic bonds – *3)* hydrogen enhanced localized plasticity (HELP), which has been shown by GAVRILJUK et al. [80-82] to be dominant. The localization of plasticity corresponds to a lowering of the activation stress of dislocation source and to an increase of the dislocation mobility. Electronic structure calculations as well as experimental spectroscopy measurements show that in austenite hydrogen interstitial increases the density of states at the Fermi level. The observed improvement resulting from nitrogen is therefore suggested to correspond to a competitive effect of H and N interstitials on the electronic structure. The "saw-tooth" relief of the fracture surface which is the consequence of hydrogen embrittlement appears to belong to a generic morphology also observed in other cases, duplex H.N.S. [60] and, super alloys [83].

The increasing demand of steels designed for high temperature resistance stimulates studies on austenitic as well as martensitic H.N.S. likely to provide improved oxidation resistance above 500°C. Experimental results [69] show that a significant improvement can be reached thanks to moderate concentrations (0.12 – 0.2 wt % N) in synergy with additions of elements such as Al, Si, Y, etc.

A rather concise conclusion about the role of nitrogen on corrosion can sound: that very often an optimized concentration of nitrogen lead to a significant improvement of the resistance. The complete identification of the mechanisms involved is missing because the corrosion properties are among the most difficult to analyse and to model, therefore the optimization of the concentration is based on corrosion tests, electrochemical experiments, characterisation and analysis and the expertise of corrosion scientists and engineers who have to decyfer the corrosion complexity.

V. CONCLUSION

H.N.S. constitutes a specific class of alloys. As interstitial solute, or involved in substitutional - interstitial clusters or forming ordered distributions, combined in nitrides and carbonitrides precipitates, nitrogen originates unambiguous beneficial effects on mechanical and corrosion resistance. Meanwhile the existence of an optimum composition, which constitutes a corollary of any alloying, has been obtained for H.N.S. more or less empirically according to the properties demanded by applications. Many other fruitful grades virtually exist, they have to be explored. Further breakthrough will be based on revisited and updated R&D work supported by deepening of theoretical concepts, exploitation of classical and new tools of characterization, more powerful means of simulation and calculation. The economical driving force for this future exists : the cost of raw materials increases and the price of nitrogen is not yet an object of speculation.

VI. REFERENCES

1. *High Nitrogen Steels*, HNS'88, Proceedings of the 1st International HNS Conference, Lille France, Edited by Foct J. and Hendry A., The Institute of Metal, (1988).
2. *High Nitrogen Steels*, HNS'90, Verein Deutscher Eisenhüttenleute (VDEh) and Deutsche Gesellschaft für Metallkunde e.V., Edited by Stein G. and Witulski H., Aachen, Germany, (1990).
3. *High Nitrogen Steels*, HNS'93, Proceedings of the 3rd International Conference, Edited by Gavriljuk V. G. and Nadutov V. M., **Part I and Part II**, Kiev, Ukraine, (1993).
4. *Special Issue on High Nitrogen Steels*, Proceeding of 4th HNS Conference, Kyoto (Japan), Edited By Kikuchi M. and Mishima Y., ISIJ International, The Iron and Steel Institute of Japan, Vol. 36, Number 7, (1996).
5. *High Nitrogen Steels*, HNS'98, Proceedings of the 5th International Conference on High Nitrogen Steels, Editors Hänninen H., Hertzman S., Romu J., Trans. Tech. Publications LTD, Espoo, Finland, and Stockholm, Sweden, (1998).
6. *High Nitrogen Steels*, Part A, Part B, Edited by Baldev Raj, Speidel M., Kumachi Mudali U., Srinivasan V. S., Trans. Indian Inst. Met., Vol. 55, No 4 and No 5, Chennay, Inde, (2002).
7. *High Nitrogen Steels*, HNS'2003, Edited by Speidel M. O., Kowanda C. and Diener M., Institute of Metallurgy, ETH Zürich, Switzerland, (2003).
8. *High Nitrogen Steels*, HNS'2004, Edited by Akdut N., De Coomand B. C. and Foct J., International Conference, GRIPS media, Ostend, Belgium, (2004).
9. Speidel M. O. und Uggowitzer P. J., Ergebnisse der Werkstoff - Forschung, Band 4, Verlag der Schweizerischen Akademie der Werkstoff-Wissenschaften, (1999).
10. Kunze J., *Nitrogen and Carbon in Iron and Steel*, Akademie - Verlag Berlin, (1990).
11. Grigorova N., *Carbonitrides and High Speed steels - Chemical phase analysis*, Interlsoft, Sofia, (1995).
12. Rashev T., *High Nitrogen Steels, Metallurgy Under Pressure*, Publishing House of the Bulgarian Academy of Sciences, Sofia, (1995).
13. Murata T. and Sakamoto M., *Nitrogen-Alloyed Steels - Fundamentals and Applications* - Edited by Imay Y., AGNE Publishing Inc., (1997).

14. Gavriljuk V. G. and Berns H., *High Nitrogen Steels*, Springer-Verlag Berlin Heidelberg, (1999).
15. *High Nitrogen Steels and Stainless Steels*, Edited by Kamachi Mudali U. and RAJ B. Manufacturing, Properties and Applications, Narosa Publishing House, Chennai, Inde, (2004).
16. Physical Metallurgy and the design of steels, Pickering F. B., Apllied Science Pub., (1983).
17. Stainless steel, Lula R. A., ASM, (1989).
18. Perrot P. and Foct J., Techniques de l'Ingénieur, Paris, M4 275, pp. 1-23, (2003).
19. Goldschmidt H. J., Interstitial Alloys, Butterworth, Pub. U.K., (1967).
20. Speidel H. J. C. and Speidel M. O., in [7], pp. 101-112, (2003).
21. Stein G., Menzel J. and Dörr H., in [I], pp. 32-38, (1988).
22. Romu J. and Hänninen H., in [5], pp. 673-680, (1998).
23. Medovar B. I., Marinskii G. S., Latash Yu. V., Tikhonovskii A. L. and Torkhov G. F., Special Electrometallurgy, Vol. 1, No 5, Sov. Tech. Rev. C, (1989).
24. Foct J. and Mastorakis A., Solid State Phenomena, 25/26, pp. 581-590, (1992).
25. Foct J., Journal of Mat. Sci., 39, pp. 5011-5017, (2004).
26. Speidel H. J. C., Dissertation ETH Zürich CH, (2002).
27. Bleck W., Trip High Strength Ferrous alloys, Edited by De Cooman B. C., pp. 13-23, (2002).
28. Westbrook J. H. and Fleischer R. L., Intermetallic Compounds, Vols. I and II, Edited by Wiley J., (1995).
29. Jargelius-Petterson R. F. A., Z. Metallkd., 89, pp. 177-183, (1998).
30. Hertzman S., Scand. J. Metallurgy, 24, pp. 140-146, (1995).
31. Jack K. H., Proc. Roy. Soc., A208, PP. 216-224, (1951).
32. Rochegude P. and Foct J., Phys. Stat. Sol., A88, pp. 137-145, (1985).
33. Foct J. and Le Caër G., Annal. Chim. Sci. Mat., 22, p. 387, (1997).
34. Vanderschaeve F., Taillard R. and Foct J., J. Mater. Sci., 30, pp. 6035-6046, (1995).
35. Jack K. H., in [1], pp. 117-135, (1988).
36. Evans D. and Jack K. H., Acta. Cryst., 10, p. 762, (1957).
37. Vanderschaeve F., Taillard R. and Foct J., Steel Research, 64, pp. 221-227, (1993).
38. Anthamatten B. R., Uggowitzer P. J., Solenthaler Ch. And Speidel M. O., in [2], pp. 434-441, (1990).
39. Domain C., Becquart C. S. and Foct J., Phys. Rev. B, 144112 – 1, 12, (2004).
40. Becquart C. S., Domain C. and Foct J., Phil. Mag. 85, 4-7, pp. 533-540, (2005).
41. Mathew M. D. and Srinivasan V. S., in [15], pp. 182-204, (2004).
42. Keichel J., Gottstein G. and Foct J., Mat. Sci. Forum, Vol. 318-320, pp. 785-792, (1999).
43. Stein G. and Diehl V., in [8], pp. 421-426, (2004).
44. Shilling, Advanced Materials & Processes, A.S.M., Vol. 164, N) 1, pp. 41-43, (2006).
45. Shanina B. D., Gavriljuk V. G., Konchitz A. A., Kolesnik S. P. and Tarasenko A. V., Phys. Stat. Sol., a149, pp. 711-722, (1995).
46. Taillard R. and Foct J., in [1], pp. 387-390, (1988).
47. Degallaix S., Taillard R. and Foct J., Fatigue E.M.A.S. 1, Edited by Beevers et *al.*, pp. 45-49, (1988).
48. Diener M., in [8], pp. 539-543, (2004).
49. Grujicic M., Owen W. S., Acta. Met. Mat., 43, pp. 4201-4211, (1995).
50. Norström L. A., Metal. Sci., 11 (6), pp. 208-212, (1977).
51. Köstler H. J. and Sidan H., Z. Wirtsch Fert, 72(10), pp. 785-788, (1977).
52. Keichel J., Foct J. and Gottstein G., ISIJ International, 43, No 11, pp. 1781-1787, (2003).

53. Keichel J., Foct J. and Gottstein G., ISIJ International, 43, No 11, pp. 1788-1794, (2003).
54. Bata V. and Pereloma E. V., Acta. Mater., 52, pp. 657-663, (2004).
55. Gavriljuk V. G., Scripta. Mater., 52, pp. 951-953, (2005).
56. Akdut N., Keichel J. and J. Foct, Mat. Sci. Forum, Vol. 318-320, pp. 769-776, (1999).
57. Seeger A., Phil. Mag., 46 (382), pp. 1194-1217, (1955).
58. Nyilas A., Obst B in [1], pp. 194-198, (1988).
59. Müllner P., Sollenthaler C., Uggowitzer P. J. and Speidel M., Acta. Metall. Mater., 42, pp. 2211-2217, (1994).
60. Foct J. and Akdut N., Scripta Metall., 29, pp. 153-158, (1993).
61. Vogt J. B., Foct J., Robert G. and Dhers J., Metall. Trans., 22A, pp. 2385-2392, (1991).
62. Vogt J. B., Saadi B. A. and Foct J., Z. Metallkd. 90, pp. 323-328, (1999).
63. Foct J., Stolarz, Baffie N., Massol K. and Vogt J. B., Trip-aided High Strength Ferrous Alloys, Edited by De Cooman, GRIP'S print, pp. 333-339, (2002).
64. Balachandran G., in [15], pp. 40-93, (2004).
65. Foct J., Akdut N. and Gottstein G., Scripta Metall. Mater., 27, pp. 1033-1038, (1992).
66. Truman J. E., in [1], pp. 225-239, (1988).
67. Grabke H. J., in [4], pp. 777-786, (1996).
68. Hänninen H., in [05], pp. 479-488, (1998).
69. Kamachi Mudali U. and Ningshen S., in [15], pp. 131-181, (2004).
70. Osozawa K. and Okato N., USA – Japan Seminar, NACE, p. 135, (1976).
71. Jargelius R., Quoted by [67], Report Swed. Inst. For Metal., IM 2179, (1986).
72. Wen-Ta Tsai, Reynders B., Stratmann M. and Grabke H. J., Corros. Sci., 34, p. 1647, (1993).
73. Clayton C. R. and Martin K. G., in [1], pp. 256-260, (1988).
74. Eckstein C. and Spies H. J., in [8], pp. 545-548, (2004).
75. Speidel M., in [7], pp. 1-8, (2003).
76. Rawers J., Govier D. and Cook D., in [4], pp. 958-961, (1996).
77. Normand B., in [8], pp. 509-520, (2004).
78. Leinonen H., Virkkunen I. and Hänninen H., in [8], pp. 555-561, (2004).
79. Magnin T. and Lardon J. M., Mater. Sci. Engi., A104, pp. 21-25, (1988).
80. Shivanyuk V. N., Foct J. and Gavriljuk V. G., Mat. Sci. Engi., A300, pp. 284-290, (2000).
81. Shivanyuk V. N., Shanina B. D., Tarasenko A. V., Gavriljuk V. G. and Foct J., Scripta. Mater., 44, pp. 2765-2773, (2001).
82. Gavriljuk V. G., Shivanyuk V. N. and Foct J., Acta. Mater., 51, pp. 1293-1305, (2003).
83. Foct F., Bouvier O. and Magnin T., Metall. Mater. Trans. A, 31, pp. 2025-2032, (2000).

CHAPTER 12

Biomaterials and Corrosion

Noam Eliaz

Biomaterials and Corrosion Lab, School of Mechanical Engineering & The Materials Science and Engineering Program, Tel-Aviv University, Ramat Aviv, Tel-Aviv 69978, Israel.

A. INTRODUCTION

The field of *Biomaterials Science* deals with the study of the structure and properties of biomaterials, the mechanisms by which they interact with biological systems, and their performance in clinical use. *Biomaterials* are commonly defined as "nonviable materials intended to interface with biological systems to evaluate, treat, augment or replace any tissue, organ or function of the body" [1]. Before a new biomaterial can be introduced to the market, various issues are considered, including its designated anatomic location, toxicology and biocompatibility, healing, mechanical and other property requirements, ethics, standardization and regulation [2]. To prevent inadequately tested materials and devices from coming to the market, a complex national regulatory system has been erected by the United States government through the Food and Drug Administration (FDA). International organizations, mainly the International Standards Organization (ISO) and The American Society for Testing and Materials (ASTM), issue material, device and procedure standards.

In general, a biomaterial should not be *toxic*, unless it is specifically engineered for such requirements (e.g. a smart drug release system that seeks out cancer cells and destroys them). *Biocompatibility* is an essential requirement of any biomaterial, which implies the ability of the material to perform with an appropriate host response in a specific application [3]. There are two main factors that determine the biocompatibility of a material: the host reactions induced by the material, and the degradation of the material in the body environment. Often, both factors should be considered. By referring to *healing*, it is realized that special processes are invoked when a material (or a device) heals in the body; an injury to a tissue will stimulate an inflammatory reaction sequence that leads to healing.

Since about 4,000 years ago, humans have been using artificial materials to repair fractured and diseased tissues and organs [4]. In the early ages, the Greeks and Egyptians transplanted bones from animals in humans. Only in 1546 was a synthetic material, gold (Au) plate, used to repair a cleft palate. The development of advanced biomaterials is related to the development of modern medicine and new materials. The first alloy which was developed specifically for implantation was the "vanadium steel," in the early 1900's [5]. In the early 1960's, Sir John Charnley made the first attempt to assemble together a stainless steel hip prosthesis and a high-density polyethylene with a methacrylate bone cement. This may be considered the beginning of modern orthopedics, in which the development of better materials plays a central role. Nowadays, biomaterials are made of metals, ceramics, polymers, or combinations of them. As representative examples, one can mention the vascular stents made of stainless steel and coated with organic biomaterials (a four billion dollars annual market in the US alone), or the implantable teeth and artificial bones made of physiologically-compatible bioceramics. Polymers, on the other hand, are most common in drug delivery systems and tissue engineering. *Corrosion* is an important factor in the design and selection of alloys for service *in vivo*. Because toxic species might be released to the body during corrosion processes on one hand, and because various corrosion mechanisms can lead to implant loosening and failure on the other hand, biomaterials are often required to be tested for corrosion and/or solubility before they are approved by regulatory organizations.

The objective of this Chapter is to introduce the reader with the basics of biomaterials corrosion. First, principles of biocompatibility are presented in section B, because this term is often used in corrosion-related reports. Next, the body environment is described in section C. As shown, this environment is harsh and introduces several challenges with respect to corrosion control. Subsequently, a brief review of corrosion mechanisms *in vivo* is provided in section D. Section E presents the major metals and alloys currently used in biomedical applications. Finally, section F focuses on corrosion control strategies *in vivo*.

B. BIOCOMPATIBILITY

A *biocompatible material* disrupts the normal body function as little as possible. There are many factors which influence implant biocompatibility; for example, implant size, shape, material composition, surface wettability, surface roughness and charge. The biomaterial must not alter plasma proteins (including enzymes) so as to trigger undesirable reactions; must not cause thrombus formation, adverse immune response or cancer; must not destroy or sensitize the cellular elements of blood, produce toxic or allergic responses, or deplete electrolytes. In turn, the environment should not cause degradation (e.g. biological or mechanical) or corrosion of the biomaterial that would result in loss of physical and mechanical properties. In practice, no synthetic material is completely harmonious with the living environment; however, materials do have different levels of inertness.

Most artificial materials, once implanted in the human body, induce a cascade of reactions with the biological environment through interaction of the biomaterial with body fluids, proteins and various cells. The sequence of local events often leads to the classic *foreign-body reaction* (FBR) and the formation of a fibrous tissue capsule around an implant. Clearly, a major factor influencing the reaction of the body is the *biomaterial surface*, because it is the one that the body "senses" first. The specific reactions at the surface determine the FBR, the path and speed of the healing process, and the long-term development of the biomaterial/body interface. The chemical composition, structure and morphology of a surface are all important with this respect. They regulate the type and degree of the interactions that take place at the interface (e.g. adsorption of ions and biomolecules such as proteins, formation of calcium phosphate layers, interaction with different types of cells such as macrophages, bone marrow cells, osteoblasts, etc.). Hence, the nature of the initial interface that is established between an artificial material and the attached tissue determines the ultimate success or failure of the implant.

Synthetic biomaterials are generally not immunogenic. However, they typically elicit the FBR, which is a special form of nonimmune inflammation. The most prominent cells in the FBR are macrophages, which presumably attempt to phagocytose the material, with much difficulty. Activated macrophages elaborate cytokines that may stimulate inflammation or fibrosis. Multinucleated giant cells in the vicinity of a foreign body are generally considered evidence of a more severe FBR, in which the material is particularly irritating. This reaction is frequently called a foreign-body granuloma. The more biocompatible the implant, the more quiescent the ultimate response is. When the implant is a source of wear debris from articulating joint surfaces, particles may be seen in the extracellular matrix between the cells and within macrophages. The inability of inflammatory cells to adhere to without phagocytose particles larger than a critical size can lead to release of enzymes (exocytosis) to the extracellular environment. Thus, inflammatory cell products that are critical in killing micro-organisms in typical inflammation can damage tissue adjacent to foreign bodies. For most inert biomaterials, the late tissue reaction is encapsulation by a relatively thin fibrous tissue capsule, composed of collagen and fibroblasts. The nature of the reaction is largely dependent on the chemical and physical characteristics of the implant.

The response of individual cells to material can be considered to be dependent on how well the material mimics the natural, extracellular environment of the cell. The physical structure of the surface may have an inferior influence on the biological response of the material, which is normally non-toxic and does not release any biologically active substance. Osteolysis, bone resorption and the formation of a thick fibrous layer between the implant and bone reflect poor biocompatibility. Also, micro-particles of certain size of normally non-toxic materials may trigger an inflammatory response. These particles cause an irritation of phagocytic cells and activate them to produce and release cytokines, proteinases, growth factors and other proinflammatory factors, finally leading to chronic inflammation, fibrosis, osteolysis and porosis in bone. In the case of aseptic loosening of the prosthesis, wear particles are expected to lead to the formation of a poorly vascularized, synovial-like interface membrane between the prosthesis and bone. The formation of necrotic focuses, granulomas and osteolysis may finally result in loosening of the prosthesis. The increase of metallic wear increases the surface of the metal material and the quantity of metal ions. The porous surface increases the surface area, but also particular wear.

One of the issues that arise from the release of corrosion products into the body is systemic and remote effects. In animals and patients with either stainless steel or cobalt-base orthopedic total joint replacement components, corrosion and wear produce longer-term changes in blood composition, primarily in its metal content. These include elevations of metallic content in tissue (at both local and remote sites) and of metal-bearing ion concentrations in serum and urine. In patients with total joint replacements, large elevations of chromium levels in serum occur in the early postoperative period, significant elevations may persist for more than a decade, and accumulation of 10 to 100 times the normal chromium and nickel levels is possible in tissues remote from the implanted hip. By themselves, metal ions lack the structural complexity required to challenge the immune system. However, when combined with proteins, such as those available in the skin, connective tissues and blood, a wide variety of metals induce immune responses, and thus must be considered harmful. Cobalt, chromium, and nickel are included in this category, with nickel perhaps the most potent; at least 10% of a normal population will be sensitive by skin test to one or more of these metals, at some threshold level. The most typical response of a metal-sensitized individual to a challenge is delayed type IV hypersensitivity [6]. The principal mechanisms by which metal dissolution products can damage cells include the inhibition of enzymes, prevention of diffusion through the cell membranes or at the periphery of the cell, and breakdown of lysosomes. The corrosion products are stored in lysosomes which, after accumulation of a certain amount of the foreign material, undergo activation and release the digestive enzymes. In turn, the enzymes lyse adjacent cells. The release of further activated lysosomal enzymes initiates a self-stimulating process. The enzymes might degrade either the extracellular matrix or the cells, to provoke loosening of an implant [7]. In contrast to the discussion above, immune responses to polymers in clinical use have not been reliably reported, whereas immune responses to ceramics are highly unlikely owing to the extremely low solubility of these materials [6].

Before proceeding to describe the body environment in more detail, it should be mentioned that tissue interactions can be modified by changing the chemistry of the surface (e.g. by adding specific chemical groupings to stimulate adhesion or bone formation in orthopedic implants), inducing roughness or porosity to enhance physical binding to the surrounding tissues, incorporating a surface active agent to chemically bond the tissue, or using a bioresorbable component to allow slow replacement by tissue to simulate natural healing properties.

C. THE BODY ENVIRONMENT

The water content of the human body ranges from 40% to 60% of its total weight. Functionally, the total body water can be subdivided into two major fluid compartments, namely the extracellular and the intracellular fluids. *Extracellular fluids* (ECFs) consist of the plasma found in the blood vessels, the interstitial fluid that surrounds the cells, the lymph, and transcellular fluids (e.g. cerebrospinal fluid and joint fluids). *Intracellular fluid* (ICF) refers to the water inside the cells. Both the amount and the distribution of body fluids and electrolytes are kept normal and constant, by a mechanism known as *homeostasis*.

The normal pH range for blood plasma is 7.35 to 7.45. A decrease in blood pH below normal is known as acidosis, whereas an increase in blood pH above normal is known as alkalosis. There are two major types of mechanisms that control the body pH - chemical and physiological. The rapid-acting *chemical buffers* (e.g. bicarbonate, phosphate and protein buffer systems) immediately combine with any added acid or alkali that enters the body fluids, thus preventing drastic changes in hydrogen ion concentration and pH. If the immediate action of chemical buffers cannot stabilize the pH, the *physiological buffers* (i.e. respiratory and urinary response systems) serve as a secondary defense against harmful shifts in pH [8]. Electrolytes play a major role in body functionality. Among various functions, they take part in metabolism, determine the cell membrane potentials and osmolarity of body fluids, etc. Major cations include hydrogen, sodium, potassium, calcium, and magnesium ions. Major anions include hydroxide, bicarbonate, chloride, phosphate and sulphate ions. Under normal conditions, body fluids have a temperature of 37°C.

From the perspective of corrosion, the most important characteristics of body fluids are the chloride, dissolved oxygen and pH levels. Body fluids may seem to be slightly less aggressive than seawater, as reflected by the lower pitting resistance equivalent number (PREN) of 26 and greater, recommended to prevent *in vivo* pitting corrosion of stainless steels, in comparison to the value of 40 usually required for stagnant seawater [9]. However, the dissolved oxygen levels in blood are lower than in artificial solutions exposed to air atmosphere due to combination with hemoglobin, which is the main component of red blood cells. The partial pressure of oxygen in blood varies between 100 to 40 mmHg for arterial and venous blood, respectively. On the other hand, the corresponding value in air is 160 mmHg. Because most biomaterials rely on oxygen to repassivate, repassivation of metal surfaces is more difficult under conditions of low dissolved oxygen concentration. Thus, deaeration of the solution with high-purity nitrogen gas to maintain low O_2 concentration was found to be more appropriate to predict the *in vivo* performance of biomaterials [10]. Bicarbonate levels are about twenty times higher in blood than in seawater [11]. Other components in body fluids (e.g. phosphates, cholesterols and phospholipids) are usually thought to either play no role in the corrosion process, or exist at inconsequential levels. Therefore, most *in vitro* experiments have been conducted in either saline or standard isotonic solutions such as Ringer's or Hank's, in which the presence of bicarbonate and calcium chloride is the main difference compared to saline. Compositions of selected body fluids and simulated body fluids (SBFs) are provided in Tables 1 and 2, respectively. Phosphate buffered saline (PBS) is mostly recommended because it maintains the pH almost constant throughout *in vitro* experiments [12]. A review by Solar [13] concluded that inorganic solutions based on diluted NaCl were indeed satisfactory substitutes for human body fluids when studying the behavior of passive metals. Thus, many use the simple saline solution (0.9 wt% NaCl in DI water) for *in vitro* experiments. However, usually no attempt is made in the *in vitro* experiments to lower the dissolved oxygen content of the isotonic solution to that of veinal blood; this has been proposed as an explanation for some of the differences observed in the *in vivo* and *in vitro* corrosion behavior of implant materials [11,14]. Furthermore, the minor components in blood have occasionally been blamed for the accelerated *in vivo* corrosion. For example, it has been postulated that sulfur present in amino acids may enhance crevice corrosion of stainless steels [15].

When the implant is inserted into the body, the disturbance of the blood supply to the bone is often accompanied by severe pathological infections that might affect the healing and cause electrochemical variations in the equilibrium state [16]. On surgical insertion of the implant, the pH of the body fluid drops from the normal value of 7.4 to 5.5, and in the course of 10 to 15 days regains neutrality. However, bacterial infection at the tissue site results in a variable pH from acidic to alkaline (4.0 to 9.0, respectively) in the vicinity of the implant. Laing [17] reported that the pH around a newly inserted surgical implant can drop to as low as 4.0 due to the build-up of haematomas, a condition that could last for several weeks. The lowering of pH in vicinity to the implant implies severe localized corrosion of the implant. Hydrogen peroxide may also be generated during the initial stages of the inflammatory response, following insertion of an implant [18,19]. The level of these pathological changes depends on the biological activity of any corrosion products released from the implant, and also on the implant size and shape. The extent of the pathological changes may vary across the surface of the

Table 1 Compositions of various body fluids [10]

Component	Interstitial fluid (mg/L)	Synovial fluid (mg/L)	Serum (mg/L)
Na^+	3,280	3,127	3,265
K^+	156	156	156
Ca^{2+}	100	60	100
Mg^{2+}	24	—	24
Cl^-	4,042	3,811	3,581
HCO_3^-	1,892	1,880	1,648
HPO_4^{2-}	96	96	96
SO_4^{2-}	48	48	48
Organic acids	245	—	210
Protein	4,144	15,000	66,300

Table 2 Compositions of various simulated body fluids (SBFs) [10]

Component	PBS (g/L)	Ringer's (g/L)	Hank's (g/L)
NaCl	8.00	8.60	8.00
$CaCl_2$	—	0.33	0.14
KCl	0.20	0.30	0.40
$MgCl_2 \cdot 6H_2O$	—	—	0.10
$MgSO_4 \cdot 7H_2O$	—	—	0.10
$NaHCO_3$	—	—	0.35
Na_2HPO_4	1.15	—	0.0476
KH_2PO_4	0.20	—	0.06
Phenol red	—	—	0.02
Glucose	—	—	1.00

implant, which could lead to the development of electrochemical cells [20]. Variations in the local pH on titanium alloys have also been observed during *in vitro* experiments, which could also generate the potential gradients required to drive localized corrosion [21]. The amount of ion leaching out might be very high and lead to allergic and carcinogenic effects to the patient.

With respect to dental applications, the environment within the oral cavity is not well defined [22]. Although there are several recipes for artificial saliva, the most popular is that of Fusayama [23], i.e. 0.400 g/dm^3 NaCl, 0.400 g/dm^3 KCl, 0.795 g/dm^3 $CaCl_2 \cdot H_2O$, 0.690 g/dm^3 $NaH_2PO_4 \cdot H_2O$, and 0.005 g/dm^3 $Na_2S \cdot 9H_2O$, at pH 5.5. Yet, in reality the make-up of human saliva varies considerably between individuals, especially in the sulfide content, which can cause tarnishing of both silver- and gold-based amalgams. Many foodstuffs are acidic, with high chloride levels, and are thus far more corrosive than saliva. In addition, oral hygiene has a strong effect on the corrosiveness of the oral environment; what rots the teeth is likely to corrode the amalgams and dental fixtures. Finally, many dental products and solutions contain fluoride, with some of the special varnishes used by dentists containing over 2 wt% fluoride [24]. Thus, although fixtures in the oral cavity are readily accessible for repair, there is a concern that the galvanic cells and the toxicity of the metals leaching out might cause oral cancer [25,26].

Biological macromolecules can influence the rate of corrosion by interfering in different ways with the anodic or cathodic reactions. Proteins and lipids from the ECF adsorb onto the surface of the implant material and might trigger changes in its chemical properties through oxidation and/or hydrolytic reactions. First, proteins can bind to metal ions and transport them away from the implant surface. This will upset the equilibrium across the charged double layer and allow further dissolution of the metal. Second, proteins can affect the electrode potential due to their electron-carrying capabilities, whereas bacteria can alter the pH of the local environment through generation of acidic metabolic products. Third, the adsorption of proteins onto the surface of biomaterials could limit the diffusion of oxygen to certain regions of the surface, thus causing preferential corrosion of oxygen-deficient regions and breakdown of the passive layer. Finally, bacteria in the vicinity of an implant could consume hydrogen that is released in cathodic reactions, thus accelerating the corrosion process [2]. In addition, cells may release strong oxidizing agents and enzymes that are targeted at decomposing the implant material. Furthermore, relative motion between tissues and the implant might cause wear of both surfaces, thereby promoting chronic inflammation and establishing an even harsher chemical environment [7]. Figure 1 shows an adapted potential-pH (Pourbaix) diagram that illustrates the range and complexity of conditions which may be experienced by biomaterials *in vivo* [27]. Pourbaix diagrams are derived from the Nernst equation, the solubility of the degradation products, and the equilibrium constants of the reaction. In general, they are very useful in determining the regions of corrosion, passivity and immunity based on thermodynamic considerations. The upper dashed line in Fig. 1 (oxygen evolution) represents the upper limit of water stability; it is associated with oxygen-rich solutions or electrolytes near oxidizing materials. In the human body – saliva, intracellular fluid, and interstitial fluid occupy regions near the oxygen line because they are saturated with oxygen. The lower dashed line in Fig. 1 (hydrogen evolution) represents the lower limit of water stability. In the human body – urine, bile, the lower gastrointestinal tract, and the secretions of ductless glands occupy a region somewhat above the hydrogen line. Thus, different parts of the body have different pH values and oxygen concentrations. Consequently, a metal which performs well in one part of the body may suffer an unacceptable amount of corrosion in another part.

As mentioned above, corrosion of biomaterials *in vivo* rises two major concerns: (1) its effect on the lifetime of the medical device, and (2) will the metal ions that leach out of the device concentrate to levels sufficient to cause the development of tumors or other medical complications. Such toxic levels could occur even at corrosion rates that are insignificant with respect to the physical performance of the implant. In many industrial applications, metal corrosion is controlled by: (1) changing the chemistry of the environment, (2) changing the pH, (3) lowering the temperature, or (4) adding inhibitors. Unfortunately, neither of these strategies can be applied to reduce the corrosion rate of surgical implants *in vivo*. Coatings are of only limited use for protecting implants because many of them are subjected to wear. Thus, corrosion control *in vivo* is limited mainly to proper material selection during implant design. A new challenge that will confront corrosion scientists in the near future results from a desire to make extended use of a number of advanced materials, such as shape memory alloys (SMAs),

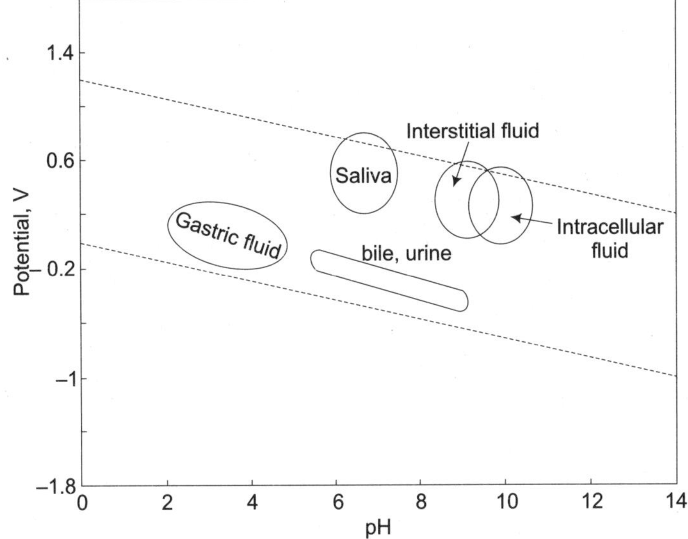

Figure 1 Potential-pH diagram that illustrates the diverse service conditions for biomaterials *in vivo* [27].

porous materials, composites and rare earth magnets. Within the scope of this chapter, we shall move on and review the major mechanisms of corrosion *in vivo*.

D. CORROSION-RELATED FAILURE MECHANISMS

Failures of implants are usually classified as either mechanical, electrochemical, biological or combinations of these. Mechanical failure mechanisms include micromotions, overload, fatigue and wear. Electrochemical failures are mainly related to different forms of corrosion. Biological failures result from infection, inflammation, enzymatic degradation, calcification, etc. Failures may also result from synergistic effects, for example – stress corrosion cracking (SCC), corrosion fatigue (CF), and fretting corrosion.

The importance of such biodegradation processes is paramount. Firstly, they might lower the structural integrity of an implant. Secondly, they may lead to periprosthetic bone loss. With this respect, one example is osteolysis resulting from formation of small polyethylene particles during wear of artificial joints. Another example is focal osteolysis, periosteal reaction and cortical thickening due to fretting corrosion of modular femoral intramedullary nails made of 316L stainless steel [28]. Thirdly, metal ions, which are released as degradation products, are transported by body fluids to remote tissues where they may elicit an adverse biological reaction (such as cytotoxicity, allergy, or even cancer).

Many authors have reported increased concentrations of local and systemic trace metals in association with metal implants. At the screw/plate junctions of internal fixation devices made of stainless steels, the membrane often contains macrophages, foreign-body giant cells, and a variable number of lymphocytes in association with two types of corrosion products: iron-containing hemosiderin-like granules, and microplates which consist of relatively larger particles of chromium compound [29]. Hallab *et al.* [30] have reviewed several concepts of metal sensitivity in patients with orthopedic implants. Dermal hypersensitivity to metal is common, affecting about 10-15% of the population. However, the prevalence of dermal sensitivity is substantially higher in patients with failed metallic implants. Formation of metal ions during corrosion may activate the immune system by forming complexes with endogenous proteins. Metals known as sensitizers include nickel, cobalt and chromium, but occasionally even tantalum, titanium and vanadium. Nickel is the most common metal sensitizer in humans, followed by cobalt and chromium.

Merrit and Brown [31] reported the ability of metallic implants to stimulate metal sensitivity reactions upon degradation. It is apparent that the presence of metal ions in sensitive animals or humans may elicit an inflammatory response and have an adverse effect on the performance of the implant, with pain, swelling and tissue necrosis at the site. It has been found that metal ions, which are released from implants *in vivo*, mostly bind to albumin; their ability to bind to red and white cells varies, hexavalent chromium cations binding most strongly. The binding of certain metal ions to tissues and proteins may be altered by slight increase in pH around the tissue during inflammatory response or infection. Studies have indicated that metallic ions released during corrosion of stainless steels accumulate in the liver and kidneys and are responsible for morphological changes in these organs [32]. Hence, it is necessary to keep the corrosion and number of failures to a minimum by developing materials with improved properties for a specific body environment.

Different failure processes may prevail in the human body. Mudali *et al.* [33] conducted a survey of 50 failures of stainless steel orthopedic implants that had been retrieved from patients. Those implants were sorted based on the reported causes of removal, type of device, anatomical location, implant lifetime, and number of components in the device. Ten cases were selected for thorough failure analysis in order to determine the mechanism and cause of failure. Fatigue-related failures were encountered in three cases. Several cases were related to conjoint action of two failure mechanisms (e.g. fatigue and intergranular corrosion attack in a total knee prosthesis, fatigue and pitting corrosion in a compression bone plate and screws fixation device, and a pit-induced SCC in an intramedullary nail). In one case, of a Sherman bone plate, failure was attributed to the combined action of pitting corrosion, crevice corrosion and CF.

The ability to isolate wear particles from body fluids may become a powerful tool in remnant life prediction, failure analysis and optimization of implants. Bio-ferrography is a method for particles isolation on a glass slide, based upon the interaction between an external magnetic field and the magnetic moments of the particles suspended in a flow stream, while non-magnetic components of the fluid flow into disposable syringes. The principle of this method is illustrated in Fig. 2a. By quantifying the number and size of captured particles and determining their chemical composition and surface morphology, the origin, mechanism and level of degradation may be determined. At Tel-Aviv University, we recently applied this method to isolate particles suspended in synovial fluids for diagnostics of natural joint chondropathies [34] and artificial hip and knee joints performance [35]. In addition, it has been found extremely sensitive and efficient in capturing nano-particles of carbon from ethanol [36] and minerals containing Ti, Zr and Hf embedded in social wasp comb cells [37].

In Ref. 35, synovial fluid aspirates and prosthesis compartments removed by revision surgery from 14 patients were analyzed. Results showed that metallic (namely, Ti-, Co- and Fe-based alloys), polymeric (namely, UHMWPE, POM and PMMA) and bone particles were suspended in synovial fluids. The formation of metal, PMMA and bone particles seemed to accelerate further the wear of certain prostheses. Figure 2b provides macroscopic view of a failed hip prosthesis. This type of cementless isoelastic prosthesis was designed to reduce

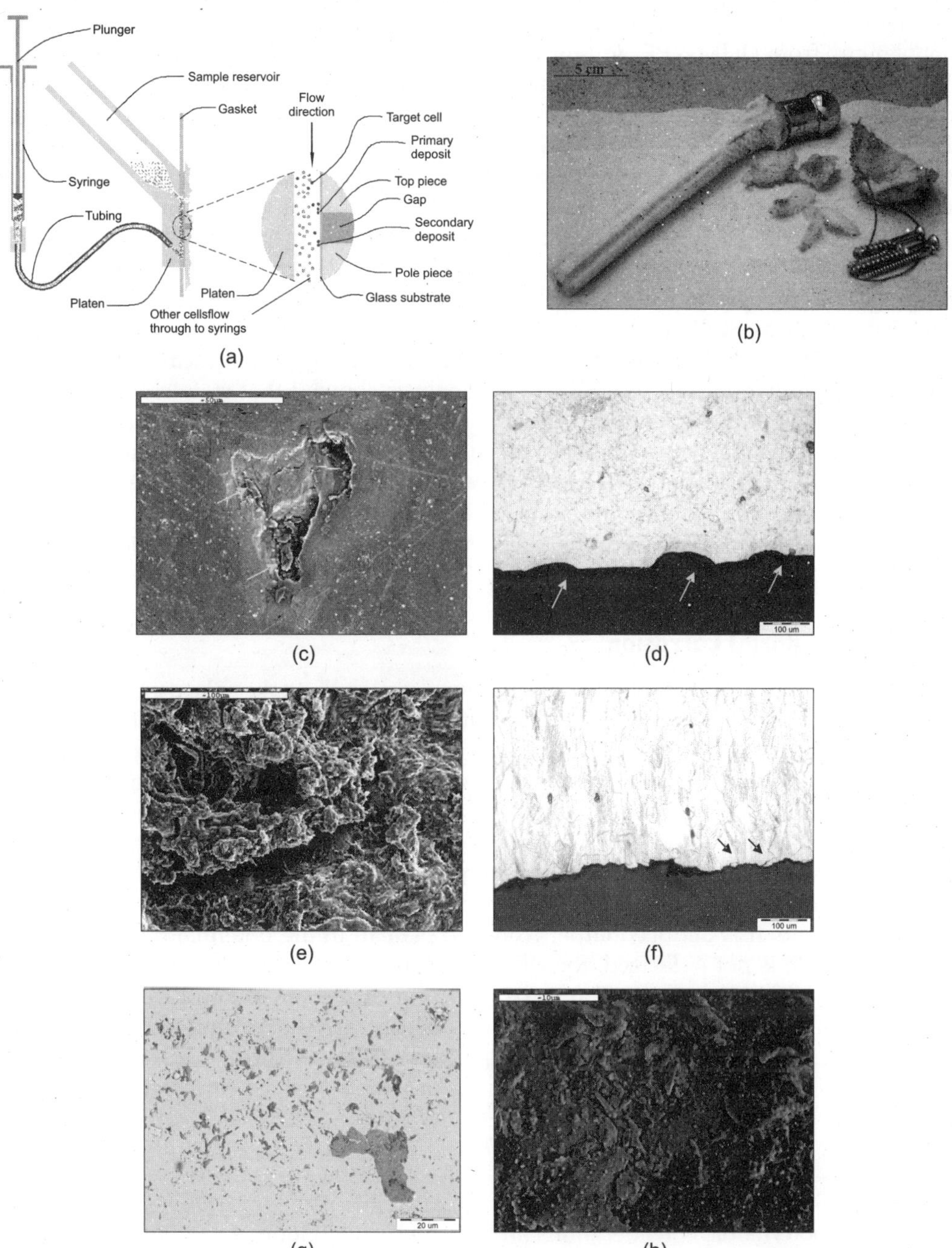

Figure 2 The application of bio-ferrography in the study of biodegradation of artificial hip joint. (a) The principle of particle isolation by bio-ferrography; (b) A retrieved hip joint; (c,d) SEM and optical microscope images of pits on the neck surface; (e,f) Transgranular SCC of a stainless steel screw; (g,h) isolated stainless steel particles as seen under an optical microscope with bichromatic illumination and by SEM.

stress shielding of the proximal femur. The stem is made of polyoxymethylene (POM), the acetabular cup from UHMWPE, and the ball and neck component from 316L stainless steel. In addition, four bone screws made of 316L stainless steel and a wire made of Ti-6Al-4V are noticed in Fig. 2b. One of the screws fractured *in vivo*. Failure analysis revealed ductile tearing of the UHMWPE, as well as crazing and micro-void coalescence in the POM component. Pitting and wear were noticed in the neck component (Figs. 2c and 2d). EDS analysis revealed traces of chloride in these pits. Transgranular stress corrosion cracking and wear were identified around the fracture surface of the failed screw (Figs. 2e and 2f). The exposure of grain boundaries to the outer surface of the screw may indicate that the threads were fabricated by machining, and not by plastic forming. This could have made them susceptible to failures by mechanisms such as SCC and fatigue. The ferrogram (i.e. microscope slide with isolated particles) revealed numerous metallic wear particles under an optical microscope with bichromatic illumination (Fig. 2g). SEM/EDS analysis showed that stainless steel, POM and bone particles were all suspended in the synovial fluid. The stainless steel particles were mainly in the form of platelets with a smooth surface and no striation marks (Fig. 2h). Thus, the failure was attributed to the synergistic effect of corrosion and wear. This example may demonstrate the potential attractiveness of bio-ferrography in studying biodegradation processes. The following paragraphs review corrosion-related failure mechanisms most relevant to the human body environment.

1) General (Uniform) Corrosion

General corrosion may be described as a corrosion reaction that takes place uniformly over the surface of the material, thereby causing a uniform thinning of the device. For a successful implant material, the long-term uniform corrosion rate should typically fall to less than 1 μm per year. Although this rate may be considered insignificant in industrial applications, it is high enough to introduce nickel, chromium and cobalt into surrounding tissues in levels which are five times normal values [15]. Therefore, experiments are often carried out *in vitro* to determine the solubility and/or corrosion potential and corrosion current density in SBFs. The corrosion potential, E_{corr}, is the potential of a corroding surface in an electrolyte, relative to a reference electrode. It is deduced either from the plateau in the potential transient when the working electrode is not polarized, for which it is also known as rest potential (E_r) or open-circuit potential (OCP), or from Tafel extrapolation of the anodic and cathodic curves in potentiodynamic polarization curves. The current density at the corrosion potential, i_{corr}, is also deduced from potentiodynamic polarization curves and is directly proportional to the corrosion rate. The higher E_{corr} and the lower i_{corr} are, the better the corrosion performance of the material is.

2) Galvanic Corrosion

Based on thermodynamic considerations and the Nernst equation, a scale of the reactivity of different metals, known as the electromotive force (emf) series, can be constructed (see Table 3). This scale ranks the equilibrium potential from most positive (most noble) to most negative (most reactive). Metals such as Au and Pt are very noble, i.e. they have low driving force for oxidation in aqueous solutions; hence, they tend to maintain their metallic form *in vivo*. Other metals at the bottom of the emf series, including titanium, have high driving force for

Table 3 Standard electromotive force series for selected reduction half-cells

	Reaction	E^0 (V *vs.* SHE)
Most noble	$Au^{3+} + 3e^- \rightarrow Au$	+1.498
	$O_2 + 4H^+ + 4e^- \rightarrow 2H_2O$ (pH 0)	+1.229
	$Pt^{3+} + 3e^- \rightarrow Pt$	+1.200
	$O_2 + 2H_2O + 4e^- \rightarrow 4OH^-$ (pH 7)	+0.820
	$Ag^+ + e^- \rightarrow Ag$	+0.799
	$O_2 + 2H_2O + 4e^- \rightarrow 4OH^-$ (pH 14)	+0.401
	$Cu^{2+} + 2e^-$ '! Cu	+0.337
	$Ti(OH)^{3+} + H^+ + e^- \rightarrow Ti^{3+} + H_2O$	+0.060
	$2H^+ + 2e^- \rightarrow H_2$	0.000
	$Fe^{3+} + 3e^- \rightarrow Fe$	-0.040
	$Ni^{2+} + 2e^- \rightarrow Ni$	-0.250
	$Co^{2+} + 2e^- \rightarrow Co$	-0.277
	$Fe^{2+} + 2e^- \rightarrow Fe$	-0.440
	$Cr^{3+} + 3e^- \rightarrow Cr$	-0.744
	$Zn^{2+} + 2e^- \rightarrow Zn$	-0.763
	$2H_2O + 2e^- \rightarrow H_2 + 2OH^-$	-0.828
	$TiO_2 + 4H^+ + 4e^- \rightarrow Ti + 2H_2O$	-0.860
	$Ti^{2+} + 2e^- \rightarrow Ti$	-1.630
	$Mg^{2+} + 2e^- \rightarrow Mg$	-2.363
Most active	$Na^+ + e^- \rightarrow Na$	-2.714

oxidation. Yet, it is well known that titanium and its alloys serve very well *in vivo*. This is because they become passive (i.e. essentially inert) under most service conditions due to the spontaneous, rapid formation of a dense, fully covering and well adhered oxide layer that serves as a kinetic barrier to the transport of metal ions and electrons. Other alloys that rely on the formation of a passive film to prevent oxidation are based on iron, cobalt, nickel, chromium, etc.

When two dissimilar metals are electrically connected in a conducting electrolyte, an electrochemical cell is established. An electric potential (voltage) exists between the two electrodes, in accordance with the two half-cell potentials as in the Table 3. The magnitude of this potential can be determined if a voltmeter is connected in an external circuit. The higher the potential of the overall cell, the higher the driving force for non-equilibrium reactions is. Dissimilarity of electrodes may result also from a non-uniform chemical composition of the electrode material, local changes in solution chemistry or dissolved oxygen concentration, different processing routes (e.g. wrought versus cast Co alloys), and surface defects.

Galvanic corrosion is an accelerated corrosion of a relatively active metal (anode) when it is brought in electrical contact with a more noble metal (cathode) in an electrolyte. This form of corrosion may be either uniform or localized. Contact between dissimilar metals immersed in an electrolyte is common in orthopedic, dental and other biomedical applications. Examples include hip prostheses with ball made of 316L stainless steel and socket made of Ti-6Al-4V, a

CoCrMo femoral head in contact with a Ti-6Al-4V femoral stem, and a gold crown coupled to an amalgam core in the oral cavity. In principal, strategies for prevention of galvanic corrosion include selection of materials with as similar electrode potentials as possible, use of insulators between dissimilar metals, and use of coatings or special designs to limit the cathode area relative to the anode area. When titanium- and cobalt-based alloys are coupled together *in vivo*, it may be anticipated that the passive titanium alloy would become the cathode while the less passive cobalt alloy would undergo accelerated corrosion. In practice, however, since the kinetics of the oxygen and water reduction reactions are slow on titanium surfaces, and because the passive current of titanium is virtually independent of potential so it is easily polarized, titanium is a poor cathode. This means that the extent of accelerated corrosion caused to any metal from coupling to titanium should be small. Thus, titanium-cobalt combinations have been found stable both *in vitro* and *in vivo*, at least as long as no relative motion (fretting) occurs [38-40]. On the other hand, 316L stainless steel is susceptible to pitting corrosion when it is coupled to either Ti- or Co-based alloys [41]. European Standard 12010 [42] defines acceptable and non-acceptable combinations of materials for either articulating or non-articulating contacting surfaces of implants.

3) Localized Corrosion – Pitting Corrosion and Crevice Corrosion

Pitting corrosion is a highly localized corrosion of a metal surface that is confined to a small area and takes the form of cavities. This is typically a process of local anodic dissolution, for example at local breakdowns of the passive layer, where metal loss is exacerbated by the presence of a small anode and a large cathode. Pitting corrosion was a common problem with the early 304 stainless steel implants. However, the addition of 2-3 wt% Mo in 316L stainless steel has greatly reduced the number of failures due to pitting corrosion [22]. Mudali *et al.* [43] reported that alloying annealed 316L stainless steel with 0.05-0.22 wt% nitrogen significantly increased the pitting corrosion resistance in a 0.5 M NaCl electrolyte. A synergistic effect of nitrogen alloying and cold working of up to 20% provided an improved pitting resistance. However, at higher cold working levels, the pitting resistance decreased, the effect being more pronounced at higher nitrogen contents. These synergistic effects were attributed to the role of nitrogen in increasing the density of fine deformation bands. Cobalt-based alloys have been found resistant to pitting corrosion under static conditions [44-46], but exposed to pitting corrosion under cyclic loads or following severe cold work [40]. Pure titanium is immune to pitting corrosion in any *in vivo* environment. Although titanium alloys may be less resistant due to discontinuities in the protective oxide film, *in vivo* pitting-related failures have not been reported.

The risk of pitting corrosion in the oral cavity is much higher due to the availability of oxygen and acidic foodstuffs. However, the development of ultraclean grades, such as 316LVM, and/or nitrogen additions have reduced this risk for stainless steels. On the other hand, *in vitro* experiments have shown that titanium alloys might suffer from pitting at high potentials in saline or in the high fluoride solutions used in dental cleaning procedures [22]. Pure Ti exposed to various static immersion tests has also shown a significant increase in ion release (by approximately four orders of magnitude) in the presence of fluoride [47].

Crevice corrosion is a form of localized corrosion occurring at locations where easy access to the bulk environment is prevented, such as the mating surfaces of metals or assemblies of metal and non-metal. It usually occurs in small areas of stagnant solution in crevices, joints and under corrosion deposits. Crevice corrosion of stainless steel implants is a very serious problem even in the Mo-containing 316L grade. In 1959, Scales *et al.* [48] reported that 24% of the 316 stainless steel bone plates and screws removed from patients showed evidence of crevice corrosion. The low non-metallic inclusion type 316LVM stainless steel [49] and an austenitic microstructure that is free of δ-ferrite [50] have been shown to reduce the extent of crevice corrosion, but not to eliminate it [22]. Syrett *et al.* [51] found no crevice corrosion on CoCrMo specimens that had been implanted in dogs and rhesus monkeys for two years. Likewise, Galante and Rostoker [52] found no crevice corrosion on implants removed from rabbits after one year, although the single pits that were observed may indicate on a very early stage of crevice corrosion. As with pitting corrosion, titanium alloys would be less resistant to crevice corrosion than pure titanium. Crevice corrosion of titanium in neutral chloride environments has only been reported at temperatures above 70°C. However, Blackwood *et al.* have shown that at a temperature of 45°C, the protective oxide film on titanium will slowly dissolve if the environment is anaerobic and pH<2. [22]. Because a porous matrix provides already-made crevices, it is not surprising that porous titanium [53] and porous CoCrMo alloys [54] have been shown to exhibit much higher corrosion rates compared to their solid counterparts. Guindy *et al.* [55] studied six dental implants whose late failure was related to suprastructure metal corrosion at the marginal gap. The suprastructures (crowns) were made of porcelain fused to a gold alloy that contained also Pt, Pd, Ag, Co, In and Sn. Extensive corrosion lesions and areas of oxidation were detected on all implants and inner crown surfaces. Bone tissue from around five implants showed higher contents of metal ions in comparison to physiologic baseline values. It was concluded that corrosion was initiated by the bonding oxides, which are necessary for fusing porcelain to gold, and rapidly propagated at the gap crevices. The pH in these regions is locally reduced due to both decrease of oxygen flow and bacteria colonization at the marginal gap spaces. To avoid recurrence of failure, the authors suggested designing a single-unit implant-abutment as well as crowns made of a homogeneous, biocompatible Ti. In addition, all conditionally removable suprastructures should be cemented or sealed to avoid bacterial colonization and possible crevice corrosion.

The most common procedure for testing susceptibility of small implants, in their final form and finish, to localized corrosion is described in ASTM F2129-04 [10]. This procedure is based upon construction of cyclic potentiodynamic polarization curves, after monitoring the OCP for 1 h. In this technique, the potential of the test specimen is controlled and the corrosion current measured by a potentiostat. The potential is first scanned in the positive (forward) direction until a predetermined potential (or current density), usually within the transpassive region, is reached. Then, the scan is reversed until the specimen repassivates or the potential reaches a preset value. Several parameters are defined in these experiments, as drawn in Fig. 3. The corrosion potential (E_{corr}) and corrosion current density (i_{corr}) are determined through Tafel extrapolation of the straight line portion (usually occurring at more than 50 mV from the OCP). The ASTM standard also defines the zero current potential, E_{zc}, as the potential at which the current reaches a minimum during the forward scan. The primary passivation potential, E_{pp}, is the potential corresponding to the maximum active current density of an electrode that exhibits an active-passive transition region. The critical current density, i_{cc}, is the corresponding maximum anodic current density. The passive current density, i_{pas}, is proportional to the charge transferred under passivation conditions. The breakdown (or critical pitting) potential, E_b, is the least noble potential at which pitting or crevice corrosion or both will initiate and propagate. An increase in the resistance to pitting corrosion is associated with an increase in E_b. The vertex potential, E_v, is a preset potential at which the scan direction is reversed. Similarly, the threshold current density, i_t, is a preset current density at which the scan direction is reversed. The protection potential, E_p, is the potential at which the reverse scan intersects the forward scan at a value that is less noble than E_b. The protection potential cannot be determined if there is no breakdown (see Fig. 3). The absence of a hysteresis loop indicates repassivation or oxygen evolution. While pitting will occur on a pit-free surface above E_b, it will occur only in the range of potentials between E_p and E_b if the surface is already pitted. The severity of crevice corrosion susceptibility increases with increasing hysteresis of the polarization curve, i.e. the difference between E_b and E_p. Therefore, a higher value of E_p reflects higher resistance to crevice corrosion. If the metal does not repassivate until a potential below E_v is reached, then it is very susceptible to crevice corrosion. It should be noted that the scan rate (typically, 0.167 or 1 mV/s) may affect the E_b value and the shape of the passive region. In addition, deaeration of the solution with nitrogen gas before and during the test is recommended. It is also recommended to run experiments under the same conditions on reference devices that have a history of good corrosion resistance *in vivo*, for comparison. To avoid intensive hydrogen absorption during the cathodic portion of the curve, it is required to start the polarization only 100 mV below E_r.

4) Stress Corrosion Cracking (SCC)

SCC is caused by the simultaneous effects of tensile stress and a specific corrosive environment. Stress may be due to applied loads, residual stresses from the manufacturing process, or a combination of both. Although Blackwood has claimed [22] that SCC had not been observed on recovered surgical implants, we did observe it in several cases of orthopedic implants made of stainless steels [33,35]. Industrial uses of stainless steels in saline environments also support possible susceptibility to SCC [2].

5) Corrosion Fatigue (CF)

CF occurs as a result of the combined action of a cyclic stress and a corrosive environment. For a given material, the fatigue strength generally decreases in the presence of an aggressive environment. Many medical devices are subjected to low-frequency loads, for example, normal walking results in a hip implant being subjected to cyclic loads at about 1 Hz. Yet, CF seems to hold only a minor percentage of total fatigue failures of implants [22]. It occurs mainly when specifications of materials and processes are not followed, or after long implantation periods [9,56]. On the other hand, Morita *et al.* [11] reported that the *in vivo* fatigue strengths of 316 stainless steel and CoCrNiFe alloy were considerably lower than the equivalent *in vitro* values. These authors suggested that this was due to the low dissolved oxygen concentration in body fluids. Piehler *et al.* [57] tested hip nail plates and found that large plates had better CF resistance than small ones, and that Ti-6Al-4V performed better than type 316L stainless steel. Hughes *et al.* [14] found that the CF resistance of titanium was almost independent of pH over the range 2 to 7, whereas the fatigue strength of stainless steel declined rapidly below pH 4.

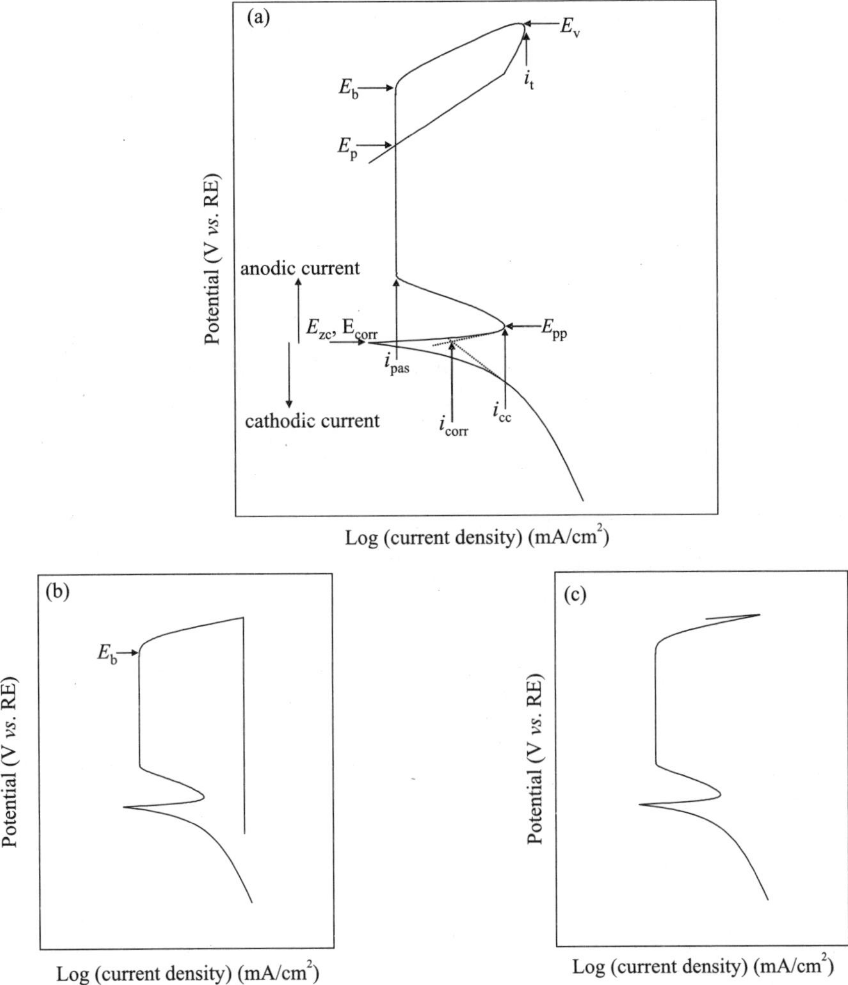

Figure 3 Schematics of cyclic potentiodynamic polarization curves of: (a) a metal that exhibits a protection potential, (b) a metal that does not exhibit a protection potential, and (c) a metal that repassivates [10].

This is consistent with the finding [58] that pitting corrosion facilitates the initiation of CF in stainless steels. In the latter work, it was also reported that the CF resistance of Ti-6Al-4V could be enhanced by nitrogen implantation and heat-treatments to produce fine grain sizes [22].

6) Fretting Corrosion

The term fretting corrosion refers to wear resulting from vibration or slip, which is enhanced by corrosion. The motion abrades surface oxide films off the metal surface, exposing the reactive metal to increased oxide formation. The effect is compounded by the oxide debris, which acts as an additional abrasive between the contacting surfaces. Fretting corrosion appears as pits or grooves in the metal, surrounded by corrosion products. Damage increases with normal load on the contacting surfaces, and with the amplitude of motion. There is no known minimum amplitude below which fretting stops [59,60].

Hallab and Jacobs [61] reviewed several aspects of fretting corrosion of orthopedic implants. Fretting corrosion at modular junctions is produced by relatively small-scale (1- 100 μm) motion between implant components induced by cyclic loading. Fretting of modular implant components commonly occurs in total hip and knee replacements (THR and TKR, respectively); for example, tibial inserts in total knee replacements have been shown to undergo over 100 μm of micromotion at low loads (<100 N) when "locked" into place, creating an ideal site for the generation of fretting corrosion products. This motion results in increased rates of corrosion produced by the repeated fracture and reformation of oxide layers (repassivation), which form over metal surfaces of implants [61]. Corrosion at joints can be a serious problem as it not only results in metal loss but also increases the dimensions of the joint, causing fixation problems. Mechanical wear at joints can also lead to loss of the surrounding cement or bone which, apart from being a serious problem in itself, increases the amount of movement of the implant, thereby increasing the likelihood of corrosion fatigue [22]. Mechanically-assisted crevice corrosion radically alters the chemistry within the crevice solution. This process is driven by the reduction in free energy associated with metal-oxide formation that results in the sequestration of O_2 from any available source (primarily H_2O). Repassivation of metal surfaces also results in a buildup of H^+ and Cl^-, thus lowering the pH within the crevice environment [62].

Orthopedic alloys rely entirely on the formation of passive films to prevent significant oxidation from taking place. Since the potentials across the body fluid/metal solution interface for these reactive metals is typically 1-2 V and the distances are so small, the electric field across the oxide is very high, on the order of 10^6-10^7 V/cm [61]. According to the theory of Cabrera and Mott [63], oxide film growth depends on the electric field across the oxide. If the potential across the metal-oxide/solution interface is decreased, the film thickness will decrease by reductive dissolution processes at the oxide, to make the electric field strength constant. By contrast, increasing the voltage will increase the thickness of the film. If the interfacial potential of an implant interface is made sufficiently negative, or the pH of the solution is made low enough, then these oxide films will no longer be thermodynamically stable and will undergo reductive dissolution [61].

As early as the late 1980's there were reports of fretting corrosion occurring in the modular junctions of TJR components. These studies demonstrated that both mixed and similar metal couples were capable of undergoing this type of corrosion attack, although there was a greater incidence of modular junction fretting corrosion in mixed metal connections. Testing of ceramic-to-metal modular junctions was found to produce significantly less fretting and soluble metal debris than geometrically similar metal-to-metal modular junctions. In order to mitigate fretting corrosion of orthopedic implants, it is required to pay attention to several factors, including: (1) metallurgical processing variables, (2) tolerances of modular connections, (3) surface processing modalities, and (4) appropriate material selection [61].

All three major classes of prosthesis implant materials (stainless steels, Co-based alloys, and Ti-based alloys) suffer from fretting corrosion [64]. The situation is made worse by the fact that the corrosion products accumulate locally as particles, e.g. black titanium oxide debris, which cause further abrasion of the implant. The cause of the shearing micro-movements that eventually lead to fretting corrosion appears to be the large differences between the elastic moduli of the solid metallic implants and the surrounding bone or PMMA cement. The poor fretting corrosion resistance of Ti-6Al-4V is a serious drawback of this alloy. Therefore, different approaches have been explored to minimize this drawback, including anodizing or formation of nitride coatings [22]. An alternative approach is to improve the binding between the implant and its surroundings (either bone or PMMA cement). This can be done through plasma spraying of titanium coating onto Ti-6Al-4V [65], design of the implant material to provide either in-growth of tissue or enhanced on-growth of mineralized bone [66], or application of strongly adhered hydroxyapatite coatings [22,67].

In vitro studies of modular junction fretting have generally quantified the degree of fretting corrosion through electrochemical measurement, or directly through measurement of weight loss of the implant. Current techniques for studying fretting corrosion testing of modular junctions generally employ OCP measurements or metal ion release testing (i.e. immersion tests) while mechanically loading the device [61]. The following section will discuss the corrosion performance of specific materials in more detail.

E. COMMON BIOMATERIALS AND THEIR CORROSION PERFORMANCE

Metals and alloys have a large range of applications, including devices for fracture fixation, partial and total joint replacement, external splints, braces and traction apparatus, dental amalgams, etc. The high modulus and yield stress coupled with the ductility of metals make them suitable for bearing high loads without leading to large deformations and permanent dimensional changes. The compositions most commonly used for load-bearing applications include stainless steels, cobalt-based wrought or cast alloys, and titanium-based alloys. In addition, the NiTi shape memory alloy (SMA) has attracted much attention due to its ability to reproduce its original shape upon exposure to body temperature and its pseudo-elastic properties (E_{eff} ~ 40 GPa) which allow, for example, the processing of low-stiffness, high springback, orthodontic wires. Dental implants are often made of commercially pure (CP) titanium or titanium alloys, amalgams and precious metals (e.g. Au). Noble metals such as Pt and Pt-Ir also find use as electrodes in cardiac pacemakers and other neuromuscular stimulatory devices. Copper is used in contraceptive intrauterine devices (IUDs). Small metallic parts may be used in a wide range of other implants, including skin and wound staples, vascular endoprostheses, filters and occluders. Although metals exhibit high strength and toughness, they have several drawbacks, mainly high modulus of elasticity (which causes stress shielding of bone) and susceptibility to chemical and electrochemical degradation. This susceptibility is usually increased by the action of applied forces and wear. In this section, the main metals and alloys currently used as biomaterials will be discussed. Obviously, within the page limitation of this chapter and the advanced materials introduced to the market every year, it is impossible to discuss all of them. Yet, similar approaches for corrosion characterization and control may be applied to other metals and alloys too.

I) Stainless Steels

The first stainless steel, which was developed specifically for implantation in the early 1900's, was the "vanadium steel" [5]. The early successful devices were fracture fixation plates. However, the surgeons quickly learnt that these devices failed due to mechanical, corrosion and poor biocompatibility reasons. Iron and steel were found to dissolve rapidly *in vivo* and caused erosion of the adjacent bone. Thus, a stainless steel, which contained 18 wt% Cr and 8 wt% Ni, and exhibited improved strength and corrosion resistance, was introduced to the market in 1926. Later that year, Mo was added to this steel in order to improve its localized corrosion resistance. The new alloy became known as 316 stainless steel. During the 1950's, the carbon content in this stainless steel was reduced from 0.08 wt% to 0.03 wt%, thus improving both the corrosion resistance and weldability of the material. This alloy is known as 316L stainless steel (UNS S31673). Standards of stainless steels include ASTMs F138 (bar and wire) [68], F139 (sheet and strip) [69], F745 (cast) [70], F899 [71], F1586 (nitrogen alloyed) [72], and ISO 5832-1 [73]. Chemical compositions of certain stainless steels currently or potentially used as biomaterials are provided in Table 4.

Austenitic stainless steels possess an fcc (γ) structure, unless they undergo a phase transformation due to severe plastic deformation. Chromium soluted evenly within the microstructure allows the formation of a thin (typically 10-50 Å thick), amorphous chromium oxide (Cr_2O_3) layer on top of the steel. The ionic bonds in this layer protect the surface from electrochemical degradation. Hence, stainless steels are often treated in nitric acid to promote the growth and thickening of this passive layer. Austenitic stainless steels are used either in the annealed state or in the cold work state. The latter improves the tensile strength and fatigue properties, with some tradeoff in corrosion resistance. Advantages of stainless steels include high strength, availability, cost effectiveness, and good formability (namely, good cold work, machinability and weldability). Shortcomes, on the other hand, include possible release of toxic ions (namely, Cr and Ni) as a result of the combined action of corrosion and wear, and stress shielding of adjacent bone due to high modulus of elasticity (E ~ 190 GPa) which is about ten times that of bone. Consequently, stainless steels are used nowadays mainly as temporary medical devices or in elder patients.

Hanawa [74] discussed briefly the chemical composition and reconstruction of the oxide films on austenitic stainless steels. The composition was found to consist of Fe and Cr, small amounts of Mo, but no Ni. After mechanical polishing in water, the surface of 316L consisted of iron and chromium oxides containing small amounts of Ni, Mo and Mn oxides. The surface oxide also contained a large amount of OH^-. Following immersion in Hank's solution and incubation with cells, calcium phosphate was formed on and within the film. Sulfate was also adsorbed on the surface of the oxide film, and was reduced to sulfite and/or sulfate in cell culture medium.

Test procedures and evaluation criteria for the corrosion resistance of surgical instruments fabricated from stainless steels and intended for reuse in surgery are described in ASTM F1089 [75]. Figure 4 illustrates a typical cyclic potentiodynamic polarization curve for 316L stainless steel in deaerated neutral Hank's balanced salt solution (HBSS) at 37°C [76]. The corrosion potential is typically more active than titanium alloys, and the corrosion rates higher. The absence of a true potential-independent passive region is evident. In addition, the passive region is small, and a quick transition to a transpassive region occurs. The material does not exhibit E_p, and the large hysteresis loop indicates susceptibility to localized corrosion.

The corrosion performance of stainless steels has been investigated both *in vitro* and *in vivo* in numerous papers. For example, von Fraunhofer *et al.* [77] studied the effect of antibiotics additions to saline on the corrosion potential of the major surgical alloys, including stainless steels. It was found that only one antibiotic, oxytetracycline, exerted a significant effect on the electrochemical behavior, producing an anodic shift of 120-250 mV in E_{corr}. Shih *et al.* [78]

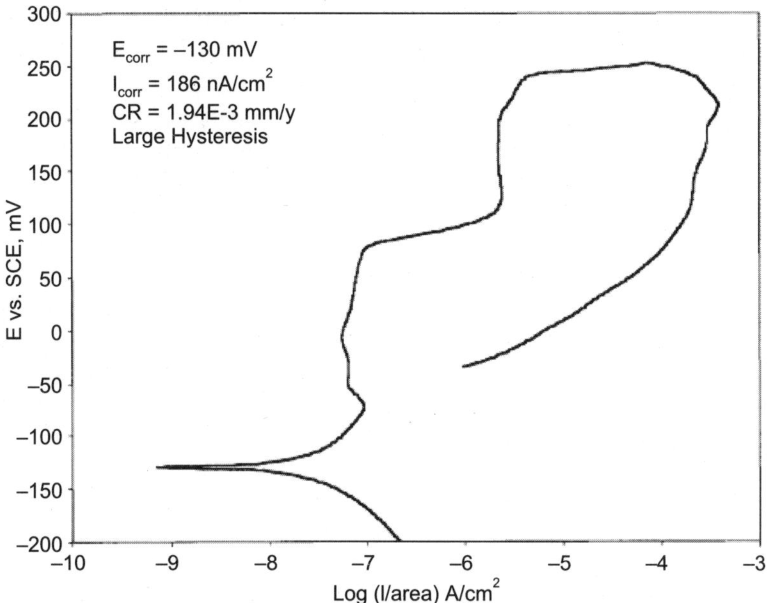

Figure 4 A typical cyclic potentiodynamic polarization curve for a 316L stainless steel in deaerated neutral HBSS at 37°C [76].

showed that while the amorphous oxides on either 316L stainless steel or nitinol were resistant against localized corrosion both *in vitro* and *in vivo*, their counterpart polycrystalline oxides experienced severe pitting or crevice corrosion. Sutow *et al.* [79] showed that the *in vitro* crevice corrosion of cold worked 316LVM steel could be reduced following passivation in 30% HNO_3. Bundy *et al.* [80] conducted *in vitro* tests and showed that cyclic anodic polarization tests with highly loaded fracture mechanics samples resulted in lowering of E_b and disruption of the passive films. The current density increased by more than an order of magnitude in the presence of plastic deformation, compared to loading to below yield stress. Brown and Merritt [81] studied the fretting corrosion of stainless steel plates *in vitro*. The results showed a tenfold decrease in fretting corrosion when 10% solution of fetal calf serum was added to saline. In a later publication [82], proteins were found to increase the corrosion rate of 316L in the static mode, while decreasing the corrosion rate in the fretting mode. The presence of proteins appeared to cause an increase in the anodic Tafel constant and a decrease in the cathodic Tafel constant of this material.

During the last decades, the role of sulfide inclusions in initiating pitting corrosion in stainless steels was recognized [83], leading to the development of type 316LVM stainless steel. This steel is vacuum melted (VM) to reduce the non-metallic inclusions content. Because biomaterials are more prone to localized corrosion in dental applications, a further improvement has been made, forming ultraclean high nitrogen austenitic stainless steels [49]. As mentioned in section C, a PREN of 26 and greater has been recommended to prevent *in vivo* pitting corrosion of stainless steels, in comparison to the value of 40 usually required for stagnant seawater [9]. Additions of nitrogen to 316L stainless steel increase the PREN of this alloy to above 26, and also increase the ultimate tensile strength, although the tradeoff is some loss of ductility (as observed by elongation to fracture). A typical composition (UNS S31675) is shown in Table 4.

Mudali *et al.* identified different failure mechanisms in stainless steel orthopedic devices, and subsequently suggested routes to improve the corrosion behavior of these steels [33]. Alloy modification was carried out through additions of titanium and nitrogen as alloying elements. A super-ferritic stainless steel (Sea-Cure®, UNS S44660, ASTM A268 [84], see Table 4), that was commercially introduced in 1979 to provide high strength and high resistance against localized corrosion and SCC in seawater, was also characterized for comparison. A second comparison was made to SAF 2205 (UNS S31803) duplex stainless steel [85], see Table 4. This steel acquires a ferrite/austenite two-phase microstructure, and is characterized by good weldability and an attractive combination of high strength, high toughness and excellent corrosion resistance (PREN = 35; high resistance to SCC in chloride-bearing environments, as well as to erosion corrosion and to CF). Another stainless steel tested was the 317L (UNS S31703) [86], see Table 4, which is expected to have a significantly higher resistance to pitting and crevice corrosion at ambient temperatures than 316L. Nitrogen was added in contents of 680-1,600 ppm. Figure 5 summarizes several corrosion parameters that were obtained from cyclic potentiodynamic polarization curves in Hank's solution. It is evident that austenitic stainless steels with higher nitrogen content exhibit increased E_b values, which indicates an improved pitting corrosion resistance. The E_b value for the duplex stainless steel is nobler than E_b of the currently used type 316L stainless steel. The super-ferritic stainless steel showed immunity to pitting and crevice corrosion attack and is thus not included in Fig. 5. The mechanisms by which each of the above steels attains its pitting corrosion resistance are summarized elsewhere [33]. Another parameter shown in Fig. 5 is the critical crevice potential, E_{cc}. Mudali and Dayal [87] have defined this parameter in anodic polarization curves as the potential above which significant crevice attack occurred on the surface of the specimen. Hence, the higher E_{cc}, the better would be the resistance of the steel against crevice corrosion. In practice, E_{cc} was identified as the potential at which a monotonic increase in the anodic current exceeding 25 μA was noticed. Figure 5 shows that 316L with high nitrogen contents and duplex stainless steel exhibit higher values of E_{cc}. Finally, the potential range (from E_{corr} to E_p) which is safe against localized corrosion attack is also marked in Fig. 5. Based on the aforementioned results, the resistance of the different stainless steels to pitting corrosion was ranked [33] as follows: super-ferritic > duplex > 316L with 1,600 ppm nitrogen > 317L with 1,410 ppm nitrogen > 317L with 880 ppm nitrogen > 316L with 680 ppm nitrogen > Ti-modified 316L > reference 316L. With respect to crevice corrosion resistance, however, the order slightly changed, as follows: super-ferritic > duplex > 317L with 1,410 ppm nitrogen > 316L with 1,600 ppm nitrogen > 317L with 880 ppm nitrogen > 316L with 680 ppm nitrogen > Ti-modified 316L > reference 316L.

The advantages of surface nitrogen alloying should be emphasized, with respect to biomedical applications. Surface barriers and coatings often have only limited use in protecting implants due to their abrasion and wear, especially in orthopedic applications. Ion implantation is a versatile surface alloying technique, which produces novel metastable solid solution surface alloys of no composition limits as those normally imposed by equilibrium phase diagrams. Nitrogen ion implantation can be carried out on finished orthopedic devices as the process does not create any dimensional changes at the surface after implantation. In addition to corrosion resistance, it also imparts excellent wear resistance to the modified surfaces [33].

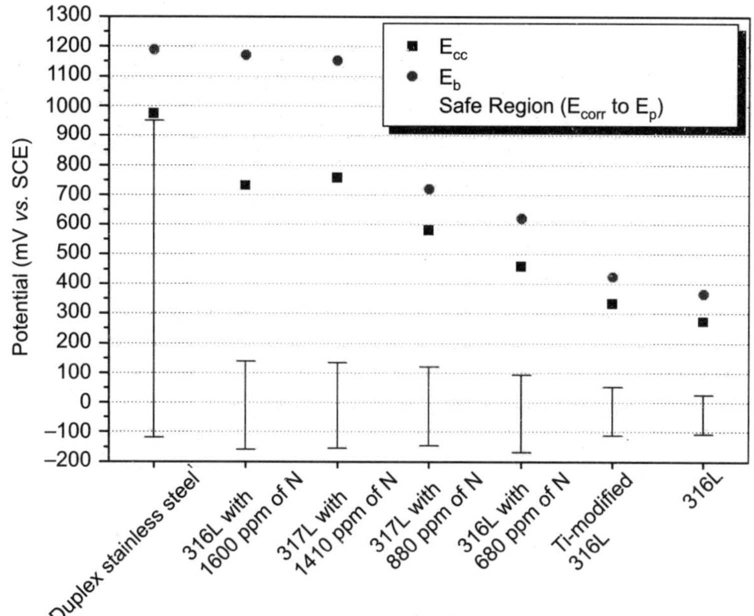

Figure 5 A chart comparing the corrosion resistance characteristics of the various stainless steels studied under simulated body fluid conditions. See text for further details [33].

Advanced stainless steels for surgical implants include the low-nickel (or nickel-free) austenitic stainless steels. One example is UNS S29108 [88], known also as BioDur® 108 from Carpenter Technology Corporation (see Table 4). This alloy was designed to minimize problems associated with Ni toxicity, and is produced by the electro-slag remelting (ESR) process to assure its microstructural integrity and cleanness. It has superior mechanical properties (both static and fatigue) and localized corrosion resistance compared to type 316L alloy.

The corrosion performance of medical devices made of stainless steels can also be improved by electrochemical polishing (electropolishing). In this process, the part to be polished is immersed in an electrolyte and, while an electric current (or voltage) is applied, a thin metal layer is removed. Thus, the surface becomes smooth, free of contaminants and internal stresses, and more passive. The improvement in corrosion resistance may be attributed to preferred dissolution of Fe and Ni, thereby forming a Cr-rich oxide layer. The decreased surface roughness results in an increased resistance against bacteria growth and a reduced protein adsorption, which is helpful in preventing ingrowth of tissue and complications (e.g. restenosis in stent applications). If the final process is electropolish, it is no longer necessary to etch the device in nitric acid in order to regain passivation. Figure 6 shows an example of a commercial medical device from Optonol Ltd., made of 316LVM stainless steel. The fabrication of this miniature glaucoma implant, called Ex-PRESS™, involves processes such as electro-erosion (EDM) and laser welding. Due to the complex geometry and surface inhomogeneities (both chemical and microstructural) resulting from the fabrication processes, it is very difficult to achieve good results using the standard electropolishing process. Therefore, delegate

Table 4 Chemical compositions of stainless steels and cobalt-based alloys currently (or potentially) used as biomaterials

Alloy	Fe	Co	W	Al	Cr	Ni	Mo	Mn	Si	C	P	S	N	Nb	Ti
UNS S31673	Bal.	—	—	—	17.00-19.00	13.00-15.00	2.25-3.00	max 2.00	max 0.75	max 0.03	max 0.025	max 0.01	max 0.10	—	—
UNS S31675	Bal.	—	—	—	19.50-22.00	9.00-11.00	2.00-3.00	2.00-4.25	max 0.75	max 0.08	max 0.025	max 0.01	0.25-0.50	0.25-0.80	—
UNS S31703	Bal.	—	—	—	18.00-20.00	11.00-15.00	3.00-4.00	max 2.00	max 0.75	max 0.03	max 0.045	max 0.03	max 0.10	—	—
UNS S31803	Bal.	—	—	—	21.00-23.00	4.50-6.50	2.50-3.50	max 2.00	max 1.00	max 0.03	max 0.03	max 0.02	0.08-0.20	—	—
UNS S44660	Bal.	—	—	—	25.00-28.00	1.00-3.50	3.00-4.00	max 1.00	max 1.00	max 0.03	max 0.04	max 0.03	max 0.04	(Ti + Nb) = 0.20-1.00 and 6 × (C + N) min	
UNS S29108	Bal.	—	—	—	19.00-23.00	max 0.05	0.50-1.50	21.00-24.00	max 0.75	max 0.08	max 0.03	max 0.01	0.85-1.10	—	—
UNS R30075	max 0.75	Bal.	max 0.20	max 0.10	27.00-30.00	max 0.50	5.00-7.00	max 1.00	max 1.00	max 0.35	max 0.02	max 0.01	max 0.25	—	max 0.10
UNS R30605	max 3.00	Bal.	14.00-16.00	—	19.00-21.00	9.00-11.00	—	1.00-2.00	max 0.40	0.05-0.15	max 0.04	max 0.03	—	—	—
UNS R30035	max 1.00	Bal.	—	—	19.00-21.00	33.00-37.00	9.00-10.50	max 0.15	max 0.15	max 0.025	max 0.015	max 0.01	—	—	max 1.00
UNS R30563	4.00-6.00	Bal.	3.00-4.00	—	18.00-22.00	15.00-25.00	3.00-4.00	max 1.00	max 0.50	max 0.05	—	max 0.01	—	—	0.50-3.50
UNS R30003	Bal.	39.00-41.00	—	—	19.00-21.00	14.00-16.00	6.00-8.00	1.50-2.50	max 1.20	max 0.15	max 0.015	max 0.015	—	—	—
UNS R31537	max 0.75	Bal.	—	—	26.00-30.00	max 1.00	5.00-7.00	max 1.00	max 1.00	max 0.14	—	—	max 0.25	—	—

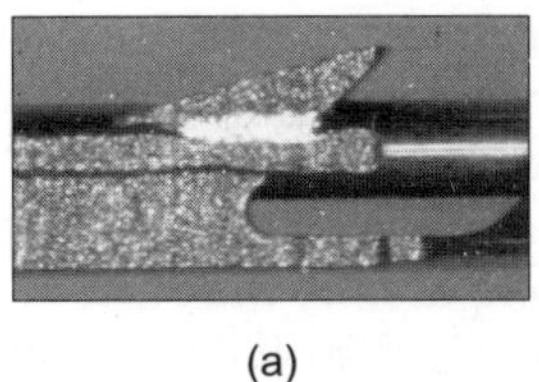

(a)

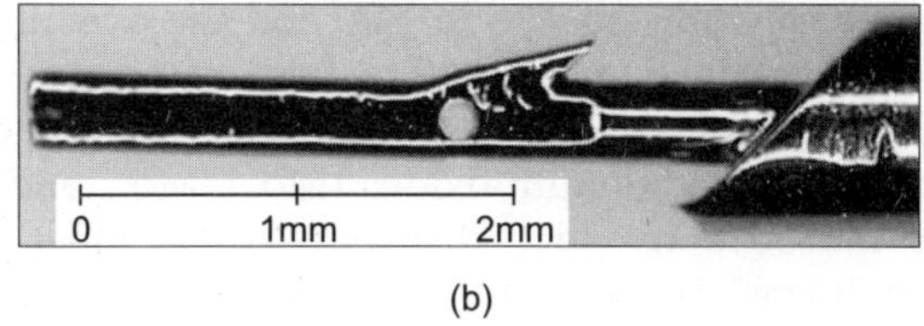

(b)

Figure 6 A miniature glaucoma implant (Ex-PRESS™, Optonol Ltd.) made of 316LVM steel: (a) A close-up after mechanical grinding; and (b) A device after novel electropolishing [89].

mechanical polishing (each implant separately) is currently conducted. However, we recently developed in our lab a novel electropolishing process which yielded excellent results [89], as evident in Fig. 6. It is anticipated that further improvements in this area will allow polishing of other miniature implants as well, thus improving their performance *in vivo*.

One last note on stainless steels: Surgeons and manufacturers of implants should pay attention not to bring in contact different grades of stainless steels (or stainless steels with other alloys) because this might result in galvanic corrosion. As an example, Jedwab *et al.* [90] reported failure when just one of a group of screws holding a 316L fracture plate in position was fabricated from the lower grade 304L.

2) Cobalt-Chromium Alloys

There are two major types of cobalt-chromium alloys – cast CoCrMo and wrought CoNiCrMo alloys. While the former has been used for years in dentistry, and more recently in orthopedics (for TJR), the latter is used for manufacturing hip stems and knee joints. The abrasive wear properties of the wrought alloy are similar to those of the cast alloy. However, the superior fatigue and ultimate tensile strength of the wrought alloy make it suitable for applications which require long service life without fatigue or fracture. The CoCrMo alloy is particularly susceptible to work-hardening; hence, the normal fabrication procedures used with other metals cannot be employed. Standards of cobalt-chromium alloys include ASTMs F75 (cast UNS R30075 alloy), F90 (wrought UNS R30605 alloy), F562 (wrought UNS R30035 alloy), F563 (wrought UNS R30563 alloy), F799 and F1537 (forged UNS R31537, R31538 and R31539 alloys), and F1058 (wrought UNS R30003 and R30008 wire and strip). ISO standards include 5832-4 (cast CoCrMo alloy), 5832-5 (wrought CoCrWNi alloy), 5832-6 (wrought CoNiCrMo alloy), 5832-7 (forgeable and cold-formed CoCrNiMoFe alloy), 5832-8 (wrought CoNiCrMoWFe alloy), and 5832-12 (wrought CoCrMo alloy). Selected compositions are given in Table 4.

Cobalt-based alloys are highly resistant to corrosion, including chloride-induced crevice corrosion. Galvanic corrosion is also of less concern compared to stainless steels. Cobalt-based alloys are quite resistant to fatigue and to environment-induced cracking. They have reasonable toughness, and elongation of more than 8% to fracture. However, wear processes can lead to release of toxic Cr, Co and Ni ions into the body. Another drawback of these alloys is their high modulus of elasticity, which enhances stress shielding compared to other alloys, including stainless steels and titanium-based alloys.

Hanawa [74] described the composition of the surface oxides on CoCrMo alloys. Typically, these are oxides of Co and Cr, without Mo. However, after mechanical polishing in DI water, there were also oxidic species of Mo, and the overall film thickness was about 2.5 nm. The surface film contained also high OH^- content, i.e. it was hydrated or oxyhydroxidized. After dissolution in Hank's solution, the surface oxide consisted of chromium oxide (Cr^{3+}) which contained also molybdenum oxides (Mo^{4+}, Mo^{5+} and Mo^{6+}). Chromium and molybdenum were found to be more widely distributed in the inner layer compared to the outer layer of the oxide film. It was also argued that in body fluids, Co is completely dissolved, and the surface oxide changes into chromium oxide containing a small amount of molybdenum oxide. Calcium phosphate is also formed on the top surface.

Figure 7 illustrates typical cyclic potentiodynamic polarization curves for cast and wrought Co-based alloys in deaerated neutral HBSS at 37°C [76]. It can be noted that the cast alloy has inferior corrosion behavior relative to the wrought alloy. Both alloys do not exhibit an active-passive transition. Their passive regions are not as potential independent as those of titanium and its alloys. They also exhibit a secondary peak at about 500-700 mV, corresponding to the oxidation reactions of Cr at a higher valence level (this is more pronounced for the wrought alloy). There is a true transition from passive to transpassive behavior, with a clearly defined E_b. For the cast alloy, the reverse scan does not cross over the forward scan immediately, but actually traces a loop in doing so. The wrought alloy does not exhibit a hysteresis compared to the cast alloy, indicating that it can repair damage to its oxide layer faster (or better). The finer grain size and more uniform distribution of carbides in the wrought alloy contribute to its chemical homogeneity and improved corrosion behavior.

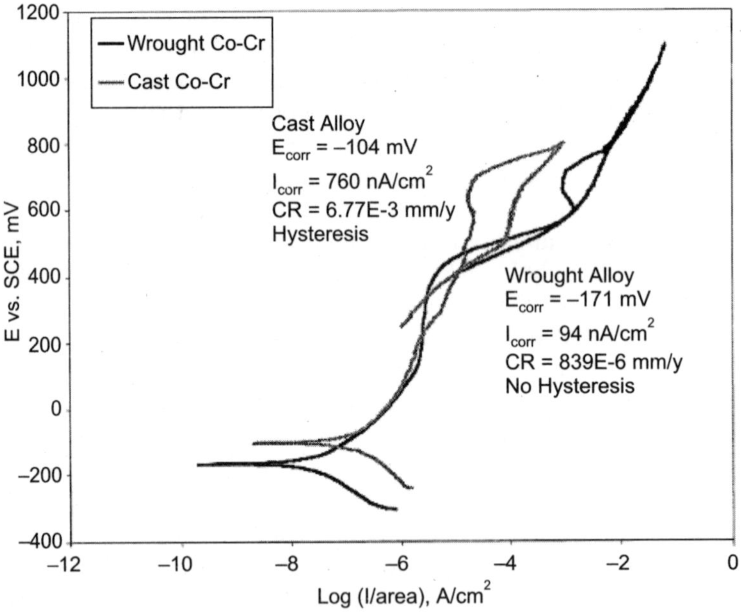

Figure 7 Typical cyclic potentiodynamic polarization curves for cast and wrought cobalt-based alloys in deaerated neutral HBSS at 37°C [76].

3) Titanium and its Alloys

Titanium and its alloys are nowadays the most widely used materials for medical implants [91]. Titanium is light (ρ = 4.51 g/cm^3), biologically and chemically inert, biocompatible, and has low electrical and thermal conductivity {$\sigma = 2.3 \times 10^6$ S/m at 22°C, k = 22 W/(m·K) at 27°C, respectively} in comparison to other metals. Titanium does not warm significantly as a result of exposure to the magnetic field involved in magnetic resonance imaging (MRI) testing. In addition, because it has a coefficient of thermal expansion which is similar to that of the bone, MRI testing complications due to thermal expansion and distortion are minimized. Since titanium is weakly paramagnetic, image interference in MRI or computed tomography (CT) scans is not observed. The modulus of elasticity of titanium and its alloys is closer to that of bone, compared to stainless steels and cobalt-based alloys, thus the concern over stress shielding is reduced. The oxide layer on the surface of titanium and its alloys is formed due to the high affinity of titanium to oxygen, provides corrosion resistance, and allows for histological osseointegration. This oxide layer, typically several nanometers thick, is thermodynamically stable and repairs itself rapidly as long as there is low concentration (several ppm) of oxygen or water in the environment. However, because osseointegration of titanium implants takes a long time (typically three to six months), they are often coated with hydroxyapatite (HAP). This bioactive ceramic is more osteoconductive (i.e. better guides bone formation on material surface in a bony environment) than the titanium surface and can form direct bonds with adjacent hard tissues, leading to an earlier fixation of the implant [92,93].

Relevant ASTM standards for titanium and its alloys include F67 for unalloyed titanium (UNS R50250, R50400, R50550 and R50700) [94], F136 for wrought Ti-6Al-4V ELI (UNS R56401) [95], F1472 for wrought Ti-6Al-4V (UNS R56400) [96], F1713 for wrought Ti-13Nb-13Zr (UNS R58130) [97], F1295 for wrought Ti-6Al-7Nb (UNS R56700) [98], F1813 for wrought Ti-12Mo-6Zr-2Fe (UNS R58120) [99], and F1580 for titanium and Ti-6Al-4V powders for coatings [100]. ISO standards include 5832-2 for unalloyed titanium, 5832-3 for wrought Ti-6Al-4V alloy, 5832-10 for wrought Ti-5Al-2.5Fe alloy, and 5832-11 for wrought Ti-6Al-7Nb alloy. Typical chemical compositions of titanium and its alloys used as biomaterials are provided in Table 5.

Titanium and its alloys are used in a variety of applications, including fracture fixation devices (plates and screws), spinal fixation devices, total hip replacement femoral components, ligament anchorage screws, dental implants and maxillofacial surgery, heart pacemaker housings and artificial heart valves, and components in high-speed blood centrifuges. Generally, alloys are used for joint replacement components because of their superior mechanical properties in comparison to commercially pure (CP) titanium. The Ti-6Al-4V alloy has a dual microstructure, which is consisted of a fine-grained hcp (α) phase and a distributed bcc (β) phase. The presence of elements such as Al, O, N, Ga and C stabilizes the α phase, whereas transition metals such as V, Nb, Ta and Mo stabilize the β phase. The microstructure and mechanical properties of Ti-6Al-4V are highly dependent on the thermo-mechanical processing treatments. If the material is cooled too slowly, the β phase becomes more prominent and lowers the strength and corrosion resistance of the alloy. Interstitial atoms such as oxygen, nitrogen and carbon increase the strength significantly, nitrogen having an almost double effect per atom.

Drawbacks of titanium and its alloys include high price, high susceptibility to friction and wear (as reflected by a high coefficient of friction), and low shear strength. Machining of titanium is not cost-effective. Although precision casting reduces the manufacturing costs, cast titanium alloys exhibit low fatigue strength and low elongation compared to wrought alloys due to the coarse microstructure of the cast alloys. Therefore, supplement microstructural refinement is often required to improve the mechanical properties while preserving the shape of the product. Thermomechanical processes with post heat treatment have found to increase the strength, elongation and fatigue strength of α+β titanium alloys. Another cost-effective process is powder metallurgy (PM). This process is very useful in producing more homogenous β alloys which contain alloying elements with high melting point such as Nb and Ta [101].

Wear and corrosion processes result in release of vanadium and aluminum from Ti-6Al-4V into the body, thus causing hypersensitivity and other clinical complications. The potentially toxic aluminum has been removed in the new Ti-50Ta alloy (an α+β alloy), which has shown excellent corrosion resistance and high tensile strength *in vitro* [102]. The presence of Zr and Nb in titanium has been found to reduce significantly the modulus of elasticity as well as to improve the corrosion resistance *in vivo*. In the middle of the 1980's, the dual phase Ti-6Al-7Nb was introduced into clinical use, realizing that Nb is both cheaper and more biocompatible than V. This alloy has been shown to possess enhanced mechanical properties and workability [103]. Another (near-β) alloy, Ti-13Nb-13Zr [97], exhibits higher adhesion of osteoblasts and lower bacterial adhesion than CP-Ti and Ti-6Al-4V. This alloy is useful in designing cardiovascular implants because the ZrO_2 passive film is thrombogenically compatible with blood. The corrosion resistance of this alloy is also enhanced due to the presence of ZrO_2 and Nb_2O_5 which strengthen the TiO_2 film formed on the surface. Hence, many modern titanium-based alloys are β stabilized in order to reduce the elastic modulus towards that of bone, although the tradeoff is typically lower wear resistance and fatigue strength. Typical mechanical properties of selected Ti-based alloys are summarized in Table 6, in comparison to some other common biomaterials.

Recently, Hong *et al.* [92,93] designed and synthesized by three-dimensional printing (3DP™) advanced Ti-5Ag and Ti-5Ag-35Sn alloys. The 3DP process offers several significant advantages over other processes for customized fabrication of prostheses. First, it can form parts of almost any geometry; e.g., structures with complex geometry and/or small dimensions can be produced with significantly more precision. In addition, it can be scaled-up both in size and rate through the use of multiple nozzle printheads on a raster machine, and is adaptable to different material systems. In the above study, liquid-phase sintering and liquid-tin infiltration were used, respectively, to densify the printed alloys. The corrosion behavior of these alloys as well as of pure titanium control samples was studied by means of short- and long-term OCP and potentiodynamic measurements during immersion in non-deaerated saline at body temperature, as well as by AES. Figure 8 shows typical potentiodynamic curves for each of the five materials. Although the curves for the pure titanium control sample and for the Ti-Ag alloy sintered at 1300°C are similar in shape, the Ti-Ag alloy exhibits a less active E_{corr}. This was attributed to the higher exchange current density, i_o, for hydrogen/oxygen reduction on Ag than on Ti. Typically, coupling of Ti with noble metals such as Pt, Au, or Pd results in spontaneous passivation, since these noble metals act as good catalysts for the hydrogen and oxygen reduction reactions. The Ti-Ag alloy sintered at 1150°C exhibits a less stable passivation (or metastable pitting) at high potentials.

Table 5 Chemical compositions of titanium and its alloys currently in use as biomaterials

	Ti	Al	V	Ta	Nb	Mo	Zr	O	N	C	H	Fe	Ni	Co	Cu
UNS R50250	Bal.	—	—	—	—	—	—	max 0.18	max 0.03	max 0.08	max 0.015	max 0.20	—	—	—
UNS R50400	Bal.	—	—	—	—	—	—	max 0.25	max 0.03	max 0.08	max 0.015	max 0.30	—	—	—
UNS R50550	Bal.	—	—	—	—	—	—	max 0.35	max 0.05	max 0.08	max 0.015	max 0.30	—	—	—
UNS R50700	Bal.	—	—	—	—	—	—	max 0.40	max 0.05	max 0.08	max 0.015	max 0.50	—	—	—
UNS R56401	Bal.	5.50-6.50	3.50-4.50	—	—	—	—	max 0.13	max 0.05	max 0.08	max 0.012	max 0.25	—	—	—
UNS R58130	Bal.	—	—	—	12.50-14.0	—	12.50-14.0	max 0.15	max 0.05	max 0.08	max 0.012	max 0.25	—	—	—
UNS R56700	Bal.	5.50-6.50	—	max 0.50	6.50-7.50	—	—	max 0.20	max 0.05	max 0.08	max 0.009	max 0.25	—	—	—
UNS R58120	Bal.	—	—	—	—	10.0-13.0	5.0-7.0	0.008-0.28	max 0.05	max 0.05	max 0.02	1.5-2.5	—	—	—
NiTi	Bal.	—	—	—	max 0.025	—	—	max 0.05	—	max 0.07	max 0.005	max 0.05	54.5-57.0	max 0.05	max 0.01

Table 6 Mechanical properties of selected biomaterials

Material	UNS designation	σ_{UTS} (MPa)	σ_{YP} (MPa)	*E* (GPa)	ε (%)
316L, annealed	S31673	min 490	min 190	190	min 40
316L, cold worked	S31673	min 860	min 690	190	min 10
BioDur® 108 stainless steel, annealed	S29108	min 827	min 517		min 30
Co-28Cr-6Mo, as-cast	R30075	min 655	min 450	210	min 8
Co-28Cr-6Mo, forged	R31537	min 1,172	min 827	210	min 12
Co-20Cr-15W-10Ni, annealed	R30605	min 860	min 310	210	min 30
Co-35Ni-20Cr-10Mo, solution annealed	R30035	793-1,000	241-448	232	min 50
CP-Ti, grade 1	R50250	min 240	min 170	110	min 24
CP-Ti, grade 4	R50700	min 550	min 483	110	min 15
Ti-6Al-4V ELI, annealed	R56401	min 825	min 760	116	min 8
Ti-6Al-7Nb, annealed	R56700	min 900	min 800	114	min 10
Ti-13Nb-13Zr, capability aged	R58130	min 860	min 725	75	min 8
Ti-12Mo-6Zr-2Fe, solution-annealed	R58120	min 932	min 897	74-85	min 12
NiTi, annealed	—	min 551		40	min 10
Cortical bone	—	70-150	30-70	10-30	0-8

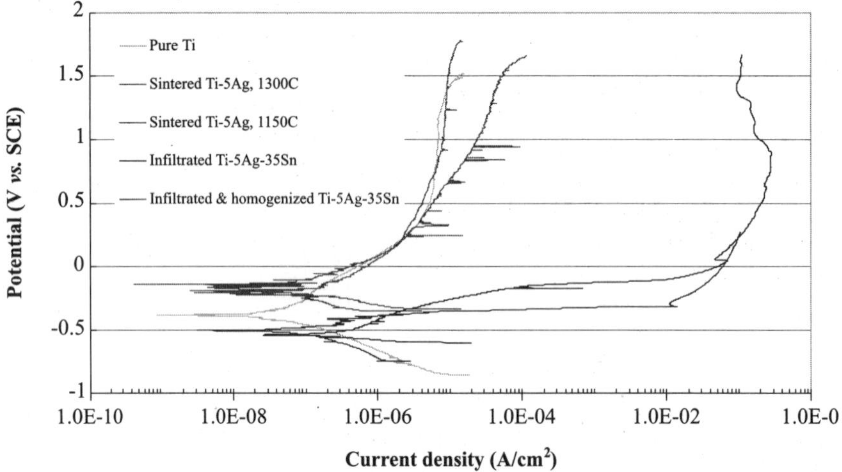

Figure 8 Typical potentiodynamic polarization curves of pure Ti as well as of Ti-5Ag and Ti-5Ag-35Sn alloys processed by 3DP™ [92].

This difference was related to the more even distribution of Ag and the lower porosity in the alloy sintered at 1300°C. With regard to i_{corr}, the Ti-Ag alloys sintered at either 1300 or 1150°C exhibited higher values compared to the titanium control samples, although of the same order of magnitude. This difference could be explained in terms of the real surface area in contact with the electrolyte, which was increased due to porosity. The sintered Ti-Ag alloys were also found to exhibit an exceptionally high hardness, which may be beneficial in providing enhanced wear resistance. The combination of good electrochemical and mechanical properties, and the presence of an anti-bacterial material (Ag), make these alloys attractive candidates to both orthopedic and dental applications.

The Ti-Ag-Sn alloys also exhibit a galvanic couple behavior. However, this behavior is much different from the one observed for the Ti-Ag system, as evident from the different shape of potentiodynamic curves in both cases (Fig. 8). At a constant potential, the cathodic polarization curve of Sn typically exhibits a lower current density, while the anodic polarization curve of Sn exhibits a much higher current density, compared to Ti. Therefore, the corrosion potential is likely determined by Sn anodic reaction and Ti cathodic reaction. Consequently, the Ti-Ag-Sn alloys don't exhibit any passivation before the dissolution of Sn is significantly increased. Therefore, these alloys cannot be used as biomaterials.

It is well known that CP-Ti and single-phase titanium-based alloys exhibit enhanced corrosion properties compared to most other metallic biomaterials. Figure 9 illustrates typical cyclic potentiodynamic polarization curves for CP-Ti or titanium alloy in deaerated neutral HBSS at 37°C [76]. The E_{corr} is nobler than that of 316L stainless steel and Co-based alloy (see Figs. 4 and 7, respectively), and the material translates directly into a stable passive behavior from the Tafel region, without exhibiting an active-to-passive transition (note that no E_{pp} or i_{cc} exist). Moreover, neither E_b nor E_p is evident, indicating that this material has a very protective oxide layer. Consequently, only minimal release of ionic or by-product residue into the periprosthetic tissue occurs, and this material may be classified as bioinert in the whole range of clinically relevant potential-pH combinations.

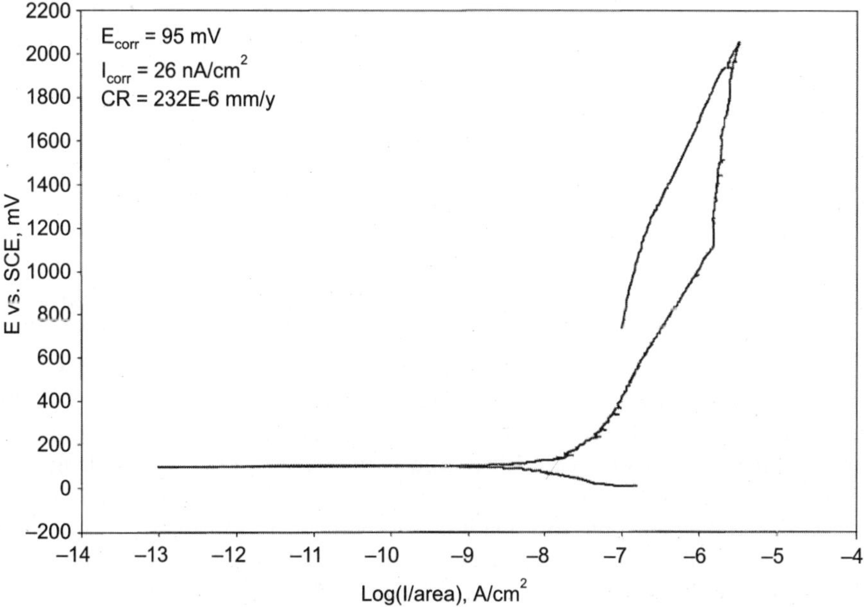

Figure 9 Typical cyclic potentiodynamic polarization curve for titanium alloy generated in deaerated neutral HBSS at 37°C [76].

In order to better understand the corrosion behavior of titanium and its alloys, the Pourbaix diagram for the Ti-water system at 25°C is shown in Fig. 10. Although titanium is thermodynamically reactive $\left(E^{0}_{\mathrm{Ti/Ti+2}} = -1.63\ \mathrm{V}\ vs.\ \mathrm{SHE}\right)$, it is highly corrosion resistant due to the instantaneous formation of a hard, tightly adherent, passive film which is stable at all pHs in oxidizing potentials. It corrodes only at low pH in solutions without oxidizers. The natural passive film formed on titanium, when exposed to different synthetic physiological solutions, is typically a few nanometers thick, and contains a relatively high concentration of oxygen vacancies. This titanium oxide layer may be described as an n-type semiconductor containing anion vacancies. It is known that the kinetics of titanium corrosion in neutral solutions is controlled by migration of oxygen vacancies across the oxide film. Thus, a thick TiO_2 film containing a low concentration of oxygen vacancies will lead to a very slow mass transport rate across the film [92]. When the oxide is sufficiently thin (0.4-3 nm), electron exchange occurs between the redox electrolyte and the underlying metal by direct tunneling or resonance tunneling via intermediate states. A freshly abraded titanium surface immediately passivates to form a low-crystalline rutile and/or anatase oxide layer. The titanium oxide gradually decreases in oxygen content from TiO_2 at the surface, to Ti_2O_3 and TiO as it approaches the metal/oxide interface. Depending on the environment, this oxide may be covered with an amorphous or hydrated surface oxide, thus exhibiting a bilayer oxide structure. The oxide may be thickened in the presence of oxidizing agents, through anodization or thermal oxidation. Anodization can thicken the very thin natural oxide from ~20 Å to several thousand angstroms, depending on the applied potential [104].

The characteristics of the surface oxide layer on titanium have been studied by different researchers. Hanawa [74] has reviewed the composition, reconstruction and regeneration in biological environments of various surface oxides, including those on titanium and its alloys. The film on Ti-6Al-4V was found to be almost the same as that on CP-Ti, although containing a small amount of Al_2O_3. Vanadium contained in the alloy was not detected in the oxide layer. Calcium, phosphorus and sulfur have been observed on the surface of CP-Ti surgically implanted into the human jaw. Calcium phosphates also form on titanium and its alloys by immersion in Hank's solution and other SBFs. When the surface oxide film of a metallic material is disrupted, corrosion proceeds, and metal ions are released continuously unless the film is regenerated. Therefore, the number of released metal ions is governed by the regeneration time of the film. The regeneration time in saline solution was found to be much shorter for Ti-6Al-4V (8.2 min) than for 316L stainless steel (35.3 min) or Co-28Cr-6Mo (12.7 min). The repassivation rate was not influenced by the pH of solution, dissolved oxygen or proteins [74].

Gilbert *et al.* [105] reported the effect of potential, pH and aeration on the titanium oxide film fracture and repassivation. In a subsequent work [106,107], Bearinger *et al.* studied the morphological changes of TiO_2 oxide films upon exposure to PBS and hydrogen peroxide-modified PBS solutions. Hydrogen peroxide (H_2O_2) is an oxidizing molecule, which is secreted by cells and is associated with wound healing, and is thus present in significant quantities in fresh implant environments. The substrates were either CP-Ti or Ti-6Al-4V. The surfaces were subjected to simultaneous polarization or impedance testing and *in situ* electrochemical atomic force microscopy (EC-AFM) imaging to evaluate how the structure and properties of the passive oxide film are affected by varying potential and hydration. All samples were found

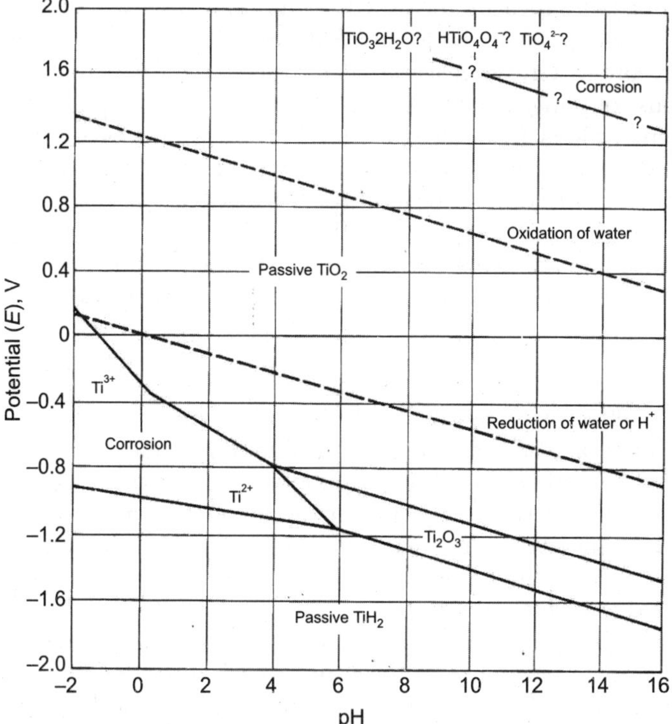

Figure 10 Pourbaix diagram of the Ti-$H_2$0 system [104].

covered with protective titanium oxide domes that grew in area and coalesced due to hydration, and as a function of increasing applied voltage and time. Reversal of dome growth did not occur upon voltage reduction, while impedance behavior was quasi-reversible, suggesting independence between structural and electrical properties. Oxide growth appeared to occur in part by lateral spreading and overgrowth of domes at the oxide/solution interface.

The corrosion behavior of various titanium alloys in different environments has been studied extensively. Nakagawa *et al.* [108] studied the corrosion behavior of Ti-6Al-4V, Ti-6Al-7Nb and Pd-containing titanium-based alloys in a wide range of pH and fluoride concentrations and found that the Ti-0.2Pd alloys were more resistant to corrosion due to surface enrichment with Pd. Khan *et al.* [109] investigated the corrosion behavior of Ti-6Al-4V, Ti-6Al-7Nb and Ti-13Nb-13Zr in PBS at various pH levels as well as in the presence of protein solutions. It was shown that the Ti-13Nb-13Zr alloy was least affected by the change in the pH level, as compared to the other two alloys. The reduction in the hardness of the surface oxides due to corrosion in protein solutions was less for Ti-13Nb-13Zr than for Ti-16Al-7Nb and Ti-6Al-4V. Williams *et al.* [82] also studied the effect of proteins on the corrosion rates of CP-Ti and Ti-6Al-4V in the static and fretting modes. It was found that proteins increased the corrosion rate of CP-Ti, but did not have an effect on Ti-6Al-4V. In the fretting mode, however, proteins did not have an appreciable effect on these two materials. Aziz-Kerrzo *et al.* [110] compared the corrosion susceptibility of CP-Ti, Ti-6Al-4V and Ti-45Ni in PBS using anodic polarization and electrochemical impedance measurements. CP-Ti and Ti-6Al-4V were found more resistant to pitting corrosion in comparison to the Ti-45Ni SMA, for which pitting potentials as low as +250 mV *vs.* SCE were recorded. Grosgogeat *et al.* [111] measured the galvanic corrosion current and potential for various couplings of either CP-Ti or Ti-6Al-4V with seven different Au-, Ag-, Pd-, or Co-Cr-based dental alloys. In all cases, the level of corrosion was found low. Manivasagam *et al.* [101] provided a comprehensive review of physical metallurgy of titanium alloys, their corrosion behavior, and surface modification approaches to improve the corrosion behavior *in vivo.* We shall now proceed and review the corrosion performance of other metallic biomaterials. Because high contents of titanium exist in SMAs such as nitinol, the following paragraphs will be dedicated to the corrosion behavior of these materials.

4) Nitinol

The shape memory effect, superelasticity, and good damping properties make the nickel-titanium SMA (Nitinol or NiTi) an attractive material for surgical applications, such as self-locking, self-expanding and self-compressing implants. The term *shape memory effect* reflects the ability of an alloy to revert its original shape following deformation at low temperatures and subsequent heating above its *transition temperature.* While NiTi is soft and easily deformable in its lower temperature phase (martensite), it resumes its original shape and rigidity when heated to its higher temperature phase (austenite). The shape memory effect is based on this temperature-dependent austenite-to-martensite phase transformation on an atomic scale. The temperature range for the martensite-to-austenite transformation that takes place upon heating is somewhat higher than that for the reverse transformation upon cooling. This difference in the transition temperatures upon heating and upon cooling is called *hysteresis.* In practice, an alloy designed to be completely transformed by body temperature upon heating (A_f < 37°C) would require cooling to about 5°C to fully retransform into martensite (M_f). *Superelasticity* (pseudoelasticity) means that the alloy specimen is able to return to its original shape upon unloading after a substantial deformation. Within a given temperature range, NiTi can be strained several times more than conventional alloys without being plastically deformed (it can reach elastic deformations of 10-12%) [6,112,113]. Relevant ASTM standards include F2005 (terminology related to SMAs) [114] and F2063 (wrought alloys) [115].

The shape memory effect was discovered in an equiatomic alloy of nickel and titanium in the early 1960's by Buehler and his co-workers at the U.S. Naval Ordnance Laboratory. This alloy was thus named Nitinol (Nickel-Titanium Naval Ordnance Laboratory). The first efforts to exploit the potential of NiTi as an implant material were made by Johnson and Alicandri in 1968. However, the use of NiTi for medical applications was first reported only in the 1970's. Although in the early 1980's some orthodontic and orthopedic applications were already available on market, it was only in the mid 1990's that the first widespread commercial stent applications made their breakthrough in medicine [6].

There are some exceptional properties of NiTi that might be useful in surgery. NiTi has an ability to be highly damping and vibration-attenuating below A_s. From an orthopedic point of view, this property could be useful in, for example, dampening the peak stress between the bone and the articular prosthesis. The low elastic modulus of NiTi ($E \cong 40$ GPa, which is much closer to that of bone compared to any other metallic biomaterial) may also be beneficial. NiTi possesses enhanced fatigue and ductility properties, as well as very high wear resistance. It is non-magnetic; therefore, MRI imaging is possible [6].

The mechanical properties of NiTi alloys are especially sensitive to the chemical composition and the individual thermal and mechanical history. A typical chemical composition is provided in Table 5, whereas typical mechanical properties are given in Table 6. The surface of NiTi consists mainly of TiO_2, smaller amounts of NiO and Ni_2O_3, and metallic Ni, while nickel-titanium constitutes the inner layer. The thickness of the oxide layer varies within 2-20 nm. During implantation, this oxide layer grows and takes up minerals (e.g. calcium phosphates) and other constituents of biofluids, resulting in remodeling of the surface. Several surface treatments have been introduced to improve the corrosion properties of NiTi, including titanium nitride coating prepared by an arch ion plating method, chemical modification with human plasma fibronectin via aminosilane and glutaraldehyde as coupling agents, plasma-polymerized tetrafluoroethylene (PPFTE) coating, laser surface treatment, electropolishing, and nitric acid passivation [6].

Most of the knowledge on the corrosion behavior of NiTi is from studies of dental arch wires and *in vitro* conditions. Although the corrosion resistance of NiTi in SBFs has yet to be fully assessed, its resistance to pitting corrosion appears to be similar to (or better than) 316L stainless steel [116]. Ryhänen [6] reviewed the literature and added his own data to demonstrate that despite its higher initial nickel dissolution, NiTi induces no toxic effects, decrease in cell proliferation, or inhibition in the growth of cells in contact with the metal surface. The muscular tissue response to NiTi was clearly non-toxic and non-irritating, as were also the neural and perineural responses. The overall inflammatory response and the presence of immune cells, macrophages and foreign body giant cells were similar compared to 316LVM stainless steel, CP-Ti and Ti-6Al-4V. After eight-week implantation, histomorphometry showed that the encapsulating membrane of NiTi was thicker than that of stainless steel, but at 26 weeks the membrane thicknesses were equal. Determination of trace metals in several distant organs showed no statistically significant differences in Ni concentration between the NiTi and 316LVM stainless steel.

5) Dental Amalgams

Dental amalgams, in widespread use for over 150 years, are among the oldest materials used in oral healthcare. They are formed by mixing liquid mercury (45-55 wt% Hg) and a powder made of Ag, Sn and Cu (sometimes – also smaller amounts of Zn, Pd, In, or Se). Mixing occurs via mechanical vibration and results in a putty-like material which is easy to manipulate (e.g. in filling cavities). Amalgams possess high compressive strength and high dimensional stability; however, shrinkage and corrosion remain of concern. Because dental amalgams are multiphase alloys, there is a risk of localized, galvanic or intergranular corrosion between the different phases. In the conventional silver amalgams, the phases that contain Ag are nobler, and accelerated corrosion attacks the γ_2 phase (Sn_7Hg), thus releasing toxic mercury into the body. On the other hand, although high copper amalgams are also susceptible to intergranular corrosion, the most susceptible phase is η (Cu_6Sn_5), which does not release Hg into the body when it corrodes [117]. Yet, because the high copper content shifts E_{corr} to more positive values, high Hg release may still be observed. Joska *et al.* [118] found that the Hg release rates from conventional silver amalgams and high copper amalgams were very similar, with the method of preparation being critical. Furthermore, it is has been shown that for most people the major source of Hg uptake is food, with less than 10% coming from dental amalgams [119]. High copper amalgams are prepared from either a mixture of silver-tin and silver-copper alloys or from a ternary silver-copper-tin alloy. The high copper amalgams have been reported to have superior clinical properties, with a higher resistance to corrosion [120]. However, too much copper can cause a positive shift in the corrosion potential, leading to increased corrosion rates [118].

Dental alloys with high contents of Au and other precious metals, used for crowns and inlays amalgams, are highly corrosion resistant in nearly all oral environments. The exception may be when high fluoride levels are introduced into the mouth during some dental cleaning procedures. However, in some patients, both Ag- and Au-based alloys can suffer from tarnishing, in which a thin black layer (probably a sulfide) develops across the surface. Although tarnishing does not dramatically affect the performance of the amalgam or the release rates of Hg, it is unsightly and is thus of concern. The solution to the problem is more likely to be eliminating the source of the sulfide, e.g. changing the patient's diet, rather than replacing the amalgam with a more corrosion resistant material [22].

6) Gold

Gold and its alloys are used in applications such as electrode materials in medical devices and dentistry due to their durability, immunity to corrosion, desired electrical and thermal properties, and esthetic appearance. Alloying of gold with copper or platinum results in increased strength, while silver is added to compensate for the color of copper. Although gold is common in dental cast restoration applications, it is not suitable for orthopedic applications because of its high density, insufficient strength, and high cost.

Under most environments, gold is immune to corrosion. The Pourbaix diagram for gold in aqueous solutions free from complexing substances at 37°C shows that, except of a small regime at pH < 1, gold is immune to corrosion over the entire domain of water stability. However, the introduction of a small amount of chloride ion into the system allows for regions of stability of soluble gold(I)-chloride complexes, thus making it prone to corrosion. In the presence of 0.1 M Cl^- (and 0.6 mM Au^+), the Pourbaix diagram shows that the domain of gold corrosion extends significantly to pH as high as 8, although at $6 < pH < 8$ it does not overlap the region of water stability, and thus requires the application of voltage perturbation [121]. With respect to the valence of gold ions, Frankental and Siconolfi [122] found that gold dissolves as a monovalent ion at $E < 0.8$ V and as a trivalent ion at $E > 1.1$ V. Gold(I)-chloride is extremely unstable in aqueous environments. Analysis of antibody sensitization indicates that AuCl quickly decomposes *in vivo* into $AuCl_3$ [123]. From kinetics point of view, gold exhibits a typical passive metal behavior in the presence of chloride. Gold corrosion rate in the presence of chlorides is a strong function of both chloride concentration and solution velocity. These effects are attributed primarily to a depletion of chloride ions at the anodic surface, leading to a diffusion-limited current density. This suggests that the corrosion rate is unlikely to be significantly affected by cathodic reactions [121].

Rosenberg [121] studied the electrochemical performance of gold membranes covering drug-filled wells in implantable drug delivery devices. Comparisons were made between behaviors in PBS, calf serum and a rat model. Results showed that PBS solutions can match the thermodynamic environment of biological media, but not the kinetics. The kinetics of gold corrosion in calf serum was shown to be limited significantly by modification of the gold surface by the media. Gold samples treated in serum and subsequently corroded in PBS exhibited similar kinetics to those corroded in serum alone. It was suggested that the kinetic variation was primarily due to protein modification of the gold surface. Gold samples corroded in serum exhibited protein hydrogels. Gold chloride was shown to prevent serum from modifying the gold surface.

F. CORROSION CONTROL STRATEGIES IN-VIVO

As explained before, most strategies commonly used in industry to mitigate corrosion are not applicable in the body environment. These include: (1) changing the chemistry of the environment, (2) controlling the oxygen level, (3) adding inhibitors, (4) changing the pH, (5) lowering the temperature, and (6) applying anodic or cathodic protection techniques. In addition, service *in vivo* raises unique challenges due to the negative effects of proteins, enzymes and other body matter on corrosion processes, and the action of wear and high loads on load-bearing implants. Hence, corrosion control *in vivo* is currently limited mainly to careful design (e.g. to prevent galvanic couplings or crevices), proper material selection, and surface modification. From section E it could be realized that the list of potential materials is currently somehow limited because of mechanical and biocompatibility requirements. Therefore, much attention has been paid to surface modification approaches.

Due to page limitation, it is not possible to discuss these strategies in detail herein. Thus, the reader is encouraged to refer to reviews of surface modification approaches provided elsewhere [7,33,66]. Relevant processes include laser-induced surface modification, ion implantation, sintering, plasma spraying, pulsed laser deposition, sputtering, physical or chemical vapor deposition (PVD or CVD, respectively), precipitation from solution, sol-gel wet chemistry, electrophoretic deposition (EPD), electrodeposition, etc. In addition to the well known hydroxyapatite (HAP) and other calcium phosphate coatings which provide enhanced osseointegration and fixation of orthopedic and dental implants, diamond-like carbon coatings (DLCs) have recently emerged to provide superior wear resistance.

Before concluding, few sentences should be dedicated to the widely used HAP coatings. The term *apatite* refers to a variety of calcium phosphate minerals with the general formula $Ca_{10}(PO_4)_6X_2$ where X is a monovalent anion such as chloride, carbonate, fluoride, or hydroxyl ion. When X is replaced by OH, the compound is known as HAP. In biological systems, apatite or apatitic calcium phosphates occur as the principal inorganic constituent of vertebrate bone and tooth. Wet cortical bone, for example, is composed of 69 wt% inorganic (mineral) constituent, 22 wt% organic matrix, and 9 wt% absorbed water. Several types of synthetic apatites are now commercially available for use in bone repair, bone augmentation, bone substitution, and as coatings on dental and orthopedic implants. HAP-based bioceramics are considered promising for osteo-implants and as a means of aiding the regeneration of bone. When implanted *in vivo*, they are able to bond to the host tissue by simulating a specific biological response at the host/biomaterial interface. Considerations in the production of calcium phosphate coatings are that: (1) the interfacial bond between the coating and the substrate should be strong enough to prevent delamination and long-term failure; (2) the coating/body interface should enhance tissue ingrowth and implant fixation; and (3) the coating should be reproducible and cost-effective. To achieve such coatings, the interest has shifted in recent years from the traditional plasma sprayed HAP to electrodeposited HAP coatings. Electrochemical processes may allow good control of the microstructure, formation of hybrid/composite systems, and incorporation of biological species such as growth factors during processing. Hence, we have been studying in detail the microstructure evolution and properties of HAP electrodeposited on CP-Ti, Ti-6Al-4V and 316L stainless steel at Tel-Aviv University [7,124-128]. Comparisons were made to EPD-HAP [7,127] and plasma-sprayed HAP [126]. It is anticipated that this research will eventually lead to processing of better HAP coatings for biomedical applications. Comprehensive data is still to be published.

REFERENCES

1. D.F. Williams, J. Black and P.J. Doherty, "Consensus report of second conference on definitions in biomaterials," in *Biomaterial-Tissue Interfaces*, Vol. 10, eds. P.J. Doherty, R.L. Williams, D.F. Williams and A.J.C Lee, Elsevier, Amsterdam (1992), 525-533.
2. B.D. Ratner, A.S. Hoffman, F.J. Schoen and J.E. Lemons (eds.), *Biomaterials Science – an Introduction to Materials in Medicine*, Academic Press, San Diego, CA (1996), 1-8; 260-267.
3. D.F. Williams (ed.), *Definitions in Biomaterials – Proc. Consensus Conf. European Soc. Biomaterials*, Vol. 4, Elsevier, New York (1987).
4. D.C. Ludwigson, "Requirements for metallic surgical implants and prosthetic devices," *Metals Engineering Quarterly*, **5**(3) (1965), 1-6.
5. J.D. Bronzino (ed.), *The Biomedical Engineering Handbook*, 2nd Ed., Vol. I, CRC Press & IEEE Press (2000).
6. J. Ryhänen, "Biocompatibility evaluation of Nickel-Titanium shape memory metal alloy," http://herkules.oulu.fi/isbn9514252217/html/index.html
7. T.M. Sridhar, N. Eliaz, U. Kamachi Mudali and Baldev Raj, "Electrophoretic deposition of hydroxyapatite coatings and corrosion aspects of metallic implants," *Corrosion Reviews*, **20**(4-5) (2002), 255-293.
8. G.A. Thibodeau and K.T. Patton, *Anatomy and Physiology*, 4th Ed., Mosby Inc., MO (1999), 851-887.
9. H. Zitter, "Case histories on surgical implants and their causes," *Werkstoffe und Korrosion*, **42** (9) (1991), 455-466.
10. ASTM F2129-04: Standard test method for conducting cyclic potentiodynamic polarization measurements to determine the corrosion susceptibility of small implant devices, ASTM, West Conshohocken, PA, USA (2004).
11. M. Morita, T. Sasada, H. Hayashi and Y. Tsukamoto, "The corrosion fatigue properties of surgical implants in a living body," *Journal of Biomedical Material Research*, **22**(6) (1988), 529-540.
12. R.A. Corbett, *Laboratory Corrosion Testing of Medical Implants*, Corrosion Testing Laboratories Inc., Newark, Delaware, USA (2004).
13. R.J. Solar, "Corrosion resistance of titanium surgical implant alloys, a review. Corrosion and degradation of implant materials," in *ASTM STP 684: Corrosion and Degradation of Implant Materials*, eds. B.C. Syrett and A. Acharya, ASTM, Baltimore (1979), 259-273.
14. A.N. Hughes, B.A. Jordan and S. Orman, "The corrosion fatigue properties of surgical implant," *Eng. Med.*, **7** (3) (1978), 135-141.
15. M. Traisnel, D. Le Maguer, H.F. Hildebrand and A. Iost, "Corrosion of surgical implants," *Clinical Materials*, **5** (1990), 309-318.
16. G.D. Walker, *Journal of Biomedical Material Research*, **5** (1974), 11.
17. P.G. Liang, *Orthopedic Clinics of North America*, **4** (1973), 249.
18. P. Thomsen and L.E. Ericson, "Inflammatory cell response to bone implant surfaces," in *The Bone-Biomaterial Interface*, ed. J.E. Davis, University of Toronto Press, Toronto (1991), 153-164.

19. P. Tengvall and I. Lundstrom, "Physico-chemical considerations of titanium as a biomaterial," *Clinical Materials*, **9** (1992), 115-134.
20. N.D. Green, in *ASTM STP 859: Corrosion and Degradation of Implant Materials: 2nd Symp.*, eds. A.C. Fraker and C.D. Griffin, ASTM, Philadelphia (1985), 5.
21. S. Ciolac, E. Vasilescu, P. Drob, M.V. Popa and M. Anghel, *Rev. Chim. (Bucharest)*, **51** (2000), 36.
22. D.J. Blackwood, "Biomaterials: past successes and future problems," *Corrosion Reviews*, **21**(2-3) (2003), 97-124.
23. T. Fusayama, T. Katayori and S. Nomoto, "Corrosion of gold and amalgam placed in contact with each other," *Journal of Dental Research*, **42** (1963), 1183-1197.
24. S. Joyston-Bechal and E.A.M. Kidd, *Dental Update*, **21** (1994), 366-371.
25. I.M.C. Lundstrom, *International Journal of Oral Surgery*, **12** (1982), 1.
26. J. Bànòczy, B. Roed-Petersen, J.J. Pindborg and J. Inovay, "Clinical and histological studies on electrogalvanical induced oral white lesious," *Oral Surgery, Oral Medicine, Oral Pathology*, **48** (4) (1979), 319-323.
27. J. Black, *Biological Performance of Materials – Fundamentals of Biocompatibility*, 2nd Ed., Marcel Decker Inc., New York, NY (1992), 38-59.
28. D.M. Jones, J.L. Marsh, J.V. Nepola, J.J. Jacobs, A.K. Skipor, R.M. Urban, J.L. Gilbert and J.A. Buckwalter, "Focal osteolysis at the junctions of a modular stainless-steel femoral intramedullary nail," *The Journal of Bone & Joint Surgery*, **83A**(4) (2001), 537-548.
29. J.J. Jacobs, J.L. Gilbert and R.M. Urban, "Corrosion of metal orthopedic implants," *The Journal of Bone & Joint Surgery*, **80A**(2) (1998), 268-282.
30. N. Hallab, K. Merritt and J.J. Jacobs, "Metal sensitivity in patients with orthopedic implants," *The Journal of Bone & Joint Surgery*, **83A**(3) (2001), 428-436.
31. K. Merritt and S.A. Brown, "Effect of proteins and pH on fretting corrosion and metal ion release," *Journal of Biomedical Material Research*, **22** (1988), 111-120.
32. E.J. Sutow and S.R. Pollack, in *Biocompatibility of Clinical Implant Materials*, ed. D.F. Williams, Vol. II, CRC Press, Boca Raton, FL (1981), 45-98.
33. U. Kamachi Mudali, T.M. Sridhar, N. Eliaz and Baldev Raj, "Failures of stainless steel orthopedic devices - causes and remedies," *Corrosion Reviews*, **21**(2-3) (2003), 231-267.
34. K. Mendel, *Application of Bio-Ferrography to the Study of Osteoarthritis and other Human Joint Diseases*, M.Sc. Thesis, Tel-Aviv University, Israel (2004).
35. A. Evron, *Bio-Ferrography and Failure Analysis of Artificial Hip and Knee Joints*, M.Sc. Thesis, Tel-Aviv University, Israel (2004).
36. N. Parkansky, B. Alterkop, R.L. Boxman, G. Leitus, O. Berkh, Z. Barkay, Yu. Rosenberg and N. Eliaz, "Magnetic properties of carbon nano-particles produced by a pulsed arc submerged in ethanol," Carbon, 46(2008), 215-219.
37. J.S. Ishay, Z. Barkay, N. Eliaz, M. Plotkin, S. Valynchick, and D.J. Bergman, "Gravity orientation in Social Wasp comb cells (Vespinae) and the possible role of embedded minerals," Naturwissenschaften, 46(2008), 215-219.
38. D.C. Mears, *Journal of Biomedical Material Research*, **6** (1975), 133.
39. H. Jackson-Burrows, J.N. Wilson and J.T. Scales, "Excision of tumours of humerus and femur, with restoration by internal prostheses," *The Journal of Bone & Joint Surgery*, **57B** (1975), 148-159.
40. L.C. Lucus, R.A. Buchanan, J.E. Lemons and C.D. Griffin, "Susceptibility of surgical cobalt-based alloy to pitting corrosion," *Journal of Biomedical Material Research*, **16** (1982), 799-810.
41. W. Rostoker, C.W. Pretzel and J.O. Galante, "Coupled corrosion among alloys for skeletal prostheses," *Journal of Biomedical Material Research*, **8** (1974), 407-419.
42. BS EN 12010: Non-active surgical implants – Joints replacement implants – Particular requirements, BSI (1998).
43. U. Kamachi Mudali, P. Shankar, S. Ningshen, R.K. Dayal, H.S. Khatak and Baldev Raj, "On the pitting corrosion resistance of nitrogen alloyed coled worked austenitic stainless steels," *Corrosion Science*, **44** (2002), 2183-2198.

44. C.O. Clerc, M.R. Jedwab, D.W. Mayer, P.J. Thompson and J.S. Stinson, "Assessment of wrought ASTM F1058 cobalt alloy properties for permanent surgical implants," *Journal of Biomedical Material Research*, **38**(3) (1997), 229-234.
45. H.J. Mueller and E.H. Greener, "Polarization studies of surgical materials in Ringer's solution," *Journal of Biomedical Material Research*, **4** (1970), 29-41.
46. B.C. Syrett and S.S. Wing, *Corrosion*, **34A** (1978), 138-145. Pitting resistance of new and conventional orthopedic implant materials-effect of metallurgical condition.
47. R. Strietzel, A. Hösch, H. Kalbfleisch and D. Bush, "In vitro corrosion of titanium," *Biomaterials*, **19** (1998) 1495-1499.
48. J.T. Scales, G.D. Winter and H.T. Shirley, *The Journal of Bone & Joint Surgery*, **41B** (1959), 810-820. Corrosion of orthopaedic implants: Screws, plates and fomoral nail-plates.
49. J. Pan, C. Karlén and C. Ulfvin, "Electrochemical study of resistance to localized corrosion of stainless steels for biomaterial applications," *Journal of the Electrochemical Society*, **147**(3) (2000), 1021-1025.
50. J.A. Disegi and L.D. Zardiackas, in ASTM STP 1361 (1999), 49.
51. B.C. Syrett and E.E. Davis, in *ASTM STP 684: Corrosion and Degradation of Implant Materials*, eds. B.C. Syrett and A. Acharya, ASTM, Baltimore (1979), 229.
52. J. Galante and W. Rostoker, "Corrosion related failures in metallic implants," *Clinical Orthopedics and Related Research*, **86** (1972), 237-244.
53. K.H.W. Seah, R. Thampuran, X. Chen, S.H. Teoh, "A comparison between the corrosion behaviour of sintered and unsintered porous titanium," *Corrosion Science*, **37**(9) (1995), 1333-1340.
54. B.S. Becker, J.D. Bolton and M. Youseffi, "Production of porous sintered Co-Cr-Mo alloys for possible surgical applications. Part 2: Corrosion behaviours," *Powder Metallurgy*, **38** (1995), 305-313.
55. J.S. Guindy, H. Schiel, F. Schmidli and J. Wirz, "Corrosion at the marginal gap of implant-supported suprastructures and implant failure," *The International Journal of Oral & Maxillofacial Implants*, **19**(6) (2004), 826-831.
56. M.F. LeClerc, in *Corrosion*, eds. L.L. Shrier, R.A. Jarman and G.T. Burstein, Vol. 1, 3rd Ed., Butterworth Heinemann, Oxford (1994), 164.
57. H.R. Piehler, M.A. Portnoff, L.E. Sloter, E.J. Vegdahl, J.L. Gilbert and M.J. Weber, in *ASTM STP 859: Corrosion and Degradation of Implant Materials: 2nd Symp.*, eds. A.C. Fraker and C.D. Griffin, ASTM, Philadelphia (1985), 93.
58. J. Yu, Z.J. Zhao and L.X. Li, "Corrosion fatigue resistances of surgical implant stainless steels and titanium alloy" *Corrosion Science*, **35** (1-4) (1993), 587-597.
59. D.A. Jones, *Principles and Prevention of Corrosion*, Macmillan Publishing Company, New York, NY (1992), 349-350.
50. M.G. Fontana, *Corrosion Engineering*, 3rd Ed., McGraw-Hill Inc., New York, NY (1987), 104-109.
61. N.J. Hallab and J.J. Jacobs, "Orthopedic implant fretting corrosion," *Corrosion Reviews*, **21**(2-3) (2003), 183-213.
62. J.L. Gilbert and J.J. Jacobs, "The mechanical and electrochemical processes associated with taper fretting crevice corrosion: a review," in *ASTM STP 1301: Modularity of Orthopedic Implants*, Philadelphia, PA (1997), 45-59.
63. N. Cabrera and N.F. Mott, "Theory of the oxidation of metals," *Reports on progress* In *Physics*, **12** (1948), 163-184.
64. J. Rieu, L.M. Rabbe and P. Combrade, in *Proc. 8th Inter. Conf. on Surface Modification Technology*, Institute of Materials, London (1995), 43.
65. B. Normand, F. Renaud, C. Coddet and F. Tourenne, in *Proc. 9th National Thermal Spraying Conf.*, ASM International, Materials Park, OH (1996), 73.

66. J.E. Lemons, "Surface modifications of surgical implants," *Surface and Coatings Technology*, **103-104** (1998), 135-137.
67. K. Hayashi, T. Mashima and K. Uenoyama, "The effect of hydroxyapatite coating on bony ingrowth into grooved titanium implants," *Biomaterials*, **20**(2) (1999), 111-119.
68. ASTM F138-03: Standard specification for wrought 18Cr-14Ni-2.5Mo stainless steel bar and wire for surgical implants (UNS S31673), ASTM, West Conshohocken, PA, USA (2003).
69. ASTM F139-00: Standard specification for wrought 18Cr-14Ni-2.5Mo stainless steel sheet and strip for surgical implants (UNS S31673), ASTM, West Conshohocken, PA, USA (2000).
70. ASTM F745-00: Standard specification for 18Cr-12.5Ni-2.5Mo stainless steel for cast and solution-annealed surgical implant applications, ASTM, West Conshohocken, PA, USA (2000).
71. ASTM F899-02: Standard specification for stainless steel for surgical instruments, ASTM, West Conshohocken, PA, USA (2002).
72. ASTM F1586-02: Standard specification for wrought nitrogen strengthened 21Cr-10Ni-3Mn-2.5Mo stainless steel bar for surgical implants (UNS S31675), ASTM, West Conshohocken, PA, USA (2002).
73. ISO 5832-1: Implants for surgery - metallic materials - Part 1: Wrought stainless steel, International Organization for Standardization, Geneva, Switzerland (1997).
74. T. Hanawa, "Reconstruction and regeneration of surface oxide film on metallic materials in biological environments," *Corrosion Reviews*, **21**(2-3) (2003), 161-181.
75. ASTM F1089-02: Standard test method for corrosion of surgical instruments, ASTM, West Conshohocken, PA, USA (2002).
76. R. Venugopalan and J. Gaydon, "A review of corrosion behaviour of surgical implant alloys," Technical Review Note 99-01, Princeton Applied Research (1999).
77. J.A. von Fraunhofer, N. Berberich and D. Seligson, "Antibiotic-metal interactions in saline medium," *Biomaterials*, **10**(2) (1989), 136-138.
78. C.C. Shih, S.J. Lin, K.H. Chung, Y.L. Chen and Y.Y. Su, "Increased corrosion resistance of stent materials by converting current surface film of polycrystalline oxide into amorphous oxide," *Journal of Biomedical Material Research*, **52**(2) (2000), 323-332.
79. E.J. Sutow, D.W. Jones and E.L. Milne, "In vitro crevice corrosion behaviour of implant materials," *Journal of Dental Research*, **64**(5) (1985), 842-847.
80. K.J. Bundy, M.A. Vogelbaum and V.H. Desai, "The influence of static stress on the corrosion behaviour of 316L stainless steel in Ringer's solution," *Journal of Biomedical Material Research*, **20**(4) (1986), 493-505.
81. S.A. Brown and K. Merritt, "Fretting corrosion in saline and serum," *Journal of Biomedical Material Research*, **15**(4) (1981), 479-488.
82. R.L. Williams, S.A. Brown and K. Merritt, "Electrochemical studies on the influence of proteins on the corrosion of implant alloys," *Biomaterials*, **9**(2) (1988), 181-186.
83. J. Stewart and D.E. Williams, "The initiation of pitting corrosion on austenitic stainless steel: on the role and importance of sulphide inclusions," *Corrosion Science*, **33** (3) (1992), 457-463.
84. ASTM A268/A-04a: Standard specification for seamless and welded ferritic and martensitic stainless steel tubing for general service, ASTM, West Conshohocken, PA, USA (2004).
85. ASTM A789/A-04a: Standard specification for seamless and welded ferritic/austenitic stainless steel tubing for general service, ASTM, West Conshohocken, PA, USA (2004).
86. ASTM A240/A-04a: Standard specification for chromium and chromium-nickel stainless steel plate, sheet, and strip for pressure vessels and for general applications, ASTM, West Conshohocken, PA, USA (2004).
87. U. Kamachi Mudali and R.K. Dayal, "Influence of nitrogen addition on the crevice corrosion resistance of nitrogen-bearing austenitic stainless steels," *Journal of Materials Science*, **35** (2000), 1799-1803.
88. ASTM F2229-02: Standard specification for wrought, nitrogen strengthened 23Mn-21Cr-1Mo low-nickel stainless steel alloy bar and wire for surgical implants (UNS S29108), ASTM, West Conshohocken, PA, USA (2002).

89. N. Eliaz and O. Nissan, "Innovative processes for electropolishing of medical devices made of stainless steels," *Journal of Biomedical Materials Research: Part A*, **83** (2) (2007), 546-557.

90. J. Jedwab, F. Burny, R. Wollast, G. Naessens and P. Opdecam, *Acta Orthop. Belg.* **40** (1974), 877-886.

91. D.M. Brunette, P. Tengvall, M. Textor and P. Thomsen (eds.), *Titanium in Medicine*, Springer, Heidelberg, Germany (2001).

92. S.-B. Hong, N. Eliaz, E.M. Sachs, S.M. Allen and R.M. Latanision, "Corrosion behavior of advanced Ti-based alloys made by three-dimensional printing (3DP™) for biomedical applications," *Corrosion Science*, **43**(9) (2001), 1781-1791.

93. S.-B. Hong, N. Eliaz, G.G. Leisk, E.M. Sachs, R.M. Latanision and S.M. Allen, "A new Ti-5Ag alloy for customized prostheses by three-dimensional printing (3DP™)," *Journal of Dental Research*, **80**(3) (2001), 860-863.

94. ASTM F67-00: Standard specification for unalloyed titanium for surgical implant applications (UNS R50250, UNS R50400, UNS R50550, UNS R50700), ASTM, West Conshohocken, PA, USA (2000).

95. ASTM F136-02: Standard specification for wrought Ti-6Al-4V ELI (Extra Low Interstitial) alloy for surgical implant applications (UNS R56401), ASTM, West Conshohocken, PA, USA (2002).

96. ASTM F1472-02a: Standard specification for wrought Ti-6Al-4V alloy for surgical implant applications (UNS R56400), ASTM, West Conshohocken, PA, USA (2002).

97. ASTM F1713-03: Standard specification for wrought Ti-13Nb-13Zr alloy for surgical implant applications (UNS R58130), ASTM, West Conshohocken, PA, USA (2003).

98. ASTM F1295-01: Standard specification for wrought Ti-6Al-7Nb alloy for surgical implant applications (UNS R56700), ASTM, West Conshohocken, PA, USA (2001).

99. ASTM F1813-01: Standard specification for wrought Ti-12Mo-6Zr-2Fe alloy for surgical implant (UNS R58120), ASTM, West Conshohocken, PA, USA (2001).

100. ASTM F1580-01: Standard specification for Ti and Ti-6Al-4V alloy powders for coatings of surgical implants, ASTM, West Conshohocken, PA, USA (2001).

101. G. Manivasagam, U. Kamachi Mudali, R.Asokamani and Baldev Raj, "Corrosion and microstructural aspects of titanium and its alloys as orthopedic devices," *Corrosion Reviews*, **21**(2-3) (2003), 125-159.

102. E.A. Trillo, C. Ortiz, P. Dickerson, R. Villa, S.W. Stafford and L.E. Murr, "Evaluation of mechanical and corrosion biocompatibility of TiTa alloys," *Journal of Materials Science: Materials in Medicine*, **12** (2001), 283-292.

103. P.F. Barbosa and S.T. Button, "Microstructure and mechanical behavior of the isothermally forged Ti-6Al-7Nb alloy," *Proceedings of the Institution of Mechanical Engineers, Part L: Journal of Materials-Design and Applications*, **214** (2000), 23-32.

104. J. Been and J.S. Grauman, "Titanium and titanium alloys," in *Uhlig's Corrosion Handbook*, 2nd Ed., ed. R.W. Revie, John Wiley & Sons Inc., New York, NY (2000), 863-885.

105. J.L. Gilbert, C.A. Buckley and E.P. Lautenschlager, "Titanium oxide film fracture and repassivation: The effect of potential, pH and aeration," in S.A. Brown and J.E. Lemons (eds.), *ASTM STP 1271: Medical Application of Titanium and its Alloys: The Material and Biological Issues*, American Society for Testing and Materials (1996), 199-215.

106. J.P. Bearinger, *The Electrochemistry of Titanium/Titanium Oxide in the Biological Environment*, Ph.D. Thesis, Northwestern University, Evanston, IL (2000).

107 J.P. Bearinger, C.A. Ormec and J.L. Gilbert, "In situ imaging and impedance measurements of titanium surfaces using AFM and SPIS," *Biomaterials*, **24**(11) (2003), 1837-1852.

108. M. Nakagawa, S. Matsuya and K. Udoyh, "Corrosion behaviour of pure titanium alloys in fluoride-containing solutions," *Dental Materials Journal*, **20**(4) (2001), 167-305.

109. M.A. Khan, R.L. Williams and D.F. Williams, "In vitro corrosion and wear of titanium alloys in the biological environment," *Biomaterials*, **17**(22) (1996), 2117-2126.

110. M. Aziz-Kerrzo, K.G. Conroy, A.M. Fenelon, S.T. Farrell and C.B. Breslin, "Electrochemical studies on the stability and corrosion resistance of titanium-based implant materials," *Biomaterials*, **22** (2001), 1531-1539.

111. B. Grosgogeat, L. Reclaru, M. Lissac and F. Dalard, "Measurement and evaluation of galvanic corrosion between Ti/Ti-6Al-4V implants and dental alloys by electrochemical techniques and auger spectrometry," *Biomaterials*, **20** (1999), 933-941.

112. D.J. Rabkin, E.V. Lang and D.P. Brophy, "Nitinol properties affecting uses in interventional radiology," *JVIR*, **11**(3) (2000), 343-350.

113. S.A. Shabalovskaya, "Surface, corrosion and biocompatibility aspects of Nitinol as an implant material," *Bio-Medical Materials and Engineering*, **12** (2002), 69-109.

114. ASTM F2005-00: Terminology for nickel-titanium shape memory alloys, ASTM, West Conshohocken, PA, USA (2000).

115. ASTM F2063-00: Standard specification for wrought nickel-titanium shape memory alloys for medical devices and surgical implants, ASTM, West Conshohocken, PA, USA (2000).

116. J. Ryhänen, E. Niemi, W. Serlo, E. Niemelä, P. Sandvik, H. Pernu and T. Salo, "Biocompatibility of nickel-titanium shape memory metal and its corrosion behavior in human cell cultures," *Journal of Biomedical Materials Research*, **35**(4) (1997), 451-457.

117. B.M. Eley, "The future of dental amalgam: a review of the literature. Part 1: Dental amalgam structure and corrosion," *British Dental Journal*, **182** (7) (1997), 247-249.

118. L. Joska, L. Bystrainsky and P. Novak, in *Proc. 15th Intern. Corrosion Congress*, Granada, paper 347, (2002).

119. T. Newton, *Chem. Br.*, **38**(10), (2002), 24.

120. N.K. Sarker and C.S. Eyer, *J. Oral. Rehabil.* **14** (1987), 27.

121. A.D. Rosenberg, *In-Vitro Electrochemical Testing of a Microchip-Based Controlled Drug Delivery Device*, M.Sc. Thesis, M.I.T., Cambridge, MA (2001).

122. R.P. Frankental and D.J. Siconolfi, "Anodic corrosion of gold in concentrated chloride solutions," *Journal of the Electrochemical Society*, **3** (1969), 465-470.

123. D. Schuhmann, M. Kubicka-Muranyi, J. Mirtschewa, J. Gunther, P. Kind and E. Gleichmann, "Adverse immune reactions to gold. I. Chronic treatment with an Au(I) drug sensitizes mouse T cells not to Au(I), but to Au(III) and induces autoantibody formation," *Journal of Immunology*, **145**(7) (1990), 2132-2139.

124. N. Eliaz, T.M. Sridhar and Yu. Rosenberg, "Electrodeposition of hydroxyapatite on titanium for implants," in *Proc. of the 10th World Conf. on Titanium – Ti-2003*, Vol. V, eds. G. Lütjering and J. Albrecht, Wiley-VCH, Weinheim, Germany (2004), 3299-3306.

125. N. Eliaz and M. Eliyahu, "Electrochemical processes of nucleation and growth of hydroxyapatite on titanium supported by real-time electrochemical atomic force microscopy," *Journal of Biomedical Materials Research: Part A*, **80** (3) (2007), 621-634.

126. H. Wang, N. Eliaz, Z. Xiang, H.-P. Hsu, M. Spector and L.W. Hobbs, "Early bone apposition in vivo on plasma-sprayed and electrochemically deposited hydroxyapatite coatings on titanium alloy," *Biomaterials*, **27** (23) (2006), 4192-4203.

127. N. Eliaz, T.M. Sridhar, U. Kamachi Mudali and Baldev Raj, "Electrochemical and electrophoretic deposition of hydroxyapatite for orthopedic applications," *Surface Engineering*, **21** (3) (2005), 238-242.

128. N. Eliaz, W. Kopelovitch, L. Burstein, E. Kobeyashi and T. Hanawa, "Electrochemical processes of nucleation and growth of calcium phosphate on titanium supported by real-time quartz crystal microbalance measurements and XPS analysis," *Journal of Biomedical Materials Research: Part A*, in press.

CHAPTER 13

Corrosion Behaviour of Nanostructured Surfaces

Anita Toppo, P. Shankar, H. Shaikh and A.K. Tyagi
Metallurgy and Materials Group, Indira Gandhi Centre for Atomic Research, Kalpakkam 603102, India

The word *nano* refers to a Greek prefix meaning dwarf or something very small and depicts one billionth (10^{-9}) of a unit. Nanomaterials, therefore, refer to the class of systems, having atleast one of its dimensions in nanometric range. Nanostructured materials can in general be classified into four categories according to their dimensionality: 0D-nanoclusters; 1D-multilayers; 2D-nanograined layers; 3D-equiaxed bulk solids.

Nanocrystalline materials have a grain size of the order of 1 - 100 nm, and, are therefore, a few thousand times smaller than conventional grain dimensions. However, compared to the size of an atom (0.2-0.4 nm in diameter), nanocrystalline grains are still significantly big. For example, a 10 nm big nanocrystal contain a few thousands of atoms, large enough to exhibit quantum properties. However, as the dimensions reduce to below 50-100 nm, they can no longer be treated as infinite systems and the resultant boundary effects, manifest extremely fascinating and useful properties, which can be exploited for a variety of structural and non-structural applications.

It is known that nearly all properties like hardness, strength, ductility, elastic modulus, melting point, density, thermal conductivity, thermal expansion coefficient, diffusivity etc, changes as we transit to nanomaterials [1]. The strength, hardness and ductility of nanomaterials are known to increase with decreasing grain size. However, below a critical grain size, strength decreases with decreasing grain size and this effect is known as inverse Hall-Petch relation. Melting point, elastic modulus and thermal conductivity decreases with decreasing grain size. Ceramics, which are conventionally brittle, can be made even superplastic under certain deformation conditions, when they are synthesised in nanocrystalline form. Nanometer sized grains contain only few thousands of atoms each. As the grain size decreases, there is a significant increase in the volume fraction of grain boundaries or interfaces. With increase in defect density, or in other words, when the fraction of atoms residing at defect cores, such as grain boundaries, dislocations, triple junctions, porosities, etc., become comparable to that residing in the core of matrix, the properties of the material are bound to be governed to a large extent on defect configurations, dynamics and interactions.

Nanomaterials have mechanical properties which are in most cases better than their microcrystalline counterparts. Table 1 shows the mechanical behaviour of nanocrystalline Ni which was made using electrodeposition technique. It shows that the mechanical properties are improved once we go from micro to nanomaterials.

Table 1 Improved properties of nanocrystalline Ni

Properties	Descriptions	Reference
Hardness	5 times harder.	2, 3, 4, 5
Wear Resistance	170 times increase.	6
Friction	Decreased by 50%	6
Corrosion Resistance	reduce or stop localized corrosion.	7, 8, 9
Strength	3 to 10 times stronger.	10
Magnetic	lower coercivity, resistivity 3 times increased, M_s reduced by 5%.	11, 12, 13, 14
Hydrogen Diffusion	higher hydrogen diffusion.	15, 16
Electrocatalysis	improved electrocatalytic activities for hydrogen evolution and hydrogen oxidation reactions.	17

Although the physical and mechanical properties of nanocrystalline solids are reasonably well understood, their influence on corrosion behaviour is still unclear. The corrosion behavior of polycrystalline metal or amorphous counterparts [18] is well understood both in terms of electrochemical polaristaion behaviour and surface morphology. However, the properties of nanocrystalline materials cannot be predicted from those usually observed for their conventional polycrystalline materials. This is due to the unique structure of nanocrystalline materials, which contain considerable volume fractions of grain boundaries and triple junctions. Extensive research has been reported on the corrosion behaviour of nanocrystalline materials. However, the findings are often conflicting and are found to depend on several other factors. The influence of nanocrystalline structure of an alloy on its passivation behaviour and properties of its passive film is yet to be unambiguously clarified. Therefore, it is necessary to characterize nanomaterials in terms of their physical and chemical properties before they can be considered for large scale industrial applications. This chapter details the corrosion behaviour of nanocrystalline materials and the variables which affect their corrosion behaviour.

1.0 FACTOR AFFECTING CORROSION PROPERTIES OF NANOCRYSTALLINE MATERIAL

Several reports suggest the high density of grain boundaries in nanocrystalline alloys to be responsible for its significantly changed properties [19]. The corrosion investigations of nanocrystalline materials showed a more intensive active anodic dissolution, compared with conventional materials [7]. This difference in the behavior can be explained by the small grain size, which results in a larger fraction of high energy grain boundary defects in the material. Such high energy grain boundary sites act as preferred anodic dissolution sites and can thus result in increased corrosion rates of nanocrystalline alloys. However, higher defect densities are not always disadvantageous. Grain boundaries being easy diffusion paths, can facilitate, enhanced surface diffusion resulting in the early formation of a well-adherent, dense oxide layer, thereby passivating the surface. Some authors showed that the grain size does not affect the thickness of the passive film [20]. Electrochemical polarization tests on nanocrystalline Ni (grain sizes 500 nm and 20 nm) indicated that they exhibited higher passive current densities than their coarser- grained conventional Ni counterparts [20]. XPS investigations of the passive film on these materials showed no difference in the thickness of the passive film between the nanocrystalline material and conventional material. These passive films on the nanocrystalline specimens were, however, more defective than that on conventional Ni. The defect density of the film was found to increase with decreasing grain size, which in turn allowed for easier Ni cation diffusion through more defective film leading to higher current densities in the passive potential range for the nanocrystalline Ni specimens. The amount of impurities per unit grain boundary area is therefore lower in nanomaterials than larger grained materials with the same bulk impurity concentration because grain boundary area increases. This purification of the grain boundaries has been associated with more uniform corrosion morphology and higher intergranular corrosion resistance for nanomaterials as compared to larger grained materials. One more factor which affects the corrosion behaviour of nanocrystalline material is the porosity of the coating. If the coating is less porous in nature, it will show high corrosion resistance [21]. Contribution of triple junctions to the structure and properties of nanocrystalline materials may also affect the corrosion properties of these nanocrystalline materials [22]. A large increase in volume fraction of triple junction is expected with decreasing grain size in the nanocrystalline range. Table 2 lists a few possible effects of nanostructured surfaces on their electrochemical behaviour. The following sections give a detailed overview on the corrosion behaviour of nanomaterials.

Table 2 Possible effects of nanocrystalline materials on its corrosion behaviour

Characteristic	Beneficial Effects	Deleterious Effects
Enhanced diffusion	• Faster surface diffusion of ions to enhance kinetics of passive film formation	• Faster diffusion can result in higher corrosion currents
Stability of passive film	• Faster grain boundary diffusion can aid healing of passive film in service	• Porous oxide film
Effect of large grain boundary area	• Lower dilution effects	• Increased anodic sites for nucleation of corrosion damage
High temperature effect	• Better thermal shock resistance.	• Rapid grain growth limits high temperature application

2.0 LOCALISED CORROSION OF NANOCRYSTALLINE MATERIALS

The properties of nanocrystalline materials cannot typically be predicted from those observed for their conventional polycrystalline counterparts. This is due to the unique structure of nanocrystalline materials which has a large number of grain boundaries. It has been shown that the structure-property relationship of bulk nanocrystalline materials differ considerably from those observed for thin films nanocrystalline materials. [5,23]. The differences are mainly attributed to microstructural variations including crystallographic texture, porosity, impurities, grain boundary and triple junctions.

2.1 Corrosion Behaviour of Nanocrystalline Nickel and its Alloys

Several studies indicate the decrease in the corrosion resistance of the surface with nanocrystallisation of Ni films. Rofagha et al. [7] investigated the electrochemical behaviour of nanocrystalline nickel which had a grain size of 32 nm as compared to cold rolled annealed nickel with a grain size of 100 mm in 2N H_2SO_4. The potentiodynamic curve is shown in Fig 1. It was observed that, although, the corrosion potential shifts to nobler direction with decrease in grain size, the open circuit current density increases, indicating higher corrosion rates with decreasing grain size. This was because the smaller grained nickel provided more sites (grain boundaries and triple junctions) for reduction to occur, thus increasing its corrosion rate vis-a-vis annealed nickel. It was shown that the high defect density in nanocrystalline Ni catalysed the hydrogen reduction (increasing the cathodic current), reduced the kinetics of passivation and compromised the stability of the passive film.

Similar effect on the corrosion behaviour of nanocrystalline Ni was reported by Wang et al. [24]. The results of their polarization studies, carried out in neutral 3% by mass sodium chloride solution, is shown in Fig. 2. It is seen in Fig. 2 that the current density in the passive range was higher for nanocrystalline Ni than for polycrystalline material. Mishra and Balasubramaniam [25] observed that the zero corrosion potential shifts to nobler direction with decrease in grain size of Ni film. This was attributed to the enhanced cathodic reaction kinetics., ie., increase in hydrogen reduction rate due to increase in catalytic site provided by nanocrystalline (NC) Ni surface compared to microcrystalline surface. However, all nanograin sized deposits exhibited higher passive current density compared to coarse- grained nickel in sulfuric acid medium. This indicated the defective nature of the passive film that formed on nanocrystalline nickel. But quite surprisingly, Mishra et al., found that the breakdown potential was much higher for NC Ni grains compared to coarse grained film. They reported that the corrosion rate of freshly exposed nanocrystalline Ni was lower compared to bulk Ni, indicating higher hindrance to anodic dissolution from the nanocrystalline Ni surface.

In sharp contrast to earlier reports, Wang et al [26] reported a beneficial effect on corrosion behaviour of surfaces by decreasing the grain size to nanometric dimensions. They studied grain size effect on the corrosion behaviour of nanocrystalline Ni coatings with grain sizes of 3 μm, 250 nm, 54 nm and 16 nm by pulsed electrodeposition. Nanocrystalline Ni was prepared by electro deposition and potentiodynamic polarizations tests were carried out in alkaline solution. Impedance measurements were also carried out. They observed that nanocrystalline Ni coatings had wider passive range and lower passive current densities. This indicated that NC Ni coatings have a high density of nucleation sites for passive film, which leads to

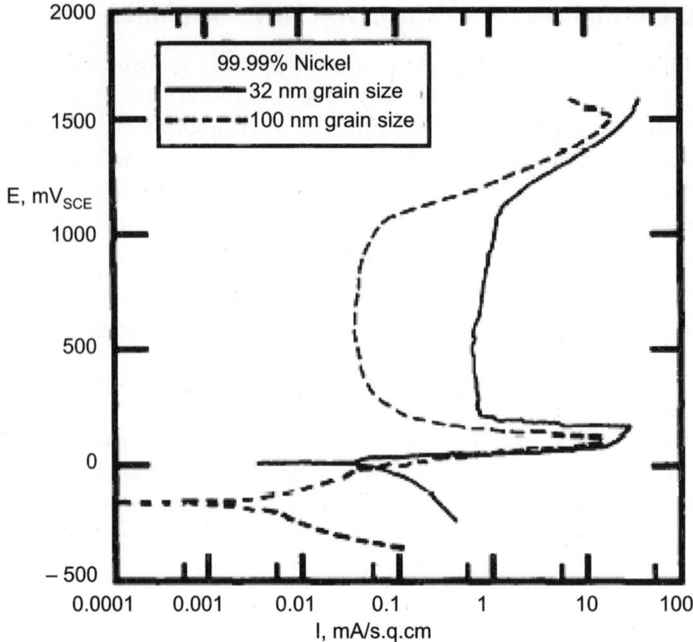

Figure 1 Electrochemical behavior of Ni in the nanocrystalline and annealed conditions [7].

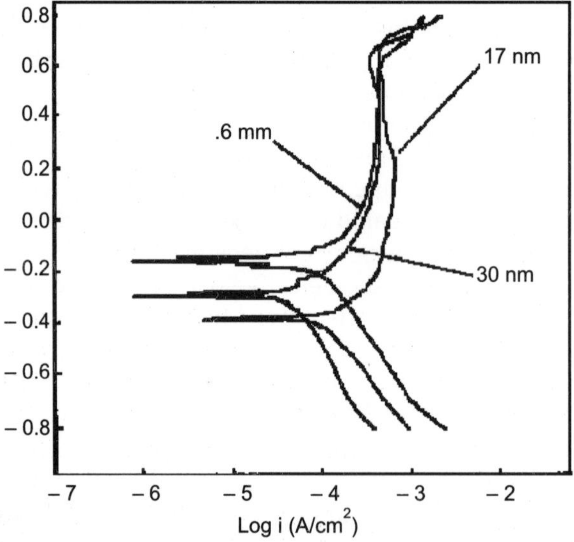

Figure 2 Potentiodynamic polarization curves of nanocrystalline and polycrystalline Ni in 3 wt% NaCl solution at ambient temperature [24].

continuous and adherent passive film and thus lower passive current densities. Microcrystalline Ni coating had the highest corrosion current density, and hence, the worst corrosion resistance compared with NC Ni coatings. Impedance studies indicated that NC Ni coatings possessed higher polaristaion resistance than the coarse grained Ni, which suggested a much lower corrosion rate of NC Ni coatings. Thus, corrosion resistance of Ni coatings in NaOH solution increased with the reduction of grain size and the NC Ni coatings exhibited better protection against corrosive media.

On the contrary, Rofagha et al. [27] showed deteriorating properties for nanocrystalline Ni-P alloys. They studied the effect of grain size and phosphorous content on the corrosion characteristics of nanocrystalline Ni-P alloys (grain sizes were 8.4 and 22.4 nm) by anodic polaristaion technique in 0.1M H_2SO_4. The results were then compared with that of the amorphous Ni-P and high purity conventional polycrystalline Ni. They found that nanocrystalline Ni-P alloys were nonpassivating in the medium similar to their amorphous Ni-P counterpart, whereas conventional Ni displayed typical active- passive behaviour. One more interesting point was that the nanocrystalline Ni-P alloys showed higher dissolution rates than the conventional Ni over the entire range. On comparing the grain sizes, the 22.6 nm specimen exhibited lower dissolution rates than the 8.4 nm material, with the difference in dissolution rate decreasing with increasing potential. This difference in dissolution rate was attributed to the lower intercrystalline volume fraction of 22.6 nm. The difference in dissolution rates decreased with increasing potential in the regime where the dissolution rates were high and the contributions of the intercrystalline content became less significant.

2.2 Corrosion Behaviour of Nanocrystalline Aluminum

Studies on corrosion behaviour of Al alloys have not been extensive. Nanocrystalline Al alloys produced by severe plastic deformation processing (SPD), showed unusual properties, such as substantially increased yield strength and ultimate strength. This behaviour of nanocrystalline Al alloys was attributed to the extremely fine grains and the high grain boundary area associated with them. While some benefits could be derived from large grain boundary areas, certain problems could also arise such as accelerated grain boundary diffusion and intergranular corrosion.

Kus et al., [28] investigated the corrosion behaviour of nanocrystalline and conventional Al 5083 samples by using pitting scans and electrochemical impedance spectroscopy. Susceptibility to intergranular corrosion was investigated using ASTM G 67 test. Pitting scans were carried out in three solutions with different chloride concentrations. Pitting scan results showed that, repassivation was more difficult in all the three solutions for nanocrystalline Al, while coarse grained Al 5083 showed more resistance to pitting. However, nitric acid mass loss tests (NAMLT) according to ASTM G 67 [29] showed that the nanocrystalline aluminium was less susceptible to intergranular corrosion than coarse grain Al 5083, in which intermetallic precipitates such as Mg_2Al_3 are attacked. Thus, it was concluded that NC samples have somewhat similar corrosion behaviour to those of coarse grained Al alloys. Hence they can also be used in marine industries, because of their superior yield strength and ultimate tensile strength. Sikora et al. [30], investigated the corrosion behaviour of a series of nanocrystalline Al-Mg alloys and then compared it with that of conventional AA 5083. The electrochemical characterization of the alloy was carried out by electrochemical noise measurements, potentiodynamic polarization experiments in neutral solutions with various chloride concentrations, and intergranular corrosion and exfoliation corrosion testing. Results showed not much difference in polarisaton behaviour in terms of passive current density or breakdown potential, when compared to microstructured conventional AA 5083. However, the open circuit potentials for the nanocrystalline alloys were typically lower than that of the conventionally processed alloy and the pits observed on nano alloy surfaces were smaller than those present on the conventional AA 5083. SEM observation of corroded surfaces revealed that corrosion attack for both the conventionally processed and the nanocrystalline alloys initiated at inclusions. In the case of the nanostructured materials, these inclusion were smaller and less numerous than those noted on the conventional alloy. Results of ASTM G-67 intergranular corrosion tests revealed that the nanocrystalline alloys were susceptible to intergranular corrosion while the conventional AA 5083 alloy was resistant. However, ASTM G66 exfoliation corrosion testing showed that the nanocrystalline alloys possessed excellent exfoliation corrosion resistance, superior to that of conventional alloys.

2.3 Corrosion Studies of Nanocrystalline Copper

Barbucci et al. [31] studied the corrosion behaviour of nanostructured $Cu_{90}Ni_{10}$ alloys in neutral solutions with different chloride concentrations. The corrosion behaviour of Cu-Ni-Fe alloy is strongly related to the ability of its passive film to protect the underlying metal. The structure of the protective layer is generally ascribed to an outer $CuO/Cu(OH)_2$ layer and an inner Cu_2O layer in the alloy when exposed to aggressive environments. The formation of the protective layer is highly influenced by the metallurgical structure of the alloy. Studies by Barbucci et al. showed weaker passivity of the nanostructured alloy, compared to the coarse grained alloy. Potentiodynamic polarisation studies showed that the breakdown potential was more negative for nanostructured alloy with respect to coarse grained alloys. Hence, the passive oxides which grew on the nanostructured metal surface were not as compact as in coarse grained alloys. Two important factors, viz., the amount of grain boundary and the presence of sintering defects govern the effectiveness of the protective properties of the passive oxide. Vnogradov et al. [32] have also shown that the corrosion behaviour of nanocrystalline Cu does not change significantly in comparison with coarse- grained polycrystalline Cu.

2.4 Corrosion of Nanocrystalline Cobalt

Kim et al., [33] examined the corrosion response of electrodeposited nanocrystalline Co (12 nm) with that of the conventional polycrystalline Co (8 μm). Corrosion studies were carried out in 0.25 M sodium sulphate solution at a pH of 6.5, using anodic polaristaion technique. In general, cobalt shows no passivation behavior [34], though exceptions have been reported. Studies carried out by Kim et al. showed that the potentiodynamic polaristaion curve for nanocrystalline Co, was very similar to that of polycrystalline Co, as shown in Fig. 3 showed only a slightly enhanced anodic dissolution current was observed for nanocrystalline Co. The corrosion behaviour of Co was not greatly affected by reducing the average grain size to the nanocrystalline range.

SEM images for both rolled and annealed electrowon Co samples surface showed homogenous surface attack with signs of grain boundary etching due to weak orientation

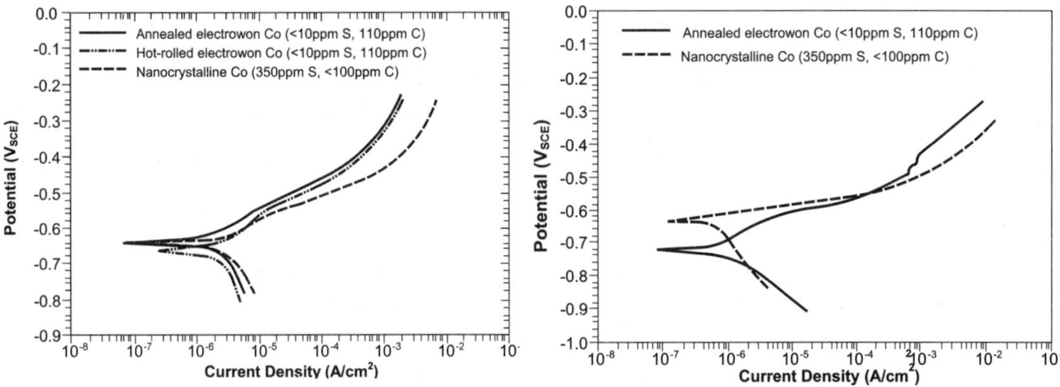

Figure 3 a and b Potentiodynamic polarization curves obtained from (a) as-ground and (b) electropolished Co surfaces in deaerated 0.25 M Na_2SO_4 at pH=6.5. [33]

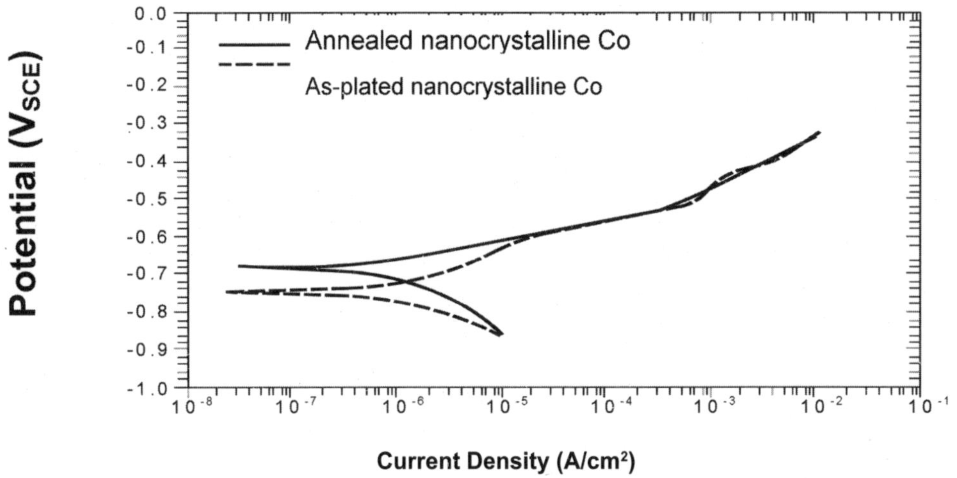

Figure 3c Potentiodynamic polarization behaviour of as- plated nanocrystalline Co (containing 440ppm S and 290 ppm C) and annealed nanocrystalline Co at 950°C for 15 min. [33]

dependant surface dissolution. Whereas in case of nanocrystalline Co, corrosion morphology was different from electrowon Co samples. Here surface showed grinding marks which was due to sulphur accumulation. To study the effect of microstructure on the corrosion behaviour of Co, electrodeposited nanocrystalline Co was annealed. Fig 3c is the potentiodynamic polarization curve of annealed nanocrystalline Co.

Although the potentiodynamic polarization curves are similar, SEM investigation of the corrosion tested specimens revealed the difference. In as- plated nanocrystalline Co, sulphur impurities were evenly dispersed throughout the nanocrystalline structure and hence uniform dissolution of surface was seen. In contrast, for the annealed nanocrystalline Co, preferential attack along the grain boundaries was seen due to segregation of Sulphur.

The improved corrosion resistance of NC Co in certain environments was reported by Wang et al. [35]. They compared the electrochemical corrosion behaviour of nanocrystalline Co coating with coarse grained Co (CG) in different corrosive media i.e. in 10wt % NaOH and 10 wt % HCl solutions. NC Co showed typical active- passive- transpassive- active behaviour in alkaline solutions, while only active behaviour without passivation is seen in chloride environments. In the NaOH solution, corrosion current density of nanocrystalline Co decreased significantly while the polarization resistance increased as compared with coarse grained Co, thus indicating improved corrosion resistance of the former. These results indicated that nanostructure enhanced both the formation of passive film and its stability on Co coatings. This enhancement in corrosion resistant properties was due to higher grain boundary density in these materials. To verify the potentiodynamic polarisation results, impedance tests were also carried out in 10% NaOH solution. The charge transfer impedance increased with the reduction in grain size to nanocrystalline, thus indicating much higher corrosion resistance of NC Co coatings. Also, the double layer capacitance of NC Co increased compared with coarse grained Co. This indicated the smooth and protective nature of passive film formed on NC Co coatings. However, in 10% of HCl solution, no passive behaviour was observed. This was because, higher grain boundary density in NC Co coating accelerated corrosion by providing high- density of active sites for preferential attack when NC Co coatings were exposed to such corrosive environments. Thus, it was concluded that nanocrystalline Co coatings behaved differently in various corrosive media due to their high-density network of grain boundaries.

2.5 Corrosion Behaviour of Nanocrystalline Titanium

Balyanov et al [36] carried out the corrosion tests on commercially pure titanium with ultrafine-grains (300 nm) and coarse grains (7 μm). The corrosion tests were carried out in 1M, 3M and 5M H_2SO_4 and HCl solutions. It was observed that corrosion potential for ultra fine grains (UFG) Ti shifts to noble direction with decreasing grain size, in HCl as well as in H_2SO_4. Higher corrosion resistance of UFG Ti was attributed to reduced segregation of impurities at the grain boundaries compared to coarse grained (CG) Ti. Also, the UFG Ti formed passive surface layer more readily than CG Ti. This was because UFG Ti would have had a high density of nucleation sites for passive film formation, thereby reducing the corrosion rates.

2.6 Corrosion of Nanocrystalline Zinc

Youssef et al., [37] compared the corrosion behaviour of nanocrystalline zinc produced by pulse electrodeposition with that of conventional electrogalvanised (EG) steel. The electrochemical studies were carried out in deaerated 0.5 N NaOH solution by potentiodynamic polarization and AC impedance techniques. It was concluded from potentiodynamic studies that the nanocrystalline zinc exhibits a lower corrosion rate than the electrogalvanised steel as shown in Fig. 4. This was attributed to the oxide film formed on the nanocrystalline zinc surface, which was more protective than that on electrogalvanised steel. The film formed on NC zinc deposit showed lower leakage current density and therefore had more protective characteristics than that on EG steel.

The oxide film formation on the surface is diffusion controlled, and it is reported that diffusion of elements in NC materials is much faster than that in polycrystalline materials.

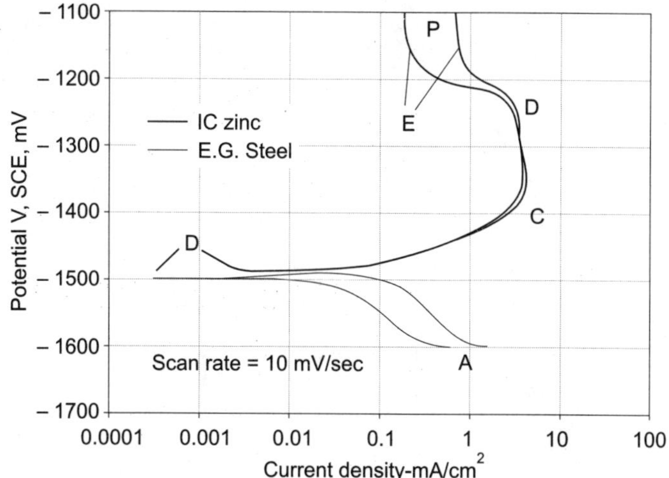

Figure 4 Polaristaion curves of nanocrystalline Zinc and EG steel in deaerated 0.5 N NaOH. [37]

Therefore, corrosion resistance of NC zinc, which was due to the formation of passive film, was higher than that of EG steel. These results were confirmed by the ac impedance technique which showed lower average capacitance value for NC zinc than for EG steel [37].

2.7 Corrosion of Nanocrystalline Fe and its Alloys

Meng et al., [38] investigated the corrosion behaviour of nanocrystalline coatings of Fe- 10 Cr having a grain size of 20- 30 nm. These corrosion studies were carried out in 0.05 mol/L H_2SO_4 + 0.25 mol/L Na_2SO_4 solution and 0.05 mol/L H_2SO_4 + 0.5mol/L NaCl solution. These tests were followed by EIS and Mott- Schottky analysis. The active dissolution of the Fe-10Cr nanocrystalline coating was accelerated due to higher dissolution of Cr as evidenced by Cr getting easily enriched in the passive film. This was due to the large fraction of grain boundaries that provided diffusion paths for Cr to the surface resulting in easier passivation. The passive films were n- type semicondutors in the acidic solution without Cl^- and p- type semiconductors in acidic solution with Cl^-. The lower breakdown potential for both materials in the solution with Cl^- was related to the p- type passive film formed on them. Lower donor density and increased Cr in the passive film on the Fe- 10Cr nanocrystalline coating were responsible for its higher chemical stability. Thorpe et al [39] found that the corrosion resistance of $Fe_{32}Ni_{36}Cr_{14}P_{12}B_6$ was greater for the nanocrystalline alloy than the amorphous form. This was because of greater Cr- enrichment in the surface film via rapid interphase boundary diffusion for nanocrystalline material. Zeiger et al [40, 41] studied the corrosion resistance of nanocrystalline Fe-8% Al by potentiodynamic polarization experiments in Na_2SO_4 solution by changing the pH of the solution. As shown in Fig 6, open circuit potential in strongly acidic medium was more negative as compared to that in a weakly acidic medium. Similarly, current density was high in strong acid as compared to that in weak acid. Increased anodic dissolution of nanocrystalline $FeAl_8$ than polycrystalline $FeAl_8$ (Fig. 5) was attributed to a higher defect density.

Considerable attention is also given to nanomaterials for use at high temperatures. Such materials are required in the aerospace industry for the applications as compressor blades and

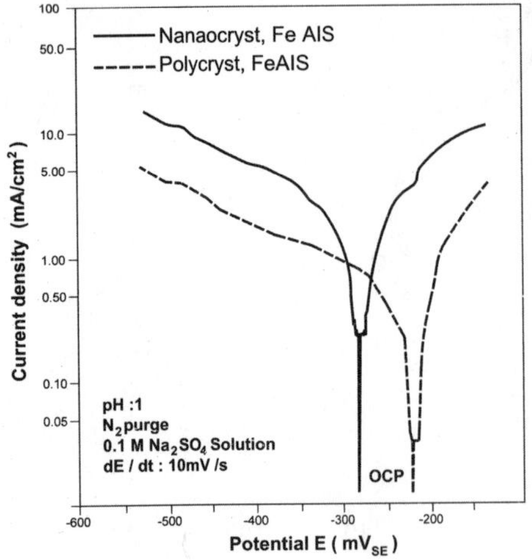

Figure 5 Current-density/ potential curve of nano- and polycrystalline $FeAl_8$ In pH: 1

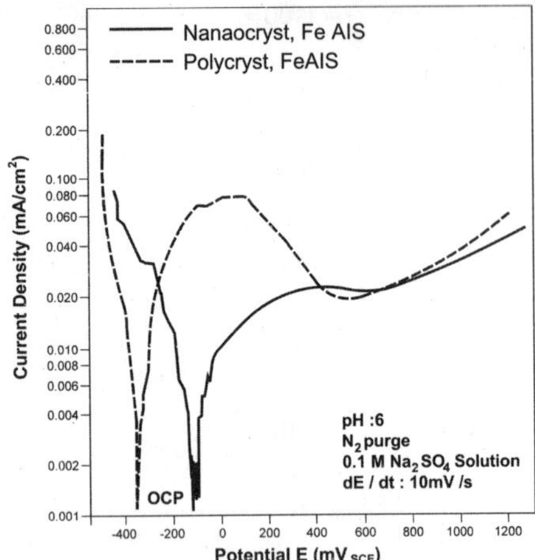

Figure 6 Current-density/ potential curve of nano- and polycrystalline $FeAl_8$ in pH: 6 [40]

missile skins. One of the examples of this is iron aluminide. El Kedim [42] investigated corrosion behaviour by potentiodynamic polarization measurements of nanocrystalline iron aluminide obtained by mechanically activated field activated pressure assisted systems (MAFAPAS) Nanocrystalline iron alumnide showed higher corrosion resistance parameters than the bulk conventional Fe- 40 Al microcrystalline material. Some heat treatments were also given to these nanocrystalline materials. Investigating the effect of grain size altered by suitable heat treatments, it was found that the corrosion resistance increased with decreasing grain size.

2.7.1 *Corrosion behaviour of nanocrystalline stainless steels*

Kwok et al., [21] succeeded in nanocrystallising the surface of 316L stainless steel by cavitation annealing. This process involves severe plastic deformation of surface by cavitation shotless peening followed by recrystallisation annealing to obtain NC grains at surface. It is known that stainless steel is prone to pitting in chloride environments. However, it was reported by Kwok et al., that nanograined 316L surfaces exhibit superior pitting resistance as seen from the shifting of pitting potential from 240 to 390 mV. The repassivation behaviour is also seen to improve by surface nanocrystallisation , as seen by increase in protection potential from -66 mV (42 μm grain) to 220 mV (91 nm). In their studies, as shown in Fig. 7, the improvement in pitting behavior was because a high grain boundary density in NC materials could promote the diffusion of Cr to the surface, thus forming a more uniform passive film containing more Cr and also results in a passive layer which is more uniform. Hence nanocrystalline 316 L showed uniform corrosion, where as a coarse grained material showed localized type of attack.

Nanomodified surface also resulted in a decrease in the corrosion current density, implying lower corrosion rates of nanograined steel surface. This was attributed to the higher electron work function (EWF), i.e. the minimum energy required to remove an electron from the interior of a solid to a position just outside of nanocrystalline stainless steel surface[43]. Similar observations on improvement in passivation capability of 304L stainless steel on

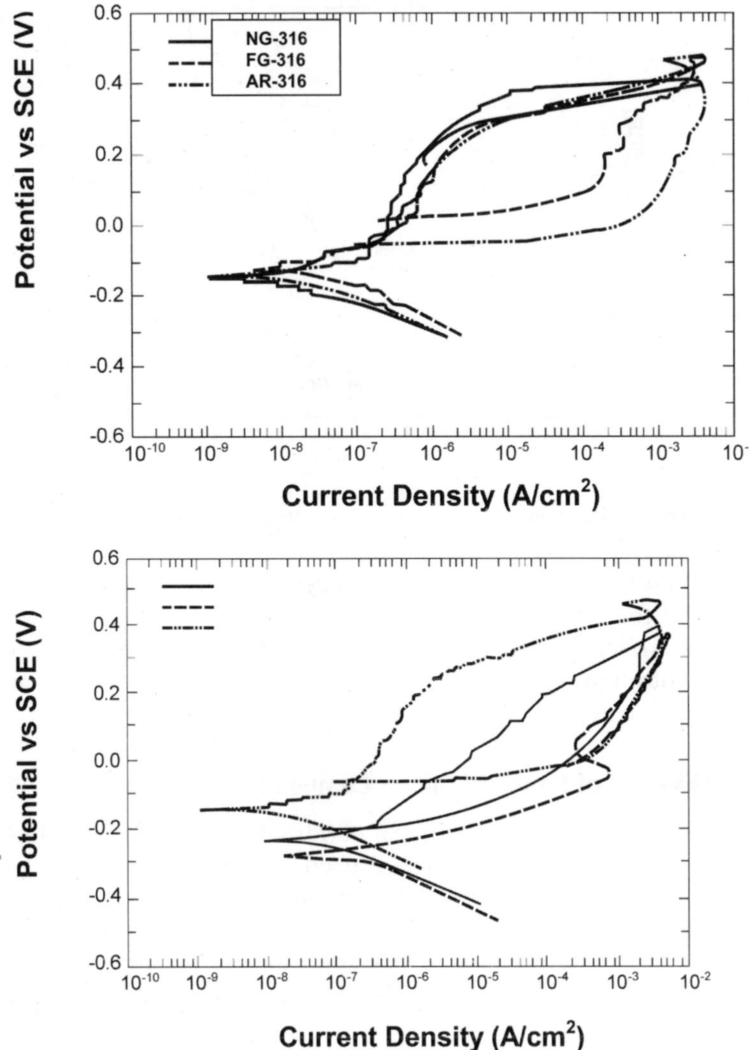

Figure 7 Potentiodynamic polarization curves of various specimens with (a) fine grains and (b) coarse grains in 0.9 % NaCl at 25°C. [21]

surface nanocrystallisation by sandblasting and annealing have been reported by Wang and Li [44].

Inturi et al [45] also performed investigations on the electrochemical behaviour of type 304 stainless steel with varying grain sizes. Conventional AISI type 304 Stainless steel (grain size 30 microns) was compared with nanocrystalline 304 stainless steel with grain size of 25 nm. Anodic polarization was carried out in a 0.3 wt% NaCl solution. It is well known that chloride ions can cause problems by breaking down the passive film that naturally forms on stainless steel surface leading to localized corrosion. Both the nanocrystalline and conventional materials showed pitting and crevice corrosion after polaristaion. Figure 8 shows that there is a wide difference in the pitting potential and passive current density between the nanocrystalline and conventional stainless steel.

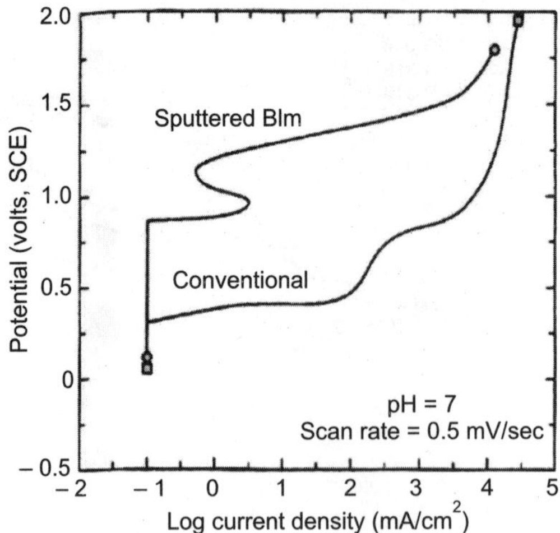

Figure 8 Anodic polarization of Type 304 stainless steel in 0.3 wt% NaCl solution [45]

The nanocrystalline type 304 stainless steel outperformed its conventional grain sized counterpart because of the large number of defects on the surface allowed the Cl^- to spread more evenly over the surface, thus leading to a decrease in Cl^- at any one point on the surface. With a decrease concentration of Cl^- ions on the surface the necessary driving force of localized corrosion is increased.

2.8 Corrosion Behaviour of Nanocrystalline Composite Materials

Yuzhan Li et al [46] studied the electrochemical performance of nanocrystalline $Li_3V_2(PO_4)_3$/ carbon composites. These are lithium batteries used as power sources for wide variety of applications such as cellular phones, notebook computers, camcorders, electric vehicles, etc. Cyclic voltammetry tests were carried out using electrochemical analyzer. The results showed that on reducing the size of particles discharge capacity of the materials improved. This increase in the discharge capacity was attributed to the decrease in crystallite size and improved homogeneity of the carbon dispersion.

2.9 Nano Hydroxyapatite Biocompatible Coatings

It is well known that hydroxyapatite is extremely biocompatible and hence integrates well with human body. Thus it opens a wide possibility for biomedical applications. However, the poor mechanical properties and toughness are limiting factors in their application. Thus hydroxyapatite is often used as coatings on prosthesis made of other biocompatible materials for improved performance. Several methods including plasma spray, solgel, dip coating etc., have been employed to obtain hydroxyapatite coatings over Ti alloy implants. However one of the main limitation of the plasma sprayed HAP coatings is that it suffers from various problems such as high degree of porosities, poor bond strength, non-stoichiometric composition and amorphous structure. Recently nano-HAP coated Ti specimens have been developed by electrophoretic deposition [47]. It was observed that the nano-hydroxyapatite coatings possessed significant improvement in corrosion resistance compared to the uncoated Ti and the micron-sized HAP coated Ti as reflected by the noble shift in corrosion potential, wider passive range, and reduction of corrosion and passive current densities. Polarization current from corrosion of electrophoretically deposited nano HA in simulated body fluid was 300 fold less than that of a thermally sprayed micron sized HA coating. In addition the bond strength of the coating was found to be much higher than the conventionally sprayed HAP coatings. After in-vitro testing, the bond strength of the n-HA coatings remained constant, (> 60 MPa), while the plasma sprayed and chemical deposited HAp coatings exhibited significant bond strength reductions. Nano-hydroxyapatite coatings are expected to result in reduced integration time with human tissues due to their higher bioactivity resulting in accelerated osteointegration. Also the implant lifetimes are expected to be higher with the use of nano coatings for biocompatible applications.

2.10 CVD Nanodiamond Coatings for Electrochemical Applications

With the development of techniques to synthesize diamond films by chemical vapour deposition methods, it has become possible to utilise many of the superior properties of diamond for engineering applications. Apart from possessing high elastic modulus and highest hardness, they are also biocompatible and possess extreme tolerance to chemical and electrochemical attack. Hence diamond electrode coatings are also being considered as potential candidates for several electrochemical applications including, electroanalysis, electrocatalysis, spectroelectrochemistry, and bioelectrochemistry.

It is known that diamond is a wide band gap material and hence they exhibit nearly insulating properties. To achieve the required surface electrochemical conductivity, diamond films are often doped with boron. This is most often accomplished by adding controlled amounts of diborane or trimethylboron to the source gas mixture. Boron-doped microcrystalline and nanocrystalline diamond thin-films possess a number of important and practical electrochemical properties, unequivocally distinguishing them from other commonly used sp2-bonded carbon electrodes, such as glassy carbon, pyrolytic graphite, and carbon paste. These properties are (i) a low and stable background current, leading to improved signal-to-background ratio (SBR) and signal-to noise (SNR) ratios; (ii) a wide working potential window in aqueous and non-aqueous media; (iii) superb microstructural and morphological stability at high temperatures and current densities (iv) good responsiveness for several aqueous and non-aqueous redox analytes without any conventional pretreatment; (v) weak adsorption of polar molecules, leading to improved resistance to electrode deactivation and fouling; (vi) long-term response stability (e.g. months during air exposure); and (vii) optical transparency in the UV-Vis and IR regions of the electromagnetic spectrum, useful properties for spectro-electrochemical measurements [48].

One of the first demonstrations of diamond's usefulness in electroanalysis was the oxidative detection of azide anion in aqueous media, for application in automotive industries, to provide environmental safety controls from release azide anions of due to its toxic nature. Diamond coated electrochemical electrodes also hold promise in the nuclear waste reprocessing plants for electrochemical processing of nitrate solutions, by virtue of their superior electrochemical stability and good irradiation resistance.

For electrocatalytic applications boron doped diamond thin-films have been reported to be better catalyst support host materials, than presently employed sp2-bonded carbon. This is because of the formers high electrical conductivity , dimensional stability, and the fact that catalyst particles can be stably anchored into the surface microstructure during deposition. Two technologies requiring dimensionally stable catalytic electrodes are fuel cells and electrolytic reactors for water disinfection and decontamination. Electrically conducting diamond is a also finding newer application as optically transparent electrodes (OTE), due to the fact that diamond exhibits transparency to electromagnetic radiation over a wide range of wavelength, from the near-UV to the far-IR. The wide optical window, coupled with the interesting electrochemical properties, make diamond a viable, new OTE. Diamond OTEs offer a number of advantages over traditional materials, such as indium-doped tin oxide (ITO): (i) stability in acidic media and chlorinated organic solvents, (ii) the ability to withstand cathodic polarization, (iii) a reasonably well-defined and stable surface chemistry, (iv) transparency in the UV, visible and IR regions of the electromagetic spectrum, (v) a wide working potential window in excess of 3 Vin aqueous media, and (vi) a low and stable background current, which leads to improved SBR in electroanalytical measurements. These materials hold great promise, for example, in bioelectrochemical research, due to the low background current, wide working potential window, and broad range of optical transparency [49, 50].

3.0 SUMMARY

The discussions above indicate that the effect of nanostructuring on corrosion behaviour is highly system dependent. Not much effect of nanocrystallisation is observed in Cu and Co on improving their corrosion behavior. In sharp contrast, NC zinc coatings show significantly improved resistance to chemical attack compared to conventional electrogalvanised steel. Similarly NC Ti film is observed to exhibit improved passivation behaviour compared to coarse Ti.

The presence of a huge fraction of grain boundaries can result in contrasting effects on the response of the NC surface to chemical environment. The net effect can be to either increase or decrease the corrosion rate, depending on the metal system, environmental parameters and various other conditions. In general classical corrosion theory would suggest that corrosion rate should increase for nanocrystalline surface due to the formation of a large number of microelectrochemical cells between matrix and grain boundaries. However, the same grain boundary defects in nanocrystalline systems can provide as diffusion paths for faster formation of passive films to protect surfaces. The large density of grain boundary defects also provide as sinks for segregation of certain elements like sulphur that can deteriorate corrosion behaviour in some systems. Such a intrinsic cleaning effect will contribute to decrease in propensity for localized attack.

A multitude of contrasting effects makes it more difficult to provide a universal mechanism for the corrosion behaviour of nanocrystalline surfàces. Hence, the corrosion behaviour is also sensitive to the corrosive environment, defect nature of surface etc. Hence, the corrosion behaviour of NC films also depends on the synthesis technique apart from grain size. Different deposition techniques result in varying surface strains, which could be either tensile or compressive. The presence of surface strains as a result of the deposition process can also have equally binding influence on the corrosion behaviour of surfaces. Also, from the limited corrosion data available in the literature for nanocrystalline alloys; it is difficult to predict the electrochemical behaviour of nanocrystalline metal from the known properties of their coarse-grained polycrystalline analogs. Hence a detailed correlation of corrosion behaviour with microstructural state (grain size, defect types and distribution etc) and surface residual stress is essential to evolve a comprehensive understanding of the fundamental mechanisms of corrosion in nanocrystalline systems. More detailed research on a variety of systems and corrosion environments is essential for better fundamental understanding of the processing in nanocrystalline systems.

However it is known that for most materials the strength and toughness are significantly improved by decreasing grain size to nano dimensions. Hence even if the difference in corrosion behaviour is not significant, nanomaterials are being preferred for applications in certain corrosive environments by virtue of the benefit they offer due to their improved mechanical properties.

4.0 ACKNOWLEDGEMENTS

The authors are grateful to Dr Baldev Raj, Director, IGCAR for his constant support and motivation for all the surface engineering and nanotechnology research activities in the Centre. The authors also thank Dr R.K.Dayal, Head, Corrosion Science and Technology Division, for useful guidance and suggestions.

5.0 REFERENCES

1. H. Gleiter, Prog. Mat. Sci, 33, 223 (1989)
2. A. M. El-Sherik, U. Erb, G. Palumbo and K.T. Aust, Scripta Metall. et Mat., 27 (1992), 1185.
3. C. Cheung, G. Palumbo and U. Erb, Scripta Metall. et Mater., 32 (1994), 735.
4. L. Wong, D. Ostrander, U. Erb, G. Palumbo and K.T. Aust, Nanophases and Nanostructured Materials, R.D. Shull and J.M. Sanchez (eds.), TMS (1994), 85.
5. U. Erb, G. Palumbo, R. Zugic and K.T. Aust, in "Processing and Properties of Nanocrystalline Materials", C. Suryanarayana, J. Singh and F.H. Froes, TMS (1996), 93.
6. U. Erb, Nanostr. Mat. 6 (1995), 533.
7. R. Rofagha, R.Langer, A.M.El-Sherik, U.Erb, G.Palumbo, K.T.Aust,, Scripta Metallurgica et Materillia, Vol.25, 1991, pp. 2867-2872
8. R. Rofagha, R. Langer, A.M. El-Sherik, U. Erb, G. Palumbo and K.T. Aust, Mat. Res. Soc. Symp. Proc., 238 (1992), 751.
9. S. Wang, R. Rofagha, P.R. Roberge and U. Erb, Electrochem. Soc. Proc., 95-8 (1995), 244.
10. N. Wang, Z. Wang, K.T. Aust and U. Erb, Acta Metall. et Mat., 43 (1995), 519.
11. M. J. Aus, B. Szpunar, U. Erb, A.M. El-Sherik, G. Palumbo and K.T. Aust, J. Appl. Phys. 75 (1994), 3632.
12. M. J. Aus, B. Szpunar, A.M. El-Sherik, U. Erb, G. Palumbo and K.T. Aust, Scripta Metall. et Mat., 27 (1992), 1637.
13. C. Cheung and U. Erb in "Novel Techniques in Synthesis and Processing of Advanced Materials", J. Singh and S.M. Copley (eds.), TMS (1995), 244.
14. M. J. Aus, B. Szpunar, U. Erb, G. Palumbo and K.T. Aust, Mat. Res. Soc. Symp. Proc., 318 (1994), 39.

15. D. M. Doyle, G. Palumbo, K.T. Aust, A.M. El-Sherik and U. Erb, Acta Metall. et Mat., 43 (1995), 3027.
16. G. Palumbo, D.M. Doyle, A.M. El-Sherik, U. Erb and K.T. Aust, Scripta Metall. et Mat., 25 (1991), 177.
17. S. Wang, J.K.Lewis, P.R. Roberge and U. Erb, www.corrosionscience.com/ events/intercorr/ techsess/papers/session4/abstract/wang.html
18. S.J.Thorpe, B.Ramaswami, A.T. Aust, J. Electrochem.Soc, 135, 2162 (1993)
19. W. Zeiger, M. Schneider, H. Worch, Trans Tech Publications, Switzerland, ISMANAM- 97, pp 833-836
20. R. Rofagha, S.J. Splinter, U.Erb, N.S. McIntyre, Nanostructured Materials, Vol. 4, No. 1, pp. 69-78, 1994
21. C.T. Kwok, F.T. Cheng, H.C. Man, W.H. Ding, Materials Letters 60 (2006) 2419- 2422.
22. G. Palumbo, S.J. Thorpe, K.T. Aust, Scripta Matallurgica et Materialia, Vol. 24, pp 1347- 1350, 1990
23. U.Erb, G. Palumbo, R. Zugic and K.T. Aust, in "Processing and properties of Nanocrystalline Materials", C.Suryanarayana (ed.), TMS, 93 (1996)
24. C. Cheung, D. Wood and U.Erb, in "Processing and properties of Nanocrystalline Materials", C.Suryanarayana (ed.), TMS, 479 (1996)
25. R.Mishra, R.Balasubramaniam, Corrosion Science 46 (2004) 3019-3029
26. Liping Wang, Junyan Zhang, Yan Gao, Qunji Xue, Litian Hu and Tao Xu, Scripta Materialia 55 (2006) 657- 660
27. R. Rofagha, U. Erb, D. Ostrander, G. Palumbo, K.T. Aust, Nanostructured Materials, Vol. 2, pp 1-10, 1993
28. E.Kus, Z.Lee, S.Nutt and Mansfeld, Corrosion Vol 62, No2, 2006 P 152- 161
29. ASTM G 67-99, " Standard Test Method for Determining the Susceptibility to Intergranular Corrosion of 5XXX Series Aluminium Alloys by Mass Loss after Exposure to Nitric Acid (NAMLT Tets)" (West Conshohocekn, PA: ASTM International, 1999).
30. E. Sikora, X.J. Wei, and B.A. Shaw, Corrosion Vol.60, No.4, 2004 P 387- 389
31. A. Barbucci, G.Farne, P.Matteazzi, R.Riccieri, G.Cerisola, Corrosion Science 41 (1999) 463-475
32. A. Vinogradov, T. Mimaki, S. Hashimoto, RZ. Valiev, Scripta Mater 1999; 41:319
33. S.H.Kim, K.T. Aust, U.Erb, F. Gonzalez, G. Palumbo, Scripta Materialia 48 (2003) 1379-1384
34. Ball GR, Prayer JH. 12th Int Proc Cor Congress 1984 ; 3A : 1132
35. Liping Wang, Yimin Lin, Zhixiang Zeng, Weimin Liu, Qunji Xue, Litian Hu, Accepted in Scripta Materialia (2006)
36. A. Balyanov, J. Kutnyakova, N.A. Amirkhanova, V.V. Stolyarov, R.Z. Valiev, X.Z. Liao, Y.H. Zhao, Y.B. Jiang, H.F. Xu, T.C. Lowe, Y.T. Zhu, Scripta Materialia 51 (2004) 225- 229
37. Kh.M.S. Yousef, C.C. Koch, P.S. Fedkiw, Improved corrosion behaviour of nanocrystalline zinc produced by pulse-current electrodeposition, Corros. Sci. 46 (2004) 51
38. Guozhe Meng, Ying Li, Fuhui Wang, Electrochemica Acta 51 (2006) 4277- 4284
39. S.J. Thorpe, B. Ramaswami, K.T. Aust, J. Electrochem. Soc. 135 (1988) 2162
40. W. Zeiger, M. Schneider, D. Scharnwber, H. Worch, Corrosion behaviour of a nanocrystalline FeAl8 alloy, Nanostruct. Mater. 6 (1995) 1013.
41. W. Zeiger, M. Schneider, H. Worch, Materials Sci. Forum, 269- 272, 833 (1998)
42. O.El Kedim, S. Paris, C. Phigini, F. Bernard, E. Gaffet, Z. A. Munir, Materials Science and Engineering A369 (2004) 49- 55

43. A.K. Neufeld, N.D. Mermin, Solid State Physics, Saunder, 1976, 354pp.
44. X.Y.Wang and D.Y.Li, Electrochem. Acta, 47 (2002) 3939
45. Inturi R.B., Szklarska-Smialowski Z, Corrosion, May 1992, pp. 398-403.
46. Yuzhan Li, Zhen Zhou, Manman Ren, Xueping Gao, Jie Yan, Electrochemica Acta 51 (2006) 6498-6502
47. http://www.inframat.com/hydro2.htm
48. Matt Hupert, Alexander Muck, Jian Wang, Jason Stotter, Zuzana Cvackova, Shannon Haymond, Yoshiyuki Show, Greg M. Swain; Diamond and Related Materials 12 (2003) 1940–1949
49. J.K. Zak, J.E. Butler, G.M. Swain, Anal. Chem. 73 (2001) 908.
50. J. Stotter, J. Zak, Z. Behler, Y. Show, G.M. Swain, Anal. Chem. 74 (2002) 5924.

CHAPTER 14

NDE Techniques for Assessment of Corrosion Damage in Materials and Components

Baldev Raj, T. Jayakumar and G. K. Sharma
Indira Gandhi Centre for Atomic Research, Kalpakkam – 603 102, India

1.0 INTRODUCTION

Metal that has been extracted from its primary ore (metal oxides or other free radicals) has a natural tendency to revert to that state under the action of oxygen and water. The action is called rusting and most common example is rusting of steel.

A number of corrosion related failures and catastrophes led to loss in production, leakages and contamination, in addition to loss of human lives. While every attempt is made in selection of materials, proper design and operating environment, anticipated and unanticipated corrosion related degradation of components is inevitable. Therefore, it becomes essential to monitor the performance of the components in service for assessing the progress of corrosion related degradation to be within the expected and acceptable limits and also to detect any accelerated/unanticipated corrosion related degradation. Such an approach is aimed at avoidance of failures, adoption of methodologies for prevention/retardation of degradation and life assessment and extension of industrial components. Application of non destructive evaluation (NDE) techniques provides an opportunity to detect and monitor corrosion related damage in plant components and structures thus ensuring their reliable, safe and economic performance.

There are a host of types of corrosion such as uniform corrosion, pitting corrosion, crevice corrosion, stress corrosion cracking, hydrogen embrittlement, oxidation, corrosion fatigue and erosion corrosion. The physical manifestation of these types of corrosion is different and hence their detection and evaluation demand employment of appropriate NDE techniques. The use of various types of materials also demands the employment of different NDE techniques. The service conditions such as operating temperature, environment and access for inspection also decide the type of NDE technique to be employed for a given case. In this chapter, the NDE techniques useful for detection and assessment of corrosion related damage are given. NDE techniques useful in laboratory for basic understanding of corrosion processes in addition to exploratory studies for widening the application base are also given.

Many of the developments in NDE are aimed at imaging and quantitative evaluation of corrosion damage, thus providing the required information for taking decisions on continued use, next inspection schedule and replacement of components. The common NDE techniques applied for corrosion damage assessment include visual examination, penetrant testing, ultrasonic thickness measurements and eddy current testing. Some of the advancements in NDE techniques include ultrasonic time of flight diffraction, ultrasonic back-scatter, ultrasonic C-scan, ultrasonic echo dynamic technique, ultrasonic crack tip echo technique, ultrasonic based internal rotary inspection, guided wave ultrasonics, remote field eddy current testing, pulsed and low frequency eddy current testing, dynamic real time radiography, infrared thermography, magnetic flux leakage, field signature mapping (FSM), alternating current field measurement (ACFM), videoendoscopy, magnetic Barkhausen emission, microwave testing technique and implementation of signal processing and imaging approaches to NDE signals. Intelligent pigs for inspection of long pipelines have been developed and are being used routinely. Multi technique, multi sensor and multi parametric approaches are also used for comprehensive evaluation of corrosion related damage. Table 1 gives the list of NDE techniques generally applied for monitoring different types of corrosion damage [1]. While the techniques like eddy current, ultrasonic and radiography are used to detect the static damage, acoustic emission technique offers the possibility for evaluation of dynamic corrosion damage in real-time.

Visual inspection is the simplest of all NDE techniques. Surface imperfections invisible to the eye may be revealed by penetrant or magnetic methods. Except for some types of hydrogen related damage, all the corrosion related degradation starts from the surface. Hence, unless serious surface damage is found, there is often little point in proceeding further to more complicated examination of the interior by other methods like ultrasonics or radiography. In multilayered structures, corrosion can occur in the hidden layers and appropriate NDE

Table 1 NDT techniques applied for monitoring different types of corrosion damage

Type of Corrosion	NDE Techniques
Uniform	MFL, UT, ECT, RT, OPTICAL/ LASER, IRT
Localized	UT, ECT, RT, OPTICAL/ LASER, IRT
Stress Corrosion Cracking	PT, AE, UT, ECT
Corrosion Fatigue	PT, AE, UT, ECT
High Temperature Oxidation	AE, ECT, UT, IRT
AE: Acoustic Emission	ECT: Eddy Current Testing
IRT: Infrared Thermography	MFL: Magnetic Flux Leakage
PT: Penetrant Testing	RT: Radiographic Testing
UT: Ultrasonic Testing	

techniques are necessary for detection of such hidden corrosion. The following sections provide the details of various NDE techniques for detection and evaluation of corrosion related damage.

2.0 VISUAL TESTING

The most valuable NDT tool is the human eye because of its excellent visual perception. Visual testing (VT) provides a means of detecting and examining various types of corrosion, as all forms of corrosion starts from the surface. The characteristics of surface features also play a vital role in corrosion resistance of materials and components. Hence visual examination is an important tool for examining metals and components for surface flaws such as corrosion, contamination, surface finish, coatings, oxide layers and surface discontinuities on joints (ex. welds, seals and adhesive joints) etc. Visual inspection provides great help for taking decision on proceeding with other inspection techniques. The basic procedure used in visual NDT involves illumination of the test specimen with light, usually in the visible region. Under ordinary conditions, eye is most sensitive to yellow green light with a wavelength of 5560 Angstrom. For visual inspection, adequate lighting of about 800 to 1000 lux is important. The time period during which a person is permitted to inspect is also limited to maintain adequate sensitivity. The specimen is then examined with eye or with the aid of the optical devices. Visual inspection is often the most cost-effective method.

2.1 Visual Inspection with Aids

Some of the equipment to aid visual inspection include Borescope, Endoscope and Telescope.

2.1.1 Borescope

A borescope is an instrument designed for an observer to inspect the inside of a narrow tube, bore, or chamber. Borescope consists of precision built-in illumination system having a complex arrangement of prisms and plain lenses through which light is passed to the observer with maximum efficiency. The illumination system is either an incandescent lamp located at distal end or a light guide bundle made from optical fibres. Rigid borescopes are generally limited to applications with a straight line path between the observer and the area to be observed (Fig. 1(a)).

2.1.2 Flexible fiber optic borescope (Flexiscope)

The Flexiscope is an instrument which is very useful when inspection has to be done at regions of difficult access. Flexible fibre optic borescope can be used around corners and through passages with several directional changes. The two types of flexible borescopes are flexible fiberscopes and videoscopes with a CCD image sensor at the distal tip (Fig.1(b)).

The experienced inspector can obtain a lot of information from the appearance of the surface. In most cases, corrosion penetration can not be ascertained, but often an indication of the corrosion can be obtained by measuring pit depth. The experienced eye can find defects such as general corrosion, pitting corrosion, stress corrosion cracking etc, which are then confirmed and exactly sized with the help of the other NDT techniques [2].

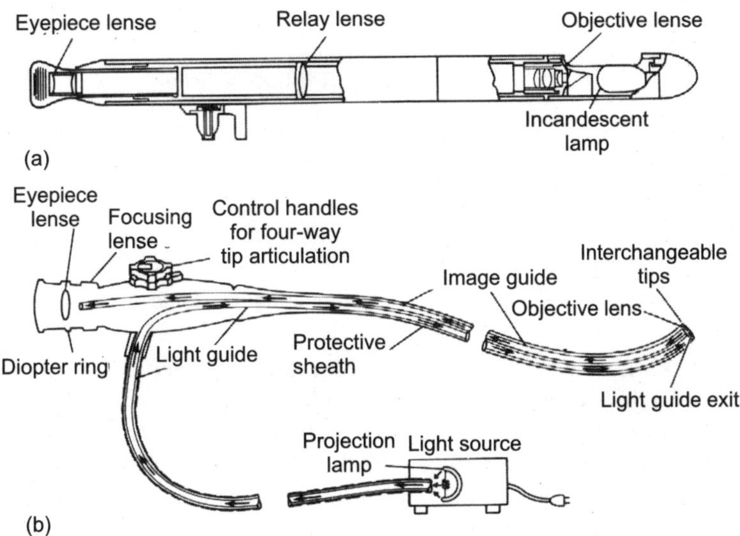

Figure 1 (a) Rigid borescope with a lamp at the distal end. (b) A flexible fiberscope with a light source [2].

3.0 ULTRASONIC TESTING

The foundation of the ultrasonic technology is closely related to the discovery of the phenomenon of Piezoelectricity by Pierre Curie in 1880. As per this discovery, asymmetrical crystals such as quartz and Rochelle salt (potassium sodium tartrate) generate an electric charge when mechanical pressure is applied. Conversely, mechanical vibrations are obtained by applying electrical oscillations to these crystals. This provided the key for ultrasonic inspection. Ultrasonic waves are generated and detected by using piezoelectric crystals (Fig. 2). A number of methods are available for ultrasonic inspection of component based on the geometry and material. Various parameters are important like ultrasonic velocity, attenuation, acoustic impedance etc and these parameters has to be correlated with material properties to characterize problem with the components. To characterize the defect or corrosion various scanning methods are available and the information can be obtained either by received echoes or images. The most widely used method used for inspection is pulse echo in which the ultrasonic wave will be generated and received by a single transducer. Three echoes are shown in Fig. 2, The first echo is from front surface, second is from back wall and a third in between is from a defect (cavity). Sound wave reflects from the defect due to acoustic impedance difference between the surroundings and the defect. Ultrasonic testing (UT) provides a sensitive detection capability for corrosion damage when access is not available to the surface exposed to corrosion. The reflecting surface that is offered by typical pitting corrosion is often poor for ultrasonic purposes. The A-scan measurement can give information about the wall loss and morphology of signals can be a tool to detect corrosion. For the detection of corrosion, A-Scan presentation is preferred and can be complimented by B-Scan and C-Scan facilities which give information in image form.

Ultrasonic testing is used for monitoring several forms of corrosion. Uniform corrosion is monitored by measuring the thickness of the component using ultrasonic thickness gauge.

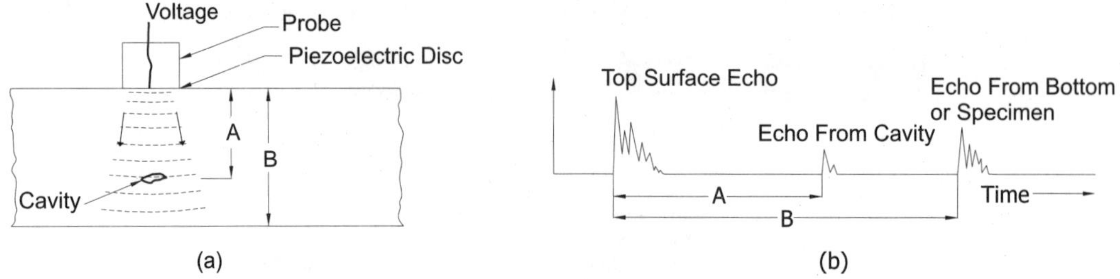

Figure 2 Principle of flaw detection using ultrasonic testing.

Stress corrosion cracking occurs by production of a new interface within the material, which causes ultrasonic reflections earlier than those from the back surface. Pitting and inter granular corrosion cause scattering of the ultrasound and can be detected by the use of shear waves in an angular incidence as well as using advanced ultrasonic imaging systems. In addition, such scattering also results in attenuation of longitudinal waves commonly referred to as loss in back-surface signal. This phenomenon serves as a means of corrosion detection in thick structures. For detection of intergranular stress corrosion cracking (IGSCC) in austenitic stainless steel welds using ultrasonic methods, it is important to identify the nature of the echoes from the root of the weld and that from the IGSCC. Advanced signal analysis methods such as split spectrum processing are useful for evaluation of IGSCC, that cause low ultrasonic reflectivity.

A number of methods are available to measure crack depth by ultrasonic testing. These are: (i) defect echo height method, (ii) decibel drop method, (iii) edge echo method, (iv) scattering method and (v) composite aperture method. However, in the case of IGSCC, the cracks tend to branch out and therefore depth measurement becomes difficult.

Corrosion pits are not ideal reflectors and hence special ultrasonic probe design is required for special applications. Dual element transducers are preferred for this. Figure 3 shows the pulse echo principle and a twin crystal probe used for detection of pitting corrosion. In this configuration, one crystal acts as a transmitter and the other as a receiver. The transmitter is isolated from the receiving circuit so that the A-Scan display is free from the presence of a transmitting signal. As a result, the transmission pulse does not obscure the first back wall echo even when testing relatively thin walled plate or pipe. Only in shaded region, where the two fields of view coincide, echoes can be obtained. Outside this region, either the transmitter doesn't interrogate the spot, or the receiver doesn't catch the echo. Within the shaded area, the significant zone is the triangular portion from the test surface to the beam path range at which the transmitter beam completely crosses the receiver viewing angle. Close to the scanning surface (the apex of that triangle), not much of the transmitter beam reflecting from that range is received by the receiver. The proportions increase until they maximize at the crossover range. In this triangle, even very large reflectors will give only small echo amplitudes. The closer to the scanning surface (i.e. the thinner the wall), the smaller would be the echo. If the reflector is within this region, the gain should be adjusted accordingly.

The easiest pits to be detected are the 'lake' type because, in the deepest region, they are relatively parallel to scanning surface and can be expected to give reasonable reflectivity. On

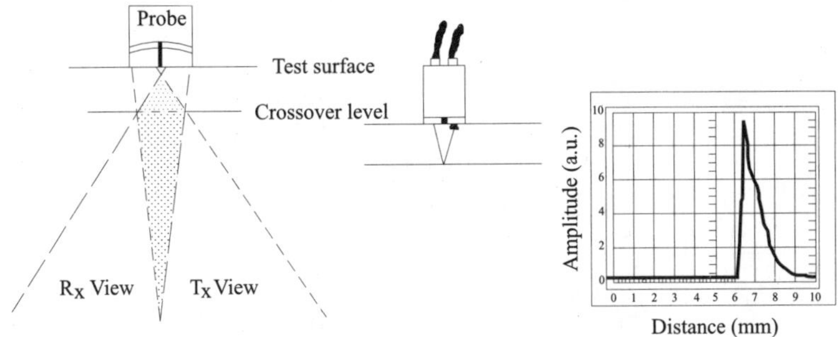

Figure 3 Twin crystal probe and effective beam for evaluation of pits [3].

the other hand, the 'conical' pits tend to reflect sound away from the receiver and the centre of the pit is often too small in area to give strong signal. Hence, for this type of pits, the probability of detection is less [3].

3.1 Ultrasonic Internal Rotary Inspection System (IRIS) for Inspection of Tubes

Ultrasonic internal rotary inspection system (IRIS) based technique has been developed as a complementary technique to eddy current testing for inspection of tubes in heat exchangers and steam generators. Conventional eddy current testing is not sensitive enough for detection of localized pitting (corrosion) type defects, which are isolated, and also defects present under support plates. Figure 4 shows the plan view of the ultrasonic IRIS system. In this system, an ultrasonic transducer lies axially along the tube and the pulse emitted impinges upon a mirror angled at 45 degrees to the axis of the tube. The ultrasonic pulse is deflected to penetrate the wall of the tube in a radial direction. Ultrasonic echoes are returned from both the inside and the outside surfaces. The signal is displayed on a computer screen. The mirror rotates 360° and successive indications from each pulse are displayed sequentially on the screen. All the data from one full circumferential rotation is displayed on the screen, at a time. As the probe is withdrawn at a controlled rate from the tube, the overlapping footprint of the ultrasonic scan covers every 25 mm square of the tube circumferentially and longitudinally, giving a complete recording of the tube's condition. The technique has the capability of detecting wall thinning and pitting due to corrosion in tubes. The advantages of this technique are: (i) it can measure the remaining wall thickness up to 500 mm of the tubes and pipes, (ii) it can indicate the reduction in wall thickness that has taken place either from outer surface or inner surface of the tubes/pipes and (iii) it also reveals the circumferential position of the defects such as localized pitting and defects under support plate. Figures 5a and 5b show the CRT patterns obtained by IRIS from a good and a corroded tube [4]. In these patterns, the left boundary of the display represents the condition of the inner surface of the tube and the right extreme trace represents the condition of the outer surface. The height of the display represents the circumference, and the wall thickness of the tube is represented by the width of the display, as shown in Fig. 5a.

3.2 Hydrogen Related Damage Assessment

Hydrogen attack is produced in steels exposed to high pressure hydrogen at high temperatures. Decarburisation of the steel due to hydrogen attack reduces the material

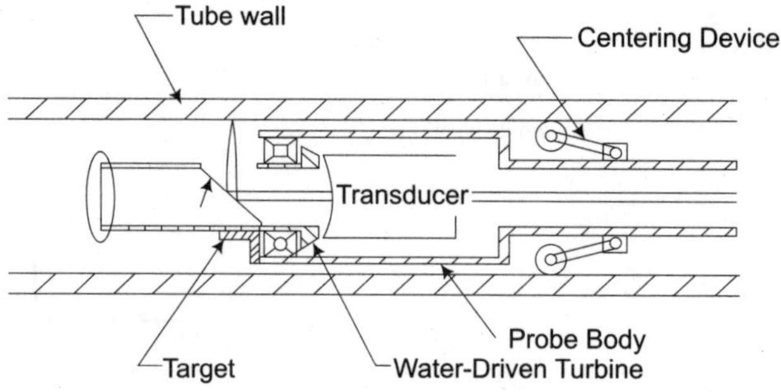

Figure 4 Ultrasonic Internal Rotary Inspection System (IRIS) [4].

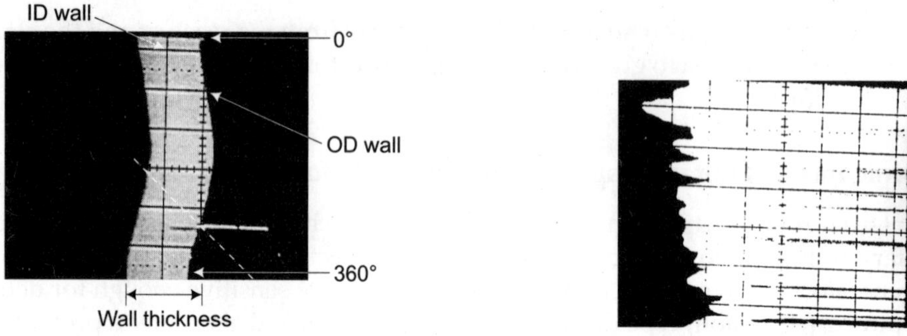

Figure 5a CRT Pattern from a good tube [4].

Figure 5b CRT Pattern from a Corroded Tube [4].

toughness without causing any reduction in the thickness. This loss of structural strength from hydrogen attack has been known to produce several failures in petroleum and fossil fuel plants. Ultrasonic parameters used for evaluation of hydrogen damage include velocity, attenuation and back-scatter amplitude [5]. It has been seen that hydrogen attack decreases ultrasonic wave velocity and increases attenuation and back-scattering. Ultrasonic back-scatter is a method in which the ultrasonic waves scattered from within a material are analysed. Back-scattering occurs from the grains because of impedance mismatch at the grain boundaries. Back-scattering increases with frequency. Velocity reduction is caused by an overall decrease in the modulus of elasticity due to microcracking. Such a decrease in bulk modulus is produced when a large number of microcracks are present, in other words, the material has undergone extensive hydrogen attack. Ultrasonic time of flight diffraction (TOFD) technique has also been employed for detection of hydrogen induced cracking (HIC) in underground pipelines [6]. A TOFD image is shown in Fig. 6.

Attempts have been made to assess hydrogen damage by analyzing the backwall echoes and backscattered signal in frequency domain, the second order moments of the the frequency spectra, which emphasize the higher frequency components have been attempted for assessment of hydrogen damage [7]. The second order moments showed variability for hydrogen attacked materials.

Figure 7 gives the methodology followed for obtaining the second order moment for the echoes. The sensitivity of this technique to hydrogen attack is also evident by the higher

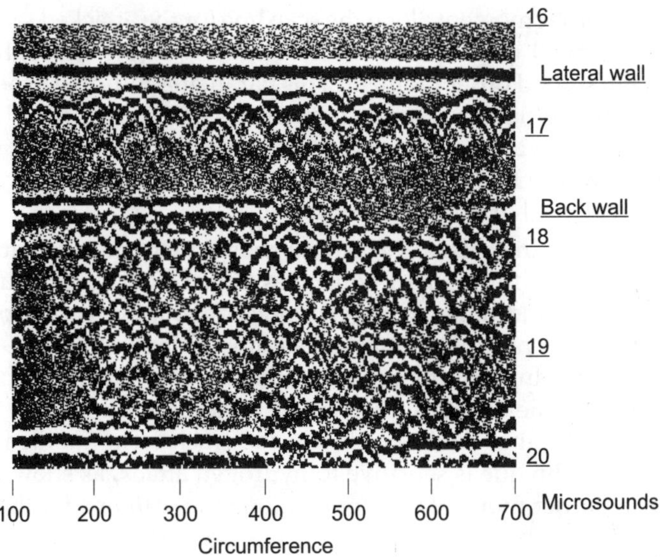

Figure 6 TOFD image shows back wall signal losses due to HIC masking the back wall [6].

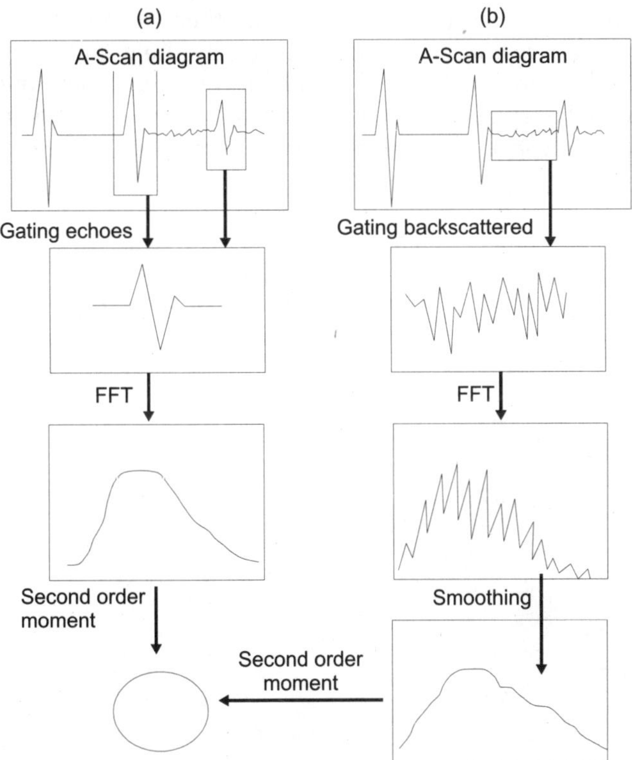

Figure 7 Signal getting and processing of (a) Echo and (b) backscattered signal for second momentum analysis [7].

variation of the second order moment along the scan line for the attacked samples, in relation to the reference. The backwall echo spectra obtained along a line through the sample were processed to obtain the second order moment. This parameter indicates whether the frequency distribution tends to shift to higher or lower frequencies (it depends mainly on attenuation of higher frequencies) and this can be compared with the level of hydrogen damage. A fall in the amplitude of the back wall echo for higher frequencies is observed when an ultrasonic pulse propagates through regions that contain cracks.

The ultrasonic spectral analysis of the echoes shows that hydrogen attack has produced higher attenuation especially for higher frequencies (Fig. 8). This is attributed to the fact that hydrogen damage acts as a additional scatterer. Optical micrograph of a typical stepwise crack of the analysed surface of a hydrogen attacked sample is shown in Fig. 9. This technique is related to the presence of many scatters present in the material, such as debonds at inclusions, in the material (Fig. 10). The spectral analysis of backscattered signals also identifies the hydrogen attacked samples, but could not locate the cracks. The moment for backscattered signals shows that this technique is sensitive to hydrogen attack, as shown in Fig. 11. Higher variation in the second order moment was found along the scan line for the attacked samples.

3.3 Microbiologically Influenced Corrosion

The deterioration by way of corrosion of materials influenced by micro-organisms is termed as microbiologically influenced corrosion (MIC). Another phenomenon related to MIC is biofouling, which is the physical obstruction caused by the accumulation of inorganic and organic deposits on a component surface by biological means. Macrofouled layers of several millimetres thick formed over a period of time affect the operation of the component in various ways, such as reduced efficiency of heat transfer and reduction in flow rate due to chocking.

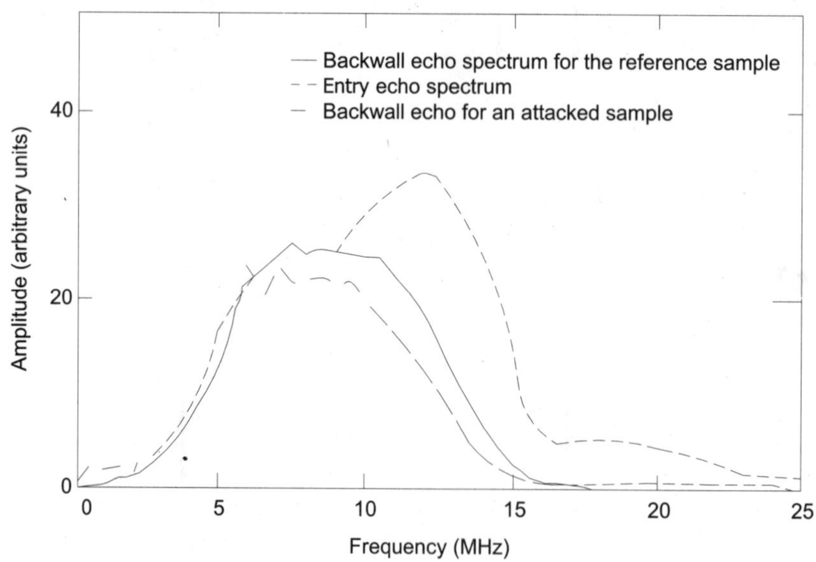

Figure 8 UT signal amplitude variation with respect to hydrogen attack [7].

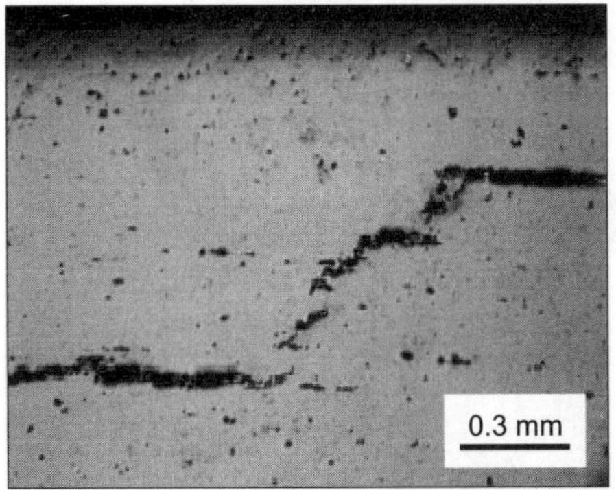

Figure 9 Optical micrograph of a typical stepwise crack of the analysed surface of a hydrogen attacked sample (Without etching) [7].

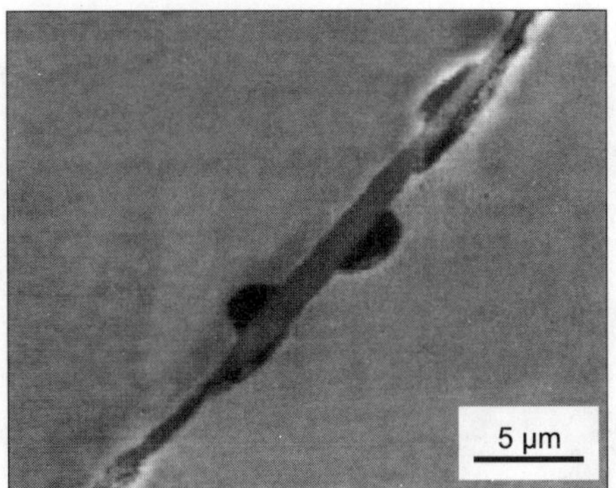

Figure 10 Hydrogen attacked sample showing disbonding at the inclusion sites [7].

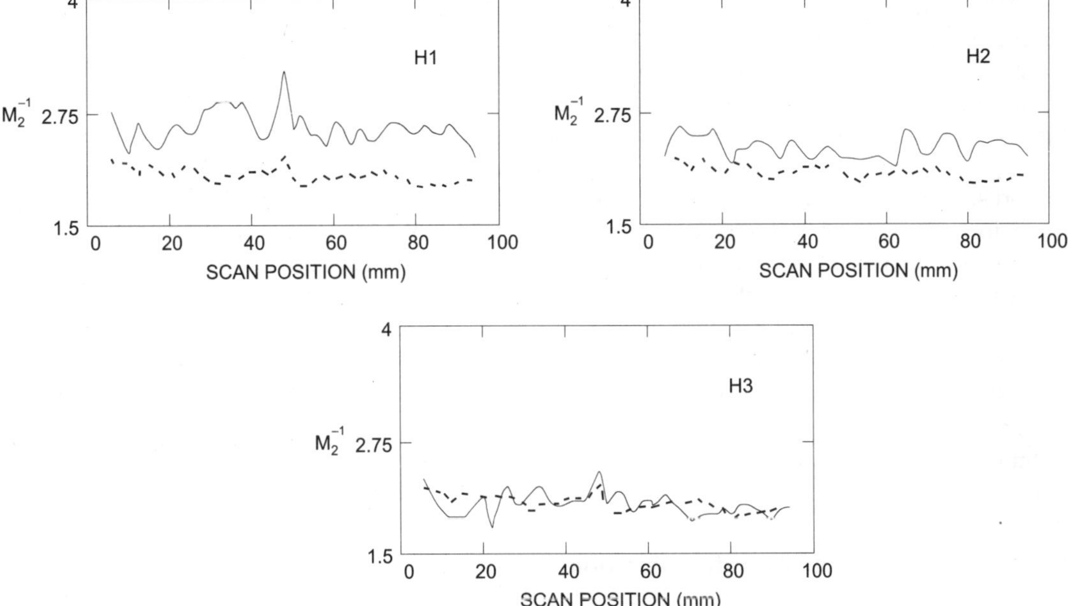

Figure 11 Inverse of second order moment (M2-1) of the backscattered signal spectrum for the attacked sample (solid line) and reference sample (dotted line) [7].

Presence of tubercles (biological corrosion products) formed at the inside surface of carbon steel pipes (Fig. 12a) poses difficulty in using conventional ultrasonic techniques for wall thickness measurements, essentially due to heavy attenuation taking place at the inside surface-corrosion layer interface. For monitoring the wall thickness of these biologically corroded pipes, a new technique based on ultrasonics has been attempted [8]. In this technique, two angle beam probes are used, one probe for transmitting the sound energy into the pipe and the other probe for receiving the refracted sound energy from the inside surface of the pipe

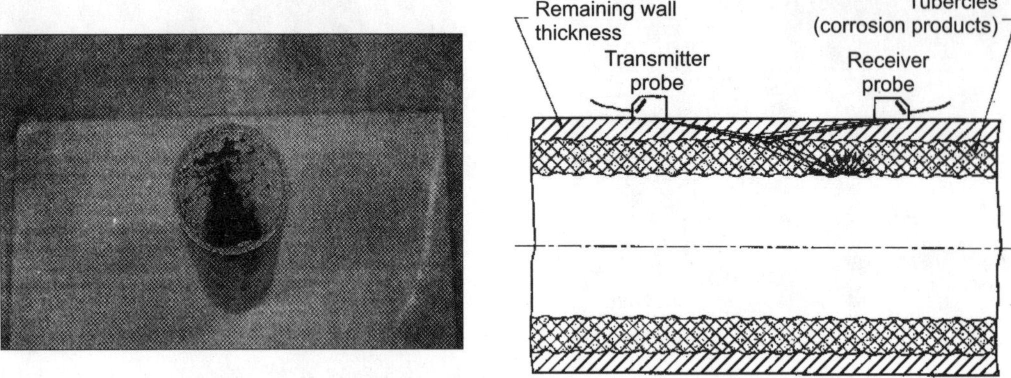

Figure 12 (a) Carbon steel pipe containing tubercles on inner surface and (b) Ultrasonic probes arrangement for wall thickness measurement [8].

(Fig. 12b). When the sound energy is transmitted in to the material, the energy is partitioned at the region of back surface of the test wall/corrosion layer and gets attenuated due to various mechanisms. However, a part of the energy transmitted from the probe gets refracted at the remaining wall thickness and gets skipped. The refracted beam is received at the outside surface either at one skip or two skip distance depending upon the probe size and the thickness of the pipe. Using this technique, wall thickness measurements were carried out in a corroded carbon steel pipe of 3 mm nominal wall thickness, 50 mm OD and containing 5-10 mm thick corrosion deposits of varying degrees. The transducer pair used for the wall thickness measurements is 70°, 4MHz, shear wave. Scanning was carried out by positioning the transducers axially on the outer surface at 2 skip distances apart, at different locations. At each measurement point, the beam path for maximum amplitude position and corresponding skip distance were recorded. From these values, the wall thickness of the pipe was determined. Measurements were carried out at various points along the circumference.

To validate the remaining wall thickness values obtained by this developed ultrasonic technique, wall thickness profiles were measured using optical metallography. Using this technique, the wall thickness could be determined to an accuracy of the order of 0.1 mm.

3.4 Techniques Based on Ultrasonic Guided Waves

3.4.1 Inspection of aircraft structures

The use of ultrasonic guided waves for non destructive inspection falls into two categories depending on the distance of propagation, short range and long range applications. For the short range applications, guided waves are used to obtain information that cannot be readily obtained by more conventional means. There applications include determination of elastic properties of materials, detection of defects near to interfaces such as in the inspection of adhesive joints and air coupled ultrasonic inspection of thin specimens. In these cases, sensitivity is of key factor and generally this is the main criterion for selecting a suitable guided wave mode. The effect of dispersion is relatively unimportant as the propagation distances are small.

The dispersion of ultrasonic guided waves causes wave packets to spread out in space and time as they propagate through a structure. This limits the resolution that can be obtained in a long range guided wave inspection system. Duration of wave packet increases linearly with propagation distance. Duration of wave packet after a given propagation distance can be minimized by optimizing the input signal.

The detection of corrosion in insulated pipes is of major importance to the oil and chemical industries. Current method involving point by point inspection is expensive because of the need to remove the insulation. An alternative method which involves the propagation of guided waves in the wall of the pipes. The guided waves are transmitted and received using a single transducer unit.

In the long range guided wave testing application, the guided waves are excited by a short burst of energy (the input signal) applied by a suitable transducer at one location on a structure. The excitation causes the packet of guided waves (the wave packet) to propagate away from the transducer into the surrounding structure. Then either the same transducer or a second transducer is used to detect signals caused by reflections from the surrounding structural features or defects. Multiple modes of guided wave propagation are possible in most structures and these modes are generally dispersive (i.e. their velocities are frequency and thickness dependent). In order to obtain useful data from a guided wave inspection system, it is necessary to selectively excite and detect a single guided wave mode while suppressing coherent noise due to other modes of guided wave propagation [9].

Guided waves or Lamb waves are created in thin plates and long pipes when oblique incident angles are used. Selection of modes can be optimised for a particular thickness of material by the calculation and analysis of dispersion curves. Defects such as corrosion or disbond in long pipe lines (insulated or uninsulated) and also between two layers are seen as thickness change and therefore are detected.

Practical application of this approach for inspection of large areas requires probes that will provide a good signal to noise ratio, and a stable scanning mechanism that will ensure a constant coupling of the probes. Among several probe types and scanning configurations examined in the laboratory using simulated and real aircraft lap joint specimens, the easiest to implement configuration is the pitch-catch configuration (Fig. 13) that involves placing transducers on either side of the joint and scanning the transducer pair along the length of the joint. This is relatively faster than raster scanning methods. When there is no disbond or corrosion, most of the sound energy is transferred through the adhesive bond into the second layer resulting in a strong RF signal. However, when a disbond or corrosion exists between the two bonded layers, the transfer of energy is reduced and the signal intensity drops.

The use of piezoelectric transducers has the advantage of easy selection of wave mode by adjusting source frequency and the angles of oblique incidence and reception. However, when conventional piezoelectric wedge probes are used for guided wave inspection, the overall signal amplitude can vary greatly due to the variation of the amount of the fluid used for transducer coupling, as well as probe alignment or the pressure applied. These problems can be resolved by using non-contact EMAT or air-coupled ultrasonic probes. EMAT and air-coupled probes have been used for the detection of disbands and corrosion in simple lap joints and encouraging results have been obtained. Although the ultrasonic guided waves using either contact wedge probes or non-contact EMAT or air-coupled sensors have shown

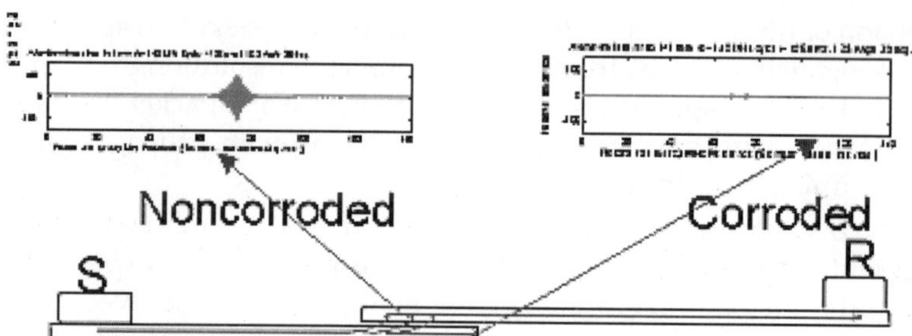

Figure 13 Ultrasonic guided wave inspection of a lap joint in pitch catch configuration and corresponding signals for corroded and non-corroded areas.

promising results in the laboratory for detecting corrosion or disbond in aluminium lap joints, their usefulness for field applications is yet to be demonstrated.

Economic inspection of aircraft fuselage or wings requires methods that can cover large areas. Attempts have been made for detection of hidden corrosion in aircraft aluminum structures using a noncontact laser based ultrasonic guided wave technique. A short laser pulse focused to a line spot is used as a broadband source of ultrasonic guided waves in an aluminum alloy 2024 sample cut from an aircraft structure and prepared with artificially corroded circular areas on its back surface (Fig. 14). The out of plane surface displacements produced by the propagating ultrasonic waves are detected with a heterodyne Mach–Zehnder interferometer.

Back surface corrosion could be detected by the dissipative effect that the associated surface roughness has on Lamb modes near their cutoff frequencies (Fig. 15). A variant of this method that uses laser based generation and detection of Lamb waves, is based on the efficient generation of a Lamb mode near its cutoff frequency, to detect the strong attenuation of that mode due to the surface roughness associated with the presence of corrosion. Naturally occurring corrosion at the back surface of an aircraft aluminum structure can be effectively detected using this method. Thickness loss of 5% (0.1 mm in a 2 mm plate) could be detected by Lamb wave technique. Since the method detects the change in the surface roughness due to

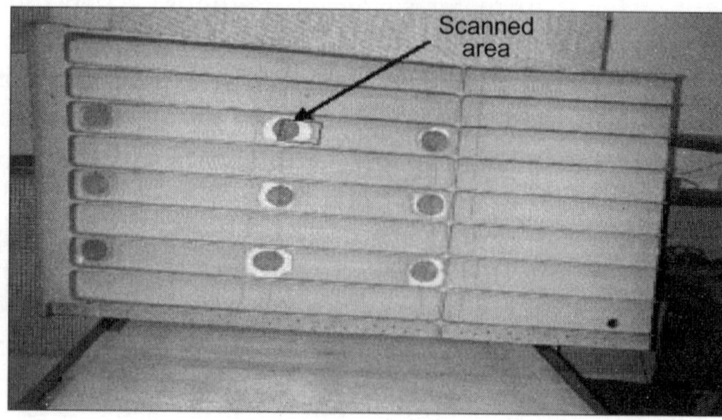

Figure 14 General view of the sample back surface showing the nine artificially corroded zones and the scanned area [10].

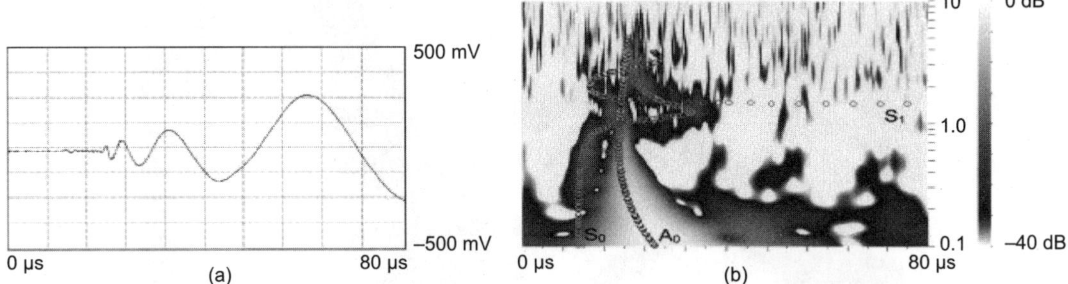

Figure 15 Ultrasonic signal measured for a 60 mm propagation distance that penetrates ~6 mm over a corroded region (0.1 mm loss in a 2 mm plate thickness). (a) Time-domain signal. (b) Relative magnitude of the CWT of the signal, in dB scale, superposed with the theoretical dispersion curves for modes *A*0, *S*0 and *S*1. Frequencies, in the vertical axis, are in MHz [10].

corrosion, and not the material loss or thinning of the plate, it is particularly suited for early corrosion detection. For this same reason, to evaluate the material loss due to corrosion using this method, it will be necessary to consider some other feature of the Lamb wave dispersion curves, such as the time of arrival of some mode at a selected frequency or the increase in the cutoff frequency (ies) due to plate thinning [10].

3.4.2 Long range inspection of chemical plant pipelines using ultrasonic guided waves

Corrosion in pipelines is a major problem in oil, chemical and other industries. Many pipes are insulated which means that even external corrosion cannot be seen without removing the insulation. The removal of full insulation and inspection is prohibitively expensive. The problems become severe at road crossings where the pipe cannot be inspected without excavation. Propagation of ultrasonic guided waves in the pipe wall and analyse this response provides a solution to this problem because they can be excited at one location on the pipe and will propagate many metres along the pipe, returning echoes indicating the presence of corrosion or other pipe features. It is shown that propagation distances of over 25 metres for pipe diameters from 50 to 600 mm can be obtained using a dry coupled piezoelectric transducer system [11].

The use of cylindrical guided waves propagating along the pipe wall is potentially a very attractive solution to this problem since they can propagate a long distance under insulation and may be excited and received using transducers positioned at a location where a small section of insulation has been removed as shown in Fig. 16.

It has been reported that corrosion defects could reliably be detected. However, echoes will also be seen from welds since the weld caps are not generally removed so the weld presents a change in acoustic impedance. This makes it difficult to identify defects at welds and also introduces the possibility of a weld being incorrectly identified as a defect. This problem can be overcome by measuring the extent of mode conversion produced by a reflector.

If an axially symmetric mode is incident on an axially symmetric feature in the pipe such as a flange, square end or uniform weld, then only axially symmetric modes are reflected. However, if the feature is non axially symmetric such as a corrosion patch, non axially symmetric waves will be generated. These propagate back to the transducer rings and can be

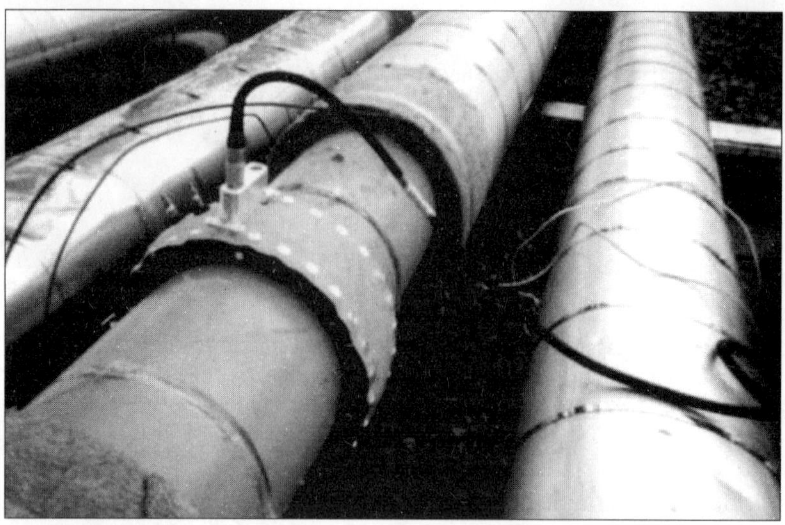

Figure 16 The transducer assembly for long pipe line inspection [11].

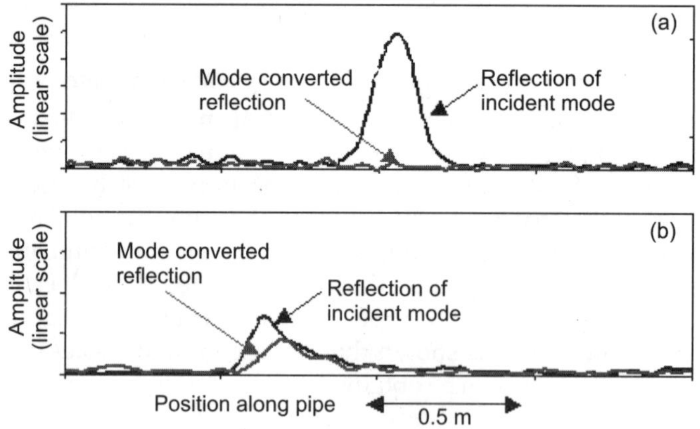

Figure 17 Typical signals from (a) axisymmetric feature e.g. weld; (b) corrosion [11].

detected. Figure 17 shows typical reflections from symmetric and asymmetric features; the increase in the mode converted signal can clearly be seen in the asymmetric case and this is a key element for the defect identification [12]

Figure 18 shows a pipe was paritally wrapped as it passed through a short earth wall. Figures 18 (a) shows small echoes at the point where the pipe exits the wall. The somewhat non-symmetric nature of these echoes (as well as their shape) indicates that there is likely to be corrosion at this location (beginning at the location marked -F3). From this side, the corrosion at the beginning of the earth wall is very evident as the large echo with large asymmetric component, marked +F2 in Fig. 18 (b). A rough approximation of the size of the corrosion can be obtained by using the DAC curves. A photograph of the corrosion is shown in Fig. 18 (c).

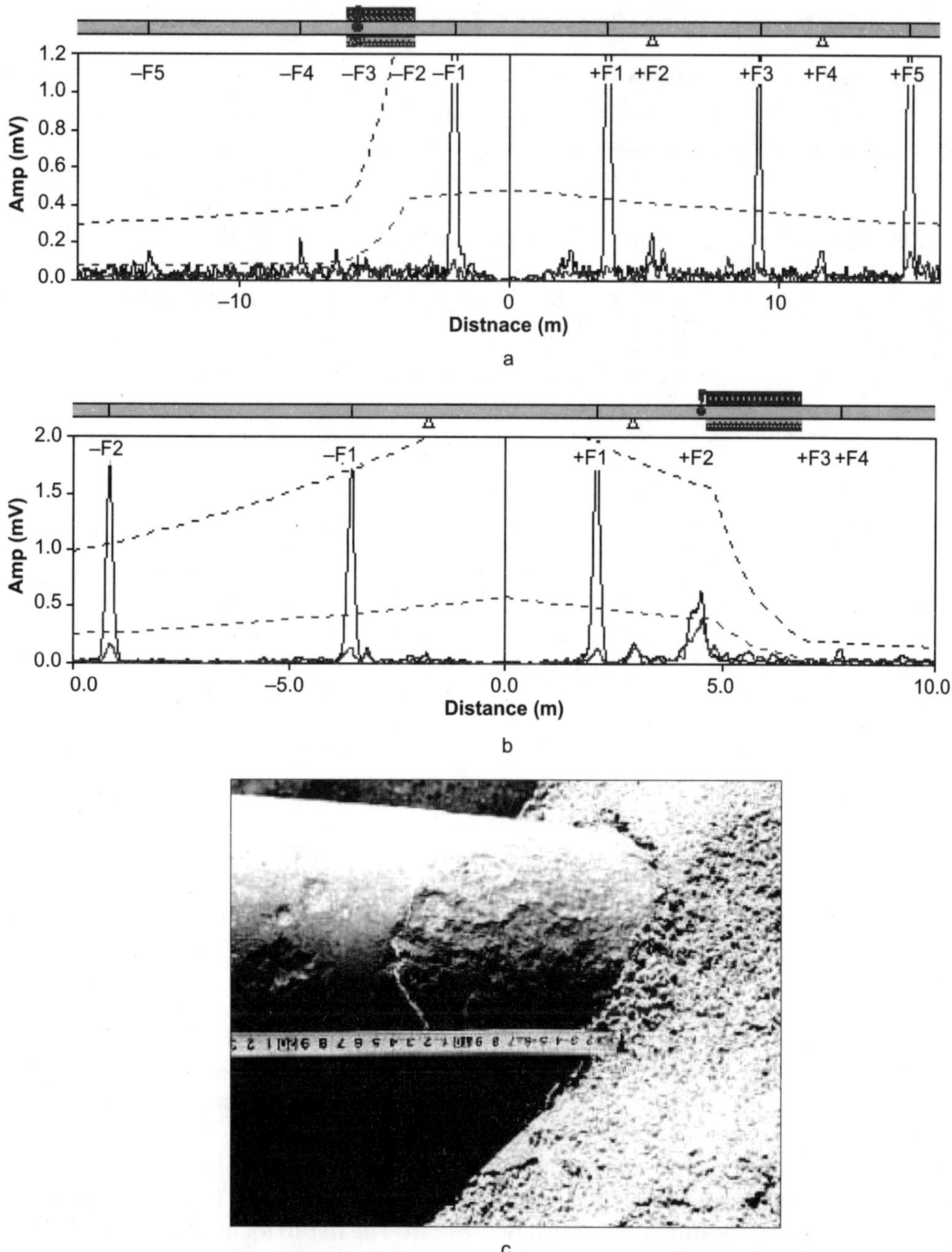

Figure 18 Inspection of a pipe passing through earth wall. (a) From the visible corrosion shown in (c) (b) From other side of wall (c) photograph of corrosion [12].

3.5 Corrosion Mapping using Ultrasonic C-Scan

3.5.1 Reinforced concrete structures

The problem of durability in reinforced concrete structures is most prevalent in chloride induced corrosion of reinforcing steel. The process of corrosion that takes place in reinforced concrete is depicted pictorially in Fig. 19. Traditional corrosion monitoring techniques such as half cell potential and corrosion rate measurements often fail when used in this type of structures and standard nondestructive testing methods such as Impact echo have not been found reliable. A new method called C-scan imaging to evaluate grouted post tensioned tendoms is found useful. The investigations carried out on laboratory specimens have shown promising results for the technique [13].

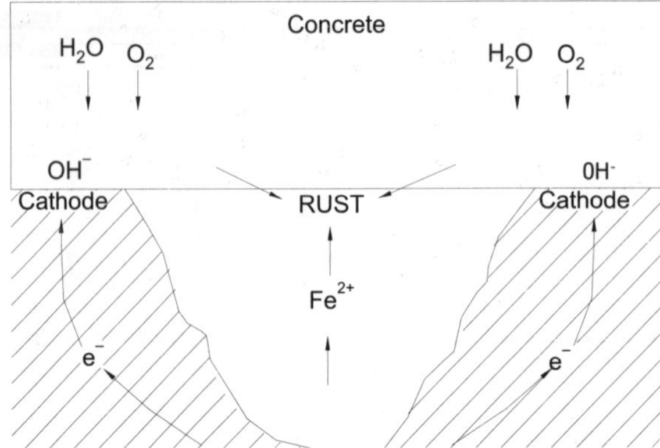

Figure 19 Corrosion process of steel in concrete [13].

The steel used in pre and post tensioned concrete is prone to corrosion attacks due to chloride induced corrosion, and are also susceptible to stress corrosion cracking or hydrogen embrittlerment. Chloride induced corrosion is the most common form of corrosion in reinforced concrete. The concrete surface is stained due to the rust produced which also causes cracking and spalling.

Pitting type of corrosion of the reinforcing element is very common in reinforced concrete structures and can be quite destructive due to the concentrated pits of corrosion causing large reduction in cross sectional area with a small amount of overall metal loss. This process is a self propagating one. As the pits grow, the surface can be undercut resulting in a deceivingly large pit making detection of pits difficult even on the bare steel.

In a study undertaken [13], the field representative samples are placed at the focus of the acoustic beam and than is received by a second receiving transducer. The grouted sample with tendon is mechanically scanned creating a raster scan. The depth focus of the transducer used is very small and hence allows observations of corrosion or void patterns in different layers by adjusting the transducer up and down in the longitudinal direction. The C-scan images of plain grouted and fully corroded specimens are given in Fig. 20, indicating that the C-scan imaging can be a promising tool for detection of corrosion in reinforced concrete. In Fig. 20, the images corresponding to different thickness levels (Gates G1 to G8) are shown in a sequence. Figure 21

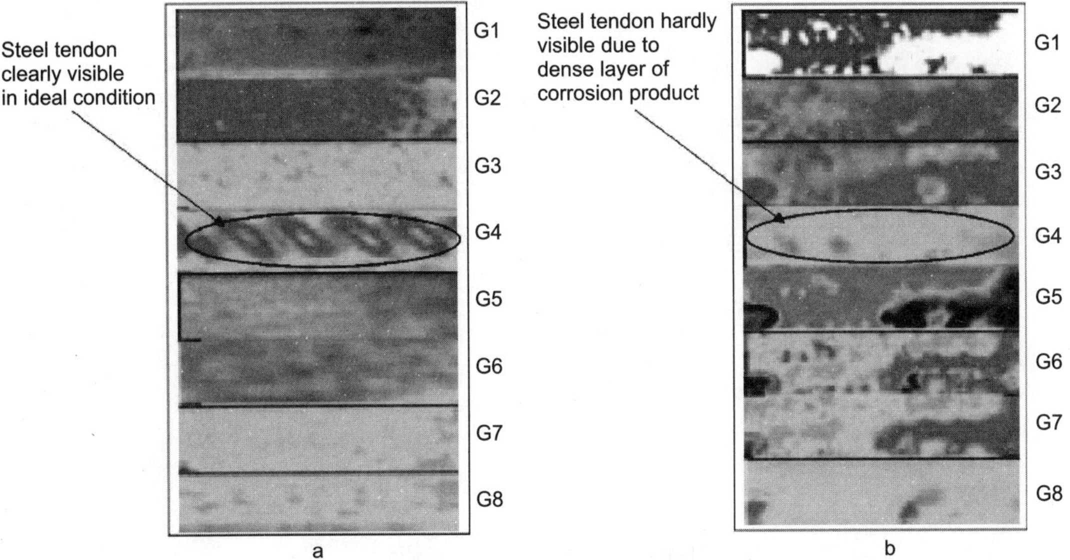

Figure 20 Images from all gates for (a) Plain grouted and (b) Fully corroded specimens [13].

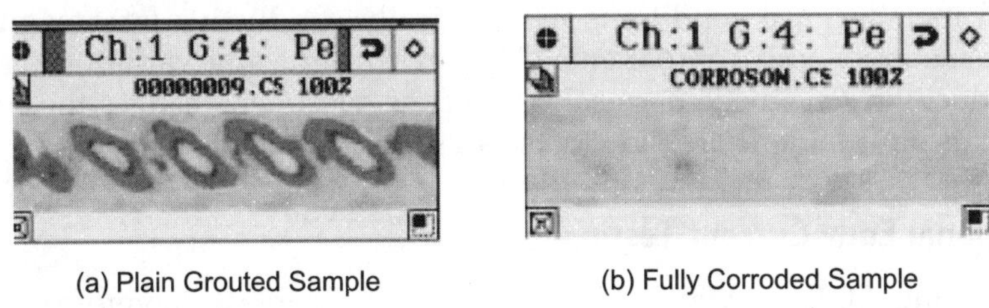

Figure 21 Comparison of different level of corrosion [13].

provides the signals from the gate 4 only. Figure 21(a) is the C-scan image from the plain grouted specimen and Fig. 21 (b) is from the fully corroded specimen.

4.0 EDDY CURRENT TESTING

4.1 Principle

Eddy current is based on the principles of electromagnetic induction. When sinusoidal excitation of a particular frequency is given to eddy current probe, primary magnetic field is generated around the coil (Ampere's law), and eddy currents are generated in conducting material (Faraday's law of induction). Secondary magnetic field is generated due to the presence of eddy currents in the material (Faraday's law of induction). Resultant magnetic field (Primary + Secondary) changes the impedance of the eddy current probe which can be detected using eddy current equipment. The principle is explained pictorially in Fig. 22.

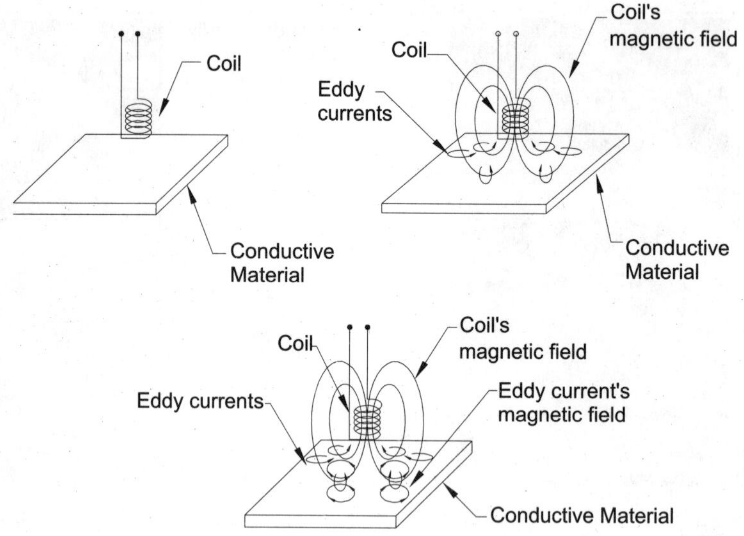

Figure 22 Eddy current principle

Eddy current inspection used to identify or differentiate among a wide variety of physical, structural and metallurgical conditions in electrically conductive ferromagnetic and non ferromagnetic materials and components. The presence of any variation in the property (electrical conductivity/magnetic permeability) of the specimen changes the coil impedance and based on this information corrosion, defects can be characterized.

4.2 Conventional Eddy Current Testing

Eddy current testing (ECT) is a very popular technique for periodic monitoring of corrosion of tubes in heat exchangers (HX) and steam generators (SG), because of its ease of operation, versatility and reliability. Heat exchanger tubes used in various industries including chemical, petro-chemical, refineries, thermal and nuclear power plants, undergo corrosion of various kinds, since they are subjected to continuous flow of high temperature fluids and steam, and operation under aggressive environments. The deterioration of the tube material leads to the possibility of leak thus affecting the integrity of the tubes and the heat exchanger. The significant problems faced include wall thinning, pitting, stress corrosion cracking (SCC), hydrogen embrittlement, carburisation, decarburisation, denting and deposition of crud.

The eddy current coil is placed in the tube and balanced at the defect free region. The probe is pulled inside the tube, and variations in coil impedance are recorded. The impedance change is related to the type and size of a defect. Conventional eddy current inspection can be performed at speeds up to 600 mm/sec [14].

Requirements for inspection of aging steam generator (SG) tubes in various industries are becoming increasingly stringent throughout the world. The dimensions and material composition of tubes of steam generators can greatly affect design features of the probes. The steam generator tubes are composed of either ferromagnetic or non ferromagnetic material. In the case of ferromagnetic tubes, powerful permanent magnets need to be integrated into eddy current probe designs to magnetically saturate the tube material. Magnetic saturation is required to ensure adequate depth of penetration of eddy current, in order for internal probes to detect defects that initiate from the outer surface of the tube. It is also needed to eliminate probe signal distortions from magnetic permeability variations that can obscure defect signals.

One important and unique characteristic of some of the SG and HX tubes is that deposition of magnetite on the internal or external surfaces. These magnetic layers partially shield the tube walls from the probe's electromagnetic fields, thereby weakening probe responses to defects. In addition, variations in the magnetic permeability and thickness of the deposits can cause distortions in the signal background that obscure defect signals. The in-service inspection of HX and SG tubes is mostly carried out using bobbin type coil eddy current probes. These probes consist of coils of wire that are coaxial with inspected tubes as shown in Fig. 23.

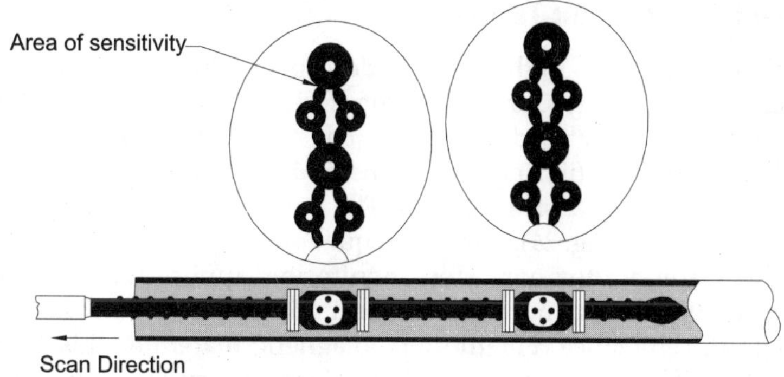

Figure 23 Eddy current probe

Conventional eddy current testing is very sensitive to uniform wall loss, pits and cracks, but is limited to non-ferromagnetic materials such as austenitic stainless steel, Admiralty brass, copper-nickel alloys and titanium. The conventional eddy current inspection is done in two modes: differential and absolute. Differential mode detects small defects such as pitting and cracking while absolute mode detects tube wall loss. A mixer channel with multifrequency approach detects defects under the support plate/baffle plate. This channel is a mixer of two differential channels that are added to cancel out the tube support plate signal.

Eddy current techniques can be used to perform a number of dimensional measurements. The ability to make rapid measurements without the need for couplant or, in some cases even surface contact, makes eddy current techniques very useful. The types of measurements that can be made include:

- thickness of thin metal sheet and foil, and of metallic coatings on metallic and nonmetallic substrate
- cross-sectional dimensions of cylindrical tubes and rods
- thickness of nonmetallic coatings on metallic substrates

4.2.1 Corrosion thinning of aircraft skins

One application where the eddy current technique is commonly used to measure material thickness is in detection and characterization of corrosion damage of the skins of aircraft. Eddy current technique can be used to carry out spot checks or scanners can be used to inspect small areas. Eddy current inspection has an advantage over ultrasound in this application because no mechanical coupling is required to get the energy into the structure. Therefore, in multi-layered areas of the structure like lap splices, eddy current technique can determine if corrosion thinning is present in multiple layers.

Eddy current inspection has an advantage over radiography also because only single sided access is required to perform the inspection. To place a piece of film on the back side of the aircraft skin might require removing interior furnishings, panels, and insulation which could be very costly and time consuming. Advanced eddy current techniques are being developed that can determine thickness changes down to about 3% of the skin thickness such as swept frequency and pulsed eddy current techniques [15].

4.3 Remote Field Eddy Current Testing

One of the recent advancements in the area of eddy current inspection of tubes for corrosion monitoring is Remote Field Eddy Current Testing (RFECT). The advantages of this technique are: (i) ability to inspect tubular products with equal sensitivity to both internal and external metal loss or other anomalies, (ii) linear relationship between wall thickness and measured phase lag and (iii) absence of lift-off problems. The output of the remote field varies linearly with increase in the wall loss (Fig. 24). This technique can be applied to both ferromagnetic and non-ferromagnetic materials for corrosion monitoring applications. In the case of non-ferromagnetic materials, cracks and other discontinuities due to corrosion can be detected because of change in reluctance. For the ferromagnetic materials, this technique is more effective for detection of wall thickness loss due to large difference in the permeability of air gap/ corrosion products and the base metal. Detection of wall thickness loss down to 15% has been established using RFECT [16].

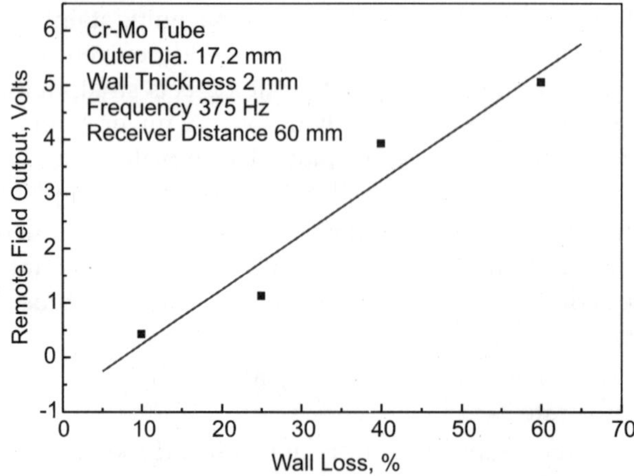

Figure 24 Variation in output of the remote field EC signal with wall loss [16].

4.4 Intergranular Corrosion in Austenitic Stainless Steels

Feasibility of detection and accurate characterisation of inter granular corrosion (IGC) in stainless steel by eddy current non-destructive method has been examined [17]. The eddy current setup for the study is shown in Fig. 25. AISI type 316L stainless steel specimens subjected to ageing treatment at 973K for 1, 20, 50, and 200 hours duration and exposed to Strauss test have been examined using EC method. The EC amplitude has been noted to change by a small amount of 0.25 volts with ageing for different durations. On the contrary, the amplitude has been observed to increase monotonously to about 3 volts, after the Strauss test. This increase in the EC voltage has been attributed to the changes that have taken place during the Strauss test i.e. dissolution of Cr depleted regions and formulation of grain boundary grooving. In order to bring out the relative changes clearly, ratio of EC amplitudes of Strauss test exposed to unexposed conditions has been evaluated for each specimen and the same is plotted and shown in Fig. 26 and 27. As can be observed, when the ageing duration is more than 20 h, the ratio of EC amplitude is more than 5 and is well delineated from the specimens with ageing duration of 1 h and without ageing. Bend tests on Strauss test exposed specimens have confirmed that sensitisation has taken place in specimens aged for duration of 20 hours and above. The study has been recently extended to SS 316 and it has been found that sensitization of specimens that leads to failure by Strauss test can be detected based on the EC amplitude, even without the Strauss test [18-21]. The degree of sensitization can be classified into four categories as seen from the bend test results: (1) No fissures (2) Microfissures (3) Macrocracks (4) Breakage. The videomicroscope pictures of these four categories of specimens classified after bend test are shown in Fig. 28.

4.5 Corrosion Under Insulation

The problem of corrosion under insulation (CUI) is found to be severe in carbon steel components and moderately severe in those made of stainless steel. This problem is found to be severe in pipes less than 300 mm diameter. The CUI occurs in the case of a variety of insulations including asbestos, calcium silicate, fiberglass, mineral fibre, wool, flexible rubber,

Figure 25 Eddy current set up

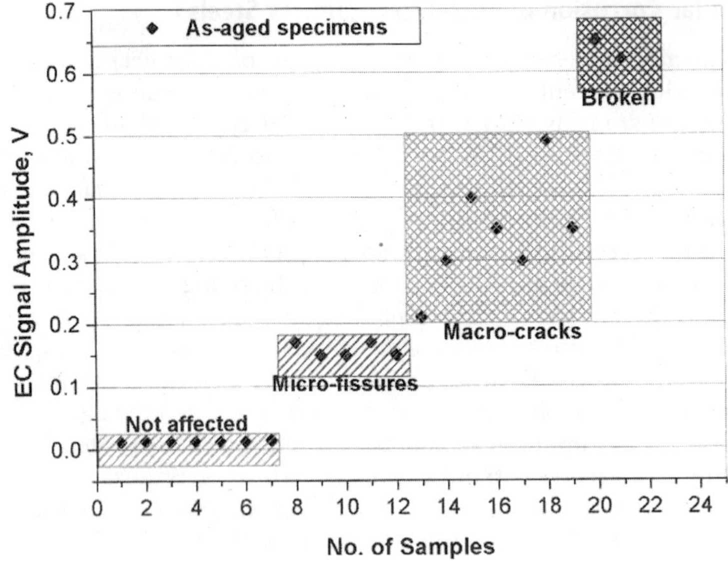

Figure 26 EC responses from the as-aged condition specimens are in good agreement with the four categories of specimens classified after bend test [18].

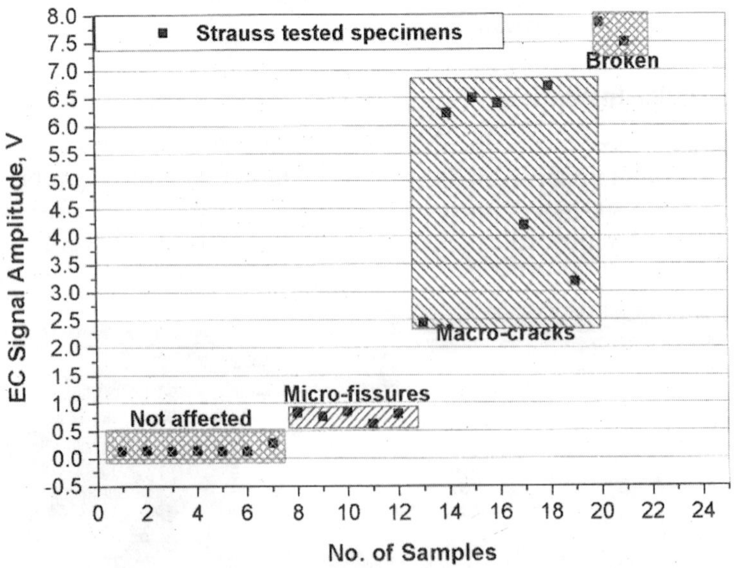

Figure 27 Comparison of EC responses from the Strauss tested specimens with severity of the cracks developed after bent test [18].

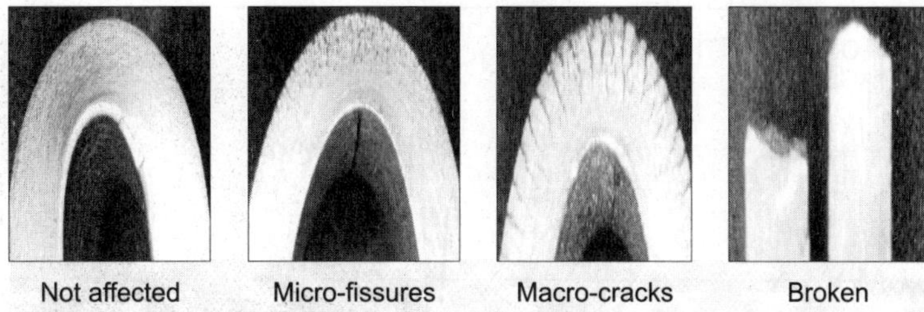

Figure 28 Four categories of specimen classified after bend test. They are a) Not-affected, b) Micro-fissures, c) Macro-cracks and d) Broken [18].

polyisocyanurate, polyurethane and other foams. Real time portable X-radiography devices have been developed to view CUI using a tangential X-ray beam. However, this method is very slow to use in routine inspection. One relatively new technique for detection of CUI is the pulsed eddy current testing technique. The pulsed eddy current technique is not a scanning technique; the probe head must be left in place for a few seconds to tens of seconds depending on the wall thickness of the component being inspected. However, it has been found that low frequency ECT (LFECT) technique that has the capability for rapid scanning in comparison to pulsed eddy current testing, can be used for detection of CUI [22]. The method is very sensitive to localized damage, such as cracks and pits, and can detect severe wall thinning. The primary limitation of this technique is its limited lift-off capabilities, which restrict the insulation thickness to a few tens of centimeters. LFECT technique uses frequencies less than 50 Hz to penetrate up to 50 mm of insulation in addition to the metallic weather cover (overcap) on the component. The technique has the capability for detection of CUI in areas as small as 50 mm diameter. Smaller areas of CUI can be detected with thinner insulation. The LFECT systems have the following capabilities: (1) For piping with up to 50 mm of insulation and smooth metallic weather covers, detection of CUI in areas as small as 20 cm^2 with a 25% loss of wall. (2) The system is capable of discriminating between CUI and other indications that provide spurious signals associated with weather cover overcaps and straps, carbon steel insulation retaining wires, heating lines, access plugs and weld joints. Figure 29 shows the response of a LFECT system, i.e. trace of voltage (Y-axis) vs. time or position (X-axis) from a pipe section covered with 50 mm insulation and an aluminum cover (overcap). The signal corresponding to the CUI (later confirmed by removing the insulation) along with the signal due to the overcap is seen in the top trace. The bottom trace, that is generated simultaneously along with the top trace, contains no signal from the CUI based on the selection of phase angle for the display. The signal due to the overcap is also shown in the impedance plane display, as given in the middle portion of Fig. 29.

4.6 Eddy Current Impedance Imaging

Application of computerised eddy current impedance imaging (ECII) technique that has the advantage for imaging corrosion pits, cracks etc. has been developed at the authors' laboratory [16]. The technique is used to scan the object surface and creates EC impedance images in the form of gray levels or pseudo colours. The system consists of a PC controlled X Y scanner which scans the component point by point in a raster fashion with an eddy current probe, acquires

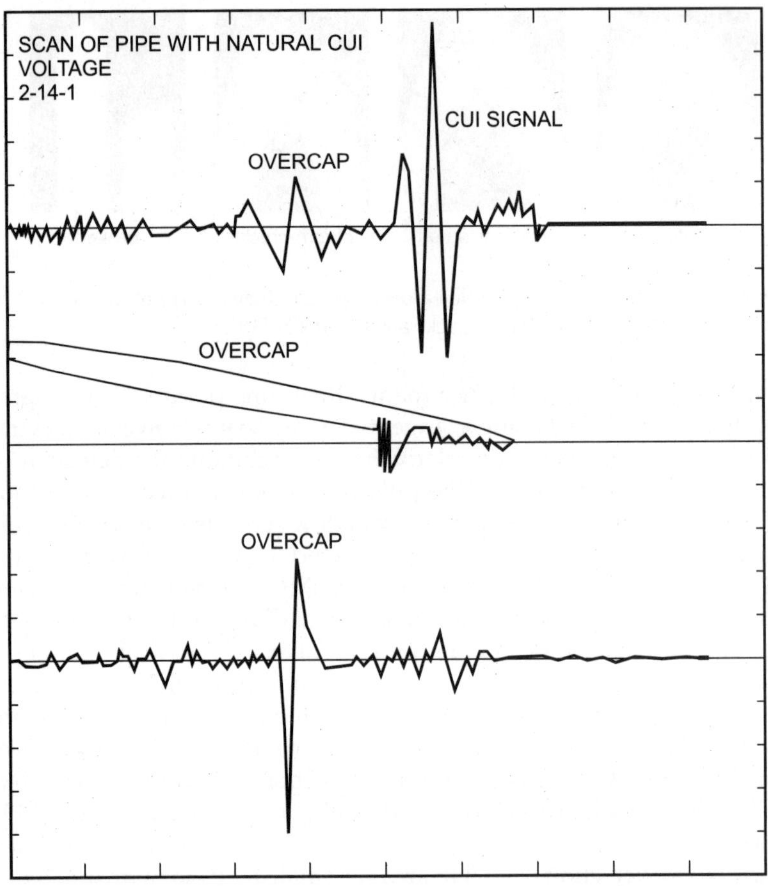

Figure 29 System Response to CUI Damage on 350 mm Pipe through 50 mm of Insulation and Weather Cover [22].

data using a 12 bit analog-to-digital converter, and finally processes and displays data in the form of images. Using ECII, it is possible to obtain images of defects. Fatigue cracks, corrosion pits, electro-discharge machined (EDM) notches and other types of defects have been imaged and characterized. Figure 30 shows eddy current impedance images of three corrosion pits in a stainless steel plate. These images are produced using an absolute surface probe operating at 150 kHz.

4.7 Pulsed Eddy Current Technique

It is very difficult for detecting and quantifying small metal loss (less than 10% of one plate thickness) in case of corrosion in multi-layered structures. Combining a multi-frequency eddy current technique with sliding probe technology, external amplification, digitization, and robotics provides, the ability to quantify small amounts (3% of one plate thickness) of material loss in sub-layers not otherwise detectable. This has been achieved by designing a pulsed eddy current (PEC) system.

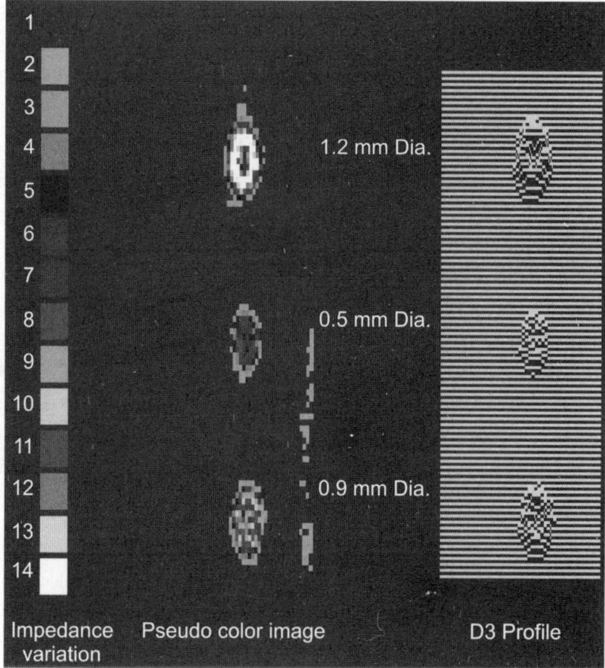

Figure 30 Eddy current Images of three corrosion pits in stainless steel plate

The traditional eddy current technique excites the probe with a sinusoidal waveform of a set frequency whereas in pulsed eddy current technique, the probe is excited with a step voltage function. This step contains a continuum of frequencies. As a result, the response to several different frequencies can be measured with just a single step. Since the depth of penetration is dependent on the frequency of excitation, information from a range of depths can be obtained with just one step function.

The measurement of pulsed eddy current response in time domain is needed for the analysis. The response of the coil in an area which is expected to have no corrosion, is digitized and stored as a null trace. All subsequent responses are then subtracted digitally from this null trace. Information about the characteristics of corrosion can be extracted from this difference curve, for example, the depth and amount of corrosion. With pulse methods, the frequencies are excited over a wide band, the extent of which varies inversely with the pulse length. As found with ultrasonic testing, the total amount of energy dissipated within a given period of time is considerably less for pulsed waves than for continuous waves having the same intensity. For example, with pulses containing only one or two wavelengths and generated 1000 times per second, the energy produced is only about 0.002 times of that for continuous waves having the same amplitude. Thus considerably higher input voltages can be applied to the exciting coil for pulsed operation than for continuous wave operation [23].

The transient current through the coil by pulse excitation induces transient eddy currents in the test piece, associated with highly attenuated magnetic pulses propagating through the material. At each probe location, a series of voltage-time data plots are produced as the induced field decays, analogous to ultrasonic inspection data. Physically, the pulse is broadened and delayed as it travels deeper into the highly dispersive material. Therefore, flaws or other anomalies close to the surface will affect the eddy current response earlier in time than deep flaws.

It has been demonstrated that PEC has very high potential for applications in corrosion detection and characterization. The time-based signals are more intuitive and can be displayed in various formats presently used in ultrasonic NDT. The A-, B- and C-scans serve well as complimentary methods, for presenting data when simple variations in test parameters occur; these displays can become very complex and confusing, however, when combinations of such variations occur simultaneously. The PEC system (Fig. 31) has demonstrated a sensitivity of at least 5% total metal loss, or 0.05 mm thinning, in the second layer. Smaller levels of corrosion and deeper penetration may be achieved with optimized probe designs and test parameters, such as excitation current shape and frequencies [24]. The amplitude C-scan image of the corroded specimen shown in Fig. 32, reveals a region of intense corrosion activity. Figure 33 shows another pulsed eddy current C-scan image, revealing the exfoliation corrosion in an aircraft specimen.

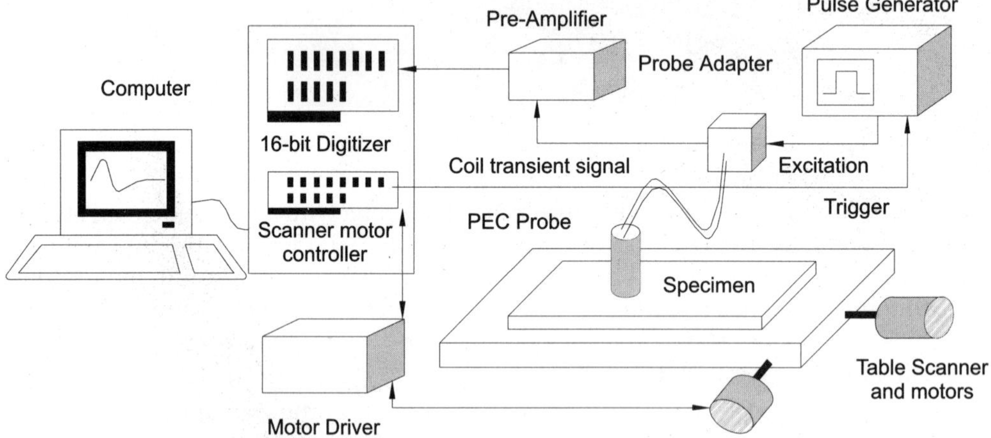

Figure 31 Pulsed eddy current testing set up [24].

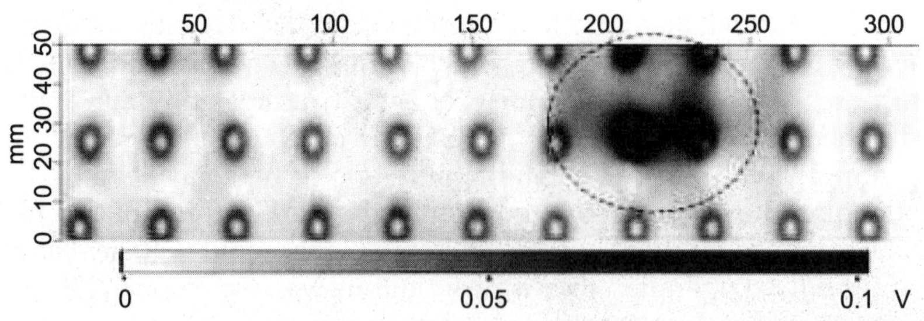

Figure 32 Amplitude c-scan of a aircraft fuselage lap splice with natural corrosion [24].

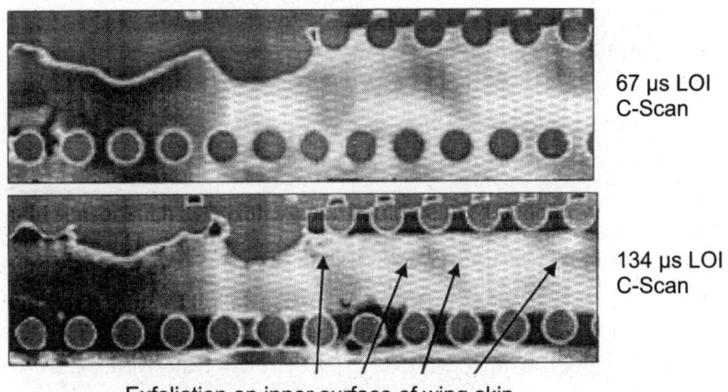

Figure 33 Exfoliation detection by pulsed eddy current testing [24].

5.0 MAGNETIC FLUX LEAKAGE TESTING

Magnetic Flux Leakage (MFL) inspection is a fast and reliable screening method to detect local corrosion in low alloy carbon steel components, such as pipelines and storage tanks and vessels. The basic principle of the MFL method involves a magnet, mounted on a carriage, induces a strong magnetic field in the plate or pipe wall. In the presence of a corrosion pit, a magnetic flux leakage field forms outside the plate or pipe wall. An array of sensors is positioned between the magnet poles to detect this flux leakage. This method is limited to thickness of 12 mm [25].

An improved MFL method has been developed and is called Saturation Low Frequency Eddy Current (SLOFEC) Technique. The SLOFEC technique has an upper thickness limit of at least 30 mm. In addition, it is suitable for application on stainless steel materials. In this improved method, instead of 'passive' Hall sensors, it uses 'active' eddy current sensors to detect flux variations. These sensors detect changes in flux density inside the component instead of flux leakage. The passive Hall sensor detects flux leakage just a few millimeters above the surface. This explains the higher sensitivity of eddy current sensors for variations of magnetic flux. The extended thickness capability of new tools using the SLOFEC principle makes them suitable for a much wider range of applications, not only for inspection of thickness of up to 30 mm but also for thinner walls covered with thick non-metallic layers such as glass fibre reinforced epoxy coatings on floor of (oil) storage tanks. Moreover, the SLOFEC method is able to differentiate near-surface defects from back-wall defects .

In MFL technique, magnetisation can be achieved using electro-magnets or permanent magnets. There are several types of sensors that can be used in MFL. These include Coils, Hall effect sensors, Magnetostrictive and similar devices. Because the MFL method responds to both far side (FS) and near side (NS) corrosion, it is necessary to introduce a strong magnetic field into the component wall. The closer this field becomes to saturation for the component, the more sensitive and repeatable the method becomes. For typical steels used in bulk liquid storage tanks, this value is generally between 1.6 and 2 Tesla. In this range, any residual magnetism from previous scans or operations will be eliminated during subsequent scans so that the resulting flux leakage signals remain relatively constant and repeatable. Working below the 1.6 T level will still detect pitting on the first scan, but residual magnetism tends to cause a progressive deterioration of the signal amplitude during subsequent re-scannings unless alternate scans are made from opposite directions.

For a given magnet system, the flux density achieved in the component will depend on the thickness and permeability of the material. For most storage tanks, the steels used are in the mild steel range equivalent to the old EN 2 grade and these steels have similar permeability. So the factor controlling flux density becomes one of plate thickness. There will be an upper thickness limit for each given magnet system above which flux density will be too low to give adequate sensitivity to defect/pitting. One advantage of using an electro-magnet is that the magnetising current can be increased to cater to a wider range of plate thicknesses than can not be achieved with a given permanent magnet. However, this will be at the expense of size, weight, and the use of an independent (battery) power supply

Centered between the poles of the Magnet Bridge and stretching the full scanning width of the system is an array of Hall effect sensors. These are spaced at 7.5 mm between centres to give optimum resolution and coverage. The sensing range of each sensor is sufficient to allow overlap with its neighbour. The arrangement is illustrated in Fig. 34.

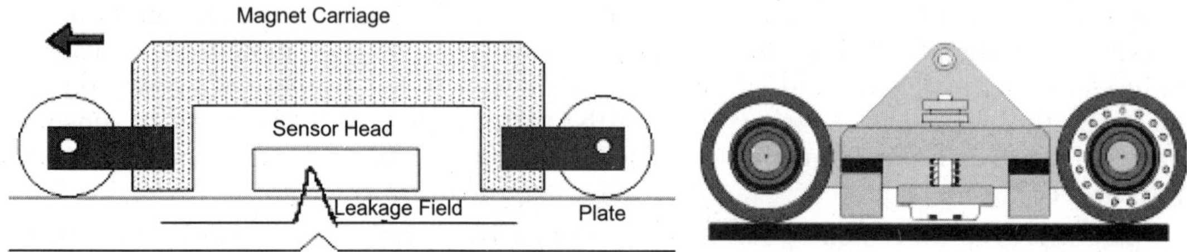

Figure 34 MFL Sensor coil and set up [25].

Figure 35 shows the field patterns for sound and pitted material. Hall effect sensors give a voltage signal proportional to the flux density of the field passing through the sensing element. For quantitative evaluation of flux leakage signals from pits, the pit morphology needs to be taken into account. The shape, profile, aspect ratio, and hence volume of a corrosion pit influences the MFL signals. It is possible to rationalize the general characteristics of the pits in to three types as illustrated in Fig. 36. These three types include (i) Lake or dish, (ii) Cone and (iii) Pipe.

These three categories refer to the profiles of the pits, along a section through the plate thickness. In plan view, pits tend to be circular or elliptical in general outline. As corrosion sites grow, these basic shapes tend to merge to give more complex outlines such as those shown in Fig. 36. A shallow 'lake' type may progress in stages, tending towards 'cone' like and ending up in a 'pipe' like pit, as shown in Fig. 36. We need to identify the types of pits expected to form in a given component and operating environment and appropriate correlation between the pit size and MFL signal should be used for evaluation.

As with other NDT methods, the effectiveness of an inspection using MFL is influenced by cleanliness and surface preparation. MFL is more tolerant of general dust and debris than ultrasonics, and it is not essential for grit blasting of floor surfaces in all cases. Nor does the

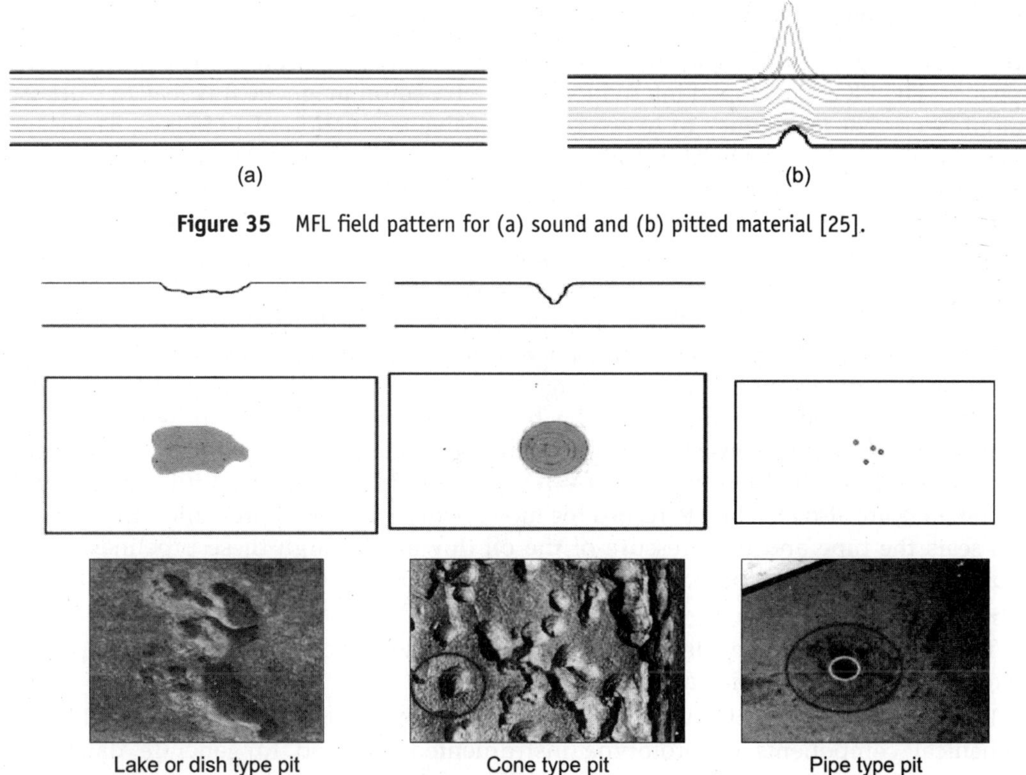

Figure 35 MFL field pattern for (a) sound and (b) pitted material [25].

Figure 36 Quantitative evaluation of flux leakage signals from different types of pits [25].

method rely on either wet or dry surfaces although any water on the surface should not be deep enough to submerge the sensor head. Presence of ferromagnetic debris is of a concern because, under certain circumstances it may lead to spurious signals.

MFL is a useful tool that allows the rapid monitoring of large surface areas for both near side and far side corrosion. It is capable of giving information on both the location and relative severity of pitting so that the amount of ultrasonic inspection is minimised and directed to very specific locations. Although an apparently simple method, it does require careful setting up and use.

The demand for integrity assessment of oil storage tanks, both for the wall and the floor of the tanks, is increasing and the early detection of defects that are potential leak locations is preferred. A recent technique, working on the principles of magnetic flux leakage, has been implemented in a professional tool called FLOORSCANNER to detect local corrosion. The use of Intelligent pigs based on magnetic flux leakage (MFL) and/ or ultrasonics are employed for inspecting pipelines for corrosion defects [26]. The pigs are useful device to carry NDT sensors inside the pipelines for inspection/ structural integrity assessment. A new development for the detection and sizing of internal corrosion in small diameter, heavy wall pipelines is the use of high frequency eddy current pigs.

5.1 Development of Instrumented Pipeline Inspection Gauge

Pipe inspection gauges (PIGs) are used for inspection of developed for the inspection of long oil pipelines. The PIGS developed internationally are based on magnetic flux leakage, ultrasonic and eddy current techniques to have comprehensive evaluation of pipelines. Thousand of kilometers of oil pipelines exist or transporting crude and oil products from different places allover the world. For example Indian Oil Corporation (IOC), owns approximately 6000 kms of these pipelines of sizes ranging from 200 mm to 70 mm. These pipelines sometimes pass through thickly populated area carrying highly inflammable and costly products hence their integrity is very important. This calls for a regular monitoring of their health. In this context, the development of an on-line pipe inspection gauge taken by BARC, Mumbai is discussed.

The Instrumented Pipe Inspection Gauge (IPIG) works on the principle of saturating the carbon steel oil pipe section with high magnetic flux and monitoring the leakage flux on the inner surface of the pipe line with the help of 64 numbers of Hall sensors over full circumference. The IPIG consists of magnetic module, data acquisition system (DAS) module, battery module, residual sensors with DASII module and PIG locator module. Three numbers of odometers are also attached to record the movement of IPIG. Polyurethane cup mounted on IPIG seals the pipe and the pressure of the oil flowing through these pipelines gives the required propelling force for its movement. Generally the front cups are sealing cups and other cups are supporting cups.

There are eight segments in the magnetic module, which covers almost ninety percent of the total periphery of the pipe as shown in Fig. 37.

Linear pull through rig was (Fig. 38) commissioned at BARC for dry evaluation of mechanical components of prototype instrumented pig and to generate database for characterization of defects. The rig consists of 25m of straight length of pipe formed by joining eight flange-joined sections. These pipe sections were retrieved from 300 mm commercial pipeline. There is a semi-cylindrical SS launching/retrieving tray connected to a reducer at either end of the rig. A motor-gear box assembly winches the pig through the linear pull through rig with the help of a steel rope. The prototype pig made about 50 successful runs in linear pull through rig. All the features in the rig e.g. reducers (entry and exit), flange joints, weld affected zones near the flange joints, support pillars, edges of the cut section with defects and straps to hold the cut section, could be identified from signal analysis of repeated runs. A set of metal loss defects were machined on a cut section of pipe line of a linear pull through rig specially constructed for calibration of the developed IPIG. The data from the runs were downloaded from the on-board computer to permanent storage. The pig recorded coherent and repeatable data in several runs over the same defects. The gray level image of the pipe section is given in Fig. 39 showing cut edges and defects. The system has been successfully used for inspection of actual pipelines of IOC [27].

6.0 LASER BASED TESTING

Laser based techniques are finding increasing applications in specialised areas because of their high sensitivity and non contact nature of measurements. These techniques are also employed for assessment of in-service degradation, for example, early stage deformation, fatigue damage, pitting corrosion and intergranular corrosion (IGC). Laser scattering technique has

Figure 37 Actual Magnet Module Assembly [27].

Figure 38 Linear Pull-through rig [27].

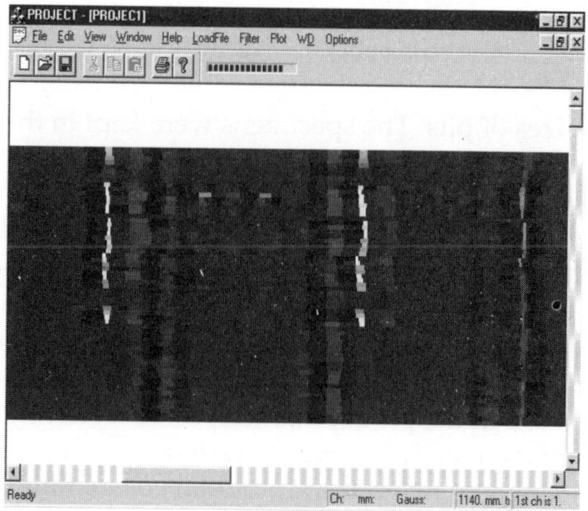

Figure 39 Grey image showing cut section and defects in linear pull through rig [27].

been explored [28] for studying pitting and intergranular corrosion in stainless steels [28]. A set up used for corrosion studies using laser scattering technique is shown in Fig. 40. A 5 mW He-Ne laser operating at 6328 Å in TEM_{00} mode is used. A beam steerer is used to raise or lower the beam or to change the direction of the beam. A specially designed specimen holder is used to keep the specimen in the same plane so that the angle of incidence is constant for all the specimens, for comparison purpose. The scattered radiation falls on a translucent screen held at a distance of 600 mm from the specimen. The screen is placed perpendicular to the specular direction of the beam. A CCD camera is placed in line with the specular direction behind the screen. The camera is interfaced with the computer through an image acquisition card. The acquired image is integrated over 20 frames to minimise the electronic noise. The image of the scattering pattern is processed using a suitable software IMAGER 2 PLUS Ver. 3.0.

AISI type 316 stainless steel specimens in solution annealed condition were used for pitting corrosion investigations. Potentiodynamic anodic polarisation of the specimens polished up to 1 mm finish was carried out in a solution containing 0.5 M NaCl and 1 N H_2SO_4 to produce

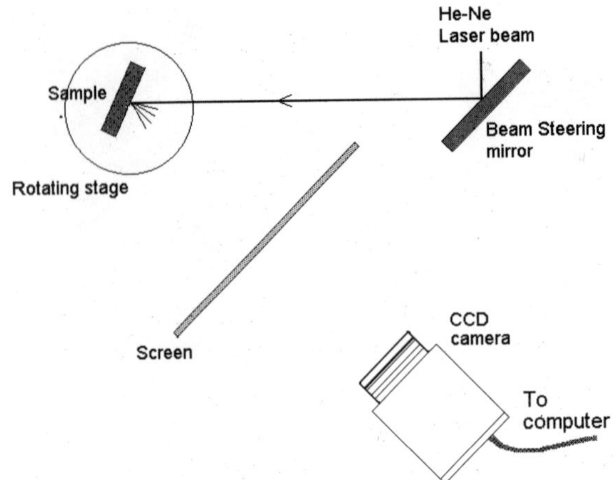

Figure 40 Schematic of optical configuration using CCD camera [28].

specimens with varying sizes of pits. The specimens were kept in the pitting range for known periods of time in order to produce pits of sizes of 35, 60, 80, 100, 225, 250 and 275 µm. AISI type 316 stainless steel specimens sensitised at 923K for 1 h were used for inter granular corrosion as per ASTM A262 practice A. The sensitised specimens were polished to a mirror finish up to 1 µm and were electrolytically etched in 10% ammonium persulphate solution at 6 V for various durations of 10, 20, 30, 40, 50, 60, 70, 80 and 90 s.

The laser scattered patterns in specimens with pits consisted of mainly intense 'caustic' and faint 'speckle' background. The spread of the caustic pattern has increased with the increase in the pit size. With the increase in the pit depth, the speckle scattering has increased in the off-specular direction. If the pit is circular, the caustic scattered pattern is symmetric, otherwise it is asymmetric. In some cases, where two pits are simultaneously illuminated by the laser beam, interference fringes are observed. Figure 41 shows the area of scatter as a function of average pit size. The area of the scatter increases with increasing pit size. The perimeter of the scattered intensity distribution is found to vary linearly with the size of the pit.

For the specimens with intergranular corrosion, the diffuse scattering component increases and the specular scattering component decreases with increase in the electrochemical etching duration. This is attributed to the increase in the roughness of the surface. Figure 42 shows the distribution of specular scatter component for three specimens etched for durations of 20, 50 and 80 seconds. The microstructures corresponding to these specimens are shown in Fig. 43. A good correlation between the extent of grain boundary attack and the change in the distribution of the sepcular scatter could be observed. The variation in intensity (Fig. 44(a)) and full width at half maximum (FWHM) (Fig. 44(b)) of the specular scatter distribution with etching duration shows that the intensity decreases and FWHM increases with increase in the etching time, or in other words with the extent of inter granular corrosion. These studies demonstrate the usefulness of laser scattering techniques for characterisation of pitting and intergranular corrosion.

The technological advances in the area of optics and electronics enabled the development of a system based on principles of optical triangulation called Laser Optic Tubing Inspection

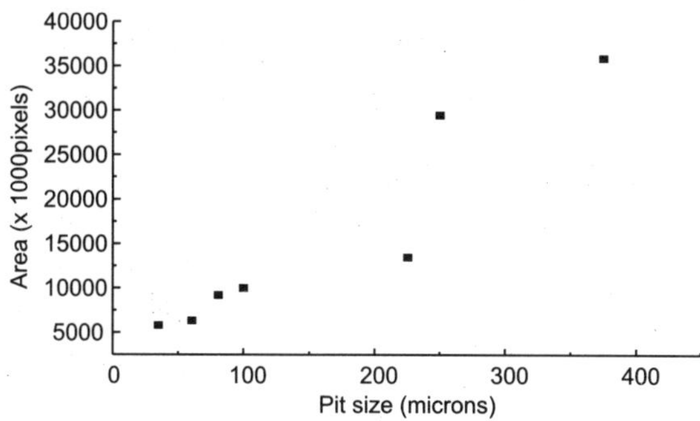

Figure 41 Area of binary image as a function of pit size [28].

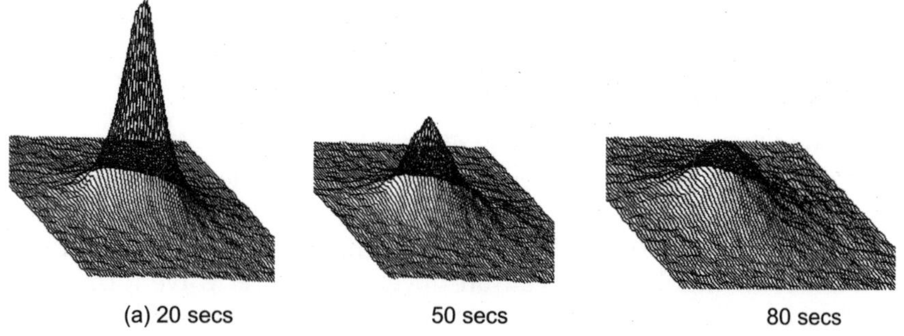

Figure 42 Intensity distribution of scattering pattern from surfaces of specimens with different etching times. The noise in the intensity distribution is removed using a Gaussian smoothening [28].

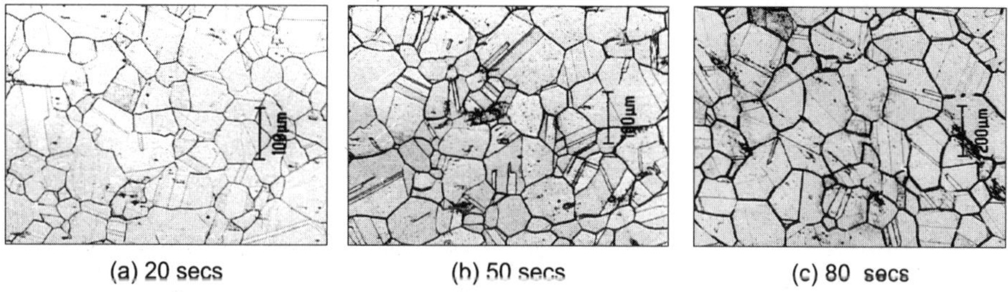

Figure 43 Micrographs of etched surface with different etching times. Grain boundary grooving increases with increasing etching time [28].

System (LOTIS) for inspection of corrosion inside the tubes. This system essentially consists of three articulated modules combined to form a probe viz. a laser source, optics and photo detector; rotational drive system; and associated electronics. In this system, a 40 m diameter laser beam is projected at near normal incidence onto the tube inner surface using an accurate rotational drive system, and the receiving optical system images this spot of light onto a single axis lateral effect photo detector.

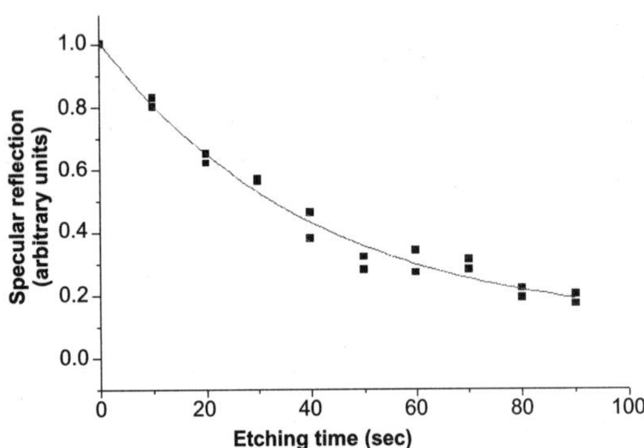

Figure 44 (a) Variation in intensity of specular reflection with etching time [28].

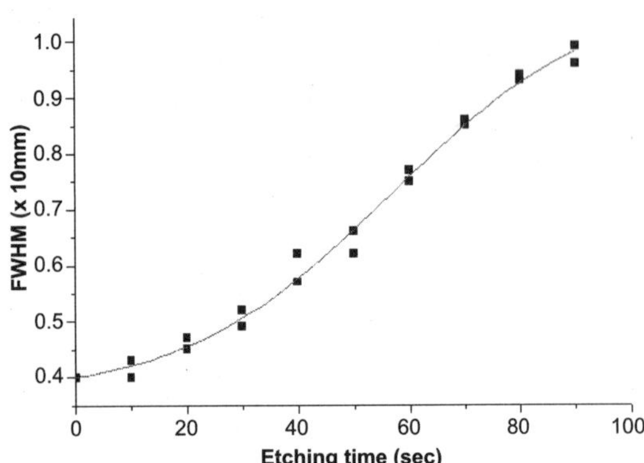

Figure 44 (b) Variation in FWHM of specular reflection with etching time [28].

7.0 ACOUSTIC EMISSION TESTING

Acoustic emission (AE) is defined as the class of phenomenon whereby transient elastic waves in the range of 20 kHz and 20 MHz are generated by the rapid release of energy from localized sources within a material (Fig. 45 (a)). Usually, AE occurs in the form of a release of a series of short impulsive packets of energy, which travel as spherical wavefront to be picked up from the material surface by highly sensitive transducers. A typical wave form picked up by a transducer is shown in Fig. 45 (b). The wave thus picked up is converted into electrical signal, which on suitable processing and analysis can reveal valuable information about the source causing the energy release (Fig. 46). In this way information about the existence and location of possible sources is obtained. Processing and analysis of the transducer output can reveal valuable information about the dynamics of AE sources. In metals, the sources of AE include generation and propagation of cracks, generation and movement of dislocations, formation and growth of twins, decohesion and fracture of inclusions, phase transformations and corrosion phenomena. There are also secondary or pseudo sources, which include leaks and cavitation, friction, realignment and growth of magnetic domains (Barkhausen effect), liquefaction and solidification. Metallic corrosion involving chemical reactions results in redistribution of energy. Any corrosion process thus, is a potential source of acoustic emission. Initiation and growth of stress corrosion cracks produce a variety of distinct and discernible signals such as electrochemical noise, electrochemical current transients, acoustic emission and load drops [1]. The sources of AE release during stress corrosion cracking process include: breakdown of thick surface - oxide film, hydrogen gas evaluation, hydrogen embrittlement etc. (Fig. 47).

Acoustic emission can be detected online and the progress of pitting corrosion, crevice corrosion and stress corrosion cracking can be assessed. Corrosion of fuel tanks of aircrafts often leads to leakage of the tanks and subsequently severe accident. Although pitting corrosion takes place in a few individual locations, they can develop in due course and become through holes, which are potentially the most dangerous structural faults. The following investigation points to a new method of using acoustic emission technique to detect corrosion initiation [29].

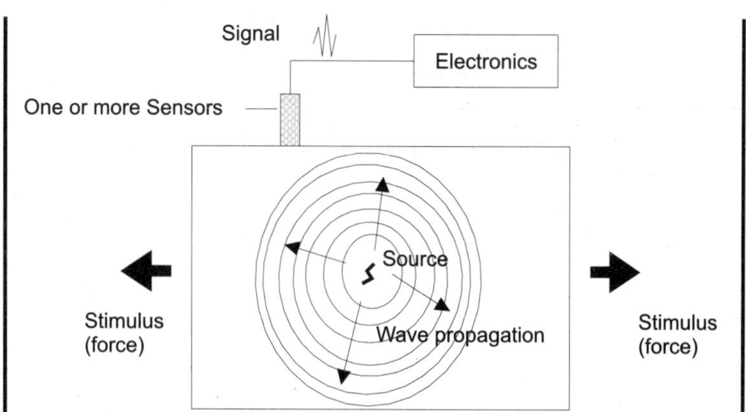

Figure 45 (a) Principle of acoustic emission technique

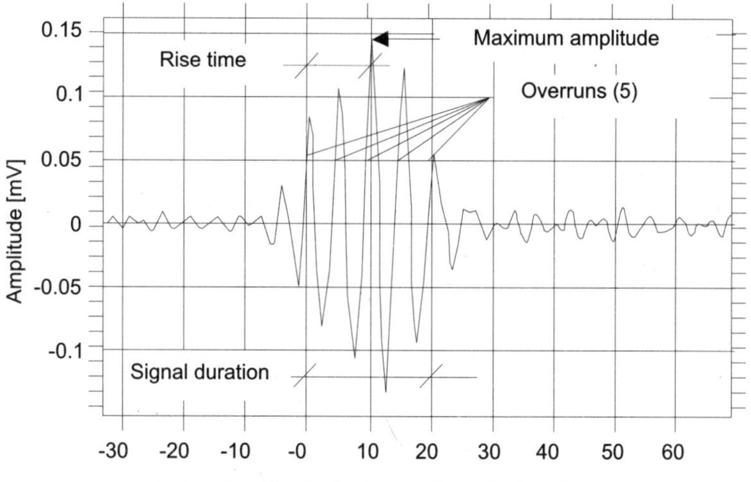

Figure 45 (b) Typical acoustic emission signal

The source mechanism of acoustic emission from pure corrosion activity is much debatable, but it is generally accepted that the most plausible source of AE activity is the nucleation of hydrogen bubbles form solution, or the breakage of a local passivation film. It is generally accepted that pitting process is caused by the breakage of passivation film. Pitting and exfoliation both produce out of plane force and hence they both produce AE signals. Specimens used in this demonstration test are aluminum alloy of LY12, cut from wing skin of an aircraft. It is easy to differentiate between corrosion related AE signal and background noise by using the combination of parameters and real time waveforms. Waveforms of typical noise and typical pitting corrosion are given in Fig. 48 and Fig. 49 respectively. The background noise and AE can be distinguished by using AE amplitude distribution. In general, noise shows a distribution below 40dB, whereas the corrosion related AE has a rather wide range of amplitudes, extending to 60dB.

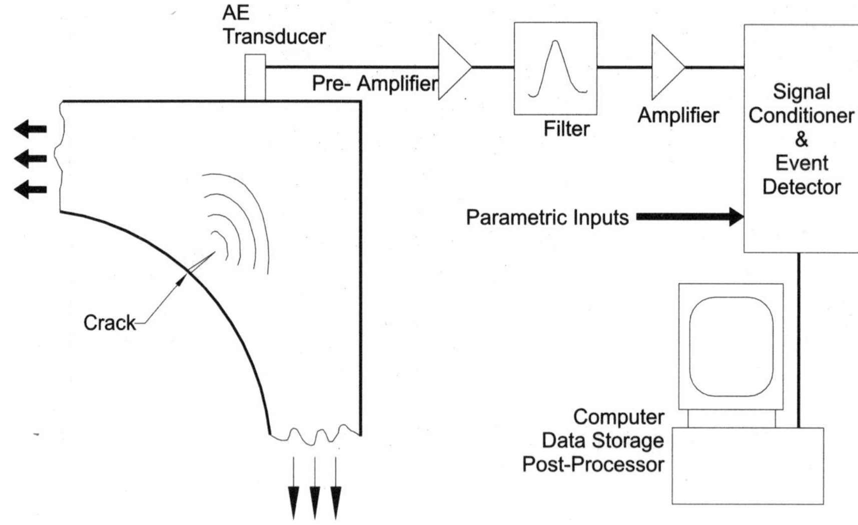

Figure 46 Acoustic emission set up.

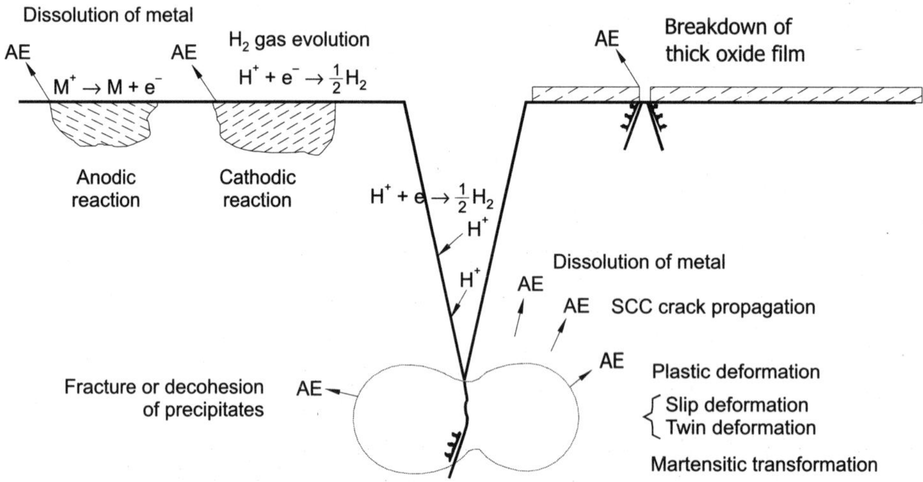

Figure 47 Possible sources of acoustic emission during stress corrosion cracking process.

AE characteristics during SCC of sensitized AISI type 304 stainless steel has been studied [30]. It was found that the AE characteristics during SCC may be divided into three stages (Fig. 50): The first stage corresponds to non-stationary state mainly related to the elevation of the specimen temperature. The AE detected during this stage is attributed to thermal expansion of the specimen. The generation of AE signals in the first stage is found to be most active. The second and third stages correspond to the stationary state of the test. In the second stage, AE activity is found to be very weak. The total Energy (E) and the total emission number (N) in this stage are only several percent of the maximum values in the first stage. In the third stage, AE activity is found to be relatively high. The values of energy and ringdown count in

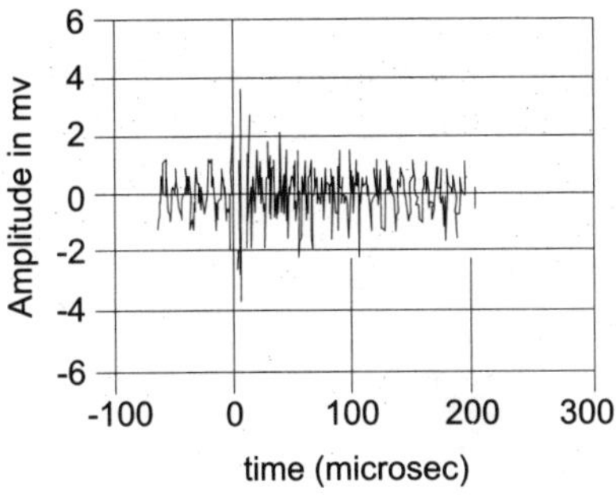

Figure 48 Background noise [30].

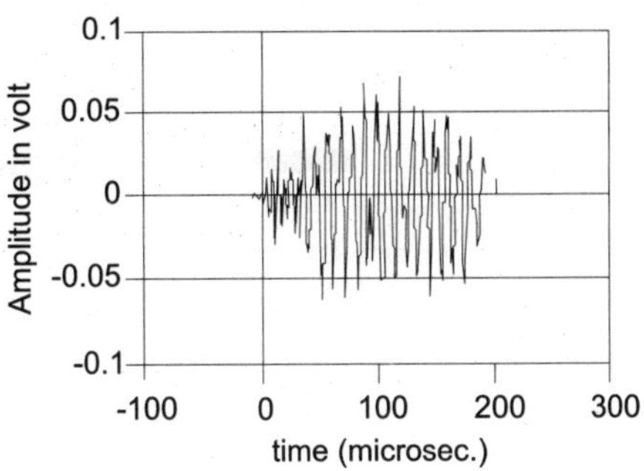

Figure 49 AE Signal from corrosion pits [30].

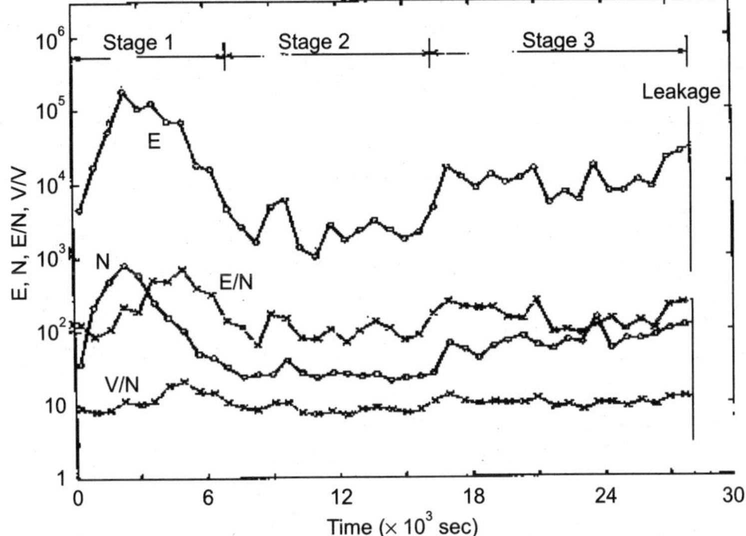

Figure 50 Variation in AE parameters with time, during SCC Test [30].

the third stage are several times larger than that in the second stage. The most probable AE source in the third stage is attributed to the phenomenon associated with inter-granular stress corrosion cracking during the propagation period. It is felt that the thermal expansion that takes place during the first stage would lead to a change in stress distribution in the specimen and this will be reflected in a distinct acoustic emission activity. Since the SCC behaviour is strongly dependent on the stress developed, it would be possible to predict the time to failure due to SCC by using the characteristics of AE generated during the first stage.

7.1 Detection of Exfoliation Corrosion

Exfoliation corrosion of aluminum alloys is a form of localized corrosion which affects many industries, specially aeronautics. The studies on this corrosion mode using only electrochemical technique is not fully efficient for on-line detection and control of this

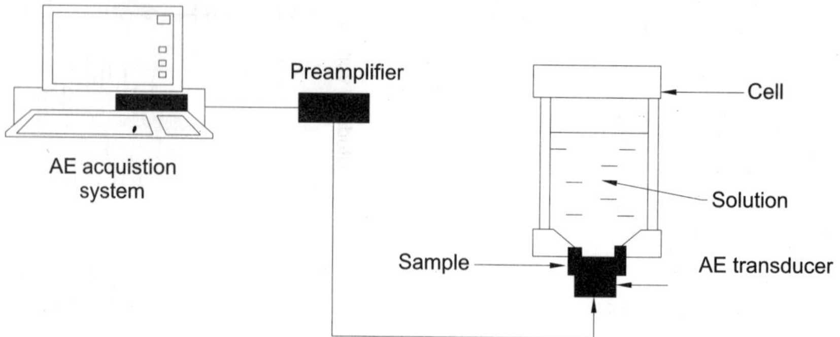

Figure 51 Experimental device [31].

phenomenon. Two aluminum alloy, samples were immersed for 4 days in the saline solution. The experimental set up used for monitoring AE during exfoliation corrosion shown in Fig. 51. A resonant piezoelectric transducer (100-300 kHz) is used to detect the acoustic signals and then filtered and amplified.

The curves representative of the number of bursts versus time are plotted with a given threshold (Fig. 52). The shapes of the curves are quite similar irrespective of the alloy. They can be divided into four steps. The first step corresponds to a latency period of time before any acoustic activity occurs. The duration of this step depends on the nature of the tested alloy. Then a second step of weak acoustic activity appears. It is related to starting of corrosion phenomenon. The third step, characterized by a strong acoustic activity, can be attributed to acceleration of this phenomenon. During the fourth period, it again shows a weak acoustic activity. This period can be related to the formation of the corrosion products which lead to a reduction in the actual anodic area despite the growth of the surface of the corroded area [31]. Observations further showed that the exfoliation resistivity of alloy 7449 T6 is below than alloy 7449 T7 which exhibits only the presence of small and non-occluded pits. Very severe exfoliation has occurred on Al 2024 T3 only after a longer immersion time or in a more acidic

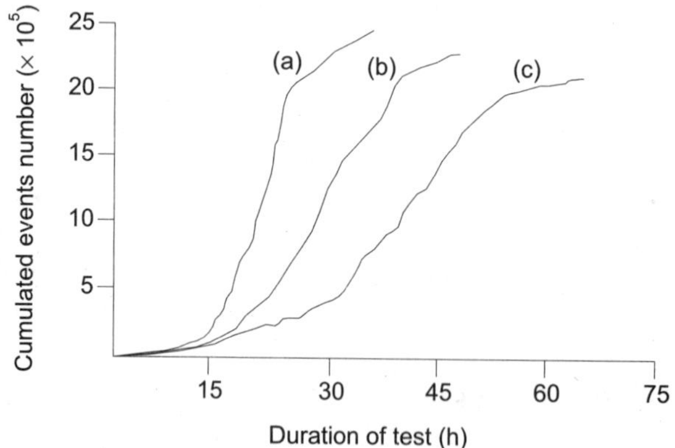

Figure 52 Cumulated AE events number versus time for (a) Al 7449 T7 (b) Al 7449 T6 and Al2024 T3 in EXCO inhibited solution [31].

solution. The acoustic emission activity shows a correlation between AE activity and the exfoliation corrosion rate.

8.0 INFRARED THERMOGRAPHY

8.1 Principle

All objects warmer than "absolute zero" (-273°C) radiate energy which cannot be detected by the naked eye. This energy lies in the range of infrared region of electromagnetic spectrum. Thermography uses this energy to identify defects. Thermography is a NDT tool and the information is extracted using infrared rays emitted by test objects. The wavelength of the infrared radiation lies in the range of 1-10 microns approximately.

Infrared rays are invisible component of the light spectrum that can only be seen with special imaging equipment. Thermography simply views objects in the infrared spectrum and remodulates the image into the visible light range so that we can see it. The image is captured by Infrared Camera (Fig. 53). When viewing photographs taken with an infrared camera, dark areas represent cooler temperatures and bright areas indicate warmer temperatures. By analyzing temperature loss and excess, infrared thermography can be used to identify problems that would otherwise go undetected until a critical failure takes place.

Figure 53 Infrared thermography camera

In recent years, infrared thermography (IRT) has emerged as a method for nondestructive testing. Thermography offers fast, noncontact, wide area detection of subsurface defects. Infrared thermography can be used to detect wall thinning in pipes. which in turns provides information about the corrosion in piping. The technique provides a qualitative alternative to commonly used techniques such as ultrasonics, X-ray, and inspection via borescope. Infrared imaging (thermography) is a non-contact optical method where an accurate two- dimensional mapping of steady or transient thermal effects is constructed from the measurement of infrared energy emitted by the target. Recent advances in infrared technology, specifically development of high-density imaging sensors have opened up a new level of applications unreachable prior to the availability of this technology. Real-time infrared image acquisition and processing allows implementation of advanced thermographic test methods. Two approaches are used in IRT: the passive approach and the active approach. The passive approach tests materials and structures which are naturally at a different (often higher) temperature than ambient, while in the case of the active approach, an external stimulus is necessary to induce relevant thermal contrasts. The interest of these thermal methods is obvious since they allow to probe under the surface in order to reveal potential subsurface defects. Once a normal thermal signature is obtained and understood, any deviation from this normal signature indicates possible damage [32]. Thermal imaging techniques have been used for the detection of wall thinning in stainless steel pipes. Wall thinning of the order of 30% could be detected [32].

8.2 Pulse Thermography

Pulse thermography (PT) is one of the most popular thermal stimulation methods in IR thermography. One reason for this popularity is the quickness of the inspection relying on a thermal stimulation pulse, with duration going from a few milliseconds for high thermal conductivity material inspection (such as metals parts) to a few seconds for low thermal conductivity materials (such as plastics and graphite epoxy). Such quick thermal stimulation allows direct deployment of IRT in plants with convenient heating sources. Moreover, the brief heating prevents damage to the component (heating is generally limited to a few degrees above the initial component temperature). Basically, PT consists of briefly heating the material and then recording its temperature decay curve. Qualitatively, the phenomenon is as follows: The temperature of the material first rises during the pulse. After the pulse, it then decays because the energy - the thermal front - propagates by diffusion into the material. Later, the presence of a subsurface defect (examples: a disbond or corrosion) reduces the diffusion rate so that when observing the surface temperature, such a subsurface defect appears as an area of higher temperature with respect to the surrounding sound area. In fact in such a case, the reduced diffusion rate caused by the subsurface defect translates into "heat accumulation" and hence higher surface temperature just over the defect. Moreover, such phenomenon occurs in time so that, deeper defects are observed later and with a reduced "diluted" or "spread" thermal contrast. One typical set up employed is given in Fig. 54.

In the case of high thermal conductivity materials such aluminum aircraft skin, quick heating schemes are necessary. Figure 55 shows the results of experiments conducted on 1 mm thick aluminum sheet in which blind holes of 20 mm diameter were drilled simulating corrosion. Flash heating was used. In the C_{max} image, the defects exhibit a different apparent size due to the spreading of the thermal front. This can be also explained by multiple internal reflections the thermal waves experience inside the specimen. It is believed that this phenomenon is not present in the φ_{max} image because only the maximum phase is concerned at each pixel location (instead of the cumulative effects of all the various frequencies), and hence a better definition of defect shape is thus noted [33].

An important hint for variation in process to process or from time to time within a process, is the temperature change in the piping systems which carry fuels, by-products or any other item for the process. Process piping may be referred to as "the arteries" of the production sequence. Without the efficient transportation of the various gasses or liquids, many processes will suffer from quality, throughput, or may even require shutdown. One of the commonly associated fluctuations is the removal of heat from the process or media. While this function is

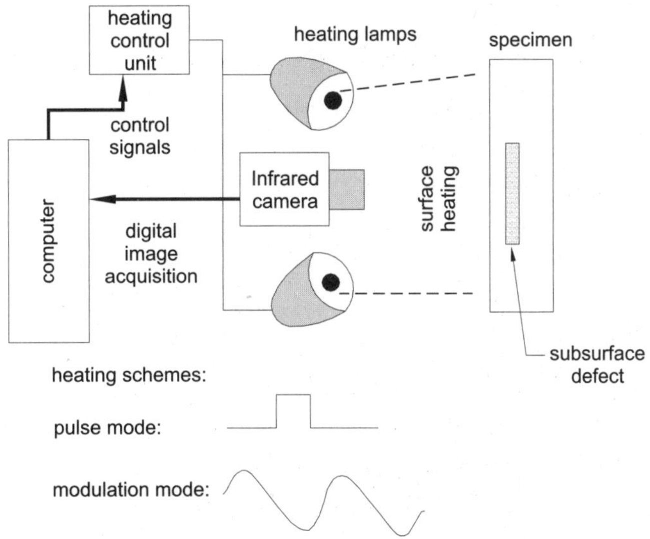

Figure 54 Experimental set-up for infrared thermography experiments (radiative heating scheme, in reflection) [33].

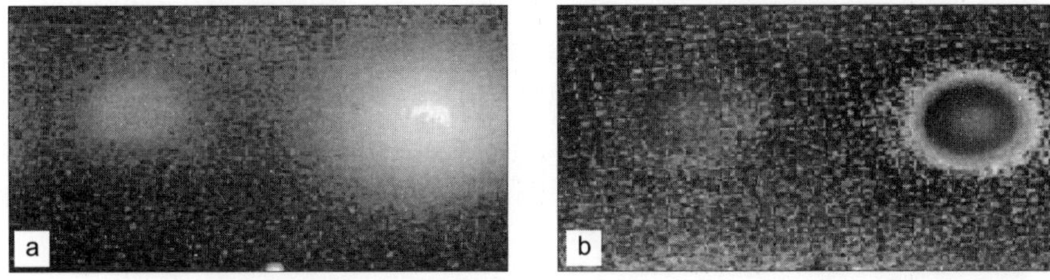

Figure 55 Aluminum specimen: simulated corrosion (a) Cmax contrast and (b) phase image [33].

commonly acknowledged, one of the functions not normally associated with process piping is the transfer of gasses or liquids while maintaining the temperature at a predetermined cooling rate. Infrared thermogarphy is used to determine restrictions in flow and thermal leakage throughout the system due to corrosion/erosion/ageing. Figure 56 shows a heat exchanger with cracked tubes. The cracked and leaking tubes were found during a pre-shutdown inspection with the help of infrared thermography. This allowed for efficient planning and scheduling of tube repairs/replacement.

8.3 Inspection of Surface Coatings

An effective application of the surface-tolerant coating requires an accurate characterization of the initial coating condition, quantitative assessment of surface preparation and evaluation of the integrity of the applied coating. It is very costly affair to remove old coatings from old bridges and recoat. It is shown that infrared thermography is most useful for quantitative location and measurement of subsurface coating failures such as disbands and blisters and visual (color) imaging is the most applicable for global coating inspection based on the changes in the coating pigmentation and visual appearances of the corroded areas.

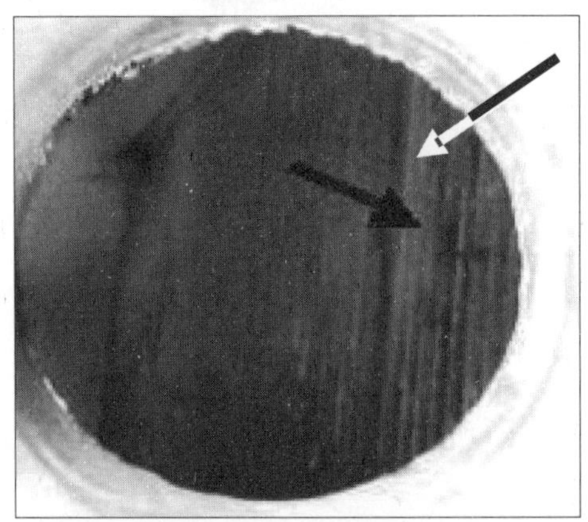

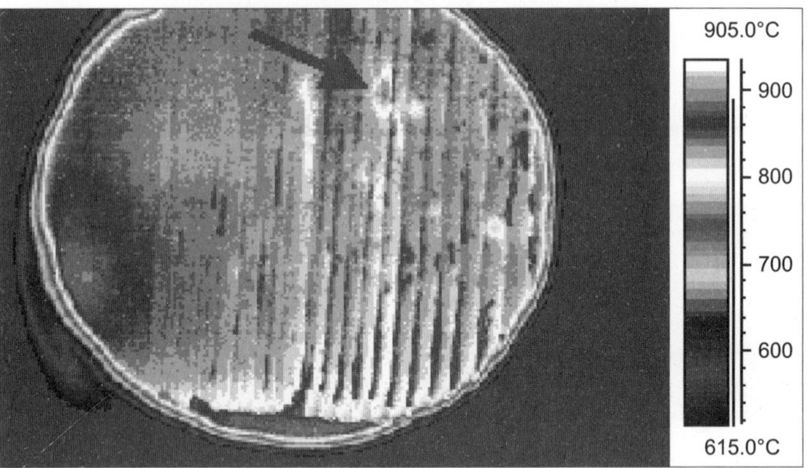

Figure 56 Thermal imaging of high temperature heat exchanger tubes - Arrows indicate split tubes as "cold" areas [34].

One of the methods selected for the coating inspection is an active or transient thermography. Depending on the type of defect and thermal characteristics of a target, an external heating or cooling is applied in the form of the short energy pulses. The energy pulses can be generated by using quartz flash lamps or hot-air heating. Thermal perturbation is then followed by a differential time-resolved infrared image analysis. Coating defects such as blistering and sub-surface corrosion spots can be detected in infrared images as a result of the differences in the thermal diffusivity of the defective and nondefective areas Fig. 57. Often a fraction of a degree difference in temperature is adequate for reliable detection and identification of the thermal signature of the defect. The same technique can potentially be used for rapid assessment of the variations in the coating thickness [34].

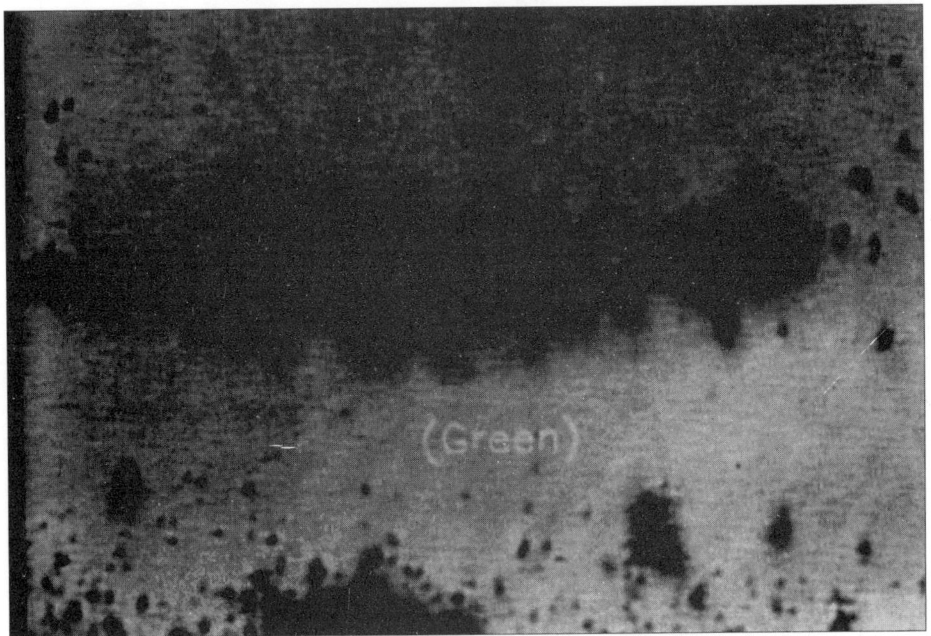

Figure 57 Color image of girder section with rusted area [34].

8.4 Inspection for Metal Thinning

IR techniques can be used to detect material thinning of relatively thin structures since areas with different thermal masses will absorb and radiate heat at different rates. In relatively thin and thermally conductive materials, heat will be conducted away from the surface faster by thicker regions. By heating the surface and monitoring its cooling characteristics, a thickness map can be produced [Fig. 58]. Thin areas may be the result of corrosion damage on the backside of a structure, which is normally not visible. The image shows corrosion damage and disbonding of a tear strap/stringer on the inside surface of an aircraft skin. This type of damage is costly to detect visually because a great deal of the interior of the aircraft must be disassembled. With IR techniques, the damage can be detected from the outside of the aircraft [35].

9.0 OTHER NDT TECHNIQUES FOR CORROSION DETECTION

9.1 Radiography

Radiographic wall-thickness measurements of insulated piping in chemical and petrochemical industry are carried out by exposing film-cassettes with Ir-192 source. Accurate wall thickness measurements are carried out by special techniques using accurately calibrated reference blocks. In film density variation technique, comparison is made between the images of the known stepped wedge and the corroded wall of the pipe.

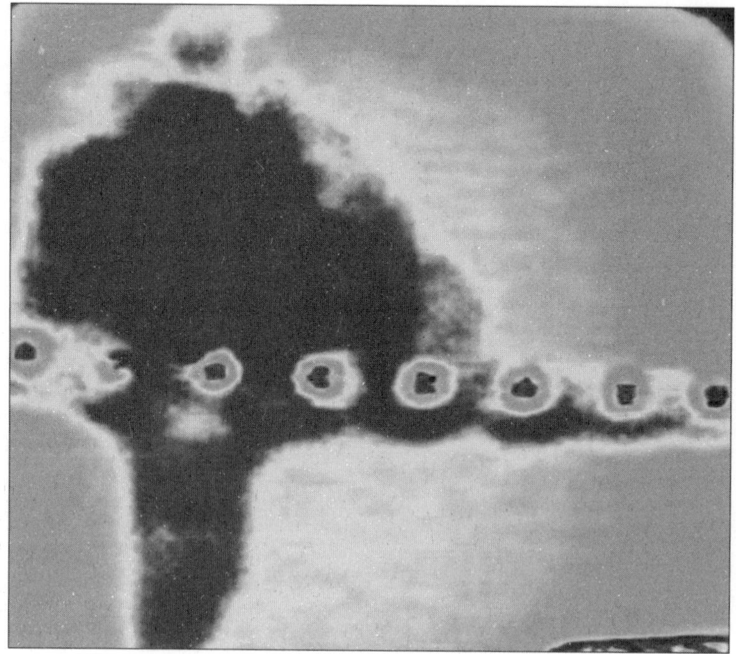

Figure 58 Thermographic image of corroded area [35].

9.2 Field Signature Method

An efficient non intrusive and rugged method called Field Signature Method (FSM) has been developed for monitoring localised corrosion, erosion or abrasion and cracking in steel and metal structures, pipelines, pipe bends and vessels [36]. FSM is based on feeding an electric direct current through the selected regions of the structures to be monitored, and sensing the pattern of the electrical field by measuring small potential differences generated on the surface of the monitored structure. Potential areas for corrosion are identified and the FSM sensing electrodes are mounted and the potential difference is monitored periodically / continuously. Typical defects that can be monitored and located by this method include: general corrosion, CO_2 corrosion, localised corrosion or pitting; bacterial corrosion; cracking in welds/HAZ and erosion.

9.3 Microwave Testing

Inspecting for corrosion under paint is an important step, but is problematic. Most NDT systems capable of detecting corrosion under paint (for example Zinc Chromate primer) in aircraft industry are expensive and complicated, requiring highly trained technician-operators. The first sign of corrosion on an aircraft surface is a function of oxide on the metal substrate. Though invisible to the eye under the paint system, these oxides are the precursor to metal loss and future structural damage, and present a different dielectric signature along the microwave signal path. Microwave Corrosion Detector (MCD) senses these dielectric changes by

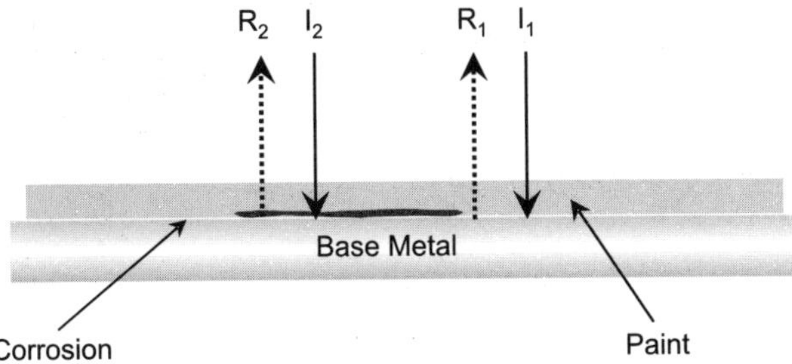

I and R are Incident Reflected microwave signals respectively. Once corrosion is initiated, $R_2 \neq R_1$ due to differences in dielectric property

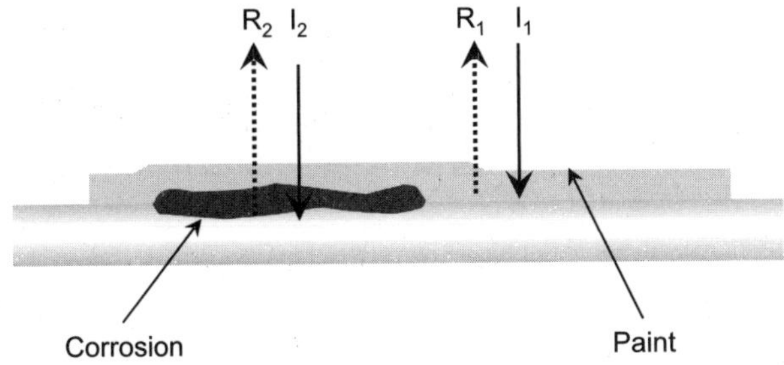

Detection of differences in distance to base metal and in turn the extent of wastage caused by corrosion

Figure 59 Corrosion detection using microwaves [37].

comparing the incident and reflected microwave signals as they pass through the paint system and reflect off the aircraft's metal substrate. Where corrosion is more advanced and metal loss has already taken place, the MCD senses both the oxides and the metal wastage Fig. 59 [37].

9.4 Magnetic Barkhausen Emission

The typical magnetization curve, the change in flux density with applied field appears to be a continuous function of H, the magnetic field. However if we examine the loop more closely we find that the B-H curve consists of small, discontinuous changes of B as H varies. These discontinuous changes are a result of the Barkhausen effect. This corresponds to small magnetization 'jumps' due to domain walls becoming pinned and released from microstuctural obstacles such as grain boundaries, dislocations, second phase particles, and non-metallic inclusions.

Each abrupt jump produces a brief burst of magnetic 'noise' as shown in Fig. 60 which can be detected and analyzed using special MBN probes that respond to high frequency signals. The characteristics of this magnetic Barkhausen emission (MBE) signal can be related to a number of different microstructural features like stress, microstructural features etc

The feasibility studies on use of a relatively new NDE technique called magnetic Barkhausen Emission (MBE) for evaluation of carburisation depth in reformer tubes have been carried out [38]. Carburisation takes place in service in ferritic steel reformer tubes used in

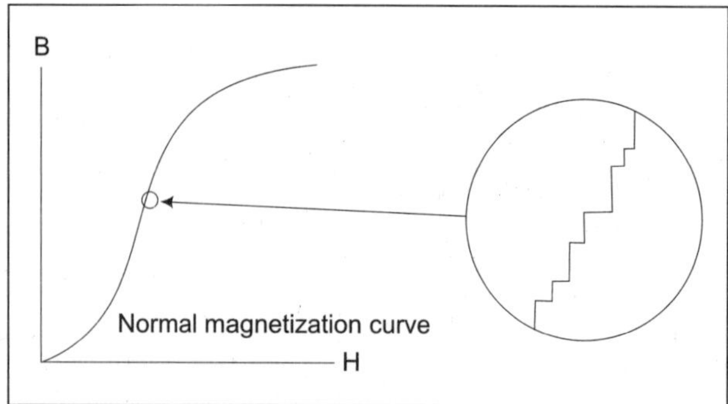

Figure 60 Magnetic barkhausen principle [38].

petrochemical industries. The depth of carburised layer decides the continued use/ replacement of these tubes. MBE measurements have been carried out on samples from service exposed 0.5Cr- 0.5Mo ferritic steel tubes at different depths (cross section) from the carburised ID surface, to simulate the variation in carbon concentration gradient within the case depth of MBE with increasing time of exposure of the tubes to carburisation in service. It has been observed that the MBE level increases with increasing depth of measurement (Fig. 61). An inverse relation between MBE level and carbon content / hardness value has been observed. This study suggests that the MBE measurements on the carburised surface can be correlated with the concentration gradient within the case depth of the MBE, which in turn would help in predicting the appropriate depth of the carburised layer with proper prior calibration.

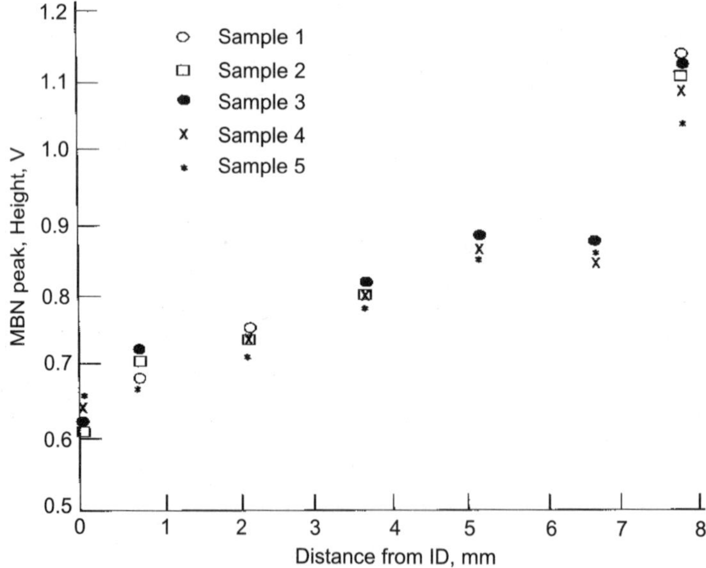

Figure 61 Variation in MBE peak height with distance from inside surface [38].

9.5 In-situ Metallography

All the NDT techniques measure change in parameters with respect to reference to find out corrosion damage. In-situ metallography reveals microstructure, and hence inspector can visualize the corrosion damage.

Portable equipment can be used to metallographically prepare very small area of component by grinding and polishing. Portable microscope can be used conveniently to obtain the microstructure from the component. The technique also helps in establishing root cause analysis for corrosion related failures to enable taking remedial measurements to prevent recurrence of such failures. A case history is given to demonstrate the usefulness of in-situ metallograpy for failure analysis.

9.5.1 Failure investigation on turbine lube oil cooler tubes

A failed turbine lube oil cooler tube of a nuclear power plant was investigated for failure analysis. The tubes of 15 mm in outer diameter are made of grade 90:10 cupro- nickel alloy, and have 1 mm wall thickness. These tubes carry raw water in the tube side and lube oil in the shell side. The oil make up rate was observed to be on the rise during operation of the plant, and hence it was suspected that the same of cooler tubes might be leaking. The presence of leak was confirmed by the plant personnel by analyzing the cooling water for oil content. The failed sample was received for investigation.

The two tubes received from the plant are shown in Fig. 62 and Fig. 63. The sectioned tube in Fig. 62 shows extensive pitting on the inner surfaces. The diameter of these near circular pits varies approximately from 0.7 mm to 1.5 mm. The inner oxide scale in this particular tube (Fig. 62) was removed by the plant personnel for identifying the nature of the scale by chemical analysis. Figure 63 shows a partially sectioned tube with the oxide layer formed inside during service. A through and through hole is also clearly seen (as indicated by arrow). The oxide coat is found to be impaired at a few locations as can be seen in Fig. 63 (indicated by arrow).

Figure 64 shows the microstructure of the cross section of the failed tube at a location away from a pit. Slight grain boundary attack and grain dislodgement at the ID can be seen. The figure also shows that the OD of the tube has not undergone any degradation. Figure 65 shows the normal microstructure at the mid wall. Figure 66 shows the microstructure of a cross section at a region close to a corrosion pit. Extensive wall thinning at the pitting region can be

Figure 62 As received cut cross-section of the failed 90:10 cupro nickel tube from lube oil cooler, Kaiga-I. Extensive pitting attack indicated by arrows [39].

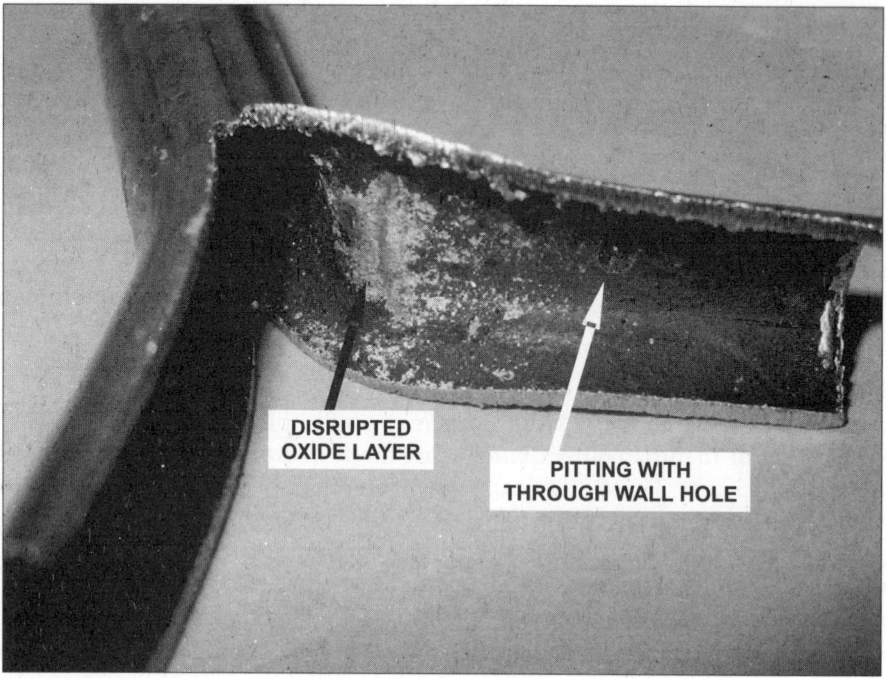

Figure 63 Magnified view of the inner portion showing pitting attack and disruption of oxide scales. A through wall hole is also visible [39].

Figure 64 Cross section microstructure revealing mild grain boundary attack at the ID region. OD is free from defects [39].

Figure 65 Mid-wall microstructure of the tube [39].

observed. Fig.67 shows the on set of pitting attack close to the major pit. From the above microstructural observations, it is confirmed that the pitting has originated from the inner side of the tube carrying the raw water. [39]

Figure 66 Extensive intergranular pitting attack on the inner surface of the tube [39].

Figure 67 Onset of pitting attack on the inner surface of the tube [39].

10.0 CONCLUSION

Applicability of various non destructive evaluation techniques for detection and assessment of corrosion related damage in various materials and components has been given. A few case studies have also been addressed to elucidate the capabilities of these techniques to assess the corrosion damage.

REFERENCES

1. T. Jayakumar and Baldev Raj, "Corrosion Monitoring by NDT Methods", First Asian - Pacific Conference and 6th National Convention on Corrosion, Bangalore, 2001.
2. ASM Handbook on non destructive evaluation and quality control (Vol. 17), Materials Park, Ohio, (Edition 2001).
3. http://www.silverwing.ltd.uk/en/technical/paper03.html
4. C.V.Subramanian, A. Joseph, A.S. Ramesh, T. Jayakumar, P.Kalyanasundaram and Baldev Raj, "Wall thickness measurements of tubes by internal rotary inspection system (IRIS) - a comparative study with metallography", Proc. of 14th World Conf. on NDT, New Delhi, India, 1996, pp. 985-988.
5. A.S.Birring, M.L.Birtlett and K.Kawano, Corrosion, 45, 1989, p. 259
6. G.D.Reid, Proc. First Int. Conf. on NDT in Gulf, University of Qatar, Doha, Qatar, 1999, pp. 153-174.
7. S.E. Krüger, J.M.A. Rebello, P.C. de Camargo, "Hydrogen damage detection by ultrasonic spectral analysis", NDT&E International, 32, 1999, pp. 275-281.
8. M.Thavasimuthu, A. Joseph and T. Jayakumar, "Monitoring of biologically influenced corroded pipes by ultrasonic technique", Proceedings of the National Conference on Non-destructive Evaluation for Life Extension, Hyderabad, Dec. 1997, pp. 218-223.
9. P. Wilcox, M. Lowe and P. Cawley, " The effect of dispersion on long-range inspection using ultrasonic guided waves", NDT&E International, 34, 2001, pp 1-9
10. M.Z. Silva, R. Gouyon, F. Lepoutre, "Hidden corrosion detection in aircraft aluminum structures using laser ultrasonics and wavelet transform signal analysis" Ultrasonics, 41, 2003, pp. 301-305.

11. Alleyne, D.N. and Cawley, P. 'The excitation of Lamb waves in pipes using dry coupled piezoelectric transducers', J NDE, Vol 15, pp11-20, 1996.
12. Alleyne, D.N. and Cawley, P. 'Long range propagation of Lamb waves in chemical plant pipework', Materials Evaluation, Vol 55, pp504-508, 1997.
13. Shivprakash Iyer, Andrea J. Schokker, Sunil K. Sinha, "Ultrasonic imaging - A novel way to investigate corrosion status in post - tensioned concrete members", J. Indian Inst. Sci., 82, Sept. - Dec. 2002, pp. 197-217.
14. http://www.nde.com/paper54.htm
15. http://www.ndt-ed.org/EducationResources/CommunityCollege/EddyCurrents/Applications/thicknessmeasurements.htm.
16. B.P.C. Rao C.B. Rao and Baldev Raj, Proc. of Seminar on Non Destructive Evaluation (NDE 93), ISNT, Madras, 1993.
17. B. Sasi, Hasan Shaikh, B.P.C. Rao and T.Jayakumar, Proc. National Seminar on Non Destructive Evaluation (NDE 2001), Lunawala, 2001.
18. H. Shaikh, N. Sivaibharasi, B. Sasi, T. Anita, R. Amirthalingam, B. P. C. Rao, T. Jayakumar, H. S. Khatak and Baldev Raj, Use of eddy current testing method in detection and evaluation of sensitisation and intergranular corrosion in austenitic stainless steels, Corrosion Science, 48(2006), pp 1462-1482.
19. H. Shaikh, B. P. C. Rao, S. Gupta, R.P. George, S. Venugopal, B. Sasi, T. Jayakumar, H. S. Khatak, "Assessment of Intergranular Corrosion in AISI Type 316L Stainless Steel Weldments", British Corrosion Journal, Vol. 37, No. 2, 2002, pp. 129-140.
20. R. Amirthalingam, H. Shaikh, N. Sivaibharasi, B. Sasi, T. Anita, B. P. C. Rao, T. Jayakumar, and H. S. Khatak, "Detection and quantification of sensitization and intergranular corrosion in Austenitic stainless steels by eddy current testing technique", Proc.of 12th National congress on corrosion control at Vishakapatnam, September, 2004.
21. B. Sasi, Hasan Shaikh, B. P. C. Rao and T. Jayakumar, "Characterisation of intergrannular corrosion in austenitic stainless steel using eddy currents", Proc. of National Seminar on NDE-2001, December 2001, Lonavala, pp.108.
22. D.M.Stevens and H.L.Whaley, Non Destructive Evaluation (ASME), Vol54, 1993, pp. 43-47.
23. http://www.public.iastate.edu/~engeltra/NDE/Education/College/EddyCurrents/AdvancedTechniques/ec_5_2.htm
24. B.A.Lepine Pulsed Eddy Current Method Development for Hidden Corrosion in Aircraft Structure, NDT.net - January 1999, Vol.4 No.1
25. http://www.silverwing.ltd.uk/en/technical/paper02.html
26. H.J.M. Jansen and M.M. Festen, Insight, 37, 1995, pp. 421-425.
27. Manjit Singh, V.M. Joshi, V.H. Patankar, S. Vaidyanathan, B.P.C. Rao, T. Jayakumar, G.P. Srivastava, K.V. Kasiviswanathan, C. Rajagopalan and Baldev Raj. Insight 45(1) 2003, 73-86.
28. C.Babu Rao, "Studies on Laser Scattering to Characterise Surfaces", Ph D Thesis, University of Madras, Chennai, 2001.
29. http://www.tms.org/pubs/journals/JOM/9811/Huang/Huang-9811
30. H.Kusanagi, H.Nakasa, T.Ishihara and S.Ohashi, Proc. of the 4th Acoustic Emission Symp., Tokyo, 1998.
31. F. Bellenger, h. Mazille, H. Idrissi, "Use of acoustic emission technique for the early detection of aluminum alloys exfoliation corrosion", NDT& E International, 35, 2002, 385-392,
32. Jeong-Seb Han, Jin - Hwan Park, Detecion of corrosion steel under and organic coating by infrared photography, Corrosion Science ,46, 2004, 787-793.

33. Allan D. Hines, Infrared Thermography Applications at Dofasco Inc., Proc. 10th APCNDT, Brisbane, http://www.ndt.net/article/apcndt01/papers/920/920.htm
34. http://www.iti.northwestern.edu/publications/technical_reports/tr10.html
35. http://www.ndted.org/EducationResources/CommunityCollege/Other%20 Methods/IRT/IR_Applications.htm
36. R.D.Strommen, H.Harold and K.R.Wold, Proc. NACE Annual Conf., 1992, pp. 712-716.
37. http://www.systemsandmaterials.com/nde/projects/aircraft.html
38. S.Vaidyanathan, V.Moorthy, T.Jayakumar and Baldev Raj, "Magnetic Barkhausen emission analysis for assessment of microstructure and damage", Materials Evaluation, 56, 1998, pp. 449-452.
39. N.Raghu, N.G. Muralidharan and K.V. Kasiviswanthan, "Failure Investigaion on Turbine Lub Oil Cooler Tubes, Kaiga I". (Internal Report of IGCAR)

CHAPTER 15

Scanning Kelvin Probe and Scanning Kelvin Probe Force Microscopy and Their Application in Corrosion Science

Michael Rohwerder
Max-Planck-Institut für Eisenforschung, Düsseldorf, Germany

I. INTRODUCTION

As the name already suggests, the Kelvin Probe technique itself is not a new technique. It was invented by W. Thomson [1]. In the classical set-up sample surface and the Kelvin probe plate form a parallel plate capacitor, where the lateral dimensions of the capacitor plates are several orders of magnitude larger than the distance between sample and probe plate. If the capacitance between probe plate and sample surface is known, the work function difference between probe (reference material) and sample can be calculated, although the measurement of the charge is not easily performed. This problem can be overcome by measuring the discharge current when the distance between probe and sample is varied, first introduced by Lord Kelvin. This classical set-up is in fact one of the oldest techniques for measuring the work functions of materials, see also Kohlrausch [2]. By manually changing the probe-sample separation and measuring the resulting current flow, the work function difference can be calculated [1]. Zisman developed the technique further to the vibrating capacitor technique, in which the probe plate vibrates periodically, thus causing a steady ac current to flow [3]. The vibrating capacitor technique is the approach commonly used nowadays.

The scanning version of the classical capacitor setup was first introduced by Parker and Warren [4], who obtained in their studies on lateral variations of the work function on gold and graphite a resolution of several millimetres. Since then, the Scanning Kelvin Probe (SKP) technique has been steadily improved, and a resolution of several tens of μm can be easily obtained [5-6]. Mäckel et al. prepared probe tips with a flattened surface of just a few microns in diameter and by controlling the distance between tip and sample surface to about 50 nm they achieved a spatial resolution of 5 μm [7-8]. An even more impressive lateral resolution of 100 nm was reported by Nabhan et al. [9], who also use active distance control.

A high resolution technique that is much easier to use is the Scanning Kelvin Probe Force Microscope (SKPFM), which is based on atomic force microscopy (AFM) and which was first introduced by Nonnemacher et al. (10, 11) for microelectronic applications.

While since decades Kelvin probe techniques are successfully used in surface physics to study adsorption of molecules or even reconstruction processes of single crystal surfaces (see e.g. [12]), which both cause a change in the surface or dipole potential, their application in corrosion science is since just very recent. The reason for this is that in contrast to measurements in UHV or at ambient conditions corrosion experiments usually need to be carried out at high humidity. In regard of the extremely sensitive pre-amplifiers required for the measurement of the ultra-low displacement currents in SKP measurements at high humidity are a challenge not easily met.

Stratmann et al. [13-22] were the first who introduced SKP to corrosion science. As will be discussed in the following, SKP is unique insofar as it allows a non-contact measurement of corrosion potentials. It can be used instead of electrochemical reference electrodes and allows measuring electrode potentials through insulating dielectric media such as air or polymeric films. It is mainly used where standard electrochemical techniques, which require a finite electrical resistance between working and reference electrodes, will fail.

II. THE SCANNING KELVIN PROBE TECHNIQUE AS A NATURAL METHOD FOR THE MEASUREMENT OF ELECTRODE POTENTIALS

When we talk about electrode potentials, we usually refer to the electrical double layer, i.e. the potential drop at the metal/solution interface. However, this potential difference is not directly measurable. This is why electrode potentials are referred to a standard, such as e.g. the standard hydrogen electrode. In the discussion on the theory of electrochemistry the concept and the physical meaning of an "absolute" electrode potential is a long-standing question, that experienced a revival in 1970s by Bockris et al. [23- 25] and Gileadi [26], but it was essentially S. Trasatti who formed the modern concept of an absolute electrode potential [27-33]. The following discussion of the absolute electrode potential is inspired by his ideas presented in [34].

First we consider the definition of the work function:

The work function is defined as the minimum work that is required to extract an electron from within the sample to a position just outside the sample (far enough to eliminate contributions from image forces). It is comprised of a part that takes into account the chemical work and a part that takes into account the electrostatic work to transport the charged electron through the dipole layer of the sample's surface:

$$\Phi = -\mu_e + e \cdot \chi = -(\mu_e - e \cdot \chi), \quad \text{...(1)}$$

where μ_e is the chemical potential (i.e. the chemical work to transfer the electron from the infinity into the sample, i.e. the other way round, hence the minus sign) and χ is the so called dipole or surface potential, i.e. the potential drop between just inside the bulk and just outside of it.

The surface potential χ is usually positive as an electrode cloud hovers over the surface, causing a positively charged bulk. It is significantly affected by factors such as crystal face orientation and step density etc., as these directly affect this electron cloud just outside the surface.

The use of the phrase "just outside the sample" for the definition of the work function is unimportant for the case of an uncharged sample, where the energetic position just outside the sample is the same as infinitely far away, i.e. it is equivalent with the vacuum level. But in the general case the sample may be charged and then an additional work component is required to transfer the electron from just outside the surface to a position infinitely far away: $e \cdot \psi$ (ψ being the Volta potential, which is equivalent for the potential drop from the infinity to a position just outside the surface, i.e. which is due just to the surface charge, i.e. the external charge, of the sample; see fig.1).

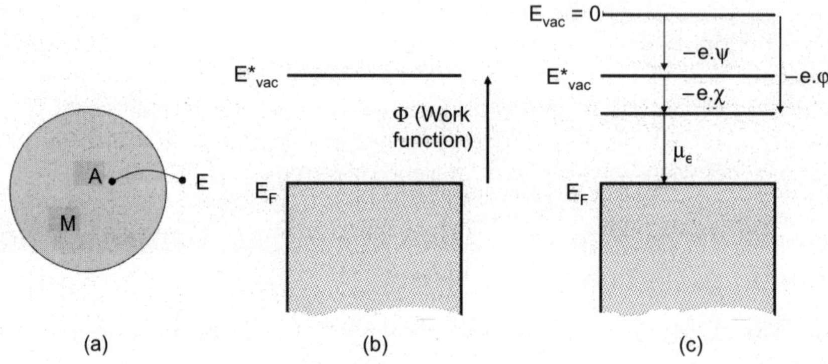

Figure 1 a) the work function is defined as the minimum work that is required for the removal of an electron from within the sample (i.e. at the Fermi level, since only then it is the minimum work) to a point just outside the sample (far enough to eliminate contributions from image forces). For uncharged samples the energy level at this position is equivalent to the vacuum level (b)). For charged samples (with the external potential (Volta potential) $\psi \neq 0$) this is still $-e \cdot \psi$ away from the absolute vacuum level E_{vac} (c)). The energy level E^*_{vac} just outside the sample is conveniently used in semiconductor physics.

Combined the dipole potential χ and the Volta potential ψ give the Galvani potential φ, i.e. the potential drop between the bulk and the vacuum level infinitely far away from the surface:

$$\varphi = \chi + \psi$$

The absolute electrode potential E_{abs} of a sample covered by a liquid layer in analogy to the definition of the work function coincides as the minimum work to extract an electron from the Fermi level, through the solid/liquid interface (with the potential drop $\Delta\varphi_M^{El}$), through the liquid and through the surface layer of the liquid to a position just outside the liquid, see fig. 2 a) and b).

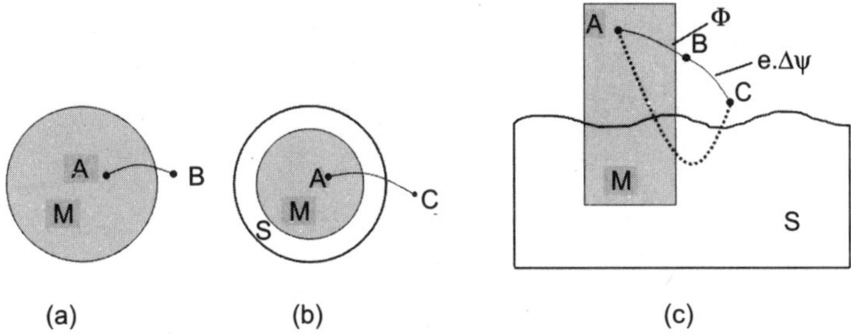

Figure 2 In analogy to the definition of the work function Φ (see a))a more general work function F* may be defined where the electron is also transferred through the liquid layer covering the sample (see b). The absolute electrode potential then is defined as $e \cdot E_{abs} = \Phi^*$. c) Both paths for the electron extraction are energetically equivalent: directly AC and ABC.

Hence, E_{abs} is directly obtained from eq.(1) by adding the additional term $\Delta\varphi_M{}^{El} = \varphi_M - \varphi_{El}$ and exchanging χ_S with χ_M:

$$E_{abs} = \frac{1}{e} \cdot \Phi^* = \left(-\frac{\mu_e}{e} + \Delta\varphi_M{}^{El} + \chi_s \right) \qquad ...(2)$$

where $\Delta\varphi_M{}^{El}$ is the potential drop at the metal/liquid interface (the superscript "El" referring to "electrolyte" in reference to the importance of this term for electrochemistry) and χ_S the surface potential of the surface of the liquid.

Obviously, the following relation holds (see fig. 2 c)):

$$E_{abs} = \Phi/e + (\Psi^M - \Psi^S), \qquad ...(3)$$

where ψ^M and ψ^S denote the Volta potentials outside the metal and outside the solution.

Now let us consider the case where a metallic probe KP is positioned near the surface of the electrolyte as sketched in fig. 3.

Then the following equations results:

$$E_{abs} = \frac{1}{e} \cdot \Phi^* = \left(-\frac{\mu_e}{e} + \Delta\varphi_M{}^{El} + \chi_s \right) = \Phi^M/e + (\psi^M - \psi^S) = \Phi^{KP}/e + (\psi^{KP} - \psi^S) \qquad ...(4)$$

as no work is required for an electron to go from the metal M into the probe KP.

Hence, if the work function of the probe KP is known, the measurement of the Volta potential difference ($\psi^{KP} - \psi^S$) directly provides the absolute electrode potential. Now, the measurement of such Volta potential differences is exactly what the Kelvin Probe technique provides, as will be discussed in the next section. In this way the Kelvin Probe technique may be described as the natural reference electrode, which in fact -as we will see- has many advantages over conventional reference electrodes.

The absolute electrode potential is directly correlated to the electrode potential vs. the [SHE] standard hydrogen electrode by the simple equation:

$$E_{SHE} = E_{abs} - E^0(H_2)_{abs},$$

where $E^0(H_2)_{abs}$ is the potential of the standard hydrogen electrode (SHE) in the absolute electrode potential scale. For the case of a pure water layer adsorbed on the metal surface, i.e. in

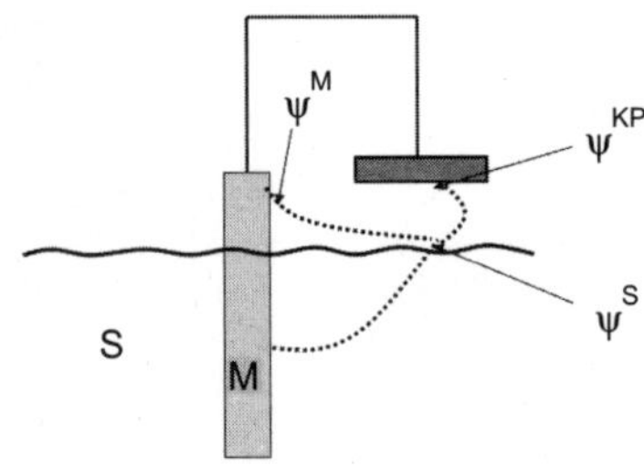

Figure 3 For a probe KP electrically connected to the sample M the work for transferring an electron from the bulk of M to a point near the surface of the solution phase S is the same for all three indicated transfer paths, and hence: $E_{abs} = \frac{1}{e}\cdot F^* = \left(-\frac{\mu_e}{e} + \Delta\varphi_M^{El} + c_S\right) = \Phi^M/e + (\psi^M - \psi^S) = \Phi^{KP}/e + (\psi^{KP} - \psi^S)$ Thusly, the Volta potential difference between the Kelvin Probe and the electrolyte surface is directly correlated to the electrode potential of the immersed electrode M.

the absence of supporting electrolyte, the following correlation between the modified work function Φ^* and the potential of zero charge can be made [32]:

$$\Phi^* = e\cdot E^M(\sigma = 0)_{abs} = e\cdot(E^M(\sigma = 0)_{SHE} + E^0(H_2)_{abs}) \quad ...(5)$$

i.e., the measurement of the work function of the water layer covered metal provides directly $E^0(H_2)_{abs}$, if the potential of zero charge vs. SHE is known. Hence, the determination of $E^0(H_2)_{abs}$ should not be a problem. However, the values found in literature scatter in the range from 4.44 V to 4.85 V [32].

III. THE KELVIN PROBE TECHNIQUE

III.I The Classical Needle Sensor SKP set up

Having introduced the Scanning Kelvin Probe as the natural technique for measuring electrode potentials, we will now have a closer look on how the Kelvin Probe technique works. In principle, the Kelvin Probe consists of a metallic reference electrode, which is separated from the sample by a dielectric medium and connected to the sample by a metallic wire (see fig.4). Thusly, the reference electrode and the sample form a capacitor.

Now, if the work function of the sample is given as Φ_{sample} and the work function of the reference electrode as Φ_{KP}, then:

$$\Phi_{sample} = -\mu_e^{sample} + e\cdot\chi^{sample}$$

and

$$\Phi_{KP} = -\mu_e^{KP} + e\cdot\chi^{KP}$$

As after connection of the two metals the electrochemical potential $\tilde{\mu}_e$ of the electrons within both phases will be identical, a charging of one sample with respect to the other causing a Volta-potential difference between the two will be observed (see fig.5):

$$\tilde{\mu}_e^{sample} = \mu_e^{sample} - e\cdot\varphi^{sample} = -\Phi^{sample} - e\cdot\psi^{sample} =$$

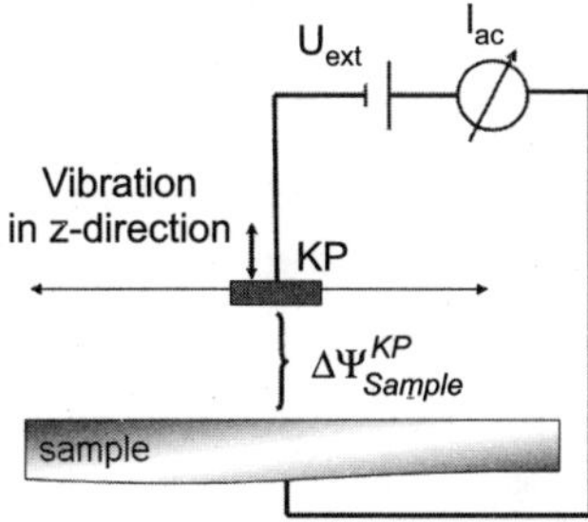

Figure 4 Schematic set-up of a Scanning Kelvin Probe

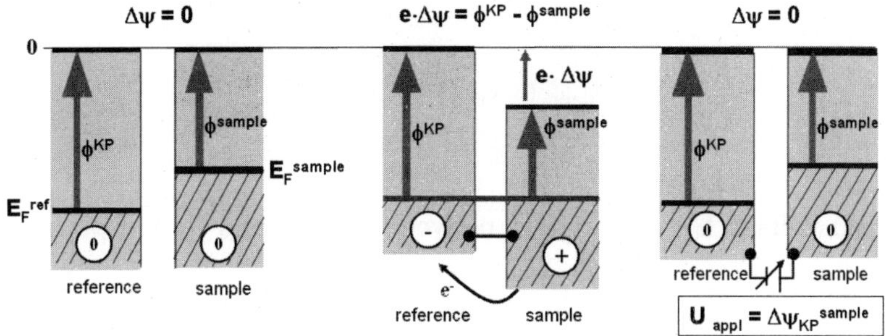

Figure 5 When sample and a reference, the Kelvin probe KP, are brought into contact the equilibration of the Fermi levels results in a Volta potential difference due to the correlated charge transfer (middle). If an external bias is applied between sample and Kelvin probe so that the Volta potential difference is compensated, both surfaces are uncharged again.

$$\tilde{\mu}_e^{KP} = \mu_e^{KP} - e\cdot\varphi^{KP} = -\Phi^{KP} - e\cdot\psi^{KP}$$

and hence:

$$e\cdot(\psi^{sample} - \psi^{KP}) = e\cdot\Delta\psi^{KP}_{sample} = \Phi^{KP} - \Phi^{sample} \quad ...(6)$$

This means that due to the difference of the work functions of sample and reference upon contact an equivalent Volta potential difference is established. The difference in work functions is reflected in non-contact in the difference in Fermi level positions, the energetic position of which in respect to the absolute vacuum level is given by the electrochemical potential.

A Volta potential difference between reference and sample is directly correlated to a corresponding charging of the capacitor formed by Kelvin Probe and the sample. Now, if the reference, i.e. the Kelvin probe, is vibrated over the surface of the sample, i.e. the sample-reference distance is periodically modulated, $d = \bar{d} + \Delta d\cdot\sin(\omega t)$, then this induces a displacement current to flow.

In the simplified expression for the capacity C of an ideal parallel plate capacitor we can write:

$$C = \varepsilon\cdot\varepsilon_0\cdot\frac{A}{(\bar{d} + \Delta d\sin\omega t)}$$

with: ε : dielectric constant of the medium

ε_o : electric field constant

ω : frequency of vibration for the reference metal

A : surface area of the reference plate

and the induced ac-current in the external circuit is:

$$i_{ac} = \Delta\Psi^{KP}_{sample} \cdot \frac{dC}{dt}$$

$$= \Delta\Psi^{KP}_{sample} \cdot (\varepsilon \cdot \varepsilon_0 \cdot A \cdot \Delta d \cdot \omega) \frac{\cos(\omega t)}{(\bar{d} + \Delta d \sin(\omega t))^2}$$

$$\approx \Delta\Psi^{KP}_{sample} \cdot (\varepsilon \cdot \varepsilon_0 \cdot A \cdot \Delta d \cdot \omega) \frac{\cos(\omega t)}{(\bar{d})^2}$$

for $\bar{d} >> \Delta d$

If an external voltage U_{appl} is applied between sample and Kelvin probe (see fig. 4 and fig.5), then:

$$i = (\Delta\Psi - U_{appl}) \cdot \frac{dC}{dt}$$

with: $i = 0$ for $\Delta\Psi = U_{appl}$

In the conventional nulling technique, the voltage U is changed until the current i vanishes, and for this condition $\Delta\Psi^{KP}_{sample} = U_{appl}$ is measured. The accuracy of the measurement is determined by the sensitivity S:

$$S = \frac{di_{rms}}{dU} \approx \frac{\varepsilon \cdot \varepsilon_0 \cdot A \cdot \Delta d \cdot \omega}{(\bar{d})^2}$$

Therefore, for small areas of the reference plate A, as it is the case in Scanning Kelvin Probe microscopy (SKP), sensitivity problems will arise. This can be overcome by decreasing the sample-reference separation, which then allows high resolution SKP.

For decreasing the distance between tip and sample surface to very small values in the range of a few micrometers or even below it is obviously necessary to be able to actively and permanently control this distance in order to prevent tip crashes on asperities of the sample surface. For this the general approach is to maintain the capacity C between tip and sample constant. This is most conveniently done by applying in addition to the externally applied dc bias U_{appl} a small harmonically oscillating ac bias $U_{ac} = U_{ac0} \cdot \sin(\omega_U t)$.

In this case we have

$$i = (\Delta\Psi - U_{appl} + U_{ac}) \cdot \frac{dC}{dt} + C \cdot \frac{dU_{ac}}{dt} = (\Delta\Psi - U_{appl} + U_{ac}) \cdot \frac{dC}{dt} + C \cdot U_{ac0} . \omega_U . \cos\omega_U t$$

and since U_{ac} is negligible in the first term:

$$i = (\Delta\Psi - U_{appl})\cdot\frac{dC}{dt} + C\cdot\frac{dU_{ac}}{dt} = (\Delta\Psi - U_{appl})\cdot\frac{dC}{dt} + C\cdot U_{ac0}\cdot\omega_U\cdot\cos\omega_U t$$

$$\approx (\Delta\Psi - U_{appl})\cdot(\varepsilon\cdot\varepsilon_0\cdot A\cdot\Delta d\cdot\omega)\cdot\frac{\cos(\omega t)}{(\bar{d})^2} + C\cdot U_{ac0}\cdot\omega_U\cdot\cos\omega_U t$$

Hence the amplitude of the discharge (or displacement) current demodulated at frequency ω with a lock-in amplifier is:

$$\hat{I}_\omega = (\Delta\Psi - U_{appl})\cdot(\varepsilon\cdot\varepsilon_0\cdot A\cdot\Delta d\cdot\omega)\cdot\frac{1}{(\bar{d})^2},$$

nulling of which by the external bias U_{appl} directly provides the Volta potential difference, while the amplitude demodulated by a second lock-in amplifier at frequency w_U gives:

$$\hat{I}_{\omega U} = \omega_U\cdot C\cdot U_{ac0}$$

which is at given ac amplitude directly proportional to the capacity *C* between sample and tip, and thus is directly proportional to the inverse distance. If the vertical tip position is adjusted via high precision motors and/or piezo elements to maintain the capacity constant, i.e. the distance constant, a topographical mapping is simultaneously obtained together with the Volta potential mapping of the scanned sample area.

Note that the measurement of the Volta potential difference is independent of the value of C and therefore of the distance between tip and sample. In fact, the measurement of the dependence on distance is a measure for the quality of the experimental set-up: the higher the range of d where the signal is basically independent of the distance, the better the quality of the experimental set-up and the higher the reliability of the obtained data. A high dependence in *d* usually signals the contribution of stray capacities or other stray signals.

The problematic of non-tip area related signal contributions becomes clear from fig.6: capacity contributions from the sides to the overall capacity between tip and sample may be regarded in a first approximation as a sum of small individual contributions to the total capacity C_{tot}:

$$C_{tot} = \sum_n C_n$$

and the signal (i.e. the current amplitude demodulated at ω) becomes

$$i = \sum_n (\Delta\Psi_n - U_{appl})\cdot\frac{\partial C_n}{\partial d}\cdot\dot{d}$$

and nulling requires

$$0 = \sum_n (\Delta\Psi_n - U_{appl})\cdot\frac{\partial C_n}{\partial d} \Rightarrow U_{appl} = \frac{\sum_n \Delta\Psi_n\cdot\frac{\partial C_n}{\partial d}}{\sum_n \frac{\partial C_n}{\partial d}} \qquad ...(7)$$

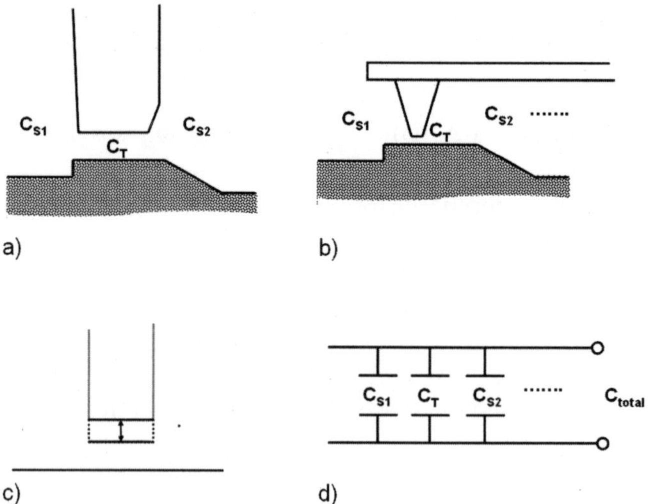

Figure 6 The totally effective capacitance C_{tot} between KP sensor and sample depends strongly on sensor and sample morphology: a) classical metal needle of standard SKP and b) AFM tip of a SKPFM. Especially in the latter case it is obvious that for increased tip-sample distance the contribution of the cantilever will become more and more important. For the classical needle set-up in the ideal case of exactly parallel oscillation to the needle axis only the lower part of the tip contributes to $\partial C/\partial d$ and hence to the signal. d)The contribution of the different capacities can be regarded as an assembly of parallel capacities.

Only if $\frac{\partial C_{tip}}{\partial d}$ is the dominating term, nulling results $U_{appl} = \Delta\psi_{tip}$, i.e. the Volta potential difference between tip and the sample surface right beneath it.

For the case of a perfect cylindrical SKP tip and tip vibration exactly parallel to the cylinder axis, the $\frac{\partial C_n}{\partial d}$ contribution from the shaft is zero except for the contributions at the lower end of the tip, as only there capacity changes with contributions from the shaft occur (see fig.6).

Hence, although it might be assumed that the contribution of the tip shaft becomes more and more important as the distance is increased, in this ideal case the shaft does not contribute to the discharge current. Thusly, high resolution in the order of the tip area is routinely achieved, if the distance between tip and sample is of the same order or smaller.

However, if the tip size is decreased to submicron dimensions the ideal case is difficult to fulfil. In this case it is necessary to shield the shaft by an electrostatic guard in order to obtain reliable high resolution. This can be achieved by first coating the needle with e.g. a thin insulating varnish, followed by a deposition of a thin metal film (by evaporation) and then removing the coatings at the tip by electric sparking through high bias (see Nabhan et al. [9, 35]).

III.2 Scanning Kelvin Probe Force Microscopy (SKPFM)

The principle and technical details of SKPFM are described in detail in other references [10, 11, 36-38]. Shortly summarized, it involves scanning the surface in the AFM tapping mode to determine the topography on a line-by-line basis. The metal-coated or doped silicon cantilever is then lifted a fixed distance from the surface, typically 100 nm, and the tip is rescanned across the surface at this fixed height in "lift mode". On the rescan, the tapping piezo is turned off, and an ac voltage $U_{ac} \cdot sinwt$ is applied to the tip, which stimulates oscillations of the cantilever, as tip and cantilever carry a charge in the presence of the Volta potential difference $\Delta\psi$ over the capacitor formed by tip/cantilever and sample. The magnitude of the oscillations at the stimulating frequency ω, monitored by the AFM detection scheme, is nulled on a point-by-point basis during the lift mode rescan by adding to the tip a dc voltage U_{appl} that balances the Volta potential difference $\Delta\psi$, and thusly the charge.

Hence, unlike to the standard SKP here it is not the displacement current that is nulled but the force exerted by an electric ac field on the charged cantilever/tip. However, as we will see everything discussed above for the case of the classical SKP remains valid.

The electric energy stored in the capacitor formed by tip/cantilever and sample is:

$$W_{cap} = \tfrac{1}{2} \cdot V^2 C \quad \text{and the electric force is} \quad F_e = -\frac{dW_{cap}}{dz} = \tfrac{1}{2} \cdot V^2 C' .$$

With $V = \Delta\psi - U_{ac} + U_{ac} \cdot sin\omega t$ and with $\sin^2 x = \tfrac{1}{2} \cdot (1 - \cos 2x)$ this gives for the total electric force:

$$F_e = \frac{1}{2} \cdot \left\{ (V_{dc} - \Delta\Psi_{sample}^{\mathrm{Re}f})^2 + \frac{1}{2} V_{ac}^2 \right\} \cdot C' + C' \cdot (V_{dc} - \Delta\Psi_{sample}^{\mathrm{Re}f}) \cdot V_{ac} \cdot \sin(\omega t) + \frac{1}{4} \cdot C' \cdot V_{ac}^2 \cdot \cos(2\omega t)$$

Hence the force modulation at frequency ω becomes zero for $V_{dc} = \Delta\Psi_{sample}^{\mathrm{Re}f}$. Thus, nulling of this force component by the applied dc potential indeed yields the sought for Volta potential difference.

More precisely, taking into account the individual contributions of tip and different parts of the cantilever one has to sum up the resulting different force contributions, which again results in equation (7).

Obviously, in SKPFM the signal is expected to strongly depend on tip-sample distance as the contribution of the cantilever becomes more and more important as the distance between sample and tip is increased (see fig. 6). In addition, the measured potential differences on a sample do not represent the real potential differences as the cantilever contribution is difficult to exclude. The passage of the cantilever over substrate areas far from the tip, with inhomogeneous topographic and compositional features, can significantly influence the Kelvin signal and result in incorrect results. Hochwitz et al. report deviations of the measured from the real work function differences between two surface features by a factor of two or more, depending on the tip/cantilever used even an inversion of the potential contrats was observed [39-40]. This fact is important, as the wide-spread availability of SKPFM, which usually is available for most commercial AFM systems, has led to an intense use of SKPFM in Corrosion Science for the characterisation of potential differences on alloy surfaces, which are thought to be an indicator for possible galvanic element formation when the sample is immersed in a corrosive solution. This point will be discussed in more detail further below.

Another important difference to the classical SKP is that in SKPFM the field between tip and sample is not free of electric filed, as a high field ac bias remains (only the dc field is nulled), as the SKPFM generates a rather large ac voltage modulation. High electric ac fields are active, in the order of 5V/10 nm. These extremely high fields might have an effect on the physical conditions of the surface, especially for semiconducting samples, where a quantitative determination of the Volta potential difference can be significantly influenced by this high electric field of the applied ac voltage.

For instance, Sommerhalter et al. [41] found a strong dependence of the measured Volta potential difference on a nondegenerate p-type WSe_2 ($p \approx 10^{16}/cm^{-3}$) single crystal surface. Due to the layered structure, the (0001) van der Waals surface of WSe_2 represents an ideal semiconductor surface free of cleavage-induced defects and surface states. They clearly observed a decrease of the measured Volta potential difference with increasing ac voltage U_{ac}. They contribute this effect to the penetration of the electric field into the semiconductor, which induces band bending and the ac bias is partially rectified at the semiconductor surface, resulting in an apparent reduction or increase of the measured Volta potential difference for p- or n-type doped samples, respectively. Since this effect should strongly depend on the tip-sample distance, the change of the Volta potential difference for decreasing distance was measured in addition. And indeed, a further decrease of the Volta potential difference with decreasing tip-sample distance was observed. Measurements on metallic HOPG did not show this effect which confirms that this effect only occurs on semiconducting samples.

As the surfaces of most metals and alloys are covered by a semiconducting oxide layers, depending on their thickness and charge carrier density in the oxide this effect could also play a role on metal surfaces that emerge from corrosion experiments.

III.3 Applying Kelvin Probe Techniques on layered Systems

Before coming to examples for the application of Kelvin probe, it is helpful to consider what information Kelvin probe techniques will provide on layered systems.

Let us consider the case depicted in fig.7a): a three layer system consisting of metals A and B and the semiconducting coating C. Starting first with the metals A and B, fig.7b) shows that upon establishment of the electronic equilibrium between A and B, i.e. Fermi levels alignment, the Volta potential difference measured with the Kelvin probe technique will be equivalent to the work function difference between the surface of metal B and the Kelvin Probe tip. Electronic equilibrium with the additional semiconducting (here p-type) layer C requires band bending near the interface, as the required charge transfer for the Fermi level alignment of a low work function metal with a higher work function semiconductor causes the formation of a depletion zone in the semiconductor (Mott-Schottky, see e.g. [42]). The lower the charge carrier density in the semiconductor, the more extended the space charge region. Since most organic semiconducting materials have a lower charge carrier density than inorganic semiconductors typically have, much thicker films are often necessary to achieve bulk Fermi level alignment, i.e. for thin films the measured work function might deviate from the real work function of the coating material (see e.g. [43]). In such cases the measured work function is usually referred to as "apparent" work function.

However, if bulk Fermi level alignment is fulfilled, the Volta potential difference measured by the Kelvin probe on the ABC system will be just the work function of the topmost layer C.

For organic semiconductor/metal interfaces often the formation of interfacial dipole layers is observed [see e.g. 43-44]. As shown in figure 8, the Kelvin probe technique is insensitive to

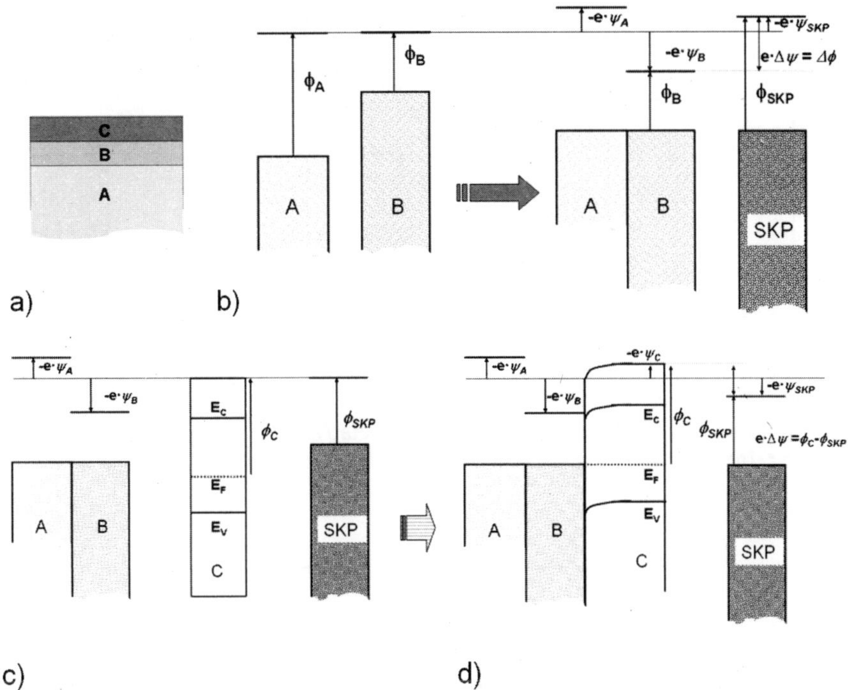

Figure 7 **a)** Consider a metal A coated by a layer of metal B and semiconductor C. **b)** The charge transfer required for the Fermi level alignment between A and B causes shift of the quasi vacuum levels vs. the absolute vacuum level (thin line), i.e. Volta potentials develop. Fermi level alignment of the SKP with metal A (as the SKP electrically contacts the layered system through A, see figs. 4 and 5) causes an according shift of the quasi vacuum level of the SKP. As a result the measurement of the Volta potential difference between the surface of B and the SKP corresponds to the according work function difference. **c)** Equilibration with the Fermi level of the semiconducting layer C (p-type) results in **d)**. If layer C is thick enough so that band bending establishes the bulk Fermi level position, the SKP will measure the correct work function of C. If the layer is thinner than that, a too low apparent work function will be measured.

these interfacial dipoles, if Fermi level alignment is achieved and the coating thickness is large enough to establish bulk Fermi position, since then the effect of the dipole layer just causes a different band bending.

For organic semiconductor coatings of very low conductivity, i.e. with a very low charge carrier density, the band bending across the thin coating will not reach the bulk Fermi level alignment and the apparent work function will with decreasing film thickness increasingly approach the value of the work function of the underlying metal, modified by Δ of the dipole layer.

This is fully the case if the Fermi level alignment is not achieved. Then the topmost layer is practically invisible to the Kelvin probe, and the measured Volta potential difference corresponds to the difference in work function between the underlying metal and the Kelvin probe, modified by the potential drop across the interfacial dipole layer Δ/e, because the Fermi level of the Kelvin probe is aligned with the Fermi level of the metal and the quasi vacuum level

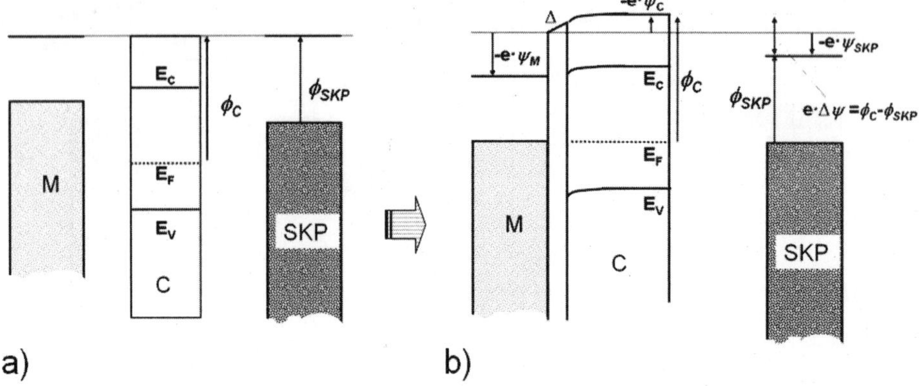

Figure 8 The formation of a dipole layer at the interface metal/organic semiconductor just causes a modified band bending. The Volta potential difference does not detect the dipole layer at the buried interface.

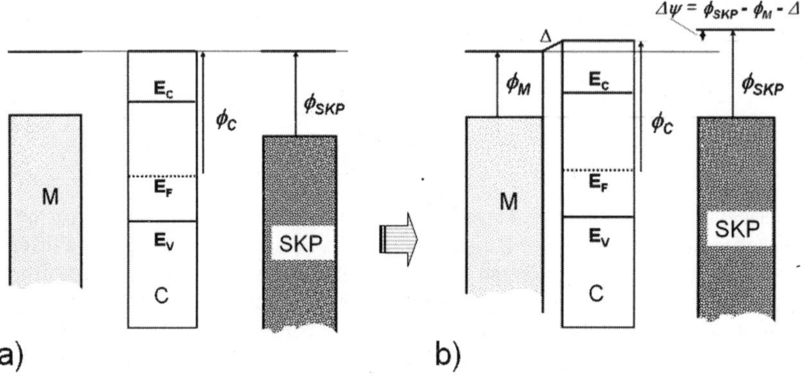

Figure 9 If Fermi level alignment of the organic semiconductor and the metal is inhibited, the SKP will measure the work function difference to the underlying metal, modified by the dipole layer. In the general case there will be also an additional dipole potential χ_C at the surface of the coating, not shown here and in fig.7-8.

near the surface of the semiconducting layer just differs by Δ from the one of the metal (see figure 9).

In such cases the apparent work function measured with the Kelvin probe depends strongly on the underlying metal.

For insulating coatings this is the general case, because there are no charge carriers available for the alignment of Fermi levels. Hence, for insulating coatings the case described in fig. 8 merges into the one described in fig. 9.

IV. COMPARISON OF SKP AND SKPFM FOR APPLICATION IN CORROSION SCIENCE

It is useful to consider the main advantages of SKP and SKPFM over conventional reference electrodes.

The following advantages of SKP are obvious:

i. The electrode potentials of surfaces which are covered by extremely thin electrolyte layers can be studied. This is not possible with conventional reference electrodes, which disturb the composition and thickness of the electrolyte film at the point of measurement. This is of importance for studies on atmospheric corrosion, fuel cells or all other electrodes of electro-catalytic importance, and corrosion of electronic materials.

ii. Since the Kelvin probe technique is characterized by the separation of reference probe and sample by a dielectric, local corrosion and electrochemical processes in general beneath insulating films, such as organic or inorganic coatings, can be studied.

iii. Hence, localised corrosion phenomena on surfaces covered by highly resistive films or electrolytes can be monitored. This property may be used to detect defects such as inclusions in metallic matrices or to analyze corrosion in the presence of such defects.

iv. SKP can also be applied in dry environments for determining changes in the surface potential of a sample.

However:

The Kelvin Probe shows no improvement with respect to conventional reference electrodes, if the surface is covered by low resistive films or thick electrolyte layers. Under these circumstances, potential coupling between different surface sites will result and the Kelvin probe will loose its spatial resolution and detect only averaged electrode potentials as any other reference electrode.

The following advantages of SKPFM are obvious:

v. The extremely high resolution which is in principle given by the tip-samples distance, i.e. in the range of 20-30 nm.

vi. Its availability for most AFMs

However:

SKPFM is not applicable for measurements on samples covered by liquid films, because the large ac voltages applied to the tip may cause Faradaic reactions in solution and AFM on thin liquid layers is very difficult to perform due to the high capillary forces.

Also, long-term measurements at high humidity are a problem for the piezo scanner, as the insulation my slowly fail and current creep may occur, significantly shortening the scanner lifetime.

Another problem with SKPFM is the limitation of the field of view. AFM piezoelectric scanners typically are limited to 100 mm in range or smaller. Relevant electrochemical processes, especially in delamination of organic coatings from metal, sometimes take place on a larger scale more appropriate for a standard SKP instrument.

The tips are only pseudo-references since their potential may vary with changes in their surface oxide. Although this is in general also true for the classical SKP needle, the latter ones are usually by far more stable and it is possible to prepare SKP needles with extremely high potential stability, e.g. based on Ag/AgCl (see Ehahoun et al. [45]).

Also, errors arising from cantilever contributions and as a consequence of the high ac fields can lead to faulty measurements and even contrast inversion (as discussed above).

Summarizing, the range of applications is by far more versatile for SKP than for SKPFM. In fact, in Corrosion Science SKPFM is nearly exclusively applied for potential mappings of alloy surfaces at low humidity.

V. APPLICATION OF SKP AND SKPFM IN CORROSION SCIENCE

The measurement of $\Delta\psi^{KP}_{sample}$ has many important applications. As we have seen above (see eq. (4)):

$$(\psi^{KP} - \psi^{S}) = E_{abs} - \frac{1}{e}\cdot\Phi^{KP} = \left(-\frac{\mu_e}{e} + \Delta\varphi_M{}^{El} + \chi_s\right) - \frac{1}{e}\cdot\Phi^{KP}, \qquad ...(8)$$

i.e. for an electrolyte covered sample the measurement of the Volta potential difference between Kelvin Probe and electrolyte surface directly provides the absolute electrode potential, if the work function of the Kelvin Probe is known. Even more convenient is to directly calibrate the Kelvin Probe to a standard reference electrode, which will be discussed in the next section. Then scanning over an electrolyte film covering the sample of interest will directly provide a local mapping of the corrosion potential distribution beneath it and thusly information about local galvanic elements, e.g. at active inclusions. This interpretation, of course assumes that χ_S remains constant, which is usually the case in a typical corrosion experiments.

In surface physics, to the contrary, the change of the Volta potential difference between sample and Kelvin probe provides information about adsorption of molecules or surface reconstruction, which both will cause changes in the surface potential χ of the sample.

$$\psi^{KP} - \psi^{S}) = \left(-\frac{\mu_e}{e} + \chi\right) - \frac{1}{e}\cdot\Phi^{KP}, \qquad ...(9)$$

As can be seen from these two examples, depending on the kind of experiments either buried interfaces, as e.g. in the case of an electrolyte covered sample the solid/liquid interface, or the surface itself, i.e. $\Delta\varphi_M{}^{El}$ resp. χ, determine the changes in the measured Volta potential difference.

Hence, the interpretation of the measured data requires a thorough understanding of the system that is investigated. For some important examples this will be discussed in more detail in the following.

V.I Examples for the Application of SKP in Corrosion Science: Measurement of Corrosion Potentials

A) The wet surface

In Corrosion Science the potential of the freely corroding interface is called corrosion potential E_{corr}, and this term is also used for the open cell potential of a passive surface, which are not (or practically not) corroding.

Since in practice the measured electrode potential E_{corr} (as in Corrosion Science the potential of the non is referenced to a standard reference electrode, we may write:

$E_{corr} = E_{abs}^{corr} - E_{abs}^{standard}$, where E_{abs}^{corr} and $E_{abs}^{standard}$ denote the corresponding potentials in the absolute electrode potential scale.

This gives (see eq. 8):

$$(\psi^{KP} - \psi^{S}) = \left(-\frac{\mu_e}{e} + \Delta\varphi_M^{El} + \chi_s\right) - \frac{1}{e}\cdot\Phi^{KP} = E_{abs}^{corr} - \frac{1}{e}\cdot\Phi^{KP} = E_{corr} + E_{abs}^{standard} - \frac{1}{e}\cdot\Phi^{KP} \quad ...(10)$$

Since $E_{abs}^{standard}$ and Φ^{KP} are constant, measured changes in $(\psi^{KP} - \psi^{S})$ are directly correlated to E_{corr} vs. a given standard reference electrode, i.e.:

$$E_{corr} = (\psi^{KP} - \psi^{S}) + \text{const.(standard)}, \quad ... (11)$$

where the constant depends on the chosen standard reference electrode.

The value for const.(standard) can be readily obtained by calibrating the Kelvin probe by means of a couple of reference systems of well known corrosion potential, as e.g. the Cu/Cu^{2+}, Zn/Zn^{2+} etc. electrodes, or by direct calibration with a chosen reference electrode as shown in fig.10.

The Kelvin probe cannot only be applied for measurements of electrode potentials of metals covered by ultrathin electrolyte layers, but also for an active potentiostatic control of this surface and hence for performing electrochemical experiments on such samples, which is of great significance for fundamental studies of atmospheric corrosion [46].

B) The dry surface

On dry metal or oxide surfaces the measurement of the Volta potential difference is just the measurement of the work function difference (eq. (6)):

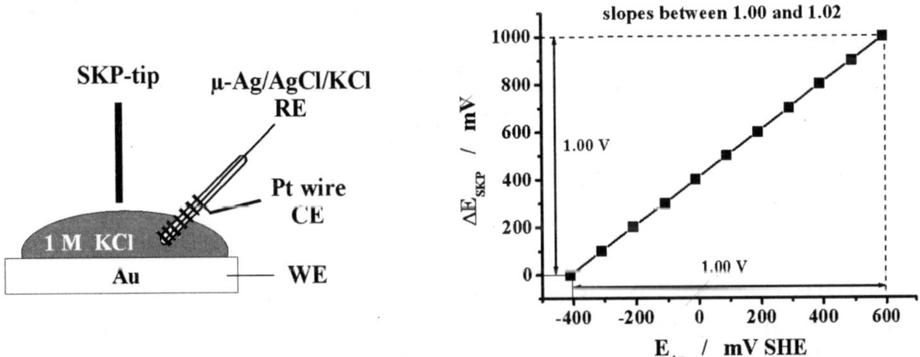

Figure 10 A convenient method for the calibration of the Kelvin probe is to position it over the surface of a droplet. The potential of the surface beneath it is adjusted to different values by a standard electrochemical set-up consisting of a micro reference electrode, a small counter electrode and a potentiostat. A typical calibration curve is shown on the right. As can be seen, there is a linear relation between measured Kelvin probe signal and the potentiostatically applied electrode potential. In practice, it is not necessary to perform such a calibration very often. Once the correct functioning of a SKP set-up has been checked this way, it is sufficient to calibrate the offset value (here it is about 400 mV). This is done by positioning the SKP over a small Cu beaker filled with $CuSO_4$ solution of well known concentration (usually saturated) and referencing the SKP signal to the corresponding Cu Cu^{2+} potential [45].

$$e\cdot(\psi^{sample} - \psi^{KP}) = e\cdot\Delta\psi^{KP}_{sample} = \Phi^{KP} - \Phi^{sample}$$

However, if the Kelvin probe has been calibrated to SHE, the measured work function will be referenced as electrode potential vs. the standard hydrogen potential, low work function materials (such as zinc) will be correlated with a more negative electrode potentials than high work function materials (such as gold). This may feel unusual, but it is very concenient for the discussion of corrosion mechanisms, as will be discussed further below.

The surfaces of reactive metals are covered by an oxide layer, which may be electronically conductive, semiconducting or insulating or consist of layers of different conductivity. During oxide growth an electric field forms in the oxide. An overview about the different growth mechanisms and the resulting field distribution can be found in [47-49].

The potential drop over the oxide is correlated to different chemical potentials in the oxide at the internal interface (x = 0) and at the surface (x = d), since the electrochemical potential need be constant:

$$\mu_e^{ox}(0) - e\cdot\varphi(0) = \mu_e^{ox}(d) - e\cdot\varphi(d)$$

$$\Rightarrow \quad \Delta\varphi_{ox} = \varphi(d) - \varphi(0) = 1/e\cdot(\mu_e^{ox}(d) - \mu_e^{ox}(0))$$

The different chemical potential of the electron at different positions in the oxide is due to the different composition of the oxide at different distances from the metal. Since the electrochemical potential is constant across the oxide.

Graphically, the oxide layer may be considered as consisting of a series of thin oxide slabs, with gradually changing Fermi level (i.e. electrochemical potential). Bringing them together causes band bending as schematically shown in fig. 11.

However, the Kelvin probe only provides information about the position of the Fermi level at the surface of the oxide as long as there is electronic equilibrium is established across the metal/oxide interface and across the oxide layer (see fig. 11).

The position of the Fermi level in the band gap depends on the density of donor and acceptor sates in the band gap. For an n-conducting semiconductor with clearly predominating donor state density the position of the Fermi level measured from the conduction band edge can be calculated from (see e.g. [42]):

$$E_C - E_F = kT\cdot\ln\left(\frac{Nc}{N_D}\right) \qquad ...(12)$$

where N_C is the so called effective density of states and N_D is the donor density.

For instance, in the passive film on iron the Fe^{2+} states are the donor states, i.e. the concentration of Fe^{2+} will directly determine the position of the Fermi level at the surface. Now, as we have seen in the discussion above electrode potentials and work functions are naturally linked. Hence, the position of valence and conduction band edges, as well as the position of the Fermi level, can not only be referred to the vacuum level but also to the standard hydrogen electrode (SHE). The position of E_C and E_V for a number of practically important oxides referenced to the vacuum level as well as SHE can e.g. be found in Schultze and Hassel [49].

In this context, the resemblance of equation (12) with the Nernst equation tells us that we may consider the oxide as a solid state electrode where oxidized states (e.g. Fe^{3+}) are in equilibrium with reduced states (e.g. Fe^{2+}).

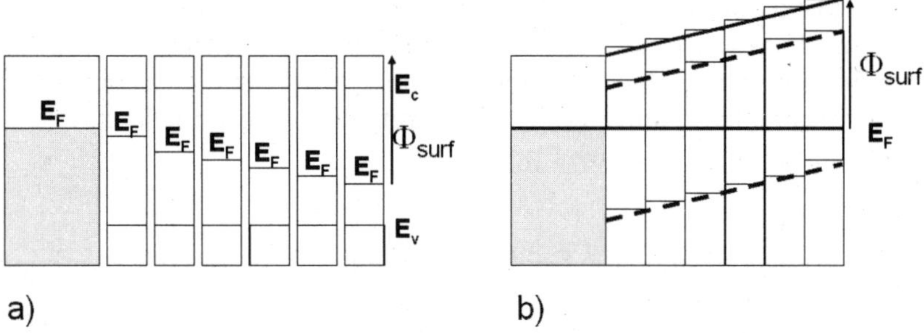

Figure 11 a) The oxide scale on the metal can be considered to be composed of slabs of different donor/acceptor ratio, i.e. of different position of the Fermi level within the band gap between conduction band and valence band. **b)** In electronic equilibrium the Fermi level in the metal and everywhere in the oxide is the same. The work function that will be measured by a Kelvin probe is Φ_{surf}, the work function of the topmost oxide layer, i.e. in other words the position of the Fermi level in the topmost oxide layer is measured. The Fermi level at the surface is always lower than at the metal/oxide interface, corresponding to a higher oxidation stage. The exact shape of the band structure across the oxide depends on the growth mechanism of the oxide.

Indeed, with eq. (8):

$$(\psi^{KP} - \psi^{S}) = E_{abs} - \frac{1}{e}\cdot\Phi^{KP} = \left(-\frac{\mu_e}{e} + \Delta\varphi_M^{\ El} + \chi_s\right) - \frac{1}{e}\cdot\Phi^{KP}$$

and with the equilibrium potential of Fe^{2+} to Fe^{3+} in the iron oxide

$$E_{abs}\left(Fe^{2+}/Fe^{3+}\right) = -\frac{\Delta\mu^0_{Fe^{2+}/Ee^{3+}}}{F} + \frac{RT}{F}\ln\frac{[Fe^{3+}]}{[Fe^{2+}]} + \chi_{ax}$$

$(\psi^{KP} - \psi^{S})$ becomes:

$$\Delta\Psi^{KP}_{ax} = -\frac{\Delta\mu^0_{Fe^{2+}/Fe^{3+}}}{F} + \frac{RT}{F}\ln\frac{[Fe^{3+}]}{[Fe^{2+}]} + \chi_{ax} - \frac{\Phi_{KP}}{F}$$

$$\approx const. + \frac{RT}{F}\ln\left(\frac{[Fe^{3+}]}{[Fe^{2+}]}\right) \equiv const! + \left(\frac{E_C - E_F}{e}\right) = const! + \frac{kT\cdot\ln\left(\frac{N_C}{N_D}\right)}{e}$$

where the concentrations refer to the surface concentrations.

Such a correlation between the oxidation state in iron oxide and the work function of the oxide was indeed experimentally found [50].

In a simple understanding of this equation the oxide surface on iron will be oxidized in air until the oxidation level and therefore the Volta potential is high enough to prevent any further oxygen reduction.

In other oxides where the metal cation has nominally only one oxidation state N_C/N_D will be determined by defects concentrations (vacancies, impurities etc..).

For many passive oxides formed at room temperature the defect concentration in the oxide is quite high. As most of these oxides are n-semiconductors this means that the Fermi level will be near the conduction band edge, i.e., since in most cases the conduction band edge shifts upwards with increasing band gap (see Schultze and Hassel [48]), the larger the band gap, the higher E_F, which corresponds to a lower potential.

C) The wet oxide surface

If the metal is oxide covered, then the Galvani potential difference between the metal and the electrolyte is composed of three terms:

$$\Delta\varphi_M{}^{El} = \Delta\varphi_M{}^{ox} + \Delta\varphi_{ox} + \Delta\varphi_{ox}{}^{El} \qquad ...(13)$$

taking into account the potential drops at the two interfaces and the potential drop across the oxide phase itself. However, as long as only the relation between the corrosion potential and the Volta potential difference is concerned, all potential drops will contribute to the electrode potential and the Volta potential difference in the same way and therefore equation (10) still holds. Neither the electrode potential nor the Volta potential difference can distinguish between an oxide-covered and an oxide-free surface.

D) Polymer covered metal (oxide) surfaces

In the application of SKP for studying delamination the tip will scan over the surface of a organic coating with already delaminated, delaminating and still intact interface beneath it. Also here the measured signal is in all cases referenced to a standard electrode potential, conveniently SHE, although only the values measured on the coating at the delaminated and delaminating interfaces would be thought to have a direct association with electrochemistry. This may seem a little bit confusing, but in fact one may also ask where there should be the difference between the electrical double layer at the metal/solution interface and the potential drop at the metal/polymer interface?

As discussed above (see fig. 9), scanned above the surface of an insulating organic coating SKP will measure the work function of the metal beneath it, modified by a dipole layer at the metal/organic interface and the surface potential of the organic coating.

As E_{corr} of a solution covered surface is (eq. (10)):

$$(\psi^{KP} - \psi^{S}) = \left(-\frac{\mu_e}{e} + \Delta\varphi_M{}^{El} + \chi_s\right) - \frac{1}{e}\cdot\Phi^{KP} = E_{corr} + E_{abs}^{standard} - \frac{1}{e}\cdot\Phi^{KP}$$

By replacing the "El" designating electrolyte by "Pol" for polymer (and keeping the "S" in χ_s, denoting "surface") we obtain:

$$(\psi^{KP} - \psi^{S}) = \left(-\frac{\mu_e}{e} + \Delta\varphi_M{}^{Pol} + \chi_s\right) - \frac{1}{e}\cdot\Phi^{KP} \qquad ...(14)$$

At the delaminated interface a thin layer of electrolyte is formed (see fig. 12). In that case an additional term has to be added, $\varphi_{El} - \varphi_{Pol} = \Delta\varphi_{El}{}^{Pol}$, the potential drop at the electrolyte/polymer interface.

This potential drop is determined by the so called Donnan potential $\Delta\varphi_D$: $\Delta\varphi_D = \varphi_{Pol} - \varphi_{El} = -\Delta\varphi_{El}{}^{Pol} = f(c)$, which depends on the concentration of fixed charged

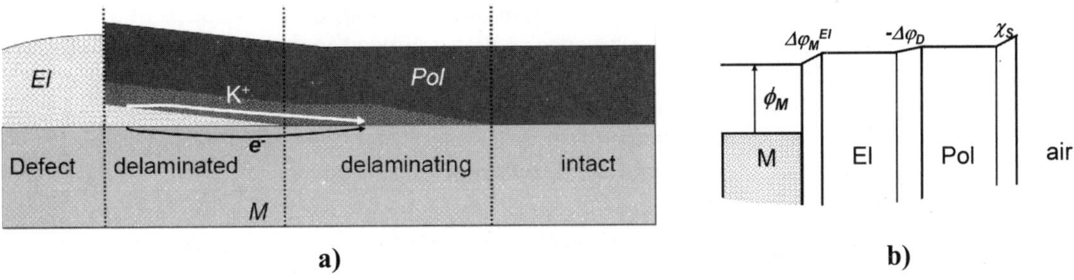

Figure 12 a) Polymer coated metal sample with a defect covered by electrolyte. At the defect site the metal is corroding: $Me \rightarrow Me^{z+} + ze^-$. Migration of cations along the interface allows a corresponding flow of electrons to the delaminating interface, thusly enabling oxygen reduction. This causes interfacial degradation which turns the interfacial polymer gradually into a gel [20]. **b)** At the delaminating site three interfaces are crossed when going from the metal bulk to the air above the surface. At the delaminating site the degrading polymer may considered as a kind of electrolyte (of very low ionic strength near the intact interface).

groups in the polymer and on the concentration of the electrolyte (see below). Adding this to eq.14 leads to:

$$(\psi^{KP} - \psi^{S}) = \left(-\frac{\mu_e}{e} + \Delta\varphi_M{}^{El} - \Delta\varphi_D + \chi_s\right) - \frac{1}{e}\cdot\Phi^{KP} \qquad ...(15)$$

$$= E_{corr} + E_{abs}^{standard} - \frac{1}{e}\cdot\Phi^{KP} - \chi_{El} - \Delta\varphi_D + \chi_s$$

$$\text{with } \Delta\varphi_D = \frac{zRT}{F}\ln\left(\frac{x}{2} + \sqrt{1 + \left(\frac{x}{2}\right)^2}\right) \approx \frac{zRT}{2F}\cdot x \text{ for } x << 1 \text{ and} \approx \frac{zRT}{F}\cdot \ln x \text{ for } x >> 1 \qquad ...(16)$$

where E_{corr} is the corrosion potential referenced vs. a standard, e.g. SHE, at the buried interface referenced vs. a standard, e.g. SHE, $E_{abs}^{standard}$ is the absolute electrode potential of this standard according to eq.(4), z is the valence of the ions in the assumed (1:1) electrolyte, and x is the ratio between the fixed charge carrier density in the polymer and c the concentration of the electrolyte [51-52].

Since Φ^{KP}, χ_{El} and χ_S are constants, a suitable calibration of the measured Volta potential difference will directly derive the electrode potential at the buried interface. For this, a film of the respective polymer is placed on the surface of an electrolyte covering a metal surface. The electrolyte is connected to standard electrochemical cell. Thus the electrode potential of the metal in the electrolyte and the Volta potential of the polymer surface can be measured simultaneously by a standard reference electrode in the electrolyte and Kelvin probe over the polymer surface (see Leng et al [20]). With this procedure it is possible to assess the susceptibility of a special organic coating for developing Donnan potentials, i.e. whether it contains a non negligible amount of fixed charge carriers.

Leng et al. [20] could show that the Donnan potential of the investigated polyfunctional epoxylester cured with polyamidoamines (which contained groups of -OH, -COOH, -CONH and -NH-) did not lead to significant Donnan potentials, although these groups in the coating caused a considerable swelling in the presence of water. Since the delamination behaviour of this coatings investigated by SKP was found to be quite similar to those of commercial nonpigmented clearcoats with different organic chemistry, it can be assumed that this is also valid for most coating systems. This means that the electrolyte concentration is much larger than the density of fixed charge carriers in the polymer, i.e. $x << 1$.

However, during delamination very alkaline electrolytes are formed at the metal/polymer interface and radicals such as $OH^{\bullet}$ are generated as side products of the oxygen reduction at the interface. Under these circumstances the interfacial polymer is destroyed, via e.g. hydrolysis of ester groups and oxidative attack by the radicals, and new functional groups such as COO^- may be formed which may give rise to a locally high density of fixed ionic groups. Then the Donnan potential may become significant if the electrolyte concentration is low. Leng et al. [20] found for the investigated polymer a Donnan potential development from about 0 mV to 80 mV caused by increasing the pH from pH 9 to pH 14, indicating such an increase in fixed charge carriers (of net negative charge) in the polymer, obviously caused by the pH increase.

Hence, at intermediate stages of interfacial degradation (when the ionic concentration of the electrolyte at the interface may still not be very high) x may be high enough to cause a certain contribution. However, in practical application of SKP for investigating delamination usually there is no indication for substantial contribution from Donnan potentials. Since χ_S and χ_{El} are both not very high (the surface potential of polymer is usually about 50 mV [50]) and are about the same magnitude, $\chi_S - \chi_{El}$ in eq. (15) will in most cases also be very small.

Consequently, according to eq. (15) the measurement of the Volta potential difference between the surface of the polymer and the Kelvin probe will directly provide the corrosion potential at the buried interface. In this case it is sufficient to calibrate the Kelvin probe over electrolyte/metal (see fig. 10).

Fig.13 shows the typical delamination behaviour of a coated zinc sample. During the scan from the defect over the polymer coated metal the tip of the SKP passes different sites at the buried interface: the defect (first dry, then corroding), corroding area at the delaminated interface near the defect (very low potential around -900 mV_{SHE}), delaminating interface where the potential makes a steep increase towards the potentials of the intact interface, and the intact interface.

Although for the case of coated zinc both anodic and cathodic reactions occur at the interface, the delamination has more or less the character of a cathodic delamination process, as at the front the process is predominantly cathodic [53-55]. This cathodic front is followed by an anodic one, because the zinc oxide dissolves at the very high pH caused by the oxygen reduction at the interface.

For coated iron, the delamination process is purely cathodic, because its oxide is stable at these high pHs [20-22, 56-59]. The mechanism of cathodic delamination is described in detail elsewhere [20-22, 56-59]. The principle is sketched above in fig. 12: the oxygen reduction at the interface is enabled by the migration of cations along the buried interface. This migration of cations determines the possible electron flow into the intact interface (as the coating at the intact and nearly intact interface prevents metal dissolution all electrons for the oxygen reduction have to come from the defect), i.e. it closes the electric circuit. This migration of

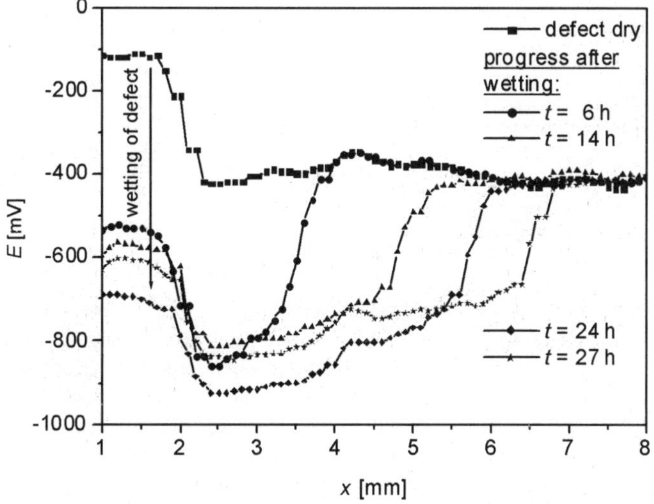

Figure 13 Potential profiles measured with SKP above a coated zinc sample. The first line was measured before wetting the defect (between 1-2 mm). The potential at the intact interface is -400 mV_{SHE}, the typical value for freshly prepared native zinc oxide. At the dry defect site the potential higher because here a thicker oxide layer has formed due to ageing (at the intact interface ageing is inhibited by the coating). Upon wetting the defect potential decreases to the potential of corroding zinc. The potential decrease at the interface shows the progress of the delamination (see Rohwerder et al., Galvatech (2001) 585-592).

cations is directed from the negative corrosion potential to the more positive ones, because the Galvani potential in the polymer (or the gel like electrolyte it is transformed to by the degradation) becomes more negative in this direction (see fig. 14):

$$E_{corr}(x_2) - E_{corr}(x_1) = \Delta\varphi_M^{Pol}(x_2) - \Delta\varphi_M^{Pol}(x_1) = (\varphi_M - \varphi_{Pol}(x_2)) - (\varphi_M - \varphi_{Pol}(x_1))$$
$$= -(\varphi_{Pol}(x_2) - \varphi_{Pol}(x_1))$$

This is of great importance for the development of improved coating systems: if it were possible to invert the potential difference between intact interface and defect, the fast cathodic delamination is not possible any more. Then delamination can only occur via the much slower anodic delamination mechanism [60].

This was for the first time successfully achieved on coated zinc-magnesium alloys, namely on the intermetallic $MgZn_2$ (see fig. 15 [59]). For this intermetallic the potential of the passive oxide surface is more negative than the corrosion potential of the NaCl electrolyte covered surface. This is unusual as the potential where passivation of the metal occurs is always at more anodic potentials than the corrosion potential (see fig. 14a).

As discussed above, for the corrosion scientist it is convenient to reference the work function, or more accurately the position of the Fermi level in the band gap, to a standard reference electrode, such as SHE, because that is where the Kelvin probe is usually calibrated to. When the passivated metal is emersed from the solution at a potential within the passive potential range, the measured potential (i.e. work function) will be quite similar to the potential of the electrode it had when it was immersed within the passive potential range. $\Delta\varphi_{Ox}^{El}$ in the

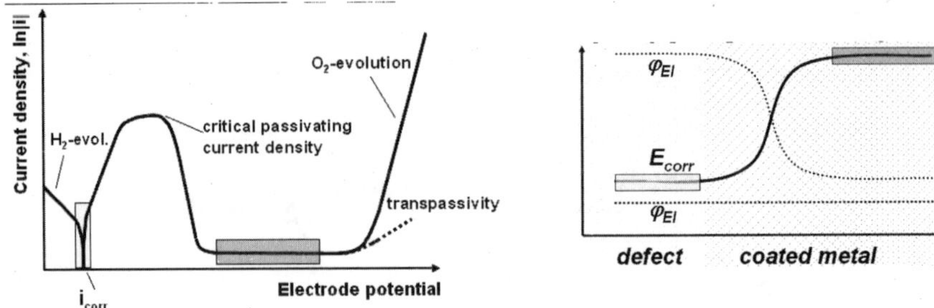

Figure 14 **a)** Current-potential curve for the passivation metal: the passive oxide prevails at potentials more positive than the active corrosion of the metal. **b)** The potential at the polymer/metal (oxide) interface is determined by the work function of the passive oxide (or its equivalent potential value), while that at the defect is determined by the active corrosion of the metal (see also the text below). When delamination couples the potential at the delaminating interface with the much lower on in the defect, the typical potential profile observed with SKP develops. The Galvani potential in the degrading polymer gel is lower where the potential is higher (see test below), thus providing the bias required for the migration of cations into the interface. If the potentials were inverted, no cation migration from the defect could occur).

passive range is small (at the isoelectric point) and hence according to eq. 13 $\Delta\varphi_M^{El} \approx \Delta\varphi_M^{Ox} + \Delta\varphi_{ox}$, i.e. the potential drop between metal and electrolyte is more or less the potential drop between metal and oxide and across the oxide, which determines the work function of the oxide).

Now the passive oxide of $MgZn_2$ is dominated by low work function magnesium oxide and hydroxide, while the corrosion potential in the defect is determined by the corrosion of the zinc, because the less noble magnesium is quickly dealloyed from the defect [59]. Thusly, the corrosion potential in the defect is determined by the nobler component, i.e. Zn, and the potential at the intact interface by the less noble element, i.e. Mg. Only in this way it is possible to invert the potential difference between intact interface and defect. Then no cation transport is possible (see fig. 14 b).

If however the defect is polarized artificially below the potential of the intact $MgZn_2$/ polymer interface, for example by placing magnesium in the defect (which corrodes in NaCl solution around -1400 mV_{SHE}) a fast cathodic delamination is observed (see fig. 16 [60]).

Hence, in the development of novel alloys for improved corrosion protection both, the potential of the passive oxide and the corrosion potential of the metal have to be carefully tailored.

Hence, a positive potential between intact interface and corrosion potential in the defect bears intrinsically the driving force for cathodic delamination. As was shown one approach is to try to invert this potential difference as it was successfully achieved for $MgZn_2$. Another approach is to use the potential decrease at the delaminating polymer/metal interface as trigger signal for intelligent inhibitor release.

Such an intelligent self-healing can be achieved by adding particles of inhibitor anion doped polypyrrole to the organic coating [61]. When corrosion at a defect occurs and a delamination front starts to precede into the intact interface, the pulling down of the electrode potential at the interface by coupling with the more negative corrosion potential at the defect causes reduction of the polypyrrole: $PPy^+A^- + e^- \rightarrow PPy + A^-$.

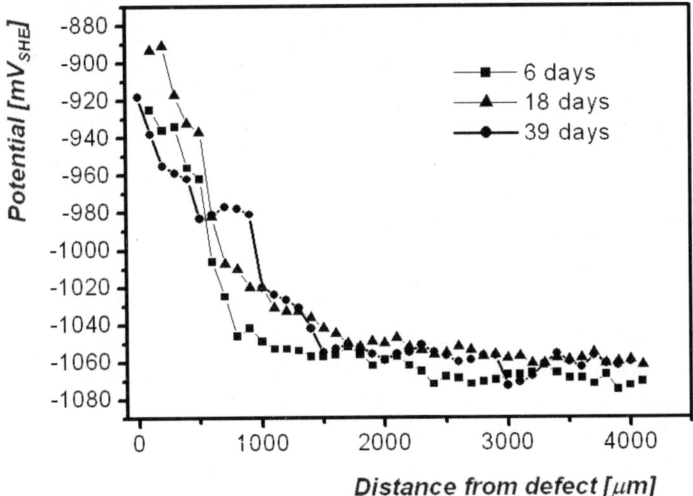

Figure 15 Potential profile measured over the surface of polymer coated $MgZn_2$. Only next to the defect (d < 1000 μm) the interface seems to be affected by the corrosion in the defect (d < 0). Even after more than a month no significant changes can be observed. Hence, the delamination progress is very slow and as interfacial analysis shows of definitely anodic character [60].

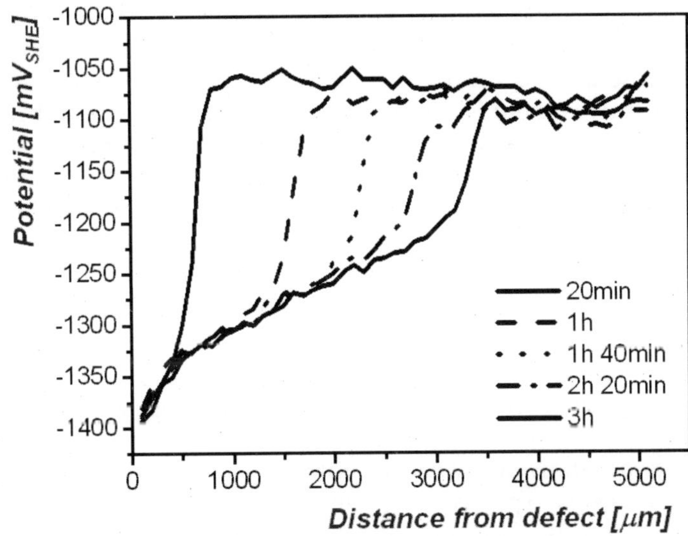

Figure 16 Potential profile measured over the surface of polymer coated $MgZn_2$, with corroding Mg in the defect. Now cathodic delamination is possible and, indeed, very fast cathodic delamination is observed in this case [60].

The anions migrate along the interface to the defect (just as cations migrate in the opposite direction) and passivate the defect, which results in an increase of the corrosion potential. Thusly, the potential difference between intact interface and defect decreases and the delamination is stopped [61]. This just shows how the corrosion protection by conducting polymers can work. However, conducting polymers can also promote corrosion (e.g. a coating of pure conducting polymer in contact with a larger defect becomes quickly reduced and hence in that case does not provide any corrosion protection) and great care has to be taken for tailoring coating systems based on corrosion protection by conducting polymers [61, 62].

V.2 Examples for the Application of Scanning Kelvin Probe Force Probe Force Microscopy (SKPFM) in Corrosion Science

In principle SKPFM can be applied to the same problems as SKP and should also provide the same information as the standard Kelvin probe. However, as already discussed in section *IV: " Comparison of SKP and SKPFM for application in Corrosion Science"*, SKPFM has several limitations (e.g. it is not applicable over thin electrolyte films, is difficult to use for long term measurements at high humidity and has a scan width). Also, it should be kept in mind that it may deliver faulty contrast due to cantilever contributions and the effect of the high ac fields.

But on the other hand, standard Kelvin probe does not allow a high enough resolution necessary for the study of microscopic aspects of delamination. However, such information is necessary for a real fundamental of delamination processes. For instance, one important question concerning cathodic delamination is how exactly the process is initiated at the delamination front. For Filiform corrosion on aluminium alloys another important question is in how far the distribution of microscopic intermetallic particles determines the growth dynamics of the Filiform filaments. In both cases scanning Kelvin probe force microscopy (SKPFM) can play an important role in deriving new information on delamination processes on the submicroscopic scale. Just like it is the case for all scanning Kelvin probe (SKP) techniques also for SKPFM the resolution is strongly dependent on the distance between tip and surface. Since the surface of interest here is the interface between polymer and metal surface, this requires the preparation of special model samples that are characterised by ultrathin polymer coatings[1] and specially prepared defects that show a very sharp borderline to the intact coating. One step even further is the preparation of model samples with molecularly highly ordered interfaces between polymer and metal. This allows studying the correlation between the detailed processes of delamination and the molecular structure at the buried interface and will hopefully open the possibility for a future computational simulation of the involved processes.

But because of the involved difficulties, only very few attempts to use SKPFM for the study of delamination have were made so far. The first was the work of Rohwerder et al. [64] where the delamination behaviour of ultra-thin plasma polymer coatings with and without organized molecular monolayers at the interface was studied. SKPFM and correlated ToF-SIMS studies of the delaminated interface showed that the delamination zone can be as small as a few microns and that delamination seems to proceed step-wise in discrete microscopic steps [64].

Another example for the application of SKPFM to delamination problems is the investigation of delamination initiated by nanoscopic salt impurities at the buried interface (see fig. 17, see [65]).

---

[1] Although experiments with a height controlled SKP have shown that the capacity between Kelvin probe and sample surface is the one between polymer surface and tip, and not between metal surface and tip [63], experience shows that the resolution is determined by the distance between tip and metal surface. Furthermore, it is affected by the kind of polymer applied to the metal [59].

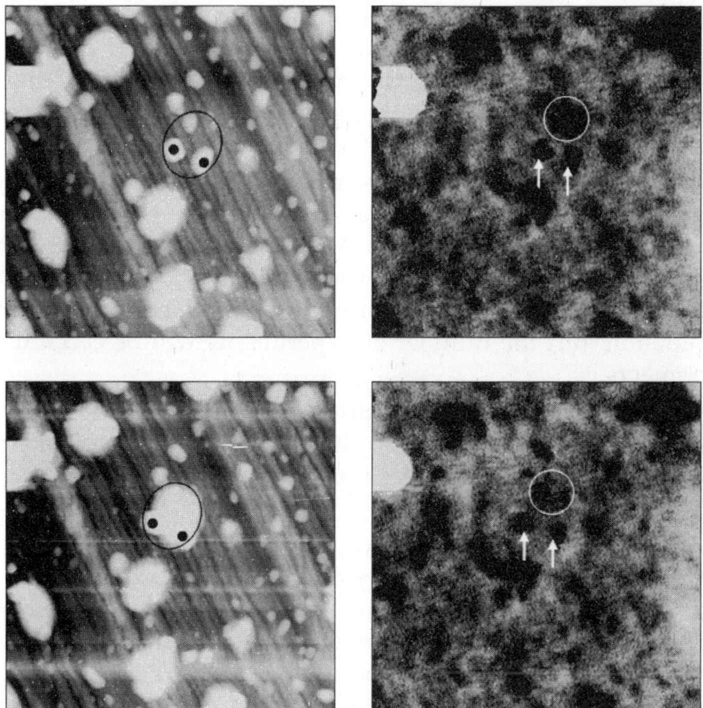

Figure 17 Simultaneous height and potential mappings measured by SKPFM scanned over the surface of contaminated iron coated with ultra-thin plasma-polymer (about 20 nm) for two subsequent scans (lower row recorded about 20 min after the first row). Size: 10 μm × 10 μm. Left column: topography; Δz = 500 nm. Right column: Volta potential, $\Delta\psi$ = 200 mV. Thehe two droplets in the middle of the first scan (see black dots and white arrows) fuse together in one huge, much bigger droplet (second scan). The Volta potential map in the first scan shows already activity (dark colour) in the area of the latter centre of the big droplet, even before it is formed, indicating galvanic coupling to the droplet [65].

Salt contamination originating from atmospheric exposure (e.g. during transport of steel sheet) is often found to be distributed in submicroscopic crystals. SKPFM was applied for monitoring delamination initiating at such nanoscopic defects, which were prepared by depositing nanoscopic KCl crystals by thermophoretic deposition. At high humidity the salt takes up water and high concentrated droplets of salt solution form at the buried interface, causing topographical protrusions at the surface of the ultra-thin polymer.

The area covered by these highly saline droplets was observed (by AFM and SKPFM) to be significantly larger than the area originally covered by the deposited salt crystallites: the uptake of water obviously causes a de-adhesion of the polymer from the surface in the vicinity of the salt particles. This initial delamination is obviously driven by mechanic tension at the interface caused by the swelling droplets. After this initial droplet formation even after prolonged exposure to high humidity only very seldom electrochemically driven delamination initiating at these droplets was observed. An example for a cathodic delamination process is shown in fig. 17. Interestingly, the delamination from nanoscopic defects seems to be much slower than from larger, external defects.

Schmutz et al. applied SKPFM for the investigation of Filiform delamination [66]. As the scan window was still too small even for the filaments observed on their model samples covered by an ultra-thin polymer coating, they had to stitch several 100 mm scans together in order to obtain a full Volta potential mapping of the filament´s head region. They found indications for oxygen diffusion pathways to the filament head along the sides of the filament.

But because of all the involved problems (the need for ultra-thin coatings, the too small scan window and also the requirement of a high humidity which is a problem for most commercial devices (see Rohwerder et al. how it can be done [63])) SKPFM is usually not applied for the study of delamination. Nevertheless, these two examples [64] show that SKPFM can and will take an important role in uncovering still open questions concerning microscopic aspects of delamination. But the number of experiments will be certainly limited to well defined key experiments.

However, SKPFM still has found a widespread application in Corrosion Science, although not for delamination studies, but for routine Volta potential mapping of alloy surfaces. Frankel et al. were the first to perform such studies on aluminium alloy surfaces [67-71], and many others were soon to follow [72-81].

In general, microstructural heterogeneities in alloys are a major driving force for pitting corrosion [82]. For instance, in Al alloys, alloying elements are added for increased strength. They are often found to have segregated to and be enriched in intermetallic particles [83-84]. Such particles can be large constituent phases on the order of tens of microns in size, or be precipitated as hardening particles of just a few nanometres in size. Localized corrosion typically initiates at the larger particles (micron size or larger), but the extent of the particle´s activity in the localized corrosion process depends on the exact particle type.

Clearly, it is of interest to be able to understand the exact role of these intermetallic particles in the localized corrosion process on Al alloys. Owing to their small size, techniques with high spatial resolution are required to do so. A number of standard techniques with sub-micron resolution exist, such as SEM, EDS, AES, and TEM. The advantage of SKPFM is that it provides additional simultaneous information about topography and Volta potential.

Assuming that there is a correlation between the Volta potential measured in dry air and the corrosion potential in solution, the measurement of the Volta potentials of the different particles in respect to the matrix should provide direct information about a possibly galvanic activity between these particles and the matrix during corrosion. Of course, during immersion in the electrolyte the particles and the matrix will have basically the same potential, because the ohmic resistance in the electrolyte is very low. However, at the mixed corrosion potential the corrosion potential of the less noble metal will be pulled up, increasing the anodic reaction, the potential of the nobler metal will be pulled down, increasing the cathodic reaction.

Indeed, for a series of metals such a correlation between corrosion potential and Volta potential was found by Frankel et al. [67, 69]: the corrosion potentials in deionized water, 0.5 M NaCl and 0.1 M Na_2SO_4 solution corresponded well with the Volta potentials after emersion. A linear relation was found. Note that the Volta potential measurement by SKPFM was carried out after emersion of the samples not prior. This is important, because the oxide layers formed during corrosion in different oxides may differ significantly from the one of the native oxide before immersion.

That such a correlation is found for many metals is not surprising, if the above discussion of fig. 14a) is considered: the potential of the passive oxide is always more positive than the potential of active corrosion. Hence, a rough correlation of corrosion potential and Volta potential of the passive oxide is to be expected for pure metals. For alloys, however, the different stability regimes of the alloying elements may lead to significant deviations. Indeed, even for non-alloys this correlation is not generally fulfilled [34].

Unfortunately furthermore, upon extended immersion the formation of the corrosion layers result in smearing out the potential contrasts. That is why very often the Volta potential mapping is carried out on the fresh sample surface, although this may often lead to wrong interpretations, especially for alloys when dealloying occurs during corrosion.

While the application of SKPFM for mapping Volta potentials on alloy surfaces is therefore not a real corrosion application (i.e. not a really electrochemical application), monitoring the the evolution of corrosion potentials during exposure at high humidity is.

For instance, Scanning Kelvin probe force microscopy (SKPFM) was used to study the initial stages of atmospheric corrosion of an AlMg alloy (AZ91D) and of a model sample prepared by physical vapor deposition (PVD) of Al dots of 2 μm in diameter on a layer of pure Mg. The latter system served as a model for a two-phase AlMg alloy (AZ91D) [85].

Since this experiment was carried out in humid atmosphere, a nanoscopic thin layer of electrolyte will have covered the alloy surface. Such a nanoscopic electrolyte layer will not cause capillary forces high enough to negatively interfere with the tapping mode scanning procedure. The ohmic resistance of this ultra-thin electrolyte layer will be very high and thusly the potential contrast between the different phases is not affected, if the involved galvanic currents very low. Since the observed build up of corrosion products amounted to just a few nanometres even after more than 150 hrs, this assumption is certainly correct.

It was observed that the corrosion products accumulated primarily on the magnesium substrate between the aluminium islands. The aluminium islands were surrounded by corrosion product halos. AES showed that these corrosion products were rich in aluminium. The islands, originally about 20 nm thick, decreased to about 12 nm thickness after exposure. Between the halos and the Al islands narrow trenches were observed. Further away from the Al islands the corrosion products were magnesium-rich. In addition, there were scattered Al-rich corrosion product accumulations that increased in number and size as the exposure continued (see fig. 18, [85]).

The difference in Volta potential between the aluminium islands and the magnesium substrate at 85% RH was about 70 mV as measured with SKPFM, while the Volta potential difference between pure Al and pure Mg at the same relative humidity as measured by SKP is about 600 mV. It is suggested that this discrepancy has to be attributed to the small size of the islands; even though their shape is clearly resolved in the images, the correct measurement of the full potential difference requires larger patterns (see discussion on cantilever contribution for SKPFM). It is assumed that usually with SKPFM not the correct potential differences are measured.

The localized corrosion seen in the absence of CO_2 is interpreted in terms of the formation of electrochemical corrosion cells on the sample surface. The cathodic reaction occurring on the Al islands gives rise to a local increase in pH. As a result, the passive film on aluminium is attacked and aluminium corrosion occurs, being redeposited again after reaching less alkaline pH (at the halos) [85].

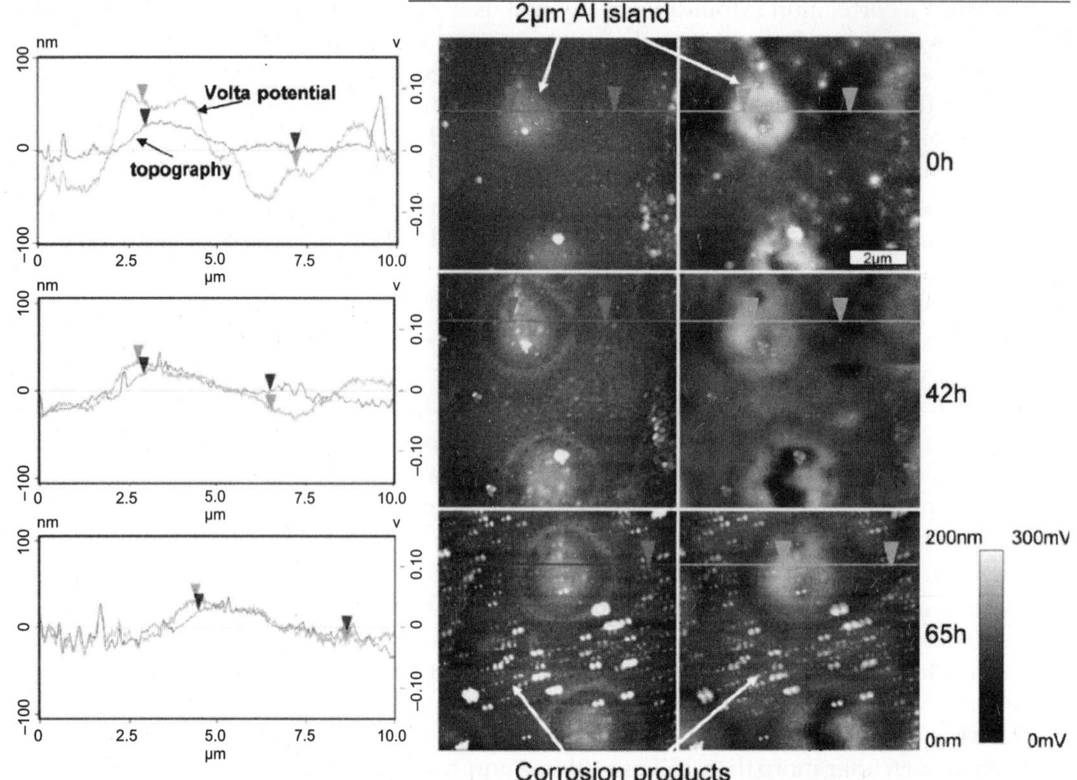

Figure 18 Time resolved in situ AFM and SKPFM images of PVD-deposited pure Al on pure Mg in the absence of CO_2. The RH was 85% and the temperature was 22°C. The images in the left column show the topography by tapping mode AFM while the right column shows Volta potential maps obtained by using the Kelvin mode of the AFM. The brighter areas are protruding from the surface in the topography images and have a higher relative potential in the Volta potential images. The height range is 200 nm and the Volta potential range is 300 mV [85].

REFERENCES

1. W. Thomson, *Philos. Mag.* 46 (1898) 82
2. I.F. Patai and M.A. Pomerantz, *J. Franklin Inst.* 252 (1951) 239
3. W.A. Zisman, *Rev. Sci. Instrum.* 3 (1932) 367
4. J.H. Parker, R.W. Warren, *Rev. Sci. Instruments* 33 (1962) 948
5. H.Baumgärtner, H. D. Liess, *Rev. Sci. Instruments* 59 (1988) 802
6. A. Broniatowski, W. Nabhan, B. Equer, G. de Rosny, *Mater. Sci. Forum* 207-209 (1996) 149
7. R. Mäckel, H. Baumgärtner, J. Ren, *Rev. Sci. Instruments* 64 (1993) 694
8. H.-D. Liess, R. Mäckel, J. Ren, *Surf. Int. Anal.* 25 (1997) 855
9. W. Nabhan, B. Equer, A. Broniatowski, G. de Rosny, Rev. Sci. Instruments 68 (1997) 3108.
10. M. Nonnenmacher, M. P. O´Boyle, H. K. Wickramasinghe, *Appl. Phys. Lett.* 58 (1991) 2921
11. M. Nonnenmacher, O. Wolter, J. Greschner and R. Kassing, *J. Vac. Sci. Technol.* B9 (1991) 1358
12. M. Rohwerder, C. Benndorf, *Surf. Sci.* 307-309 (1993) 789

13. M. Stratmann, K. T. Kim, H. Streckel, *Zeitschrift für Metallkunde* 81 (1990) 715
14. S. Yee, M. Stratmann, and R. A. Oriani, *J. Electrochem. Soc.* 138 (1991) 55
15. M. Stratmann and H. Streckel, *Corros. Sci.*, 30 (1990) 681
16. M. Stratmann and H. Streckel, *Corros. Sci.*, 30 (1990) 697
17. M. Stratmann, H. Streckel, K. T. Kim and S. Crockett, *Corros. Sci.*, 30 (1990) 715
18. A. Leng and M. Stratmann, *Corros. Sci.*, 34 (1993) 1657
19. M. Stratmann, *Bull. of Electrochem.* 8 (1992) 30
20. A. Leng, H. Streckel and M. Stratmann, *Corros. Sci.* 41 (1999) 547
21. Leng, H. Streckel and M. Stratmann, *Corros. Sci.* 41 (1999) 579
22. A. Leng, H. Streckel, K. Hofmann and M. Stratmann, *Corros. Sci.* 41 (1999) 599
23. J.O'M. Bockris and S.D.Argade, *J. Chem. Phys.* 49 (1968) 5133
24. J.O'M. Bockris, *Energ. Conv.* 10 (1970) 41
25. J. O'M. Bockris, *J. Electroanal. Chem.* 36 (1972) 495
26. E. Gileadi and G. Stoner, *J. Electronal. Chem.* 36 (1972) 492
27. S. Transatti and B. Damaskin, *J. Electroanal. Chem.* 52 (1974) 313
28. S. Trasatti, *J. Electroanal. Chem.* 66 (1975) 155
29. A. De Battisti and S. Transatti, *J. Electroanal. Chem.* 79 (1977) 251
30. S. Transatti, *J. Electroanal. Chem.* 139 (1982) 1-13
31. S. Transatti, *Electrochimica Acta* 3 (1990) 269-271
32. S. Trasatti, *Electrochimica Acta* 36(11/12) (1991) 1659-1667
33. S. Trasatti, *Surf. Sci.* 335 (1995) 1-9
34. M. Rohwerder and F. Turcu, Electrochimica Acta 53 (2007) 290
35. Walid Nabhan, Alexandre Broniatowski, Gilles de Rosny and Bernard Equer, *Microsc. Microanal. Microstruct.* 5 (1994) 509-517
36. H. Jacobs, P. Leuchtmann, O. Homan, A. Stemmer, *J. Appl. Phys.* 84(3) (1998) 1168
37. H. Jacobs, H. Knapp, A. Stemmer, *Rev. Sci. Instrum.* 70(3) (1999) 1756
38. F. Robin, H. Jacobs, O. Homan, A. Stemmer, W. Bachtold, *Appl. Phys. Lett.* 76(20) (2000) 2907
39. T. Hochwitz, A. K. Henning, C. Levey, C. Daghlian, J. Slinkman, *J. Vac. Sci. Technol.* B 14(1) (1996) 457
40. A. K. Henning, T. Hochwitz, J. Slinkman, J. Never, S. Hoffmann, P. Kaszuba, C. Daghlian, *J. Appl. Phys.* 77(5) (1995) 1888
41. Ch. Sommerhalter, Th. Glatzel, Th.W. Matthes, A. Jager-Waldau, M.Ch. Lux-Steiner, *Applied Surface Science* 157 (2000) 263–268
42. S.M. Sze, Semiconductor Devices-Physics and Technology, John Wiley & Sons, 1985
43. H. Ishii, N. Hayashi, E. Ito, Y. Washizu, K. Sugi, Y. Kimura, M. Niwano, Y. Ouchi, and K. Seki, *phys. stat. sol.* 201(6) (2004) 1075–1094
44. Y. Harima, K. Yamashita, H. Ishii, and K. Seki, *Thin Solid Films* (2000) 237
45. H. Ehahoun, M. Stratmann, M. Rohwerder, *Electrochim. Acta* 50 (2005) 2667–2674
46. M. Stratmann, H. Streckel, K. T. Kim and S. Crockett, *Corros. Sci.*, 30 (1990) 715
47. J.W. Schultze, M.M. Lohrengel, *Electrochim. Acta* 28(7) (1993) 973-984
48. J.W. Schultze, M.M. Lohrengel, *Electrochim. Acta* 45 (2000) 2499–2513
49. J.W. Schultze, A.W. Hassel, Passivity of metals , alloys and semiconductors, in: Encyclopedia of Electrochemistry, vol.4 (G.S. Frankel, M. Stratmann, eds.), (2004) p. 234

50. G. Grundmeier, K.-M. Jüttner and M. Stratmann in M. Schütze (ed.) *Corrosion and Environmental Degradation*, Vol. I, Wiley-VCH, Weinheim, 2000, pp.285
51. K. Doblhofer, *Bulletin of Electrochemistry* 8 (1992) 96
52. W. Kutner, *Electrochimica Acta* 37 (1992) 1109
53. W. Fürbeth, M. Stratmann, *Corrosion Science* 43 (2001) 207
54. W. Fürbeth, M. Stratmann, *Corrosion Science* 43 (2001) 229
55. W. Fürbeth, M. Stratmann, *Corrosion Science* 43 (2001) 243
56. M. Rohwerder, M. Stratmann, *MRS Bulletin* 24 (1999) 43-47
57. M. Rohwerder, G. Grundmeier, M. Stratmann, *"Corrosion Prevention by Adsorbed Organic Monolayers and Ultrathin Plasma Polymer Films"*, Corrosion Mechanisms in Theory and Practice, ed. P. Marcus, Marcel Dekker, New York ch. 14 (2002) 479-528
58. G.S. Frankel, M. Rohwerder, *" Electrochemical Techniques for Corrosion"*, in *Encyclopedia of Electrochemistry*, eds. A.J. Bard and M. Stratmann, Vol.4 *"Corrosion"*, pp.687-723 (eds. M. Stratmann and G.S. Frankel), Wiley-VCH, Weinheim 2003
59. M. Rohwerder, G.S. Frankel, P. Leblanc, M. Stratmann: *"Application of Scanning Kelvin Probe in corrosion science"*, in: Methods for Corrosion Science and Engineering, F. Mansfeld, P. Marcus (eds.), Marcel Dekker New York, 2006, p. 605-648
60. R. Hausbrand, M. Stratmann, M. Rohwerder, *STEEL RESEARCH INTERNATIONAL* 74(7) (2003) 453-458
61. G. Paliwoda, M. Stratmann, M. Rohwerder, K. Potje-Kamloth, Y. Lu, A.Z. Pich, H.-J. Adler, *Corros. Sci.* 47 (2005) 3216-3233
62. A. Michalik, M. Rohwerder, *Z. Phys. Chemie* 219, 11 (2005), 1547-1560
63. K. Wappner, B. Schömnberger, M. Stratmann, G. Grundmeier, *J. Electrochim. Acta* **152**(3) (2005) E114-E122
64. M. Rohwerder, E. Hornung, M. Stratmann, *Electrochim. Acta* 48(9) (2003) 1235-1243
65. Final report ECSC project 7210-PR/320
66. P.P Leblanc and G.S. Frankel, *J. Electrochem. Soc.* 151(3) (2004) B105-B113
67. P. Schmutz and G.S. Frankel, *J. Electrochem. Soc.* 145(7) (1998) 2285-2295
68. P. Schmutz and G.S. Frankel, *J. Electrochem. Soc.*, 145(7) (1998) 2295-2306
69. V. Guillaumin, P. Schmutz, G.S. Frankel, *J. Electrochem. Soc.* 148(5) (2001) B163-B173
70. P. Schmutz and G. S. Frankel, *J. Electrochem. Soc.*, 146(12) (1999) 4461-4472
71. P. Leblanc and G. S. Frankel, *J. Electrochem Soc.*, 149 (6) (2002) B239-B247
72. M.L. Zheludkevich, K.A. Yasakau, S.K. Poznyak, M.G.S. Ferreira, *Corr. Sci.* 47(12) (2005) 3368-3383
73. J.X.Jia, A. Atrens, G. Song, T.H. Muster, *Mater. and Corr.* 56(7) (2005) 468-474
74. X. Zhang, W.G. Sloof, A. Hovestad, E.P.M. van Westing, H. Terryn, J.H.W. de Wit, *Surf. & Coat. Tech.* 197(2-3) (2005) 168-176
75. B.S.Tanem, G. Svenningsen, J. Mardalen, *Corr. Sci.* 47(6) (2005)1506-1519
76. S.Mato, G. Alcala, T.G. Woodcock, A. Gebert, J. Eckert, L. Schultz, *Electrochim. Acta* 50(12) (2005) 2461-2467
77. J.H.W. de Wit, *Electrochim. Acta* 49(17-18) (2004) 2841-2850
78. F. Andreatta, H. Terryn, J.H.W. de Wit, *Electrochim. Acta* 49(17-18) (2004) 2851-2862
79. F. Andreatta, M.M. Lohrengel, H. Terryn, J.H.W. de Wit, Electrochim. Acta 48 (2003) 3239-3247
80. M. Femenia, C. Canalias, J. Pan, C. Leygraf, *J. Electrochem. Soc.* 150(6)(2003) B274-B281

81. P. Campestrini, E.P.M. van Westing, H.W. van Rooijen, J.H.W. de Wit, *Corros. Sci.* 42 (2000) 1853-1861
82. Z. Szklarska-Smialowska, *Pitting Corrosion of Metals* (NACE, Houston, 1986)
83. R. G. Buchheit, R. P. Grant, P. F. Hlava, B. McKenzie and G. L. Zender, *J. Electrochem. Soc.*, 144 (1997) 2621
84. V. Guillaumin and G. Mankowski, *Corr. Sci.* **41** (1999) 421-438
85. D.B. Blucher, J.E. Svensson, L.G. Johansson, M. Rohwerder, M. Stratmann, *J. Electrochem. Soc.* 151(12) (2004) B621-B626

CHAPTER 16

Corrosion Monitoring in Industry

Alec Groysman
Oil Refineries Ltd., Haifa, Israel.
ORT Braude College, Mechanical Engineering Department, Karmiel, Israel.

INTRODUCTION

I wake up every morning, wash my face and look in the mirror at myself. This is primary *monitoring* of myself. Then I go out my house and check the wheels of my car before driving. This is a simple *monitoring*. One can give many examples of *monitoring* in everyday life. Everything that relates to metals and safety of people, must be *monitored*. Every enterprise wants to operate for longer periods between scheduled shut-downs, reduce incidents resulting in hazard or injury to both plant personnel and the environment. The cost of production loss and unplanned shutdowns which arise due to corrosion can be very high, and, with the added desire for excellence in plant operation and safety, the drivers for effective online corrosion monitoring can be appreciated. About 70 to 90% failures in different industries are corrosion-related, resulting in unplanned shutdowns, damage to the environment and production penalties [1-3]. Most corrosion damages occur during relatively short periods of time, usually during process startup, changes, upsets or shutdowns [4]. Therefore corrosion monitoring methods must supply the realtime (immediate) information. To use the metallic constructions without corrosion monitoring is similar to live and work with closed eyes and ears. *Corrosion monitoring* is an integral part of *anti-corrosion management* in all industries [5]. It is impossible to reduce the cost of corrosion to zero, because we have to use corrosion control methods and *corrosion monitoring*, which have a definite cost. The books on corrosion science and engineering published during the last 10 years contain chapters about *corrosion monitoring* [6, 7]. Special book on *corrosion monitoring* was published in 2000 [8]. Many corrosion conferences contain the session "*Corrosion Monitoring*" (CM). Most information about news in CM is dispersed in the periodical literature. This chapter describes the principles and achievements in CM in industry during the last 10 years. We can emphasized that this period was marked by spread *on-line real-time CM* methods in different industries, especially concerning monitoring and determination of localized corrosion phenomena (pitting, stress corrosion cracking, crevice, etc.). *"On-line"* means CM application during equipment operation, not requiring shut-down. *"Real-time"* means measuring instantaneous corrosion rate.

There is no strong division of CM methods. They range from non-direct/non-intrusive to direct/intrusive techniques [9]. Sometimes one method (for instance, pH measurement) may be intrusive (on-line pH meter), or non-intrusive (if pH is measured for sample obtained through an existing valve). Sometimes CM techniques are divided on direct (weight loss - coupons and electrical resistance probes; corrosion current - linear polarization probes; remaining wall thickness - ultrasonic or Eddy-current technique), and indirect (radiography and chemical analytical methods). An intrusive technique requires entry into the process stream (coupons, electrochemical sensors, and pH-meters). Non-intrusive technique does not require entry into the stream (Non-DestructiveTechnique - Eddy current, acoustic emission; hydrogen flux probes). CM of anti-corrosion coatings and cathodic protection is beyond the scope of this article.

WHY HAVE WE TO MONITOR CORROSION?

It is impossible to prevent corrosion of many metals and alloys, because they are thermodynamically unstable in the environment. If corrosion of metals is inevitable, we want to know how long equipment or structure will serve under particular conditions safely and efficiently. Various electronic devices are used in industries. Their reliable functioning depends on the quality of electric contacts made of different metals and alloys. If corrosive species are present in the environment, electric contacts may corrode, reliability of the data receiving by these devices diminishes, and corrosion may result in malfunctioning and unpredicted disasters. Therefore we have to monitor corrosive aggressiveness of the environment where electronic or other devices are in contact with it. The third reason of CM requirement relates to preventive anticorrosion measures. Various methods are used for corrosion control. Such chemicals as corrosion inhibitors, neutralisers, anti-scaling inhibitors, biocides, and demulsifiers are used in order to prevent or to diminish corrosion. Oxygen scavengers are injected in the boiler feed water in order to diminish water corrosiveness and thus to prevent corrosion in boilers, superheaters, and other equipment. Corrosion inhibitors and neutralisers are injected in the overhead of the distillation columns at the oil refineries in order to neutralize acid gases and thus to prevent dew point corrosion in the pipelines, air coolers, and heat exchangers. We must know the efficiency of these "anticorrosive" chemicals. When cathodic protection is used for the corrosion control of the underground and under water equipment and structures we must know its efficiency. If coatings are used for corrosion control, we have to know their efficiency and protective properties. In other words, CM helps us to define the efficiency of the corrosion control methods. To sum up, there are three important questions in industry related to CM. The first question is a "qualitative" one. Does corrosion occur? Is there any equipment or structure suffered from corrosion? Many side - issues exist. What is the corrosion form? Is the corrosion uniform or localized? Here is wide spectrum of corrosion phenomena: pitting, crevice, stress corrosion cracking (SCC), galvanic corrosion, dealloying, intergranular corrosion, erosion, cavitation, microbiologically induced corrosion (MIC), fretting corrosion, etc. We have to define corrosion form in order to take correct preventive measures. The second question relates to the previous one. What are the reasons and factors influencing corrosion? Certainly this question is connected to selection of anticorrosion measures, type of corrosion resistant materials, coating systems, inhibitors and other chemicals used for corrosion control. The third question is a "quantitative" one. What is the corrosion level? What is the corrosion rate value? *The corrosion rate is the corrosion of material occurring in the unit of time.* This is not always the corrosion product quantity. Corrosion may change with time and with distance on the metal surface. Even if we know the corrosion level, that is if we received the corrosion rate 0.01 mm/year, or 0.1 mm/year, or 1 mm/year, the following questions remain. Is this value low or high? Is this situation dangerous or can we exist with such value of corrosion rate? May we continue to use the equipment? The replies are beyond of this article. We shall concentrate on

I. CORROSION MONITORING METHODS

Three main factors influence corrosion: a metal type, an environment, and conditions at a border between a metal and the environment. Therefore, we can use the properties of a metal, of an environment, and of a border metal - environment for CM. In the light of this, CM methods can be divided into three groups: control of a metal condition, of the environment, and of a border metal - environment. Control of a metal condition is based on the physical properties of metals: mass, thickness, and electrical resistance of metal sample. Control of the environment is based on the chemical, physical and microbiological properties and conditions of the environment. Control of the border metal - environment is based on the detection of its physico - chemical properties.

I.1 Control of Physical Properties of a Metal (Physical Methods)

Historically they were the first: man's eyes or optical devices were used for the control of properties of corroded metal surface. Optical microscopes and boroscopes are used for visual analysis of corroded metal surface. Video cameras are used for inspection of inner surfaces (including coating assessment) of tanks and voids on the ships [10]. Different devices based on physical phenomena allow to measure the changes in metal thickness of metallic equipment (ultrasonic and Eddy-current methods), and weight loss (WL) of the metal (WL "coupon" method and electrical resistance measurements of a corroding metal sensor). Acoustic emission method is used for determination of local damage of metallic and concrete constructions (cracks, pits, and holes) [11]. X-ray radiographic methods are used for online wall thickness measurements of insulated pipes, detection of intergranular corrosion, and flow-accelerated corrosion.

I.2 Control of the Environment (Chemical Analytical, Physico-Chemical, Physical, and Microbiological Methods)

These methods are based on chemical analysis of media which is in contact with a metal equipment. It is impossible to use the WL method in the case of corrosion of copper alloys or stainless steels in aqueous solutions because of very low mass changes of coupons made of these alloys. One can use various sensitive physico-chemical analytical methods for the determination of concentration of metal cations in media as a result of corrosion of metallic equipment. On the other hand, analytical methods are used for the determination of aggressive components in media: pH, Cl^-, SO_4^{2-}, O_2, CO_2, H_2S, NH_3, CN^-, Fe^{3+}, Cu^{2+}, hardness, alkalinity, other dissolved and suspended solids, microorganisms influencing corrosion, the presence of contaminants inducing erosion and cavitation in water and steam; and some physico-chemical properties of media: electrical conductivity and redox potential. For organic liquids (fuels, solvents, etc.), it is very important to measure water content, dissolved oxygen concentration and electrical conductivity. For boiler feed water, it is important to examine concentration of dissolved oxygen, copper and iron cations, silica, and total dissolved solids. For steam and high temperature gases, it is very important to examine temperature in order to eliminate water condensation and, as a result, cavitation and dew point corrosion. For the determination of the aggressiveness of atmosphere, we have to measure the relative humidity, the temperature and its changes, the period during which a metal surface is covered by a film of electrolyte that is capable of causing atmospheric corrosion (time of wetness), sulfur oxides (SO_x), nitrogen oxides (NO_x), ammonia and its derivatives, ozone, hydrogen sulfide, and dust in atmosphere. The determination of hydrogen gas in some processing (Hydrogen Flux Monitoring) relates to the analytical methods. For the determination of corrosiveness of soil, we have to measure its electrical conductivity. In some cases it is preferable to identify the microorganisms, the humidity and pH of soil [12]. Microbiological analysis of water, soil, atmosphere, crude oil and fuels is very important for the control of potentiality of MIC occurrence.

Control of process (technological) parameters include the measuring of flow rate, temperature, pressure, dew point temperature, heat transfer resistance, presence of solid particles in streams, etc.

The analysis of residue concentrations of corrosion inhibitors, neutralizers, oxygen scavengers, biocides in technological streams is very important for control of efficiency of anti-corrosion treatment, as well for ecology.

I.3 Control the Border Metal – Environment (Physico – Chemical Methods)

These methods are based on the physico-chemical properties of the border metal - environment, and may be divided into electrochemical methods, identification of corrosion products and deposits, examination of morphology of metal surface after corrosion.

Among electrochemical methods are corrosion potential measurements, polarization measurements (potentiodynamic polarization), linear polarization resistance (LPR), galvanic probes (ZRA - Zero Resistance Ammeter), electrochemical noise measurements (ENM), electrochemical impedance spectroscopy (EIS), and Harmonic Distortion Analysis (HDA). Radioactive method "Thin Layer Activation" is used for the determination of corrosion resistance (corrosion rate) of some alloys under specific conditions [8]. Microbiological analysis of deposits (sessile bacteria) on metal surface is also very important. Scanning Electron Microscopy (SEM), Scanning Tunneling Microscopy (STM), and Atomic Force Microscopy (AFM) are used for the examination of metal surface morphology [13]. Energy Dispersive Spectroscopy (EDS), X-ray Photoelectron Spectroscopy (XPS), also known as Electron Spectroscopy for Chemical Analysis (ESCA), Auger Electron Spectroscopy (AES), Mössbauer Spectroscopy, Secondary Ion Mass Spectroscopy (SIMS), Fourier Transform Infrared Spectroscopy (FTIR), Raman spectroscopy, UV-Visible reflectance studies, and Ultraviolet Photoelectron Spectroscopy (UPS) are used for the identification of chemical content of corrosion products [14].

We may receive very useful information from the color of corrosion products formed on the metal equipment, but there is some uncertainty. Firstly, there is no objectivity in the visual color determination. Secondly, the same color can belong to several compounds. For instance, black color may be attributed to the presence of various sulfides (FeS, Ag_2S, CuS) or oxides (Fe_3O_4, CuO); white – the presence of AlOOH, $Zn(OH)_2$, or $CaCO_3$, etc. In any case, this is very useful preliminary information about corrosion and its reasons. Slippery surface of brown, orange, black, green, or intermediate colors recognizes usually the presence of biofouling.

2. PHYSICAL METHODS OF CORROSION MONITORING.

We shall concentrate on two physical methods widely used both in laboratory and in industry: weight loss (WL) method and electrical resistance (ER) measurements, both based on a loss of corroded metal.

2.1 Weight Loss (WL) Method.

Rectangular or circular in shape metal specimen of different dimensions called *"coupons"* (Figure 1) made of a metal or an alloy representative of the material of the equipment (or another one that we want to know corrosion resistance and corrosion rate), can be inserted into a process stream by means of retractable or retrievable holders (Figure 2).

Corrosion rate of a coupon is calculated from the WL measurements according to formulas (1) and (2).

$$k = (M_i - M_f) / A \cdot T \quad ...(1)$$

k – corrosion rate of a metal, gram/(cm^2·hour); M_i and M_f – initial and final mass of a coupon, gram; A – area of a coupon surface, cm^2; T – period of immersion of a coupon in a solution, hours.

$$k_p = (k \cdot 10 \cdot 8760) / d \quad ...(2)$$

k_p – penetration of corrosion, mm/year; k – corrosion rate defined in the equation (1), gram/(cm^2·hour); d – density of a metal, gram/cm^3; 10 – conversion of cm into mm; 8760 – conversion of hours into years.

The coupons' surface must be carefully prepared before installation. The dimensions and initial weight of coupons must be defined with high precision. Every coupon must be stored in a special envelope in a dessicator. Then they are installed for some period depending on

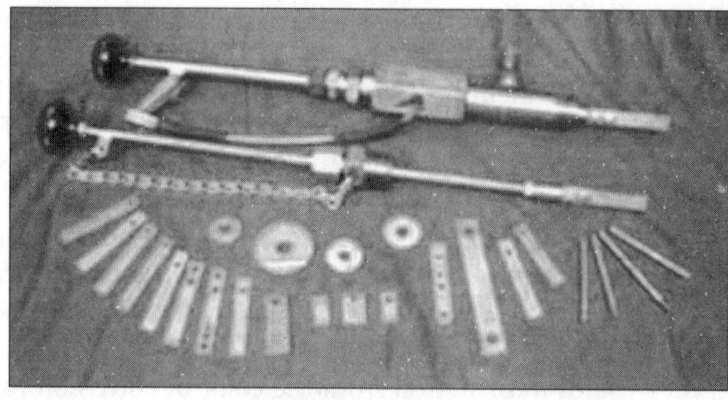

Figure 1 Various "coupons" for corrosion examination (Courtesy of Metal Samples Company).

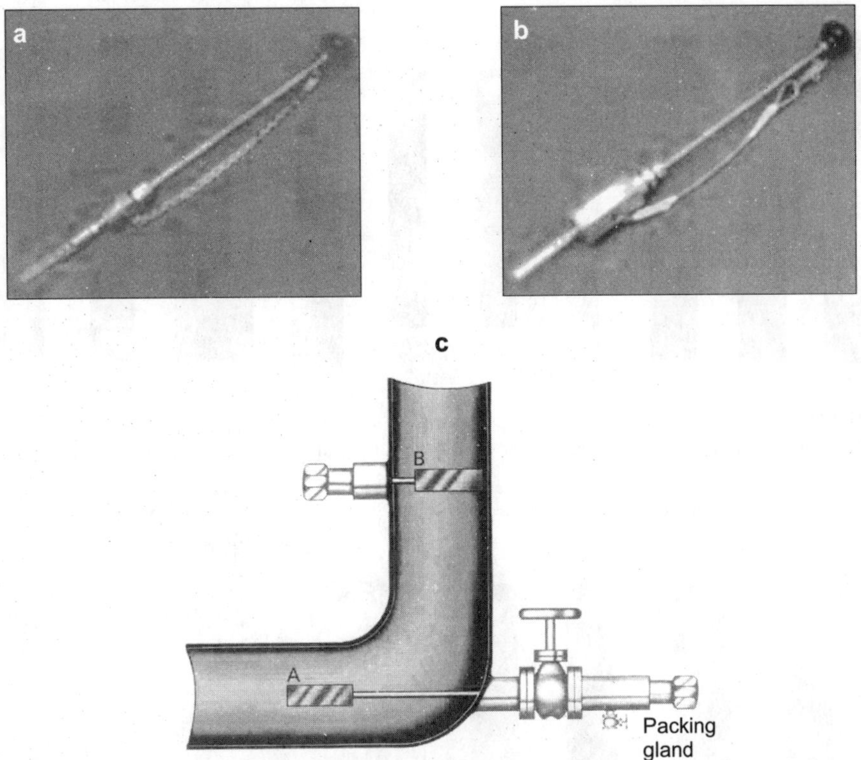

Figure 2 Holders (a, b) and their installation in a process stream (c): a – retrievable coupon holder (up to 8 atm.); b – high pressure or packing gland holder (up to 105 atm.). A and B – coupons. (Courtesy of Metal Samples Company).

corrosiveness of a process stream and a coupon's material type. The coupons' immersion time must be determined for every particular case. Usually 30 to 40 days are recommended for the determination of corrosion rate of coupons in water systems. This period is correct for carbon steel. Copper alloys, stainless steel and titanium are resistant to water. Therefore immersion period for these alloys must be increased to 180 days and even more. Sometimes period of 30 to 40 days for carbon steel coupons is unacceptable. If water is not treated by chemicals (corrosion inhibitors, anti-scaling agents and biocides) corrosion rate of carbon steel may be high (about 0.6 to 1 mm/year), and we have to change coupons every 5 to10 days. On the other hand, if chemicals are injected in cooling water, corrosion rate of carbon steel is low (about 0.001 to 0.05 mm/year and less), and you may change coupons every 90 to 120 days, and even more. If corrosion rate of carbon steel is measured in two phase system, for example, "hydrocarbons – water" in the overhead of the crude oil distillation column, the immersion period of coupons' depends on the inhibitor's efficiency and may change from 40 to 180 days, and even more. One has to keep in mind that pitting corrosion is very dangerous phenomenon in any streams, and some initiation period (above 60 days) is needed for the determination of pitting corrosion. After some period of immersion that you choose according to corrosiveness of a stream and goals defined, the coupons must be taken out. You may take their pictures with, and then without corrosion products, and compare with the original coupon before installation. Some

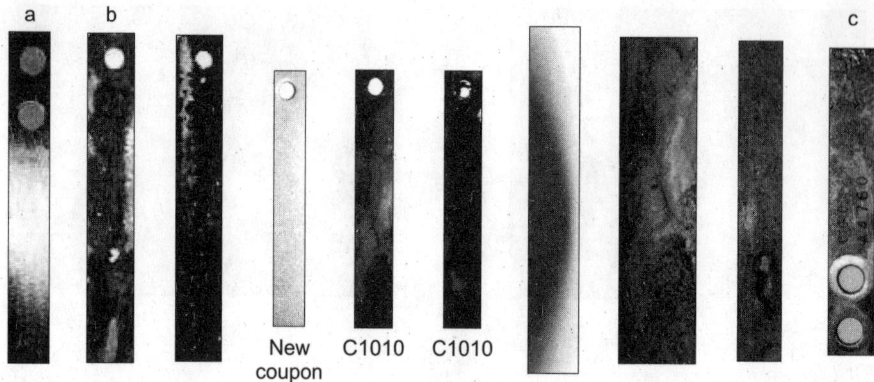

Figure 3 The coupons after being exposed to cooling water: a – original coupon; b – the coupons with corrosion products after immersion in cooling water for 60 days;c – coupon after cleaning from corrosion products.

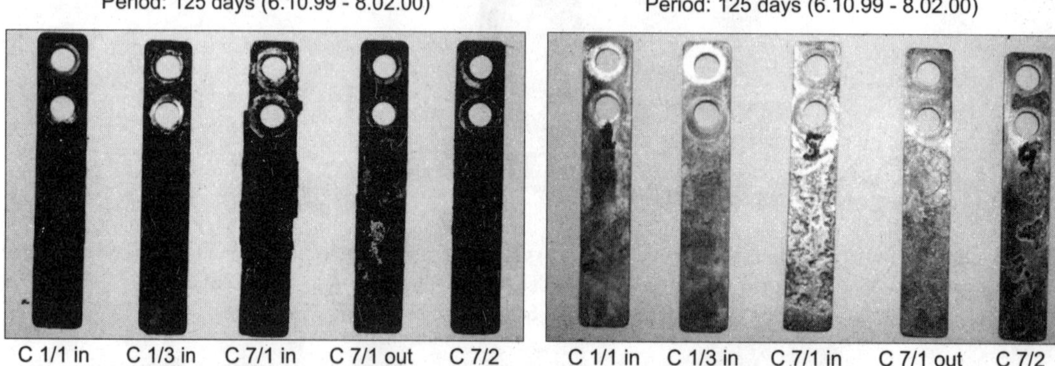

Figure 4 The coupons after being exposed to the process stream in the overhead of the crude distillation column: a – before cleaning; b – after chemical cleaning.

examples of the coupons after immersion in some industrial streams are shown at Figures 3 and 4. Color identification of corrosion products formed on the coupons' surface may give preliminary information about the presence of corrosive species in the streams caused corrosion. If dense films of corrosion products are formed on the coupons' surface, (for instance, iron oxides or iron sulfides) their thickness can be measured. Then you have to remove corrosion products by means of knife made of some hard polymeric material (Teflon, polyethylene, etc.) for their subsequent chemical identification. Then coupon's (carbon steel and copper alloys) surface must be cleaned mechanically with eraser or chemically with hydrochloric acid (5 to 8 wt% HCl) inhibited with 0.2% of organic corrosion inhibitor (usually liquid amines or their mixtures) for decrease of loss of clean metal during chemical cleaning. A procedure is not simple and is needed experience. Here are some recommendations regarding correct installation of coupons into equipment at the plant.

1. A coupon must be installed against flow, but in such manner that to prevent any erosion or impingement.
2. A bypass with three and more places for coupons' installation is often used in cooling water systems (Figure 5). A bypass should be connected to the return riser to the cooling tower or at the hottest part of the cooling water system. It is very important to install copper alloys' coupons after carbon steel and aluminum ones. Do not use copper and galvanized pipes as well as brass valves for the connection of the bypass to the riser. Bypass is usually made of polymeric material (polyvinylchloride, etc.).
3. The coupons must be mounted closely (but not to be in contact!) to the surface of equipment that you examine corrosion rate.
4. Flow rates 1 to 2 m/s of stream are adequate.
5. You have to remember that a coupon gives information about corrosion situation only in the place of its installation. Corrosion situation may change with moving away from the place of a coupon's installation. Therefore the more coupons are installed, the better *"corrosion situation map"* may be received.

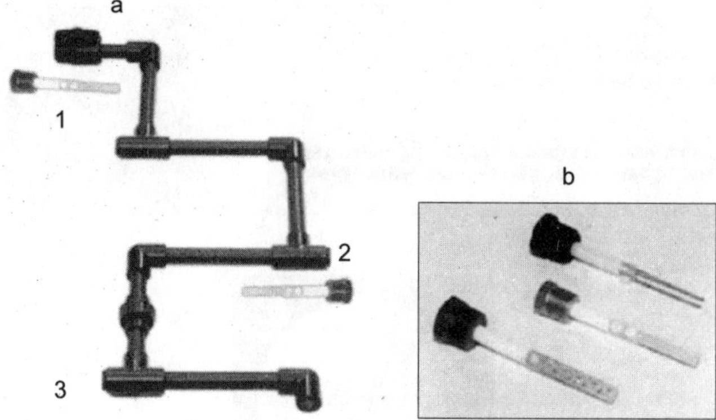

Figure 5 Bypass piping assemblies (a) and coupons (b) installed inside. 1, 2, 3 – places for insert of coupons. (Courtesy of Metal Samples Company).

The WL method is standardized [15].

The benefits of the *WL method*:

1. The corrosion rate calculated by this method is very precise: 3 to 5 percent of reproducibility. The reliability of this method is high. It is recommended to compare other methods of CM (for instance, electrochemical ones) with the *WL* method.
2. It is possible to measure the corrosion rate in any media (without any relationship to its electrical conductivity): in gases, in liquids, and in solids, and under wide process conditions: up to 450°C and to 350 Bar in the case of retrievable holders, and without restrictions in the case of fixed probes.
3. This is the only method for the quantitative information about pitting corrosion rate. The electrochemical noise measurements give qualitative information about pitting tendency.
4. It is possible to examine coupon's surface and corrosion products, and then to identify the corrosion mechanism and corrosion type (phenomenon).

5. You may use the "*scale coupons*" for the examination of a scale formation; the "*disc coupons*" - for the crevice corrosion examination; the "*stressed coupons*" - for the stress corrosion cracking examination; the "*rod coupons*" - for erosion examination, the "*welded coupons*" – for study the variance between corrosion behavior between welded and non-welded metals (Figure 6), and special gauze coupon for collection of biofouling (see Figure 21).
6. The coupon itself is usually has low cost.

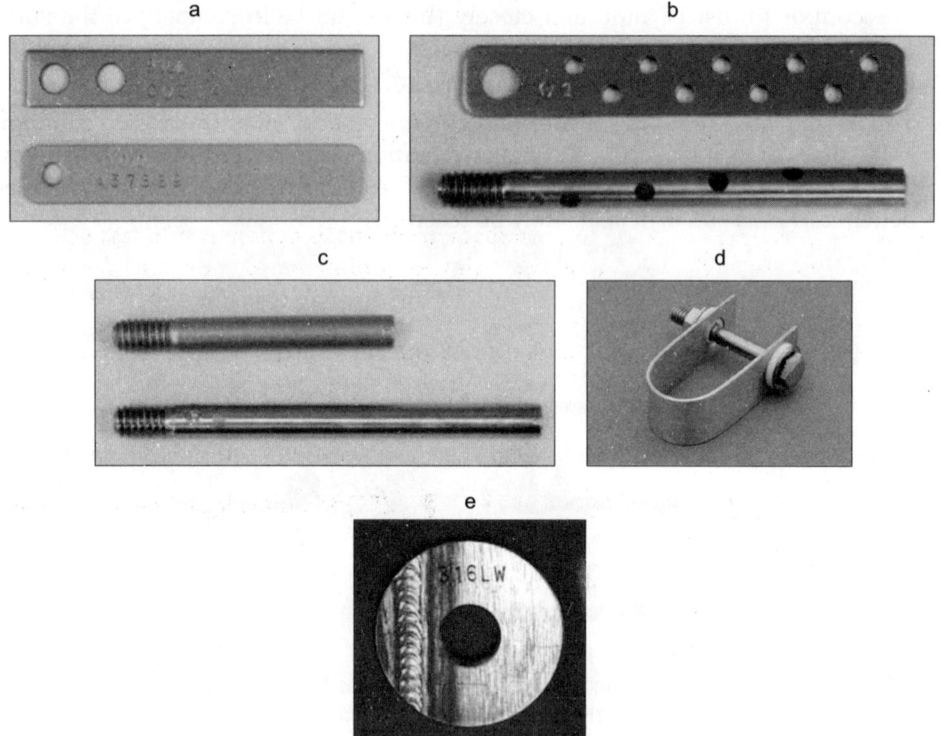

Figure 6 Various coupons' forms for the examination of different corrosion phenomena: a – flat coupons for the examination of general and pitting corrosion; b – "scale coupons" for examination of scale formation; c – rod (cylindrical) coupons for erosion examination; d – stressed coupons for stress corrosion cracking examination: e – welded coupons. (Courtesy of Metal Samples Company).

The limitations of the *WL method*:

1. In order to receive reliable data, in many cases we need to wait for a long time: at least 30 to 40 and more days. Corrosion rates can be calculated only after removal of a coupon from the system. If corrosion rate is low (about one thousandth mm/year), we need to wait longer: about 4 to 9 months. In order to receive information about pitting corrosion, we need to wait at least more than 60 days.
2. The coupons provide general and averaged data about corrosion over some period (Figure 7). Environmental conditions may change every moment, and it is impossible to know what happened during such upsets.

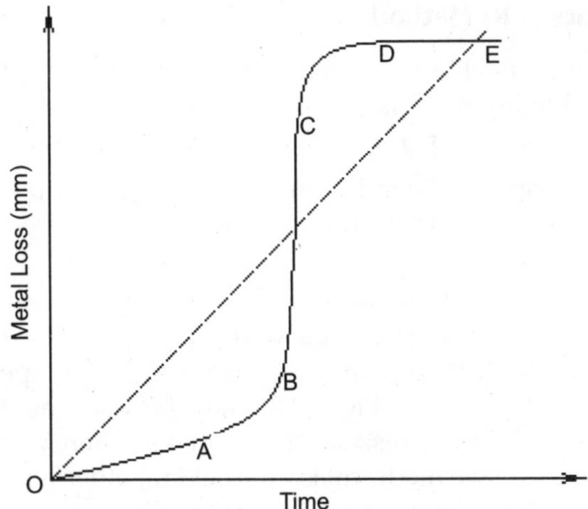

Figure 7 A typical plot of metal loss versus time. OE – Average corrosion rate over total time period. OA and DE – Corrosion rate for non-upset conditions. BC – Corrosion rate at process upset.

3. The *WL method* is time-consuming. We have to treat the coupons: to clean, to weigh, and to take into account the *cleaning factor*. When we clean coupon's surface with inhibited acid from corrosion products, some WL of clean metal occurs. We may estimate this mistake in percent (called *cleaning factor*) in the WL definition. In other words, *cleaning factor* shows the influence of chemical cleaning on the accuracy of WL method. In order to define the *cleaning factor*, you may immerse a clean coupon in the same acid solution with inhibitor for similar period that treated coupon and define its WL. The ratio of the WL determined for the clean coupon to the WL determined for the treated coupon in percent is the *cleaning factor*. If the WL of treated coupon is relatively high (10 to 100 mg), cleaning factor is not high: about 1 to 2 percent. If the WL of treated coupon is low (0.5 to 1 mg), cleaning factor may be 50 to 100 percent.

Sometimes *WL method* does not reply regarding corrosion of non-ferrous alloys, for example, copper alloys in water and selection of corrosion inhibitors for them. Sensitivity and accuracy of WL of copper coupons are low in water in the presence of corrosion inhibitors. In this case Atomic Absorption and ICP (Ion Couple Plazma) methods are used for the determination of concentration of copper ions appeared in water because of the corrosion of copper alloys.

The main drawback of the *WL method* is that it is not always suitable to use it in industry because of difficulties of installation and change of coupons every month or every several months. For example, if there are dangerous processing conditions, such as high temperature and pressure, harmful substances (hydrogen sulfide, hydrogen chloride, etc.), it is difficult or even impossible to come to the points needed for measuring. The *WL method* does not permit real-time monitoring. Therefore we will discuss the methods that overcome partly the limitations of the *WL method*.

2.2 Electrical Resistance (ER) Method.

If you take a wire, a strip or a cylindrical metal specimen (Figure 8), you may use following formula known for a conducting material from physics:

$$R = \rho \cdot L / A \qquad ...(3)$$

R – the electrical resistance of conducting material, Ohm; ρ – the specific electrical resistivity of conducting material, Ohm · m; L – the length of a specimen, m; A – the cross-sectional area of a specimen, m^2.

If a piece of a metal to immerse in some corrosive media (in a liquid, in a gas or in a solid), corrosion may occur, and as a result the cross-sectional area A of a specimen will decrease. According to the formula (3), the electrical resistance of a metal specimen would increase. Thus, a metal specimen (element) at Figure 8, may be used as a corrosive ER sensor. Temperature influences the electrical resistance of a metal. Therefore an ER-probe (a sensor element contacting with a corrosive medium) is mounted together with a reference element for temperature compensation, that is not contacting physically with a medium but being at the same temperature. Thus, a reference element is influenced by temperature by the same way as a sensor, but is not influenced by corrosive media. Practically, we measure the ratio of the electrical resistance of a sensor exposed to environment to the electrical resistance of a reference element not exposed to environment at the same temperature (Figure 9). Since temperature changes effect the electrical resistance of both the exposed and reference element equally, measuring the resistance ratio minimizes the influence of changes of the ambient temperature.

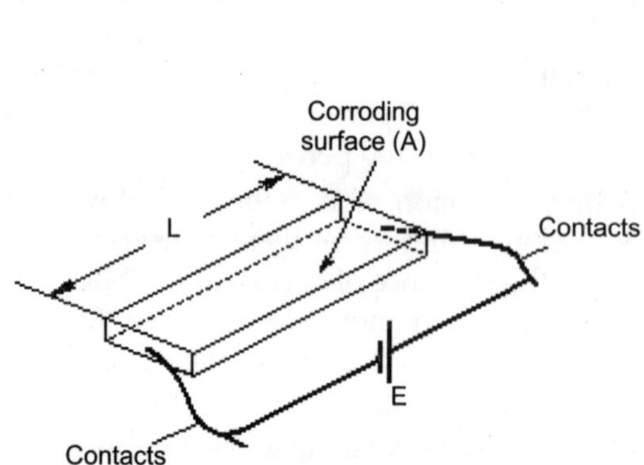

Figure 8 A piece of a conducting material (metal) and the operating principle of the electrical resistance (ER) probe. L – distance; E – Voltage, Volt.

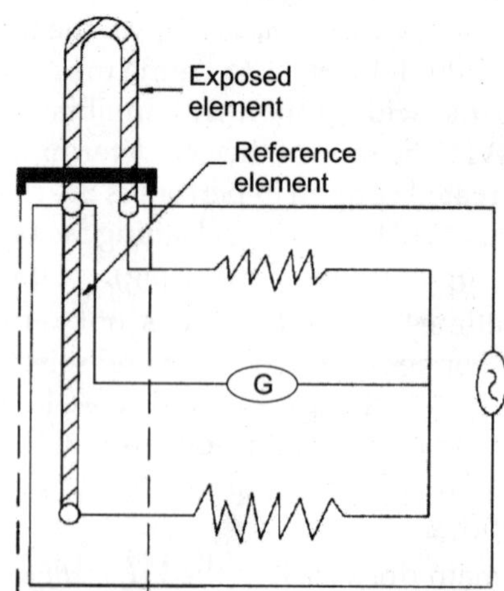

Figure 9 Exposed element (sensor) and a reference element in the ER-probe. (Courtesy of Metal Samples Company).

Sensing elements in the ER-probes are made in a variety of geometric configurations, thickness and alloy materials (Figure 10). The selection of a particular element form depends on sensitivity, response time and element (probe) "lifetime" required. Response time is the minimum period in which a measurable change takes place, governs the speed with which useful results can be obtained. A probe "lifetime", or a period required for the effective

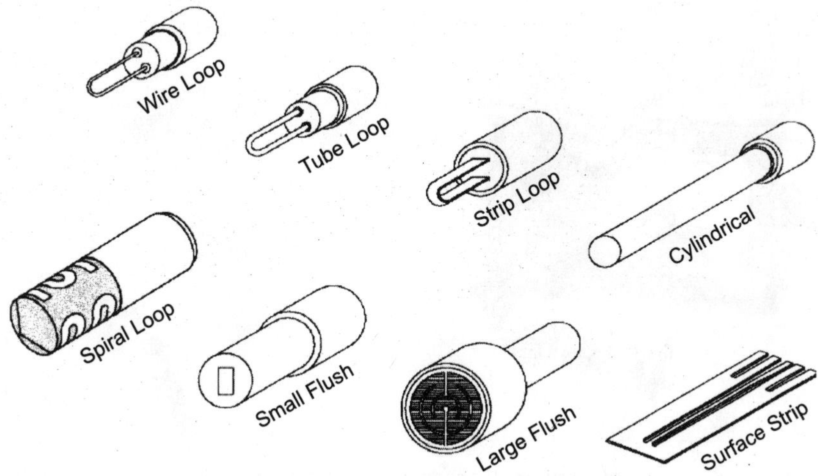

Figure 10 Various types of the ER sensing elements (Courtesy of Metal Samples Company).

thickness of the exposed element to be consumed, governs the probe replacement schedule. The less element thickness, the more its sensitivity, and the less its "lifetime". Since probe`s "lifetime" and response period are directly proportional each other, a selection of an element is a compromise between sensitivity, data frequency, and probe replacement frequency. Detection levels as low as 1 nm are achieved with elements with thicknesses of several millimeters [1].

A portable device may be used for registration of changes of sensor's electrical resistance (Figure 11, a). In this case you must go every several days to measure the electrical resistance changes of the sensor mounted in a stream. You may continually monitor by means of a transmitter connected to a distributed control system (Figure 11, b).

An increase of an electrical resistance of a metal specimen (sensor) is equivalent to decrease of a metal thickness. Thus, a slope of a curve "dial reading of ER-probe – exposure time" at the Figure 11 shows that a metal corrodes with time, and as a result a metal thickness and a metal quantity diminish. One can calculate corrosion rate according to slope of such curve according to formula:

$$k = |slope| \cdot 0.001 \cdot 365 \text{ [mm/year]} = \Delta L/\Delta t \cdot 0.001 \cdot 365 \text{ [mm/year]} \quad ...(4)$$

$\Delta L/\Delta t$ – ratio of diminishing of thickness of sensor (change of the dial reading ER-probe) in microns to the measured period in days, microns/day; 0.001 mm/μm and 365 days/year – conversion factors. The more a slope of a line "dial reading of ER-probe versus time", the more corrosion rate of a metal (Figure 12).

This method is used for erosion control [1], corrosion monitoring in district heating systems [16], in the overhead of crude distillation units in the oil refining industry [5, 17, 18], for monitoring the efficiency of cathodic protection of fuel storage tank bottoms [19], for inhibitor selection and optimization [20], internal corrosion monitoring of subsea production flowlines [21]. The design of ER-probes permits operation up to 160°C and 700 atm. [1].

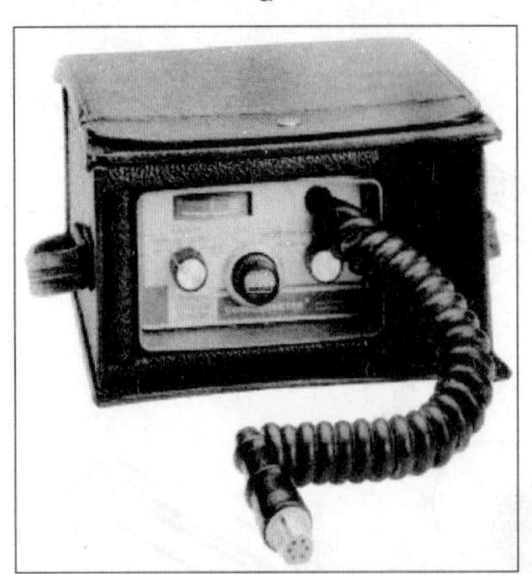

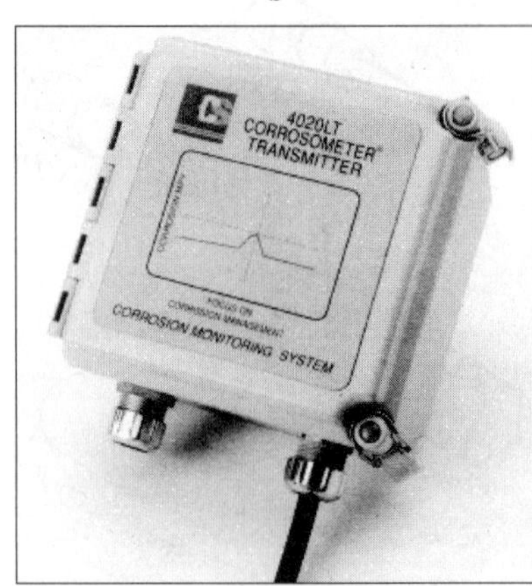

Figure 11 A portable device (a) and a transmitter connected to a distributed control system (b) for registration of electrical resistance of sensors (Courtesy of Metal Samples Company).

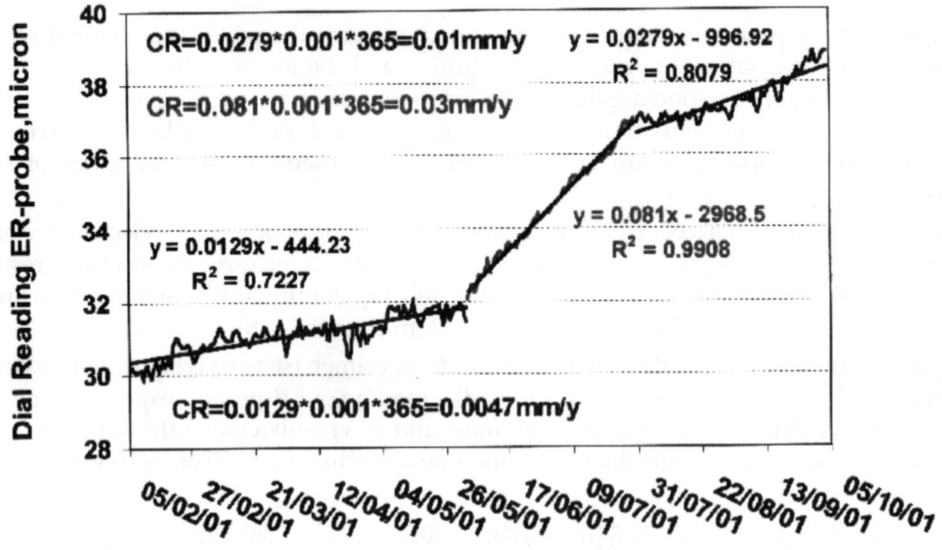

Figure 12 The ER measured in the corroded thickness of an ER-probe versus time. CR – Corrosion Rate for three various periods.

The *benefits* of the ER-method:

1. One can observe from the curve "ER versus time", that real time corrosion rate may be measured.
2. Because the measurements are based on the WL of metallic elements, one can measure the ER, and as a result the corrosion rate of metals in media both of high electrical conductivity (solutions of electrolytes) and low one (gases, hydrocarbons, two-phase electrolyte-nonelectrolyte systems, and other non-conducting solvents). ER-method is

a little bit similar to the WL method, and is contrary to the electrochemical methods. The latter ones may be used only in the media of relatively high electrical conductivity.

3. The lifetime and sensitivity of the ER-probes depend on geometrical dimensions (form) of a sensor. Therefore, one can select a sensor of any lifetime and sensitivity required. Very high corrosion rates can be quickly detected (for several hours).
4. It is possible to monitor corrosion without removing sensors and at any distance. Of course, we must replace a sensor after its full corrosion. You may select a sensor of such configuration that replacement will be needed once a year or once in two years.

The *limitations* of ER-method:

1. ER-method allows measuring only uniform corrosion. It is impossible to identify localized corrosion phenomena. Pitting may result in wrong results. Therefore, we have to examine the surface of a sensor if sudden changes in ER data occur.
2. Some salts (FeS_2, PbS) and oxides (Fe_3O_4, MnO_2) formed on the sensor's surface are electrically conductive and give rise to wrong results.
3. Some deposits (fouling) formed on the surface of a sensor also give rise to wrong results.
4. If there is no preliminary information about corrosion rate values in a particular medium, it is very difficult to select a correct sensor of a suitable sensitivity and a lifetime. Therefore, it is important to define corrosion rate values by means of a WL method in a particular medium at the beginning, and then to select an ER-probe of a suitable sensitivity and a lifetime.
5. Usually several days are needed for an accurate measurements by ER-method.

We have to remember to change sensing elements when they reach 90 to 95 percent their nominal life. For instance, if the ER-probe with the thickness of 250 microns of sensing element is used, we must change it when 240 microns will corrode. However, sensing elements must be immediately changed if localized corrosion has occurred.

In order to receive reliable data, we must compare the ER data with the WL method. How are these methods used in practice? What useful information may we receive by means of these methods? The following is an example of online corrosion monitoring using ER probe in an oil refinery.

2.3 On-line Corrosion Monitoring at the Overhead of the Crude Distillation Unit at the Oil Refinery.

Crude oil distillation is a process of crude oil separation of the petroleum distillates (fuels) proceeding in a distillation tower (also called an atmospheric column, because a distillation, or a rectification, of crude oil is carried out at the atmospheric pressure). The light petroleum fractions (gaseous methane, propane, butane, pentane, and others), water vapors, hydrogen chloride (HCl) and hydrogen sulfide (H_2S) at 120 to 130°C go out an overhead of a distillation column (Figure 13). All gases going out of the distillation column are cooled below 100°C in the air cooler and condensers (heat exchangers), usually made of carbon steel. If temperature goes down 100°C acid aqueous solutions of HCl and H_2S are formed, and cause severe corrosion of carbon steel equipment. Neutralizers and corrosion inhibitors are injected in the overhead system in order to diminish or prevent acid corrosion. In order to follow the efficiency of neutralizers and corrosion inhibitors injected in the overhead, as well as corrosion situation in

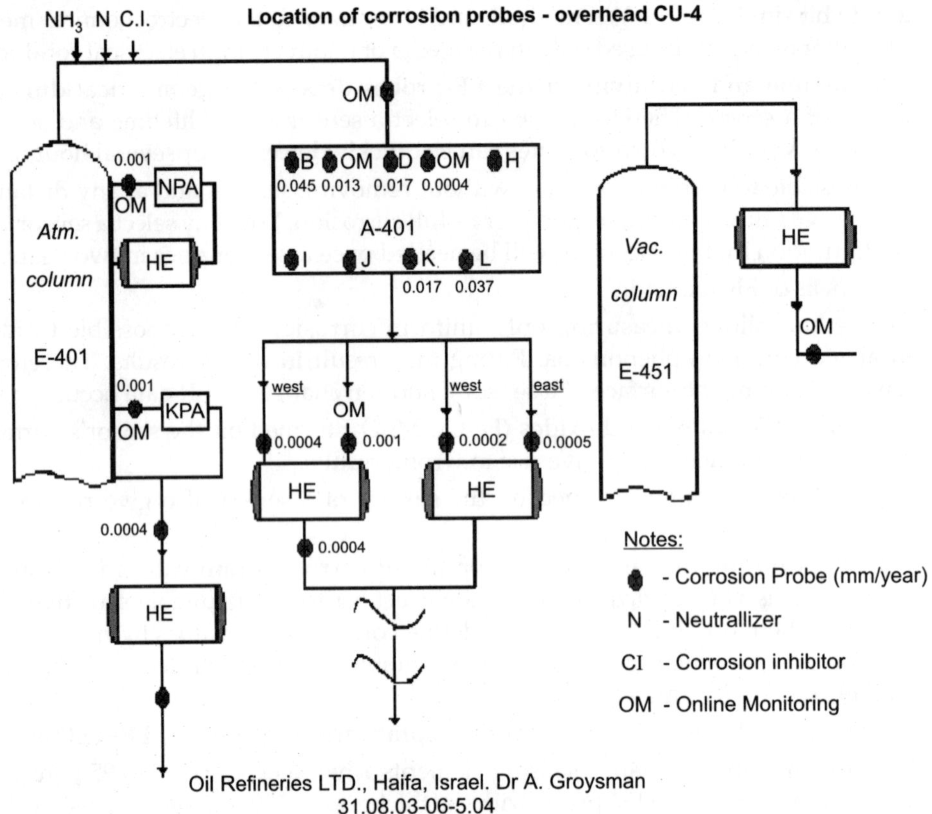

Figure 13 Corrosion monitoring at the overhead of the crude oil distillation column at the oil refinery. OM – On-line monitoring (ER-probes). B, D, H, I, J, K, L – WL coupons.CU-4 – Crude Unit. NPA – Naphtha Pump-Around. KPA – Kero Pump-Around. HE – Heat Exchanger. Corrosion rates are shown near the read points.

the air cooler and condensers, both WL coupons and ER-probes are mounted in all possible places (see Figure 13). Usually heat exchangers have two inlets and two outlets. It is very important to mount WL coupon parallel to ER-probes in such inlets and outlets. The philosophy of on-line ER real-time corrosion monitoring includes the data collecting system, their treatment (performing the calculation), utilization the data, their interpretation, and comparing every 30 to 120 days with the results of the WL coupons [5, 17, 18]. The analog signals (4 to 20 mA) sent from the ER-probes are proportional to the uncorroded metal (sensor) left and are calibrated to give an indication of total corrosion of the carbon steel sensors in microns. The outputs of these probes are monitored via the existing distributed control system (Honeywell/TDC 3000) and are also sent to the Plant Information system (PI) where they are analyzed. Figure 14 shows these data as collected and stored within the PI system.

The PI system enables easy access to the history data base with any PC which is connected to the local net or by dialing-in from remote PC's. This means that you may receive information about corrosion in any place you are with the computer.

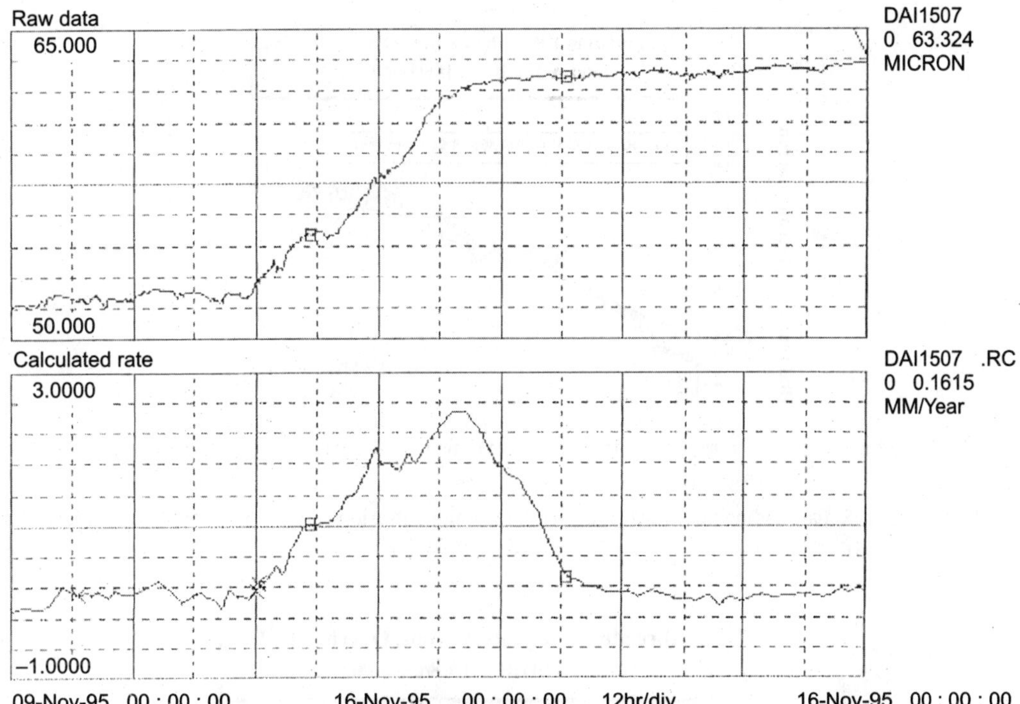

Figure 14 Dial reading and calculated corrosion rate stored in the PI system. The upper curve shows the changes (diminishing) of sensor's thickness (in microns) with time as a result of its corrosion; the lower curve shows the corrosion rate in mm/year (differential of the upper curve).

Performing the calculations. Based on the field instrument readings corrosion rates are calculated by means of differentiation of received data regarding given period. The simplest way is to take all readings that were collected within the Information system and find the regression line which best fits the data (Figures 15 and 16).

Another technique which is more suitable for on-line calculations is filtering by averaging over periods of time and making the corrosion calculations between separated averages. There are two parameters to adjust in the application of this algorithm: the length of the period to be averaged to give representing values and the time to separate between these two values (Figure 17).

The calculated corrosion rates depend on measured period, and these data may be presented for various periods, for example, 2 hours – 2 days, 5 hours – 5 days, 7 hours – 7 days, etc. (Figure 18). These data show how many hours are averaged and what is the distance in days between these averages. These calculations are continuously done and the results are stored in the system story.

Long periods (usually above 30 days) give integral corrosion rate values which we must compare with the WL coupon's data. One can observe that ER-data coincide with the data of WL method (Figure 19), but the latter does not allow to follow changes in corrosion as it happens during this period (real-time).

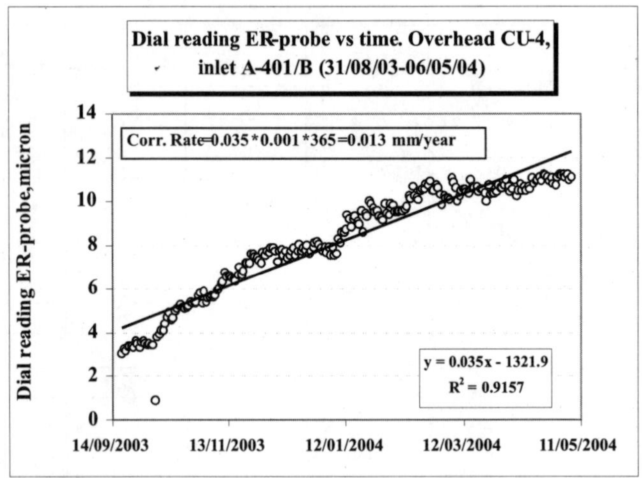

Figure 15 ER dial reading and linear regression (average corrosion rate for all period). CU – Crude Unit. A-401/B – Air Cooler.

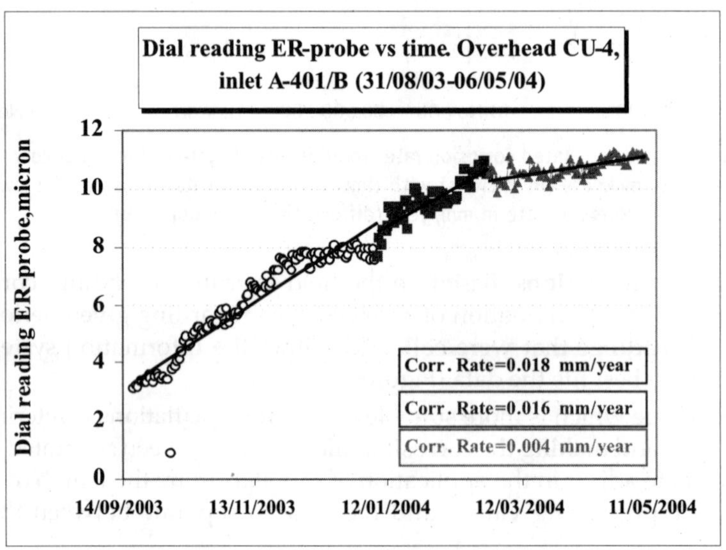

Figure 16 ER dial reading and linear regression. Period of 8 months is divided for three periods with various corrosion rate (compare with Figure 15).

On-line ER data allow analyzing during various small periods (several hours). These data reflect process changes as it reflects the corrosion and enables the experts to find the reasons for the upsets and to try to improve or eliminate the causes and minimize the corrosion damages (Figure 20). One can define, that high content of chlorides caused high corrosion rate and high content of iron in condensed water in the overhead at the crude oil distillation column.

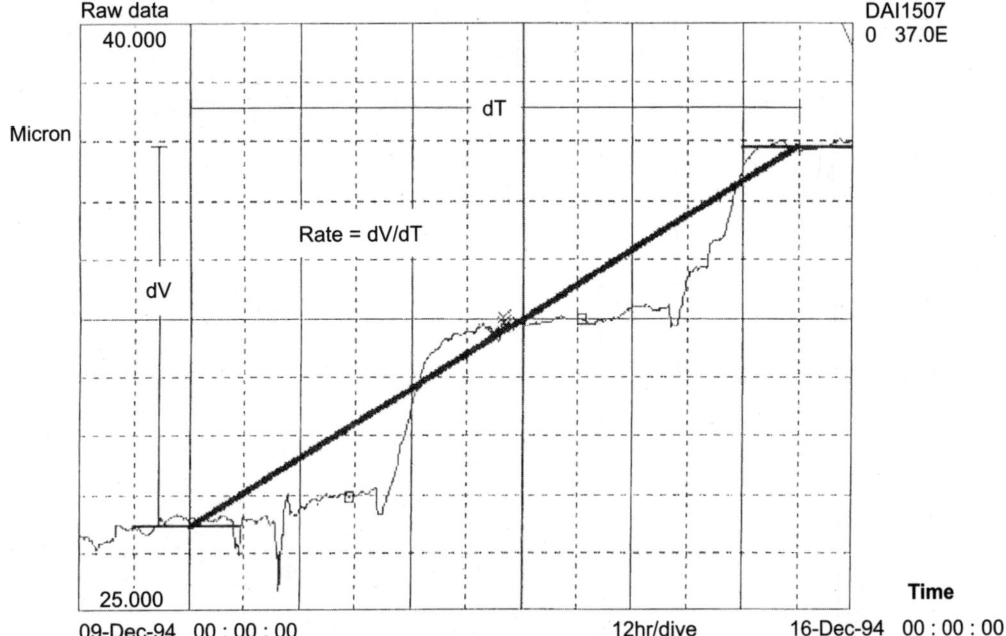

Figure 17 On-line corrosion rate calculation.

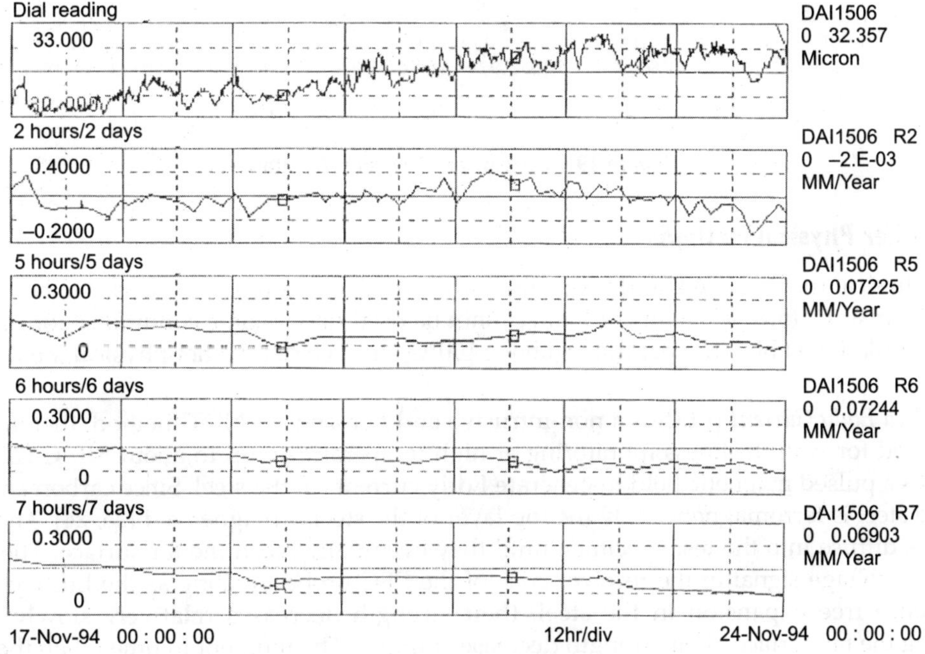

Figure 18 Averaged corrosion rates versus time base.

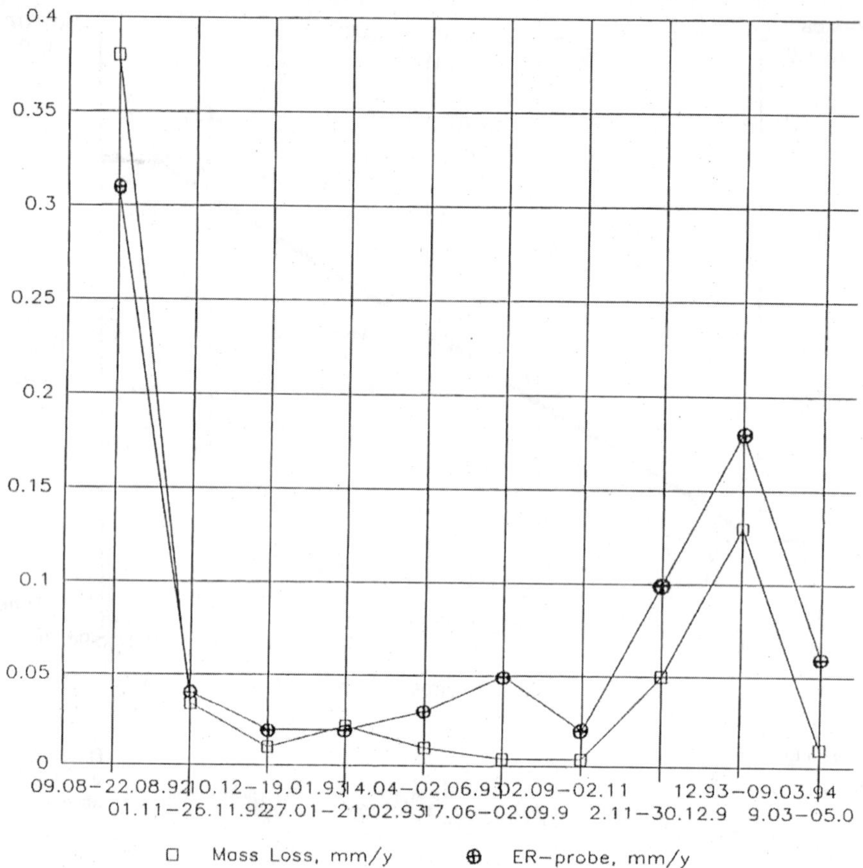

Figure 19 Comparison of ER and WL data.

2.4 Other Physical Methods

Some physical methods (WL and ER-probes) are intrusive and require a sensor to be inserted into the stream. However, many of them cannot be used under some conditions, for instance, under insulation at high temperatures (above 100°C). In such cases other physical methods are used.

Pulsed Eddy Current (*PEC*) is a non-intrusive and non-contact NDT method, therefore can be applied for wall thickness monitoring at high temperatures, up to about 540°C [22]. PEC employs a pulsed magnetic field to generate Eddy currents in the steel. Since carbon and low-alloyed steel is ferromagnetic, only the top layer of the steel is magnetized [23, 24]. The Eddy currents diffuse into the test specimen until they eventually reach the far surface. Then they induce a voltage signal in the receiver coils of the PEC probe. As long as the Eddy currents experience free expansion in the steel, their strength decreases relatively slowly. Upon reaching the far surface, their strength decreases rapidly. The moment in time when the Eddy currents first reach the far surface is indicated by a sharp decrease in the PEC signal. The onset of the sharp decrease point is a measure of wall thickness. An earlier onset of this sharp decay

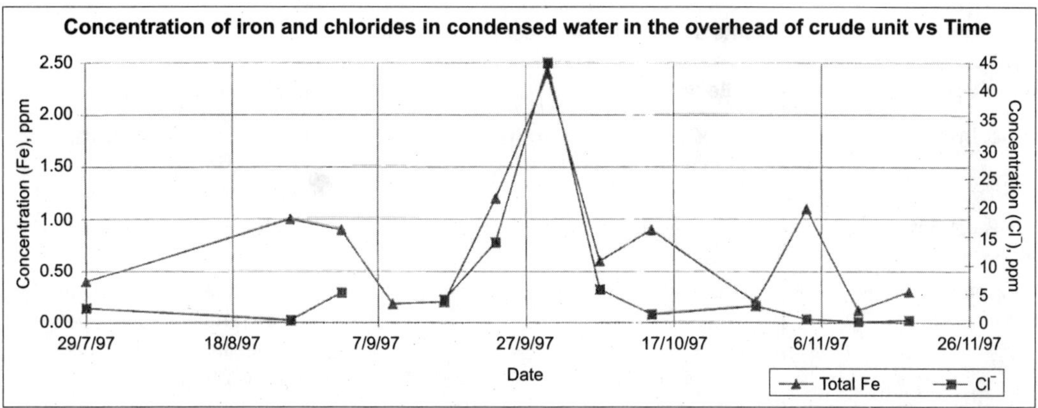

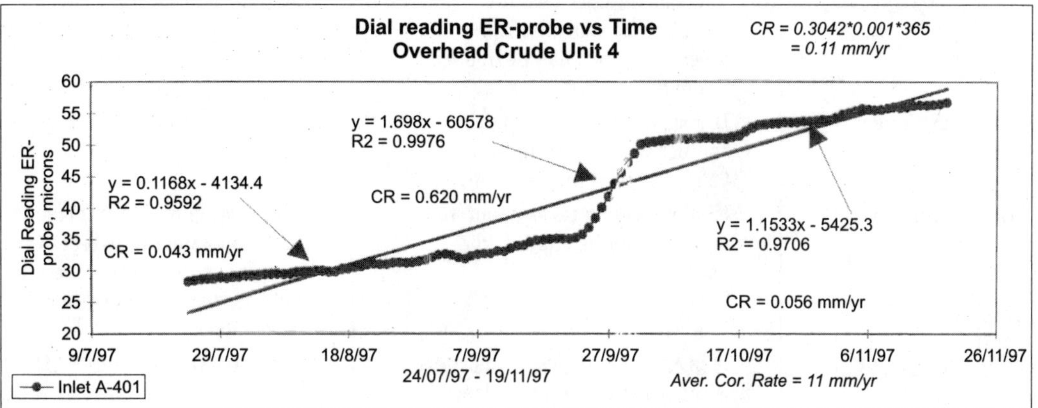

Figure 20 The ER data (lower graph) and the chemical analysis results of condensed water after heat exchangers (upper graph) in the overhead distillation column.

of one PEC signal compared to a reference signal indicates wall loss. PEC has a much better reproducibility than ultrasonic wall thickness measurements and has been applied to monitor wall thickness in piping of refineries and oil production platforms. Physical methods (NDT techniques) used for CM under thermal insulation are summarized in Table 1.

3. CONTROL OF THE ENVIRONMENT (CHEMICAL ANALYTICAL, PHYSICO-CHEMICAL AND MICROBIOLOGICAL METHODS).

Chemical analytical and physico-chemical methods probably were the second one after visual analysis in CM methods and include chemical and microbiological analysis of the environment (content of dissolved ions, gases and other compounds, microorganisms), and its physico-chemical parameters (see chapter 1.2). These methods allow to detect both corrosion level (sometimes also corrosion rate) and corrosiveness of the environment. They must be defined for specific streams at every enterprise: cooling water, condensed water, two-phase systems water + hydrocarbons, boiler feed water, etc. All chemical treatment programs in the cooling

Table 1 NDT for CM under thermal insulation [25]

Technique	Benefits	Limitations
Visual Inspection	It is possible to quantify the areas of corrosion.	Needs to remove insulation. It is impossible to measure remaining wall thickness.
Dye Penetrant	It gives indication for pitting or SCC.	Needs to remove insulation. Only for stainless steels.
Ultrasonic Thickness Measurement	It gives remaining wall thickness.	Needs to remove insulation. It needs surface cleaning before measurements. It is impossible to detect SCC.
Guided Wave Ultrasonic	The probe is applied at large intervals of pipes (~ 6 m). Period of measurements is several minutes. Can be applied in operation.	Needs to remove insulation. It does not provide for localised corrosion, gives percent wall thickness loss, only for pipes. It is impossible to detect SCC.
Pulsed Eddy Current	It does not need to be in contact with metal surface. Can be applied in operation.	It is impossible to detect pitting and SCC.
Flash Radiography*	Film processing takes about 15 minutes. Can be applied in operation.	It is used up to 1 m in diameter. It is impossible to detect SCC. Contrast and resolution are not good as that for conventional radiography.
Profile Radiography*	It gives remaining wall thickness without removing insulation.	It is impossible to detect SCC. It is used only for pipes of small sections.
Real-Time Radiography*	It gives remaining wall thickness without removing insulation.	It is impossible to detect SCC. It gives only the video images or recorded using a standard for evaluation.
Digital Radiography*	It gives remaining wall thickness without removing insulation.	It is impossible to detect SCC.
Infrared	It gives difference temperature information for defining if moisture or water are present in insulation. Can be applied in operation.	It does not detect corrosion under insulation.
Neutron Backscatter	It gives accurate information about the presence of moisture or water in insulation. Can be applied in operation.	It does not detect corrosion under insulation.

*Cautions must be taken for radiation safety.

water systems require particular water quality, namely chemicals (corrosion inhibitors, dispersants and biocides) work only under specific chemical and physico-chemical parameters of water. These methods are described in literature [26-28], and their analysis is beyond the scope of this chapter.

Photo-colorimetric, spectroscopic (IR – Infra-Red, UV – Ultra-Violet, visible light), Atomic Absorption, Ion Plazma Couple (IPC), liquid and gas chromatography, ion-selective electrodes, fibre-optic electrodes, and pH-meters are used in the chemical analytical methods.

Many crude oil refineries combine WL and ER methods with chemical analytical ones of streams (measuring of iron, copper, pH, chlorides and sulfates) [5, 17, 18, 29]. The chemical analysis of iron in oil and gas products is standardized [30]. Analytical measurements of iron and nickel in crude oil and gasoil are used for evaluation of corrosiveness of these organic liquids in the presence of naphthenic acids [31]. Test methods for measuring pH of soil and soil resistivity are standardized [32-34]. The question is what are the dangerous values of analytical parameters (pH, iron, copper, chlorides, microorganisms, etc.) determining corrosion level (high or low) in the system? Usually allowable values of some analytical parameters in water, fuels, other streams, and soil are recommended which show the possible danger of corrosion in the system (Table 2).

Table 2 Effect of chlorides, sulfates and pH on corrosion of burried steel pipelines [12]

Parameter	Unit	Concentration	Degree of corrosivity
Chlorides	> 5000	ppm	Severe
	1500 to 5000		Considerable
	500 to 1500		Corrosive
	> 500		Threshold
Sulfates	> 10000		Severe
	1500 to 10000		Considerable
	150 to 1500		Positive
	0 to 150		Negligible
pH	< 5.5	–	Severe
	5.5 to 6.5		Moderate
	6.5 to 7.5		Neutral
	> 7.5		None (alkaline)

These values depend on specific system (type of materials and media used). For instance, if equipment is made of iron and copper alloys, total iron and copper content in cooling water systems must not exceed 1 and 0.05 ppm respectively. For condensed water in overhead of the crude oil distillation columns at the oil refineries recommended values are: pH = 5.5 to 6.5; 0.5 ppm total iron, and 20 ppm chlorides. Otherwise, corrosion occurs. In many cases, it is important to monitor not absolute values, but their trend. That is sudden changes of these values can show beginning of corrosion (see Figure 20). Sometimes hydrogen sulfide and ammonia are measured in condensed water at the overhead, but these data do not give useful information. Chemical analytical methods are very sensitive and allow to measure corrosion rates in the cases when WL method is not applicable. For instance, copper alloys are very stable in cooling water in the presence of copper corrosion inhibitors, and WL method does not give reliable results, but atomic absorption allows to measure dissolved copper in solution. Any analytical method of determination of metallic cations in water or any other medium after immersion of coupon in this medium, allows to calculate the corrosion rate of the metal (supposing that all coupon's area dissolves and metallic ions are in the solution):

$$k = C \cdot (1/A) \cdot V \cdot \rho_s \cdot (1/\rho_m) \cdot (1/t) \cdot 87.6 \qquad ...(11)$$

k – corrosion rate, mm/year; C – concentration of dissolved metal in solution, determined by analytical method, ppm (mg/kg); A - area of coupon, cm^2; V – volume of solution, Liter; ρ_s – density of solution, kg/liter; ρ_m – density of metal, g/cm^3; t – period of immersion of coupons in solution, hours; 87.6 – coefficient of inversion to mm/year. If corrosion products are formed on coupon's surface, they must be removed and dissolved in the same solution.

Dissolved oxygen plays important role in corrosion of boilers and its concentration is strictly restricted [35, p. 10]. The same concerns the content of iron, copper, total hardness, silica, total alkalinity and electrical conductance of boiler feed water and boiler blowdown [35, p. 98].

The benefits of the *chemical analytical* methods:

1. High sensitivity to all metals dissolved in liquids.
2. Estimation of corrosion rate of all metals in liquids under specific conditions.
3. The trend of analytical parameters in the environment can show sudden corrosion (see Figure 20), or any danger for corrosion occurrence.
4. Determination of concentrations of dangerous metals (Cr, Co, Pb and Zn) for the environment can decide the problem of maximum contaminant levels for regulated drinking water suppliers.

The disadvantages of the *chemical analytical* methods:

1. *Quantitative* estimation of corrosion rate is applicable for measuring only in liquids. The presence of corrosion products on metallic surface can give rise wrong results.
2. Chemical content of the environment (water, soil and atmosphere) gives *qulitative* estimation of corrosion situation.
3. These methods require sometimes much time for receiving results.

3.1 Monitoring of Microbiological Activity Towards Metals.

Microorganisms can be responsible for corrosion in many systems: cooling water, drinking water, crude oil, fuel, and water storage tanks, heat exchangers, pipes, and other systems. Different methods of microbiological analysis in water, soil, crude oil, fuels, and sludge are desrcibed in [27, 36-39]. There is no one method of MIC monitoring. Following techniques are used for biofilm monitoring: collection of samples on exposed studs or special coupons and afterward examination (Figure 21), electrochemical devices, or the measurements of the change of heat transfer over a tube array arising from the influence of biofouling, scale and corrosion products.

Microbiological examination includes identification of microbial activity, that is various kinds of *planktonic* (free-floating bacteria whose movements are controlled by water movement – not attached to surfaces) and *sessile* (attached to solid surfaces) microorganisms by means of

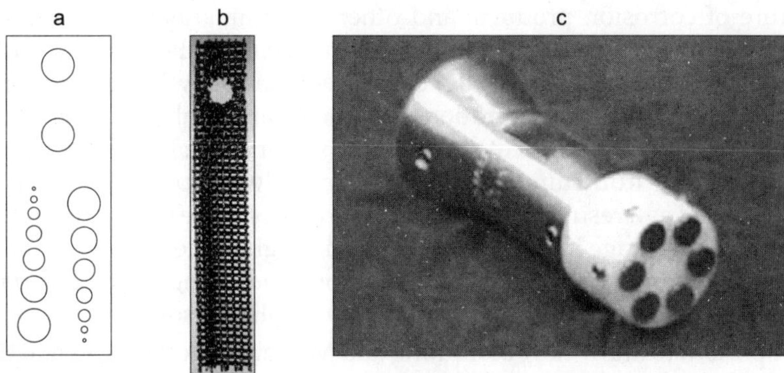

Figure 21 Special metallic coupons with holes (a), Ni-Cr screen gauze (b) and" bioprobe" (c) for biofouling collection and afterward examination (Courtesy of Metal Samples Company).

"kits", biosensors, and conventional methods of microbiological analysis (bacteria growth on special artificial media) [40, 41]. Direct visualization includes the usage of the fluorescent dyes that cause cells to light up under ultraviolet radiation. Use of commercial "kits" allows quickly to assay the presence of enzymes associated with microorganisms suspected to cause corrosion. The most common way to assess microbial populations in various industrial streams is use of growth artificial media for particular microorganisms associated with corrosion. In this case one has to wait for the results more time than in the case of kit`s use. Microbial activity may be determined by measuring the amount of adenosine triphosphate (ATP) in industrial streams. This metabolite drives many reactions in the cells. Detection of other metabolites such as organic acids, some esters, hydrogen sulfide or methane by analytical methods can indicate biological activity in the biofilm. Identification of proteins (for example, lipopolysaccharide) or nucleic acids may also show quickly the microbial activity. It is described on-line real-time determination of total biological activity with fluorescent bioreporters [42]. It is recommended to determine the presence of microorganisms in deposits by means of mixing with 2 wt% OsO_4 solution [43]. Microorganisms intensively absorb osmium. The further determination of osmium in the deposits by means of one of physico-chemical methods (for example, EDS - energy dispersive spectroscopy) may point out the presence of microorganisms. Some physico-chemical methods, such as scanning vibrating probe, allow examining electrochemical conditions at the metal surface when microorganisms are present on it.

The question is what is the allowable value of microorganisms for usage of metallic equipment out of MIC danger? It is recommended that total bacterial count (TBC) in cooling water and fuels must not exceed 10^4 bacteria/ml, and sulfate reducing bacteria (SRB) must not exceed 10 bacteria/ml. Otherwise, severe MIC occurs or can occur. It is very important to carry out the trend of measurments' results. Recommended allowable values of microorganisms depend on specific system. For instance, if there are not favorable conditions for the proliferation of microorganisms in some industrial systems, the TBC values 10^4 to 10^5 bacteria/ml are out of danger. But ... if leakage of oil occurs into water system, or there is no enough flow rate of water, or water phase appears in the fuel, even the TBC = 100 bacteria/ml may immediately cause severe growth of the microorganisms in water and then give rise to corrosion during several months and even weeks. Chemical and physico-chemical analysis of deposits (mixture of corrosion products and other contaminants with microorganisms and their metabolitic products) may give the important information about microbiological activity. If the quantity of inorganic matter in the deposits is less than 90 wt%, organic contamination is present, and MIC is possible. Concentration of organic carbon above 20 wt% in the deposits also shows possible MIC. Sometimes the presence of sulfur >1 wt% in deposits may show SRB activity. The presence of iron sulfides in deposits not always points out SRB activity. Iron sulfides may be formed as a result of presence of hydrogen sulfide in the environment: in crude oil and fuels, and their leakage into the cooling water. High concentration of chlorides (Cl^-) in aqueous solutions shows the possible activity of iron- and manganese-bacteria, especially if high concentrations of iron and manganese ions are revealed in solution.

It is very important to know how biofouling and MIC are developing in real time. The most spread method which found wide application in industry is the method of measuring of deposit accumulation on the metal surface.

3.1.1 Deposit accumulation test ("Heat transfer resistance method").

This method is based on the principle that microorganisms attaching inside of heat exchanger tubes (if water flows inside) result in decrease of heat transfer. We have to measure the resistance to heat transfer from outer surface of tube inside [38, pp.137-150; 44]. It is simple to do this, if we measure the inlet and outlet water temperature and flow rate of water in heat exchanger. Some companies manufacture such devices (for instance, corrosion and deposit monitoring system CorrDATS of Rohrback Cosasco Systems, Inc.) similar to small heat exchanger that allow carrying out measurements needed and to calculate the "heat transfer resistance" values (Figure 22).

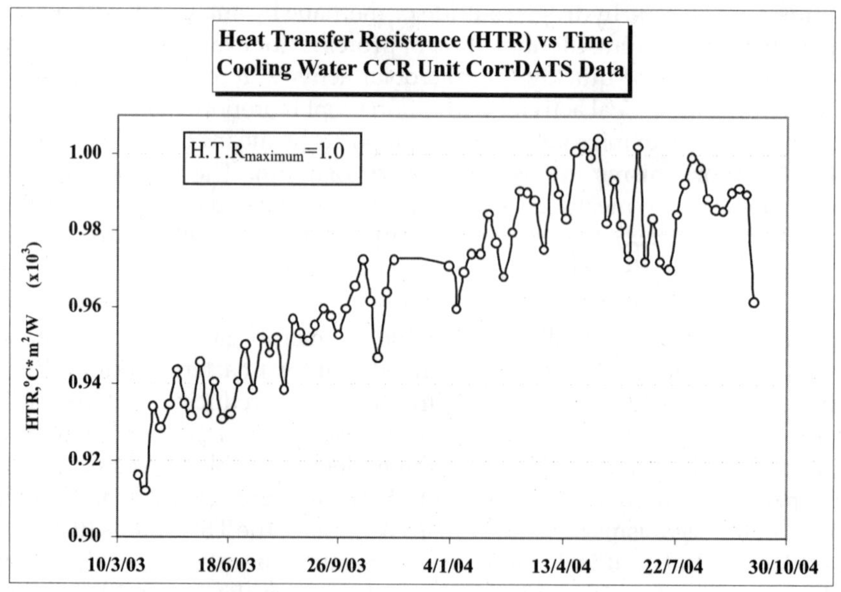

Figure 22 Heat transfer resistance (HTR) vs time (cooling water system).

This value shows the formation of all possible organic and inorganic deposits inside of heat exchanger tube: microorganisms, oil and other hydrocarbons, scale of carbonates, phosphates, silicates, corrosion products, silt, dirt, etc. Thus, we do not know the exact contribution of microorganisms in the "heat transfer resistance". Even if we would be able to define only the biofouling formation, we would not be able to know which part and which kind of microorganisms is responsible for MIC. In spite of this substantial drawback, the "heat transfer resistance" method is used together with microbiological analysis and corrosion monitoring. Certainly, we have to wait for biofouling formation much time, and we would not know what happened during this period. Electrochemical methods found application for the MIC on-line monitoring.

Electrochemical methods of MIC monitoring in real time are based on the principle, that MIC is an electrochemical process, and therefore influences such electrochemical parameters of corrosion as electric current, electric potential, and electric resistance of metals. Linear polarization resistance (LPR), electrochemical impedance spectroscopy (EIS), and electrochemical noise measurements (ENM) are used for this purpose [38, pp.265-287; 45, 46]. Electrochemical methods give general information about all electrochemical corrosion processes occurring on the metal surface. In other words, the changes of the measured electrochemical parameters occur as a result of all corrosion processes, even which do not relate to MIC, and it is impossible to differentiate the only MIC from other electrochemical corrosion processes.

3.2 Hydrogen Monitoring

Some corrosion reactions can produce hydrogen (atomic H and molecular H_2), which can penetrate into metals and alloys and cause blistering, and hydrogen induced cracking (HIC). This hydrogen can be collected either in an intrusive or non-intrusive devices called *hydrogen probes* [9, 47]. Intrusive probes (often called "finger" probes) inserted into a vessel or pipe. They consist of a steel sensing element which have a hollow space inside connected to a pressure devise which monitors the build up (or vacuum loss) of hydrogen pressure. Non-intrusive hydrogen probes utilize an externally applied cell, patch or gauge to monitor the rate of hydrogen egress from the outer surface of a vessel or pipe wall. Electrochemical cell is also used where hydrogen reacts electrochemically producing a current signal. This method is standardized [48]. Hydrogen monitoring is used in wet sour (H_2S) environments (oil, gas, refining and petrochemical industries).

The benefits of *hydrogen monitoring*:

1. It provides information about potential hydrogen attack.
2. In some cases, the method provides on-line real-time monitoring of hydrogen permeation.

The limitations of *hydrogen monitoring*:

1. It provides qualitative information.
2. Hydrogen probe data usually do not correlate with other CM methods, for instance, WL method.
3. The hydrogen probes can be subjected to error or delays in reading process transients, since many processes with hydrogen participation can be prone to rapid changes in hydrogen charging severity with plant operating conditions.

4. ELECTROCHEMICAL METHODS

Electrochemical corrosion monitoring methods are based on the established fact, that corrosion of metals and alloys in solutions of electrolytes is an electrochemical process. This means that electrochemical methods are suitable only for corrosion of metals in media with highly enough electrical conductivity. Thus, if corrosion reactions have electrochemical nature, electrical potential, electrical current and electrical resistance (as derivative of the two first parameters) formed on the metal surface are main characteristics of corrosion. If we measure these three parameters, we are able to characterize and monitor corrosion of metals in conductive media. Electrical potential of a metal is a thermodynamic characteristic of corrosion process, and one can use the formula for its definition: $\Delta G = - nFE$ (n - number of moles of electrons involved in the metal dissolution; F – Faraday constant, E - difference of electromotive forces for anodic and cathodic reactions). We must be careful, as usually we measure the *corrosion potential* and not a *reversible electrical potential* used in this formula. The electrical current shows the kinetics characteristic of corrosion and is connected with a physical loss (m) of a metal by Faraday law:

$$I \cdot t = (n \cdot F \cdot m/M) \qquad \text{...(12)}$$

I – a corrosion current; t – a period of current flow; n – a number of moles of electrons involved in the metal dissolution; F – Faraday constant; M – a molecular weight of the metal; m – a mass of corroded metal.

Corrosion rate of a metal is proportional to electrical, or corrosion, current. Based on three electrochemical parameters *(electrical potential, electrical current* and *electrical resistance*), electrochemical methods may be divided into qualitative and quantitative ones. Qualitative methods include the identification of corrosiveness of media (determination of *redox potential*) and determination of *corrosion (electrode) potential* of corroded metal surface. Quantitative methods include the identification of *corrosion current* on a metal surface. Any electrochemical method that gives rise to the calculation of a *corrosion current* and as a result a *corrosion rate* may be related to a quantitative method.

The advantages of the *electrochemical methods*:

1. The electrochemical methods allow receiving the quickest dynamic information about corrosion rate in real time. Therefore we can follow the fast changes in the corrosiveness of a system (a system includes a metal and an electrolytic environment).
2. From the engineering point of view, the electrochemical methods are convenient as they give electrical signal which allowing to react quickly and to store the data in the electronic devices.
3. Some electrochemical methods use polarization of metal in the environment. Such methods allow receiving useful information about active and passive regions at the metal surface, mechanism of corrosion and inhibition, as well as a corrosion rate.

The limitations of the *electrochemical methods*:

1. Electrochemical methods are based on measurement of electric current flowing through solution. Therefore there are limitations for the electrochemical methods usage in solution with low electrical conductivity: in non-electrolytes (organic solvents, fuels), two-phase systems electrolyte – non-electrolyte (with large quantity of the latter), or in very pure water.

2. The corrosion rate measurements assume uniform (general) corrosion.
3. If used in multi-phase systems, non-conductive phases may fairly quickly isolate the metal electrodes, and invalidate the readings until the electrodes are cleaned again.
4. Non corrosive electrochemical reactions occurring on metal surface result in mistakes. For example, hydrogen sulfide (H_2S), hydrogen peroxide (H_2O_2), or other species in solutions take part in the oxidation – reduction reactions. Such non corrosive reactions give rise to additive electrical current non referring to a corrosion current.
5. Not always all metallic surface corrodes, but calculations relate to a current density (A/cm^2), and to a weight loss of all metallic surface ($gram/cm^2$). Therefore we do not receive a real corrosion rate.
6. So many factors influence the electrochemical processes on a metal surface, and they measure so small changes in the electrical current and potential, that the electrochemical methods are not so reliable as WL method, and their repeatability is bad. But … "trend" of the electrical parameters may give useful information about corrosion behavior of metals.

The results of the electrochemical methods must be compared with the WL data. Electrochemical methods use analysis of the interface metal – electrolyte with direct current (DC), alternating current (AC), and without electrical current. The measurement of oxidation / reduction potential and electrode (corrosion) potential relate to the technique without electrical current. DC technique uses the linear polarization resistance (LPR), zero resistance ammeter (ZRA), and potentiodynamic polarization. AC technique uses the electrochemical impedance spectroscopy (EIS), the electrochemical noise (ENM), and harmonic distortion analysis (HDA). We shall describe some of them.

4.1 The Measurement of Oxidation / Reduction (Redox) Potential (ORP).

Atoms of different elements may be present in solutions in various oxidized states, for instance, Fe^{2+} and Fe^{3+}, Mn^{2+} and Mn^{4+}, Co^{2+} and Co^{3+}, Cu^{2+} and Cu^{+}, S^{2-} and SO_4^{2-}, O_2 and OH^-, Cl_2 and Cl^-, etc. If such ions and molecules are present in solution in significant quantities, and when electron exchange with the electrode is sufficiently fast, oxidation / reduction (or redox) equilibrium is established at some inert electrode (for example, platinum), giving it a well – defined potential, or reversible *oxidation / reduction potential* (ORP):

$$E = E^o_{Ox/Red} + (RT/nF)\ ln(a_{Ox}/a_{Red}) \qquad ...(13)$$

$E^o_{Ox/Red}$ – standard electrode potential (at $a_{Ox} = a_{Red} = 1$); a_{Ox} and a_{Red} – activities of oxidized and reduced forms; $R = 8.314\ J{\cdot}mol^{-1}{\cdot}K^{-1}$, gas constant; T – temperature in Kelvin; n – number of electrons participating in the redox reaction; $F = 96{,}500\ C{\cdot}mol^{-1}$ (Faraday constant).

The redox potential E is a voltage which is proportional to the ratio of concentrations of oxidized to reduced states of specific substances in liquid solutions and is measured on the platinum electrode in relation to some reference electrode (calomel, for example). Atoms of different elements are present in various oxidized states (oxidizers and reducers) in electrolytic media, may take part in cathodic and anodic corrosive reactions, and may influence microbiological activity. Therefore an ORP is a qualitative parameter for an identification of corrosiveness of a medium. It is useful to monitor the "trend" of ORP in order to define the changes in the corrosiveness of an environment. The presence of O_2, Fe^{3+}, and other oxidizing agents in a solution results in a deficiency of electrons on a metal surface and will attempt to acquire electrons. Likewise, the presence of H_2S, Fe^{2+}, Mn^{2+}, and other reducing agents in a solution results in the situation that electrons are available and will attempt to give up electrons. In an oxidizing environment, a higher ORP values exist, whereas a lower ORP values exist in a more reducing environment. The polarity and value of ORP depend on the concentration of oxidizing and reducing species in the environment. Here are not so many benefits of the ORP method:

1. It is very simple, quick, and can be used for control the injection of oxidizing biocides or other additives.
2. It is used for the control of microbiological activity.
3. It can serve as a rough indication of the corrosiveness of the solution.

There are more limitations of the ORP method. The main drawback is that the measured ORP is referred to the bulk conditions in the solution and is different from the ORP formed at the surface of equipment. Some ions which are present in various oxidized states, do not take part in corrosive anodic and cathodic reactions, but influence the ORP. Deposits (for example, biofilm) may be formed on the platinum electrode, and do not allow to some aggressive species (for example, oxygen molecules, chloride anions, etc.) to penetrate to the electrode's surface. Such situation may give rise to mistakes in the determination of real ORP. Therefore this method is applicable for corrosion control in boiler feed water, and microbiological growth in cooling water. This is the only electrochemical method concerned with the determination of corrosive parameter of the environment. Other electrochemical methods deal with corrosive parameter of the border metal - environment. Now we shall consider the electrochemical methods designated to the determination of corrosive properties of metal equipment in contact with solutions of electrolytes, that is connected with the border metal - environment. These methods are the measurement of corrosion potential, LPR, ENM, ZRA, EIS, and HDA.

4.2 The Measurement of a Corrosion Potential of Metallic Equipment

Corrosion potential is an electrical (electrode) potential of a metal surface formed in contact with an electrolyte solution when an electrical current does not flow through a metal. The voltmeter with high inner electrical resistance (above 10^6 Ohm) is used for this purpose. Corrosion potential measured at open electrical circuit is called an *open circuit potential* (OCP), and is used as a *qualitative* characteristic of a corrosion process on a metal surface. The *corrosion potential* is measured in comparison with reference electrodes such as calomel, silver/silver chloride, or copper/copper sulphate electrodes. This method is similar to the redox potential measurement, but instead of inert platinum electrode, the metallic equipment or some metallic specimen made of the same equipment metal is used. If a corrosion potential of a metal increases with time (changes in a positive direction), passivation occurs, and corrosion slows down. But pitting corrosion may occur! If a corrosion potential of a metal decreases with time (changes in a negative direction), corrosion process intensifies. But ... it is impossible to answer how quick corrosion may be decreased or intensified. This general rule is not always correct, as many factors can shift the potential values towards more positive or negative values, but these shifts may not necessarily be related to the occurring of severe corrosion of steel or other alloys. For example, decrease of oxygen concentration at the metallic surface causes more negative corrosion potential values, but they may not necessarily be associated with a high probability of steel corrosion. The presence of cathodic (or mixed anodic and cathodic) inhibitors causes the corrosion potential to shift towards a more negative value with a corresponding reduction in the severity of the steel corrosion. It is used to assist prediction of corrosion behavior by comparison with polarization data obtained from laboratory or site polarization scan [49]. This method is used to control cathodic and anodic protection systems, which hold the metal at controlled immune or passive potential respectively [8]; and standardized for measurement of the corrosion potential of rebars in concrete [50-52]. This method provides general guidelines for qualitative evaluating corrosion in concrete structures (Table 3). The measurements of potentials of epoxy (or other paints) coated and galvanized rebars, as well the structures under cathodic protection are not suitable.

Table 3 Probability of corrosion according to half-cell readings

Half-cell potential reading vs Cu/$CuSO_4$	Corrosion activity
Less negative than – 0.200 V	90% probability of no corrosion
Between – 0.200 V and – 0.350 V	An increasing probability of corrosion
More negative than – 0.350 V	90% probability of corrosion

There are the same advantages of *corrosion potential* measurements as for the *redox* method: it is very simple, fast and can be used for an on-line monitoring. There are more drawbacks:

1. Changes in corrosion potential with time do not allow defining which process, cathodic or anodic, is responsible for increase or decrease of a metal corrosion.
2. It is impossible to identify the corrosion phenomenon type, such as a pitting corrosion, a crevice corrosion, etc.
3. The method is a qualitative one and does not give information about corrosion kinetics.

Another electrochemical method, linear polarization resistance (LPR) is widely used for the quantitative determination of corrosion rate of metallic equipment in aqueous solutions of electrolytes.

4.3 Linear Polarization Resistance (LPR) method.

This method is based on the measurement of electric current formed between two similar electrodes when electric potential between them is about 10 to 30 mV (Figure 23). The propensity of the metal ions of an alloy to pass into solution, or corrode, is inferred by the relationship between a small change in electric potential (typically 10 to 30 mV) across the corrosion interface and the increase in current density, which result from a consequent change in the flow of metal ions into solution.

If we take two metal identical electrodes as at Figure 23, put into solution of electrolytes, begin to change electric potential from minus 30 mV regarding the open circuit potential (E_{ocp} = E_{corr}) of a corroding electrodes with constant rate to plus 30 mV, we receive nearly straight line "electric potential versus electric current" (Figure 24) [53]. In other words, when the corroding electrode is polarized within a few millivolts around E_{corr}, the measured

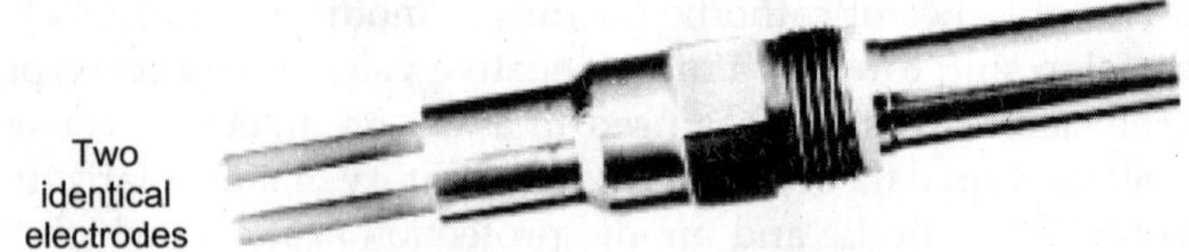

Figure 23 Linear polarization resistance probe (Courtesy of Metal Samples Company).

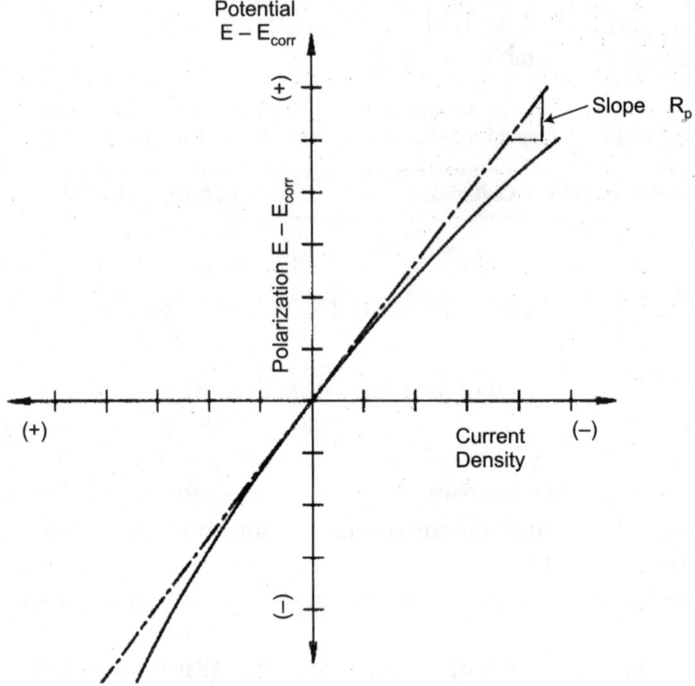

Figure 24 Electric potential versus electric current between two similar electrodes (linear polarization resistance method).

current ΔI is proportional to the corrosion current I_{corr}. The ratio of potential change (ΔE) to current change (ΔI) at the point of zero current ($i_o = 0$) gives the "polarizarion resistance" R_p to corrosion reaction, that is the slope of the linear curve corresponds to the polarization resistance:

$$R_p = (\Delta E/\Delta I)_{io = 0} \quad ...(15)$$

Milton Stern noted [54], that this is not a "resistance" in the usual sense. Qualitatively, we may emphasize that the more polarization resistance the less corrosion rate. For example, the efficiency of corrosion inhibitors in aqueous solutions of electrolytes may be evaluated according to R_p values.

The theory of this method was developed by M. Stern and A.L. Geary [54, 55] which showed in 1957 that the corrosion current I_{corr} (mA/cm^2) can be calculated if we know polarization resistance R_p:

$$I_{corr} = B / R_p \qquad ...(16)$$

Where B is the Stern-Geary constant (mV) depended on type of a metal and an environment.

The principle of the technique has a basis only in the evaluation of steady-state uniform (general) corrosion. The B values are generally taken to be in the range 26 to 30 mV for most metal-environment systems [56].

The use of the "polarization resistance" for measuring corrosion rates has one particularly important advantage. The potential range investigated is close to the corrosion potential, and the applied currents are generally smaller than the corrosion current. Thus, the nature of the corroded surface is not changed significantly, and the reactions which proceed during polarization are those which actually occur during the corrosion [54].

If the electrical conductivity of solution above 100 μS/cm ($R < 10^4$ Ohm · cm), two electrodes are used for the LPR measurements. If the electrical conductivity of solution between 1 and 100 μS/cm ($10^4 < R < 10^6$ Ohm · cm), three electrodes are used for the LPR measurements. The third electrode is used for the compensation of IR-drop and for the measurement of real electrical potential of the electrode. LPR measurements can be made both manually and automatically to produce data from 30 minutes to 24 hours [9].

LPR method is used for determination of internal corrosion monitoring of subsea production flowlines [21], corrosion rate of reinforcement in concrete [57-59], in furnaces [60], a nuclear waste tanks [61], for evaluation of the efficacy of corrosion inhibitors in oil and gas pipelines [62, 63], for measuring corrosion rates of outer surfaces of pipes at a dig sites [64] and water cooling systems [65]. Guidelines for on-line in-plant CM using electrochemical techniques are described in [66].

Using Faraday's Law, one can convert the results of electrochemical measurements (i_{corr}, mA/cm^2) to rates of uniform corrosion (CR, mm/yr) [67]:

$$CR = K \cdot i_{corr} \cdot \rho^{-1} \cdot EW \qquad ...(17)$$

CR – Corrosion Rate, mm/yr; $K = 3.27 \times 10^{-3}$, mm·g/μA·cm·yr; ρ - density of metal, g/cm^3; i_{corr} – corrosion current, μA/cm^2; EW – equivalent weight; EW = W/n (W – the atomic weight of the element; n – the valence of the element).

The benefits of the *LPR* method:

1. It allows on-line real-time measuring corrosion rate of uniform corrosion. The method provides corrosion rate data directly in a few minutes.
2. The LPR method is particularly well suited to aqueous electrolyte solutions' applications so that process upsets or other corrosion conditions can be detected very quickly to enable remedial action to be taken almost immediately. The method is used for the fast estimation of corrosion inhibitors performances in conductive environment.
3. Some systems combine this method with the electrochemical current noise measurements, to provide a determination of an initiation of localized corrosion or an indication of the stability of the inhibitor film on the metal's surface.

The limitations of the *LPR* method:

1. It is impossible to measure localized corrosion. We shall show later that two-electrode system shown at Figure 25 is used for the measuring of pitting tendency by means of electrochemical noise measurements.

2. Scale and other deposits formed on the electrode surface give rise to increase of the electrical resistance that non referred to the change of corrosion rate of the metal.
3. The redox reactions, different from corrosion reactions, contribute in the increase of corrosion current, and can result in mistakes.
4. The constant B in (16) may vary with time, and this change may result in wrong results. The method requires the prior knowledge of Tafel constants for the calculation of corrosion rates.
5. It is very important to mount correctly the electrodes in a flow stream and to prevent a "shade" of one electrode by another (Figure 25).
6. The main disadvantage is that the LPR method does not work in solutions of low conductivity, and it is impossible to monitor localized corrosion. Electrochemical noise measurements (ENM) have been introduced into practice in order to overcome the above mentioned drawbacks.

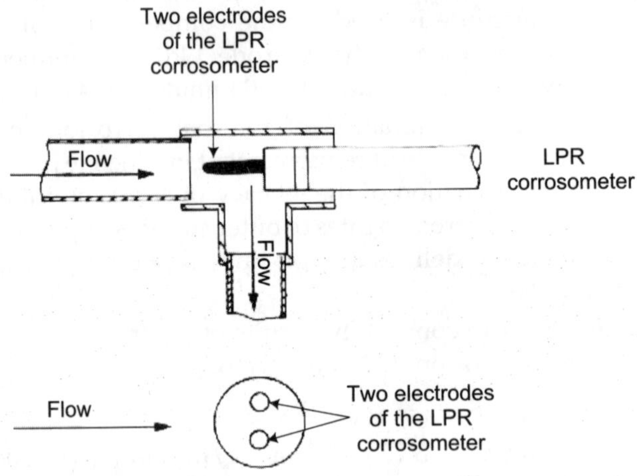

Figure 25 Correct installation of the LPR electrodes.

4.4 Electrochemical Noise Measurements (ENM)

Analysis of corrosion failures shows that 70 to 90% of them result from localized corrosion. Our data in oil refining industry show that these values are between 40 to 65%. Ninety per cent of these failures occur in 10% of time of equipments' service [68]. ENM allow monitoring localized corrosion failures. First of all we have to explain what does this mean "*electrochemical noise*", a term coined in 1979. In an "ideal world of corrosion", the electrical potential and current would be constant, metals would be corroded uniformly, and all measurements would be easy. In the real world of corrosion, metals tend to corrode non-uniformly, corrosion potentials and currents vary with time. All this is connected with the "*electrochemical noise*". All in the universe is under *noise*. *Noise* is everywhere, from atom to planetary orbits. All particle trajectories exhibit chaotic, fractal and turbulent behavior. *Noise* is opposite to silence. As it is impossible to reach the *absolute zero temperature*, it is impossible to reach the *absolute silence*. *Noise* in many cases is undesired phenomenon. People learned to use *noise* in some cases. If a metal is "ill", hole or crack is formed on the metal surface or inside, high energy goes out (leaves) the active region, and we may "listen" the energy flow effects. Acoustic emission method is based on this principle. Please, tear a piece of a paper or a tree. You will listen to some *noise* - crash, crackle. This means that we listen how the bonds between atoms and molecules break. Similar situation is in a metallic corrosion. Metal ions leave their places in a crystal and get across into a solution in the anodic process. A metal corrodes, that is some bonds are destroyed, and some bonds are formed. Some cations return into a crystal from a solution. Corrosion potential and corrosion current are not constant. Their fluctuations occur. "*Electrochemical noise*" in corrosion refers to the naturally occurring fluctuations in corrosion potential and corrosion current. Adsorption/desorption processes, pitting and crack initiation, etc., may be the sources of *noise*. If corrosion is an electrochemical phenomenon, it occurs according to a random or stochastic process. *Electrochemical noise* may be explained by the uncertainty principle of quantum mechanics. Some uncertainties or fluctuations exist in the positions and velocities of the particles (ions, electrons and molecules) taking part in the electrochemical process. Up to 1970s, scientists thought that the sources of the "*electrochemical noise*" were bias and error diminishing the reliability of the electrochemical results. Therefore scientists did not use the "*electrochemical noise*" information. Warren P. Iverson was the first who used in 1968 the *electrochemical potential noise* measurements for the analysis of the metallic corrosion.

The *electrochemical noise* technique is defined as the analysis of the spontaneous fluctuations of the electrical current and electrical potential of a corroding electrode. Two techniques are used for measuring of the electrochemical potential and the electrical current noise. In the first technique, two identical electrodes are connected through zero resistance ammeter. Changes of the electrical potential and current between two "identical" electrodes are measured regarding to the reference electrode with time. It is explicit that it is impossible to create two ideally "identical" electrodes. Therefore another technique was developed with one working electrode which is connected with a counter electrode (usually made of platinum) and a reference electrode of the potentiostat. This technique is called the "*potentiostatic electrochemical noise technique*" where both potential noise and current noise are measured at the same working electrode at an open-circuit potential. Current noise relates to current variations between working and counter electrodes, whereas potential noise refers to the variations in potential between a working electrode and a reference electrode. Various software programs were developed for mathematical treatment of measured values of potential noise and current noise.

Electrical resistance noise can be calculated as a derivative of a potential and a current noise according to the Ohm's law, and these values are concordant with the linear polarization resistance data [69]. The electrical resistance noise values are very important for analysis of efficiency of protective coatings on the metals.

Both electrochemical noise techniques with two "identical" and one electrode are used for monitoring of both general and localized corrosion, especially pitting and crack initiation. The main benefit of this method is that it is used for on-line measuring of all types of localized corrosion in real time. The theory and practice of ENM is described in [70, 71]. The pitting index (PI) or pitting factor is calculated according to [71]:

$$PI = \sigma_I / I_{corr} \qquad ...(18)$$

where σ_I is the curent recorded from ECN (Electrochemical Current Noise) (mA); I_{corr} is the corrosion current recorded by LPR (mA). Mixed quantitative-qualitative assessment is used:

a) PI < 0.01 (general corrosion);

b) 0.01 < PI < 0.1 (intermediate zone, but still predominantly general corrosion);

c) PI > 0.1 (localized corrosion).

The industrial devices are built on this principle for the *pitting tendency* corrosion monitoring, that is transform raw electrochemical noise data into a process parameter that can be understood and used by operation personnel just like a temperature or pressure reading (Figure 26). ENM are unique among other electrochemical monitoring techniques in that it can be used to detect and distinguish between different localized corrosion phenomena such as pitting, crevice corrosion and SCC [72-75]. EN probes were used for tanks containing high-level radioactive waste [76, 77], sour oil processing facilities [78], pitting tendency in cooling water systems [79].

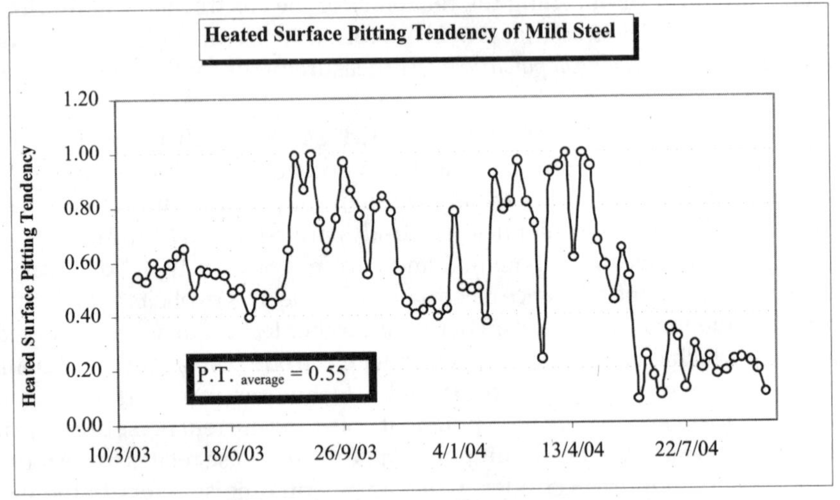

Figure 26 Pitting Tendency of heated carbon steel surface vs time in cooling water (measured by means of CorrDATS system of Rohrback Cosasco).

4.5 Zero Resistance Ammetry (ZRA)

ZRA is a current-to-voltage converter, which provides a voltage that is proportional to the current flowing between the input terminals while imposing a "zero" voltage drop to the external circuit. In the ZRA technique, a macro cell current is measured between two corroded sensor elements in an electrolytic environment. ZRA is used for measuring the galvanic coupling current between two dissimilar electrodes. The resultant galvanic current is a corrosion rate value which can be obtained through Farady`s Law (12) [80]. This method is used also for measuring the coupling current between two "identical" electrodes to monitor electrochemical current noise [70, 71]. In reality, these electrodes will be slightly different, and a small coupling current will exist. Such measurements are relevant to detecting the breakdown of passivity and the early stages of corrosion. The restriction is, if extensive corrosion is occurring on both electrodes, the macro cell current measured will not accurately reflect the severity of attack.

4.6 Electrochemical Impedance Spectroscopy

Everyone knows about the concept of electrical *resistance* – the ability of a circuit element to resist the flow of electrical current, known as Ohm's law:

$$R = E/I \quad ...(19)$$

This relationship is restricted by the ideal resistor which has the following properties: it follows Ohm's law at all current and voltage values; it's resistance value is independent of frequency; AC current and voltage signals through a resistor are in phase with each other.

The real world contains circuit elements that exhibit much more complex behavior. Instead resistance we use *impedance*, which is a more general circuit parameter. Like resistance, *impedance* is a measure of the ability of a circuit to resist the flow of electrical current. Unlike resistance, *impedance* is not limited by the simplifying properties listed above.

Electrochemical Impedance is the frequency dependent, complex valued proportionality factor, $\Delta E/\Delta I$, between the applied potential (or current) and the response current (or potential) in an electrochemical cell [81]. This factor becomes the impedance when the perturbation and response are related linearly and the response is caused only by the perturbation. The value may be related to the corrosion rate when the measurement is made at the corrosion potential.

Electrochemical Impedance Spectroscopy (EIS, also known as *AC impedance*, or the *electrochemical impedance technique*) is an electrochemical technique in which a low amplitude alternating potential (or current) wave is imposed on top of a potential (usually the corrosion potential and zero imposed current) [82]. AC means an alternating current. In other words, EIS monitors the electric response of the border metal-environment to the applied AC signal over a frequency spectrum. *Electrochemical Impedance* is normally measured using a small excitation signal. The relationship between the voltage and current is used to make a judgement about the corrosion rate and other corrosion characteristics. The theory of EIS based on generation of impedance spectra is described in literature [56, 70, 83-86]. The goal of the EIS technique is to measure the impedance as a function of frequency by proper procedures and then analyze the resulting spectrum to estimate corrosion rates and mechanism that might give rise to the spectra. The *impedance* is affected by the interaction between frquency and all of the physical and chemical processes that respond to that frequency change within the electrochemical cell and across the corroding interface. In the limit of zero frequency, the impedance approaches the DC resistance of the corroding system. Thus, the low frequency limit of EIS is the same as the LPR method.

The benefits of the *EIS*:

1. Rapid estimation of general corrosion rates (0.5 to 24 hours after immersion).
2. Estimation of corrosion rates in slowly changing systems (as long as stability is not compromised).
3. Estimation of low corrosion (less than 0.01 mpy).
4. Corrosion rate estimation in low conductivity media (organic media, soil, and concrete), namely, in concrete structures and the evaluation of cathodic protection . In

these environments the solution resistance can be separated from the actual polarization resistance.

5. Rapid assessment of corrosion inhibitors. EIS technique can be more effective than LPR method.
6. Evaluation corrosion under coatings and quality assurance of coatings (pore resistance, film capacitance, etc.).

The shortcomings of the *EIS*:

1. Measuring an EIS spectrum takes time (often many hours).
2. The system must be at a steady state. Otherwise, inaccurate results will be received. In practice, steady state can be difficult to achieve, as the cell can change through adsorption of solution impurities, growth of an oxide layer, build up of reaction products in solution, temperature changes, coating degradation, etc.
3. The analysis of the EIS data is relatively complex compared to the commonly used ER or LPR techniques. The EIS method requires benchmarking with other CM techniques.

4.7 Harmonic Distortion Analysis (HDA), or Harmonic Analysis (HA)

This method is based on measuring the fundamental, second and third harmonic currents from the current response at corrosion potential by perturbing a corroding system with a nondistorted sinusoidal signal of low amplitude [87, 88]. HDA is a low frequency impedance measurement, and similar to LPR relies on a steady state approximation, but requires more mathematical treatment than the LPR technique. The current response to a low frequency voltage sine wave is distorted due to the non-linearities of the charge transfer process. This distortion is analyzed in terms of the higher harmonics, to provide values for corrosion current, the Tafel slopes (anodic and cathodic coefficients), and hence the Stern-Geary value (B constant) [89].

This technique is related to electrochemical impedance spectroscopy (EIS), in that an alternating potential perturbation is applied to one sensor in a three element probe, with a resultant current response. Theory has been developed [87, 88, 90], whereby all kinetic parameters (including the Tafel constants) can be calculated explicitly. No other technique offers this facility. HDA is based on an evolution and improving the performance of the LPR technique. By applying a low frequency sine wave to the measurement current, the resistance of the corrosive solution can be calculated through a harmonic analysis of the resulting signals. With both the polarization resistance and the solution resistance, a more accurate general corrosion rate can be determined. Both EIS and HDA were used for measuring corrosion rates of metals in acids [91], neutral solutions of electrolytes [92], and concrete [93]. The advantage of HDA is that mathematical data treatment facilitates direct computation of the Tafel constants and the corrosion rate. Stern-Geary value (B - constant) is received by HDA and is stored in the transmitter [94].

The restrictions of *EIS and HDA*:

1. As in the LPR technique, the assumption of uniform corrosion has to be made.
2. If localized corrosion is taking place, the data are of a qualitative nature, indicating the breakdown of passivity and the possibility of localized attack.

5. CM IN THE ATMOSPHERE

We must know in many cases corrosiveness of the atmosphere at the plant or in rooms with electronic and electric equipment. Many aggressive variables, such as sulfur oxides, nitrogen oxides, chlorine, ammonia, hydrogen sulfide, ozone, water vapours, salts, dust, etc., contained in air, influence corrosion of metals. Failure may occur because of corrosion of electric contacts made of gold, silver, copper, zinc, or their alloys. The WL and ER methods are used. The WL method allowed to classify the corrosive level of the atmosphere (Table 4).

Table 4 Corrosiveness of the atmosphere regarding to carbon steel and zinc [95]

Corrosivity	Carbon steel		Zinc		Typical environment
	First year	Steady state	First year	Steady state	
Very low	< 1.3	< 0.1	< 0.1	< 0.05	Dry indoors
Low	1.3 - 25	0.1 - 1.5	0.1 - 0.7	0.05 - 0.5	Desert or non-polluted urban
Medium	25 - 50	1.5 - 6	0.7 - 2.1	0.5 - 2	Mild marine or mild industrial
High	50 - 80	6 - 20	2.1 - 4.2	2 - 4	Marine (calm sea)
Very high	80 - 200	20 - 90	4.2 - 8.4	4 - 10	Marine (surf beach)

Note: Corrosion rates are given in µm/year.

ER method allows on-line monitoring and to learn corrosiveness of the atmosphere in real time. Sometimes film thickness of corrosion products formed on copper surface, corresponds to the corrosive level of the atmosphere (Table 5). Electrochemical methods are used for measuring film thickness formed on copper surface.

Table 5 Classification of the corrosive level of atmosphere according to thickness of corrosion products on copper surface (30 days' exposure)

Film thickness of corrosion products on copper, Angströms	Corrosive level of the atmosphere
< 300	Mild
300 - 1,000	Moderate
1,000 - 2,000	Harsh
> 2,000	Severe

"Metal piece test kit" was developed [96], in which several metal strips are placed in the atmosphere for 1 month. Corrosion products are defined by X-ray fluorescence and corrosiveness of the atmosphere is determined according to presence sulfur, chloride and oxygen.

6. ON-LINE, REAL-TIME CM

Some main suppliers of equipment, devices and accessories for CM are shown in the Appendix 1. On-line, real-time CM systems manufactured by them which found application in industry are shown in the Appendix 2.

The ER and LPR methods are standardized and used for on-line CM of metallic equipment in plants [66]. LPR technique entered in use for corrosion monitoring in industry in 1960's. The problem is that the B values in the equation (16) are generally taken to be in the range 26 to 30 mV for most metal-environment systems, and are regarded by most manufactures of LPR instrumentation to be a device constant that is configured into an instrument at the factory [89]. The B values, however, not constant, change from 1 to more than 100 mV for various systems, can vary from time to time, and even within the same system [97, 98]. Therefore, for accurate corrosion rate measurements, variable B values should be used. HDA technique overcomes this limitation and allows to determine the Stern-Geary B constant. Usually an average B value is analyzed with time, and then post-correct the LPR corrosion rate. It was shown good correlation between the B value corrected corrosion rate and WL method and ultrasonic thickness measurements [99].

On-line, real-time CM must be integrated with the process control system, that is we have to monitor technological parameters together with corrosion ones [100]. CM allows quickly determine changes in water or other process stream quality, chemical changes (content of phases), or other parameters (temperature, pressure, flow rate, chemical feed rate), and inhibitor performance. The pertinent operating and analytical data are entered into the risk matrix analysis where the data is used to develop the Corrosion Potential Index, to determine the relative risk and the consequence of the corrosion [101]. CM technology can determine general and localized corrosion (pitting, crack, etc.), even when the general corrosion rate is low. CorrDATS (Rohrback Cosasco) works well as CM of general corrosion, pitting tendency and heat transfer resistance (criterion of biofouling) in cooling water systems (see chapter 3.1.1) [79].

Three electrochemical methods (LPR, HDA and ENM) described in [102-106] are realized in SmartCET Technology (Honeywell International Inc.) and are used for on-line, real-time CM general and localized corrosion in the oil refining industry [4, 97, 107], influence of microbiological activity [97, 108, 109], in district heating systems [16], in a water injection systems and suitable optimization of corrosion inhibitors selection [89] and for determining inhibitor optimal dosage requirements [97, 110], in a hot organic streams with small concentrations (1 to 2%) of water [97, 102], in multiphase oil-water systems with corrosive gases in liquid and vapor condensing phases [111], in natural gas transmission pipelines [111]. Measurements using three electrochemical techniques take only a few minutes [97].

The Resistance Corrosion Monitoring (RCM) device was developed for continuous monitoring of pipe wall thickness in high temperature areas where corrosion is suspected [112-115]. The RCM operates on the same underlying principal as traditional ER probes except it utilizes the pipe wall as the active sensor element, and it provides much more accurate and precise data. The RCM is an array of pins welded directly onto the pipe, covering approximately 1 m^2 area of pipe to be monitored.

Field Signature Method (FSM) developed by CorrOcean (Norway) is used on subsea pipelines, nuclear power plants, and storage tanks [116]. The FSM method is based on feeding an electric direct current through the selected sections of the structure to be monitored and sensing the pattern of the electrical field by measuring small potential differences set up on the surface of the monitored object [117]. Proper interpretation of these potential differences can lead to conclusion about wall thickness reduction. The FSM provides online information on wall thickness loss, erosion, cracking or pitting.

The *multielectrode array sensor* probe was developed for monitoring corrosion rates of carbon steels, stainless steels, Ni-Cr, Ni-Cr-Mo, and Cu-Ni alloys in both the liquid and vapor phase till temperature of 300°C and pressure of 140 atm. [118, 119]. This probe allows to measure the corrosion rate from 10^{-5} mm/yr (stainless steels in passive conditions) till localized penetration rates for these alloys up to 10 mm/yr. The *multielectrode array sensor* probe consists of a bundle of metal electrodes, all insulated from each other, but connected through a network of resistors. This probe was used for monitoring the general and localized corrosion (pitting, crevice and SCC), and coating integrity.

Concerto MK II developed by CAPSIS (UK), is used for monitoring localized corrosion in nuclear waste tanks and sour gas pipelines.

We have to take into consideration that CM is very dynamic field, new companies and devices appear every year, and Appendices 1 and 2 represent only part of them.

7. MONITORING OF CATHODIC PROTECTION

The effectiveness of cathodic protection, especially in the presence of stray currents, is difficult to measure. Monitoring of efficiency of cathodic protection of underground and underwater metallic constructions and structures is decsribed [120]; remote monitoring system to monitor the stray current drainage [121] and wireless remote monitoring [122]; and real-time monitoring the IR-free polarization (E_{off}) potentials of multiple coupon test station buried in close proximity to the monitored feeder [123].

8. CM INSIDE OF PIPELINES

The internal corrosion in pipelines is assessed by following techniques [124]:

1. Use of "intelligent" ("smart") pigs and verification by excavation and detailed examination at selected sites [125].
2. In-line inspection in unpiggable lines, based on fluid flow modeling and determination of the critical pipeline inclination angles that may be likely sites for water accumulation [126, 127].
3. Use of electrochemical probes based on LPR, ENM, HDA, and multielectrode array sensor method [97, 106, 110, 111, 128, 129].
4. Field Signature Method (FSM) [116, 117].
5. Use of carbon-silver galvanic couple thin-film sensors for indication of water accumulation possessing a corrosion risk in pipelines [124].
6. Fiber-optic sensors are used as a noninvasive tool for monitoring general and pitting corrosion in pipelines in real time [130].

The main limitation of techniques No. 3 - 5 is the need to have a prior knowledge of the optimum location for sensors. Integrity management includes all data that supports the estimation of corrosion situation in the pipelines, namely, direct data (corrosion rate) and indirect data (gas or liquid composition, microbiology, operating conditions, flow rate, etc.) [131].

CONCLUSION

1. CM is an important part of any plant anti-corrosion management, including corrosion control program. Comparison of properties and possibilities of some *CM* methods appropriate for the environmental conditions and for identification of particular corrosion phenomena are shown in Tables 6 and 7.
2. Why are there so many methods of *CM*? To follow up corrosion is more complicated than to monitor such processing parameters as a temperature, a pressure, a flow rate of streams, the concentrations of reagents, etc., because of:
   (a) There are many corrosion phenomena (types, or forms). One can see from the Table 7 that not every *CM* method is suitable for monitor all corrosion phenomena.
   (b) Corrosion may be uniform (general) or localized (pitting, or crack). Specific methods are suitable for monitoring of localized corrosion.
3. Selection of *CM* must be carried out at the stage of project. It is very important correctly select *CM* type and inspection points. Ultrasonic thickness measurements must be taken in the vicinity of the CM points. Corrosion rate even for uniform corrosion may be significantly changed at different distances. The more places we monitor, the more complete and the more real corrosion situation may be determined.

Table 6 Comparison of possibilities of CM methods used in industry

Corrosion phenomenon	Corrosion Monitoring Method					
	Ultra-sonic	Coupons (weight loss)	ER - probes	LPR	Chemical analytical methods	ENM
General (uniform) corrosion	Excellent	Excellent	Excellent	Good	Good	Good
Pitting corrosion	Fair	Excellent	NA	NA	NA	Excellent
Galvanic corrosion	NA	Excellent	NA	NA	NA	Good
MIC	Fair	Good	NA	NA	NA	Good
Erosion – corrosion	Good	Excellent	Excellent	NA	NA	Excellent
SCC	Fair	Good	NA	NA	NA	Excellent
Intergranular corrosion	NA	Good	NA	NA	NA	Excellent
Hydrogen Induced Corrosion	Fair	Fair	NA	NA	NA (Good for H_2 determination)	Excellent
Dealloying	Fair	Excellent	NA	NA	Good	Good
Crevice corrosion and underdeposit corrosion	NA	Good	NA	NA	NA	Excellent

Notes: ER – Electrical Resistance; LPR - Linear Polarization Resistance; ENM - Electrochemical Noise Measurements; MIC – Microbiologically Induced Corrosion; SCC – Stress Corrosion Cracking; NA – Non Applicable.

Table 7 Comparison of characters of some corrosion monitoring methods used in industry

Character	Corrosion Control Method					
	Ultra -sound	Coupons (weight loss)	ER – probes	LPR	Chemical analytical methods	ENM
Period of measurements needed	6 to 24 months	1 to 6 months	One to several days	Several minutes	Minutes to hours	Several minutes
Periodical inspection, on-line or off-line measurements	Inspection	Off-line measure-ments	On-line measure-ments	On-line measure-ments	On-line off-line measurements	On-line measure-ments
Real-time information ability	No	No	Yes	Yes	Yes	Yes

Notes: 1. On-line measurements are continuous control of metal loss, or corrosion rate, or other parameters in a system. Data are obtained without disrupting of monitoring process

2. Off-line measurements are periodical control of corrosion rate, or other parameter, in a system with disrupting of monitoring process.

3. Real-time information is corrosion rate data that can be received instantaniously, or with more frequency than the changes in the parameter being investigated. These data are usually received during several minutes.

4. Probes or their sensing elements should be replaced when they reach 90 to 95 percent of their nominal life. They must be changed immidiately if localized corrosion has occurred. In some cases, the sensing elements need to be cleaned if the data received appears questionable.
5. Use of only one method of *CM* gives some, but not full information. The more methods are used, the more we cover corrosion situation and as a result production, injection of chemicals, and technological parameters. It is important to use automated multi-technique monitoring capabilities.

CM should be integrated with other plant programs designed optimize the process conditions, injection of chemicals and inspection. Only in this case we can successfully manage plant operation.

REFERENCES

1. Development in Real-time Corrosion Monitoring - Recent Advances in Techniques and Management, Display and Correlation of Data. Report by Cormon, Business Briefing: Exploration & Production: The Oil & Gas Review, 2004, pp. 1-4.
2. A. Groysman and N. Brodsky, Corrosion and Quality, Proceedings of the 15th International Conference of the Israel Society for Quality, November 16-18, 2004, Jerusalem, Israel, pp. 223-230.
3. A. Groysman and N. Brodsky, Corrosion and Quality, Accreditation and Quality Assurance: Journal for Quality, Comparability and Reliability in Chemical Measurement, Publisher: Springer Berlin / Heidelberg, February 2006, Vol.10, No. 10, pp. 537-542.
4. Russel D. Kane, Dawn C. Eden and David A. Eden, Corrosion: a new process variable, Hydrocarbon Engineering, October 2005.
5. A. Groysman, Anti-Corrosion Management and Environment at the Oil Refining Industry, Proceedings of the International Conference on Corrosion CORCON2005, 28th-30th November 2005, Chennai, India, 2005, p.1 (18 p.).

6. Corrosion, Vol. 2, Editors: L.L. Shreir, R.A. Jarman & G.T. Burstein, 3rd Edition, Butterworth Heinemann, UK, 1994, pp. 19:154-19:178.
7. Pierre R. Roberge, Handbook of Corrosion Engineering, McGraw-Hill, New York, 2000, pp. 371-483.
8. Neil Rothwell and Martin Tullmin, The Corrosion Monitoring Handbook, First Edition, Coxmoor Publishing Company, UK, 2000, 180 p.
9. Russel D. Kane, Selection of Corrosion Monitoring Methods for Refinery Application, Corrosion in the Oil Refining Industry, September 26 to 27, 1996, Houston, Texas, NACE International, USA, 1996, pp. 14/1-14/17.
10. Bruce N. Nelson, Edward J. Lemieux, Paul Slebodnick, William Groeninger, Tank Coatings Assessments using the Insertible Stalk Inspection System and Corrosion Detection Algorithm, CORROSION2006, Paper No. 06323, NACE International, 2006, San-Diego, USA, 17 p.
11. Mark Lewis, Acoustic Monitoring of a Prestressed Concrete Tank, Materials Performance, Vol. 41, No. 7, 2002, pp. 56-59.
12. A.W. Peabody, Control of pipeline corrosion, Second edition, Edited by Ronald L. Bianchetti, NACE International, USA, 2001, 347 p.
13. Darrell R. Louder, Bruce A. Parkinson, An Update on Scanning Force Microscopies, Analytical Chemistry, Vol.67. No. 9, May 1, 1995, pp. 297A-303A.
14. A.K. Neufeld and I.S. Cole, Using Fourier Transform Infrared Analysis to Detect Corrosion Products on the Surface of Metals Exposed to Atmospheric Conditions, Corrosion, Vol. 53, No. 10, pp. 788-799.
15. NACE Standard RP0775-2005. Preparation, Installation, Analysis, and Interpretation of Corrosion Coupons in Oilfield Operations, 10 p.
16. Ragnheidur I. Thorarinsdottir, Lisbeth R. Hilbert, Online real-time, corrosion monitoring in district heating systems, News from DBDH 2/2004, 3 p.
17. A. Groysman, Corrosion Monitoring at the Oil Refinery, Second NACE Asian Corrosion Conference (Singapore), Paper No. 1005, 1994.
18. A. Groysman and A. Hiram, Corrosion Monitoring and Control in Refinery Process Units. Paper No. 512, CORROSION/97, New Orleans, USA, 1997, 25 p.
19. Ron A. Welsh, Jed Benfield, Environmental Protection Through Automated Remote Monitoring of Fuel Storage Tank Bottoms Using Electrical Resistance Probes, Materials Performance, Vol.45, No.3, 2006, pp. 38-40.
20. B. Ridd, T.J. Blakset, D. Queen, Field Trails for Corrosion Inhibitors Selection and Optimization, Using a New Generation of Electrical Resistance Probes, CORROSION/98, Paper No. 78, NACE International, 1998, 6 p.
21. M.W. Joosten, J. Kolts, P.G. Humble, T.J. Blakset, D.M. Keilty, Internal Corrosion Monitoring of Subsea Production Flowlines - Probe Design and Testing, CORROSION/98, Paper No. 77, NACE International, 1998, 24 p.
22. Paul C.N. Crouzen, W. Verstijnen, Ian J. Munns, Roy C. Hulsey, Application of Pulsed Eddy Current Corrosion Monitoring in Refineries and Oil Production Facilities, CORROSION2006, Paper No. 06312, NACE International, 2006, San-Diego, USA, 9 p.
23. P.W. van Andel, Eddy Current Inspection technique, US patent 6,291,992, September 2001.
24. P.C.N. Crouzen, Method for inspecting an object of electrically conducting material, US patent 6,570,379, May 2003.
25. EFC (European Federation of Corrosion) Working Party 15 Meeting: 15th September 2004, Corrosion Under Insulation, Guideline.

26. Annual Book of ASTM Standards, Water and Environmental Technology, Vol.11.01, ASTM, USA, 2000, 917 p.
27. Standard Methods for the Examination of Water and Wastewater, 21st Edition, American Public Health Association, USA, 2005.
28. BETZ. Handbook of Industrial Water Conditioning, Eighth Edition, BETZ LABORATORIES, INC., Trevose, PA, 19047, USA, 1980, pp. 349-420.
29. Ivanka Georgieva, Nedyalka Petkova, Application of inhibiting protection at the Crude Oil Atmospheric Distillation Units, Proceedings EUROCORR 2004, Nice, France, 2004, 9 p.
30. NACE Standard RP0192-98. Monitoring Corrosion in Oil & Gas Product. NACE International, USA.
31. Terry Jackson, M. Craig Winslow, Marl Wilson, Prolonged Experience Processing High Acid Crude Cross Oil & Refining Company, The European Corrosion Congress, Proceedings EUROCORR 2005, Lisbon - Portugal, 4 – 8 September, 2005, 13 p., EFC event No. 273, Book of Abstracts, p.609.
32. ASTM G0051-95R05. Test method for Measuring pH of Soil for Use in Corrosion Testing. Annual Book of ASTM Standards, Vol. 03.02, 2006.
33. ASTM G0057-95AR01. Test Method for Field Measurement of Soil Resistivity Using the Wenner Four-Electrode Method. Annual Book of ASTM Standards, Vol. 03.02, 2006.
34. ASTM G0187-05. Test Method for Measurement of Soil Resistivity Using the Two-Electrode Soil Box Method. Annual Book of ASTM Standards, Vol. 03.02, 2006.
35. Robert D. Port, Harvey M. Herro, The Nalco Guide to Boiler Failure Analysis, McGraw-Hill, Inc., New York, USA, 1991, 293 p.
36. Annual Book of ASTM Standards, Vol. 11.05, ASTM, USA, 2006.
37. Biofouling and Biocorrosion in Industrial Water Systems, Eds.: Flemming H.C., Geesey G.G., Springer-Verlag, Berlin, 1991, 220 p.
38. Biofouling and Biocorrosion in Industrial Water Systems, Eds.: Geesey G.G., Lewandowski Z., Flemming H.C., Lewis Publishers, CRC Press Inc., USA, 1994, 297 p.
39. Héctor A. Videla, Manual of Biocorrosion, Lewis Publishers CRC Press, USA, 1996, 273 p.
40. Robert D. Port, Harvey M. Herro, The Nalco Guide to Cooling Water System. Failure Analysis, McGraw-Hill, Inc., New York, USA, 1993, pp. 127-129, 157.
41. Mohammad Setareh, Reza Javaherdashti, Precision Comparison of Some SRB Detection Methods in Industrial Systems, Materials Performance, Vol. 42, No.5, 2003, pp.60-63.
42. Mita Chattoraj, Michael J. Fehr, and Steven R. Hatch, Eric J. Allain, Online Measurement and Control of Microbiological Activity in Industrial Water Systems, Materials Performance, Vol. 41, No.4, 2002, pp.40-45.
43. P.J.B. Scott, Materials Performance, Vol. 39, 2000.
44. Bennett P. Boffardi, Water Treatments, In: Corrosion Tests and Standards: Application and Interpretation, Editor: Robert Baboian, ASTM Manual Series: MNL 20, ASTM, USA, 1995, pp. 697- 704.
45. M.H. Dorsey, G.L. Licina, B.J. Saldanha, R.C. Ebersole, Monitoring for Corrosion and Microbiological Activity in a Cooling Water System, CORROSION/2002, Paper No.02009, Houston, Texas, NACE, 2002.
46. Matthew V. Veazey, Plant Uses Unique Strategy, to Fight MIC, Materials Performance, Vol. 42, No. 12, 2003, pp. 16-18.
47. Frank Dean, Continuous Active Corrosion and HIC risk Measurement, Eurocorr 2004, Working Party 15 (Corrosion in the Refining Industry), September 15, 2004, Nice, France.

48. ASTM G148-97, Standard Practice for Evaluation of Hydrogen Uptake, Permeation, and Transport in Metals by Electrochemical technique, 2003.
49. ASTM G0005-94R04, Reference Test Method for Making Potentiostatic and Potentiodynamic Anodic Polarization Measurements, Annual Book of ASTM Standards, Vol. 03.02, 2006.
50. ASTM C0876 - 91R99 (1999). Standard Test Method for Half-Cell Potentials of Uncoated Reinforcing Steel in Concrete, Annual Book of ASTM Standards, Vol. 04.02, 2006.
51. B. Elsener and H. Bohni, Potential Mapping and Corrosion of Steel in Concrete, Corrosion Rates of Steel in Concrete, ASTM STP 1065, American Society for Testing and Materials, West Conshohocken, PA., 1990, pp. 143-196.
52. J.P. Broomfield, P.E. Langford and A.J. Ewins, The Use of a Potential Wheel to Survey Reinforced Concrete Structures, Corrosion Rates of Steel in Concrete, ASTM STP 1065, American Society for Testing and Materials, West Conshohocken, PA., 1990, pp. 157-173.
53. ASTM G0059-97R03. Test Method for Conducting Potentiodynamic Polarization Resistance Measurements. Annual Book of ASTM Standards, Vol. 03.02, 2006.
54. Milton Stern, A Method for Determining Corrosion Rates from Linear Polarization Data, Corrosion, 1958, Vol. 14, No. 9, pp. 440t-444t.
55. M. Stern and A.L. Geary, Electrochemical Polarization: I. A Theoretical Analysis of the Shape of Polarization Curves, J. of the Electrochemical Society, 1957, Vol.102, No.1, pp. 56-63.
56. Electrochemical Impedance: Analysis and Interpretation, Editors: J.R. Scully, D.C. Silverman, and M.W. Kendig, ASTM, USA, 1993.
57. K.R. Gowers, S.G. Millard, J.S. Gill and R.P. Gill, Programmable linear polarization meter for determination of corrosion rate of reinforcement in concrete structures, British Corrosion Journal, 1994, Vol.29, No.1, pp. 25-32.
58. J. Flis, H.W. Pickering and K. Osseo-Asare, Assessment of data from Three Electrochemical Instruments for Evaluation of Reinforcement Corrosion Rates in Concrete Bridge Components, Corrosion, 1995, Vol. 51, No.8, August, pp. 602-609.
59. R.G. Kelly et al., Embeddable microinstruments for corrosion monitoring, CORROSION97, Paper No. 294, NACE, Orlando, USA, 12 p.
60. Temi M. Linjewile et al., Prediction and Real-Time Monitoring Techniques for Corrosion Characterization in Furnaces, Materials at High Temperature, Vol. 20, Issue 2, 2003, 15 p.
61. R.K. Shukla et al., Corrosion Monitoring of High Level Waste Storage Tank 8-D2 at the West Valley Demonstration Project, CORROSION/94, Paper no. 121, Houston, TX, NACE International, 1994.
62. S. Papavinasam, R.W. Revie, M. Attard, and A. Demoz, Corrosion, Vol.59, No. 10, 2003.
63. Jorge J. Perdomo, Marjorie Ramirez, Alfredo Viloria, Test spot inconsistencies, Oil & Gas Journal, Sept. 18, 2000, pp. 46-48.
64. Mark Yunovich, Brett M. Tossey, Conchita Mendez, Measuring Corrosion Growth Rates for Determining Integrity Verification Reassessment Intervals by In-situ LPR Technique, CORROSION2006, Paper No. 06311, NACE International, 2006, San-Diego, USA, 14 p.
65. S. Fofano, A Chaves Meto and H.A. Ponte, Determining Corrosion Rates by Electrochemical Techniques in an Industrial Water Cooling System, Materials Performance, Vol. 42, No. 2, 2003, pp. 62-65.
66. ASTM G0096-90R01E01, Guide for On-Line Monitoring of Corrosion in Plant Equipment (Electrical and Electrochemical Methods). Annual Book of ASTM Standards, Vol. 03.02, 2006.
67. ASTM G0102-89R04E01, Practice for Calculation of Corrosion Rates and Related Information from Electrochemical Measurements, Annual Book of ASTM Standards, Vol. 03.02, 2006.

68. Dawn C. Eden and Russell D. Kane, Achieving Improved Process Control and Real-time Corrosion Monitoring, Eurocorr 2004, Working Party 15 (Corrosion in the Refining Industry), September 15, 2004, Nice, France.
69. D.C. Eden and J.D. Kintz, Real-Time Corrosion Monitoring for Improved Process Control: A Real and Timely Alternative to Upgrading of Materials of Construction, CORROSION/2004, Paper No. 04238, NACE International, 2004, San-Diego, USA, 14 p.
70. Robert Cottis and Stephen Turgoose, Electrochemical Impedance and Noise, NACE International, USA, 1999, 149 p.
71. Electrochemical Noise Measurement for Corrosion Applications, Editors: Jeffery R. Kearns, John R. Scully, Pierre R. Roberge, David L. Reichert, and John L. Dawson, STP 1277, ASTM, USA, 1996, 476 p.
72. J.G. González-Rodriguez, V.M. Salinas-Bravo, E. García-Ochoa and A. Diaz-Sánchez, Corrosion, 1997, Vol.53, No. 9, pp. 693-699.
73. G.L. Edgemon, M.J. Danielson and G.E.C. Bell, J. Nucl. Mater., 2 June 1997, Vol.245, No. 2-3, pp. 201-209.
74. Y. Watanabe, T. Kondo, CORROSION/98, NACE International, USA, 1998, Paper No. 376.
75. G.L. Edgemon, P.C. Ohl, G.E.C. Bell and D.F. Wilson, CORROSION/96, NACE International, 1996, Paper No. 094.
76. J.I. Mickalonis and E.M. Tshishiku, G.L. Edgemon, Development of a Movable Electrochemical Noise Corrosion Probes for Nuclear Waste Tanks, Report WSRC-MS-2001-00734 U.S. Department of Energy, 2006, 14 p.
77. G.L. Edgemon, Electrochemical Noise Based Corrosion Monitoring at the Hanford Site: Third Generation System, CORROSION/2001, Paper no. 1140, 2001, NACE International.
78. E.M. Barr, R.B. Goodfellow, and L.M. Rosenthal, Noise Monitoring in Canada's Simonette, Sour Oil Processing Facility, COOROSION/2001, Paper no. 414, Houston, TX, NACE International, 2001.
79. Alec Groysman and Inna Schwarz, Study of Efficiency of Industrial Corrosion Inhibitors for Cooling Water Systems at Oil Refining Industry, CORROSION 2006, Paper 06097, San Diego, California, USA, 12-16 March 2006, 10 p.
80. ASTM G0071-81R03. Guide for Conducting and Evaluating Galvanic Corrosion Tests in Electrolytes. Annual Book of ASTM Standards, Vol. 03.02, 2006.
81. ASTM G0015-05. Terminology Relating to Corrosion and Corrosion Testing. Annual Book of ASTM Standards, Vol. 03.02, 2006.
82. D. C. Silverman, Primer on the AC Impedance Technique, In: Electrochemical Techniques for Corrosion Engineering (Ed. R. Baboian), NACE, USA, 1987, p. 73.
83. J. R. Macdonald, Impedance Spectroscopy, John Wiley & Sons, New York, 1987.
84. D.C. Silverman and J.E. Carrico, Electrochemicl Impedance Technique – A Practical Tool For Corrosion Prediction, Corrosion, Vol. 44, No. 5, 1988, p.280.
85. Impedance Spectroscopy: Theory, Experiment, and Applications, 2nd ed., Editors E. Barsoukov, J.R. Macdonald, Wiley Interscience Publications, 2005.
86. A.J. Bard, L.R. Faulkner, Electrochemical Methods, Fundamentals and Applications, Wiley Interscience Publications, 2000.
87. S.K. Rangarajan, J. Electroanal. Chem., 1975, Vol.62, p.62.
88. J. Devay and L. Meszaros, Acta Chim. Acad Hung., 1979, Vol. 100, p. 183.
89. Dawn C. Eden, David A. Eden, Ian George Winning, and David Fell, On-line, Real-Time Optimization of Corrosion Inhibitors in the Field, CORROSION/2006, Paper No. 06321, NACE International, 2006, San-Diego, USA, 14 p.

90. K. Darowicki, A. Krakowiak, The application of Gabor transformation in the harmonic analysis of corrosion processes, Anti-Corrosion Methods and Materials, June 2003, Vol. 50, Issue 3, pp. 193-200.
91. S. Sathiyanarayanan and K. Balakrishnan, Brit. Corros. J., 1991, Vol. 29, p. 152.
92. J.S. Gill, L.M. Callow, and J.D. Scantlebury, Corrosion, 1983, Vol.39, p. 61.
93. S. Srinivasan, G. Venkatachari, and N.S. Rengaswzmy, Proceedings 13th International Corrosion Congress, November 1996, Paper No. 210.
94. David Hohenstein, Corrosion as a Real-Time Process Variable, Control Engineering, March 1, 2005.
95. ISO 9223, Corrosion of metals and alloys - Classification of corrosivity of atmospheres, 1992, 13 p.; ISO 9224, Corrosion of metals and alloys - Corrosivity of atmospheres - Guiding values for the corrosivity categories, 1992, 5 p; ISO 12944-2:1998(E), Paints and varnishes - Corrosion protection of steel structures by protective paint systems, Part 2: Classification of environments, p. 5.
96. Tsutomu Iikawa, Yasuo Udo and Eiichi Nakajima, Assessment of specific environments by metal piece test kits, Corrosion Science, Vol. 35, No. 1-4, 1993, pp. 735-742.
97. Russel D. Kane, Dawn C. Eden and David A. Eden, Real-Time Solutions Integrate Corrosion Monitoring with process Control, Materials Performance, Vol. 44, No. 2, 2005, pp. 36-41.
98. Russel D. Kane, E. Trillo, Evaluation of Multiphase Environments for general and Localized Corrosion, CORROSION/2004, Paper No. 04656, NACE International, Houston, TX, USA, 2004.
99. M.V. Veazey, A Paradigm Shift for Process Control? Materials Performance, December 2004, pp.16-19.
100. D.A. Eden, S. Srinivasan, Real-time, On-line and On-board: The Use of Computers, Enabling Corrosion Monitoring to Optimize Process Control, CORROSION/2004, Paper No. 59, NACE International, Houston, TX, USA, 2004.
101. Neil Morgan, M. Craig Winslow, Craig Howard, Development and Implementation Strategies for Safe & Profitable Opportunity Crude Processing, Technical Paper GE Betz TP1005EN 0503, 2005, 9 p.
102. D.C. Eden and J.D. Kintz, Real-Time Corrosion Monitoring for Improved Process Control: A Real and Timely Alternative to Upgrading of Materials of Construction, CORROSION/2004, Paper No. 04238, NACE International, Houston, TX, USA, 2004.
103. D.C. Eden, et al., Making Credible Corrosion Measurements - Real Corrosion, Real Time, CORROSION/2003, Paper No. 03376, NACE International, 2003, San-Diego, USA.
104. P. Teevens, Defeating Internal Corrosion of Petroleum Pipelines in Remote Locations: Winning Strategies intergrating Real-Time Electrochemical Noise Corrosion Surveillance with Recent Advancements in Power and Communications Technologies, Corrosion and Prevention, 2000, Publication no. 064, Australasian Corrosion Association.
105. D.E. Eden, Electrochemical Noise - the First Two Octaves, , CORROSION/99, Paper No. 386, NACE International, 1999, San-Diego, USA.
106. P. Teevens, Electrochemical Noise - a Potent Weapon in the Battle Against Sour Gas Plant Corrosion: Over Three Years Operating and Turnaround Inspection Experiences in Two Canadian Plants, Corrosion and Prevention, 1998, Publication no. 038, Australasian Corrosion Association.
107. Dawn C. Eden and Russell D. Kane, Achieving Improved Process Control and Real-time Corrosion Monitoring, Eurocorr 2004, Working Party 15 (Corrosion in the Refining Industry), September 15, 2004, Nice, France.
108. Dawn C. Eden and Russell D. Kane, MIC, Eurocorr 2004, September 15, 2004, Nice, France.

109. Russel D. Kane, S. Campbell, Real-Time Corrosion Monitoring of Steel Influenced by Microbial Activity (SRB) in Simulated Seawater Injection Environments, CORROSION/2004, Paper No. 04579, NACE International, Houston, TX, USA, 2004.

110. Russel D. Kane, Dawn C. Eden and David A. Eden, Online, Real-Time Corrosion Monitoring for Improving Pipeline Integrity - technology and Experience, CORROSION/2003, Paper No. 03175, NACE International, Houston, TX, USA, 2003.

111. Bernard S. Covino, Sophie J. Bullard, Stephen D. Cramer, Gordon R. Holcomb, Margaret Ziomek-Moroz, and Russel D. Kane, Detecting Internal Corrosion of Natural Gas Transmission Pipelines: Field Tests of Probes and Systems for Real-time Corrosion Measurement, EUROCORR 2005, 2-5 September, 2005, Lisbon, Portugal, 10 p.

112. B. Lasiuk, M. Wilson, C. Winslow, Advances in Optimizing Refinery Profitability, Technical Report GE Betz AM-05-13, January 2005, 12 p.

113. M. Craig Winslow, Mark Wilson, Brian Lasiuk, Peter Allison and Collin Cross, Solutions for Processing Opportunity Crudes, ERTC (European Refining Technology Conference), 10th Annual Meeting, Vienna, Austria, November 2005, 14 p.

114. Terry Jackson, M. Craig Winslow, Mark Wilson, Prolonged Experience Processing High Acid Crude - Cross Oil & Refining Company, ERTC 9th Annual Meeting, Prague, Czech Republic, November 2004.

115. Andre Vanhove, Advances in Corrosion Monitoring while Processing High Acidic Opportunity Crudes, The 7th Israel Conference on Corrosion and Electrochemistry, May 10th-11th 2006, Bar-Ilan University, Israel.

116. H. Horn, S.T. Sivertsen, A.E. Pedersen, CORROSION/2003, Paper No. 03425, NACE International, 2003, Houston, TX, USA.

117. Roe D. Strommen, Harald Horn, Kjell R. Wold, New technique monitors pipeline corrosion, cracking, Oil & Gas Journal, Dec.27, 1993, pp. 88-92.

118. L. Yang, N. Sridhar, G. Cragnolino, Comparison of Localized Corrosion of Fe-Ni-Cr-Mo Alloys in Concentrated Brine Solutions Using a Coupled Multielectrode Array Sensor, CORROSION/ 2002, Paper No. 545.

119. L. Yang, N. Sridhar, Coupled Multielectrode Online Corrosion Sensor, Materials Performance, Vol. 42, No. 9, pp. 48-52.

120. Thomas Mierau, Remote Monitoring & Control Systems as a Solution for Corrosion Monitoring at Airports with Elevated Security Standards, CORROSION2006, Paper No. 06309, NACE International, 2006, San-Diego, USA, 12 p.

121. Paul S. Rothman, Steve Nikolakakos, Michael J. Szeliga, Innovative Approach to Stray Current Monitoring at the Site of the World Trade center in New York City, CORROSION2006, Paper No. 06310, NACE International, 2006, San-Diego, USA, 14 p.

122. Lee Blankenstein, Wireless Remote Monitoring for Oil and Gas Corrosion Protection Assets, Materials Performance, Vol. 42, No. 8, 2003, pp. 26-29.

123. M. Yunovich and N.G. Thompson, Coupon Monitoring for Cathodic Protection Optimization: Real-Time Automated Remote Control of Dynamic Stray Current, on Pipe-Type Cables. 20 p.

124. Narasi Sridhar, Garth Tormoen, C. Sean Brossia, and Ashok Sabata, Development and Application of Mobile sensor Network to Monitor Corrosion in Pipelines, CORROSION2006, paper No. 06322, NACE International, 2006, USA, 17 p.

125. Cynthia Greenwood, Pigging the Diesel Pipeline Between Hawaii's Red Hill Facility and Pearl Harbor, Materials Performance, Vol. 45, No. 3, 2006, pp. 16-19.

126. D. Burwell, N. Sridhar, O.C. Moghissi, and L. Perry, Internal Corrosion Direct Assessment of Dry Gas Transmission Pipelines - Validation, CORROSION/2004, paper No. 04195, NACE International, 2004, Houston, TX, USA.

127. O.C. Moghissi, L. Perry, B. Cookingham, and N. Sridhar, Internal Corrosion Direct Assessment of Dry Gas Transmission Pipelines - Application, CORROSION/2003, paper No. 03204, NACE International, 2004, Houston, TX, USA.
128. N. Sridhar, L.T. Yang, and F. Song, CORROSION/2006, Paper No. 06673, NACE International, 2006, San-Diego, USA.
129. B.S. Covino, S.J. Bullard, S.D. Cramer, G.R. Holcomb, and M. Ziomek-Moroz, CORROSION/2005, Paper No. 05133, NACE International, 2005, Houston, TX, USA.
130. Rod C. Tennyson, Tom Miesner, Fiber-optic monitoring focuses on bending, corrosion, Oil & Gas Journal, Feb. 20, 2006, pp. 55-60.
131. Richard B. Eckert, Bruce Cookingham, Lynsay Bensman, Optimizing Internal Corrosion Monitoring and response Through Intergration of Direct and Indirect Data, CORROSION2006, Paper No. 06307, NACE International, 2006, San-Diego, USA, 13 p.

APPENDICES

Appendix 1 Suppliers of Equipment for CM

No.	Company	Country	Address	Telephone, Fax, E-mail, and Web Site
1	Rohrback Cosasco Systems, Inc.	USA	11841 East Smith Avenue Santa Fe Springs, CA 90670, USA	Tel: 1-562-949-0123 Fax: 1-562-949-3065 rcs@rohrbackcosasco.com www.rohrbackcosasco.com
2	Metal Samples Corrosion Monitoring Systems	USA	152 Metal Samples Road P.O.Box 8Munford, AL36268	Tel: 1-256-358-4202 Fax: 1-256-358-4515 msc@alspi.com www.alspi.com
3	Cormon Ltd.	United Kingdom	Cormon Ltd. Unit 3, Robell Building, Chartwell Road, Lancing Business Park, Lancing West Sussex BN15 8TU, UK	Tel: 44 1903 854800 Fax: 44 1903 854854 www.cormon.com
4	CAPCIS	United Kingdom	CAPCIS House1 Echo StreetManchesterM1 7DP	Tel: 44 161 933 4000 Fax: 44 161 933 4001 info@capcis.co.uk www.capcis.com
5	CorrOcean ASA	Norway	Teglgården7485 Trondheim	Tel.: 47 73825000 Fax: 47 73825050 mail@corrocean.no www.corrocean.com
6	Honeywell International Inc.	USA	2500 West Union Hills Drive, Phoenix, AZ 85027	Tel: 877-466-3993 Tel: 602-313-6665 Tel: (281) 444-2282 x 32 Fax: (281) 444-0246 Mobile: (713) 458-8463 russ.kane@honeywell.com www.honeywell.com
7	Pepperl+Fuchs Inc.	USA	1600 Enterprise Parkway, Twinsburg Ohio, 44087 USA	Tel: 330-425-3555 Fax: 330-425-4607 sales@us.pepperl- fuchs.com www.pepperl-fuchs.com
8	MetriCorr ApS	Denmark	1.Produktionsvej 2 DK-2600 Glostrup 2. Glerupvej 20, Roedovre 2610	Tel: 45 72 177 410 Fax: 45 72 177 216 info@metricorr.dk www.metricorr.com Tel: 45 44240026 Fax: 45 44240015

Appendix 2 On-line, Real-time CM Systems

No.	Company	Model, or System	Corrosion Forms	Basics of measurements
1	Rohrback Cosasco Systems, Inc.	Microcor	General Corrosion	ER, LPR
2		CorrDATS	General Corrosion, Pitting Tendency, Heat Treansfer Resistance	LPR+ENM
3	CAPCIS	Concerto™ MK II	Localized Corrosion	LPR+ENM
4		Concerto™ XE	General and Localized Corrosion, temperature, pressure, and vibration	LPR, ENM
5		Concerto RCC	Corrosion rate in concrete structures	LPR, ENM
6	Honeywell International Inc.	SmartCET Corrosion Monitoring Transmitter: CET5000-G, CET5000-P	General Corrosion	LPR
7		SmartCET Corrosion Monitoring Transmitter: CET5000-M	Localized Corrosion (Pitting)	ENM + HDA
8	Pepperl+Fuchs Inc.	CorrTran	General and Localized Corrosion	LPR+ENM + HDA
9	CorrOcean ASA	Field Signature Method	General Corrosion	ER

Notes: 1 Localized corrosion is defined according to Pitting Factor (Pitting Index, or Pitting Tendency) in the default range 0.001-1.0. Low Pitting: 0.001-0.01; Average Pitting: 0.01-0.1; High Pitting: 0.1-1.0.

2 ER – Electrical Resistance; LPR – Linear Polarization Resistance; ENM – Electrochemical Noise Measurements; HDA – Harmonic Distortion Analysis.

CHAPTER 17

Nuclear and Auger Electron Spectroscopic Techniques for Corrosion Studies

G. Amarendra and R. Govindaraj
Materials Science Division, Indira Gandhi Centre for Atomic Research, Kalpakkam-603 102, T.N, India

I. INTRODUCTION

The study of corrosion of materials is of importance and useful, as corrosion plays very critical role in limiting the lifetime of many components used in a variety of industries [1],[2]. A large amount of money is lost each year because of corrosion. This loss is mainly due to the corrosion of iron and steel, although many other metals may corrode as well. In certain metals for example aluminum, oxide coating formed strongly bonds to the surface preventing the surface from further exposure to oxygen and corrosion. On the other hand, the main problem with iron as well as many other metals is that the oxide film formed by oxidation does not firmly adhere to the surface of the iron or steel and flakes off easily resulting in "pitting". Extensive pitting eventually causes structural weakness and disintegration of the metal. The chemical nature of the exposed surfaces, the environment, surface reactivity and exposure time all play determining role in influencing the corrosion rate. It is of fundamental and technological interest to study the corrosion phenomena and the surfaces of these materials of interest using a wide variety of experimental techniques. Conventionally, electrochemical and electroimpedance techniques are used to study the passivation, corrosion characteristics of the materials [3],[4]. Apart from various macroscopic techniques, the application of microscopic techniques such as electron microscopy, X-ray scattering, X-ray photoelectron spectroscopy and Auger electron spectroscopy have enriched the fundamental understanding of the corrosion processes [5].

Bulk oxidation of a metal occurs in general either by outward diffusion of metal cation or the inward diffusion of oxygen anion in to the bulk. Hence, in bulk oxidation the metal cation vacancies injected at the metal-oxide interface play a crucial role for the corrosion. If these vacancies grow in to a void of dimensions comparable to the thickness of passive oxide film, they become pore like structures exposing the underlying metal to the environment. Therefore the study of vacancies would provide a basic understanding of the process of corrosion. This naturally leads to the importance of positron annihilation spectroscopic (PAS) techniques for corrosion studies. A strong selectivity and sensitivity of positrons to annihilate at open volume defects such as vacancies and vacancy clusters makes PAS a powerful spectroscopic technique to study vacancies and their clustering and their critical role for corrosion phenomena.

Once the oxide phase is formed it becomes essential to understand the structural, magnetic and electronic aspects of these phases apart from studying their growth and stability under various environmental conditions. Hyperfine interaction techniques such as Mossbauer spectroscopy (MS) [6] and perturbed angular correlation (PAC) [7],[8] can be used to study structural, electronic and magnetic aspects of oxide phases. As the hyperfine techniques are also powerful to study impurity (probe)-vacancy interactions, the formation of oxides and the role of cation vacancies for corrosion can also be studied by these techniques. Owing to the fact that the corrosion studies are important in iron and steel, ^{57}Fe based MS is one of the most powerful techniques to study corrosion in these materials. The grown oxide phases can also be studied by electron spectroscopic techniques such as Auger Electron Spectroscopy (AES) [9]. X-ray Photoelectron spectroscopy (XPS) has been extensively used for corrosion studies, owing to their surface sensitivity as well as chemical specific nature. Shown below is Table 1 containing some of the important nuclear techniques (not exhaustive) used for studying different aspects of corrosion.

In Positron Annihilation Spectroscopy (PAS), positrons from a radioactive source are used as a probe of the medium, wherein the positron annihilation with electrons of the medium gives rise to the emission of annihilation gamma rays. By monitoring the gamma rays, information pertaining to the point defects present in the bulk state of the sample can be obtained [10]. On the other hand, variable low energy positron beams can enable the study of defects present on the surfaces and interfaces of materials. In Perturbed Angular Correlation (PAC), the angular correlation of two gamma rays emanating from a probe radioactive nucleus becomes time dependent due to the presence of local electric and magnetic fields [7],[8]. Thus, by monitoring the time dependence of angular correlation of gamma rays emitted in cascade by radioactive probe nuclei introduced in the sample, one can deduce pointed information about the local magnetic field and electric field gradients. Similarly, in Mossbauer Spectroscopy (MS), recoilless emission and absorption of gamma rays between a radioactive source and the sample under study is monitored to obtain information pertaining to the local electric and magnetic fields of the sample [11].

Corrosion occurs in the presence of moisture. Among other factors, the presence of salt in the moisture greatly enhances the rusting of metals. Iron when exposed to moist air reacts with oxygen to form rust labeled as Fe_2O_3. X H_2O. X indicates the varying amount of water complexed with the iron (III) oxide (ferric oxide). In the presence of oxygen Fe is converted in to Fe^{2+} and Fe^{3+}. In the presence of water this results in the formation of rust. The iron oxides commonly identified as corrosion products in steel are listed in Table 2. The oxy-hydrides, FeOOH are commonly observed in atmospheric corrosion, whereas the others tend to form when exposed to high temperature, high friction or aqueous environments.

Table 1 Different aspects of corrosion as studied by various nuclear and electron spectroscopic techniques

No	Technique	Probed quantity	Aspects studied	Limitations
1	Positron Annihilation spectroscopy using tunable low energy positron beams. (positron lifetime, Doppler broadening) techniques	Annihilation of positrons with electrons is probed to deduce electron density, electron momentum distribution at different thickness of the sample.	The role of point defects in corrosion. Presence of vacancies and vacancy clusters as a function of sample depth.	Large voids beyond about 10 Å, cannot be studied.
2	Mossbauer spectroscopy	Hyperfine interaction between nuclear moments of probe atoms with local electromagnetic fields. Deduced parameters are Isomer shift, Magnetic hyperfine field, Electric field gradients.	Various phases of rust formed in steel and Iron structures are identified. Especially identification of distinct phases of iron oxides of same structure but different magnetic properties. Kinetics of phase formation can be studied. Corrosion has been studied in detail by Conversion electron Mossbauer spectroscopy(MS).	Identification of phases of very low volume fraction is difficult.
3	Perturbed angular correlation	Hyperfine interaction technique. Measurable parameters are Magnetic hyperfine field, Electric field gradients	Internal oxidation of solute atoms can be studied. Bulk oxidation of metals and alloys can also be studied.	Phases of low volume fraction are difficult to be identified.
4	Auger Electron Spectroscopy	Energy of emitted Auger electrons, excited by a primary electron beam	Elemental composition, Oxidation, surface chemical changes – high surface sensitivity.	Limited to top surface layers.

Table 2 Iron oxides commonly identified as corrosion products

Oxide name	Formula	Oxide name	Formula
Ferrihydride	$5Fe_2O_3.9H_2O$	Wustite	FeO
Geothite	α-FeOOH	Hematite	α-Fe_2O_3
Akaganeite	β-FeOOH	Maghemite	γ-Fe_2O_3
Lepidocrocite	γ-FeOOH	Magnetite	Fe_3O_4
Feroxyhite	δ-FeOOH		

The present review is organized in the following way. The experimental details of various nuclear techniques such as PAS, PAC and MS as well as AES are briefly discussed in section 2. The application of these experimental techniques is illustrated through selected examples in section 3, followed by summary and conclusions in section 4.

2. EXPERIMENTAL DETAILS

2.1 Positron Annihilation Spectroscopy

In PAS studies, positrons from a radioactive source are used as a probe of the medium under study. Positrons injected in to the sample get thermalized and annihilate with electrons of the medium at various depths, giving rise to the emission of annihilation 511 keV γ-rays. The presence of defects such as vacancies, vacancy-impurity complexes and vacancy clusters act as trapping centres for thermalised positrons, altering the positron annihilation characteristics as compared to those of defect-free sample. By monitoring the outcoming gamma rays, experimental measurements such as positron lifetime, Doppler broadening and angular correlation can be carried out. These provide information [10],[12],[13] about the electronic structure of the material as well as the nature, size and concentration of defects present. Angular correlation studies are used for investigating the electronic structure of solids, while positron lifetime and Doppler broadening are more extensively used for defect studies. Figure 1 illustrates the experimental techniques. Due to the intrinsic continuous energy spectrum of positrons from the radioactive source, they get implanted at various depths in the sample. Therefore, information about defects, averaged over depths starting from the sample surface to about a few hundred microns is obtained from these experiments. Thus, defects present at a specific depth cannot be selectively probed by the conventional PAS techniques.

This shortcoming had been overcome with the advent of variable low energy positron beams [10], [12], [13]. In this scheme, a fraction of positrons from a primary radioactive source is thermalized and reemitted as thermal positrons using a moderator, giving rise to a secondary source of monoenergetic positrons. The slow positrons are selectively filtered from un-thermalized fast positrons using a velocity filter. The resultant slow positron beam can be focused, transported and accelerated to the desired energy (0 - 30 keV), to form a variable low

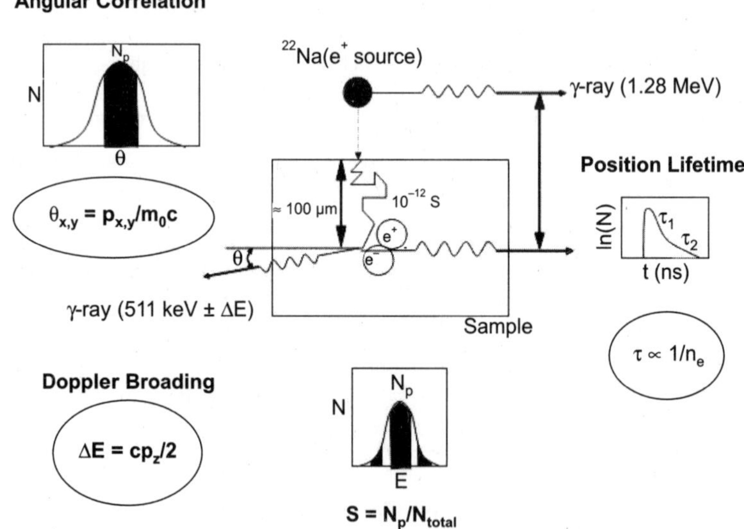

Figure 1 Positron annihilation spectroscopy comprises of Angular correlation, Doppler broadening and positron lifetime techniques. For defects studies, the latter two techniques are extensively used.

energy positron beam [13]. Beam transport in terms of electrostatic or magnetic fields is used to transmit the slow positrons from moderator to the sample. Ultra high vacuum is preferable in the beam line to preserve the surface cleanliness of the moderator so as to have optimal moderation efficiency, as well as to ensure that sample surface is not contaminated. Doppler broadening S-parameter measurements are most extensively used for positron beam studies and figure 2 defines the defect-sensitive S and W parameters. Due to the presence of vacancy-defects, the Doppler curve gets narrowed giving rise to an increase in S-parameter. On the other hand, the presence of chemical impurities gives rise to significant changes in W-parameter. While in conventional positron annihilation studies, the measured parameters are averaged over various depth regions, in variable positron beam measurements one can obtain depth-resolved parameters.

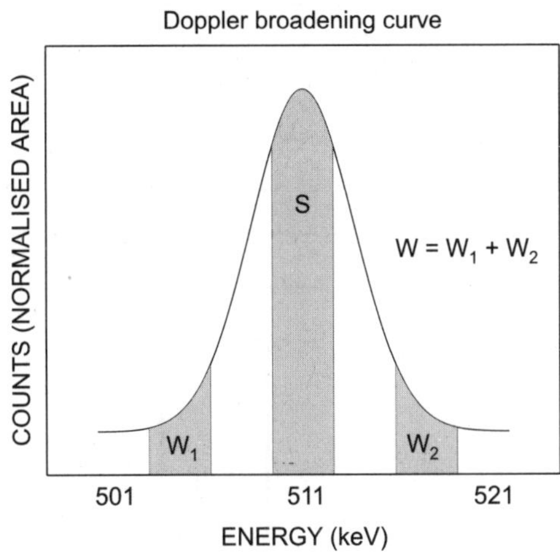

Figure 2 A typical Doppler broadening curve, measured using a Germanium detector. Definition of lineshape S and W parameters is shown.

2.2 Techniques Based upon Hyperfine Interactions

Hyperfine interaction [8] i.e., the interaction between nuclear moments with the electro magnetic fields at the site of a probe nucleus at isomeric state of spin I results in the removal of (2I+1) fold degeneracy as shown in Figure 3. This mainly comprises of electric monopole, magnetic dipole and electric quadrupole interactions. Electric monopole interaction is due to non-zero volume of the nucleus and the electron charge density due to s-electrons within the nucleus and this causes Isomer shift. Magnetic dipole interaction is due to interaction between magnetic dipole moment of the nucleus with the internal magnetic field at the site of the probe nucleus. Electric quadrupole interaction is due to interaction between nuclear quadrupole moment of the nucleus with electric field gradient (EFG) at probe nucleus.

Schematic of a hyperfine interaction induced splitting of I = 3/2 level of ^{57}Fe is shown in Figure 3.

Hyperfine interaction causes shift / splitting of spectral lines and the precession of nuclear spin. Magnitude of the splitting or the nuclear spin precession frequency is proportional to the

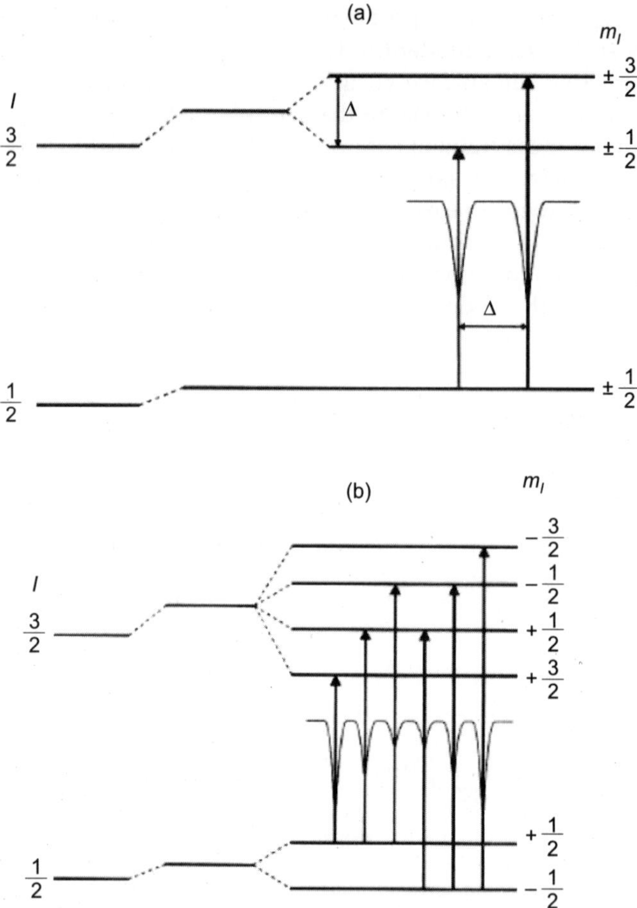

Figure 3 Shift and splitting of spectral lines due to electric monopole, electric quadrupole and magnetic dipole interactions are schematically shown above. (a) The effect on the nuclear energy levels for a 3/2 transition in ^{57}Fe, for an asymmetric charge distribution. The magnitude of quadrupole splitting Δ is seen. (b) The effect of magnetic splitting on nuclear energy levels. The magnitude of splitting is proportional to the total effective magnetic field.

strength of the hyperfine interaction. By means of measuring the hyperfine interaction one could deduce the solid state parameters such as Isomer shift, electric field gradients and the internal magnetic fields. These parameters are very useful in understanding a number of phenomena associated with the structural, electronic and magnetic properties of system concerned.

The magnitude of hyperfine interaction could be deduced either directly by means of measuring the shift/splitting of spectral lines using resonance absorption of gamma rays in the case of Mossbauer spectroscopy or by measuring the nuclear spin precession frequency using Perturbed angular correlation. In the following we will see in detail about the experimental aspects of Mossbauer and Perturbed angular correlation spectroscopy.

2.2.1 Mossbauer spectroscopy

The technique of Mossbauer spectroscopy involves the measurement of resonant absorption of gamma rays by suitable absorber atoms in the ground state introduced in the samples of interest to study the local structural, magnetic properties at the sites of absorber atoms. This is performed as follows. The source contains the parent nuclei of any suitable radioactive isotope embedded in a defect free, cubic, non-magnetic and rigid matrix. The γ-rays emitted from the source are passed through the material under investigation and those transmitted through the absorber are detected and counted. If the source and absorber nuclei are exactly in an identical environment (i.e the energy of the nuclear transition is equal in both the nuclei) the γ-rays will be resonantly absorbed with the absorption peak occurring at zero velocity. This is schematically illustrated in Figure 4. The Doppler effect produces an energy shift in the γ-ray energy enabling us to match the resonant energy level(s) in the absorber.

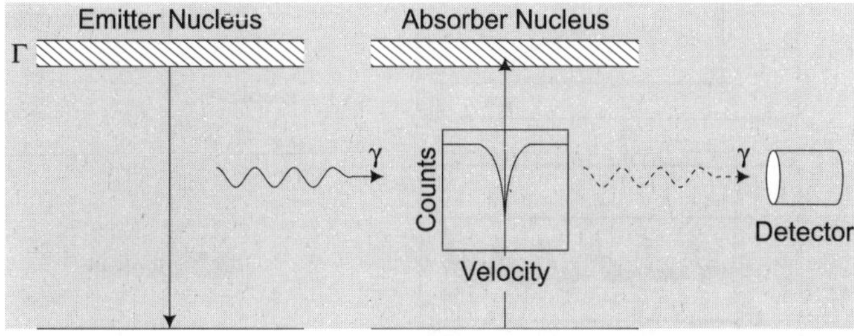

Figure 4 Example of the Mossbauer spectrum showing the simplest case of emitter and absorber nuclei in the same environment. The uncertainty in the energy of the excited state is shown exaggerated.

In the following example ^{57}Co is used as a Mossbauer source and ^{57}Fe as absorber atoms. By using ^{57}Fe having the spin I = 3/2 at its isomeric state the quadrupole interaction results in a doublet and magnetic interaction leads to a sextet in the Mossbauer spectra. Thus, using Mossbauer spectroscopy one deduces Isomer shift (IS) to measure the value of ionic charge of the Mossbauer ion. The magnitude of quadrupole and magnetic splitting as shown in Figure 3, is proportional to the electric field gradient (EFG) and magnetic hyperfine field (MHF) respectively at the site of Mossbauer nuclei. EFG is used as a parameter to deduce local structural and electronic properties, while MHF characterizes the magnetic property of the system of concern. Detailed studies of the parameters such as IS, EFG and MHF associated with different rust phases of iron or steel have provided good understanding of their corrosion properties [14] .

2.2.2 Perturbed angular correlation

Principle of a PAC method [8] is shown schematically in Figure 5. This is an indirect method of determining hyperfine parameters by means of measuring nuclear spin precession frequency. Radioactive PAC probe nuclei (selectively chosen based upon the matrix atoms) are introduced in a dilute concentration (typically of 10 μCi) in samples of interest by means of diffusion, implantation or by alloying [8]. By measuring the intensity of coincidence events of $\gamma_1 - \gamma_2$ of cascade as a function of time elapsed between them, one gets a wiggle pattern embedded in the exponentially decaying coincidence spectrum. Appearance of such wiggles is essentially due to the spin precession caused by the hyperfine interaction at the isomeric state.

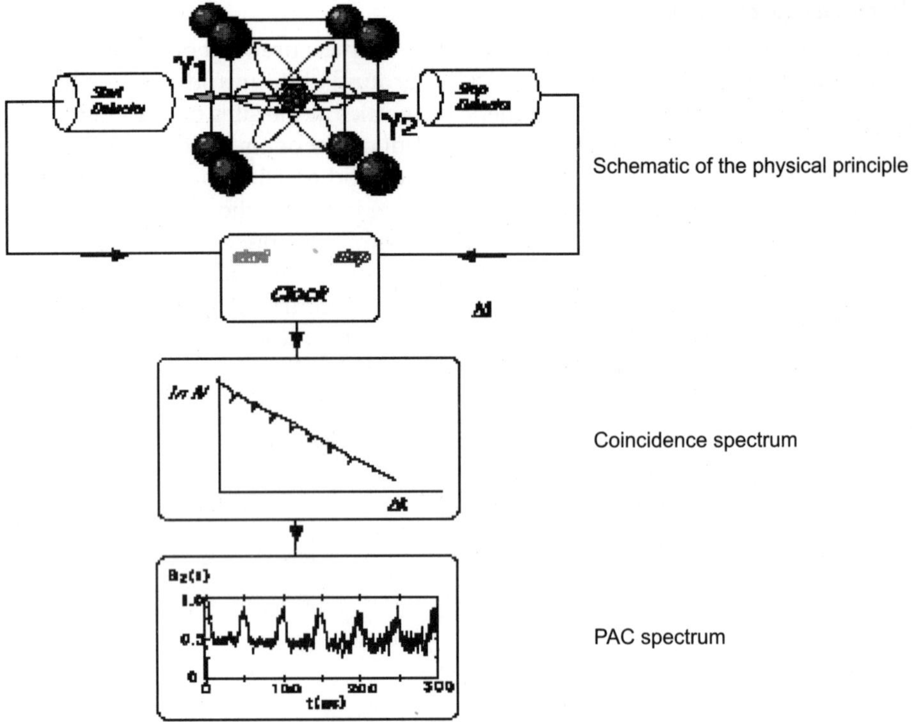

Figure 5 Schematic of the physical principle and the experimental spectra from perturbed angular correlation.

The experimental setup consists of three detectors with one set for γ_1 of the cascade while the other two detectors are set for γ_2 and are placed at 90° and 180° respectively with respect to the detector set for γ_1. Hence for this geometry of the detectors, two time dependent coincidence spectra corresponding to 90° and 180°, are obtained simultaneously. By means of obtaining the ratio of the difference and addition of these time dependent coincidence spectra, the wiggle part which is the PAC spectrum is deduced. Precession frequency is obtained by Fourier transform of the spectrum which is proportional to the magnitude of splitting as illustrated in Figure 5. The perturbation factor is analyzed for quadrupole interaction if the system is non-cubic else cubic but containing defects or impurities, to deduce electric field gradient (EFG) tensor. EFG tensor is basically represented in terms of V_{xx}, V_{yy} and V_{zz} in principal axis system. The maximum value of the tensor represented as V_{zz} is proportional to measured quadrupole frequency. Other two components are represented in terms of an asymmetry parameter which is given as $\eta = (V_{xx} - V_{yy})/V_{zz}$. If the system is cubic, defect free and magnetically ordered at the measurement temperature then the PAC spectrum is analyzed for magnetic interaction for deducing the Larmor precession frequency which is proportional to the splitting caused by the magnetic interaction. Thus, one measures the effective internal magnetic fields at the probe atoms. While both the magnetic and quadrupole interactions are involved the perturbation factor is analyzed for combined interactions to deduce both EFG tensor and magnetic fields. As the probe atoms are in the excited state, only electric field gradient and magnetic hyperfine field and not Isomer shift could be deduced by PAC. In studying problems related to defect-impurity interactions or oxidation one obtains the evolution of hyperfine parameters such as quadrupole / magnetic frequencies (ν_{Qi} / ω_{Li}), asymmetry parameter (η_i), magnitude of the fraction(s) (f_i) experiencing these parameters with annealing temperature. Details of the data analysis of the PAC spectra and hyperfine parameters are referred to the literature [7].

2.3 Auger Electron Spectroscopy

In Auger electron spectroscopy (AES), one uses energetic primary electrons to excite electronic transitions in the sample under study. The resultant Auger electrons provide identification of the elements present on the surface in terms of the characteristic electron binding energies [15]. It is one of the standard chemical characterization techniques used in semiconductor industry. AES as well as X-ray photoelectron spectroscopy (XPS) are quite useful for investigating the surface aspects related to corrosion. AES is quite fast, versatile and quantitative estimation of surface chemical species is also possible. However, the determination of the oxidation state of elemental species is rather difficult with AES, while it is possible with XPS. AES is quite surface sensitive and provides information about top surface layers up to a depth of ~ 1 nm.

AES measurements are carried under UHV conditions using cylindrical mirror analyzer (CMA) for energy analysis of Auger electrons. In view of the surface sensitivity of the technique, as-mounted samples from ambient conditions need to be cleaned of the surface contamination by Ar-ion sputtering. A primary electron beam of typically 3 keV is used for the excitation. The outcoming Auger electrons are detected using the CMA. A typical experimental arrangement of AES instrumentation is shown in Figure 6. One of the main disadvantages of AES is that insulating samples cannot be analyzed due to sample charging problems.

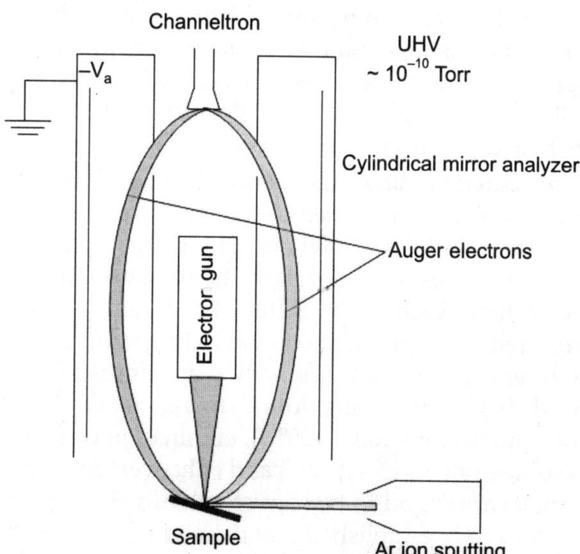

Figure 6 Experimental arrangement of AES, showing the important components.

3. RESULTS AND DISCUSSION

3.1 Positron Annihilation Spectroscopy

PAS is one of the established techniques to study electron momentum distribution as well as vacancy-defects in materials. It has been applied to a range of problems dealing with defect investigations in materials science such as vacancy clustering in metals and alloys [16], solute atom clustering in binary alloys [17] and nucleation and growth of helium bubbles [18],[19]. Radioactive source (Na-22) based fast positrons are used for investigating the bulk phenomena, by making positron lifetime or Doppler broadening measurements. When positrons are injected in to the medium (such as metals, alloys and semiconductors), it undergoes annihilation with the electrons of the medium to give rise to positron lifetimes in the range of a few hundred ps. The presence of vacancy-defects and other positron trapping defects give rise to more lifetime components. Thus, by monitoring the changes in the respective defect-specific lifetime parameters as a function of process parameters such as aging time, quenching or aging temperature or alloying elements, one can deduce important information. However, in the case of molecular and polymer media, positrons can also bind with electrons to form Positronium (Ps), which can occur in two states viz., para-Ps (p-Ps) or ortho-Ps (o-Ps). The lifetime of p-Ps is 125 ps, while that of o-Ps is 142 ns. The o-Ps particle, when trapped in low electron density regions such as polymeric holes in polymer media can undergo pick-off annihilation, thereby exhibiting lifetimes much smaller than that of vacuum value. This lifetime component and its intensity are monitored to obtain useful information about free volume hole fraction and its variation as a function of process parameters such as coating thickness, its curing cycle.

Electronic circuits on semiconductor devices are protected by polymeric coatings for insulation purposes as well as for environmental protection. The mechanism by which a polymeric coating protects a metallic material from the environment is a complex process. Subsequent to the polymeric coatings, the coated surface needs to be cured at select temperatures. This will ensure the stabilization of the microstructure of the polymer in terms of defects and polymeric hole distribution.

In a study aimed at understanding this, comprehensive studies using electroimpedance and positron annihilation spectroscopic measurements were carried out on two commercial polymeric coatings (Polymide- PI2566, Polymide-PI2610D) on a model metallic surface viz., Tungsten [20]. The coated surfaces are cured in the temperature range from 100 to 325°C for a duration of 30 min at each temperature. The deduced variation of free volume hole fraction (as deduced from the third lifetime component corresponding to o-Ps) as well as average lifetime (average of three lifetime components) are shown in Figures 7 and 8, respectively. The observed results for PI2566 coatings are found to fall in to three temperature regimes corresponding solvent evaporation (100 – 125°C), elimination of water from the reaction site leaving a small free volume cavity (125 – 250°C) and enhancement of free volume fraction (250-325°C). These observations are found to be consistent with electroimpedance measurements. These corroborative studies have established that the performance of these polymeric coatings are critically dependent on the extent of cure, which influences the free volume fraction, which facilitates the pathways for water molecules and ion species to reach the interface. Also commercial polymeric coatings on steel, consisting of vinyl ester and epoxy coatings [21] as

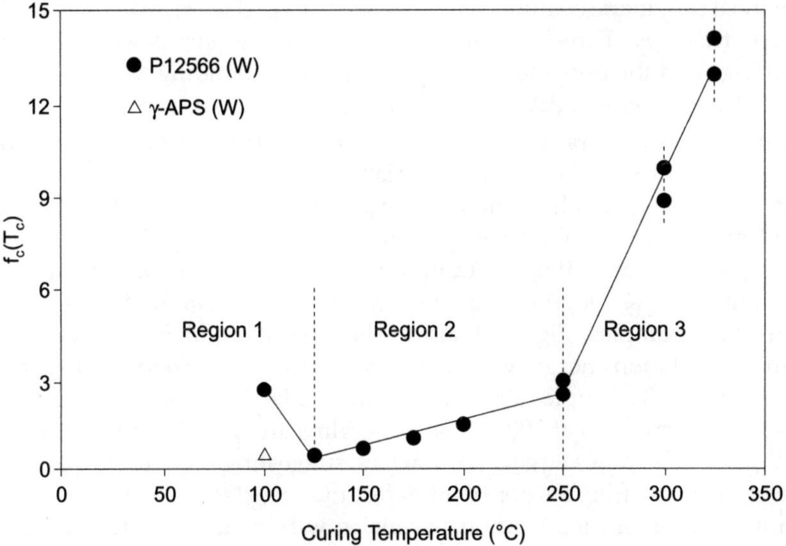

Figure 7 Free volume fraction of polymeric coating of PI2566 on W substrate monitored as a function of curing temperature. The measured lifetime in a thin film of γ-APS (used as adhesion promoter between the substrate and the polymeric coating) coated on W substrate is also shown for comparison [20].

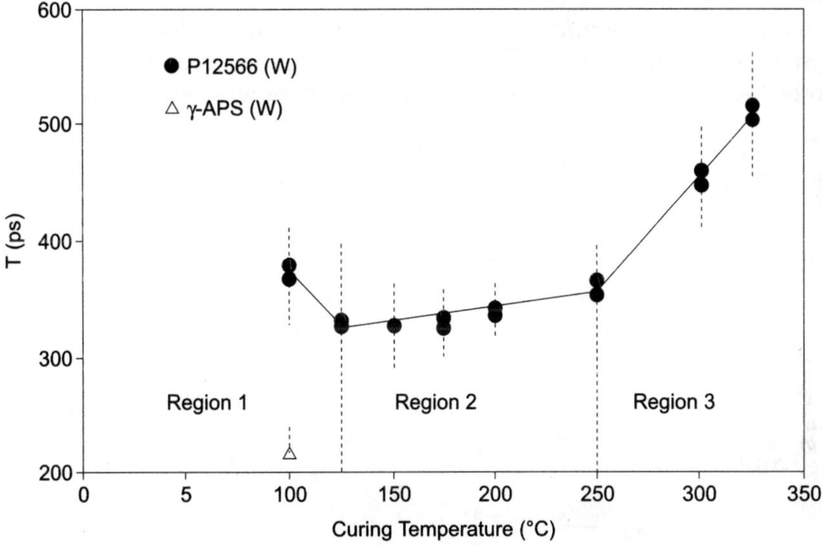

Figure 8 Average positron lifetime of polymeric coating of PI2566, monitored as a function of curing temperature [20].

well as commercial polymeric coatings on iron substrates [22] were investigated using positron annihilation spectroscopy. Three experimental lifetime components were observed in all the polymeric coatings and the o-Ps lifetime parameters corresponding to the polymeric hole are found to be influenced wet and dry exposure conditions of the coatings.

In an attempt to understand the role of vacancy-defects in corrosive breakdown of oxide passive films grown on metallic surfaces, variable low energy positron beam studies were carried out on Ti samples [23]. The samples were prepared by anodic polarization of a titanium substrate in sulphuric acid. The foils were put into a 0.1 M H_2SO_4+0.1 M KBr corroding solution and anodically polarized at a 10 mV/s ramp rate to ultimate potential ranging from 1.5 – 8.0 polarization voltage (V_{SCE}). Depth-resolved defect-sensitive S-parameter measurements were carried out on these samples. Figure 9 shows the experimental S-parameter as function of positron beam implantation energy (which probes various sample depths). In comparison with the S-parameter of pure Ti sample (observed around 2.5 keV of positron implantation energy), the surface S-parameters (seen at 500 eV) of oxide films are found to be different. Further, the sample with V_{SCE} = 8.0 V shows large S-parameters, suggesting the presence of vacancy-defects at the surface of the oxide film. As per point defect model (PDM), it is proposed that passivity breakdown initiates when metal cation vacancies within the film, formed at the surface, migrate to the metal/film interface and cannot be annihilated by the oxidative injection of metal cations into the vacancies. The excess vacancies then necessarily condense at the metal/film interface and result in localized detachment of the barrier layer. The present results indicate clearly the existence of vacancy-defects in oxide films and the polarization voltage is seen to considerably influence these.

Anodic oxidation of aluminum was investigated using variable low energy positron beams [24]. Positron annihilation measurements were carried out on as-received pure Al (99.985 %) foil and another Al foil treated in NaOH. Caustic treatment was carried out by immersion of

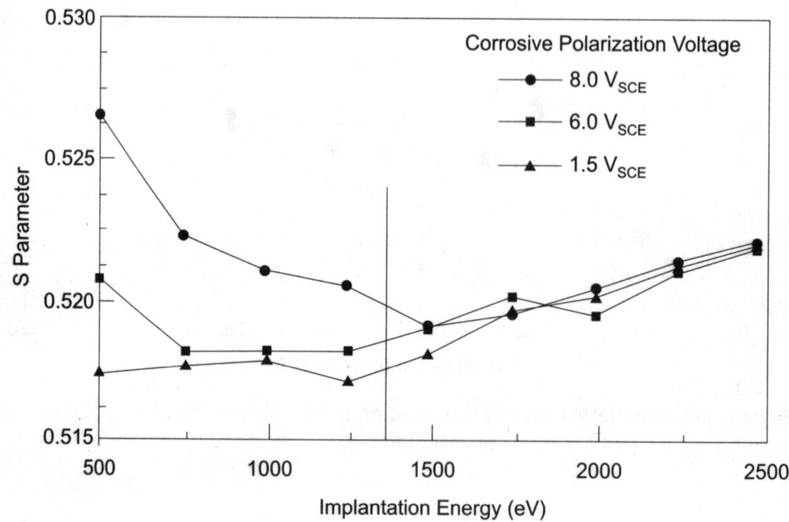

Figure 9 Variation of defect-sensitive S-parameter as a function of positron beam implantation energy for oxide films grown on Ti samples for various polarization voltages [23].

foil in aqueous 1 N NaOH solution for 15 min at room temperature. Anodic oxidation of this sample was carried out in a borate buffer solution at room temperature, at a constant applied current density of 2.5 mA/cm^2, until attaining voltages of 27, 53, 80 and 106 V.

The results in terms of S-parameter as a function of positron beam energy are shown in Figure 10 for (a) as-received and (b) NaOH treated Al samples. As can be seen, the S-parameter corresponding to the oxide layer at the surface (until 2 keV) exhibits a lower value as compared to the bulk Al value (seen beyond 15 keV of positron beam energy). It is found that the plateau region corresponding to the oxide layer increases as the anodic potential is increased, indicating that the oxide film thickness is increasing. Further, a small increase in S-parameter is seen around 3 keV energy range, indicating the existence of intermediate defect layer between the oxide and bulk Al regions. It may be pointed out this is more prominently seen for higher anodic potential. As compared to the as-received sample (Figure 10(a)), the S-parameter changes are larger in NaOH treated sample (Figure 10(b)).

The experimental data are analyzed in terms of positron diffusional analysis, using VEPFIT simulation program [25] so as to obtain useful information about the S-parameter and the thickness of the oxide film and also the S-parameter of intermediate defect layer. Figure 11 shows the deduced S-parameter of the defected layer for as-received and treated Al samples as a function of anodic potential. The main result is that for NaOH treated sample, the S-parameter is relatively larger compared to the as-received sample. Further, anodic potential does not seem to affect the defect S-parameter. The observed large S-parameter values (around 1.07) in NaOH treated sample indicate the presence of voids, having a size larger than 1 nm. Since, nanometer-size voids should not be mobile, the constant S-parameter has been explained based on the regeneration of voids i.e., after preexisting voids are incorporated into the film, new voids are created by oxidation at the same sites. This experimental data supports the picture that repeated formation of voids at certain "defect" sites at the metal/film interface. The effect of metallic impurities in stabilizing the interfacial defects were subsequently investigated using PAS, atomic force microscopy (AFM) and Rutherford backscattering spectroscopy (RBS) [26] .

In a study aimed at investigating the corrosion-induced defects at near-surface regions, detailed positron beam measurements have been carried out on pure Fe and stainless steel samples, subjected to anodic polarization studies in 1 N H_2SO_4 solution [27]. Figure 12 shows the variation of (a) defect-sensitive lineshape S-parameter (b) the difference parameter viz., (S(corroded) - S(as-received)), as function of positron beam energy for pure Fe foils in as-received and different anodic potentials in 1 N H_2SO_4 solution for 1 h. As compared to as-received sample, it is noticeable in Figure 12(a) that the S-parameter is larger for treated samples and further, it increases as anodic potential is increased. It is clear that the observed changes in S-parameter are larger near the surface than in the deeper regions of the sample. These changes are more clearly brought out in the difference S-parameter plot shown in Figure 12 (b). By plotting, the S-parameter together with its complementary W-parameter, it has been deduced from the S-W correlation plots that vacancy-defects are present at near-surface regions of the treated films as compared to that of as-received film. On the other hand, when similar studies are carried out on SS-304 and SS-316 it is found that relatively smaller size defects are formed at near-surface regions.

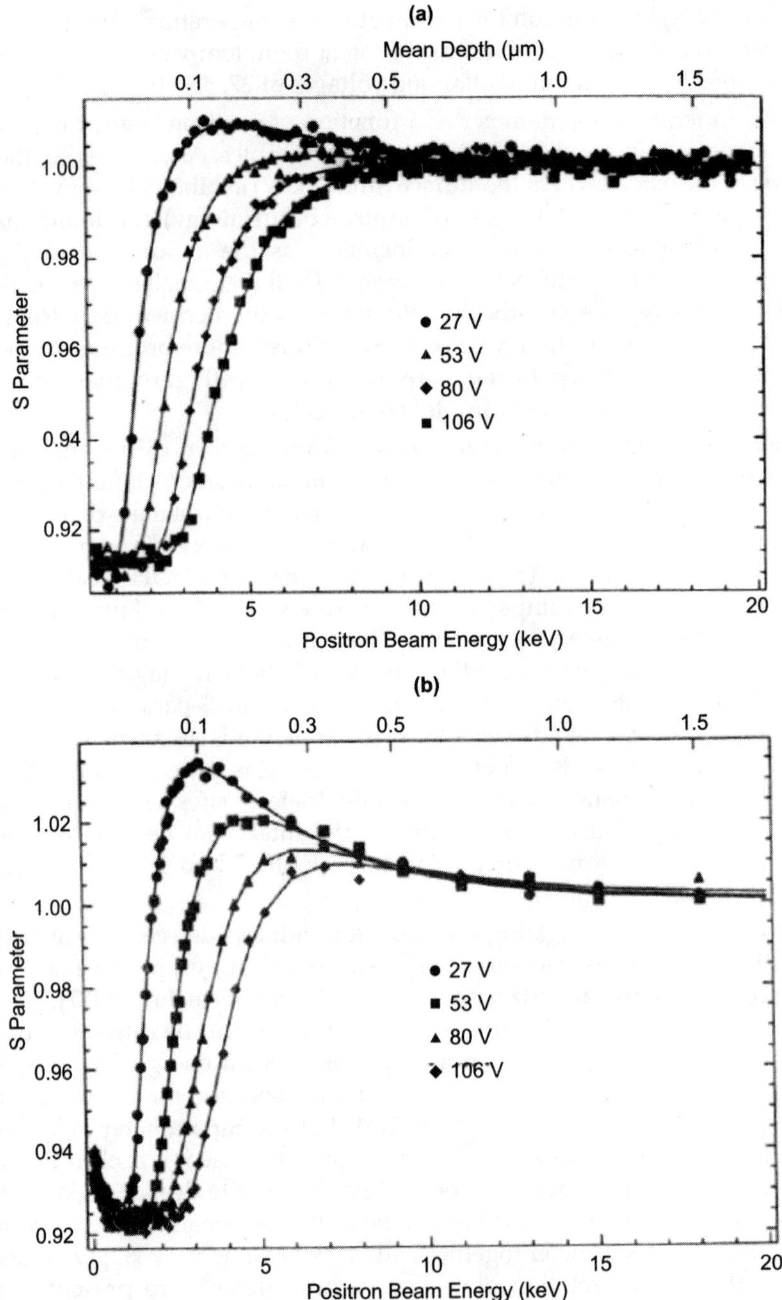

Figure 10 Defect-sensitive S-parameter as a function of positron beam energy for Al foils at various anodic potentials (with 2.5 mA/cm^2 current density) in (a) as-received (b) after 15 Min NaOH-treated conditions. The mean implantation depth ie., depth range probed positron beam energy is indicated on the top axis. Solid line through the data points is a result of model fit [24].

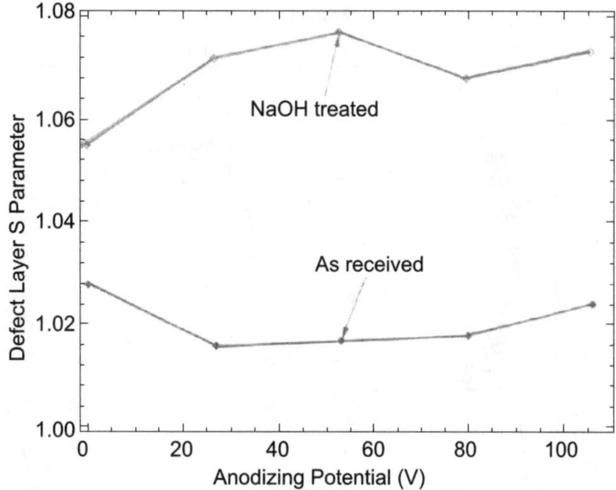

Figure 11 Variation of the S-parameter of the intermediate defect layer between the oxide film and the metal for as-received and NaOH treated Al samples [24].

3.2 Perturbed Angular Correlation (PAC) Spectroscopy

PAC spectroscopy has been widely used to study point defects, structural, magnetic phase transitions in solids [28]. With respect to the problem of corrosion it is important to understand about impurity-oxygen interaction in a material. PAC measurements have been extensively carried out on the oxidation of fcc metals [29],[30]. Foils of Ag, Al, Au, Cu, Pd and Pt of thickness around 50-100 μm were implanted with 10^{12} radioactive ^{111}In atoms to a depth of about 0.2 micron. After implantation these samples are vacuum annealed and a fully unperturbed PAC spectrum is obtained implying that the probe atoms are occupying defect-free sites in fcc matrices. Measurements were carried out subsequent to each isochronal oxidation cycle. Up to a critical oxidation temperature, the spectra remained unperturbed (except for further annealing of Pd).

The spectra, shown in Figure 13, obtained in the oxidized metals show that the quadrupole frequency is almost same in all the cases independent of the matrix, implying that this represents the internal oxidation of ^{111}In impurities. The variation of oxidized fraction (F_0) is shown in Figure 14 as a function of temperature. At higher oxidation temperature T_{ox} values, an increasing fraction F_0 evolves, characterized by a broad frequency distribution.

In the case of Ag where F_0 rises up to 100% it is shown that the oxygen trapping by the probe ^{111}Cd atoms is diffusion controlled. In the case of systems of higher melting point such as Pd and Pt, it is seen that the oxidation starts around $T_{ox} \sim 0.6\ T_{melt}$, where thermal production of vacancies, necessary for substitutional diffusion sets in. This implies that in these systems oxidation is not only controlled by fast interstitial diffusion of oxygen, but also by substitutionally diffusing indium. The formation of In_2O_3 precipitates observed in these systems require In diffusion.

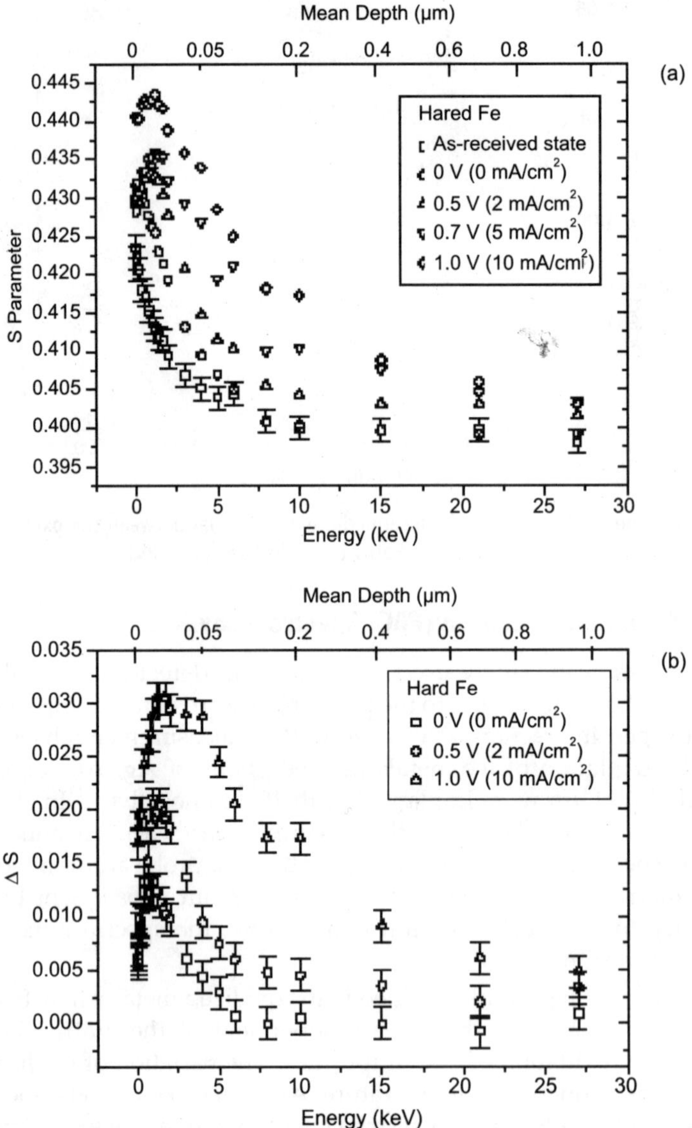

Figure 12 Variation of (a) defect-sensitive lineshape S-parameter (b) the difference parameter – (S(corroded) – S(as-received)) as function of positron beam energy for pure Fe foils in as-received and different anodic potentials in 1 N H_2SO_4 solution for 1 h. The mean implantation depth probed by the positron beam is shown on the top axis [27] .

Oxidation of Hafnium

PAC is a powerful technique to study problems ranging from impurities-vacancies binding [31], impurity-oxygen binding, internal oxidation of impurities and to bulk oxidation. Trapping of oxygen by Hf impurities in Ta matrix has been studied by ^{181}Ta PAC at different

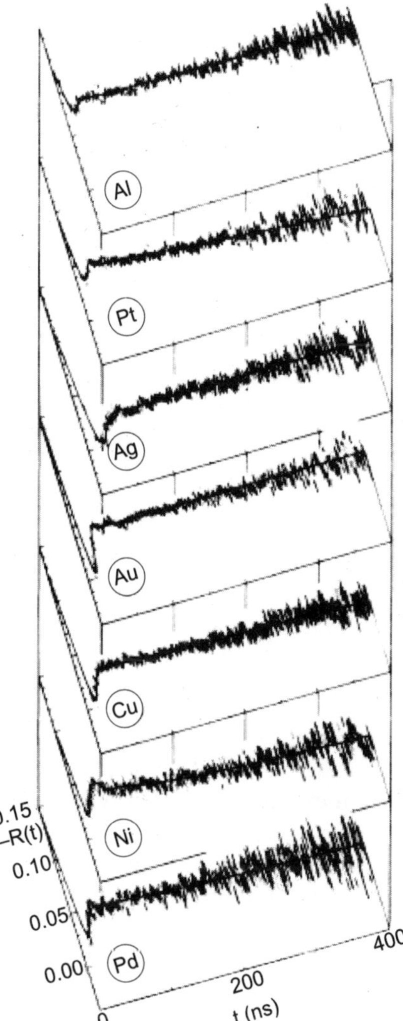

Figure 13 Comparison of perturbation functions for ^{111}In in fcc metals obtained with maximum oxidized fraction [29].

temperatures to find out the diffusion coefficient of oxygen in Ta matrix [32]. Internal oxidation of impurities in fcc matrices [29] has been studied in a detailed manner using ^{111}Cd PAC as illustrated in the previous section. Recently we have carried out ^{181}Ta PAC studies to understand the process of oxidation of hafnium at atomic scale. In this study we could delineate the existence of three regions viz., clusters of hafnium, oxygen deficient hafnium oxide and hafnium oxide based on the distinct quadrupole parameters experienced by probe atoms associated with these regions in the oxidized hafnium sample [33].

Hafnium oxide due to its high dielectric constant, large band gap and thermal stability may replace conventional gate dielectric SiO_2 in microelectronics. Oxidation of hafnium is of

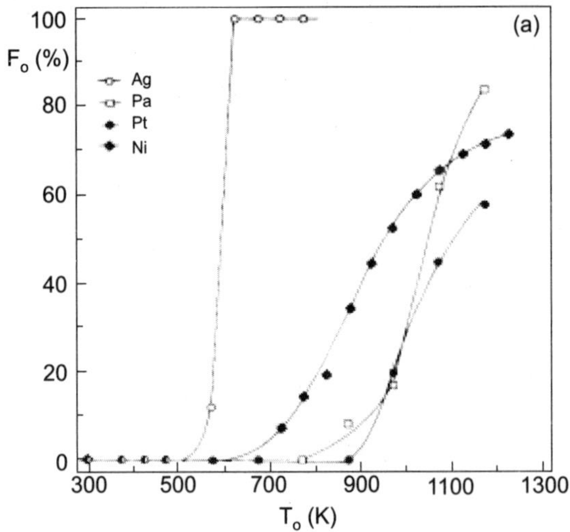

Figure 14 Oxidized fractions of ^{111}In in fcc metals observed in cumulative isochronal oxidation cycles in Ag, Pd, Pt and Ni hosts [29].

interest in semiconductor industry. Hf foil (made by Aldrich, USA) of 25 micron thickness with hcp structure as deduced by XRD was taken as the starting sample for the present measurement. The foil was irradiated with thermal neutrons to a fluence of 2×10^{22} n/m^2 to produce probe nuclei ^{181}Ta by the reaction ^{180}Hf(n,γ) ^{181}Hf $\xrightarrow{(\beta-)}$ ^{181}Ta homogeneously in the sample. The 133-482 keV γ-γ coincidences of the cascade in ^{181}Ta (separated by the isomeric state of I = 5/2 with quadrupole moment Q = 2.51b) was measured. Hf foil is subjected to a controlled annealing treatment in oxygen atmosphere at different temperatures. PAC measurement has been carried out at room temperature after each step of oxygen annealing treatment. PAC spectra obtained in the reference and oxygen annealed sample are shown in Figure 15.

In the oxygen annealed sample, three fractions experiencing distinct quadrupole frequencies are deduced. Values of quadrupole parameters corresponding to different annealing temperatures are shown in Table 3. Based on the quadrupole frequencies and asymmetry parameters, these are identified to be due to probe atoms associated with Hf, HfO_2 and oxygen depleted hafnium oxide HfO_{2-x}. Variation of the fractions (shown in Figure 16) shows that the fraction corresponding to HfO_2 increases at the cost of Hf metal part and HfO_{2-x}. Thus, PAC can be used to investigate the internal and bulk oxidation behaviour of materials.

3.3 Mossbauer Spectroscopy

Mossbauer spectroscopy (MS) is widely used for studying defect-probe interactions along with structural and magnetic phase transitions in a number of systems [34]. With respect to the problem of corrosion in steel or any iron containing metals, MS has been used to study the formation of oxide layer and its structural and magnetic properties. This knowledge is essential to tailor make a stable oxide film by different means either properly alloying or implantation of

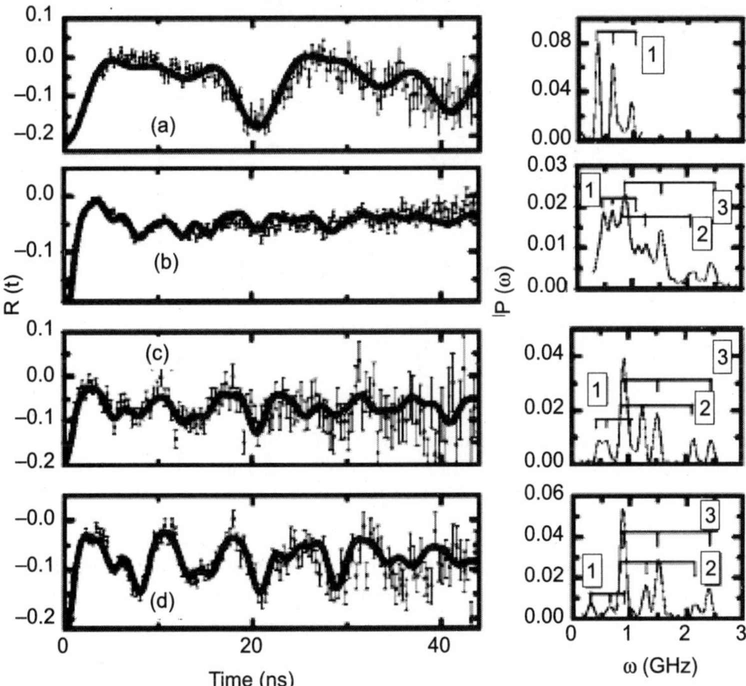

Figure 15 ^{181}Ta PAC spectra along with the Fourier transforms for Hf foil at 300 K (a) untreated (b)-(d) refer to measurements subsequent to oxygen annealing at temperatures viz., (b) 900 K, (c) 1173 K and (d) 1373 K. [33].

Table 3 Results of hyperfine parameters corresponding to a few representative annealing treatments in oxygen atmosphere

Sample treatment	ν_{Qi} (MHz)	η_i	δ_i	Identification
Hf foil (untreated)	300 ± 11	0.14 ± 0.06	0.038 ± 0.01	Substitutional probe atoms in grain of hafnium
Annealed at 900 K	348 ± 19	0.68 ± 0.08	0.058 ± 0.01	Substitutional probe atoms
	654 ± 16	0.68 ± 0.12	0.058 ± 0.01	Hf associated with HfO_{2-x}
	790 ± 34	0.48 ± 0.11	0.064 ± 0.01	Hf associated with HfO_2
Annealed at 1373 K	304 ± 12	0.14 ± 0.05	0.048 ± 0.01	Hf cluster
	688 ± 21	0.56 ± 0.11	0.05 ± 0.01	Hf associated with HfO_{2-x}
	758 ± 30	0.37 ± 0.12	0.048 ± 0.01	Hf associated with HfO_2

ions of suitable elements on the surface of the concerned metal. With respect to these studies MS has been carried out in stainless steel and iron exposed to different environments and for a varying length of time [36]. In the following, the study of oxide in steel using MS has been illustrated with a few examples.

Millscale is a thin, adherent oxide coating that forms on steel during heat treatment, hot rolling around 560°C, and forging processes. The thickness of the millscale is about 10-20 μm

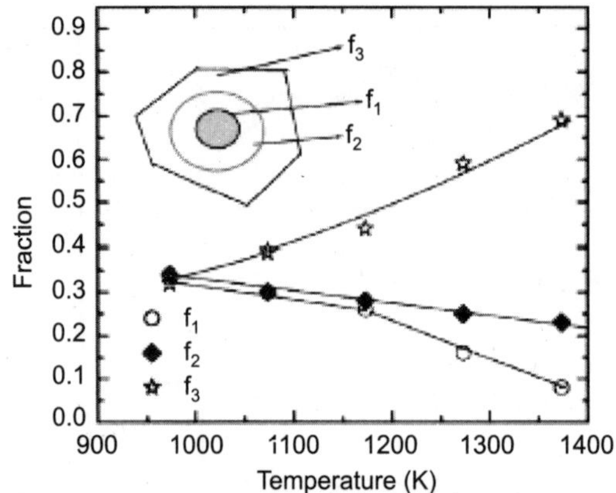

Figure 16 Variation of fractions with temperature at which Hf foil is subjected to oxygen annealing. f_1, f_2 and f_3 denote the fractions of probe atoms associated with Hf core, oxygen vacancy rich zones of oxide and hafnium oxide shell [33].

and depends upon the type of steel, its thickness and thermal, chemical and mechanical processing history. Scattering Mossbauer spectrum (XMS) of the millscale coating formed on type A-36 grade carbon steel rolled to a thickness of 0.375" at 560°C as obtained by Cook et al [35],[36] is shown in Figure 17. Spectral analysis showed that the millscale consists of three iron oxides, wustite FeO (53%), magnetite, Fe_3O_4 (33%) and hematite α-Fe_2O_3 (14%) as shown in Figure 17. Wustite is only stable at high temperature and at room temperature it may decompose to iron and magnetite. Therefore, millscale offers little long term production to atmospherically exposed stainless steel.

Shown in Figure 18 are the Mossbauer spectra obtained on weathering steel exposed for 15 years at Bethlehem both using X-ray resonant scattering and transmission geometry. XMS spectrum of Figure 18 of 40 micron protective corrosion coating , and the TMS spectrum , of the removed coating, both recorded at 300 K, show the presence of a small fraction of magnetic goethite, α-FeOOH(m) corresponding to particle size > 15 nm. Central doublet, occupying a larger area and hence larger fraction, was identified by a low temperature analysis to be lepidocrocite, γ-FeOOH, and the superparamagnetic goethite α-FeOOH(s) having particle size < 15 nm. In addition steel substrate signal is also present in each spectrum. The TMS spectrum recorded at 77 K (Cf. Figure 18 (c) shows a larger component of magnetic goethite resulting from the reduced magnetic relaxation rate at the lower temperature. This component is labeled as α-FeOOH (m + s1), and is comprised of the magnetic component and part of the super paramagnetic component seen in the doublet at 300 K.

We have seen in brief, with illustrative examples, the use of Mossbauer spectroscopy to obtain the various phases of iron oxides formed in rust. Though the studies of corrosions is very important to be understood in steel the most widely used structural material, depending upon specific applications such as space, aeronautics and medical fields certain other metals and their alloys may also be of interest. Mossbauer spectroscopy could be extended in studying corrosion in such materials by a proper choice of suitable Mossbauer source and absorber atoms.

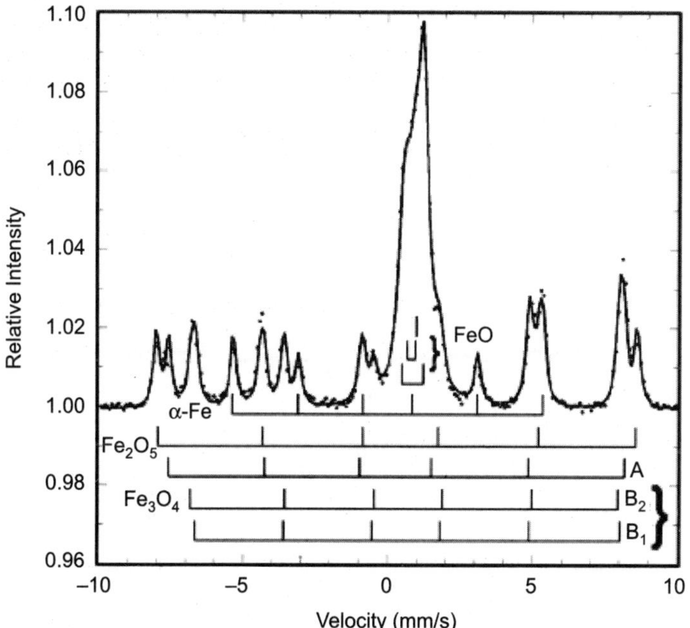

Figure 17 Scattering Mossbauer spectrum (XMS) recorded at 300K, of millscale formed on carbon steel rolled at 560°C. The composition is Wusite-53%, hematite-14% and magnetite-33%. The component labeled α-Fe is from the steel substrate [36].

3.4 Auger Electron Spectroscopy

The corrosion behavior of stainless steels is often correlated with the protective nature of the passive surface oxide film. The understanding and control of passivity has been important factor for protection against corrosion. Since, the passive film forms on the surface of the material exposed to various corroding environments, it is necessary to study the microstructure and chemical composition of near-surface regions. In view of its surface-sensitivity and ability for elemental identification, AES is quite powerful for investigating various corrosion-related phenomena in metals and alloys.

3.4.1 Hydrogen effects on the passive film formation in nitrogen containing SS316L stainless steels

The presence of hydrogen in structural materials strongly influences the physical, chemical and mechanical properties. An attempt has been made to understand the role of hydrogen in nitrogen containing stainless steels using a variety of techniques [37].

The measured AES spectra of as-received (taken as reference), passivated and hydrogen-charged samples are shown in Figure 19 (a), (b) and (c), respectively. In reference samples (Figure 19 (a)), peaks due to C, N, Cr, O, Fe and Ni are clearly seen. Further, as the nitrogen content is increased, the Auger peak due to N is found to progressively increase. With respect to passivated sample (Figure 19 (b)), the main observation is the increase in Cr and O peak heights and reduction in Fe and Ni peaks. The increase in Cr and O peaks indicates the formation of Cr_2O_3 in the passive film and consequently, the peaks due to Fe and Ni are

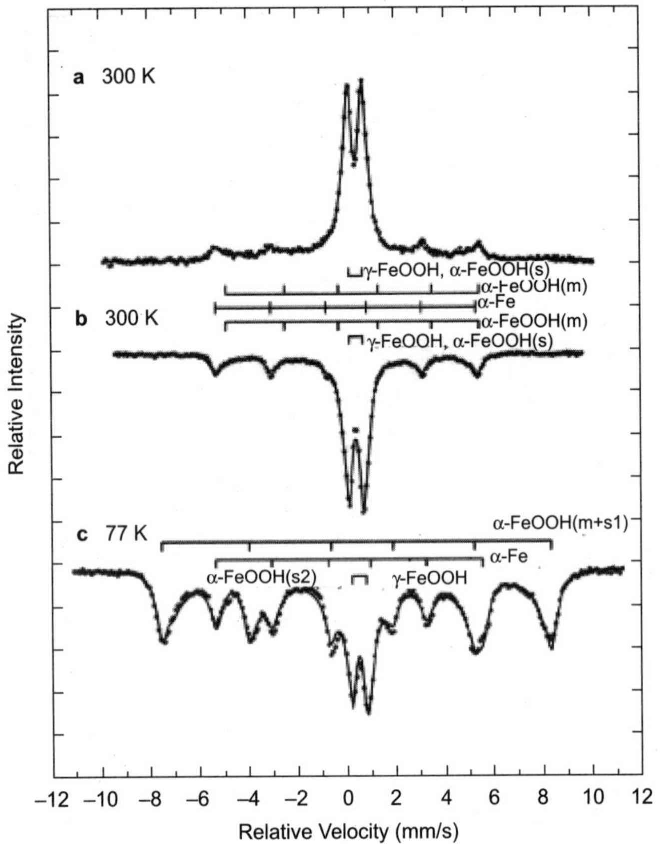

Figure 18 Mossbauer spectra recorded at (a) XMS, 300 K, (b) TMS, 300 K and (c) TMS at 77 K, of the iron oxides in the adherent and protective corrosion coating formed on weathering steel following atmospheric exposure at Bethlehem, PA for 15 yrs. The coating is comprised of 75% goethite of which 75% is nanophase of particle size < 15 nm, and has $(\alpha/\gamma) = 4.9$ and $(\alpha_m/\gamma) = 1$[36].

diminished. A small peak around 180 eV is found to be associated with chlorine, which is know to exist in the form of Cl in the passive film. The large increase in Auger peak due to carbon (272 eV) is also noticed. On the other hand, pre-charging with hydrogen, the surface composition of the passive film changes, as shown in Figure 19 (c). Perhaps, the most notable feature is enhanced uptake of Cl on the surface. It is also seen that nitrogen peak intensity decreases and marginal increase in peak intensities due to chromium and oxygen. Thus, iron dissolution and enhanced Cl uptake was more pronounced for the passive film, which has been hydrogen-charged (Figure 19 (c)) as compared to that of hydrogen-free sample (Figure 19 (b)). Thus, the presence of hydrogen seems to influence the chemical nature of passive film.

3.4.2 Effect of heat input on stress corrosion cracking behavior of weld metal of SS316N

High heat input during welding increases the risk of sensitization because the slow cooling rates lead to longer residence time in the sensitization temperature range. Reducing the carbon content and adding nitrogen to offset the loss in strength caused by the reduction in carbon

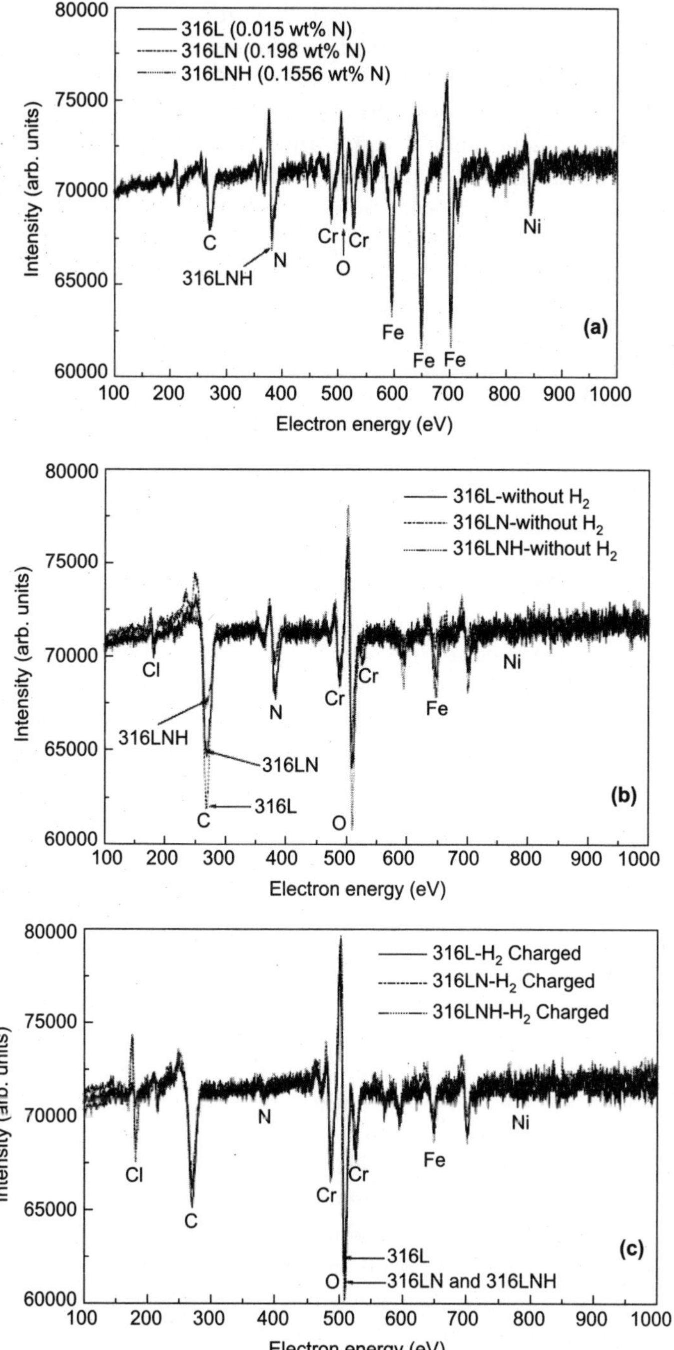

Figure 19 AES spectra of (a) as-received (b) passivated and (c) hydrogen-charged SS316 samples containing 0.015 wt% (316L), 0.198 wt% (316LN) and 0.556 wt% (316LNH) of nitrogen [37].

content can mitigate the problem of sensitization. Towards investigating this, detailed studies have been carried out to investigate the effect of heat input on the stress corrosion cracking (SCC) of AISI type 316N SS weld metal in boiling 45 % $MgCl_2$ and 5 M NaCl+0.15 M Na_2SO_4+2.5 mil/L HCI solutions for four different heat input conditions of 3.07, 4.8, 7 and 7.4 kJ/cm [38]. Detailed SCC tests, micrography and AES studies were carried out on these samples. SCC tests have indicated the formation of an adherent and protective passive film on the weld metal surface, which has necessitated corroborative evidence from AES. Figure 20 shows AES spectra of reference sample and adherent film on the fractured surfaces in different corroding solutions. Dominant Auger peaks due to Fe, Cr, Ni and oxygen are clearly seen both in reference and fractured surfaces. However, in fractured surfaces, the Auger peak height due to Fe has decreased significantly together with the enhancement of oxygen peak (arising due to oxide formation) as compared to that of reference samples. Auger peaks due to Na could not be clearly detected in view of the less relative sensitivity for Na but its presence cannot be ruled out. On the other hand, the presence of surface Cl has been clearly seen in fractured surface. The surface chemical composition is found to be marginally different between two corroding solutions. For 45% $MgCl_2$ solution tested sample, low energy AES scan has shown evidence towards the presence of Mg (seen at 48 eV) on the surface as shown in Figure 20 (c).

3.4.3 Surface segregation in D9 steel at elevated temperatures

Surface segregation of alloying elements in Ti-modified stainless steel (D9), a candidate material for fuel clad and wrapper applications in Fast Breeder Reactor (FBR) programme, has been investigated by Auger electron spectroscopy [39]. Since, this core structural material will be subjected to elevated temperatures (higher than 670 K) during operation, it is necessary to understand the surface chemical changes that can take place. D9 samples in cold-worked and solution annealed states, having Ti/C ratio of 4, 6 and 8 have been annealed at temperatures ranging from 670 to 970 K in-situ in Auger analysis chamber for a duration of 5 min and subsequently, Auger scans are obtained. Major as well is kept for time duration of 5 Min and subsequently, Auger scans are obtained. Major as well as minor elemental profiles are obtained at each temperature. Figure 21 shows the AES spectra obtained in as-received condition for 20 % CW D9 sample, after heating to 870 K and subsequently sputter cleaned. In as-received sample, Auger peaks due to all major elements such as Fe, Cr, Ni and other elements like Ti, N, Mo, Si and C could be seen. Auger peak due to Argon arises due to Ar ions used for sputtering the sample surface. When the sample is heated to 973 K, it is found that the Auger peaks due to major elements are diminished and further, enhanced Auger peaks due to minor elements like P and S are seen. So as to check whether it is surface-segregation or not, the sample has been in-situ sputtered again with Ar ions, which results in the spectrum shown in Figure 21 (C), which tallies with the as-received spectrum. This clearly establishes that due to exposure to elevated temperatures, minor light elements segregate to the surface.

Influence of pre-annealing thermo-mechanical treatments on the surface segregation behavior of various elements has also been investigated, by carrying out measurements on solution annealed sample. Figure 22 shows the relative intensities of various elements (normalized with respect to Fe intensity) for annealed and 20 % CW D9 samples. As can be seen, due to the existence of defects in cold-worked sample, the surface segregation behavior of most of the minor elements seems to be enhanced as compared to that of annealed sample. The present results are understood in terms of the formation of precipitates such as TiC and TiN at these temperatures [40].

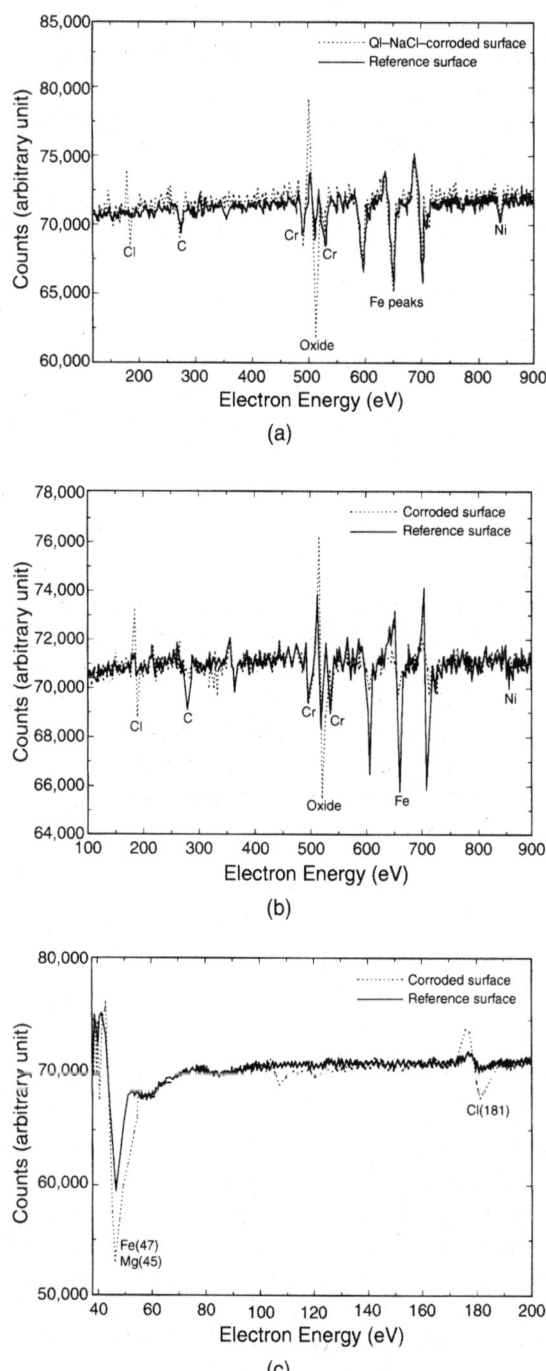

Figure 20 AES profiles of the film on the fracture surfaces tested (a) in boiling 5 M NaCl + 0.15 M Na_2So_4 + 2.5 mil/L HCl solution (b) in boiling 45 % $MgCl_2$ solution and (c) showing the presence of Mg in the film in the low energy range [38].

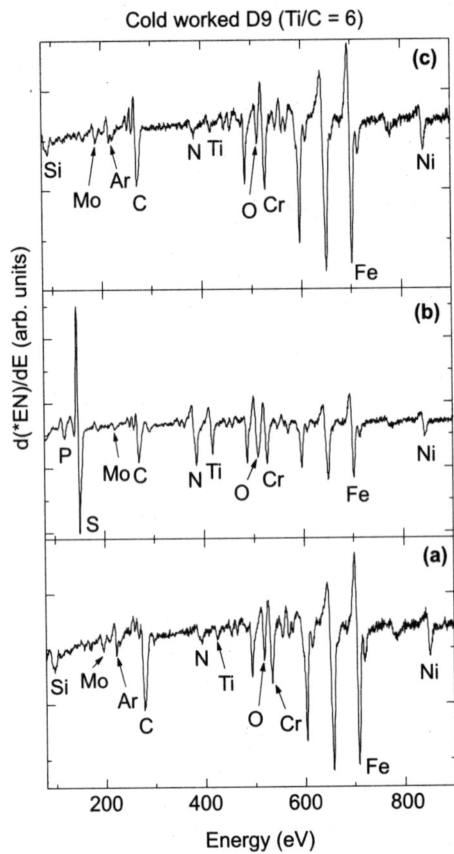

Figure 21 Auger electron spectra of 20 % cold-worked sample with Ti/C = 6 after (a) as-received and sputtered (b) annealing at 973 K and (c) sputter cleaning after 973 K anneal [39].

3.3.5 Oxygenation studies on modified 9Cr-1 Mo Steels

Modified 9Cr-1Mo steels due to their high temperature mechanical behavior and stability are used in steam generator applications in nuclear reactor applications. Since, the steel pipes get exposed to high temperatures and steam environments, it is important to study its surface chemical changes under these conditions. Towards this, modified 9Cr-1Mo steel samples have been studied using AES [40] with controlled exposure to oxygen under UHV conditions. By using a controlled leak valve, oxygen partial pressure in the AES chamber has been increased to 1×10^{-8} torr (as against the base pressure of 4×10^{-10} torr) for various time durations at different in-situ samples temperatures ranging from 300 K to 670 K. The oxygen exposure is measured in units of Langmuir (L). Figure 23 shows the surface chemical concentration of various elements as a function of oxygen exposure at room temperature. As can be seen, the most notable effect seems to be the carburization as evidenced by the increase in Auger peaks due to C and O with concomitant reduction in Fe peak. While, Auger peak due to Mo initially increases, for prolonged oxygen exposure, it settles at un-exposed values.

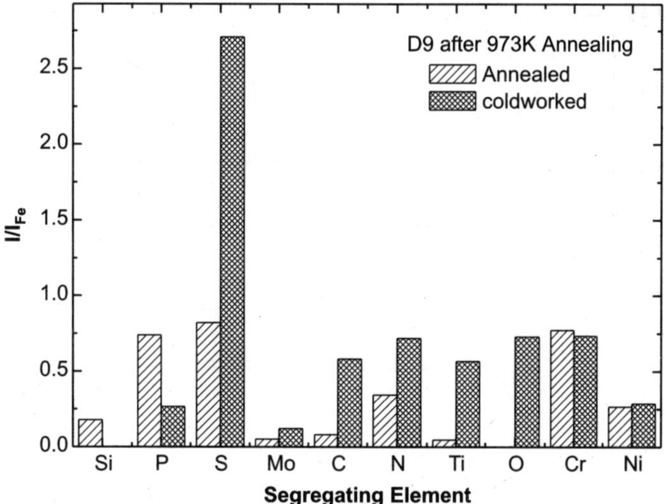

Figure 22 Elemental Auger intensities normalized with respect to that of Fe for various elements in D9 sample annealed at 973 K in 20% cold-worked and solution annealed conditions [39].

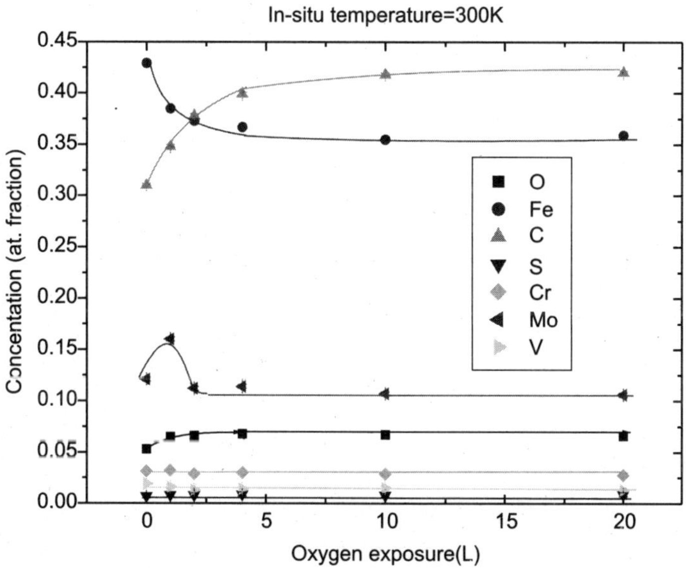

Figure 23 Surface chemical profiles as a function of in-situ oxygen exposure at room temperature obtained using AES. Respective elements are indicated by different symbols [41].

As the in-situ annealing temperature is increased noticeable changes are seen in major and minor alloying elements [41]. As a representative, the variation of iron as a function of oxygen exposure at various temperature is shown in Figure 24. At room temperature, as the oxygen exposure is increased the surface Fe concentration decreases and stabilizes beyond 10 L. This

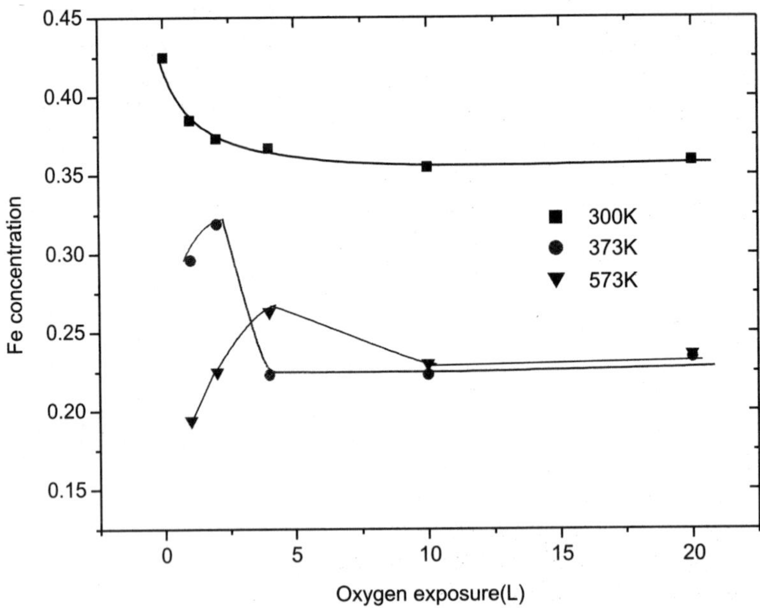

Figure 24 Surface concentration profile of Iron as a function oxygen exposure at in-situ temperatures of 300, 373 and 573 K [41].

suggests that due to oxidation and formation of iron oxides and other oxide phases, the elemental surface concentration of Fe decreases as the oxygen exposure is increased. This is also a clear indication that even at room temperature, the oxygen chemically reacts with the surface elements to form respective oxides. As temperature is increased, the relative surface Fe concentration comes down considerably compared to that of 370 K. From the variation of Fe concentration at 373 K and 573 K, it is clear that beyond 10 L of oxygen exposure there is a saturation behavior. Apart from Fe, other major and minor elemental surface concentrations have also been monitored [41], which show considerable changes for initial oxygen exposures but showing a saturation behavior beyond 10 L. These in-situ oxygenation studies have provided useful insights on the surface chemical changes of the system.

SUMMARY & CONCLUSIONS

The methodology and the application of nuclear techniques in particular positron annihilation, perturbed angular correlation, Mossbauer spectroscopy and Auger electron spectroscopy to corrosion related studies are discussed in this review. In view of their site-specific and selective nature, positron annihilation techniques can provide unique information pertaining to the existence of atomistic-sized vacancy defects or pores in the matrix. It is illustrated that oxygen impurity - probe binding and hence the details of internal and/or bulk oxidation of samples of interest can be understood using PAC. Mossbauer spectroscopy is shown to be quite useful in identifying and studying the structural and magnetic and non-magnetic aspects of the corrosion products. Auger electron spectroscopy is useful in probing the surface chemical composition of the passive layers exposed to various corrosion environments.

ACKNOWLEDGEMENTS

The authors would like to thank Ms. S. Abhaya for her assistance in the preparation of the manuscript.

REFERENCES

1. D. A. Jones in "Principles and Prevention of Corrosion", Macmillan Publishing Company, USA, 1992, pp. 568.
2. J. C. Scully in "The fundamentals of Corrosion", Second Edition, Pergamon Press Inc., Maxwell House, Fairview Park, Elmsford, New York, 1975, pp.234
3. W. J. Lorenz, F. Mansfeld, Corrosion Science 21, 647 (1981).
4. F. Mansfeld, Electrochimica Acta 35, 1533 (1990).
5. S. J. Kerber, J. Tverberg, Advanced Materials & Processes (USA) 158, 33 (2000).
6. D.C. Cook, S. J. Oh, R. Balasubramanian, M. Yamashita , Hyperfine interaction 122, 59 (1999).
7. "Perturbed angular correlation" edited by E. Karlsson, E. Maththias, K. Siegbahn, North-Holland, Amsterdam, 1964.
8. Th. Wichert and E. Recknagel, in "Microscopic Methods in Metals" edited by U. Gonser, Topics in current physics 40, springer-verlag, Heidelberg, 1986.
9. N. S. Mclntyre and T. C. Chan in "Uses of Auger Electron and Photoelectron Spectroscopies in Corrosion Science", John Wiley & Sons Ltd., Practical Surface Analysis. Second Edition. Vol. 1. Auger and X-Ray Photoelectron Spectroscopy (UK), 1990, pp. 485-579.
10. W. Brandt and A. Dupasquier (eds), "Positron Solid State Physics", North Holland, Amsterdam, 1983; K. Saarinen, P. Hautojarvi and C. Corbel, "Identification of Defects in Semiconductors ed M Stavola", Academic Press, New York, 1998.
11. "Mossbauer spectroscopy I&II", Edited by Gonser, Springer-Verlag, Berlin, 1975 and 1981.
12. P. J. Schultz and K. G. Lynn, Rev. Mod. Phys. 60, 701 (1988).
13. G. Amarendra, B. Viswanathan, G. Venugopal Rao, J. Parimala and B. Purniah, Curr. Sci. 73, 409 (1997).
14. G.S. Collins, S.L. Steven, F. Jiawen, Hyp. Int. 62, 1 (1990).
15. M. Thompson, M. D. Baker, A. Christie and J. F. Tyson, "Auger electron spectroscopy" Vol. 74, Chemical Analysis Series, John Wiley & Sons, New York, 1985; J. F. Watts and J. Wolstenhome, "An Introduction to Surface Analysis by XPS and AES", John Wiley & Sons, Chichester, 2003.
16. C. S. Sundar, "Positron Annihilation Studies of Defects in Metals and Metallic Glasses", Ph.D thesis University of Madras, 1984.
17. A. Bharathi, "Positron Annihilation Studies of Precipitation in Aluminium Alloys", Ph.D thesis, University of Madras, 1990.
18. G. Amarendra, "Positron annihilation studies of helium in metals and alloys", Ph. D thesis, University of Madras, 1990.
19. R. Rajaraman, "Positron Annihilation Studies of Light Impurities in Metals", Ph.D thesis, University of Madras, 1994.
20. M. M. Madani, H. L. Vedage and R. D. Granata, J. Electrochem. Soc., 144, 3293 (1997).
21. H. Leidheiser Jr., C. Szeles and A. Vertes, Nucl. Instr. and Met. in Physics Research A255, 606 (1987).
22. Cs. Szeles, A. Vertes, M. L. White and H. Leidheiser Jr., Nucl. Instr. and Met. in Physics Research A271, 688 (1988).

23. T. L. Dull, W.E. Frieze, D. W. Gidley, B. G. Scherer, D. J. Ellerbrock and D. D. Macdonald, Mater. Sci. Forum, 255-257, 671 (1997).
24. K.R. Hebert, T. Gessmann, K. G. Lynn and P. Asoka-Kumar, J. of Electrochemical Society 151, B22, (2004).
25. A. Van Veen, H. Schut, M. Clement, J. M. M. de Nijs, A. Krusemen and M. R. Ijpma, Appl. Surf. Sci. 85, 216 (1995).
26. R. Huang, K. R. Hebert, T. Gessmann and K. G. Lynn, J of Electrochemical Society 151, B227 (2004).
27. Y. C. Wu, R. Zhang, H. Chen, Y. Li, J. Zhang, D.M. Zhu and Y. C. Jean, Radiation Physics and Chemistry 68, 599 (2003).
28. Th. Wichert in "Point defects and defect interactions in Metals" edited by J. I. Takamura and M. Doyama, M. Kiritani, Univ of Tokyo press, Tokyo, 1982, pp 19.
29. W. Bolse, M. Uhrmacher and K. P. Lieb, Phys. Rev. B 36, 1818 (1987).
30. A. F. Pasquevich, F. H. Sanchez, A. G. Bibiloni, J. Desimoni and A. Lopez-Garcia, Phys. Rev. B 57, 963 (2006); J. Desimoni, A. G. Bibiloni, L. Mendoza-Zelis, A. F. Pasquevich, F. H. Sanchez and A. Lopez-Garcia, Phys. Rev. B 28, 5739 (1983).
31. R. Govindaraj, "Studies on defect-Hf solute interactions in fcc, bcc and hcp alloys by Time differential perturbed angular correlation and positron annihilation spectroscopies, PhD Thesis, University of Madras, 1998.
32. A. Weidinger, M. Deicher, T. Butz, Hyp. Int. 10, 717 (1981).
33. R. Govindaraj, C. S. Sundar and R. Kesavamoorthy , J. Appl. Phys. 100 , 084318 (2006).
34. A Vertes, L Korecz, K Burger in "Mossbauer Spectroscopy", Elsevier Scientific Pub. Co Amsterdam, 1979.
35. D.C. Cook, S. J. Oh, R. Balasubramanian, M. Yamashita , Hyperfine interaction 122, 59 (1999); S. J. Oh, D. C. Cook , H. E. Townsend, Corrosion science 41, 1687(1999); S. J. Oh, D. C. Cook, S. J. Kwon, H. E. Townsend, Hyperfine Interaction C4, 49(1999).
36. D. C. Cook, Corrosion science 47, 2550 (2005).
37. S. Ningshen, U. Kamachi Mudali, G. Amarendra, P. Gopalan, R. K. Dayal and H. S. Khatak, Corrosion Science 48, 1106 (2006).
38. T. Anita, H. Shaikh, H. S. Khatak and G. Amarendra, Corrosion 30, 873 (2004).
39. R. Rajaraman, P. Gopalan and G. Amarendra, J of Nucl. Mater. 349, 178 (2006).
40. P. Gopalan, R. Rajaraman and G. Amarendra, International Conference and Exhibition on Pressure Vessels and Piping "OPE 2006", Chennai, Feb 2006.
41. N. Pavan, B. Tech Project Report, Indian Institute of Technology, Chennai, June 2006 (unpublished).

Index

About the Editors

Dr. U. Kamachi Mudali, Head, Corrosion Science and Technology Section-Reprocessing Plant Materials (CSTS-RPM), is at Indira Gandhi Centre for Atomic Research, Kalpakkam, India since 1984. He is a leading corrosion specialist well known internationally for his contributions in localised corrosion, coatings and materials development for severe corrosive environments, and surface modification for corrosion protection. Dr. Mudali has published 125 papers in journals, 145 in proceedings, books and internal reports, and has 2 patents to his credit. He has contributed an article to the Encyclopedia of Electrochemistry, and has edited and published two books and two special issues in journals in his research field. He is a Professor and Faculty member of the Homi Bhabha National Institute (University), Mumbai. He has been a visiting scientist at leading institutions in Germany, Japan, UK, and Israel. Dr. Mudali's research contributions are well cited in the literature, and he has won several awards in recognition of his excellent contributions. Dr. Mudali is a Member of the Editorial Board of International Journals of: (i) Corrosion Reviews, (ii) Surface Engineering, (iii) Materials and Manufacturing Processes, (iv) Journal of Materials Science and Technology, and (v) Transactions of the Indian Institute of Metals. He is a member of several professional associations in India and abroad, including NACE (USA) and ASM International.

Dr. Baldev Raj is a Distinguished Scientist & Director, Indira Gandhi Centre for Atomic Research, Kalpakkam. His specializations include materials characterization, materials development, performance assessment and technology management. He is Fellow, Third World Academy of Sciences, German National Academy of Sciences, Indian National Science Academy, Indian National Academy of Engineering, Indian Academy of Sciences, Bangalore and The National Academy of Sciences, Allahabad. He is an AICTE-INAE Distinguished Visiting Professor in eminent institutes of engineering and technology, and is a Senior Professor at Homi Bhabha National Institute, (University), Mumbai. He has more than 700 publications in journals, and has co-authored 11 books and co-edited 29 books and special journal volumes. He has 5 Indian Standards and 20 patents to his credit. He is Editor-in-Chief of two series of books related to NDE Science & Technology and Metallurgy & Material Science. He is member, Scientific Advisory Council to Prime Minister, and Nano Science & Technology Mission of Department of Science & Technology. He has won many awards and honours, in recognition of his outstanding contributions to science and technology, including Padma Shri from Government of India. Dr. Baldev Raj has contributed significantly towards corrosion science and technology of materials used in nuclear industry, and played a key role in establishing the state of the art corrosion and NDE nlaboratories at IGCAR.

About the Editors